THE SOC SOLUTION

STUDENT RESOURCES

- Interactive eBook
- Practice Quiz Generator
- Flashcards
- Games: Crossword Puzzles and Beat the Clock

- Videos
- Trackable Quizzes
- Trackable Interactives
- Media Quizzes
- Review Cards

Students sign in at **www.cengagebrain.com**

INSTRUCTOR RESOURCES

- All Student Resources
- Engagement Tracker
- First Day of Class Instructions
- LMS Integration
- Instructor's Manual
- Test Bank
- PowerPoint® Slides
- Instructor Prep Cards

Instructors log in at **www.cengage.com/login**

Print

SOC4 delivers all the key terms and all the content for the **Introduction to Sociology** course through a visually engaging and easy-to-reference print experience.

CourseMate

CourseMate provides access to the full **SOC4** narrative, alongside a rich assortment of quizzing, flashcards, and interactive resources for convenient reading and studying.

SOC 4
Nijole V. Benokraitis

Vice President, General Manager, 4LTR Press and the Student Experience: Neil Marquardt

Product Director, 4LTR Press: Steven E. Joos

Product Manager: Clinton Kernen

Content/Media Developer: Sarah Dorger

Product Assistant: Mandira Jacob

Market Strategist: Elizabeth Rankin

Sr. Content Project Manager: Kim Kusnerak

Manufacturing Planner: Ron Montgomery

Production Service: MPS Limited

Sr. Art Director: Stacy Jenkins Shirley

Guest Art Directors: Andrius Benokraitis, Gema Benokraitis, Vitalius Benokraitis

Cover/Internal Design: KeDesign, Mason, OH

Cover Image: ©davorana/Shutterstock.com

Monitor, back cover: ©A-R-T/Shutterstock

Tablet, page i: ©tele52/Shutterstock

Intellectual Property

    Analyst: Deanna Ettinger

    Project Manager: Brittani Morgan

Vice President, General Manager, Social Science & Qualitative Business: Erin Joyner

Product Director: Marta Lee-Perriad

Library of Congress Control Number: 2014957068

ISBN: 978-1-305-09455-0

Cengage Learning
20 Channel Center Street
Boston, MA 02210
USA

Cengage Learning is a leading provider of customized learning solutions with employees residing in nearly 40 different countries and sales in more than 125 countries around the world. Find your local representative at **www.cengage.com**.

Cengage Learning products are represented in Canada by Nelson Education, Ltd.

To learn more about Cengage Learning Solutions, visit **www.cengage.com**

Purchase any of our products at your local college store or at our preferred online store **www.cengagebrain.com**

Printed in the United States of America
Print Number: 01   Print Year: 2015

BENOKRAITIS
SOC⁴

BRIEF CONTENTS

© Anelina/Shutterstock.com

CONTENTS

Stewart Cohen/Blend Images/Alamy

4 Socialization 60

5 Social Interaction in Everyday Life 82

6 Social Groups, Organizations, and Social Institutions 100

7 Deviance, Crime, and Social Control 118

8 Social Stratification: United States and Global 138

AP Images/Paul Sancy

12 Families and Aging   222

Ted Thai/Time Life Pictures/Getty Images

13 Education and Religion   244

14 Health and Medicine 270

15 Population, Urbanization, and the Environment 290

16 Social Change: Collective Behavior, Social Movements, and Technology 312

1 | Thinking Like a Sociologist

LEARNING OBJECTIVES

1-1 Explain what sociology is and how it differs from common sense and other social sciences.

1-2 Explain how and why a sociological imagination helps us understand society.

1-3 Identify and illustrate why it's worthwhile to study sociology.

1-4 Describe and explain the origins of sociology, why sociology developed, and its most influential early theorists.

1-5 Compare, illustrate, and evaluate the four contemporary sociological perspectives.

After you finish this chapter go to **PAGE 19** for **STUDY TOOLS.**

Text messaging is associated with the highest risk of car crashes, and headset cell phones aren't much safer than handheld cell phones. Almost 89 percent of American drivers consider someone talking on a cell phone while driving a serious threat to safety, and 96 percent say the same about texting while driving. However, 69 percent admit using cell phones, and 35 percent send text messages while driving (AAA Foundation for Traffic Safety, 2013; Arnold et al., 2013; see also Naumann and Dellinger, 2013).

Why is there such a disconnection between many Americans' attitudes and behavior? This chapter examines these and other questions. Let's begin by considering what sociology is (and isn't) and how a "sociological imagination" can give us more control over our lives. We'll then look at how sociologists grapple with complex theoretical issues in explaining social life.

What do you think?

Sociology is basically common sense.

1	2	3	4	5	6	7

strongly agree   strongly disagree

1-1 WHAT IS SOCIOLOGY?

Stated simply, **sociology** is the systematic study of human behavior in society. When sociologists refer to the *systematic* study of behavior, they mean that social behavior is regular and patterned, and that it takes place between individuals, small groups (such as families), large organizations (such as Apple), and entire societies (such as between the United States and other countries). But, you might protest, "I'm unique."

1-1a Are You Unique?

Yes and no. Each of us is unique in the sense that you and I are like no one else on earth. Even identical twins, who have the same physical characteristics and genetic matter, usually differ in personality and interests. One of my colleagues likes to tell the story about his 3-year-old twin girls who received the same doll. One twin chattered that the doll's name was Lori, that she loved Lori, and would take good care of her. The second twin muttered, "Her name is Stupid," and flung the doll into a corner.

Despite some individual differences, identical twins, you, and I are like other people in many ways. Around the world, we experience grief when a loved one dies, participate in rituals that celebrate marriage or the birth of a child, and want to have healthy and happy lives. Some actions, such as terrorist attacks, are unpredictable. For the most part, however, people conform to expected and acceptable behavior. From the time that we get up until we go to bed, we follow a variety of rules and customs about what we eat, how we drive, how we act in different social situations, and how we dress for work, classes, and leisure activities.

So what? you might shrug. Isn't it "obvious" that we dress differently for classes than for job interviews? Isn't all of this just plain old common sense? Before reading further, take the quiz at right.

TRUE OR FALSE?

Everybody Knows That …

1. **The death penalty reduces crime.**
2. **Women's earnings are now similar to men's, especially in high income jobs.**
3. **People age 65 and older make up the largest group of those who are poor.**
4. **Cohabitation (living together) increases the chance of having a happy and lasting marriage.**
5. **Divorce rates are higher today than in the past.**
6. **Latinos are the fastest-growing racial-ethic group in the United States.**
7. **The best way to get an accurate measure of public opinion is to poll as many people as possible.**
8. **Drug abuse is the biggest health hazard.**

The answers are on the next page.

sociology the systematic study of human behavior in society.

For example, many Americans are most concerned about street crimes, such as robbery or violent assaults by strangers. However, FBI and sociological data show that we're much more likely to be assaulted or murdered by someone we know or live with (see Chapters 7 and 12).

▶ **Common sense perceptions vary across groups and cultures.** In the United States, it's common sense to date before choosing a marriage mate. In many other societies, however, it's common sense to marry someone whom parents and relatives have selected. Thus, common sense notions about mate selection vary considerably around the world.

▶ **Much of our common sense is based on myths and misconceptions.** A common myth is that living together is a good way to find out whether partners will get along after marriage. Generally, however, couples who live together before marriage have higher divorce rates than those who don't (see Chapter 12).

Sociology, in contrast to conventional wisdom, examines claims and beliefs critically, considers many points of view, and enables us to move beyond established ways of thinking. The *sociological perspective* analyzes how social context influences people's lives. The "sociological imagination" is at the center of the sociological perspective.

1-1b Isn't Sociology Just Common Sense?

No. Sociology goes well beyond conventional wisdom, what we call common sense, in several ways:

▶ **Common sense is subjective.** If a woman crashes into my car, I might conclude, according to the conventional wisdom, statements that we've heard over the years, that "all women are terrible drivers." In fact, most drivers involved in crashes are men—especially teenagers and those age 70 and older (Insurance Institute for Highway Safety, 2013). Thus, *objective* data show that, overall, men are worse drivers than women.

▶ **Common sense ignores facts.** Because common sense is subjective, there's little room for facts that might be disturbing or challenge cherished beliefs.

Marriage without Love? No Way!

When I ask my students, "Would you marry someone you're not in love with?" most laugh, raise an eyebrow, or stare at me in disbelief. "Of course not!" they exclaim. In fact, the "open" courtship and dating systems common in Western nations, including the United States, are foreign to much of the world. In many African, Asian, Mediterranean, and Middle Eastern countries, marriages are arranged. In these societies, marriages forge bonds between families rather than individuals, and preserve family continuity along religious and socioeconomic lines. Love isn't a prerequisite for marriage in societies that value the intergenerational and community relations of a kin group rather than an individual's choices (see Chapters 9 and 12).

1-2 WHAT IS A SOCIOLOGICAL IMAGINATION?

According to sociologist C. Wright Mills (1916–1962), our individual behavior is influenced by social factors such as religion, ethnicity, and politics. Mills (1959) called this ability to see the relationship between individual experiences and larger social influences the **sociological imagination**. The sociological imagination emphasizes the connection between personal troubles (biography) and structural (public and historical) issues.

Consider unemployment. If only a small group of people can't find a job, it's a *personal trouble* that may be due, in part, to an individual's low educational attainment, lack of specific skills that employers want, failure to search for work, and so on. If unemployment is widespread, it's a *public issue* because economic problems are also the result of structural factors such as mass layoffs, sending jobs overseas, technological changes, and restrictive hiring policies (see Chapter 11). Thus, people may be unemployed regardless of skills, a college degree, and job searches.

A sociological imagination helps us understand how larger social forces affect individuals and how individuals affect society. It identifies why our personal troubles often reflect larger public issues and policies over which we have little, if any, control. A sociological imagination relies on both micro- and macro-level approaches to understand our social world.

1-2a Microsociology: How People Affect Our Everyday Lives

Microsociology examines the patterns of individuals' social interaction in specific settings. In most of our relationships, we interact with others on a micro, or "small," level (such as members of a work group discussing who will perform which tasks). These everyday interactions involve what people think, say, or do on a daily basis.

1-2b Macrosociology: How Social Structure Affects Our Everyday Lives

Macrosociology focuses on large-scale patterns and processes that characterize society as a whole. Macro, or "large," approaches are especially useful in understanding some of the constraints—such as economic forces and public policies.

Microsociology and macrosociology differ conceptually, but they're interrelated. Consider the reasons for

AF archive/Alamy

The Wire was a popular television show that aired on HBO between 2002 and 2008. It depicted the lives of poor people and blue-color workers in Baltimore, Maryland. For sociologists, *The Wire* illustrates the connection between macro-level structural constraints and micro-level individual behavior.

divorce. On a micro level, sociologists study factors such as extramarital affairs, substance abuse, arguments about money, and other everyday interactions that fuel marital tension and unhappiness, leading to divorce. On a macro level, sociologists look at how the economy, laws, cultural values, and technology affect divorce rates (see Chapter 12). Examining micro, macro, and micro–macro forces is one of the reasons why sociology is a powerful tool in understanding (and changing) our behavior and society at large (Ritzer, 1992).

1-3 WHY STUDY SOCIOLOGY?

Sociology offers explanations that can greatly improve the quality of your everyday life. These explanations can influence choices that range from your personal decisions to expanding your career opportunities.

sociological imagination the ability to see the relationship between individual experiences and larger social influences.

microsociology examines the patterns of individuals' social interaction in specific settings.

macrosociology examines the large-scale patterns and processes that characterize society as a whole.

1-3a Making Informed Decisions

Knowing some sociology can help us make more informed decisions. For example, we often hear that grief counseling is essential after the death of a loved one. In fact, 4 in 10 Americans are better off without such counseling. Grief is normal, and most people work through their losses on their own, whereas counseling sometimes prolongs the feelings of depression and anxiety (Stroebe et al., 2000).

1-3b Understanding Diversity

The racial and ethnic composition of the United States is changing. By 2025, only 58 percent of the U.S. population is projected to be white, down from 76 percent in 1990 and 86 percent in 1950 (Passel et al., 2011; U.S. Census Bureau, 2012). As you'll see in later chapters, this racial/ethnic shift has already affected interpersonal relationships as well as education, politics, religion, and other spheres of social life.

Recognizing and understand diversity is one of sociology's central themes. Our gender, social class, marital status, ethnicity, sexual orientation, and age—among other factors—shape our beliefs, behavior, and experiences. If, for example, you're a white middle-class male who attends a private college, your experiences are very different from those of a female Vietnamese immigrant who is struggling to pay expenses at a community college.

Increasingly, nations around the world are intertwined through political and economic ties. What happens in other societies often has a direct or indirect impact on contemporary U.S. life. Decisions in oil-producing countries, for example, affect gas prices, spur the development of hybrid cars that are less dependent on oil, and stimulate research on alternative sources of energy.

© Suzanne Tucker/Shutterstock.com

1-3c Shaping Social and Public Policies and Practices

Sociology is valuable in applied, clinical, and policy settings because many jobs require understanding society and research, and applying theoretical perspectives to create social change. According to a director of a research institute, sociology increased her professional contributions: "I can look at problems of concern to the National Institutes of Health and say 'here's a different way to solve this problem'" (Nyseth et al., 2011: 48).

"I'm a social scientist, Michael. That means I can't explain electricity or anything like that, but if you ever want to know about people I'm your man."

J.B. Handelsman The New Yorker Collection/CartoonBank.Com

1-3d Thinking Critically

We develop a sociological imagination not only when we understand and can apply the concepts, but also when we can think, speak, and write critically. Much of our thinking and decision making is often impulsive and emotional. In contrast, critical thinking involves knowledge and problem solving (Paul and Elder, 2007).

Critical sociological thinking goes even further because we begin to understand how our individual lives, choices, and troubles are shaped by race, gender, social class, and social institutions such as the economy, politics, and education (Eckstein et al., 1995; Grauerholz and Bouma-Holtrop, 2003). *Table 1.1* summarizes some of the basic elements of critical sociological thinking.

1-3e Expanding Your Career Opportunities

A degree in sociology is a springboard for entering many jobs and professions. A national survey of undergraduate sociology majors found that 44 percent were in administrative support or management positions, 22 percent were employed in social service and counseling, 18 percent were in sales and marketing occupations, and 12 percent were teachers (Senter et al., 2014).

TABLE 1.1 WHAT IS CRITICAL SOCIOLOGICAL THINKING?

Critical sociological thinking requires a combination of skills. Some of the basic elements include the ability to:

- rely on reason rather than emotion

- ask questions, avoid snap judgments, and examine popular and unpopular beliefs

- recognize one's own and others' assumptions, prejudices, and points of view

- remain open to alternative explanations and theories

- require and examine competing evidence (see Chapter 2)

- understand how public issues affect private troubles

What specific skills do sociology majors learn that are useful in their jobs? Some of the most important skills are being better able to work with people (71 percent), to organize information (69 percent), to write reports that nonsociologists understand (61 percent), and to interpret research findings (56 percent) (Van Vooren and Spalter-Roth, 2010). In other cases, students major in sociology because it provides a broad liberal arts foundation for professions such as law, education, and social work.

Even if you don't major in sociology, developing your sociological imagination can enrich your job skills. Sociology courses help you learn to think abstractly and critically, formulate problems, ask incisive questions, search for data in the most reliable and up-to-date sources, organize material, and improve your oral presentations (ASA Research Department..., 2013; Spalter-Roth et al., 2013).

1-4 SOME ORIGINS OF SOCIOLOGICAL THEORY

During college, most of my classmates and I postponed taking theory courses (regardless of our major) as long as possible. "This stuff is boring, boring, boring," we'd grump, "and has nothing to do with the real world." However, theorizing is, in fact, part of our everyday lives. Every time you try to explain why your family and friends behave as they do, for example, you're theorizing.

As people struggle to understand human behavior, they develop theories. A **theory** is a set of statements that explains why a phenomenon occurs. Theories produce knowledge, guide our research, help us analyze our findings, and, ideally, offer solutions for social problems.

Sociologist James White (2005: 170–171) describes theories as "tools" that don't profess to know "the truth" but "may need replacing" over time as our understanding of society becomes more sophisticated. Like hardware tools that change over the years, theories evolve over time because of cultural and technological transformations. As you'll see shortly, for example, sociological theories about behavior changed considerably after the rise of feminist scholarship during the late 1960s.

Sociology and Other Social Sciences: *What's the Difference?*

How would different social scientists study the same phenomenon, such as homelessness? Criminologists might examine whether crime rates are higher among homeless people than in the general population. Economists might measure the financial impact of programs for the homeless. Political scientists might study whether and how government officials respond to homelessness. Psychologists might be interested in how homelessness affects individuals' emotional and mental health. Social workers are most likely to try to provide needed services such as food, shelter, medical care, and jobs. Sociologists have been most interested in examining homelessness across gender, age, and social class, and explaining how this social problem devastates families and communities.

Mitchell Funk/Photographer's Choice/Getty Images

According to sociologist Herbert Gans (2005), sociologists "study everything." There are currently 43 different subfields in sociology, and the number continues to increase, because sociologists' interests range across many areas.

theory a set of statements that explains why a phenomenon occurs.

that the study of society must be **empirical**. That is, information should be based on observations, experiments, or other data collection rather than on ideology, religion, intuition, or conventional wisdom.

He saw sociology as the scientific study of two aspects of society: social statics and social dynamics. *Social statics* investigates how principles of social order explain a particular society, as well as the interconnections between institutions. *Social dynamics* explores how individuals and societies change over time. Comte's emphasis on social order and change within and across societies is still useful today because many sociologists examine the relationships between education and politics (social statics), as well as how such interconnections change over time (social dynamics).

Spencer Arnold/Hulton Archive/Getty Images

1-4b Harriet Martineau

Harriet Martineau (1802–1876), an English author, published several dozen books on a wide range of topics in social science, politics, literature, and history. Her translation and condensation of Auguste Comte's difficult material for popular consumption was largely responsible for the dissemination of Comte's work. "We might say, then, that sociology had parents of both sexes" (Adams and Sydie, 2001: 32). She emphasized the importance of systematic data collection through observation and interviews, and an objective analysis of data to explain events and behavior. She also published the first sociology research methods textbook.

Martineau, a feminist and strong opponent of slavery, denounced many aspects of capitalism as alienating and degrading, and criticized dangerous workplaces that often resulted in injury and death. Martineau promoted improving women's positions in the workforce through education, nondiscriminatory employment, and training programs. She advocated women's admission into medical schools and emphasized issues such as infant care, the rights of the aged, the

Theories = Tools

Creatas Images/Jupiter Images

Sociological theories didn't emerge overnight. Nineteenth-century thinkers grappled with some of the same questions that sociologists try to answer today: Why do people behave as they do? What holds society together? What pulls it apart? Of the many early sociological theorists, some of the most influential were Auguste Comte, Harriet Martineau, Émile Durkheim, Karl Marx, Max Weber, Jane Addams, and W. E. B. Du Bois.

1-4a Auguste Comte

Auguste Comte (pronounced oh-gust KONT; 1798–1857) coined the term *sociology* and is often described as the "father of sociology." Comte maintained

Hulton Archive/Getty Images

The Father of Sociology—Auguste Comte

empirical information that is based on observations, experiments, or other data collection rather than on ideology, religion, intuition, or conventional wisdom.

prevention of suicide, and other social problems (Hoecker-Drysdale, 1992).

After a long tour of the United States, Martineau described American women as being socialized to be subservient and dependent rather than equal marriage partners. She also criticized American and European religious institutions for expecting women to be pious and passive rather than educating them in philosophy and politics. Most scholars, including sociologists, ridiculed her ideas and dismissed them as too radical.

1-4c Émile Durkheim

Émile Durkheim (1858–1917), a French sociologist and writer, agreed with Comte that societies are characterized by unity and cohesion because their members are bound together by common interests and attitudes. According to Durkheim, however, Comte didn't show that sociology could be scientific. Like Comte, Durkheim ignored Martineau's contributions on the importance of systematic and objective data collection (Adams and Sydie, 2001).

SOCIAL FACTS

To be scientific, Durkheim maintained, sociology must study **social facts**—aspects of social life, external to the individual, that can be measured. Sociologists can determine *material facts* by examining demographic characteristics such as age, place of residence, and population size. They can gauge *nonmaterial facts*, such as communication processes, by observing everyday behavior and how people relate to each other (see Chapters 3 to 6). For contemporary sociologists, social facts also include collecting and analyzing data on *social currents*, such as collective behavior and social movements (see Chapter 16).

DIVISION OF LABOR

One of Durkheim's central questions was how people can be autonomous and individualistic while being integrated in society. **Social solidarity**, or social cohesiveness and harmony, according to Durkheim, is maintained by a **division of labor**—an interdependence of different tasks and occupations, characteristic of industrialized societies, that produces social unity and facilitates change.

As the division of labor becomes more specialized, people become increasingly dependent on others for specific goods and services. Today, for example, many couples who are getting married often contract "experts" such as photographers, florists, deejays, caterers, bartenders, travel agents (for the honeymoon), and even "wedding planners."

SOCIAL INTEGRATION

Durkheim was one of the first sociologists to test a theory using data. In his classic study *Suicide*, Durkheim (1897) relied on extensive data collection to test his theory that suicide is associated with social integration. He concluded that people who experience meaningful social relationships in families, social groups, and communities are less likely to commit suicide than those who feel alone, helpless, or hopeless. Thus, many seemingly isolated individual acts, including suicide, are often the result of structural arrangements, such as weak social ties.

Are Durkheim's findings on social integration dated? No. We typically read about the high suicide rates of teens, but suicide rates are much higher at later ages (see *Figure 1.1*). As Durkheim found, men still have higher suicide rates than women across all age groups, especially those age 75 and older, and across all racial and ethnic groups. White males age 85 and older have the highest suicide rates (National Center for Health Statistics, 2014).

Durkheim's concept of social integration is useful in explaining the high suicide rates of older men. The rates may reflect being widowed and feeling alone, a sense of hopelessness because of terminal illnesses, and not being "connected" to family, friends, community groups, and support systems that women tend to develop throughout their lives (American Association of Suicidology, 2009; see also Chapters 9 and 12).

1-4d Karl Marx

Karl Marx (1818–1883), a German social philosopher, is often described as the most influential social scientist

social facts aspects of social life, external to the individual, that can be measured.

social solidarity social cohesiveness and harmony.

division of labor an interdependence of different tasks and occupations, characteristic of industrialized societies, that produces social unity and facilitates change.

Émile Durkheim

Pictorial Press Ltd/Alamy

FIGURE 1.1 U.S. SUICIDE RATES, BY SEX AND AGE

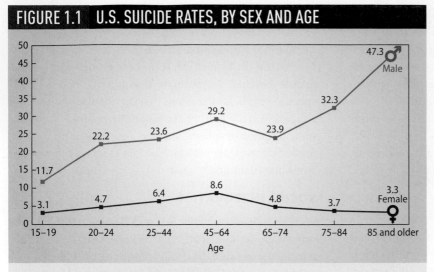

Suicide deaths per 100,000 population

Source: Based on National Center for Health Statistics, 2014, Table 35.

who ever lived. Marx, like Comte and Durkheim, tried to explain the changes that were taking place in society during the Industrial Revolution.

The Industrial Revolution began in England around 1780 and spread throughout Western Europe and the United States during the nineteenth century. A number of technological inventions—such as the spinning wheel, the steam engine, and large weaving looms—enabled the development of large-scale manufacturing and mining industries over a relatively short period. The extensive mechanization shifted agricultural and home-based work to factories in cities. As masses of people migrated from small farms to factories to find jobs, urbanization and capitalism grew rapidly.

CAPITALISM

Unlike his predecessors and contemporaries, Marx (1867/1967, 1964) maintained that economic issues produce divisiveness rather than social solidarity. For Marx, the most important social change was the development of **capitalism**, an economic system in which the ownership of the means of production—such as land, factories, large sums of money, and machines—is in private hands. As a result, Marx saw industrial society as composed of three social classes:

▸ **capitalists**—the ruling elite who own the means of producing wealth (such as factories)

▸ **petit bourgeoisie**—small business owners and owner workers who still have their own means of production but might end up in the proletariat because they're driven out by competition or their businesses fail

▸ **proletariat**—the masses of workers who depend on wages to survive, have few resources, and make up the working class

CLASS CONFLICT

Marx believed that society is divided into the haves (capitalists) and the have-nots (proletariat). For Marx, capitalism is a class system in which conflict between the classes is common and society is anything but cohesive. Instead, class antagonisms revolve around struggles between the capitalists, who increase their profits by exploiting workers, and workers, who resist but give in because they depend on capitalists for jobs.

Marx argued that there's a close relationship between inequality, social conflict, and social class. He maintained that history is a series of class struggles between capitalists and workers. As wealth becomes more concentrated in the hands of a few capitalists, he predicted, the ranks of an increasingly dissatisfied proletariat would swell, leading to bloody revolution and eventually a classless society. The Occupy Wall Street movement

Roger Viollet Collection/Getty Images

Karl Marx

Comstock/Stockbyte/Getty Images

capitalism an economic system in which the ownership of the means of production is in private hands.

Occupy Wall Street (OWS) was a protest movement against corporate greed, corruption, and influence on government. It began in mid-September, 2011, in New York City's Wall Street financial district. The OWS slogan, "We are the 99%," referred to U.S. income and wealth inequality between the wealthiest 1 percent and the rest of the population. OWS received global attention and spawned similar movements worldwide but was short-lived (see Chapter 16).

showed that thousands of Americans are very unhappy about the growing inequality between the haves and the have-nots, but there hasn't been a "bloody revolution" in the United States, unlike some countries in the Middle East.

ALIENATION

In industrial capitalist systems, Marx (1844/1964) contended, **alienation**—the feeling of separation from one's group or society—is common across all social classes. Workers feel alienated because they don't own or control either the means of production or the product. Because meaningful labor is what makes us human, Marx maintained, our workplace has alienated us "from the essence of our humanness." In plain English, instead of collaboration, a capitalistic society encourages competition, backstabbing, and "looking out for number one."

According to Marx, capitalists are also alienated. They regard goods and services as important simply because they're sources of profit. Capitalists don't care who buys or sells their products, how the workers feel about the products they make, or whether buyers value the products. The major focus, for capitalists, is on increasing profits as much as possible rather than feeling "connected" to the products or services they sell. Every year, for example, companies must recall cars, pharmaceutical items, toys, and food products that cause injuries, illness, or death.

1-4e Max Weber

Max Weber (pronounced VAY-ber; 1864–1920) was a German sociologist, economist, legal scholar, historian, and politician. Unlike Marx's emphasis on economics as a major factor in explaining society, Weber focused on social organization, a subjective understanding of behavior, and a value-free sociology.

SOCIAL ORGANIZATION

For Weber, economic factors were important, but ideas, religious values, ideologies, and charismatic leaders were just as crucial in shaping and changing societies. He maintained that a complete understanding of society requires an analysis of the social organization and interrelationships among economic, political, and cultural institutions. In his *Protestant Ethic and the Spirit of Capitalism*, for example, Weber (1920/1958) argued that the self-denial fostered by Calvinism supported the rise of capitalism and shaped many of our current values about working hard (see Chapters 3 and 6).

SUBJECTIVE UNDERSTANDING

Weber posited that an understanding of society requires a "subjective" understanding of behavior. Such understanding, or *verstehen* (pronounced fer-SHTAY-en), requires knowing how people perceive the world in which they live. Weber described two types of *verstehen*. In *direct observational understanding*, the social scientist observes a person's facial expressions, gestures, and listens to his/her words. In *explanatory understanding*, the social scientist tries to grasp the intention and context of the behavior.

If a person bursts into tears (direct observational understanding), the observer knows what the person may be feeling (anger, sorrow, and so on). An explanatory understanding goes a step further by spelling out the

Max Weber

alienation the feeling of separation from one's group or society.

Is It Possible to Be a Value-Free Sociologist?

Max Weber was concerned about popular professors who took political positions that pleased many of their students. He felt that these professors were behaving improperly because science, including sociology, must be "value free." Faculty must set their personal values aside to make a contribution to society. According to one sociologist who agrees with Weber, sociology's weakness is its tendency toward moralism and ideology:

> Many people become sociologists out of an impulse to reform society, fight injustice, and help people. Those sentiments are noble, but unless they are tempered by skepticism, discipline, and scientific detachment, they can be destructive. Especially when you are morally outraged and burning with a desire for action, you need to be cautious (Massey, 2007: B12).

There's considerable disagreement on whether sociologists can *really* be value free. Some argue that being value free is a myth because it's impossible for a scholar's attitudes and opinions to be totally divorced from her or his scholarship (Gouldner, 1962). Many sociologists, after all, do research on topics that they consider significant and about which they have strong views. If you talk to your sociology instructor, for example, you'll probably find that she or he often teaches and does research on topics that are intensely and personally interesting.

CAN SOCIOLOGISTS BE VALUE FREE—ESPECIALLY WHEN THEY HAVE STRONG FEELINGS ABOUT MANY SOCIETAL ISSUES? *SHOULD THEY BE?*

Other sociologists maintain that one's values should be passionately partisan, should frame research issues, and should have an impact on improving society (Feagin, 2001). That is, sociologists should not apologize for being subjective in their teaching and research.

Can sociologists be value free—especially when they have strong feelings about many societal issues? Should they be?

Massey, Douglas S. 2007. Categorically Unequal: The American Stratification System. New York: Russell Sage Foundation.

---

reason for the behavior (rejection by a loved one, frustration when your computer crashes, humiliation if a boss yells at you in public).

VALUE-FREE SOCIOLOGY

One of Weber's most lasting and controversial views was the notion that sociologists must be as objective, or "value free," as possible in analyzing society. A researcher who is **value free** is one who separates her or his personal values, opinions, ideology, and beliefs from scientific research.

During Weber's time, the government and other organizations demanded that university faculty teach the "right" ideas. Weber encouraged everyone to be involved as citizens, but he maintained that educators

and scholars should be as dispassionate as possible about political and ideological positions. The task of the teacher, Weber argued, was to provide students with knowledge and scientific experience, not to "imprint" the teacher's personal political views and value judgments (Gerth and Mills, 1946). The box "Is It Possible to Be a Value-Free Sociologist?" examines this issue further.

1-4f Jane Addams

Jane Addams (1860–1935) was a social worker who cofounded Hull House, one of the first settlement houses in Chicago that served as a community center for the neighborhood poor. An active reformer throughout her life, Jane Addams was a leader in the women's suffrage movement and, in 1931, was the first American woman to be awarded the Nobel Peace Prize for her advocacy of negotiating, rather than waging war, to settle disputes.

value free separating one's personal values, opinions, ideology, and beliefs from scientific research.

Sociologist Mary Jo Deegan (1986) describes Jane Addams as "the greatest woman sociologist of her day." However, she was ignored by her colleagues at the University of Chicago (the first sociology department established in the United States in 1892) because discrimination against women sociologists was "rampant" (p. 8).

Despite such discrimination, Addams published articles in numerous popular and scholarly journals, as well as many books on the everyday life of urban neighborhoods, especially the effects of social disorganization and immigration. Much of her work contributed to symbolic interaction, an emerging school of thought that you'll read about shortly. One of Addams' greatest intellectual legacies was her emphasis on applying knowledge to everyday problems. Her pioneering work in criminology included ecological maps of Chicago that were later credited to men (Moyer, 2003).

Jane Addams with a child at Hull House

1-4g W. E. B. Du Bois

W. E. B. Du Bois (pronounced Do-BOICE; 1868–1963) was a prominent black sociologist, writer, editor, social reformer, and passionate orator. The author of almost two dozen books on Africans and black Americans, Du Bois spent most of his life responding to the critics and detractors of black life. He was the first African American to receive a Ph.D. from Harvard University, but once remarked, "I was in Harvard but not of it."

Du Bois helped found the National Association for the Advancement of Colored People (NAACP) and became editor of its journal, *Crisis*. The problem of the twentieth century, he wrote, is the problem of the color line. Du Bois believed that the race problem was one of ignorance and wanted to provide a "cure" for prejudice and discrimination. Such cures included promoting black political power and civil rights and providing blacks with a higher education rather than funneling them into techical schools.

These and other writings were unpopular at a time when Booker T. Washington, a well-known black educator, encouraged black people to be patient instead of demanding

W. E. B. DuBois

equal rights. As a result, Du Bois was dismissed as a radical by his contemporaries but was rediscovered by a new generation of black scholars during the 1970s and 1980s. Among his many contributions, Du Bois examined the oppressive effects of race and class, advocated women's rights, and played a key role in reshaping black–white relations in America (Du Bois, 1986; Lewis, 1993).

All of the early thinkers agreed that people are transformed by each other's actions, social patterns, and historical changes. They and other scholars shaped contemporary sociological theories.

1-5 CONTEMPORARY SOCIOLOGICAL THEORIES

How one defines "contemporary sociological theory" is somewhat arbitrary. The mid-twentieth century is a good starting point because "the late 1950s and 1960s have, in historical hindsight, been regarded as significant years of momentous changes in the social and cultural life of most Western societies" (Adams and Sydie, 2001: 479). Some of the sociological perspectives had earlier origins, but all matured during this period.

Sociologists typically use more than one theory to explain behavior. The theories view our social world somewhat differently, but all of them analyze why society is organized the way it is and why we behave as we do. Four of the most influential theoretical perspectives are functionalism, conflict theory, feminist theories, and symbolic interaction.

1-5a Functionalism

Functionalism (also known as *structural functionalism*) maintains that society is a complex system of interdependent parts that work together

> **functionalism (*structural functionalism*)** maintains that society is a complex system of interdependent parts that work together to ensure a society's survival.

to ensure a society's survival. Much of contemporary functionalism grew out of the work of Auguste Comte and Émile Durkheim, both of whom believed that human behavior is a result of social structures that promote order and integration in society.

One of their contemporaries, English philosopher Herbert Spencer (1820–1903), used an organic analogy to explain the evolution of societies. To survive, Spencer (1862/1901) wrote, our vital organs—like the heart, lungs, kidneys, liver, and so on—must function together. Similarly, the parts of a society, like the parts of a body, work together to maintain the whole structure.

SOCIETY IS A SOCIAL SYSTEM

Prominent American sociologists, especially Talcott Parsons (1902–1979) and Robert K. Merton (1910–2003), developed the earlier ideas of structure and function. For these and other functionalists, a society is a system that is composed of major institutions such as government, religion, the economy, education, medicine, and the family.

Each institution or other social group has *structures*, or organized units, that are connected to each other and within which behavior occurs. Education structures such as colleges, for instance, aren't only organized internally in terms of who does what and when, but depend on other structures such as government (to provide funding), business (to produce textbooks and construct buildings), and medicine (to ensure that students, staff, and faculty are healthy).

FUNCTIONS AND DYSFUNCTIONS

Each structure fulfills certain *functions*, or purposes and activities, to meet different needs that contribute to a society's stability and survival (Merton, 1938). The purpose of education, for instance, is to transmit knowledge to the young, to teach them to be good citizens, and to prepare them for jobs (see Chapter 13).

Dysfunctions are social patterns that have a negative impact on a group or society. When one part of society isn't working, it affects other parts by creating

dysfunctions social patterns that have a negative impact on a group or society.

manifest functions purposes and activities that are intended and recognized; they're present and clearly evident.

latent functions purposes and activities that are unintended and unrecognized; they're present but not immediately obvious.

> # SOCIOLOGISTS TYPICALLY USE MORE THAN ONE THEORY TO EXPLAIN BEHAVIOR AND WHY SOCIETY IS ORGANIZED THE WAY IT IS.

conflict, divisiveness, and social problems. Consider religion. In the United States, the Catholic Church's stance on issues such as not ordaining women to be priests and denouncing abortion and homosexuality has produced a rift between those who espouse the papal edicts and those who question them. In other countries, religious intolerance has led to wars and terrorist attacks (see Chapter 13).

MANIFEST AND LATENT FUNCTIONS

There are two kinds of functions. **Manifest functions** are intended and recognized; they're present and clearly evident. **Latent functions** are unintended and unrecognized; they're present but not immediately obvious. Consider the marriage ceremony. The primary manifest function of the marriage ceremony is to publicize the formation of a new family unit and to legitimize sexual intercourse and childbirth (even though both might occur outside of marriage). Its latent functions include communicating a "hands-off" message to suitors, providing the new couple with household goods and products through bridal showers and wedding gifts, and redefining family boundaries to include in-laws or stepfamily members.

CRITICAL EVALUATION

You'll see in later chapters that functionalism is useful in seeing the "big picture" of interrelated structures and functions. Its influence waned during the 1960s and 1970s, however, because functionalism is so focused on order and stability that it often ignores social change. For example, many sociologists believed that functionalism couldn't explain the many rapid changes sparked by the civil rights, women's, and gay movements.

A second and related criticism is that functionalism often glosses over the inequality that a handful of powerful people create and maintain. Conflict theorists,

Blue_Cutler/iStockphoto.com

Among other manifest functions, schools transmit knowledge and prepare children for adult economic roles. Among their latent functions, schools provide many matchmaking opportunities. What are some other examples of education's manifest and latent functions?

especially, have pointed out that what's functional for some privileged groups is dysfunctional for many others.

1-5b Conflict Theory

In contrast to functionalism—which emphasizes order, stability, cohesion, and consensus—**conflict theory** examines how and why groups disagree, struggle over power, and compete for scarce resources (such as property, wealth, and prestige). Conflict theorists see disagreement and the resulting changes in society as natural, inevitable, and even desirable.

SOURCES OF CONFLICT

The conflict perspective has a long history. As you saw earlier, Karl Marx predicted that conflict would result from widespread economic inequality, and W. E. B. Du Bois criticized U.S. society for its ongoing and divisive racial discrimination. Since the 1960s, as you'll see in later chapters, many sociologists—especially feminist and minority scholars—have emphasized that the key sources of economic inequity in any society include race, ethnicity, gender, age, and sexual orientation.

Conflict theorists agree with functionalists that many societal arrangements are functional. But, conflict theorists ask, who benefits? And who loses? When corporations merge, workers in lower-end jobs are often laid off while the salaries and benefits of corporate executives soar and the value of stocks (usually held by higher social classes) increase. Thus, mergers might be functional for those at the upper end of the socioeconomic ladder, but dysfunctional for those on the lower rungs.

SOCIAL INEQUALITY

Unlike functionalists, conflict theorists see society not as cooperative and harmonious, but as a system of widespread inequality. For conflict theorists, there's a continuous tension between the haves and the have-nots, most of whom are children, women, minorities, people with low incomes, and the poor.

Many conflict theorists focus on how those in power—typically wealthy white Anglo-Saxon Protestant males (WASPs)—dominate political and economic decision making in U.S. society. This group controls a variety of institutions—such as education, criminal justice, and the media—and passes laws that benefit primarily people like themselves (see Chapters 8 and 11).

CRITICAL EVALUATION

Conflict theory explains how societies create and cope with disagreements. However, some have criticized conflict theorists for overemphasizing competition and coercion at the expense of order and stability. Inequality exists and struggles over scarce resources occur, critics agree, but conflict theorists often ignore cooperation and harmony. Voters, for example, can boot dominant white males out of office and replace them with women and minority group members. Critics also point out that the have-nots can increase their power through negotiation, bargaining, lawsuits, and strikes.

1-5c Feminist Theories

Rebecca West, a British journalist and novelist, once said, "I myself have never been able to find out precisely what feminism is; I only know that people call me a feminist whenever I express sentiments that differentiate me from a doormat." Feminist scholars agree with West and conflict theorists that much of society is characterized by tension and struggle between groups.

Feminist theories go a step further because they examine women's social, economic, and political inequality. Feminist theorists maintain that women often suffer injustice primarily because of their gender rather than because of personal inadequacies such as low educational

conflict theory examines how and why groups disagree, struggle over power, and compete for scarce resources.

feminist theories examine women's social, economic, and political inequality.

levels or not caring about success. Feminist scholars assert that people should be treated fairly and equally regardless not only of their sex but also of other characteristics such as their race, ethnicity, national origin, age, religion, class, sexual orientation, or disability. They emphasize the importance of freeing women from traditionally oppressive expectations, constraints, roles, and behavior (see Reger, 2012).

FOCUSING ON GENDER

Many feminist scholars contend that women have historically been excluded from most sociological analyses (Smith, 1987). Before the 1960s women's movement in the United States, very few sociologists published anything about gender roles, women's sexuality, fathers, or intimate partner violence. According to sociologist Myra Ferree (2005: B10), during the 1970s, "the Harvard social-science library could fit all its books on gender inequalities onto a single half-shelf." Since then, because of feminist scholars, many researchers—both women and men—now routinely include gender as an important research variable on both micro and macro levels.

Globally, except for some predominantly Muslim countries, solid majorities of both women and men support gender equality and agree that women should be able to work outside the home. When jobs are scarce, however, many women and men believe that men should be given preferential treatment (see *Figure 1.2*). Thus, even equal rights proponents place a higher priority on men's economic rights.

LISTENING TO MANY VOICES

Feminist scholars contend that gender inequality is central to *all* behavior, ranging from everyday interactions to political and economic institutions, but feminist theories encompass many perspectives. For example, *liberal feminism* emphasizes social and legal reform to create equal opportunities for women. *Radical feminism* sees male dominance in social institutions (such as the economy and

> "I myself have never been able to find out precisely what feminism is; I only know that people call me a feminist whenever I express sentiments that *differentiate me from a doormat.*"
>
> **Rebecca West, British journalist**

politics) as the major cause of women's inequality. *Global feminism* focuses on how the intersection of gender with race, social class, and colonization has exploited women in the developing world (see Lengermann and Niebrugge-Brantley, 1992). Most of us are feminists because we endorse equal opportunities for women and men in the economy, politics, education, and other institutions.

CRITICAL EVALUATION

Feminist scholars have challenged employment discrimination, particularly practices that routinely exclude women who aren't part of the "old boy network" (Wenneras and Wold, 1997). One criticism, however, is that many feminists are part of an "old girl network" that hasn't always welcomed different points of view from black, Asian American, American Indian, Muslim, Latina, lesbian, working-class, and disabled women (Lynn and Todoroff, 1995; Jackson, 1998; Sánchez, 2013).

A second criticism is that feminist perspectives tend to downplay social class inequality by focusing on low-income and minority women but not on their male counterparts. Thus, some contend, feminist theories aren't as gender balanced as they claim. Some critics, including feminists, also question whether feminist scholars have lost their bearings by focusing on personal issues such as greater sexual freedom rather than broader social issues, particularly wage inequality (Rowe-Finkbeiner, 2004; Chesler, 2006; Shteir, 2013).

1-5d Symbolic Interaction

Symbolic interaction theory (sometimes called *interactionism*) is a micro-level perspective that examines individuals' everyday behavior through the communication of knowledge, ideas, beliefs, and attitudes. Whereas functionalists, conflict theorists, and some feminist theories emphasize structures and large (macro) systems, symbolic interactionists focus on *process* and keep the *person* at the center of their analysis.

There have been many influential symbolic interactionists whom we'll cover in later chapters. In brief, George Herbert Mead's (1863–1931) proposal that the human mind and self arise in the process of social communication became the foundation of the symbolic interaction schools of thought in sociology and social psychology.

symbolic interaction theory (*interactionism*) examines individuals' everyday behavior through the communication of knowledge, ideas, beliefs, and attitudes.

FIGURE 1.2 WOMEN SHOULD HAVE THE SAME RIGHTS AS MEN. WELL, MAYBE NOT ALWAYS...

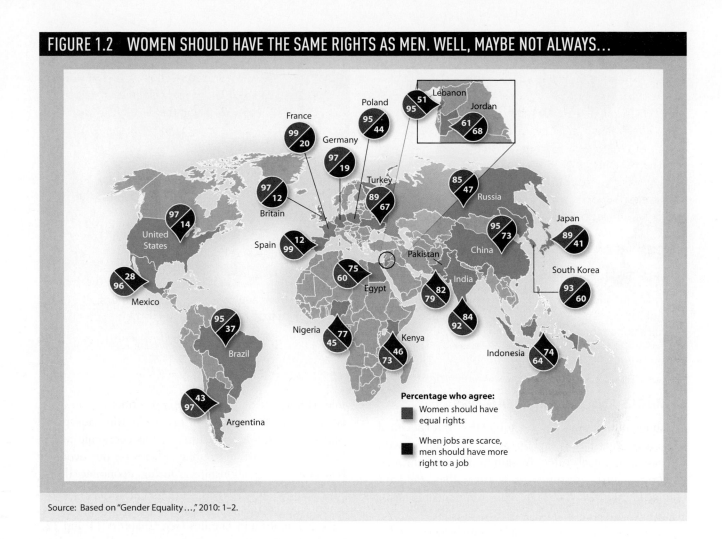

Source: Based on "Gender Equality...," 2010: 1–2.

Herbert Blumer (1900–1987) coined the term *symbolic interactionism* in 1937, developed Mead's ideas, and emphasized that people interpret or "define" each other's actions, especially through symbols, instead of merely reacting to them.

Erving Goffman (1922–1982) contributed significantly to these earlier theories by examining human interaction in everyday situations ranging from jobs to funerals. Among his other contributions, Goffman used "dramaturgical analysis" to compare everyday social interaction to a theatrical presentation (see Chapter 5).

> "Sometimes the best man for the job isn't."
> Author Unknown

CONSTRUCTING MEANING

Our actions are based on **social interaction** in the sense that people take each other into account in their own behavior. Thus, we act differently in different social settings and continuously adjust our behavior, including our body language, as we interact (Goffman, 1959; Blumer, 1969). A woman's interactions with her husband are different from those with her children. And she will interact still differently when she is teaching, talking to a colleague in the hall, or addressing an audience of colleagues at a professional conference.

For symbolic interactionists, society is *socially constructed* through human interpretation (O'Brien and Kollock, 2001). That is, meanings aren't inherent but are created and modified through interaction with others. For example, a daughter who has batting practice with her dad will probably interpret her father's behavior as loving and involved. In contrast, she'll see batting practice with her baseball coach as less personal and more goal-oriented. In this sense, our interpretations of even the same behavior, such as batting practice, vary across situations and depend on the people with whom we interact.

social interaction a process in which people take each other into account in their own behavior.

For many people, a diamond, especially in an engagement ring, signifies love and commitment. For others, diamonds represent Western exploitation of poor people in Africa who are paid next to nothing for their backbreaking labor in mining these stones.

SYMBOLS AND SHARED MEANINGS

Symbolic interaction looks at subjective, interpersonal meanings and at the ways in which we interact with and influence each other by communicating through *symbols*—words, gestures, or pictures that stand for something and that can have different meanings for different individuals.

After the 9/11 terrorist attacks, many Americans displayed the flag on buildings, bridges, homes, and cars to show their solidarity and pride in the United States. In contrast, some groups in the Middle East burned the U.S. flag to show their contempt for U.S. culture and policies. Thus, symbols are powerful forms of communication that show how people feel and interpret a situation.

To interact effectively, our symbols must have *shared meanings,* or agreed-on definitions. One of the most important of these shared meanings is the *definition of the situation,* or the way we perceive reality and react to it. Relationships often end, for example, because partners view emotional closeness differently ("We broke up because Tom wanted sex. I wanted more communication."). We typically learn our definitions of the situation through interaction with *significant others*—especially parents, friends, relatives, and teachers—who play an important role in our socialization (as you'll see in Chapters 4 and 5).

CRITICAL EVALUATION

Unlike other theorists, symbolic interactionists show how people play an active role in shaping their lives on a micro level. One of the most common criticisms, however, is that symbolic interaction overlooks the widespread impact of macro-level factors such as economic forces, social movements, and public policies on our everyday behavior and relationships. During economic downturns, for example, unemployment and ensuing financial problems create considerable interpersonal conflict among couples and families (see Chapters 11 and 12). Symbolic interaction rarely considers such macro-level changes in explaining everyday behavior.

A related criticism is that interactionists sometimes have an optimistic and unrealistic view of people's everyday choices. Most of us enjoy little flexibility in our daily lives because deeply embedded social arrangements and practices benefit those in power. For example, people are usually powerless when corporations transfer many jobs overseas or cut the pension funds of retired employees.

Some also believe that interaction theory is flawed because it ignores the irrational and unconscious aspects of human behavior (LaRossa and Reitzes, 1993). People don't always consider the meaning of their actions or behave as reflectively as interactionists assume. Instead, we often act impulsively or say hurtful things without weighing the consequences of our actions or words.

1-5e Other Theoretical Approaches

Table 1.2 summarizes the major sociological perspectives that you've just read about. However, new theoretical perspectives arise because society is always changing.

TABLE 1.2 LEADING CONTEMPORARY PERSPECTIVES IN SOCIOLOGY

THEORETICAL PERSPECTIVE	FUNCTIONALISM	CONFLICT	FEMINIST	SYMBOLIC INTERACTION
Level of Analysis	**Macro**	**Macro**	**Macro and Micro**	**Micro**
Key Points	• Society is composed of interrelated, mutually dependent parts. • Structures and functions maintain a society's or group's stability, cohesion, and continuity. • Dysfunctional activities that threaten a society's or group's survival are controlled or eliminated.	• Life is a continuous struggle between the haves and the have-nots. • People compete for limited resources that are controlled by a small number of powerful groups. • Society is based on inequality in terms of ethnicity, race, social class, and gender.	• Women experience widespread inequality in society because, as a group, they have little power. • Gender ethnicity, race, age, sexual orientation, and social class—rather than a person's intelligence and ability—explain many of our social interactions and lack of access to resources. • Social change is possible only if we change our institutional structures and our day-to-day interactions.	• People act on the basis of the meaning they attribute to others. Meaning grows out of the social interaction that we have with others. • People continuously reinterpret and reevaluate their knowledge and information in their everyday encounters.
Key Questions	• What holds society together? How does it work? • What is the structure of society? • What functions does society perform? • How do structures and functions contribute to social stability?	• How are resources distributed in a society? • Who benefits when resources are limited? Who loses? • How do those in power protect their privileges? • When does conflict lead to social change?	• Do men and women experience social situations in the same way? • How does our everyday behavior reflect our gender social class, age, race, ethnicity, sexual orientation, and other factors? • How do macro structures (such as the economy and the political system) shape our opportunities? • How can we change current structures through social activism?	• How does social interaction influence our behavior? • How do social interactions change across situations and between people? • Why does our behavior change because of our beliefs, attitudes, values, and roles? • How is "right" and "wrong" behavior defined, interpreted, reinforced, or discouraged?
Example	• A college education increases one's job opportunities and income.	• Most low-income families can't afford to pay for a college education.	• Gender affects decisions about a major and which college to attend.	• College students succeed or fail based on their degree of academic engagement.

For example, *postmodern theory* attempts to explain social life in contemporary societies that are characterized by postindustrialization, consumerism, and global communications.

Sociology, like other social sciences, has subfields. These subfields—such as socialization, deviance, and social stratification—offer specific theories that reinforce and illustrate functionalist, conflict, feminist, and interactionist approaches. No single theory explains social life completely. Each theory, however, provides different insights that guide sociological research, the topic of the next chapter.

STUDY TOOLS 1

READY TO STUDY? IN THE BOOK, YOU CAN:

☐ Check your understanding of what you've read with the Test Your Learning Questions provided on the chapter review card at the back of the book.

☐ Rip out the chapter review card for a handy summary of the chapter and key terms.

ONLINE AT CENGAGEBRAIN.COM YOU CAN:

☐ Prepare for tests with quizzes.

☐ Review the key terms with Flash Cards.

☐ Play games to master concepts.

© Anelina/Shutterstock.com

2 | Examining Our Social World

After you finish
this chapter go to
PAGE 37 for
STUDY TOOLS.

LEARNING OBJECTIVES

- **2-1** Compare knowledge based on tradition, authority, and research methods.
- **2-2** Explain why sociological research is important in our everyday lives.
- **2-3** Describe the scientific method.
- **2-4** Describe the basic steps of the research process.
- **2-5** Compare and illustrate the five most common sociological data collection methods, including their strengths and limitations.
- **2-6** Explain why ethics are important in scientific research.

Spring break is all about beer fests, wet T-shirt contests, frolicking on the beach, and hooking up, right? Maybe not. A national survey found that 70 percent of college students stay home with their parents, and 84 percent of those who throng to vacation spots report consuming alcohol in moderation (The Nielsen Company, 2008). If you suspect that these numbers are too high or too low and wonder how the survey was done, you're thinking like a researcher, the focus of this chapter.

2-1 HOW DO WE KNOW WHAT WE KNOW?

Much of our knowledge is based on *tradition*, a handing down of statements, beliefs, and customs from generation to generation ("The groom's parents should pay for the wedding rehearsal dinner"). Another common source of knowledge is *authority*, a socially accepted source of information that includes "experts," parents, government officials, police, judges, and religious leaders ("My mom says that..." or "According to the American Heart Association...").

Knowledge based on tradition and authority simplifies our lives because it provides us with basic rules about socially and legally acceptable behavior.

What do you think?

People can find data to support any opinion they have.

1	2	3	4	5	6	7

strongly agree   strongly disagree

However, the information can be misleading or wrong. Suppose a 2-year-old throws a temper tantrum at a family barbecue. One adult comments, "What that kid

needs is a smack on the behind." Another person immediately disagrees: "All kids go through this stage. Just ignore it."

Who's right? To answer this and other questions, sociologists rely on **research methods**, organized and systematic procedures to gain knowledge about a particular topic. Much research shows, for example, that neither ignoring a problem nor inflicting physical punishment (such as spanking) stops a toddler's bad behavior. Instead, most young children's misbehavior can be curbed by having simple rules, being consistent in disciplining misbehavior, praising good behavior, and setting a good example (see Benokraitis, 2014).

2-2

WHY IS SOCIOLOGICAL RESEARCH IMPORTANT IN OUR EVERYDAY LIVES?

In contrast to knowledge based on tradition and authority, sociological research is important in our everyday lives for several reasons:

1. **It challenges overgeneralizations.** A common reason for the disconnection between reality and perception is

overgeneralization, drawing conclusions about behavior or events based on limited observations. U.S. gun homicide rates dropped by 49 percent from 1993 to 2011, but 56 percent of Americans believe that the rate has gone up. This misperception—fueled by mass shootings, the media's focus on crime, and powerful lobby groups such as the National Rifle Association—is partly responsible for the growth of gun ownership in the past 20 years (Cohn et al., 2013). In effect, then, overgeneralization can change behavior.

2. **It exposes myths.** According to many newspapers and television shows, suicide rates are highest during the end of the Christmas holidays. In fact, suicide rates are lowest in December and highest in the spring and fall (but the reasons for these peaks are unclear). Another myth is that more women are victims of domestic violence on Super Bowl Sunday than on any other day of the year, presumably because men become intoxicated and abusive. In fact, intimate partner violence is common throughout the year and doesn't spike on Super Bowl Sunday (Mikkelson and Mikkelson, 2005; Romer, 2011).

research methods organized and systematic procedures to gain knowledge about a particular topic.

"Facebook Causes 20 Percent of Today's Divorces." What?!

The founder and self-described leader of the United Kingdom's online divorce site (Divorce-Online) sent out a press release titled "Facebook Is Bad for Your Marriage—Research Finds," and claimed that Facebook causes 20 percent of today's divorces. News media around the world ran stories about this press release with headlines such as "Facebook to Blame for Divorce Boom." You'll see in Chapter 12 that there's no divorce boom, so where did the 20 percent number come from?

In 2009, the managing director of Divorce-Online scanned its online divorce petition database for use of the word *Facebook*, and found 989 instances in about 5,000 petitions. Divorce-Online never said that the petitions were only those filed by members of the American Academy of Matrimonial Lawyers, who comprise a very small percentage of all divorce attorneys. Two years later, many Internet sites and blogs were still spreading the fiction that "Facebook Causes Divorce" (see Bialik, 2011). In reality, as you'll see in Chapter 12, there are a number of interrelated macro- and micro-level reasons for divorce; there is no single "cause," much less a website.

3. **It helps explain *why* people behave as they do.** A recent study predicted that older drivers, particularly those age 70 and older, would be more likely than younger drivers to have fatal crashes, but found the opposite (Braitman et al., 2011). The researchers couldn't explain why their prediction turned out to be false. Sociologists posit that older Americans are less likely to have fatal crashes because many avoid driving at night or during bad weather, and they are much less likely than younger drivers to use cell phones or text while driving (Halsey, 2010).

4. **It influences social policies.** In the past, many psychologists attributed school bullying to students with emotional and behavioral problems. Recently, however, sociologists have found that bullying is due largely to jockeying for social status rather than to mental or emotional problems (Faris and Felmlee, 2011). That is, the most aggressive bullies are trying to prevent their rivals from becoming members of a higher-ranking group. Because of these findings, some educators have enlisted the support of students at the top of the high school social ladder to prevent or decrease bullying (Parker-Pope, 2011).

5. **It sharpens critical thinking skills that affect our everyday lives.** Many Americans, especially women, rely on talk shows for information on a number of topics. During 2009 alone, Oprah Winfrey featured and applauded guests who maintained, among other things, that children contract autism from the measles, mumps, and rubella (MMR) vaccinations they receive as babies; that fortune cards can help people diagnose their illnesses; and that people can wish away cancer (Kosova and Wingert, 2009)—*all of these claims are false*.

Such misinformation can be dangerous. Because of the "MMR vaccinations can cause autism" scare, about 30 percent of U.S. parents are hesitant to vaccinate their children (Kennedy et al., 2011). Partly because of such fears, by mid-2014, the United States was experiencing the largest increase in measles cases since 1996. One or two of every 1,000 cases of measles are fatal (Gastañaduy et al., 2014; see also Gibson, 2012, and "Clueless," 2014, for other recent examples of "celebrity bogus science").

2-3 THE SCIENTIFIC METHOD

To explain behavior, sociologists rely on the **scientific method**, a body of objective and systematic techniques for investigating phenomena, acquiring knowledge, and testing hypotheses and theories. The techniques include careful data collection, exact measurement, accurate recording and analysis of the findings, thoughtful interpretation of results, and, when appropriate, generalization of the findings to a larger group. Before collecting any data, however, social scientists must grapple with a number of research-related issues. Let's begin with concepts, variables, and hypotheses.

2-3a Concepts, Variables, and Hypotheses

A basic element of the scientific method is a **concept**—an abstract idea, mental image, or general notion that represents some aspect of the world. Some examples

scientific method a body of objective and systematic techniques for investigating phenomena, acquiring knowledge, and testing hypotheses and theories.

concept an abstract idea, mental image, or general notion that represents some aspect of the world.

of concepts are "blood pressure," "climate change," and "marriage."

Because concepts are abstract and may vary among individuals and cultures, scientists rely on variables to measure (*operationalize*) concepts. A **variable** is a characteristic that can change in value or magnitude under different conditions. Variables can be attitudes, behaviors, or traits such as ethnicity, age, and social class.

An **independent variable** is a characteristic that has an effect on the **dependent variable**, the outcome. A **control variable** is a characteristic that is constant and unchanged during the research process.

Scientists can simply ask a research question ("Why are people poor?"), but they usually begin with a **hypothesis**—a statement of the expected relationship between two or more variables—such as "Unemployment increases poverty." In this example, "unemployment" is the independent variable and "poverty" is the dependent variable.

Researchers might also use control variables, such as education, to explain the relationship between unemployment and poverty. For example, people with at least a college degree generally have lower poverty rates than those with lower educational levels because the former are less likely to experience long periods of unemployment.

2-3b Deductive and Inductive Reasoning

Deduction and induction are two different but equally valuable approaches in examining the relationship between variables. Generally, **deductive reasoning** begins with a theory, prediction, or general principle that is then tested through data collection. An alternative mode of inquiry, **inductive reasoning**, begins with specific observations, followed by data collection, a conclusion about patterns or regularities, and the formulation of hypotheses that can lead to theory construction (see *Figure 2.1*).

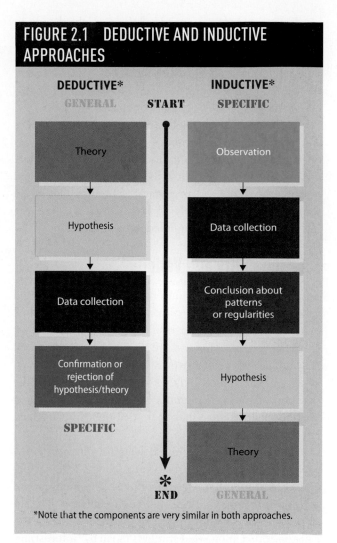

FIGURE 2.1 DEDUCTIVE AND INDUCTIVE APPROACHES

*Note that the components are very similar in both approaches.

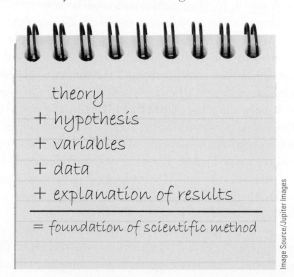

theory
+ hypothesis
+ variables
+ data
+ explanation of results
_____
= foundation of scientific method

Image Source/Jupiter Images

variable a characteristic that can change in value or magnitude under different conditions.

independent variable a characteristic that has an effect on the dependent variable.

dependent variable the outcome that may be affected by the independent variable.

control variable a characteristic that is constant and unchanged during the research process.

hypothesis a statement of the expected relationship between two or more variables.

deductive reasoning an inquiry process that begins with a theory, prediction, or general principle that is then tested through data collection.

inductive reasoning an inquiry process that begins with a specific observation, followed by data collection, a conclusion about patterns or regularities, and the formulation of hypotheses that can lead to theory construction.

Taking a deductive approach, you might decide to test a theory of academic success using the following hypothesis: "Students who study in groups perform better on exams than those who study alone." You would collect the data, ultimately confirming or rejecting your hypothesis (or theory).

Alternatively, you might notice that your classmates who participate in study groups seem to get higher grades on exams than those who study alone. Using an inductive approach, you would collect the data systematically and formulate hypotheses (or suggest a theory) that could then be tested deductively. Most social science research involves both inductive and deductive reasoning.

2-3c Reliability and Validity

Sociologists are always concerned about reliability and validity. **Reliability** is the *consistency* with which the same measure produces similar results time after time. If, for example, you ask "How old are you?" on two subsequent days and a respondent gives two different answers, such as 25 and 30, there's either something wrong with the question or the respondent is lying. Respondents might lie, but scientists must make sure that their measures are as reliable as possible.

Validity is the degree to which a measure is accurate and *really* measures what it claims to measure. Consider student course evaluations. The measures of a "good" professor often include items such as "The instructor is interesting" and "The instructor is fair." Because we don't know what students mean by "interesting" and "fair," how accurate are such measures in differentiating between "good" and "bad" professors? A study at two large public universities found that a third of the students admitted being dishonest in end-of-semester course evaluations. Some fibbed to make their instructors look good, but most lied to "punish" professors they

Are *American Idol* voters an example of a probability or nonprobability sample of the show's fans?

didn't like, especially when they received lower grades than they thought they deserved (Clayson and Haley, 2011). Such research findings raise questions about the accuracy and usefulness of student course evaluations in measuring an instructor's actual performance and, consequently, an administrator's using course evaluations to make decisions about merit pay and promotions.

2-3d Sampling

Early in the research process, sociologists decide what sampling procedures to use. Ideally, researchers would like to study all the units of the population in which they're interested—say, all adolescents who use drugs. A **population** is any well-defined group of people (or things) about which researchers want to know something. Obtaining information about and from populations is problematic, however. The population may be so large that it would be too expensive and time consuming to conduct the research. In other cases—such as all adolescents who use drugs—it's impossible even to identify the population.

As a result, researchers typically select a **sample**, a group of people (or things) that is representative of the population they wish to study. In obtaining a sample, researchers decide whether to use probability or nonprobability sampling. A **probability sample** is one in which each person (or thing, such as an email address) has an equal chance of being selected because the selection process is *random*. The most desirable characteristic of a probability sample is that the results can be generalized to the larger population because all the people (or things) have had an equal chance of being selected.

In a **nonprobability sample**, there's little or no attempt to get a representative cross section of the population. Instead, researchers use sampling criteria

reliability the consistency with which the same measure produces similar results time after time.

validity the degree to which a measure is accurate and really measures what it claims to measure.

population any well-defined group of people (or things) about whom researchers want to know something.

sample a group of people (or things) that is representative of the population researchers wish to study.

probability sample each person (or thing) has an equal chance of being selected because the selection is random.

nonprobability sample there is little or no attempt to get a representative cross section of the population.

such as convenience or the availability of respondents or information. Nonprobability samples are very useful when sociologists are exploring a new topic or want to get insights on how people feel about a particular topic before launching a larger study (Babbie, 2013).

Television news programs, newsmagazines, and entertainment shows often provide a toll-free number, a texting number, or a Web address and encourage viewers to vote on an issue (such as whether marijuana should be legal in all states). How representative are these voters of the general population? And how many enthusiasts skew the results by voting more than once?

But, you might think, if as many as 100,000 people respond, doesn't such a large number indicate what most people think? No. Because the respondents are self-selected and don't comprise a random sample, they're not representative of a population.

2-3e The Time Dimension

Researchers have two principal options to deal with the time issue: cross-sectional studies and longitudinal studies. The data can be *longitudinal* (collected at two or more points in time from the same or different samples of respondents) or *cross-sectional* (collected at one point in time). *Figure 2.2* shows a change over time in Americans' attitudes toward homosexuality; this is an example of a longitudinal study. If the researchers had collected data at only one point in time (2014, 2000, or 1995), this would have been a cross-sectional study. Cross-sectional studies provide valuable information, but longitudinal studies are especially useful in examining trends in behavior or attitudes; a researcher can compare similar populations across different years or follow a particular group of people over time.

2-3f Qualitative and Quantitative Approaches

In **qualitative research**, sociologists examine nonnumerical material that they then interpret. In a study of grandfathers who were raising their grandchildren, for example, the researcher tape-recorded in-depth interviews and then analyzed the responses to questions about financial issues and daily parenting tasks (Bullock, 2005).

In **quantitative research**, sociologists focus on a numerical analysis of people's responses or specific characteristics, studying a wide range of attitudes, behaviors, and traits (such as homeowners versus renters).

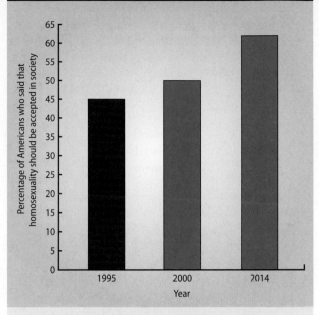

FIGURE 2.2 ACCEPTANCE OF HOMOSEXULAITY HAS INCREASED

Source: Based on Kohut et al., 2011: 133, and Dimock et al., 2014.

In one national probability study, for example, the researchers surveyed almost 7,000 respondents to understand the influence of grandparents who live with their children and grandchildren (Dunifon and Kowaleski-Jones, 2007).

Which approach should a researcher use? It depends on her or his purpose. Consider college attrition. Quantitative data provide information on characteristics such as national college graduation rates. Qualitative data, in contrast, yield in-depth descriptions of why some college students drop out whereas others graduate. In many studies, sociologists use both approaches.

2-3g Correlation Is Not Causation

Ideally, researchers would like to determine **causation**, a relationship in which one variable is the direct consequence of another. For example, the more groceries you

qualitative research examines nonnumerical material and interprets it.

quantitative research focuses on a numerical analysis of people's responses or specific characteristics.

causation a relationship in which one variable is the direct consequence of another.

buy, the more money you spend (buying groceries causes you to spend money). Most human behavior is much more complex. Because causation is difficult to prove and nothing in life (except death) is certain, sociologists and other scientists try to establish **correlation**, the relationship between two or more variables. For example, much research shows that there's an association between alcohol abuse and domestic violence. Moreover, the more frequent the alcohol abuse, the greater the likelihood of domestic violence (see Chapter 12).

Alcohol abuse and domestic violence often occur together, but this doesn't mean that one *causes* the other. Domestic violence also occurs when people don't drink, and not all people who abuse alcohol become violent or aggressive towards others. Instead, there may be other factors that affect alcohol abuse, domestic violence, or both variables (see *Figure 2.3*). As a result, sociologists rarely use the term *cause* when interpreting their results because they can't *prove* that there's a cause-and-effect relationship. Instead, a researcher might conclude that alcohol abuse is "associated (or correlated) with," "contributes to," or "increases the likelihood of" rather than "causes" domestic violence.

2-4 THE RESEARCH PROCESS: THE BASICS

Hypotheses construction, deductive and inductive reasoning, establishing reliability and validity, and sampling are some of the preliminary and often most challenging steps in the research process. *Figure 2.4* outlines the scientific

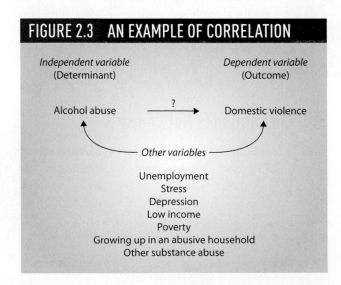

FIGURE 2.3 AN EXAMPLE OF CORRELATION

correlation the relationship between two or more variables.

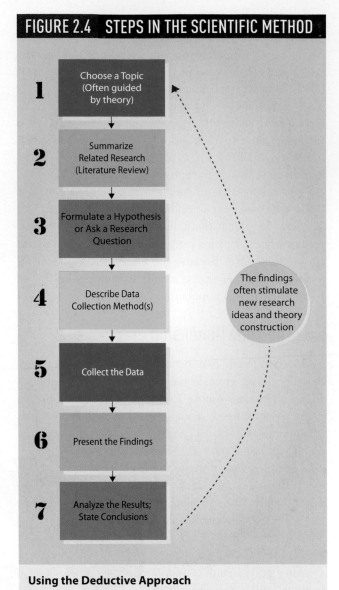

FIGURE 2.4 STEPS IN THE SCIENTIFIC METHOD

1 Choose a Topic (Often guided by theory)

2 Summarize Related Research (Literature Review)

3 Formulate a Hypothesis or Ask a Research Question

4 Describe Data Collection Method(s)

5 Collect the Data

6 Present the Findings

7 Analyze the Results; State Conclusions

The findings often stimulate new research ideas and theory construction

Using the Deductive Approach

method, using a deductive approach that begins with an idea and ends with writing up (and sometimes publishing) the results. Later in this chapter, we'll examine some studies that use an inductive approach.

1. **Choose a topic to study.** The topic can be general or very specific. Some sociologists begin with a new question or idea; others extend or refine previous research findings. A topic can generate new information, replicate a previous study, or propose an intervention (such as implementing a new substance abuse program).

2. **Summarize the related research.** In what is often called a *literature review*, a sociologist summarizes the pertinent research, shows how her or his topic is related to previous and current research, and indicates how the study will extend the body of knowledge. If the research is applied, a sociologist also explains how the proposed service or program will improve people's lives.

3. **Formulate a hypothesis or ask a research question.** A sociologist next states a hypothesis or asks a research question. In either case, she or he has to be sure that the measures of the variables are as reliable and valid as possible.

4. **Describe the data collection method(s).** A sociologist describes which method or combination of methods (sometimes called *methodology, procedure,* or *research design*) is best for testing the hypothesis or answering the research question. This step also describes sampling, the sample size, and the respondents' characteristics.

5. **Collect the data.** The actual data collection might rely on fieldwork, surveys, experiments, or existing sources of information such as Census Bureau data.

6. **Present the findings.** After coding (tabulating the results) and running statistical tests, a sociologist presents the findings as clearly as possible.

7. **Analyze and explain the results.** After analyzing the data, a sociologist explains why the findings are important. This can be done in many ways. The researcher might show how the results provide new information, enrich our understanding of behavior or attitudes that researchers have examined previously, or refine existing theories or research approaches.

In drawing conclusions about the study, sociologists typically discuss the study's implications. For instance, does a study of juvenile arrests suggest that new policies should be implemented, that existing ones should be changed, or that current police practices may be affecting the arrest rates? That is, the researcher answers the question "So what?" by showing the importance and usefulness of the study.

2-5 SOME MAJOR DATA COLLECTION METHODS

During the research process, sociologists typically use one or more of the following major data collection methods: surveys, field research, content analysis, experiments, and secondary analysis of existing data. Because each method has strengths and weaknesses, researchers must decide which will provide the most accurate information, given time and budget constraints.

"How long have you been dead? Do you have any complaints about your treatment here? Do you have any suggestions?"

Farris, Joseph/CSL, CartoonStock Ltd

2-5a Surveys

Many sociologists use **surveys** that include questionnaires, face-to-face or telephone interviews, or a combination of these techniques. Two important elements in survey research are sampling techniques and constructing a series of questions for *respondents*, the people who answer the questions.

SELECTING A SAMPLE

Random sample surveys are preferred because the results can be generalized to a larger population. Researchers can obtain representative samples through *random digit dialing*, which involves selecting area codes and exchanges (the next three numbers) followed by four random digits. In the procedure called *computer-assisted telephone interviewing* (CATI), the interviewer uses a computer to select random telephone numbers, reads the questions to the respondent from a computer screen, and then enters the replies in precoded spaces, saving time and expense by not having to reenter the data after the interview.

Two types of electronic surveys are becoming increasingly popular. The first type is a survey sent via email, either as text in the body of the message or as an attachment. The second type is the more familiar survey that is posted on a website. Electronic surveys are cost effective and can represent a broad spectrum of the general population. In addition, Web surveys provide

survey a method for collecting that data that includes questionnaires, face-to-face or telephone interviews, or a combination.

respondents with visual material, including videos, to look at and respond to (Keeter, 2009, 2010).

QUESTIONNAIRES AND INTERVIEWS

Beyond selecting subjects, a survey must have a specific plan for asking questions and recording answers. The most common way to do this is to give respondents a **questionnaire**, a series of written questions that ask for information. The questions can be *closed-ended* (the researcher provides a list of answers that a respondent chooses), *open-ended* (the researcher asks respondents to answer questions in their own words), or a combination (see *Table 2.1*). Questionnaires can be mailed, used during an interview, or self-administered (such as student course evaluations).

The **interview**, in which a researcher directly asks respondents a series of questions, is another way to collect survey data. Interviews can be conducted face-to-face or by telephone. *Structured interviews* use closed-ended questions, whereas *unstructured interviews* use open-ended questions that allow respondents to answer as they wish.

STRENGTHS

Surveys are usually inexpensive, simple to administer, and have a fast turnaround. Because the results are anonymous, respondents are generally willing to answer questions on sensitive topics such as income, sexual behavior, and drug usage (Hamby and Finkelhor, 2001).

Researchers often simplify surveys to increase response rates. During the 2010 census, for example, the Census Bureau used only a short form and slogans such as "10 Questions, 10 Minutes" to encourage people to mail back the forms. And, for the first time, the Census Bureau used six different languages for questionnaires, and hired staff who spoke over 150 different languages. Doing so generated higher response rates, reduced the need for expensive follow-up interviews, and saved millions of dollars (Groves and Vitrano, 2011; Mather et al., 2011; Wozniak and Groves, 2011).

questionnaire a series of written questions that ask for information.

interview a researcher directly asks respondents a series of questions.

TABLE 2.1 EXAMPLE OF OPEN AND CLOSED-ENDED QUESTION

Open-Ended Question
How would you describe your current financial situation?

Closed-Ended Question
How would you describe your current financial situation?
[ ] Excellent
[ ] Very good
[ ] Good
[ ] Fair
[ ] Poor
[ ] Terrible
[ ] Not sure

Telephone interviews are popular because they're a relatively inexpensive way to collect data. Face-to-face interviews have high response rates (often up to 85 percent) because they involve personal contact. People are more likely to discuss sensitive issues in an interview than via a mailed questionnaire, a phone survey, or electronic survey. If respondents don't understand a question, are reluctant to answer, or give incomplete answers, interviewers can clarify, keep respondents from digressing, or *probe* by asking respondents to elaborate on an answer (Babbie, 2013).

With the innovation of "robo-polls," the entire interview is conducted by a programmed recording that interprets the respondent's spoken answers, records them, and determines how to continue the interview. This method is cost effective because it cuts out the cost of hiring people, but respondents may be more reluctant to answer sensitive questions (Babbie, 2013).

LIMITATIONS

A major limitation of surveys that use mailed questionnaires is a low response rate, often only about 10 percent (Gray et al., 2007). If the questions are unclear, complicated, or seen as offensive, a respondent may simply throw the questionnaire away or give answers on subjects they know nothing about (Babbie, 2013).

A survey's wording can also substantially affect people's answers. According to a recent Gallup poll, for instance, 51 percent of Americans agreed that doctors should be allowed to "assist the patient to commit suicide," but a whopping 70 percent said that doctors should be allowed to "end the patient's life by some painless means." Thus, not using the word *suicide* got much more support even though the patient's outcome was the same (Saad, 2013).

Another concern is a *social desirability bias*—the tendency of respondents to give the answer that they think they "should" give or that will cast them in a favorable light. For example, the proportion of Americans who say they voted in a given election is always much higher than the actual number of votes cast (Taylor and Lopez, 2013). Respondents also underreport (or lie about) behaviors perceived negatively (such as using illicit drugs or having multiple sexual partners), but exaggerate attendance at religious services, watching the evening news, and washing their hands after using a restroom. In a recent survey, for example, 28 percent of the respondents admitted that they had lied to their doctors about smoking, drinking alcohol, using illicit drugs, not exercising, having unsafe sex with multiple partners, and other risky behavior (Prior, 2009; Zezima, 2010; Reddy, 2013).

Spurred by the rising cost of telephone surveys and plummeting participation rates, *The New York Times* and CBS News recently announced that they'll use online-only surveys for their public-opinion polling. The announcement drew sharply divided views. Opponents maintain that because the surveys are based on non-probability samples, the respondents are self-selected, and people who don't use the Internet tend to be older, poorer, and less educated than Internet users. Thus, the results of online surveys aren't accurate and nationally representative. Proponents argue, on the other hand, that the online surveys will increase the pool of people who are increasingly difficult to reach. Moreover, the potential bias from excluding non-Internet users is getting smaller—89 percent of U.S. adults were online in mid-2014, and the numbers are steadily increasing (DeSilver, 2014b; Voosen, 2014a). Future research will determine whether online surveys are as accurate as traditional telephone surveys.

Unlike questionnaires and telephone surveys, face-to-face interviews can be very expensive. They can also be dangerous. During the 2010 census, for example, a number of Americans, frustrated with the economy and government, took their anger out on census takers. Hundreds of census workers were shot at with pellet guns; hit by baseball bats; spat at; and confronted with shotguns, packs of snarling pit bulls, pickaxes, crossbows, and hammers (Morello, 2010).

Because the survey is the research approach you'll encounter most often, in sociology and many other disciplines, it's important to be an informed consumer. You can't simply assume that a survey, including a public opinion poll, is accurate or representative of a larger population. Asking a few basic questions about

TABLE 2.2 CAN I TRUST THESE NUMBERS?

To determine a survey's credibility, ask:

- Who sponsored the survey? A government agency, a political organization, a business, or a group that's lobbying for change?

- What is the purpose of the survey? To provide objective information, to promote an idea or a political candidate, or to get attention through sensationalism (Seife, 2010)?

- How was the sample drawn? Randomly? Or was it a "self-selected opinion poll" (a SLOP, according to Tanur, 1994)?

- How were the questions worded? Were they clear and objective, or loaded and biased? If the survey questions are not provided, why not?

- How did the researchers report their findings? Were they objective, or did they make value judgments?

a survey, such as those in *Table 2.2*, will help you evaluate its credibility.

2-5b Field Research

In **field research**, sociologists collect data by systematically observing people in their natural surroundings. In *participant observation*, researchers interact with the people they're studying; they may or may not reveal their identities as researchers. If you recorded interaction patterns between students and professors during your classes, you would be engaging in participant observation. In *nonparticipant observation*, researchers study phenomena without being part of the situation. For example, child psychologists, clinicians, and sociologists often study young children in classrooms through one-way mirrors. Researchers sometimes combine participant and nonparticipant observation.

Some field research studies are short term (e.g., observing, over a few weeks or months, whether and how parents discipline their unruly children in grocery stores). Others, often called *ethnographies*, require a considerable amount of time in the field. For example, Sudhir Venkatesh (2008), while a graduate student at the University of Chicago, spent more than 6 years studying the culture and members of the Black Kings, a crack-selling gang in Chicago's inner city.

Observational studies are usually highly structured and carefully designed projects in which data are recorded and then converted to quantitative summaries. These studies may examine complex communication

field research data collected by observing people in their natural surroundings.

Field researchers study a variety of topics, including how customers, cocktail waitresses, bouncers, and paid consultants often engage in deception, hustling, and bribes to help people hook up with someone at a nightclub (see Grazian, 2008, for a description of this study).

patterns, measure the frequency of acts (e.g., the number of head nods or angry statements), or note the duration of a particular behavior (e.g., the length of eye contact) (Stillars, 1991). Thus, observational studies are much more complex and sophisticated than they appear to the general public.

STRENGTHS

Unlike other data collection methods, field research provides rich detail in describing and understanding attitudes and behavior in the "real world." The researcher can explore new topics from the respondents' viewpoint using observation or in-depth interviews. Because observation usually doesn't disrupt the natural surroundings, the researcher doesn't directly influence the subjects.

Counting the homeless is an ongoing research problem. Communities must provide accurate numbers to qualify for federal funds, but census takers have difficulty getting such counts because safety rules prohibit them from entering private property (such as warehouses) or dark alleys. In one approach, the YWCA in Trenton, New Jersey, gave free haircuts at a fair for the homeless to attract the city's homeless people to a place where they could be counted (Jonsson, 2007). Is such field research unethical because it violates people's privacy? Or is it innovative in helping a community get more federal funding for the homeless?

Field research is also more flexible than some other methods. The researcher can modify the research design, for example, by deciding to interview (rather than just observe) key people after the research has started.

Validity is usually a strength of field research because researchers can determine how participants define and interpret concepts such as "success," "inequality," or "marital infidelity." Field research is also more responsive to changes that occur in the field. Studying gang behavior, for example, may require shifting the focus of the analysis if members are killed or imprisoned (see Venkatesh, 2008). Finally, field research is a powerful data collection method because it describes a phenomenon using the subjects' own words.

LIMITATIONS

Observation can be expensive if a researcher needs elaborate recording equipment, must travel far or often, or has to live in a different society or community for an extended period. Researchers who study other cultures must often learn a new language, a time-consuming task.

A second limitation is that if people know they're being observed, they might behave differently. This is known as the problem of *reactivity*. For example, a study of 392 restaurants from five different chains (such as the Olive Garden) found that employee theft decreased by 22 percent after management installed surveillance cameras (Pierce et al., 2013). In the case of participant observation, anything that the researcher does or doesn't do will affect what's being observed because the researcher's presence changes a group's dynamics, behavior, and outcomes.

A field researcher may encounter other data collection barriers. Homeless and battered women's shelters, for example, are usually—and understandably—wary of researchers intruding on their residents' privacy. Even if the researcher has access to such a group, it's often difficult to be objective while collecting and interpreting the data because domestic violence can evoke strong emotional reactions such as anxiety, anger against perpetrators, and sympathy for victims. Finally, the findings can't be generalized because the data come from nonprobability samples.

2-5c Content Analysis

Content analysis systematically examines some form of communication. Researchers can apply this unobtrusive data collection method to almost any form of written or oral communication: speeches, TV programs, online blogs, advertisements, office emails, songs, diaries, advice columns, poems, or Facebook chatter, to mention just a few.

The researcher develops categories for coding the material, sorts and analyzes the data in terms of frequency, intensity, or other characteristics, and draws conclusions about the results. For example, look at the two greeting cards in *Figure 2.5*. What messages do they send about gender roles? These are only two cards, but we could do a content analysis of all birth announcements at a particular store or from a number of stores or online sites to determine whether such cards reinforce stereotypical gender-role expectations.

Sociologists have used content analysis to examine a number of topics. A few examples are images of women and men in video games and music videos, changes in child-rearing advice in popular parenting magazines, gender and ethnic differences in yearbook photographs, and gender biases in job advertisements (Downs and Smith, 2010; Zhang et al., 2010; Wallis, 2011; Lindner, 2012; Wondergem and Friedlmeier, 2012).

STRENGTHS

Content analysis is usually inexpensive and often less time consuming than other data collection methods, especially field research. If, for example, you wanted to examine the content of television commercials during football games, you wouldn't need fancy equipment, a travel budget, or a research staff.

A second advantage is that researchers can correct coding errors fairly easily by redoing the work. This isn't the case with surveys. If you mail a questionnaire with poorly constructed items, it's too late to change anything.

Third, content analysis is unobtrusive. Because researchers aren't dealing with human subjects, they don't need permission to do the research, and don't have to worry about influencing the respondents' attitudes or behavior.

Finally, researchers can use content analysis to compare societies at single or different points in time. In one study, the researchers analyzed gender portrayals in the 101 top-grossing G-rated films in the United States and Canada between 1990 and 2005 (Smith et al., 2010; see also Joshi et al., 2011). It would be very difficult, using most other data collection methods, to analyze such material over a 15-year period.

LIMITATIONS

Content analysis can be very labor intensive, especially if a project is ambitious. In one study, the researchers examined the amount and intensity of violence in children's animated movies that were released between 1938 and 1999 (Yokota and Thompson, 2000). It took several years to code one or more of the major characters' words, expressions, and actions.

A second disadvantage is that the coding may be subjective. Having several researchers on a project can increase coding objectivity, but only one researcher often codes the content.

FIGURE 2.5 HOW DO VIEWS OF FEMALE AND MALE BABIES DIFFER?

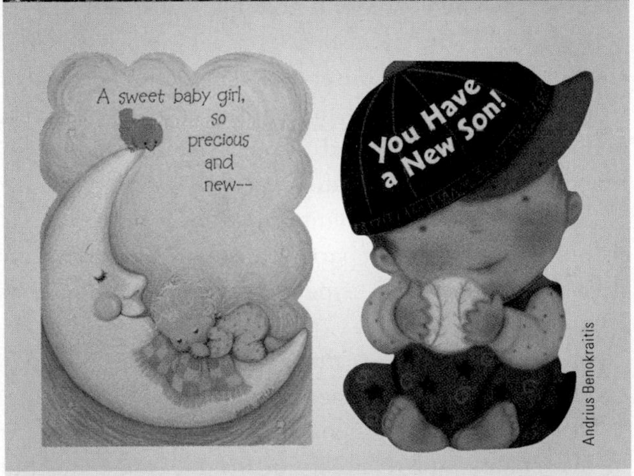

A sweet baby girl, so precious and new—

You Have a New Son!

Andrius Benokraitis

As these cards illustrate, girls and boys are viewed differently from the time they're born. In these and other birth announcements, girls, but not boys, are typically described as "sweet" or "precious." Also, the images usually show boys as active but girls as passive.

content analysis a data collection method that systematically examines some form of communication.

An Experiment That Went Awry

In a well-known 1971 experiment, psychologist Philip Zimbardo created a mock prison, using 24 undergraduate volunteers as prisoners and guards. He stopped the experiment after 6 days because some of the participants experienced intense negative reactions. For more information, go to Stanford Prison Experiment, prisonexp.org. We'll examine this study, including the ethical issues it raised, in more detail in Chapter 6.

AP Images/The Oklahoman/Chris Landsberger

A third limitation is that content analysis often reflects social class bias. Because most books, articles, speeches, films, and so forth are produced by people in upper socioeconomic levels, content analysis rarely captures the behavior or attitudes of working-class people and the poor. Even when documents created by lower-class individuals or groups are available, it's difficult to determine whether the coding reflects a researcher's social class prejudices.

Finally, content analyses can't always tell us *why* people behave as they do. We'd have to turn to the findings of studies that use different data collection methods—such as surveys and field research—to understand why, for example, people buy stereotypically feminine or masculine birth announcements or produce video games that portray men, but not women, as competitive and aggressive (see Lindner, 2012).

2-5d Experiments

An **experiment** is a controlled artificial situation that allows researchers to manipulate variables and measure the effects. The classic experimental design includes two equal-sized groups that are similar on characteristics such as gender age, ethnicity, race, and education.

Suppose our hypothesis is "Watching a film on race discrimination reduces prejudice." In the **experimental group**, the participants are exposed to the independent variable (they watch the film on race discrimination). In the **control group**, participants don't watch the film because they're not exposed to the independent variable. Before the experiment, we'll measure the dependent variable (prejudice) in both groups using a *pretest* (a prejudice scale). After the experimental group watches the film, we'll measure both groups again using a *posttest* (the same prejudice scale). If we find a difference in the scores of the dependent variable (prejudice), we might conclude that the independent variable affects the dependent variable (e.g., watching the film on race discrimination reduced or increased prejudice). *Figure 2.6* illustrates this basic experimental design.

STRENGTHS

Experiments come closer than other data collection methods in suggesting a possible cause-and-effect relationship. They're also usually less expensive and time consuming than other data collection techniques (especially large surveys and multiyear field research), and there's often no need to purchase special equipment. A second and related strength is that because researchers recruit students or other volunteers (as in medical research), participants are usually readily available and don't expect much, if any, monetary compensation.

A third advantage is that experiments can be *replicated* (repeated) many times with different participants. Such replication strengthens the researchers' confidence in the validity and reliability of the measures and the study's results. For example, doctors stopped prescribing hormone pills for menopausal women because better-designed experiments that replicated earlier studies found that the pills increased the risk for breast cancer and strokes (Ioannidis, 2005).

experiment a controlled artificial situation that allows researchers to manipulate variables and measure the effects.

experimental group the participants who are exposed to the independent variable.

control group the participants who aren't exposed to the independent variable.

FIGURE 2.6 BASIC EXPERIMENTAL DESIGN

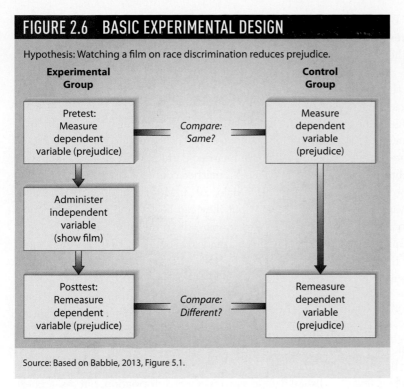

Hypothesis: Watching a film on race discrimination reduces prejudice.

Experimental Group

Pretest: Measure dependent variable (prejudice)

Compare: Same?

↓

Administer independent variable (show film)

↓

Posttest: Remeasure dependent variable (prejudice)

Compare: Different?

Control Group

Measure dependent variable (prejudice)

↓

Remeasure dependent variable (prejudice)

Source: Based on Babbie, 2013, Figure 5.1.

LIMITATIONS

One disadvantage of laboratory experiments is their reliance on student volunteers or paid respondents. Students often feel obligated to participate as part of their grade, or they may fear antagonizing an instructor who's conducting a study. Participants might also give the answers that they think the researcher expects (the social desirability bias described earlier). In the case of paid subjects, those who are the busiest, don't need the extra cash, move, or get sick may not participate fully or drop out of the study.

Another limitation is that experiments, especially those conducted in laboratories, are artificial. People *know* that they're being observed and may behave very differently than they would in a natural setting (Gray et al., 2007; see also Chapter 6 on the Hawthorne effect).

Third, the results of experimental studies can't be generalized to a larger population because they come from small, nonrepresentative, or self-selected samples. For example, college students who participate in experiments aren't necessarily representative of other college students, much less of people who aren't in college.

Fourth, experiments aren't suitable for studying large groups of people, a major focus in sociology. As you saw in Chapter 1, many sociologists examine macro-level issues (such as employment trends) that can't be observed in a laboratory or other experimental setting.

Finally, even if a researcher finds an association between variables (e.g., watching a film reduces prejudice), it doesn't mean that the former "causes" the latter. For example, the control group's responses in the posttest may have been influenced by other factors, such as communicating with the members of the experimental group (Cook and Campbell, 1979) or living in a racially-mixed neighborhood that diminishes some of their prejudices. That is, experiments have few control variables that might be explaining the relationship between the independent and dependent variables.

Because of these and other limitations of experiments, consumers are often confused and frustrated by contradictory findings and advice. For example, different experiments have concluded that caffeine can help relieve headaches *or* lead to chronic headaches, can inhibit memory *or* improve it, and can increase *or* reduce depression (Bartlett, 2011; see also Leaf, 2013).

2-5e Secondary Analysis of Existing Data

Sociologists also rely heavily on **secondary analysis**, an unobtrusive data collection method that examines information collected by someone else. The data may be historical materials (e.g., court proceedings), personal documents (e.g., letters and diaries), public records (e.g., state archives on births, marriages, and deaths), or official statistics (e.g., U.S. Bureau of Labor reports). Researchers with slightly different interests could analyze or reanalyze the data or examine only specific variables. For example, the federal government compiles information on a large number of variables—including population, age, and education—that social scientists, marketing analysts, health practitioners, and other researchers analyze for their own purposes.

© Svetlana Lukienko/Shutterstock.com

STRENGTHS

Secondary analysis is usually accessible, convenient, and inexpensive. A staggering amount of data on topics such as income and employment—for the United States and worldwide—are readily available at libraries, colleges, and online.

secondary analysis examination of data that have been collected by someone else.

Second, there's a huge savings in time. Because many of the data sets are stored electronically, researchers can spend most of their time analyzing, rather than collecting, the data.

A third advantage of secondary analysis is the high quality of the data. The federal government and nationally known survey organizations have budgets and well-trained staff equipped to address any data collection problems. Because the samples are representative of national populations, researchers can be more confident about generalizing the findings.

Fourth, data sets often contain hundreds of variables. Thus, researchers can access national, international, and historical information that they themselves can't collect (Boslough, 2007). Finally, because many of the data sets are longitudinal, researchers can look at trends and changes over time.

LIMITATIONS

Secondary analysis has several drawbacks. First, it may be difficult to gain access to historical materials because the documents are deteriorating, housed in only a few libraries in the country, or part of private collections. Determining the accuracy and authenticity of historical data may also be difficult.

Second, existing sources may not have the exact data a researcher needs. In 2000, the Census Bureau changed how it counted race and ethnicity to include the growing numbers of recent immigrants and mixed-race Americans. Doing so increased the accuracy of gauging racial and ethnic diversity but created 63 categories of possible racial-ethnic combinations. As a result, today's numbers are different from those before 2000, making comparisons on race and ethnicity across time problematic (Saulny, 2011).

A third and related limitation is that even the same source may modify its measures of variables to reflect changes in U.S. society and laws. Regarding marital and family status, for example, the Census Bureau added "separated" only in 1950, "unmarried partner" in 1990, and started categorizing same-sex married couples as families only in 2013 (Bates and DeMaio, 2013; Morello, 2014). Such changes are necessary, but they can be frustrating if, for instance, you wanted to compare the number of separated Americans between 1940 and 2015 or to determine the number of same-sex married couples since 2004, when Massachusetts became the first state to legalize same-sex marriage.

Finally, because the researcher didn't collect the data, she or he has no control over the specific variables used or how they were measured, and may not know how seriously problems such as low response rates, validity, and reliability affected the data. Most federal and other established data sets provide such information, but researchers can't simply assume that all data sets are problem-free.

Table 2.3 summarizes the data collection methods described in this chapter. Often, researchers use a combination of methods because "the social world is a multi-faceted and multi-layered reality that reveals itself only in part with any single method" (Jacobs, 2005: 4). You should also be familiar with evaluation research, an increasingly popular form of social research.

TABLE 2.3 SOME DATA COLLECTION METHODS IN SOCIOLOGICAL RESEARCH

Method	Example	Advantages	Disadvantages
Surveys	Sending questionnaires and/or interviewing students on why they succeeded in college or dropped out	Questionnaires are fairly inexpensive and simple to administer; interviews have high response rates; findings are often generalizable	Mailed questionnaires may have low response rates; respondents tend to be self-selected; interviews are usually expensive
Secondary analysis	Using data from the National Center for Education Statistics (or similar organizations) to examine why students drop out of college	Usually accessible, convenient, and inexpensive; often longitudinal and historical	Information may be incomplete; some documents may be inaccessible; some data can't be collected over time
Field research	Observing classroom participation and other activities of first-year college students with high and low grade-point averages (GPAs)	Flexible; offers deeper understanding of social behavior; usually inexpensive	Difficult to quantify and to maintain observer/subject boundaries; the observer may be biased or judgmental; findings are not generalizable
Content analysis	Comparing the transcripts of college graduates and dropouts on variables such as gender, race/ethnicity, and social class	Usually inexpensive; can recode errors easily; unobtrusive; permits comparisons over time	Can be labor intensive; coding is often subjective (and may be distorted); may reflect social class bias
Experiments	Providing tutors to some students with low GPAs to find out if such resources increase college graduation rates	Usually inexpensive; plentiful supply of subjects; can be replicated	Subjects aren't representative of a larger population; the laboratory setting is artificial; findings can't be generalized

2-5f Evaluation Research: Pulling It All Together

Evaluation research is the process of determining whether a social intervention has produced the intended result. *Social interventions* are typically programs and strategies that seek to prevent or change negative outcomes such as teenage pregnancy and substance abuse.

Evaluation research "refers to a research purpose rather than a specific method" (Babbie, 2013: 192). Thus, sociologists use a variety of data collection methods, such as secondary analysis and content analysis to examine an organization's records, surveys to gauge employee and client satisfaction, and field interviews with the staff and program recipients for in-depth information on an organization's processes.

Evaluation research is *applied* because it's intended to have some real-world effect (Weiss, 1998). Examples include *needs assessment studies* (e.g., to improve or initiate a new service such as an after-school program), *cost-benefit studies* (e.g., to determine whether funding teen abstinence programs reduces pregnancies), and *monitoring studies* (e.g., to examine the effect of new policing policies on crime rates). The findings are generally used to improve the efficiency and effectiveness of a particular policy or group.

STRENGTHS

Evaluation research is important for several reasons. First, because local and state governments have been cutting their budgets, social service agencies often rely on evaluation research to streamline their programs, and to achieve the best results at the lowest possible cost (Kettner et al., 1999).

Second, evaluation research is versatile because it includes qualitative and quantitative approaches. It can address almost any topic such as the effectiveness of driver education programs, foreign aid policies, job training programs, and premarital counseling, to name just a few. The research costs can also be low if, for example, researchers can use secondary analysis rather than collect new data.

Third, evaluation research addresses real-life problems that confront many families and communities. Because state and local resources are shrinking and there are hundreds of intervention strategies, evaluation research can be invaluable to program directors or agency heads in deciding which programs to keep, improve, or cut (Peterson et al., 1994).

LIMITATIONS

Many evaluations can address only one or a few of the many factors that affect behavior. For example, adolescent usage of illicit drugs is not only due to individual high-risk decisions, but also to multiple interrelated environments

DARE programs have been popular with schools, police departments, parents, and politicians across the country even though evaluation research has found that the curriculum doesn't reduce the incidence of drug use among students.

such as peers, school policies, and cultural values (Bandy, 2012).

Another limitation is that the social context affects evaluation research because of politics, vested interests, and conflicts of interest. Groups that fund the evaluation may pressure researchers to present only positive findings. Also, the results may not be well received if they contradict deeply held beliefs, challenge politicians' pet projects, or conflict with official points of view (Olson, 2010; Babbie, 2013).

Consider the popular DARE (Drug Abuse Resistance Education) program, introduced in 1983, to teach elementary school children about the dangers of drug use. Social scientists evaluated DARE in 1994, 1998, and 1999, and found no significant difference in the incidence of drug use among students who had and hadn't completed the DARE curriculum. When DARE funding was threatened, its promoters dismissed the research as "voodoo science" (Miller, 2001).

2-6 ETHICS AND SOCIAL RESEARCH

Researchers today operate under much stricter guidelines than they did in the past. In conducting their research, what ethical and political dilemmas do sociologists (and other social scientists) encounter?

evaluation research the process of determining whether a social intervention has produced the intended result.

2-6a What Is Ethical Research?

Because so much research relies on human participants, the federal government, university institutional review boards (IRBs), and many professional organizations have formulated codes of ethics to protect participants. Regardless of the discipline or the research methods used, all ethical standards have at least three golden rules:

▶ First, *do no harm* by causing participants physical, psychological, or emotional pain.

▶ Second, the researcher must get the participant's *informed consent* to be in a study. This includes the participant's knowing what the study is about and how the results will be used.

▶ Third, researchers must always protect a participant's *confidentiality* even if the participant has broken a law that she or he discloses to the researcher. *Table 2.4* lists some of the basic ethical principles in doing sociological research.

2-6b Scientific Misconduct

Some disciplines are more susceptible to ethical violations than others. Medical researchers, especially, have been accused of considerable scientific dishonesty. Some of the alleged violations have included the following: changing

TABLE 2.4 SOME BASIC PRINCIPLES OF ETHICAL SOCIOLOGICAL RESEARCH

In its most recent update, the American Sociological Association (1999) has reinforced its previous ethical codes and guidelines for researchers. Researchers must:

- Obtain all subjects' consent to participate in the research and their permission to quote from their responses and comments.

- Not exploit subjects or research assistants involved in the research for personal gain.

- Never harm, humiliate, abuse, or coerce the participants in their studies, either physically or psychologically.

- Honor all guarantees to participants of privacy, anonymity, and confidentiality.

- Use the highest methodological standards and be as accurate as possible.

- Describe the limitations and shortcomings of the research in their published reports.

- Identify the sponsors who funded the research.

- Acknowledge the contributions of research assistants (usually underpaid and overworked graduate students) who participate in the research project.

research results to please the corporation (usually tobacco or pharmaceutical companies) that sponsored the research; being paid by companies to deliver speeches to health practitioners that endorse specific drugs, even if the medications don't reduce health problems; allowing drug manufacturers to ghostwrite their articles that are published in prestigious medical journals (and even draft textbooks); and falsifying data. A research team examined more than 2,000 published articles that were withdrawn from some of the most prestigious medical journals such as the *Journal of the American Medical Association* (*JAMA*). The researchers found that almost 68 percent were withdrawn because of misconduct and fraud rather than errors (Project on Government Oversight, 2010; Shamoo and Bricker, 2011; Fang et al., 2012; McNeill, 2012).

In the social sciences, some data collection methods are more prone to ethical violations than others. Surveys, secondary analysis, and content analysis are less vulnerable than field research and experiments because the researchers typically don't interact directly with participants or affect them. In contrast, experiments and field research can raise ethical questions because of deception or influencing the participants' feelings, attitudes, or behavior.

In a recent experiment, Facebook and several academic researchers randomly selected 689,000 subscribers. For two of the groups, Facebook filtered the users' posts, providing either a positive or negative flow of comments, videos, pictures, and web links by other people in their social network. The study—finding that users who saw fewer positive posts were less likely to post something positive, and vice versa—concluded that similar emotional language on social networks could be contagious (Kramer et al., 2014).

Some researchers questioned the conclusion, but many criticized the experiment for breaching ethical guidelines. Facebook users weren't informed about the experiment, hadn't given their consent to participate, and didn't know that they were being manipulated. Non-researchers described the emotional manipulation as "scandalous," "spooky," and "disturbing." Even so, the university IRB had approved the study because the researchers themselves didn't collect the data (Booth, 2014; Voosen, 2014b).

All scholars face emerging ethical questions when they do research using social networks and other online environments. IRBs lack experience with Web research and sometimes unintentionally approve social science research that might violate ethics guidelines.

2-6c Do People Believe Scientific Findings?

Not always. For example, 87 percent of scientists say that humans and other living things have evolved over time, but only 32 percent of Americans agree. And although

Fisher-Price and other toy companies claim that, based on their research, their iPhone and iPad apps can teach infants as young as 9 months old to read and count (Singer, 2013). In fact, studies have found that such products do more harm than good because infants who watch the apps learn fewer words than those whose parents talk or read to them (see Chapter 12).

there has been no scientific evidence of paranormal events, 60 percent of Americans believe that people have seen or been in the presence of ghosts (Allen and Auxier, 2010; Speigel, 2013).

Why do many people reject scientific findings? One reason may be that when the media publicize the relatively rare occurrences of medical misconduct, readers generalize the few cases to all research and become more skeptical about all scientific studies.

Second, research findings often challenge personal attitudes and beliefs that people cherish. For example, a number of conservative Christians believe that global warming is a myth because only God, not humans, can affect the climate (Kahan et al., 2011; Sheppard, 2011; Smith and Leiserowitz, 2013).

Third, many Americans, especially minorities, are suspicious of the scientific community. Between 1906 and 1932, for example, medical researchers purposely infected unknowing subjects with fatal venereal diseases to test the effectiveness of penicillin. The subjects were almost always black men, prison inmates, mental patients, or soldiers (McNeil, 2010; Sanburn, 2010). Finally, many Americans know little about the scientific method, including the strengths and limitations of the research process.

STUDY TOOLS 2

READY TO STUDY? IN THE BOOK, YOU CAN:

☐ Check your understanding of what you've read with the Test Your Learning Questions provided on the chapter review card at the back of the book.

☐ Rip out the chapter review card for a handy summary of the chapter and key terms.

ONLINE AT CENGAGEBRAIN.COM YOU CAN:

☐ Prepare for tests with quizzes.

☐ Review the key terms with Flash Cards.

☐ Play games to master concepts.

3 | Culture

LEARNING OBJECTIVES

3-1 Describe and illustrate a culture's characteristics.

3-2 Explain the significance of symbols, language, values, and norms.

3-3 Discuss and illustrate cultural similarities.

3-4 Discuss and illustrate cultural variations.

3-5 Differentiate between high culture and popular culture.

3-6 Explain how and why technology affects cultural change.

3-7 Compare and evaluate the theoretical explanations of culture.

After you finish this chapter go to **PAGE 59** for **STUDY TOOLS.**

Once when I returned a set of exams, a student who was unhappy with his grade blurted out an obscenity. A voice from the back of the classroom snapped, "You ain't got no culture, man!" Is it true that people who use vulgar language "ain't got no culture"?

3-1 CULTURE AND SOCIETY

As popularly used, *culture* often means an appreciation of the finer things in life, such as Shakespeare's sonnets, the opera, and using civil language. In contrast, sociologists use the term in a much broader sense: **Culture** refers to the learned and shared behaviors, beliefs, attitudes, values, and material objects that characterize a particular group or society. Thus, culture shapes a people's total way of life.

Most human behavior isn't random or haphazard. Among other things, culture influences what you eat; how you were raised and will raise your own children; if, when, and whom you will marry; how you make and spend money; and what you read. Even people who

culture the learned and shared behaviors, beliefs, attitudes, values, and material objects that characterize a particular group or society.

What do you think?

I shape my culture; my culture doesn't shape me.

| 1 | 2 | 3 | 4 | 5 | 6 | 7 |

strongly agree   strongly disagree

pride themselves on their individualism conform to most cultural rules. The next time you're in class, for example, count how many students are *not* wearing jeans, T-shirts, sweatshirts, or sneakers—clothes that are the prevalent uniform of adolescents, college students, and many adults in U.S. society.

A **society** is a group of people who share a culture and a defined territory. Society and culture go hand in hand; neither can exist without the other. Because of this interdependence, social scientists sometimes use the terms *culture* and *society* interchangeably.

3-1a Some Characteristics of Culture

All human societies, despite their diversity, share some cultural characteristics and functions (Murdock, 1940). We don't see culture directly, but it shapes our attitudes and behaviors.

1. **Culture is learned.** Culture isn't innate but learned, and shapes how we think, feel, and behave. If a child is born in one region of the world but raised in another, she or he will learn the customs, attitudes, and beliefs of the host culture.

2. **Culture is transmitted from one generation to the next.** We learn many customs, habits, and attitudes informally through parents, relatives, friends, and the media. We also learn culture formally in settings such as schools, workplaces, and community organizations. Whether our learning is formal or informal, culture is cumulative because each generation transmits cultural information to the next one.

3. **Culture is shared.** Culture brings members of a society together. We have a sense of belonging because we share similar beliefs, values, and attitudes about what's right and wrong. Imagine the chaos if we did what we wanted (such as physically assaulting an annoying neighbor) or if we couldn't make numerous daily assumptions about other people's behavior (such as stopping at red traffic lights).

4. **Culture is adaptive and always changing.** Culture changes over time. New generations discard technological aspects of culture that are no longer practical, such as replacing typewriters with personal computers, and landlines with cell phones. Attitudes can also change over time. Compared with several generations ago, for example, many Americans now feel that premarital sex isn't immoral (see *Figure 3.1*).

Culture reflects who we are, but remember that it's people who create culture. As a result, culture changes as people adapt to their surroundings. Since the attacks on September 11, 2001, for example, many people worldwide, including Americans, don't complain at airports when we have to pass through a metal detector, when our baggage is X-rayed or searched, when items are confiscated, and when airports use full-body scanners. Thus, as the dangers in society have grown, people

society a group of people who share a culture and defined territory.

FIGURE 3.1 IS PREMARITAL SEX WRONG?

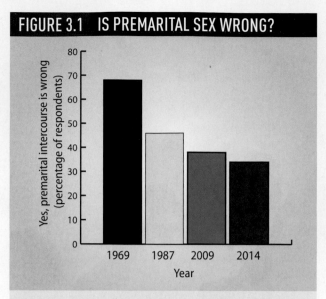

Every year since 1969, the Gallup Institute has asked Americans the following question: "Do you think it is wrong for a man and a woman to have sexual relations before marriage, or not?" Here's how attitudes have changed over the years.

Sources: Based on Saad, 2001, and Riffkin, 2014.

Communication symbols vary across societies and change over time. The earliest known map of the world (700–500 B.C.E) used cuneiform script; modern iPads are loaded with icons that provide a variety of information, including maps.

have passed laws or implemented rules that have both increased their security and decreased their privacy.

3-1b Material and Nonmaterial Culture

Cultures that people construct are both material and nonmaterial (Ogburn, 1922). **Material culture** consists of the physical objects that people make, use, and share. These objects include such things as buildings, furniture, music, weapons, jewelry, hairstyles, and the Internet. **Nonmaterial culture** consists of the ideas that people create to interpret and understand the world. Beliefs about the supernatural, customs, and rules of behavior are examples of nonmaterial culture.

Because material and nonmaterial culture are created by people, the physical objects and ideas can vary considerably across societies (e.g., the farming tools people create, how they form governments, or raise their children). Frequently, as you'll see shortly, there's a "cultural lag" because material culture (invention of the automobile and the cell phone) changes faster than nonmaterial culture (laws about pollution and texting while driving).

material culture the physical objects that people make, use, and share.

nonmaterial culture the ideas that people create to interpret and understand the world.

symbol anything that stands for something else and has a particular meaning for people who share a culture.

3-2 THE BUILDING BLOCKS OF CULTURE

Recently, 74-year-old James Davis granted his dying wife's wish to be buried in their front yard. The Alabama Supreme Court agreed with a lower court judge that family members can't bury people anywhere they like. Davis refused to budge because the town doesn't have an ordinance governing burials, and his family members and neighbors supported his decision. After four years of court hearings and appeals, a city crew exhumed the body (Reeves, 2013).

Who's right—Davis, his family and neighbors, or the judges? To answer such questions, we must understand the building blocks of culture, particularly symbols, language, values, norms, and rituals.

3-2a Symbols

A **symbol** is anything that stands for something else and has a particular meaning for people who share a culture. In most societies, for example, a handshake communicates friendship or courtesy, a wedding ring signals that a person isn't a potential dating partner, and a siren denotes an emergency. People influence each other through the use of symbols. A smile and a frown communicate different information and elicit different responses. Through symbols, we engage in *symbolic interaction* (see Chapter 1).

SYMBOLS TAKE MANY FORMS

Written words are the most common symbols, but we also communicate by tattooing our bodies, getting breast implants, and purchasing goods and services that we believe will increase our social status. Gestures (e.g., raised fists, hugs, and stares) convey important messages about people's feelings and attitudes.

SYMBOLS DISTINGUISH ONE CULTURE FROM ANOTHER

In Islamic societies, many women wear a head scarf (*hijab* or *hejab*) that covers their hair and neck, a veil that hides part of the face from just below the eyes (*nijab*), or a garment that covers the entire body (*chador* or *burqa*). Whatever form it takes, the veil is just a piece of cloth, but it means different things to people across and within cultures.

For non-Muslims, especially Westerners, veils are often symbols of women's repression by men and religious zealotry; they conjure up images of terrorism and connote hiding something. For many Muslims, on the other hand, veiling symbolizes religious commitment, women's modesty, and an Islamic identity. Other Muslims—including those who live in the Middle East or have emigrated to Western countries—view most veiling, especially *nijabs* and *burqas*, not as a religious requirement but a custom that discourages face-to-face interaction (Murphy, 2009).

SYMBOLS CAN UNIFY OR DIVIDE A SOCIETY

Symbols usually unify people culturally. About 75 percent of Americans say that they display the nation's flag in places such as their home, workplace, car, or clothing ("75%—A Nation of Flag Wavers," 2011). Every Fourth of July, many Americans celebrate the anniversary of gaining independence from Britain with parades, firecrackers, barbecues, and speeches by local and national politicians. All of this symbolic behavior signifies freedom and democracy. For this reason, many immigrants purposely choose July 4th to be naturalized.

Symbols can also be divisive. For example, some white Southerners fly the Confederate flag because they see it as a proud emblem of their Southern heritage.

The U.S. Flag Code describes the flag as a "living thing" and forbids its display on clothing, bedding, drapery, or in advertising (Luckey, 2008). Is this person disrespecting a cherished symbol? Or expressing her patriotism?

Others have abandoned the Confederate flag because it represents slavery and white domination of African Americans.

SYMBOLS CAN CHANGE OVER TIME

Symbols communicate information that varies across societies and may change over time. In 1986, for example, the International Red Cross changed its name to the International Red Cross and the Red Crescent Movement to encompass a number of Arab branches, and adopted a crescent emblem in addition to the well-known cross. Israel wanted to use a red Star of David, rejecting the cross as a Christian symbol and the crescent as an Islamic one. Red Cross officials offered a red diamond as the new shape, but some countries rejected the diamond because it represents bloody conflicts in many African countries that mine diamonds (Whitelaw, 2000). The red cross and crescent continue to be the organization's emblems because the issue hasn't been resolved.

The meaning of even simple symbols varies across groups and can change over time. The # symbol is commonly called the pound and number sign (#2 pencil). For musicians, # denotes a note that's sharp. For proofreaders, # means "insert a space." We're used to seeing # on telephone pads, but since the ascent of social media, # is a hashtag used to mark keywords or topics in a Tweet (#culture).

Left: The two emblems of the International Movement of the Red Cross and the Red Crescent. Right: The proposed red diamond emblem.

3-2b Language

Perhaps the most powerful of all human symbols is **language**, a system of shared symbols that enables people to communicate with one another. Essentially, language is an invention of human thought that communities of people have endowed with meaning. In every society, children begin to grasp the essential structure of their language at a very early age and without any instruction. Babbling rapidly leads to uttering words and combinations of words. The average child knows approximately 900 words by the age of 2 and 8,000 words by age 6 (Hetherington et al., 2006). Each word is a symbol that represents an idea, feeling, or physical object.

AP Images/Hallmark Inc.

Every year, Hallmark produces "Keepsake" Christmas tree ornaments, many of which are collectibles. In 2013, one of the new ornaments paid tribute to "Deck the Halls," a traditional Christmas carol. Hallmark changed the lyrics from "Don we now our gay apparel" to "Don we now our fun apparel" because the meaning of gay has changed since the song's publication in the 1800s (Wong, 2013). Because of some negative reactions, Hallmark apologized for the change (Murphy, 2013). Why do you think that one word generated controversy?

Anthropological linguist Edward Sapir (1929) and later his student, Benjamin Whorf (1956), theorized that thoughts and behavior are determined by language. Their theory, known as the *Sapir-Whorf hypothesis*, posited that language not only is a tool for interpersonal communication but also provides people with a framework for interpreting social reality and the world around them. Thus, as languages vary, so do interpretations of social reality.

Subsequent research found little evidence to support the Sapir-Whorf hypothesis. Some critics maintained that language affects but does not "determine" culture. Others argued that language influences some thinking processes, but not all (e.g., Carroll, 1956; Berlin and Kay, 1969; Heider and Olivier, 1972; Schlesinger, 1991). Still, the Sapir-Whorf hypothesis has generated considerable research across disciplines such as philosophy, sociology, and psychology in exploring the relationship between language, thought, and culture.

WHY LANGUAGE IS IMPORTANT

Language makes us human: It helps us understand our everyday experiences, conveys our ideas, communicates information, and influences other people's attitudes and behavior. Language directs our thinking, controls our actions, shapes our expression of emotions, and gives us a sense of belonging to a group.

Language can also spark anger and conflict. Recognizing the connection between language and behavior,

in 1990 Brazil passed a law criminalizing speech that incites racial, religious, or ethnic tensions. The penalties—including jail, fines, and community service—have been imposed on offenders such as people who produced an online video that disparaged deities at an Afro-Brazilian spiritual center, an Argentine soccer player who shouted racist slurs during a game, and a law student who sent racist comments on Twitter (Barnes, 2012).

LANGUAGE AND GENDER

Many professors routinely use phrases such as "Okay, guys, in class today…" and no one objects. One of my colleagues illustrates the linkage between language and gender, and how it affects our thinking, by saying "Okay, gals, in class today…" "I always get a reaction of gaping mouths, laughs, and bewildered looks," he says. The students react differently to "guys" and "gals" because we've internalized male terms (e.g., *guys, policeman*, and *maintenance man*) as normal and acceptable.

Language has a profound influence on how we think about and act toward women and men. Those who adhere to traditional language contend that nouns such as *fireman, chairman, mailman*, and *mankind* and pronouns such as *he* refer to both women and men, and that women who object to such usage are too sensitive. Suppose, however, that all of your professors used only *she, her*, and *women* when referring to all people. Would the men in the class feel excluded? One solution is to use sentences with plural, gender-neutral pronouns (e.g., *they, them*) and gender-neutral nouns (e.g., *firefighter, chair, police officer*, and *mail carrier*).

language a system of shared symbols that enables people to communicate with one another.

Wanted: Someone Who Can Translate Federal Gibberish into Plain English

In the United States, it's almost impossible to find a high-level government document that people can understand. To increase the readability of federal publications, in 2010 President Obama signed the Plain Writing Act, which requires federal documents to be rewritten in plain English. Here's a "before" and "after" example from the Department of Health and Human Services Handbook that assures access to essential health care for low-income HIV/AIDS patients:

Before: Title I of the CARE Act creates a program of formula and supplemental competitive grants to help metropolitan areas with 2,000 or more reported AIDS cases meet emergency care needs of low-income HIV patients. Title II of the Ryan White Act provides formula grants to States and territories for operation of HIV service consortia in the localities most affected by the epidemic, provision of home and community-based care, continuation of insurance coverage for persons with HIV infection, and treatments that prolong life and prevent serious deterioration of health. Up to 10 percent of the funds for this program can be used to support Special Projects of National Significance.

After: Low income people living with HIV/AIDS gain, literally, years, through the advanced drug treatments and ongoing care supported by HRSA's Ryan White Comprehensive AIDS Resources Emergency (CARE) Act.

In all cultures, language is supposed to help people communicate. However, the Plain Writing Act lacks enforcement, and many federal agencies, including the Internal Revenue Service, are exempted. Does a government have more control if people don't understand the government's rules and regulations?

Sources: Based on Health Resources and Services Administration (2010), and "Government Resolves…," (2011).

LANGUAGE, RACE, AND ETHNICITY

Words—written and spoken—create and reinforce both positive and negative images about race and ethnicity. Someone might receive a *black mark*, and a *white lie* isn't really a lie. We *blackball* someone, *blacken* someone's reputation, denounce *blackguards* (villains), and prosecute people in the *black market*. In contrast, a *white knight* rescues people in distress, the *white hope* brings glory to a group, and the good guys wear *white hats*.

Racist or ethnic slurs, labels, and stereotypes demean and stigmatize people. Derogatory ethnic words abound: *honky, hebe, kike, spic, chink, jap, polack, wetback*, and many others. Self-ascribed racial epithets are as harmful as those imposed by outsiders. When Italians refer to themselves as *dagos* or African Americans call each other *nigger*, they tacitly accept stereotypes about themselves and legitimize the general usage of such derogatory ethnic labels (Attinasi, 1994).

Through language, children learn about their cultural heritage and develop a sense of group identity. Spanish is the fastest-growing and most spoken non-English language in the United States today. Some demographers predict, however, that about 34 percent of Latinos will speak only English at home by 2020, up from 25 percent in 2010 (Lopez and Gonzalez-Barrera, 2013). In many Latino and other immigrant families, the native language begins to disappear after just one generation. Assimilation has positive effects, but can also lead to communication problems between Americanized children and their non-English-speaking neighbors and kin in the homeland (Tobar, 2009; see also Chapter 10).

LANGUAGE AND SOCIAL CHANGE

Language is dynamic and changes over time. The most recent edition of the New American Bible, an annual best seller, has changed the word "booty," which now has a sexual connotation, to "spoils of war." "Cereal," which many think of as breakfast food, is now "grain" to reference loads of wheat (*Huffington Post*, 2011).

U.S. English is composed of hundreds of thousands

TABLE 3.1 U.S. ENGLISH IS A MIXED SALAD

Here are a few examples of English words borrowed from other languages, showing our mixture of heritages. What words would you add to the list?

☐ Africa: apartheid, Kwanzaa, safari	☐ Spain: anchovy, bizarre
☐ Alaska and Siberia: husky, igloo, kayak	☐ Thailand: Siamese
☐ Bangladesh: bungalow, dinghy	☐ Turkey: baklava, caviar, kebob
☐ Hungary: coach, goulash, paprika	☐ France: bacon, police, ballet
☐ India: bandanna, cheetah, shampoo	☐ Japan: geisha, judo, sushi
☐ Iran and Afghanistan: bazaar, caravan, tiger	☐ Norway: iceberg, rig, walrus
☐ Israel: kosher, rabbi, Sabbath	☐ Mexico: avocado, chocolate, coyote
☐ Italy: fresco, spaghetti, piano	☐ Germany: strudel, vitamin, sauerkraut, kindergarten

of words borrowed from many countries and from groups that were in the Americas before the colonists arrived (Carney, 1997). *Table 3.1* provides a few examples of English words borrowed from other languages.

In response to cultural and technological changes, our vocabulary now includes *sexting*, *ebook*, *tweet*, *staycation*, *twerk*, *unfriend*, and *selfie*, among other new words. And to some people's dismay and others' delight, some writers are now substituting *partner* for the traditional *spouse*, *wife*, or *husband*.

3-2c Values

Values are the standards by people define what is good or bad, moral or immoral, proper or improper, desirable or undesirable, beautiful or ugly. These widely shared standards provide *general guidelines* for everyday behavior rather than concrete rules for specific situations. Honesty, for example, is an important but abstract U.S. value. Thus, faculty typically define in course syllabi what they mean by cheating, such as using material from the Internet or other sources without proper acknowledgement, signing an attendance sheet for another student, or getting help on an assignment.

MAJOR U.S. VALUES

Sociologist Robin Williams (1970: 452–500) has identified a number of core U.S. values. All are central to the American way of life because they're widespread, have endured over time, and indicate what we believe is important in life.

1. **Achievement and success.** U.S. culture stresses personal achievement, especially occupational success. Many Americans believe that if they work hard, apply themselves, and save their money, they'll be successful in the future.

2. **Activity and Work.** Americans want to "make things happen." They respect people who are focused and disciplined in their jobs, and assume that hard work will be rewarded. Journalists and others often praise those who work past their retirement age.

3. **Humanitarianism.** U.S. society emphasizes concern for others, helpfulness, kindness, and offering comfort and support. During natural disasters—such as earthquakes, floods, fires, and famines—at home and abroad, many Americans are enormously generous (see Chapter 5).

4. **Efficiency and practicality.** Americans emphasize technological innovation, up-to-dateness, practicality, and getting things done. Many American colleges now tout their programs or courses as being practical instead of emphasizing knowledge and intellectual growth.

5. **Progress.** Americans focus on the future rather than the present or the past. The next time you walk down the aisle of a grocery store, note how many products are "new," "improved," and "better than ever."

6. **Material comfort.** Americans want new products and services. Many work hard to pay for fancy new cars, large homes, and dream vacations (even if they can't afford them). Americans never have enough gadgets and always need more stuff. When the stuff accumulates, we buy stuff to stuff it into.

7. **Freedom and equality.** Countless U.S. documents affirm freedom of speech, freedom of the press, and freedom of worship. U.S. laws also tell Americans that they have been "created equal" and have the same legal rights regardless of race, ethnicity, sex, religion, disability, age, or social class.

8. **Conformity.** Most people don't want to be labeled as "strange," "peculiar," or "different." They conform because

values the standards by which people define what is good or bad, moral or immoral, proper or improper, desirable or undesirable, beautiful or ugly.

they want to be accepted, liked, hired, and promoted. As a result, we "toe the line," "go along to get along," and try to "fit in."

9. Democracy. Democracy provides the average U.S. citizen with equal political rights. It emphasizes equality, freedom, and faith in the people rather than giving power to a monarch, dictator, or emperor.

10. Individualism. U.S. culture values each person's development. Thus, we often encourage children to be independent, creative, self-directed, self-motivated, and spontaneous.

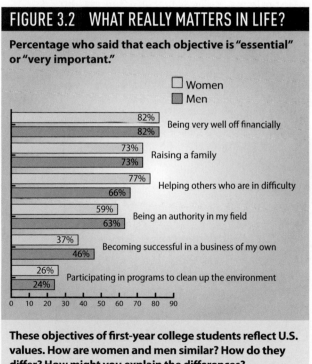

David Henderson/violinconcertono3/
iStockphoto.com

VALUES SOMETIMES CONFLICT

Williams acknowledged that these core American values are sometimes contradictory and don't always mesh with reality. For example,

▸ We value equality but are comfortable with enormous gaps in wealth and power, and continue to discriminate against people because of their ethnicity, race, sexual orientation, gender, or age (see Chapters 10 and 11).

▸ We prize individualism, but reward people who conform to a group's expectations (see Chapter 6).

▸ We encourage responsibility, but often blame popular culture rather than parenting for children's bad behavior (see Chapter 12).

▸ We rail against waste, rising food prices, and the growing number of landfills, but throw away up to 40 percent of food every year (Gunders, 2012).

VALUES VARY ACROSS CULTURES AND CHANGE OVER TIME

Cultural values can change because of technological advances, immigration, and contact with outsiders. For example, the Japanese parliament passed a law making love of country a compulsory part of school curricula. The lawmakers and their numerous supporters hope that teaching patriotism will counteract the American-style emphasis on individualism and self-expression that they believe has undermined Japanese values of cooperation, self-discipline, responsibility, and respect for others (Wallace, 2006).

In the United States, surveys of first-year college students show a shift in values. Between 1968 and 2013, for example, "developing a meaningful philosophy of life" plummeted in importance whereas "being very well off financially" surged (Eagan et al., 2014). Consistent with our humanitarian values, however, in 2013 a sizeable majority (63 percent) of the students said that "helping others who are in difficulty" is essential or very important, but women were more likely to feel this way (see *Figure 3.2*).

The share of U.S. adults ages 25 and older who have never been married is at an historic high—20 percent in 2012 compared with only 9 percent in 1960, and 25 percent may never marry. You'll see in Chapter 12 that marriage rates vary by social class, among other factors, but values also affect marital status. For example, 50 percent of Americans, another historic high, say society is just as well off if people have priorities other than marriage and children (Wang and Parker, 2014).

3-2d Norms

Values are general standards, whereas **norms** are a society's specific rules of right and wrong behavior. Norms tell us what we *should*, *ought*, and *must* do, as well as what we *should not*, *ought not*, and *must not* do: don't

FIGURE 3.2   WHAT REALLY MATTERS IN LIFE?

Percentage who said that each objective is "essential" or "very important."

☐ Women
■ Men

Objective	Women	Men
Being very well off financially	82%	82%
Raising a family	73%	73%
Helping others who are in difficulty	77%	66%
Being an authority in my field	59%	63%
Becoming successful in a business of my own	37%	46%
Participating in programs to clean up the environment	26%	24%

These objectives of first-year college students reflect U.S. values. How are women and men similar? How do they differ? How might you explain the differences?

Source: Based on Eagan et al., 2014: 66, 92.

norms specific rules of right and wrong behavior.

talk in church, stand in line, and so on. We may not like many of the rules, but they make our everyday lives more orderly and predictable. Here are some general characteristics of norms:

▶ Most are *unwritten,* passed down orally from generation to generation (use the good dishes and tablecloth on special occasions).

▶ They're *instrumental* because they serve a specific purpose (get rid of garbage because it attracts roaches and rats).

▶ Some are *explicit* (save your money "for a rainy day"), whereas others are *implicit* (be respectful at a wake or funeral).

▶ They *change* over time (it's now more acceptable to have a child out of wedlock but much less acceptable to smoke or drink alcohol while pregnant).

▶ Most are *conditional* because they apply in specific situations (slipping out of your smelly shoes may be fine at home but not on an airplane).

▶ Because they're situational, norms can be *rigid* ("You *must* turn in your term paper at the beginning of class on May 5") or *flexible* ("You can have another week to finish the paper").

Norms regulate our behavior, but can differ because of characteristics such as age, social class, race, and gender. For example, a female professor's students sometimes comment on her clothes in their course evaluations and want her to "look like a professor." In contrast, her husband, also a professor, "has yet to hear a single student comment about his wardrobe." So, she changes her businesslike outfits every day while her husband usually wears khaki pants (or jeans) and a polo shirt day in and day out (Johnston, 2005). Thus, "her" norms differ from "his."

Sociologists differentiate three types of norms—folkways, mores, and laws—that vary because some rules are more important than others. As a result, a society punishes some wrongdoers more severely than others.

FOLKWAYS

Folkways are norms that involve everyday customs, practices, and interaction. Etiquette rules are good examples of folkways: Cover your mouth when you cough or sneeze, say "please" and "thank you," knock before entering someone's office, and don't pick your ear wax or toenails in public.

We automatically practice numerous folkways because we've internalized them since birth. We often don't realize that we conform to folkways until someone violates one, such as picking one's nose or popping pimples in public, or talking loudly on a cell phone.

Folkways vary from one country to another. Punctuality is important in many Western countries but not in much of Latin America and the Mediterranean. Japan is notable for its gift giving among businesspeople, but this may be viewed as bribery in Western countries. Austrians, as well as Germans and the Swiss, consider chewing gum in public vulgar. There are also many differences in table manners: Europeans keep their hands above the table at all times, not in their laps, and Koreans frown on sniffling or blowing one's nose at the table (Axtell et al., 1997).

Folkways can change in response to macro-level changes. The growth of technology has changed many campus folkways. In the past, a syllabus was typically a few pages long, outlining course requirements and deadlines for exams and papers. Now, many syllabi—some as long as 20 pages—look more like legal documents. They

folkways norms that involve everyday customs, practices, and interaction.

YONHAP/AFP/Getty Images/Newscom

South Korea's national media criticized Microsoft founder Bill Gates for his "rude" handshake when meeting President Park Geun-Hye in 2013. As one observer tweeted, "How can he put his hand in his pocket when meeting a leader of a state?!" A one-hand shake is often seen as disrespectful in South Korea and other parts of Asia, and is normally reserved for someone younger or a good friend (Agence France-Presse, 2013; Evans, 2013). Which types of norms did Gates violate—folkways, mores, or laws?

describe, often in great detail, a variety of rules on Internet plagiarism, unacceptable online sources for papers and projects, and laptop and cell phone use during class, even forbidding videos of professors that may appear on YouTube (Wasley, 2008).

MORES

Mores (pronounced "MORE-ayz") are norms that people consider very important because they maintain moral and ethical behavior. According to U.S. cultural mores, one must be sexually faithful to one's spouse or sexual partner, must be loyal to one's country, and must not kill another person (except during war or in self-defense).

Folkways emphasize *ought to* behavior whereas mores define *must* behavior. The Ten Commandments are a good example of religious mores. Other mores include ethical guidelines (don't cheat, don't lie), expectations about behavior (don't use illicit drugs), and rules about sexual partners (don't have sexual intercourse with children or family members).

Mores, like folkways, can change. For example, until about the late 1960s, 80 percent of U.S. babies born out of wedlock were given up for adoption. This rate has dropped to about 1 percent because most unwed mothers, who are no longer stigmatized for having nonmarital babies, keep their infants (Benokraitis, 2014).

One of the most powerful types of mores are **taboos**, strong prohibitions of any act that is forbidden because it's considered to be extremely offensive. Taboos reflect social customs, religious or moral beliefs, or laws. Cannibalism, incest, and infanticide (killing infants) are examples of tabooed behavior.

Taboos define and reinforce mores, but there are many violations. For instance, 94 percent of Americans say that having an extramarital affair is morally wrong, but over a lifetime, 19 percent of men and 14 percent of women admit to having had extramarital sex (Drexler, 2012).

LAWS

Laws are formally defined norms about what is permissible or illegal. These rules for behavior are defined by a political authority that has the power to punish violators. Unlike folkways and most mores, laws are deliberate, formal, "precisely specified in written texts," and "enforced by a specialized bureaucracy," usually police and courts (Hechter and Opp, 2001: xi).

Laws change over time. In 2012 alone, states passed more than 400 laws regarding issues such as abortion, consumer protection, student bullying, and pension reform (Kelleher, 2013). So far, 23 states and the District of Columbia allow the sale of marijuana for medical use, many others are considering such legislation, and Colorado, Oregon, Alaska, and Washington have legalized marijuana for recreational use. Also, every state has legalized gambling such as a lottery, casinos, and race-track betting. Some people believe that legalizing marijuana usage and gambling violates mores, but taxing both of these activities has greatly augmented state coffers.

Laws also vary across societies. You'll see shortly and throughout this textbook that countries' laws differ considerably regarding, for example, drinking alcohol, adultery, elections, health care, religious worship, freedom of speech, and adoption.

SANCTIONS

Most people conform to norms because of **sanctions**, rewards for good or appropriate behavior and/or penalties

> **mores** norms that people consider very important because they maintain moral and ethical behavior.
>
> **taboos** strong prohibitions of any act that is forbidden because it's considered to be extremely offensive.
>
> **laws** formally defined norms about what is permissible or illegal.
>
> **sanctions** rewards for good or appropriate behavior and/or penalties for bad or inappropriate behavior.

China's new "Elderly Rights Law" allows aging parents to sue children who don't visit them "often." Pictured here, a 77-year-old woman attends a hearing against her daughter. The judge ordered the daughter to visit her mother every two months and on at least two of the country's major holidays (Makinen, 2013).

AFP/Getty Images

for bad or inappropriate behavior. Children learn norms through both *positive* sanctions (praise, hugs, smiles, new toys) and *negative* sanctions (frowning, scolding, spanking, withdrawing love) (see Chapter 4).

Negative sanctions vary in the degree of punishment. When we violate folkways, the sanctions are relatively mild (e.g., gossip, raised eyebrows, ridicule) because folkways aren't critical for a society's survival. If you don't bathe or brush your teeth, you may not be invited to parties because others will see you as crude, but not sinful or evil.

The sanctions for violating mores and some laws can be severe: loss of employment, expulsion from college, whipping, torture, banishment, imprisonment, and even execution. The punishment is usually harsh because the unacceptable behavior threatens a society's moral foundations. For example, Iran's public floggings of women who are alone with men who aren't relatives punish offenders for their "immorality" and send a warning to others to follow the rules (see Chapter 9).

Sanctions aren't always consistent, despite universally held norms. In the United States, someone who can hire a good attorney often receives no penalty or a light one for a serious crime. In some cultures, young girls and women may be lashed for engaging in premarital sex, but men aren't punished at all for the same behavior, even though it violates Islamic law (Peter, 2012). Also, laws are sometimes enforced selectively or not at all. In the United States and elsewhere, offenders who violate laws against rape, sexual harassment, and domestic violence are rarely punished (Jewkes et al., 2013; see also Chapters 7 and 12).

3-2e Rituals

Rituals are formal and repeated behaviors that unite people. They include lighting up windows during Christmas, oaths of allegiance, wedding and funeral ceremonies, worship rites, dressing up for Halloween, graduations, veterans' parades, and even shaking hands. What all rituals have in common is the transmission and reinforcement of norms that unite us with each other and strengthen relationships (Rossano, 2012).

Rituals are also outward symbols of values. During the playing of our national anthem, for example, everyone stands and faces the flag (or music). Many people

---

rituals formal and repeated behaviors that unite people.

cultural universals customs and practices that are common to all societies.

ideal culture the beliefs, values, and norms that people say they hold or follow.

place their hands over their hearts, and men remove their caps and hats. Such ritualistic behavior demonstrates respect and reinforces our shared values about patriotism and democracy.

3-3 SOME CULTURAL SIMILARITIES

You've seen that there's considerable diversity across societies in symbols, language, values, and norms. There are also some striking similarities across and within cultures because of cultural universals, ideal versus real culture, ethnocentrism, and cultural relativism.

3-3a Cultural Universals

Cultural universals are customs and practices that are common to all societies. Anthropologist George Murdock and his associates studied hundreds of societies and compiled a list of 88 categories that they found among all cultures (see *Table 3.2* for some examples).

There are many cultural universals, but specific behaviors vary across cultures, from one group to another in the same society, and over time. For example, all societies have food taboos, but they differ across societies. About 75 percent of the world's people eat insects. In Thailand, locusts, crickets, silkworms, grasshoppers, ants, and other insects are a big part of the diet (Stolley and Taphaneeyapan, 2002). Vietnamese restaurants offer cat on the menu, and many poor people in rural China eat dog meat because they can't afford poultry, pork, and beef. Increasingly, however, when individual incomes increase, many people treat their dogs as pets and family members, rather than as nutritional food sources (Masis, 2010; Wan, 2011).

3-3b Ideal versus Real Culture

The **ideal culture** comprises the beliefs, values, and norms that people say they hold or follow. In every

TABLE 3.2 SOME CULTURAL UNIVERSALS

Athletics	Games	Kin terminology	Property rights
Cooking	Gift giving	Language	Religious rituals
Courtship	Greetings	Magic	Sexual restrictions
Dancing	Hairstyles	Marriage	Status differentiation
Division of labor	Housing	Medicine	
Food taboos	Inheritance rules	Music	

Source: Based on Murdock, 1945.

society, however, ideal culture differs from **real culture**, or people's actual everyday behavior.

Consider parent-child relationships. Americans say that they love and cherish their children, but every year hundreds of thousands of children experience abuse and neglect on a daily basis. Indeed, 81 percent of the offenders are parents (U.S. Department of Health and Human Services, 2013; see also Chapter 12). In China, the 2,500-year-old Confucian ideal of *filial piety* (respecting and taking care of one's parents) still runs deep. However, nearly half of the country's older people now live apart from their children, most of whom live and work in urban areas—"a phenomenon unheard of a generation ago"—leaving elders to fend for themselves (Mencher,

About 75 percent of the world's people consume insects, which are high in protein, vitamins, and fiber, and usually low in fat. More than two out of every three American adults are overweight or obese, the highest percentage on the planet (Flegal et al., 2010). Would we be healthier if we ate insects instead of gulping down hamburgers and french fries?

Kevin Foy/Alamy

2013). Thus, ideal culture and actual behavior are often inconsistent.

3-3c Ethnocentrism and Cultural Relativism

Ethnocentrism is the belief that one's culture, society, or group is inherently superior to others. Thus, "*our* customs, *our* laws, *our* food, *our* traditions, *our* music, *our* religion, *our* beliefs and values, and so forth, are somehow better than those of other societies" (Smedley, 2007: 32). Because people internalize their culture and tend to see their way of life as the best and the most natural, they often scorn those with differing attitudes and behavior as inferior, wrong, or backward. How many of you, for example, are repelled at the thought of eating insects, cats, or dogs?

Some of my black students argue that it's impossible for African Americans to be ethnocentric because they suffer much prejudice and discrimination. *Any* group can be ethnocentric, however (Rose, 1997). An immigrant from Nigeria who believes that native-born African Americans are lazy and criminal is just as ethnocentric as a native-born African American who assumes that Nigerians are arrogant and "uppity."

Ethnocentrism can be functional. Appreciating one's own country and heritage promotes loyalty and cultural unity. When children learn their country's national anthem and customs, they have a sense of belonging. Ethnocentrism also reinforces conformity and maintains stability. Members of a society become committed to their particular values and customs, and transmit them to the next generation. As a result, life is (generally) orderly and predictable.

Ethnocentrism is also dysfunctional because viewing others as inferior generates hatred, discrimination, and conflict. Many current wars and battles—such as those in some African nations and between Israelis and Palestinians—are due to religious, ethnic, or political intolerance (see Chapter 13). Thus, ethnocentrism discourages intergroup understanding and cooperation.

The opposite of ethnocentrism is **cultural relativism**, the belief that no culture is better than

real culture people's actual everyday behavior.

ethnocentrism the belief that one's own culture, society, or group is inherently superior to others.

cultural relativism the belief that no culture is better than another and should be judged by its own standards.

another and should be judged by its own standards. Most Japanese mothers stay home with their children, whereas many American mothers work outside the home. Is one practice better than another? No. Because Japanese fathers are expected to be the breadwinners, it's common for many Japanese women to be homemakers. In the United States, in contrast, some mothers choose to stay home, but economic recessions and the loss of many high-paying jobs have catapulted many married women into the labor force to help support their families (see Chapters 10, 11, and 12). Thus, Japanese and American parenting may differ, but one culture isn't better or worse than the other.

During a trip to India, First Lady Michelle Obama chose an ensemble by designers who use traditional Indian fabrics (Givhan, 2010). How was her wardrobe decision an example of cultural relativism?

3-4 SOME CULTURAL VARIATIONS

There's considerable cultural variation not only *across* societies but also *within* the same society. *Subcultures* and *countercultures* account for some of the complexity within a society.

3-4a Subcultures

A **subculture** is a group within society that has distinctive norms, values, beliefs, lifestyle, or language. A subculture is part of the larger, dominant culture but has its own particular ways of thinking, feeling, and behaving. U.S. society contains thousands of subcultures based on a variety of characteristics, interests, or activities:

▸ **Ethnicity** (Irish, Mexican, Vietnamese)

▸ **Religion** (evangelical Christians, atheists, Mormons)

▸ **Politics** (Maine Republicans, Southern Democrats, independents)

▸ **Sex and gender** (heterosexual, lesbian, transgender)

▸ **Age** (older widows, middle schoolers, college students)

▸ **Occupation** (surgeons, prostitutes, truck drivers)

▸ **Music and art** (jazz aficionados, country music buffs, art lovers)

▸ **Social class** (billionaires, working poor, middle class)

▸ **Recreation** (mountain bikers, poker players, dancers)

subculture a group within society that has distinctive norms, values, beliefs, lifestyle, or language.

counterculture a group within society that openly opposes and/or rejects some of the dominant culture's norms, values, or laws.

As you read this list, you'll realize that most of us are members of numerous subcultures. We usually participate in subcultures without having much commitment to any of them (e.g., basketball fans, guitar players, collectors of ceramic frogs). On the other hand, members of racial or ethnic subcultures may intentionally live in the same neighborhoods, associate with each other, have close personal relationships, and marry others who are similar to themselves.

To fit in, members of most subcultures adapt to the larger society, but they may also maintain some of their traditional customs. Among Ghanaian immigrants, for instance, funerals are lavish celebrations rather than places of sadness. The funerals are "all-night affairs with open bars and window-rattling music." As in Ghana, the guests need not know the deceased or even the family. Many people attend simply to meet other Ghanaians and to "cut loose on the dance floor" (Dolnick, 2011: A1; see also Bax, 2013).

In many instances, subcultures arise because of technological or other societal changes. After the emergence of the Internet, for example, subcultures appeared that identified themselves as hackers, techies, or computer geeks.

3-4b Countercultures

A **counterculture** is a group within society that openly opposes and/or rejects some of the dominant culture's norms, values, or laws. Countercultures usually emerge when people believe they can't achieve their goals within

FIGURE 3.3 ACTIVE HATE GROUPS IN THE UNITED STATES: 2013

Source: Southern Poverty Law Center, http://www.splcenter.org/get-informed/hate-map (accessed August 9, 2014). Used with permission.

the existing society. As a result, such groups develop values and practices that run counter to those of the dominant society. Some countercultures are small and informal, but others, such as religious militias, have millions of members and are highly organized (see Chapter 11).

Most countercultures are law-abiding. Some, however, are violent and extremist, such as the 939 active hate groups across the United States (see *Figure 3.3*) that intimidate ethnic groups and gays. Some skinheads have killed gays; antigovernment militia adherents bombed a federal building in Oklahoma, killing dozens of adults and children; and antiabortion advocates have murdered physicians and bombed abortion clinics. Non-Islamic extremists are responsible for about two-thirds of terrorist acts in the United States. The FBI is particularly concerned by this violence because it's carried out by individuals who aren't affiliated with any countercultural movement and are, therefore, hard to detect (Masters, 2011).

3-4c Multiculturalism

Multiculturalism (sometimes called *cultural pluralism*) refers to the coexistence of several cultures in the same geographic area, without one culture dominating another. Many applaud multiculturalism because it encourages intercultural dialogue (e.g., U.S. schools offering programs and courses in African American, Latino, Arabic, and Asian studies). Supporters hope that emphasizing multiculturalism—especially in schools and the workplace—will decrease ethnocentrism, racism, sexism, and other forms of discrimination.

Despite its benefits, not everyone is enthusiastic about multiculturalism. Not learning the language or following the customs of the country where one lives and works, for instance, can create tension and conflict. European political leaders in Germany, Great Britain, and France have described multiculturalism, especially integrating Muslims, as a "failed policy" that's had "disastrous results." Instead of adopting the dominant culture's values and beliefs, according to critics, some ethnic and religious subcultures "build tiny nations" within the host nation, thereby jeopardizing a country's national identity (Buchanan, 2011; Marquand, 2011; Kymlicka, 2012).

3-4d Culture Shock

People who travel to other countries often experience **culture shock**—confusion, disorientation, or anxiety that accompanies exposure to an unfamiliar way of life. Familiar cues about how to behave are missing or have a different meaning. Culture shock affects some people more than others, but the most stressful changes involve differences of food, clothes, punctuality, ideas about what offends people, language, the general pace of life, and a lack of privacy (Spradley and Phillips, 1972; Pedersen, 1995).

To some degree, everyone is *culture bound* because we've internalized cultural norms and values. For example, an American journalist who works in Mexico City says that perpetual lateness is common: A child's birthday party may start 2 or 3 hours late, a wedding may begin an hour after the announced time, and interviews with top Mexico City officials may be up to 2 hours late or the official never arrives. These norms are changing because of the growth of Mexico's global economy, but the more relaxed attitude toward time in Mexico City (and elsewhere) can be a culture shock to someone from the United States who has grown up in a "clock-obsessed" culture (Ellingwood, 2009).

multiculturalism (sometimes called *cultural pluralism*) the coexistence of several cultures in the same geographic area, without one culture dominating another.

culture shock confusion, disorientation, or anxiety that accompanies exposure to an unfamiliar way of life.

Does Pop Culture Make Kids Fat?

The prevalence of obesity among American children ages 2 to 19 years has more than tripled, increasing from 5 percent in the 1960s to 17 percent in 2012 (National Center for Health Statistics, 2014). There are many reasons for the increase, but physicians and researchers lay much of the blame on popular culture, especially the advertising industry. Much of the advertising, particularly on television, uses cartoon characters such as SpongeBob SquarePants and Scooby-Doo to sell sugary cereals, cookies, candy, and

Juanmonino/iStockphoto.com

other high-calorie snacks. In contrast, many European countries forbid advertising on children's television programs. After considerable pressure from the federal government, the nation's largest food companies agreed to curb advertising, but children are watching more fast food ads every day on TV than ever before. Only six companies are responsible for more than 70 percent of all TV fast-food ads viewed by children age 18 and younger ("Food Companies Propose...," 2011; Harris et al., 2013).

3-5 HIGH CULTURE AND POPULAR CULTURE

High culture refers to the cultural expression of a society's highest social classes. Examples include opera, ballet, theatre, and classical music. In contrast, **popular culture** refers to beliefs, practices, activities, and products that are widespread among a population. Examples include television, music, magazines, radio, advertising, sports, hobbies, fads, fashions, and movies, as well as the food we eat, the gossip we share, and the jokes we pass along to others (Levine, 1998; Gans, 1999).

By definition, social class affects our participation in high culture and popular culture. Members of the upper class can pursue the fine arts and similar activities because they have **cultural capital**—resources such as knowledge, verbal and social skills, education, and other assets that give a group advantages (Bourdieu, 1986). Resources alone, however, don't determine our cultural

tastes and interests. For example, in Baltimore (as in many other cities), tickets for the symphony, opera, or ballet are usually less expensive than those for a Ravens football game or rock concert.

Cultural capital sets up boundaries between social classes and shapes our recreational activities, but our cultural expressions and products can contain elements of both popular culture and high culture. For instance, African American jazz players have inspired European classical music composers and themselves perform classical music (Salamone, 2005; Lawn, 2013). The Queen of England has conferred knighthood—one of the highest honors given to an individual in the United Kingdom—on pop singers such as Elton John and Paul McCartney. The John F. Kennedy Center for the Performing Arts in Washington, D.C., produces and presents theater, ballet, and chamber music, but since 1998 has also presented an annual Mark Twain Prize for American Humor to popular culture comedians such as Richard Pryor, Will Ferrell, and Tina Fey. Thus, the divide between popular culture and high culture isn't as rigid as we might think.

3-5a The Influence of Mass Media

People don't always believe everything they read or see on television or online. Nonetheless, many are highly influenced by a popular culture that is largely controlled and manipulated by newspapers and magazines, television,

high culture the cultural expression of a society's highest social classes.

popular culture the beliefs, practices, activities, and products that are widespread among a population.

cultural capital resources such as knowledge, verbal and social skills, education, and other assets that give a group advantages.

American fast-food restaurants are popular in many Asian countries. Because of greater competition from local fast-food chains, U.S. companies now offer a large number of items geared specifically to Asian tastes (Chu, 2010). At one of China's Pizza Huts, for example, the menu includes shrimp pizza, fried squid, and green tea ice cream. In Latin America, McDonald's sells "McMollettes," English muffins with refried beans, cheese, and salsa (York, 2012). In India, McDonald's offers an entirely meat-free menu ("Where's the Beef?" 2013).

movies, music, and ads (see Chapter 4). These **mass media**, or forms of communication designed to reach large numbers of people, have enormous power in shaping public attitudes and behavior.

Many of my students who claim that they "don't pay any attention to ads" come to class wearing branded apparel: Budweiser caps, Old Navy t-shirts, Nike footwear. Advertising affects our self-image and self-esteem. As you'll see in later chapters, many women and men spend considerable time and money to meet ideal standards of femininity and masculinity—generated by ads—that are unrealistic and impossible to attain. (We'll examine fashion and fads, two important components of popular culture, in Chapter 16.)

Because companies are driven by the pursuit of profit, considerable mass media content is basically marketing. Much of the content on television morning shows and MTV has become "a kind of sophisticated infomercial" (O'Donnell, 2007: 30). For example, a third of the content on morning shows is essentially selling something (a book, music, a movie, or another television program) that the corporation owns. One of the most lucrative alliances is between Hollywood and toy manufacturers (see Chapter 4). And, U.S. mass media exert a powerful influence abroad.

3-5b The Globalization of Popular Culture

Because the United States is the biggest producer of popular culture, many countries worldwide have become inundated with America's movies, music, television shows, newspapers, satellite broadcasts, fast food, clothing, and other consumer goods. Overseas corporations contribute to the spread of U.S. popular culture because they profit by distributing the products of popular musicians, authors, and filmmakers (The Levin Institute, 2013).

Some observers have described the spread of American popular culture as **cultural imperialism**, a process by which the cultural values and products of one society influence or dominate those of another. A great deal of cultural imperialism is voluntary. According to a British journalist, for example, much of England's population has embraced American pop music, films, television programs, architecture, and is now experiencing an "obesity epidemic" because of the "bread-potatoes-and-lard-based fast food on which America nourishes itself" (Heffer, 2010).

Other countries complain that U.S. cultural imperialism displaces authentic local culture and results in cultural loss. For instance, Iran's government has denounced Batman, Spider-Man, and Harry Potter toys as a form of "cultural invasion" that challenges the country's conservative and religious values. The curvaceous and often scantily clad Barbie dolls with peroxide-blond hair have been especially singled out as "destructive

mass media forms of communication designed to reach large numbers of people.

cultural imperialism the cultural values and products of one society influence or dominate those of another.

culturally and a social danger." However, many of the girls who watch foreign television and (illegal) satellite TV want the dolls (Peterson, 2008: 4).

3-6 CULTURAL CHANGE AND TECHNOLOGY

Why do cultures persist? How and why do they change? And what happens when technology changes faster than cultural norms, values, and laws?

3-6a Cultural Persistence: Why Cultures Are Stable

In many ways, culture is a conservative force. As you saw earlier, values, norms, and language are transmitted from generation to generation. Such **cultural integration**, or the consistency of various aspects of society, promotes order and stability. Even when new behaviors and beliefs emerge, they commonly adapt to existing ones. Recent immigrants, for example, may speak their native language at home and celebrate their own holy days, but they're expected to gradually absorb the host country's values, to obey its civil and criminal laws, and to adopt its national language. Life would be chaotic and unpredictable without such cultural integration.

3-6b Cultural Dynamics: Why Cultures Change

Cultural stability is important, but all societies change over time. Some of the major reasons for cultural change include diffusion, invention and innovation, discovery, and external pressures.

DIFFUSION

A culture may change because of *diffusion*, the process through which components of culture spread from one society to another. Such borrowing may have occurred so long ago that the members of a society consider their culture to be entirely their own creation. However, anthropologist Ralph Linton (1964) has estimated that 90 percent of the elements of any culture are a result of diffusion (see *Figure 3.4*).

Diffusion can be direct and interpersonal, occurring through trade, tourism, immigration, intermarriage, or the invasion of one country by another. Diffusion can also be indirect and largely impersonal, as in the Internet transmissions that zip around the world.

INVENTION AND INNOVATION

Cultures change because people are continually finding new ways of doing things. *Invention*, the process of creating new things, brought about products such as toothpaste (invented in 3000 B.C.E.), eyeglasses (262 C.E.), flushable toilets (the sixteenth century), fax machines (1843—that's right, invented in 1843!), credit cards (1920s), computer mouses (1964), Post-It™ notes (1980), and DVDs (1995).

Innovation—turning inventions into mass-market products—also sparks cultural changes. An innovator is someone who markets an invention, even if it's someone else's idea. For example, Henry Ford invented nothing new but "assembled into a car the discoveries of other men behind whom were centuries of work," an innovation that changed people's lives (Evans et al., 2006: 465).

Inventors and innovators are rarely isolated geniuses who work alone. Instead, they depend on collaborative relationships and social networks for ideas and assistance. Thomas Edison's laboratory, which had a staff of 35 inventors, produced the light bulb. The Human Genome Project included 110 authors from 20 universities, research centers, institutes, and hospitals. Even Michelangelo used twelve assistants to paint the renowned Sistine Chapel in Rome (Dahlin, 2011).

DISCOVERY

Like invention, *discovery* involves exploration and investigation, and results in new products, insights, ideas, or behavior. The discovery of penicillin prolonged lives, which, in turn, means that more grandparents (as well as great-grandparents) and grandchildren get to know each other. However, longer life spans also mean that many adult children and grandchildren now care for older family members over many years (see Chapter 12).

Discovery usually requires dedicated work and years of commitment, but some discoveries occur by chance, called the *serendipity effect*. For example, George de Mestral, a Swiss electrical engineer, was hiking through the woods and was annoyed by burrs that clung to his clothing. Why were they so difficult to remove? A closer examination showed that the burrs had hooklike arms that locked into the open weave of his clothes. The discovery led de Mestral to invent a hook-and-loop fastener. His invention, Velcro—derived from the French words *velour* (velvet) and *crochet*

cultural integration the consistency of various aspects of society that promotes order and stability.

FIGURE 3.4 THE 100 PERCENT AMERICAN

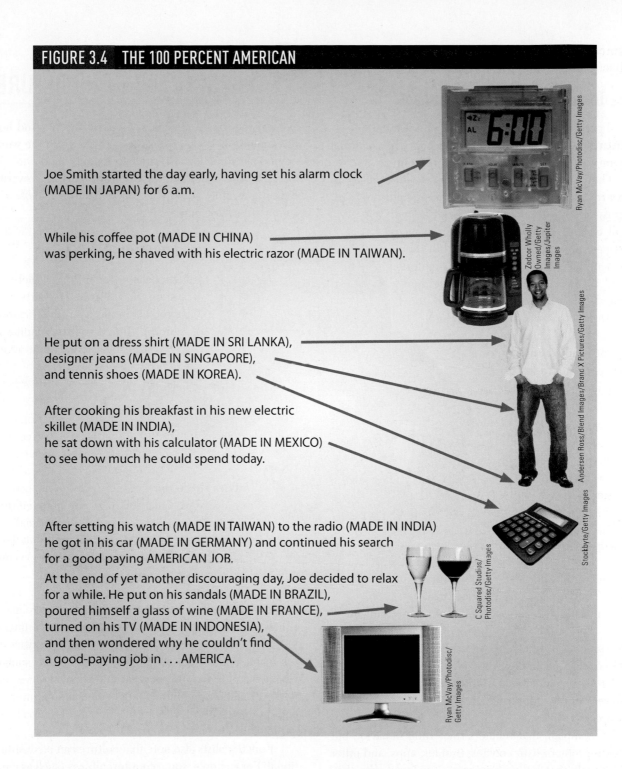

Joe Smith started the day early, having set his alarm clock (MADE IN JAPAN) for 6 a.m.

While his coffee pot (MADE IN CHINA) was perking, he shaved with his electric razor (MADE IN TAIWAN).

He put on a dress shirt (MADE IN SRI LANKA), designer jeans (MADE IN SINGAPORE), and tennis shoes (MADE IN KOREA).

After cooking his breakfast in his new electric skillet (MADE IN INDIA), he sat down with his calculator (MADE IN MEXICO) to see how much he could spend today.

After setting his watch (MADE IN TAIWAN) to the radio (MADE IN INDIA) he got in his car (MADE IN GERMANY) and continued his search for a good paying AMERICAN JOB.

At the end of yet another discouraging day, Joe decided to relax for a while. He put on his sandals (MADE IN BRAZIL), poured himself a glass of wine (MADE IN FRANCE), turned on his TV (MADE IN INDONESIA), and then wondered why he couldn't find a good-paying job in . . . AMERICA.

Ryan McVay/Photodisc/Getty Images
Zedcor Wholly Owned/Getty Images/Jupiter Images
Andersen Ross/Blend Images/Brand X Pictures/Getty Images
Stockbyte/Getty Images
C Squared Studios/Photodisc/Getty Images
Ryan McVay/Photodisc/Getty Images

(hooks)—can now be found on everything from wallets to spacecraft.

EXTERNAL PRESSURES

External pressure for cultural change can take various forms. In its most direct form—war, conquest, or colonization—the dominant group uses force or the threat of force to bring about cultural change in other groups. When the Soviet Union invaded and took over many small countries (e.g., Lithuania, Latvia,

Estonia, Ukraine, and Armenia) after World War II, it forbade citizens to speak their native languages, banned traditions and customs, and turned churches into warehouses.

Pressures for change can also be indirect. For example, some countries (e.g., Thailand, Vietnam, China, and Russia) have reduced their prostitution and international sex trafficking because of widespread criticism by the United Nations and some European countries. The United Nations has no power to intervene in a country's internal affairs but can embarrass nations by publicizing human rights violations (Farley, 2001).

3-6c Technology and Cultural Lag

Some parts of culture change more rapidly than others. **Cultural lag** refers to the gap that occurs when material culture changes faster than nonmaterial culture.

There are numerous examples of cultural lag in modern society because our nonmaterial culture (e.g., norms, values, and laws) hasn't kept up with material culture (e.g., technological advances). Texting while driving has resulted in thousands of fatal car crashes (see Chapters 1 and 2), but many states have neither banned this practice nor enforced the laws they've passed. Because technology is moving faster than laws to regulate and ensure Internet-based communication privacy, there are many breaches. The National Security Agency (NSA), one of the largest U.S. intelligence organizations, has secretly collected phone records of millions of Americans, and spied on leading U.S. technology companies (e.g., Facebook, Google, Apple, Yahoo) as well as European political heads of state who are allies (Timberg and Nakashima, 2013).

Cultural lags can create confusion, ambiguity about what's right and wrong, conflict, and a feeling of helplessness. They also expose contradictory values and behavior. U.S. technology executives have railed against the NSA's mass compilation of data, but the companies themselves collect information on users that advertisers can access (Miller, 2013; Sengupta, 2013). Some consumers complain that retailers track their behavior, but have no problem with cookies, profiles, apps, and other online tools that let e-commerce sites know who they are, how they shop, and what they purchase (Clifford and Hardy, 2013). And as you'll see in several later chapters, even when mores, values, and laws catch up with technological innovations and inventions, they can create unexpected problems such as pollution, drug shortages, and high medical costs.

cultural lag the gap that occurs when material culture changes faster than nonmaterial culture.

SOCIOLOGICAL PERSPECTIVES ON CULTURE

What is the role of culture in modern society? And how does culture help us understand ourselves and the world around us? Functionalist, conflict, feminist, and symbolic interaction scholars offer important insights in answering these and other questions about culture. *Table 3.3* summarizes these perspectives.

3-7a Functionalism

Functionalists focus on society as a system of interrelated parts (see Chapter 1). Similarly, in their analysis of culture, functionalists emphasize the social bonds that attach people to society.

KEY POINTS

For functionalists, culture is the cement that binds society. As you saw earlier, norms and values shape our lives, provide guideposts for our everyday behavior, and promote cultural integration and societal stability. Especially in countries such as the United States that have high immigration rates, cultural norms and values help newcomers adjust to the host society.

All societies have similar strategies for meeting human needs. Cultural universals, such as religious rituals, may play out differently in different countries, but all known societies have religious rituals. Critics often blame popular culture for societal problems such as violence and crime, but popular culture also serves important functions in uniting people with similar interests and producing new technologies (e.g., DVDs, YouTube) (Kidd, 2007).

Functionalists also note that culture can be dysfunctional. For example, some countercultures (such as paramilitary groups) can create chaos by bombing federal buildings and killing or injuring hundreds of people.

CRITICAL EVALUATION

Functionalism is important in showing that shared norms and values create cultural solidarity and stability. In emphasizing culture's role in meeting people's daily needs, however, functionalism often overlooks diversity and social change. For example, a number of influential functionalists have proposed that immigration should

© Edwin Verin | Dreamstime.com

TABLE 3.3 SOCIOLOGICAL PERSPECTIVES OF CULTURE

THEORETICAL PERSPECTIVE	FUNCTIONALIST	CONFLICT	FEMINIST	SYMBOLIC INTERACTIONIST
Level of Analysis	Macro	Macro	Macro and Micro	Micro
Key Points	• Similar beliefs bind people together and create stability. • Sharing core values unifies a society and promotes cultural solidarity.	• Culture benefits some groups at the expense of others. • As powerful economic monopolies increase worldwide, the rich get richer and the rest of us get poorer.	• Women and men often experience culture differently. • Cultural values and norms can increase inequality because of sex, race/ethnicity, and social class.	• Cultural symbols forge identities (that change over time). • Culture (such as norms and values) helps people merge into a society despite their differences .
Examples	• Speaking the same language (English in the United States) binds people together because they can communicate with one another, express their feelings, and influence one another's attitudes and behaviors.	• Much of the English language reinforces negative images about gender ("slut"), race ("honky"), ethnicity ("jap"), and age ("old geezer") that create inequality and foster ethnocentrism.	• Using male language (e.g., "congressman," "fireman," and "chairman") conveys the idea that men are superior to and dominant over women, even when women have the same jobs.	• People can change the language they create as they interact with others. Many Americans now use "police officer" instead of "policeman," and "single person" instead of "bachelor" or "old maid."

be restricted because it dilutes shared U.S. values, overlooking the many contributions that newcomers make to society (see Chapter 10).

3-7b Conflict Theory

Unlike functionalists, conflict theorists argue that culture can generate considerable inequality instead of unify society. Because the rich and powerful determine economic, political, educational, and legal policies for their own benefit and control the mass media, the average American has little power to change the culture—whether

it's low wages, unpopular wars, or corporate corruption (see Chapters 8 and 11).

KEY POINTS

Conflict theorists maintain that many cultural values and norms benefit some members of society more than others. We are taught to work hard, for example, but who's rewarded from the average worker's efforts? As you'll see in later chapters, U.S. taxpayers paid billions of dollars to bail out financial industries that collapsed because of greedy top executives who then used the money to give themselves and their staff higher salaries and bonuses

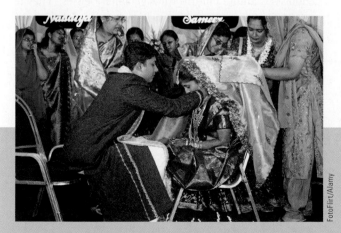

India's high-income families usually have lavish weddings that follow centuries-long traditions (left). For functionalists, the customs and rituals reinforce cultural identity and bind people. For conflict theorists, India's inequality also creates a large population of poor people (right) who are excluded from the traditions and ceremonies that reinforce social bonds.

than ever before. Very rarely, moreover, are wealthy people incarcerated for crime. Thus, those at the top of the socioeconomic ladder can violate laws and mores, including honesty, that other Americans are expected to observe.

Conflict theorists also point out that technology serves primarily the rich. For example, desperate low-income women use online sites to sell their hair, breast milk, and eggs. Selling kidneys is illegal in the United States, but the organ can fetch at least $15,000 on the Internet black market (Stillwell, 2013).

CRITICAL EVALUATION

According to some critics, conflict theorists place too much emphasis on societal discord and downplay a culture's benefits. For instance, cultural values integrate members of a society and decrease divisiveness.

Conflict theorists maintain that mass media conglomerates control popular culture and promote goods and services that most people don't need and/or can't afford (e.g., big screen TVs, the most recent smartphones, and designer clothes). Critics counter that people aren't mindless puppets but can make choices, including not living on credit. Thus, conglomerates shouldn't be blamed for people's irrational behavior.

3-7c Feminist Theories

Feminist scholars agree with conflict theorists that culture creates considerable inequality, but they focus on gender differences. Feminists are also more likely than other theorists to examine multicultural variations across groups.

KEY POINTS

Culture affects gender roles. When media portrayals of women are absent or stereotypical, we get a distorted view of reality (Neuhaus, 2010). Feminist scholars also emphasize that subcultures—for example, female students or single mothers—may experience culture differently than their male counterparts do because women typically have fewer resources, such as income and power (see Chapters 9 and 11).

Unequal access to resources often results in women having fewer choices than men, and living under laws and customs that subordinate women. In the United States, a nonmarital birth typically impoverishes women but not

men. In Japan, where rape is often kept hush-hush, police and judges rarely take the offense seriously (Makino, 2009). In many Islamic societies, men dictate women's attire and behavior (Heath, 2008; Zahedi, 2008). In many Latin American countries, violence against women is not only "firmly entrenched" but has recently spiked (Llana and Brodzinsky, 2012). In India, gang rapes are a "routine and a largely invisible crime." Also, young women who move to a city are policed by a network of males. If the women violate village moral codes, including owning a cell phone, they disappear or are quickly married to village men of their parents' choosing (Barry, 2013; Barry and Choksi, 2013; Mahr, 2013; Srivastava and Mehrotra, 2013).

FOR SALE

CRITICAL EVALUATION

Feminist analyses expand our understanding of cultural components that other theoretical perspectives ignore or gloss over. Like conflict theorists, however, feminist theorists often emphasize discord rather than how culture integrates women and men into society. Another weakness is that feminist scholars tend to downplay social class inequality by focusing on the cultural experiences of low-income and minority women but not of their male counterparts.

3-7d Symbolic Interaction

Unlike functionalists and conflict theorists, symbolic interactionists examine culture through micro lenses. They're most interested in understanding how people create, maintain, and modify culture.

KEY POINTS

Interactionists study how culture influences our everyday lives. Language, you'll recall, shapes our views and behavior. People within and across societies create a variety of symbols that may change over time. Technology has also changed many people's communication patterns. Some complain, for example, that texting dominates our lives. A much viewed YouTube clip shows the bride pulling out her cell phone from under her veil and texting with one hand as she walks down the aisle with her father (Pflaumer, 2011). On the other hand, students become more engaged when they're required to use technology, such as tweeting weekly observations that illustrate their reading assignments (Lang, 2013).

Symbolic interactionists also note that our values and norms, like other components of culture, aren't superimposed by some unknown external force. Instead, as people construct their perception of reality, they create, change, and reinterpret values and norms through interaction with others. Peer pressure, for example, can either encourage or discourage inappropriate language or behavior.

CRITICAL EVALUATION

Micro approaches are useful in understanding what culture means to people and how these meanings differ across societies. However, symbolic interactionists don't offer a systematic framework that explains *how* people create and shape culture or develop shared meanings of reality. Why, for instance, are some of us less polite than others even though we share the same cultural values and norms? Another weakness is that interactionists don't address the linkages between culture and subcultures. For example, language bonds people, but interactionists say little about how organized groups (such as the English-only movement) try to maintain control over language to promote their beliefs.

STUDY TOOLS 3

READY TO STUDY? IN THE BOOK, YOU CAN:

☐ Check your understanding of what you've read with the Test Your Learning Questions provided on the chapter review card at the back of the book.

☐ Rip out the chapter review card for a handy summary of the chapter and key terms.

ONLINE AT CENGAGEBRAIN.COM YOU CAN:

☐ Prepare for tests with quizzes.

☐ Review the key terms with Flash Cards.

☐ Play games to master concepts.

© Anelina/Shutterstock.com

4 | Socialization

LEARNING OBJECTIVES

4-1 Define and illustrate socialization and explain its importance.

4-2 Describe the nature versus nurture debate.

4-3 Compare social learning and symbolic interaction theories of socialization.

4-4 Describe and illustrate five socialization agents.

4-5 Explain how socialization changes throughout life.

4-6 Explain when and how resocialization occurs.

After you finish this chapter go to **PAGE 80** for **STUDY TOOLS.**

Two of MTV's most popular reality shows are *16 and Pregnant* and *Teen Mom*. The producers say that the goal of the programs is to decrease teen pregnancy by showing the harmful effects of not using protection or birth control, and the resulting struggles in raising a baby. Critics argue that the shows glamorize teen pregnancy and motherhood, make the teen moms instant celebrities, and ignore the long-term negative impact on the mothers' and children's healthy social development (Thompson, 2010). Both sides are talking about **socialization**, the lifelong process through which people learn culture and become functioning members of society.

4-1 SOCIALIZATION: ITS PURPOSE AND IMPORTANCE

Socialization is critical in all societies. To understand why, let's begin by looking at the purpose of socialization.

4-1a What Is the Purpose of Socialization?

Socialization—from childhood to old age—can be relatively smooth or bumpy, depending on factors such as age, sex, race/ethnicity, and social class. Generally, however, socialization has four key functions that range from

What do you think?

I'm always the same person, no matter where I am or who's around.

1	2	3	4	5	6	7

strongly agree   strongly disagree

providing us with a social identity to transmitting culture to the next generation.

SOCIALIZATION ESTABLISHES OUR SOCIAL IDENTITY

Have you ever thought about how you became the person you are today? Sociology professors sometimes ask their students to give 20 answers to the question "Who am I?" (Kuhn and McPartland, 1954). How would you respond? You would probably include a variety of descriptions such as college student, single or married, female or male, and son or daughter. Your answers would show a sense of your *self* (a concept we'll examine shortly). You are who you are largely because of socialization.

SOCIALIZATION TEACHES US ROLE TAKING

Why do you act differently in class than with your friends? Because we play different roles in different settings. A *role* is the behavior expected of a person in a particular social position (see Chapter 5). The way we interact with a parent is typically very different from the way we talk to an employer, a child, or a professor. We all learn appropriate roles through socialization.

SOCIALIZATION CONTROLS OUR BEHAVIOR

In learning appropriate roles, we absorb values and a variety of rules about how we should (and shouldn't)

interact in everyday situations. If we follow the rules, we're usually rewarded or at least accepted. If we break the rules, we may be punished. By teaching us to conform to societal expectations, socialization controls our behavior and makes life more orderly and predictable.

We conform because we've internalized societal norms and values. **Internalization** is the process of learning cultural behaviors and expectations so deeply that we accept them without question. Obeying laws, paying bills on time, and respecting teachers are examples of internalized behaviors.

SOCIALIZATION TRANSMITS CULTURE TO THE NEXT GENERATION

Each generation passes its culture on to the next generation. The culture that is transmitted, as you saw in Chapter 3, includes language, beliefs, values, norms, and symbols.

socialization the lifelong process through which people learn culture and become functioning members of society.

internalization the process of learning cultural behaviors and expectations so deeply that we accept them without question.

The Harlow Studies and Emotional Attachment

In the early 1960s, psychologists Margaret and Harry Harlow (1962) conducted several studies on infant monkeys. In one group, a "mother" made of terry cloth provided no food, while a "mother" made of wire did so through an attached baby bottle containing milk. In another group, the cloth mother provided food but the wire mother didn't. Regardless of which mother provided milk, when both groups of monkeys were frightened, they clung to the cloth mother. The Harlows concluded that physical contact and comfort were more important to the infant monkeys than nourishment. Since then, some sociologists have cited the Harlow studies to argue that emotional attachment may be even more critical than food for human infants. Do you see any problems with sociologists' generalizing the results of animal studies to humans?

EMOTIONAL ATTACHMENT

Nina Leen/Time Life Pictures/Getty Images

4-1b Why Is Socialization Important?

Social isolation can be devastating. An example of its negative effects is Genie. When Genie was 20 months old, her father decided she was "retarded," put her in a wire mesh cage, and locked her away in a back room with the curtains drawn and the door shut. Genie's father frequently beat her with a wooden stick and no one in the house was allowed to speak to her. When she was discovered in 1970 at age 13, a psychiatrist described Genie as "unsocialized, primitive, and hardly human." Except for high-pitched whimpers, she never spoke. Genie had little bowel control, experienced rages, rubbed her face frantically, and tried to hurt herself. After living in a rehabilitation ward and a foster home, she learned to eat normally, was toilet-trained, and gradually developed a vocabulary, but her language use never progressed beyond that of a 3- or 4-year-old (Curtiss, 1977).

The research on children who are isolated (such as Genie) or institutionalized (such as orphans) demonstrates that socialization is critical to our development. Behaviors such as talking, eating with utensils, and controlling our bowel movements don't come naturally. Instead, we learn to do all of these things beginning in infancy. When children are deprived of interaction with other people, they don't develop the characteristics that most of us see as normal and human.

4-2 NATURE AND NURTURE

Biologists focus on the role of heredity (or genetics) in human development. In contrast, most social scientists, including sociologists, underscore the importance of learning, socialization, and culture. This difference of opinion is often called the *nature–nurture debate* (see *Table 4.1*).

4-2a How Important Is Nature?

Those who argue that nature (biology) shapes behavior point to two kinds of evidence—developmental and

TABLE 4.1	THE NATURE–NURTURE DEBATE
Nature	Human development is . . .
	Innate
	Biological, physiological
	Due largely to heredity
	Fairly fixed
Nurture	Human development is . . .
	Learned
	Psychological, social, cultural
	Due largely to environment
	Fairly changeable

health differences between males and females, and cases of unsuccessful sex reassignment.

DEVELOPMENTAL AND HEALTH DIFFERENCES

Boys mature more slowly than girls, get sick more often, and are less likely to have mastered the self-control and fine-motor skills necessary for a successful start in school. Boys are also at greater risk than girls for most of the major learning and developmental disorders—as much as four times more likely to suffer from autism, attention deficit disorder, and dyslexia. Girls, however, are at least twice as likely as boys to suffer from depression, anxiety, and eating disorders (Eliot, 2012).

Among adults, the senses of smell and taste are more acute in women than in men, and hearing is better and lasts longer in women than in men. Women, however, have a higher risk of developing diabetes. Some conditions (such as migraine headaches and breast cancer) are more common in women, whereas others (such as hemophilia and skin cancer) are more common in men (McDonald, 1999; Kreeger, 2002a, 2002b).

UNSUCCESSFUL SEX REASSIGNMENT

Some scientists point to unsuccessful attempts at sex reassignment as another example favoring nature over nurture. In the 1960s, John Money, a highly-respected psychologist, published numerous articles and books in which he maintained that gender identity isn't firm at birth but is determined as much by culture and nurture as by hormones (see, for example, Money and Ehrhardt, 1972). His views were based on the case of David Reimer, a child who underwent sex reassignment after his penis was mutilated in a botched circumcision. His parents, upon Money's recommendation, raised David as a girl (Brenda). Money maintained that the reassignment was successful. A biologist and a psychiatrist who followed up on David/Brenda's case in the 1990's, however, concluded that the sex reassignment had not been successful (Diamond and Sigmundson, 1997).

The Case of Brenda/David

In 1963, twin boys were being circumcised when the penis of one of the infants, David Reimer, was accidentally burned off. Encouraged by John Money, a highly respected psychologist, the parents agreed to raise David as "Brenda." The child's testicles were removed, and surgery to construct a vagina was planned. Money reported that the twins were growing into happy, well-adjusted children, setting a precedent for sex reassignment as the standard treatment for 15,000 newborns with similarly injured genitals (Colapinto, 1997, 2001).

In the mid-1990s, a biologist and a psychiatrist followed up on Brenda's progress and concluded that the sex reassignment had not been successful. Almost from the beginning, Brenda refused to be treated like a girl. When her mother dressed her in frilly clothes as a toddler, Brenda tried to rip them off. She preferred to play with boys and stereotypical boys' toys such as machine guns (Diamond and Sigmundson, 1997).

When she was 14, Brenda rebelled and stopped living as a girl: She refused to wear dresses, urinated standing up, turned down vaginal surgery, and decided she would either commit suicide or live as a male. When his father finally told David the true story of his birth and sex change, David recalls that "all of a sudden everything clicked. For the first time things made sense and I understood who and what I was" (Diamond and Sigmundson, 1997: 300).

David had a mastectomy (breast removal surgery) at the age of 14 and underwent several operations to reconstruct a penis. At age 25, he married an older woman and adopted her three children. At age 38, he committed suicide. Most suicides have multiple reasons, but some speculated that David committed suicide because of the "physical and mental torments he suffered in childhood that haunted him the rest of his life" (Colapinto, 2004: 96). David's experience suggests to some scientists that nature outweighs nurture in shaping a person's gender identity.

Reuters/Landov

David Reimer was raised as a girl, Brenda, until he was 14.

4-2b How Important Is Nurture?

Most sociologists maintain that nurture is more significant than nature because socialization and culture shape human behavior. They point to successful sex reassignment cases (see Vitello, 2006) and two types of data to support their argument: cross-cultural variations in male violence, and the environment's effect on biology.

CROSS-CULTURAL VARIATIONS IN MALE VIOLENCE

We often hear that males are "naturally" more aggressive than females because their glands produce more testosterone, the dominant male hormone. If men were innately aggressive due to biology, they would be equally violent across all societies. This isn't the case. The proportion of women who have ever suffered physical violence by a male partner varies considerably: 90 percent in Pakistan, 61 percent in Peru, 36 percent in the United States, and 13 percent in Japan (Chelala, 2002; World Health Organization, 2005; Black et al., 2011; Trust-Law, 2011). All mass murderers (those who have killed a large number of people during one incident) have been men, but most of the mass murders, usually committed by white males, have occurred in the United States (Christakis, 2012; Farhi, 2012). Such variations reflect cultural laws and practices and other environmental factors (nurture) rather than biology or genetics (nature) (Chesney-Lind and Pasko, 2004).

HOW ENVIRONMENT AFFECTS BIOLOGY

Much research suggests that environment (nurture) influences children's genetic makeup (nature). For example, birth defects associated with prenatal alcohol exposure can occur in the first 3 to 8 weeks of pregnancy, before a woman even knows she's pregnant. A woman's single drinking "binge"—lasting four hours or more—can permanently damage an unborn child's brain. A man who drinks heavily may have genetically damaged sperm that also leads to birth defects. Thus, alcohol abuse (an environmental factor) by either a woman or a man can have a devastating, irreversible, and lifelong negative impact on a child (Denny et al., 2009; Warren, 2012).

In a growing field known as "fetal origins," scientists are finding that the nine months of pregnancy constitute the most consequential period of our lives, permanently influencing the wiring of the brain, the functioning of organs such as the heart and liver, and behavior. That is, the kind and quantity of nutrition you received in the womb;

Is children's obesity due to nature, nurture, or both?

the pollutants, drugs, and infections you were exposed to during gestation; your parents' health and eating habits; and your mother's stress level during pregnancy—all these factors have lasting effects in infancy, childhood, and adulthood (Begley, 2010; Paul, 2010).

Adult behavior can influence a child's biological makeup in other ways. Physical, psychological, or sexual abuse can impair a child's developing brain, and these changes can trigger disorders such as depression in adulthood (see Chapter 12). Thus, childhood mistreatment can blunt biological development and lead to behavioral and emotional problems.

Even when identical twins look alike because they have the same genetic makeup, socialization can affect their interests, personalities, lifestyles, and life outcomes.

4-2c What Can We Conclude About the Nature–Nurture Debate?

A growing number of sociologists—who are calling for a "genetically informed sociology"—maintain that the nature *or* nurture debate is becoming outdated in understanding socialization processes and outcomes (Ledger, 2009). Ignoring the impact of genetics, they argue, leads to an incomplete understanding of behavior because genes shape our lives and could help explain why there's so much variation across families and other groups.

People have at least 52 characteristics—such as aggression, leadership traits, and cognitive ability—that are partially inherited (Freese, 2008). Our social environment can enhance or dampen these genetic characteristics. For example, children who are genetically predisposed to obesity don't always become overweight if parents discourage overeating and encourage physical recreational activities (Martin, 2008). Social scientists, including sociologists, who study the relationship between genetics (nature) and the environment (nurture) maintain that research that combines both aspects enhances our understanding of how social factors affect genetic predispositions (Schnittker, 2008; D'Onofrio and Lahey, 2010; Shanahan et al., 2010).

© Lenetstan/Shutterstock.com

4-3 SOCIOLOGICAL EXPLANATIONS OF SOCIALIZATION

Functionalism provides a foundation for understanding the purposes of socialization described at the beginning of this chapter. However, functionalists don't tell us *how* socialization works on a micro level. There are well-known psychological and psychosocial theories of human development (e.g., Sigmund Freud, Erik Erikson, Jean Piaget, and Lawrence Kohlberg). In sociology, however, two influential approaches that explain how socialization works on a micro level are *social learning theories* and *symbolic interaction theories* (*Table 4.2* on page 66 summarizes these perspectives). Let's begin with social learning theories.

4-3a Social Learning Theories

The central notion of **social learning theories** is that people learn new attitudes, beliefs, and behaviors through social interaction, especially during childhood. The learning is direct and indirect, and a result of observation, reinforcement, and imitation (Bandura and Walters, 1963; Mischel, 1966; Lynn, 1969).

DIRECT AND INDIRECT LEARNING

Reinforcement occurs when we receive direct or indirect rewards or punishments for particular behaviors. A little girl who puts on her mother's makeup may be told that she's cute, but her brother will be scolded ("boys don't wear makeup"). Children also learn through indirect reinforcement. If a little boy's male friends are punished for crying, he will learn that "boys don't cry."

Children also learn through *observation* and *imitation*. Even when children aren't directly rewarded or punished for "behaving like boys" or "behaving like girls," they learn about gender by watching who does what in their families. A father who is rarely at home because he's always working sends the message that men are supposed to earn money. A mother who is always complaining about being overweight or old sends the message that women are supposed to be thin and young. Because parents are emotionally important to their children, they're typically a child's most powerful **role models**, people we admire and whose behavior we imitate. Other role models, as you'll see shortly, include siblings, grandparents, teachers, friends, and even celebrities.

LEARNING AND PERFORMING

Social learning theorists also distinguish between *learning* and *performing* behavior. Children and adults learn norms and roles through observation, but they don't always imitate the behavior. For example, children may see their friends cheat in school but don't do so themselves. Adults, similarly, may see their coworkers steal office supplies but buy their own. Social learning theorists maintain that we behave as we do, then, because our society teaches us what's appropriate and inappropriate.

CRITICAL EVALUATION

Social learning theories help us understand why we behave as we do, but much of the emphasis is on early socialization rather than on what occurs throughout life. Social learning theories don't explain why reinforcement and modeling affect some children more than others, especially those in the same family. There may be personality differences, but if learning is as effective as social learning theorists maintain, siblings' attitudes and behavior should be more similar than different. This is often not the case, even with identical twins.

social learning theories maintain that people learn new attitudes, beliefs, and behaviors through social interaction, especially during childhood.

role models people we admire and whose behavior we imitate.

TABLE 4.2 KEY ELEMENTS OF SOCIALIZATION THEORIES

Social Learning Theories	Symbolic Interaction Theories
• Social interaction is important in learning appropriate and inappropriate behavior.	• The self emerges through social interaction with significant others.
• Socialization relies on direct and indirect reinforcement.	• Socialization includes role taking and controlling the impression we give to others.
Example: Children learn how to behave when they are scolded or praised for specific behaviors.	*Example:* Children who are praised are more likely to develop a strong self-image than those who are always criticized.

Most social learning theories ignore factors such as birth order, which brings different advantages and disadvantages. Except for affluent families, larger families have fewer resources for the second, third, and later children. The first child may be disciplined and monitored more, but also has higher grades than her or his younger siblings, and is more likely to attend college. In contrast, later-born children have fewer rules and less parental monitoring regarding homework, lower grades, and are less likely to receive financial support to attend college (Conley, 2004; Hotz and Pantano, 2013).

4-3b Symbolic Interaction Theories

Symbolic interaction theories have had a major impact in explaining social development. Sociologists Charles Horton Cooley (1864–1929), George Herbert Mead (1863–1931), and Erving Goffman (1922–1982) were especially influential in showing how social interaction shapes socialization.

CHARLES HORTON COOLEY: EMERGENCE OF THE SELF AND THE LOOKING-GLASS SELF

Newborn infants lack a sense of **self**, an awareness of their social identity. Gradually, they begin to differentiate themselves from their environment and to develop a sense of self.

After carefully observing the development of his young daughter, Charles Horton Cooley (1909–1983) concluded that children acquire a sense of who they are through their interactions with others, especially by imagining how others view them. The sense of self, then, isn't innate but emerges out of social relationships. Cooley called this social self the *reflected self*, or the **looking-glass self**, a self-image based on how we think others see us. He proposed that the looking-glass self develops in an ongoing process of three phases:

▶ **Phase 1: Perception.** We imagine how we appear to other people and how they *perceive* us ("She thinks I'm attractive" or "I bet he thinks I'm fat").

▶ **Phase 2: Interpretation of the perception.** We imagine how others *judge* us ("She's impressed with me" or "He's disgusted with the way I look").

▶ **Phase 3: Response.** We experience *self-feelings* based on what we regard to be others' judgments of us. If we think others see us in a favorable light, we may feel proud, happy, or self-confident ("I'm terrific"). If we think others see us in a negative light, we may feel angry, embarrassed, or insecure ("I'm pathetic").

Our interpretation of others' perceptions (phase 2) may be totally wrong. The looking-glass self, remember, refers to how we *think* others see us rather than how they *actually* see us. However, our perceptions of other people's views—whether we're right or wrong—affect our self-image, which then affects our behavior.

Parents who encourage their children to express themselves help them develop a positive self-image.

self an awareness of one's social identity.

looking-glass self a self-image based on how we think others see us.

Cooley focused on how children acquire a sense of who they are through their interactions with others, but he noted that the process of forming a looking-glass self doesn't end in childhood. Instead, our self-concept may change over time because we reimagine ourselves as we think others see us—attractive or ugly, interesting or boring, intelligent or stupid, graceful or awkward, and so on.

Sometimes we're aware of the process shaping our looking-glass self. Consider your own self-concept: If others keep telling you what a great student you are, you'll likely see yourself as intelligent. In many settings, however—such as public places, large classrooms, and sports events—the looking-glass self is irrelevant because there's relatively little interpersonal interaction and the other people are strangers who aren't very significant in our lives.

GEORGE HERBERT MEAD: DEVELOPMENT OF THE SELF AND ROLE TAKING

Cooley described how an individual's sense of self emerges, but not how it develops. George Herbert Mead, one of Cooley's colleagues, took up this task. For Mead (1934), the most critical social interaction occurs in the family, the foundation of socialization.

Our self develops, according to Mead, when we learn to differentiate the *me* from the *I*—two parts of the self. The *I* is creative, imaginative, impulsive, spontaneous, nonconformist, self-centered, and sometimes unpredictable. The *me* that has been successfully socialized is aware of the attitudes of others, has self-control, and has internalized social roles. Instead of impulsively and selfishly grabbing another child's toys, for example, as the *I* would do, the *me* asks for permission to use someone's toys and shares them with others.

For Mead, the *me* forms as children engage in **role taking**, learning to take the perspective of others. Children gradually acquire this ability early in the socialization process through three sequential stages:

1. **Preparatory stage (roughly birth to 2 years).** An infant doesn't distinguish between the self and others. The *I* is dominant, while the *me* is forming in the background. In this stage, children learn through imitation. They may mimic daddy's shaving or mommy's angry tone of voice without really understanding the parent's behavior. In this exploratory stage, children engage in behavior that they rarely associate with words or symbols, but they begin to understand cause and effect (e.g., crying leads to being picked up). Gradually, as the child begins to recognize

others' reactions (to form a looking-glass self), he or she develops a self.

2. **Play stage (roughly 2 to 6 years).** The child begins to use language and to understand that words (such as *dog* and *cat*) have a shared cultural meaning. Through play, children begin to learn role taking in two ways. First, they emulate the words and behavior of **significant others**, the people who are important in one's life, such as parents (or other primary caregivers), siblings, and grandparents. The child learns that he or she has a self that is distinct from that of others, that others behave in many different ways, and that others expect her or him to behave in specific ways. In other words, the child learns social norms (see Chapter 3).

   In the play stage, the child moves beyond imitation and acts out imagined roles ("I'll be the mommy and you be the daddy"). The play stage involves relatively simple role taking because the child plays one role at a time but doesn't yet understand the relationships between roles. This stage is crucial, according to Mead, because the child *is learning to take the role of the other*. For the first time, the child tries to imagine how others behave or feel. The *me* grows stronger because the child is concerned about the judgments of significant others.

   Also in the play stage, children experience **anticipatory socialization**, the process of learning how to perform a role they don't occupy. By playing "mommy" or "daddy," children prepare themselves for eventually becoming parents. Anticipatory socialization continues in later years, for example, when expectant parents attend childbirth classes, job seekers practice their skills in mock interviews, and many high school students visit campuses and attend orientation meetings to prepare for college life.

3. **Game stage (roughly 6 years and older).** This stage involves acquiring the ability to understand connections between roles. The child must "not only take the role of the other ... but must assume the various roles of all participants in the game, and govern his action accordingly" (Mead, 1964: 285).

   Mead used baseball to illustrate this stage. In baseball (or other organized games and activities), the child

© OtnaYdur/Shutterstock.com

---

role taking learning to take the perspective of others.

significant others the people who are important in one's life, such as parents (or other primary caregivers), siblings, and grandparents.

anticipatory socialization the process of learning how to perform a role one does not yet occupy.

FIGURE 4.1 MEAD'S THREE STAGES IN DEVELOPING A SENSE OF SELF

STAGE 1: **Preparatory Stage** (under age 2)

No distinction between self and others; the child is self-centered and self-absorbed

Learns through observation

STAGE 2: **Play Stage** (aged 2 to about 6)

Distinguishes between self and others

Imitates significant others (usually parents)

Learns role taking, assuming one role at a time in "let's pretend" and other play that teaches **anticipatory socialization**

STAGE 3: **Game Stage** (aged 6 and older)

Understands and anticipates multiple roles

Connects to societal roles through the **generalized other**

plays one role at a time (such as batter) but understands and anticipates the actions of other players (pitcher, shortstop, runners on bases) on both teams. To successfully participate in a game of baseball, the individual must be able to anticipate the roles of others. The same is true of participating in society. Thus, as children grow older and interact with a wider range of people, they learn to anticipate, respond to, and fulfill a variety of social roles.

The game stage enables the child to understand and take the role of the **generalized other**, people who don't have close ties to a child but who influence her or his internalization of society's norms and values. The generalized other thus exerts control over the *I* and ensures some predictability in life. For sociologists, the development of the generalized other is a central feature of socialization because the *me* becomes an integral part of the self (see *Figure 4.1*).

Even after the generalized other has developed, the *me* never fully controls the *I*, even in adulthood. We sometimes break rules or act impulsively, even though we know better. When we are frustrated,

not feeling well, or angry, we may lash out against other people even though they haven't done anything wrong.

ERVING GOFFMAN: STAGING THE SELF IN EVERYDAY LIFE

Cooley and Mead described how the self and role taking emerge and develop during early socialization. Erving Goffman (1959, 1969) extended these analyses by showing that we interact differently in different settings throughout adulthood. Goffman proposed that social life mirrors the theater because we're like actors: We engage in "role performances," want to influence an "audience," and can have considerable control over the image that we project while we're "on stage." (We'll explore these concepts in greater detail in Chapter 5.)

In a process that Goffman called **impression management**, we provide information and cues to others to present ourselves in a favorable light while downplaying or concealing our less appealing characteristics. Being successful in this presentation of the self requires managing three types of expressive resources. First, we try to control the *setting*—the physical space, or "scene," where the interaction takes place. In the classroom, "a professor may use items such as chalk, lecture notes, computers, videos, and desks to facilitate a class and show that he or she is an excellent teacher" (Sandstrom et al., 2006: 105).

A second expressive resource is controlling *appearance*, such as clothing and titles that convey information about our social status. When physicians or professors use the title "Doctor," for example, they're telling the "audience" that they expect respect.

The third expressive resource is *manner*—the mood or style of behavior we display that sends important messages to the audience. For example, faculty members regularly manage their manner when interacting with students. When you email or text a professor a question about your research project, the response is usually encouraging ("Have you also thought about...") even though the professor may actually be annoyed ("Your project has nothing to do with this course; stop wasting my time").

Purestock/Getty Images

generalized other people who don't have close ties to a child but who influence her or his internalization of society's norms and values.

impression management the process of providing information and cues to others to present oneself in a favorable light while downplaying or concealing one's less appealing characteristics.

Sometimes, no matter how much we try to manage setting, appearance, and manner, we slip up in our role performances. This leads to discomfort and embarrassment for everyone involved, but the audience is typically tactful:

> *Imagine that you are giving a presentation in one of your classes and, just as you make a key point, a large droplet of saliva lands near someone sitting in the front row. Although several of your classmates cannot help but notice it, none of them are likely to shout out, "Hey, you just about spit on someone in the first row!" Instead, they will probably act as if nothing unusual or embarrassing took place (Sandstrom et al., 2006: 109).*

As we move from one situation to another, according to Goffman, we maintain self-control, for example, by avoiding emotional outbursts and altering our facial expressions and verbal tones. Thus, all of us engage in impression management practically every day.

CRITICAL EVALUATION

Symbolic interactionists have provided major insights about socialization, particularly of young children, and have shown that fitting into our social world is a complex process that continues through adulthood. Like other theories, however, symbolic interaction has its limitations.

First, it's not clear why some children have a positive looking-glass self and are successful later in life even when the cues are consistently negative, as when children grow up in abusive homes or attend low-quality schools.

Second, essential concepts such as *self, me,* and *I* are vague. Because the concepts are imprecise, it's difficult, if not impossible, to measure them.

Third, stage theories, such as Mead's, maintain that children automatically pass through play and game stages as they get older. Some critics argue that the stages aren't as rigid as symbolic interactionists claim, may overlap, or vary depending, for example, on whether parents and other primary caregivers are actively involved in the child's upbringing and provide enriching interaction. Even then, some children never develop the role of the generalized other (see Ritzer, 1992).

Fourth, some have questioned the value of the concept of the generalized other in understanding early socialization processes. Because children interact in many social contexts (home, preschool, play groups), they don't simply assume the role of one generalized other, but many. That is, as children get older, they may have several **reference groups**—groups of people who shape an individual's self-image, behavior, values, and attitudes in different contexts (Merton and Rossi, 1950; Shibutani, 1986).

Fifth, interactionists credit people with more free will than they have. Individuals don't always have the power or ability to affect others' reactions or their own lives. For example, impression management is less common among lower than higher socioeconomic groups, whose members have internalized such skills to be successful in jobs. Also, how people react to us may reflect characteristics (e.g., gender, age, or race/ethnicity) that we can't change (Powers, 2004).

Finally, a major criticism of Cooley, Mead, Goffman, and other interactionists is that they tend to downplay or ignore macro-level factors that affect our development. As the next two sections show, institutions such as the family, education, economy, and popular culture have a major impact on socialization. Let's begin by looking at socialization agents who play important roles in our lives.

4-4 PRIMARY SOCIALIZATION AGENTS

By fretting, crying, and whining, infants teach adults when to feed, change, and pick them up. Babies are actively engaged in the socialization process, but the family, friends, peer groups, teachers, and the media are some of the primary **socialization agents**—the individuals, groups, or institutions that teach us how to participate effectively in society.

© Gelpi JM/Shutterstock.com

4-4a Family

Parents, siblings, grandparents, and other family members play a critical role in our socialization. Parents, however, are the first and most influential socialization agents.

---

reference groups groups of people who shape an individual's self-image, behavior, values, and attitudes in different contexts.

socialization agents the individuals, groups, or institutions that teach us how to participate effectively in society.

TABLE 4.3 PARENTING STYLES

Parenting Style	Characteristics	Example
Authoritarian	Very demanding, controlling, punitive	"You can't borrow the car because I said so."
Authoritative	Demanding, controlling, warm, supportive	"You can borrow the car, but be home by curfew."
Permissive	Not demanding, warm, indulgent, set few rules	"Borrow the car whenever you want."
Uninvolved	Neither supportive nor controlling	"I don't care what you do; I'm busy."

HOW PARENTS SOCIALIZE CHILDREN

The purpose of socialization is to enable a child to regulate her or his own behavior and to make responsible decisions. In teaching their children social rules and roles, parents rely on several learning techniques, such as reinforcement, that you've just read about. Parents also manage many aspects of the environment that influence a child's social development: They often choose the neighborhood (which then determines what school a child attends), decorate the child's room in a masculine or feminine style, provide her or him with particular toys and books, and arrange enriching social events and other activities (such as sports, art, and music).

PARENTING STYLES

Four common parenting styles affect socialization (see *Table 4.3*). In *authoritarian parenting*, parents tend to be harsh, unresponsive, and rigid, using their power to control a child's behavior. *Authoritative parenting* is warm, responsive, and involved, but parents set limits and expect appropriately mature behavior from their children. *Permissive parenting* is lax: Parents set few rules but are usually warm and responsive. *Uninvolved parenting* is indifferent and neglectful. Parents focus on their own needs rather than those of the children, spend little time interacting with the children, and know little about their interests or whereabouts (Baumrind, 1968, 1989; Maccoby and Martin, 1983; Aunola and Nurmi, 2005).

Healthy child development is most likely in authoritative homes because parents are consistent in combining warmth, monitoring, and discipline. Authoritative parenting tends to produce children who are self-reliant, achievement-oriented, and successful in school. Adolescents in households with authoritarian, permissive, or uninvolved parents tend to have poorer psychosocial development, lower school grades, lower self-reliance, higher levels of delinquent behavior, and are more likely to be swayed by harmful peer pressure

(e.g., to use drugs and alcohol) (Hillaker et al., 2008; Meteyer and Perry-Jenkins, 2009).

Such findings vary by social class and race/ethnicity, however. For many Latino and Asian immigrants, authoritarian parenting produces positive outcomes, such as academic success. This parenting style is also effective in safeguarding children who are growing up in low-income neighborhoods with high crime rates and drug peddling (Brody et al., 2002; Pong et al., 2005).

SIBLINGS

Siblings, like parents, are important socialization agents. Older siblings may transmit beliefs that having sex at an early age, smoking, drinking, and marijuana use "aren't a big deal" (Altonji et al., 2010). On the other hand, older

AP Images/Yonhap/Shin Young-kuyn

During the last decade, numerous summer camps have sprung up in South Korea where children ages 7 to 19 live in military-style barracks, undergo drills and exercise from dawn to dusk, eat very simple meals, and have no access to computers, television, or cell phones. Most are sent by parents "who realize that their pampered offspring need more discipline to become better students and grow into conscientious adults" (Glionna, 2009: A1). Some of the kids' offenses have included getting a low grade on an important test, accidentally breaking a window in the family's apartment, and, most commonly, talking back to their parents.

siblings are positive role models when they encourage their younger brothers and sisters to do well in school, to stay away from friends who get in trouble, and to get along with people. Siblings can also have a positive impact by helping their younger brothers and sisters with their homework and protecting them from neighborhood and school bullies. During adulthood, siblings often act as confidantes and offer financial and emotional assistance during stressful times (McCoy et al., 2002; Whiteman et al., 2011; McHale et al., 2012).

GRANDPARENTS

In many ways, grandparents are the glue that keeps a family close. Some grandparents and grandchildren live far apart and rarely see each other, but the majority of grandparents see their grandchildren often, frequently do things with them, and offer them emotional and instrumental support (e.g., listening and giving them money). Many recent Asian immigrants live in extended families and are more likely to do so than any other group, including Latinos. In such co-residence, grandparents are often "historians" who transmit values and cultural traditions to their grandchildren even if there are language barriers (Benokraitis, 2014; see also Chapter 12).

Grandparents are sometimes *surrogates* because they provide regular care or raise the grandchildren. Whether custody is legal or informal, about 10 percent of all U.S. children live with a grandparent, up from 5 percent in 1990. About 2.7 million grandparents are raising their children; 33 percent of these households have no parent present (Ellis and Simmons, 2014; "National Grandparents Day…," 2014).

The most common pattern, however, is *living-with* grandparents who have the grandchild in their own home or, less often, live in the home of a grandchild's parents. Living-with grandparents take on child-rearing responsibilities either because their children haven't yet moved out of the house or because teenage or adult parents can't afford to live on their own with their young children. These living arrangements have increased the number of **multigenerational households**, homes in which three or more generations live together. In effect, then, grandparents are important socialization agents. (We'll examine the influence of religion on socialization in Chapters 7 and 9–14.)

4-4b Play, Peer Groups, and Friends

As you saw earlier, Mead believed that play and games are important in children's development. They help us to become more skilled in using language and symbols, learn role playing, and internalize roles that we

Grandparents often provide stability in family relationships and a continuity of family rituals and values. When a teacher asked her class of 8-year-olds what they liked best about a grandparent, the essays included comments such as "They don't say 'Hurry up'" and "When they take us for a walk, they slow down for things like pretty leaves and caterpillars."

don't necessarily enact (the generalized other). A **peer group** consists of people who are similar in age, social status, and interests. All of us are members of peer groups, but such groups are especially influential until about our mid-20s. After that, coworkers, spouses, children, and close friends are usually more important than peer groups in our everyday lives.

PLAY

Play serves several important functions. First, it promotes cognitive development. Whether it's doing simple five-piece puzzles or tackling complex video games such as *Sim City* and *Roller Coaster Tycoon*, play encourages children to think, formulate strategies, and budget and manage resources. From an early age, however, play is generally gender typed. Among preschoolers, when playing alone, girls prefer feminine activities (such as playing with dolls) whereas boys are more likely to play with balls and transportation toys. However, when girls and boys are in mixed-sex play groups, they're as likely to play with "masculine" as with "feminine" toys. Thus, gender-typed play breaks down and promotes a greater range of activities (Goble et al., 2012).

multigenerational households homes in which three or more generations live together.

peer group people who are similar in age, social status, and interests.

In 2010 and 2011, the top-selling toys for girls were dolls, animal toys (e.g., Zhu Zhu pets), and cooking sets. The top-selling toys for boys were video games, ride-on toys (e.g., pedal cars), "blaster" guns, Legos, and Transformers (Rattray, 2010; Rowland, 2011). Girls' sections of catalogs and toy stores are swamped with cosmetics, dolls and accessories, arts and crafts kits, and housekeeping and cookware. In contrast, boys' sections feature sports equipment, building sets, workbenches, construction equipment, and toy guns. Some family scholars propose that children's toys should be more gender neutral to encourage play for boys and girls that promotes less gender stereotyping and develops a "wider repertoire of skills" (Orenstein, 2011; Auster and Mansbach, 2012; Kapp, 2013).

The second function of play—especially when it's structured—is to keep children out of trouble and to enhance their social development. Children who devote more of their free time to structured and supervised activities—such as hobbies and sports—rather than just hanging out with their friends, perform better academically, are emotionally better adjusted, and have fewer problems at school and at home. In effect, sports and hobbies provide children with constructive ways to channel their energy and intelligence (McHale, 2001).

Third, play can strengthen peer relationships. Beginning in elementary school, few things are more important to most children than being accepted by their peers. Even if children aren't popular, belonging to a friendship group enhances their psychological well-being and ability to cope with stress (Rubin et al., 1998; Scarlett et al., 2005).

PEERS AND FRIENDS

Peer influence usually increases as young children get older. Especially during the early teen years, friends often reinforce desirable behavior or skills that enhance a child's self-image ("Wow, you're really good in math!"). Thus, to apply Cooley's concept, peers can help each other develop a positive looking-glass self.

Peers also serve as positive role models. Children acquire a wide array of information and knowledge by observing their peers. Even during the first days of school, children learn to imitate their peers at standing in line, raising their hands in class, and being quiet while the teacher is speaking. Among teens and young adults who are lesbian, gay, or bisexual, heterosexual friends can be especially supportive in accepting one's homosexuality and disclosing it to family (Shilo and Savaya, 2011).

Not all peer or friend influence is positive, however. Among seventh- to twelfth-graders, having a best friend who engages in sexual intercourse, is truant, joins a gang, and uses tobacco and marijuana increases the probability of imitating such behavior (Card and Giuliano, 2011; Haas and Schaefer, 2014). Among adolescents and college students, poor physical fitness, obesity, and eating "junk food" increases the likelihood of adopting or maintaining an unhealthy diet and developing health problems (Carrell et al., 2010; De La Haye et al., 2013).

4-4c Teachers and Schools

Oprah Winfrey often praised her fourth-grade teacher, who recognized her abilities, encouraged her to read, and inspired her to excel academically. Like family and peer groups, teachers and schools affect our socialization.

In Sweden, Top-Toy Group, a licensee of the Toys "R" Us brand, published a "gender-blind" catalog for the 2012 Christmas season. For example, "On some pages, girls brandish toy guns and boys wield blow-dryers and cuddle dolls" (Molin, 2012: D12). The United States is going in the opposite direction (see Auster and Mansbach, 2012).

Jekaterina Nikitina/Getty Images

GODONG/BSIP SA/Alamy

THE SCHOOL'S ROLE IN SOCIALIZATION

By the time children are 4 or 5, school fills an increasingly large portion of their lives. The school's primary purpose is to instruct children and enhance their cognitive development. Schools don't simply transmit knowledge; they also teach children to think about the world in different ways. Because of the emphasis on multiculturalism, for example, children often learn about other societies and customs. Even outside of classes, schools affect children's daily activities through homework assignments and participation in clubs and other extracurricular activities.

Because many parents are employed, schools have had to devote more time and resources to topics—such as sex education and drug abuse prevention—that were once the sole responsibility of families. In many ways, then, schools play an increasingly important role in socialization.

TEACHERS' IMPACT ON CHILDREN'S DEVELOPMENT

Teachers are among the most important socialization agents. From kindergarten through high school, teachers play numerous roles in the classroom—instructor, role model, evaluator, moral guide, and disciplinarian, to name just a few. Once children enter school, their relationships with their teachers are important for academic success. Kindergarten and elementary school teachers' reports of behavioral problems (such as unexcused absences) and poor grades often follow students into middle school (Hamre and Pianta, 2001; see also Chapter 13).

4-4d Popular Culture and the Media

Because of television, iPads, smartphones, YouTube, and social networking sites young people are rarely out of the reach of the electronic media (see Chapter 5). How does such technology affect socialization?

ELECTRONIC MEDIA

The American Academy of Pediatrics (AAP) advises parents to avoid television entirely for children younger than 2. The Academy also counsels parents to limit the viewing time of elementary school children to no more than 2 hours a day to encourage more interactive activities such as talking, playing, singing, and reading together "that will promote proper brain development." Still, 66 percent of children under age 2 spend, on average, an hour a day using screen media, and 16 percent of 1-year-olds have a TV in their bedrooms. Among parents of children age 8 and younger, 55 percent use media as a babysitter while they do chores around the house. Indeed, 71 percent of all children under age 18 have a TV in their bedroom (Rideout, 2013; Strasburger and Hogan, 2013).

The average young American now spends more time with electronic media than in school. The typical 8- to 10-year-old spends nearly 8 hours a day engaged with some type of electronic media (which is more time than many adults spend in a full-time job), and older children and teenagers spend more than 11 hours a day. Too much time on the Internet has been linked with violence, cyberbullying, trouble in school, obesity, lack of sleep, and other problems (Strasburger and Hogan, 2013). Generally,

Class field trips enhance children's socialization. Among other benefits, the field trips expand learning beyond the classroom, enrich students' understanding of subject matter, and encourage them to think about occupational choices and careers that they haven't considered.

TABLE 4.4 HOW DO ELECTRONIC MEDIA AFFECT CHILDREN?

Among all 8- to 18-year-olds, percentage who said that they are...

	Heavy Users (More Than 16 Hours/Day)	Moderate Users (3–16 Hours/Day)	Light Users (Less Than 3 Hours/Day)
Get good grades (A's and B's)	51	65	66
Get fair/poor grades (C's or lower)	47	31	23
Have been happy at school this year	72	81	82
Are often bored	60	53	48
Get into trouble a lot	33	21	16
Are often sad or unhappy	32	23	22

Source: Based on Rideout et al., 2010: 4.

youths who spend more time with media have lower grades and lower levels of personal contentment (see *Table 4.4*). These findings are similar for girls and boys across all categories of age, race/ethnicity, parents' social class, and single- and two-parent households (Rideout et al., 2010).

The prevalence of gun violence in PG-13 movies has more than tripled since the rating was introduced in the mid-1980s, and now surpasses the violence in R-rated movies. (PG-13-rated movies caution parents that the material isn't appropriate for children under age 13. R-rated movies require children age 17 and younger to be accompanied by a parent or adult guardian.) The researchers of this study concluded that even if youth don't use guns, "because of the increasing popularity of PG-13-rated films, youth are exposed to considerable gun violence… that may increase aggressive behavior" (Bushman et al., 2013: 1017).

There's also a growing concern that violent video games make violence seem normal. Playing violent video games (e.g., *Grand Theft Auto V*, *Call of Duty*, *Dead Space*, and *God of War 4*) can increase a person's aggressive thoughts, feelings, and behavior both in laboratory settings and in real life. Violent video games also encourage male-to-female aggression because much of the violence is directed at women (Anderson et al., 2003; Carnagey and Anderson, 2005).

Still, it's not clear why violent video games affect people differently. Many young males enjoy playing such games, but aren't any more aggressive, vicious, or destructive than those who aren't violent video game enthusiasts (Kutner and Olson, 2008; Ferguson and Olson, 2013). Some maintain, however, that only *repeated* exposure to media violence increases the likelihood of violent behavior in the long run (see Dennis, 2013).

ADVERTISING

Advertisers are targeting children at increasingly younger ages. A new form of advertising called *advergaming* combines free online games with advertising. Advergaming is growing rapidly. Sites such as Wonka.com, operated by Nestlé, and Barbie.com, operated by Mattel, attract millions of young children and provide marketers

What? Adolescents Can Buy Ultraviolent Video Games, but Not a Magazine with an Image of a Nude Woman?!

In 2011, the Supreme Court ruled (7–2 in *Brown v. Entertainment Merchants Association*) that video games, even ultraviolent ones, that are sold to minors are protected by the First Amendment's guarantee of free speech. The majority said that none of the scientific studies *prove* that violent games *cause* minors to act aggressively, and that parents, not the government, should decide what is appropriate for their children (Schiesel, 2011; Walls, 2011).

© Hot Property/Shutterstock.com

Not surprisingly, the video game industry was jubilant over the decision. Many parents and lawmakers, on the other hand, agreed with one of the dissenting justices that it makes no sense to forbid selling a 13-year-old boy a magazine with an image of a nude woman, but not an interactive video game in which the same boy "actively … binds and gags the woman, then tortures and kills her" (Barnes, 2011: A1).

Many men's health and fitness magazines routinely feature models who have undergone several months of extreme regimens, including starvation and dehydration, to tighten their skin and make their muscles "pop." The magazines also use camera and lighting tricks and Photoshop to project an idealized image of hypermasculinity that, in reality, is impossible to attain (Christina, 2011; see also Ricciardelli et al., 2010).

with an inexpensive way to "draw attention to their brand in a playful way, and for an extended period of time" (Moore, 2006: 5).

In the print media, young people see 45 percent more beer ads and 27 percent more ads for hard liquor in teen magazines than adults do in their magazines (Strasburger et al., 2006; see also Jernigan, 2010). Girls ages 11 to 14 are subjected to about 500 advertisements a day on the Internet, billboards, and magazines in which the majority of models are "nipped, tucked, and airbrushed to perfection" (Bennett, 2009: 43).

What effect do such ads have on girls' and women's self-image? About 43 percent of 6- to 9-year-old American girls use lipstick or lip gloss, 38 percent use hairstyling products, and 12 percent use other cosmetics. In addition, 8- to 12-year-old girls spend more than $40 million a month on beauty products, 80 percent of 10-year-old girls have been on a diet, and 80 percent of girls ages 13 to 18 list shopping as their favorite activity (Hanes, 2011; Seltzer, 2012). Many girls and young women believe that they have to be gorgeous, thin, and almost perfect to be loved by their parents and boyfriends (Schwyzer, 2011).

Many women, especially white women, who are unhappy with their bodies, turn to cosmetic surgery. In 2013, women had 91 percent of all cosmetic procedures; the number undergoing surgery (almost 10.4 million) has increased 471 percent since 1997. Men had more than 1 million procedures, an increase of 273 percent from 1997. Almost 79 percent of all patients were white. The most common cosmetic surgery was breast enlargement for women and liposuction for men (American Society for Aesthetic Plastic Surgery, 2014).

SOCIALIZATION THROUGHOUT LIFE

Biological aging comes naturally, but social aging is a different matter. As we progress through the life course—from infancy to death—we are expected to learn culturally approved norms, values, and roles.

4-5a Infancy and Toddlerhood

Infancy (between birth and 12 months) and toddlerhood (between the ages of 1 and 3 years) encompass only a small fraction of the average person's life span, but are periods of both extreme helplessness and enormous physical and cognitive growth. Some scientists describe healthy infants' brains as "small computers" because of their enormous capacity for learning (Gopnik et al., 2001: 142). The quality of relationships with adults and other caregivers has a profound effect on infants' and toddlers' development. Engaging infants in talk increases their vocabulary. By age 2, toddlers who know more words have better language skills that, in turn, help them control their behavior at age 3 and later (Vallotton and Ayoub, 2011).

More than half (53 percent) of children ages 1 and 2 are read to seven or more times a week by a family member ("Half of Young Children…," 2011). The biggest change is children's media use. In 2013, 38 percent of all children under age 2 were using mobile devices (e.g., smartphone, tablet), up from only 10 percent in 2011. You saw earlier that the AAP has advised parents to limit their infants' and toddlers' time in front of screens. Only 28 percent of parents say that media decreases face-to-face family time, but they often underestimate the amount of time that children under age 2 spend with electronic media (Rideout, 2013).

That's Not All . . .

There are many socialization agents other than family, play and peer groups, teachers and schools, and the media. If you think about the last few years, which groups or organizations have influenced who you are? For example, what about the military, a company you've worked for, college clubs, religious organizations, support groups, athletic memberships, community groups, or others? All affect our socialization throughout life.

Childhood—viewed as a distinct stage of development—is a fairly recent phenomenon (Ariès, 1962). Until 1938, young children made up a large segment of the U.S. labor force, especially in factories. They worked 6 days a week, 12 or more hours a day, received very low wages, and were often injured because of dangerous equipment.

not constant) praise strengthens the parent–child relationship, teaches children to overcome setbacks, and improves their problem-solving skills (Owen et al., 2012; Petersen, 2012).

Parental expectations play an important role in adolescent development. For example, families with high educational aspirations for their offspring provide more out-of-school learning opportunities, the children have more positive attitudes toward school, and they are more likely to attain a four-year college degree (Child Trends, 2010).

Most U.S. children enjoy happy and healthy lives, but many don't. Almost 3 million children younger than 18 (almost 4 percent of all children in this age group) have an incarcerated parent. Another 400,000 are in the foster care system. Almost 13 percent of all children live in households where a parent or other adult uses, manufactures, or distributes illicit drugs. And in 2012, at least 686,000 children experienced neglect or physical and sexual abuse (U.S. Department of Justice, 2011; Children's Bureau, 2013; Reilly, 2013; "Child Maltreatment," 2014). Thus, for millions of U.S. children, the socialization process is shaky at best.

4-5b Childhood Through Adolescence

Childhood and adolescence (roughly ages 4 to 12) are marked by considerable physical, emotional, social, and cognitive growth. You'll recall that as children grow older, they learn social roles through anticipatory socialization and reference groups. Parents and teachers, particularly, emphasize self-control, academic achievement, fulfilling responsibilities, and getting along with others.

Adults reinforce norms and values through rewards and punishments (see Chapter 3), but are U.S. children and adolescents being coddled? For example, children usually get trophies and prizes simply for participating in extracurricular activities, and are constantly assured that they're winners, but "collapse at the first experience of difficulty" (Merryman, 2013: A29; see also Twenge, 2007). Self-help books and self-proclaimed child experts routinely instruct parents to always praise their kids to get them to behave. Such advice is misguided. According to a recent review of 41 studies of discipline strategies targeted at children up to 11 years old, the researchers concluded that, in the long run, consistent and well-deserved (but

4-5c The Teenage Years

Ages 13 to 19 is a period of tremendous change. Most societies have distinct *rites of passage*, public rituals that mark the transition from one social position to another. Dating, the most common rite of passage in the United States, marks the end of childhood and readiness for adult responsibilities and rights (see Chapters 10 and 13 for religious and ethnic rites of passage).

Teenagers are establishing their own identity and testing their autonomy as they mature and break away from parental supervision, a healthy human development process (Erikson, 1966). As teenagers become more independent and more likely to confide in friends, parents may feel rejected and suspicious. The most difficult part of parenting adolescents, according to some parents, is dealing with their changing moods and behavior.

For many years, people attributed such dramatic changes to "raging hormones." Scientists now believe that there's a link between a teen's baffling behavior and the fact that his or her brain may be changing far more than was thought previously. During adolescence, the brain matures at different rates. Areas involved in basic functions such as processing information and controlling physical movement mature first. The parts of the brain responsible for controlling impulses, avoiding risky behavior, and planning ahead—the hallmarks of adult behavior—are among the last to mature (National Institute of Mental Health, 2011).

Parental involvement is usually beneficial in a child's development, but "helicopter parents," who hover over their kids, are hyperinvolved, intrusive, and overcontrolling. Helicopter parenting can occur at any stage, but usually refers to parents of preteens, teenagers, and young adults. Anecdotal examples include parents verbally attacking teachers about their children's low grades, demanding that their child be moved to another class before the school year has even begun, completing difficult homework assignments, showing up in the guidance counselor's office with college applications that they have filled out for their children, and haggling with college professors over the student's grade on an exam or paper (Krache, 2008; Weintraub, 2010; Rochman, 2013).

Helicopter parenting diminishes teens' and young adults' ability to develop decision-making and problem-solving skills. However well-intentioned, helicopter parents increase their own stress and decrease their children's feelings of autonomy and competence. In turn, not believing in one's ability to successfully accomplish tasks and achieve goals on one's own leads to teenagers' and young adults' feeling anxious, depressed, and

more dissatisfied with life (Schiffrin et al., 2013; Shellenbarger, 2013).

4-5d Adulthood

Socialization continues throughout adulthood, the period roughly between ages 21 and 65. Most adults adopt a series of new roles that may include work, marriage, parenthood, divorce, remarriage, buying a house, and experiencing the death of a child, parent, or grandparent. We'll examine these transitions in later chapters. Two of the most important roles in adulthood are work and parenthood, but let's begin with young adults who are reluctant to leave their parents' nest.

THE CROWDED EMPTY NEST

During the 1960s and 1970s, sociologists almost always included the "empty nest" in describing the family life course. This is the stage in which parents, typically in their 50s, find themselves alone after their children have married, gone to college, or found jobs and moved out.

Today, however, young adults are living at home longer than they did in the past. The terms *boomerang children* and *boomerang generation* refer to young adults who move back in with their parents. Many people in their 30s, 40s, and older—often with a spouse, girlfriend or boyfriend, and children in tow—are moving back into their parents' home.

Most young adults leave the parental nest by age 23, but the proportion of those ages 25 to 34 who are living with parents increased from 9 percent in 1960 to 12 percent in 2013 (Fields and Casper, 2001; U.S. Census Bureau, Current Population Survey…, 2013). Among young adults ages 18 to 34, men are more likely than women to live with their parents (see *Figure 4.2*).

Hosting an adult child who lives at home can cost between $8,000 to $18,000 a year, depending on how much parents are paying for extras such as travel and entertainment. About 26 percent of parents have taken on debt to pay for their adult children's expenses (Grind, 2013).

Macro-level factors have been the major reason for the delay in many Americans' transition to adulthood, especially living independently. In 2013, 50 percent of young adults ages 24 to 34 either lived with their parents or moved back because they were unemployed or employed part time because they couldn't find a full-time job (Jones, 2014).

Declining employment, financial insecurity, student loan debt, low wages, divorce, credit card debt,

Howard McWilliam

FIGURE 4.2 YOUNG ADULTS LIVING AT HOME WITH THEIR PARENTS, 2013

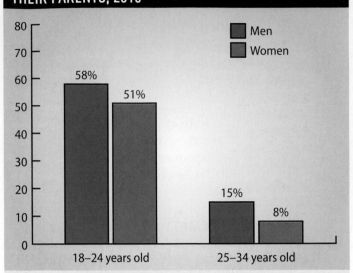

Source: Based on U.S. Census Bureau, Current Population Survey…, 2013, Table A2.

and going from job to job until they find work they enjoy have made it harder for young middle-class adults to maintain the lifestyles that their parents created. And, the transition to independence gets tougher the lower a person's occupation and education (Qian, 2012; Parker, 2012; Fry, 2013).

Economic factors aren't the only reason for living at home. On a micro level, almost 33 percent of today's parents, compared with 19 percent in 1993, say that children shouldn't be expected to be on their own financially until age 25 or later (Taylor et al., 2012). Among those ages 18 to 24, 87 percent live at home because parents "make it easy for me to stay" (Payne and Cobb, 2013). They enjoy the comforts of the pampering parental nest. According to one of my male students in his late 20s, for example, "My mom enjoys cooking, cleaning my room, and just having me around. I don't pitch in for any of the expenses, but we get along great because she doesn't hassle me about my comings and goings" (Benokraitis, 2014: 354).

Many young adults, as you'll see in Chapter 12, are delaying marriage and don't feel a need to establish their own homes. Perhaps most important, the stigma traditionally linked to young adults' living at home has faded because the practice is widespread. Among those ages 25 to 34 who live at home, 61 percent say they have friends or family members who have moved back in with their parents (Parker, 2012). The implication (and perhaps justification) of moving back home or not leaving is that "a lot of people are doing it."

WORK ROLES

The average American who was born between 1957 and 1964 has held 11 jobs from ages 18 to 46 alone; 26 percent have held 15 jobs or more ("Number of Jobs Held…," 2012). Younger generations will probably change jobs more often because of a slumping economy. Such job changing means that our occupational socialization is an ongoing process. Learning work roles is difficult because each job has different demands and expectations. Even when there's formal job training, we must also learn the subtle rules that are implicit in many job settings. A supervisor who says, proudly and loudly, "I have an open door policy. Come in and chat whenever you want," may *really* mean "I have an open door policy if you're not going to complain about something." Even if we remain in the same job for many years, we must often acquire new skills, especially technological ones, or risk being laid off.

As U.S. companies have transferred more jobs overseas, many workers have been fired. Being laid off can be stressful at any age. In the case of midlife men (generally defined as those ages 45 to 60), such changes are especially traumatic: They must learn new job-hunting skills and often adjust to as much as a 70 percent decrease in their earnings (see Chapter 11). Instead of enjoying the security of a job in their midlife years, these men must undergo occupational socialization that assaults their self-image as good employees and family providers.

PARENTING ROLES

Like workplace roles, parenting does *not* come naturally. Most first-time parents muddle through by trial and error. Family sociologists often point out that we get more training for driving a car than for marriage and parenting. For example, most couples don't realize that children are expensive. Middle-income couples, with an average income of $82,800 a year, spend about 16 percent of their earnings on a child during the first 2 years (see *Figure 4.3*). These couples will spend almost $246,000 for each child from birth through high school. Child-rearing costs are much higher for single parents and especially for low-income families if a child is disabled, chronically ill, or needs specialized care that welfare benefits don't cover. Expenses are also much higher for parents whose children attend private schools (Lino, 2014).

You'll see in Chapter 12 that arguments over finances and child-rearing strategies are two of the major

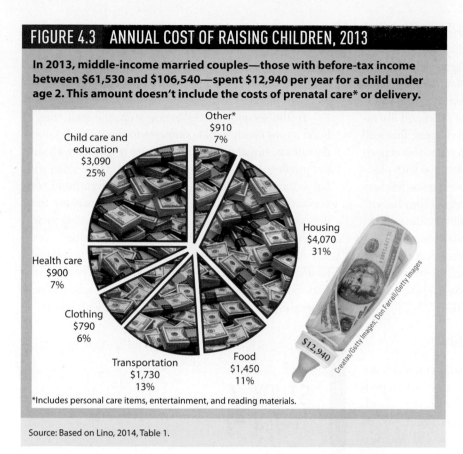

FIGURE 4.3 ANNUAL COST OF RAISING CHILDREN, 2013

In 2013, middle-income married couples—those with before-tax income between $61,530 and $106,540—spent $12,940 per year for a child under age 2. This amount doesn't include the costs of prenatal care* or delivery.

Other* $910 7%

Child care and education $3,090 25%

Housing $4,070 31%

Health care $900 7%

Clothing $790 6%

Transportation $1,730 13%

Food $1,450 11%

$12,940

Creatas/Getty Images; Don Farrall/Getty Images

*Includes personal care items, entertainment, and reading materials.

Source: Based on Lino, 2014, Table 1.

reasons for divorce. All in all, adjusting to marital and parental roles during adulthood requires considerable patience, effort, and work.

4-5e Later Life

Socialization continues in later life. People age 65 and older often remarry and must learn to play new roles such as stepparent or step-grandparent. Many others undergo unwanted rites of passage such as widowhood, divorce or re-divorce, becoming a custodial parent, and dealing with the death of an adult child or grandchild.

The adult child–aging parent relationship has mixed outcomes. Adult children who receive parental financial and emotional support are happier, not surprisingly, than their counterparts who don't receive such assistance. However, older parents often experience stress when their adult children are doing poorly and have to rely on their parents for financial and emotional support (Fingerman et al., 2012).

Because we live longer, we may spend 20 percent of our adult life in retirement. When retired people are unhappy, it's usually because of health or income problems rather than loss of the worker role (see Chapter 14).

If retirement benefits don't keep up with inflation or if a retiree isn't covered by a pension plan, people can plunge into poverty soon after retirement. Poverty is devastating at any age, but sliding down the economic ladder during one's 60s and 70s is especially traumatic because there's little chance of increasing one's income (see Chapters 12 and 14).

Older people who are in poor health and experience continuous pain sometimes welcome death. Among those who are healthy, some reenter the labor force or volunteer in various organizations, many are involved with their grandchildren, and others forge new relationships after widowhood (see Chapter 12). Thus, even in the later years, many people continue to learn new roles.

4-6 RESOCIALIZATION

Throughout life, many people undergo **resocialization**, the process of unlearning old ways of doing things and adopting new attitudes, values, norms, and behavior. Resocialization ranges from mild to intense, and can be voluntary or involuntary.

4-6a Voluntary Resocialization

Much resocialization is voluntary, as when an American wife moves to the Middle East with her Iranian husband and follows local customs about veiling, women's submissive roles, and staying out of public life. Other examples of voluntary resocialization include entering a religious order, joining a religious cult, seeking treatment in a drug abuse rehabilitation facility, or serving in the military. The process of voluntary resocialization is usually mild. First-year college students, new employees, and members of support groups, for instance, must learn new rules and live up to an individual's, a group's, or an

resocialization the process of unlearning old ways of doing things and adopting new attitudes, values, norms, and behavior.

organization's expectations. Such changes, however, are usually short-term and may even be enjoyable.

On the other hand, some voluntary resocialization can be long, difficult, and intense. New parents' lives change dramatically as they learn parenting roles. In divorce, people may experience a sense of relief and liberation, but must also cope with emotional distress, financial changes, and child custody disputes. People who experience a permanent physical disability must cope with physical limitations, psychological issues, and stigma. Soldiers who return from Iraq or Afghanistan, even if they haven't suffered physical injuries, describe coming home and living a normal life as daunting. They must reconnect with their children, adjust to spouses who have become more independent, and deal with posttraumatic stress disorders that include depression and nightmares (Jones, 2013; Taormina-Weiss, 2013; van Agtmael, 2013).

4-6b Involuntary Resocialization

Resocialization is also involuntary, as when children are sent to a foster home or a detention camp, or when people are imprisoned or committed to a psychiatric hospital. Such settings are **total institutions** where people are isolated from the rest of society, stripped of their former identities, and required to conform to new rules and behavior (Goffman, 1961).

Resocialization in total institutions is an intense two-step process. First, the staff uses *degradation ceremonies*, humiliating rituals that publicly stigmatize people (Garfinkel, 1956). The inmates are strip-searched, deloused, and fingerprinted; their heads are shaved; they're issued a uniform and are given a serial number; they have no privacy, limited access to family and friends, and little communication with the outside world; and they're told what to do and when. In the military, soldiers get short haircuts, wear matching uniforms, and follow rigid schedules. The purpose of these practices and restrictions is to destroy any sense of individualism, autonomy, and past identity, and to produce a more compliant person (Goffman, 1961; see also Chapter 7).

In the second resocialization step, the staff tries to build a new identity that conforms to the institution's expectations. Inmates who conform are rewarded with simple privileges (e.g., making phone calls, more visits with family) and an earlier release. Soldiers often bond with one another, become more self-disciplined, may use their new skills in civilian jobs, and are proud of having served in the military ("Once a Marine, always a Marine").

As this chapter shows, socialization is a powerful force in shaping who we are. Socialization doesn't produce robots, however, because people are creative, adapt to new environments, and change as they interact with others.

total institutions settings where people are isolated from the rest of society, stripped of their former identities, and required to conform to new rules and behavior.

Davincidig/iStockphoto.com; Bridget McGill/iStockphoto.com

Most resocialization occurs in institutions ranging from monastic orders to the military.

4LTR Press solutions are designed for today's learners through the continuous feedback of students like you. Tell us what you think about **SOC4** and help us improve the learning experience for future students.

YOUR FEEDBACK MATTERS.

Complete the Speak Up
survey in CourseMate at
www.cengagebrain.com

 Follow us at
www.facebook.com/4ltrpress

5 Social Interaction in Everyday Life

© Anelina/Shutterstock.com

LEARNING OBJECTIVES

5-1 Explain the importance of social interaction and its relationship to social structure.

5-2 Describe and illustrate status set, ascribed and achieved statuses, master status, and status inconsistency.

5-3 Explain how and why social roles differ, and how people cope with role conflict and role strain.

5-4 Describe and illustrate how symbolic interaction, social exchange theory, and feminist theories explain social interaction.

5-5 Describe and illustrate nonverbal communication, its importance, and cross-cultural variations.

5-6 Describe the benefits and costs of online interaction.

After you finish this chapter go to **PAGE 99** for **STUDY TOOLS.**

David Sacks/Stone/Getty Images

In mid-2011, a Boeing 767 took off from an airport in Washington, D.C., bound for Ghana, Africa. Shortly after the flight was airborne, a passenger reclined his seat, which was close to the lap of the man sitting behind him. There was a smack to the head, angry words, and a fist fight broke out. A flight attendant and another passenger jumped in between. The pilot returned to the airport to determine the scope of the problem (Halsey, 2011: A1; see also Grossman, 2011).

This incident of air rage is an example of **social interaction**, the process by which we act toward and react to people around us. The scuffle began with a typical passenger annoyance, escalated into an aggressive nonverbal reaction (a smack to the head), burst into an angry argument and fight, and quickly involved others—a third passenger, the airline attendant, and the pilot. This incident illustrates four critical components of social interaction (Maines, 2001; Schwalbe, 2001):

▶ It is central to all human social activity, and includes both nonverbal behavior (a physical smack) and words (an argument).

What do you think?

I spend more time texting and on Facebook than I do talking to my family.

| 1 | 2 | 3 | 4 | 5 | 6 | 7 |

strongly agree   strongly disagree

- ▶ People respond differently during social interaction, depending on what they think is at stake for them ("I'm not going to put up with this guy all the way to Ghana").

- ▶ People influence each other's behavior through social interaction ("No one can hit me on the head and get away with it").

- ▶ Elements of social structure, such as rules about proper behavior, affect all social interaction but can produce different personal outcomes (e.g., most airline passengers tolerate cramped seats quietly whereas others explode).

5-1 SOCIAL STRUCTURE

Social structure is an organized pattern of behavior that governs people's relationships (Smelser, 1988). Because social structure guides our actions, it gives us the feeling that life is orderly and predictable rather than haphazard or random. We're often not aware of the impact of social structure until we violate cultural rules,

formal or informal, that dictate our daily behavior—whether in class, at work, or in an airplane.

Every society has a social structure that encompasses statuses, roles, groups, organizations, and institutions (Smelser, 1988). We'll examine groups, organizations, and institutions in later chapters. Let's take a closer look here at statuses and roles, two building blocks of social structure.

5-2 STATUS

For most people, the word *status* signifies prestige: An executive, for example, has more status than a receptionist, and a physician has a higher status than a nurse.

social interaction the process by which we act toward and react to people around us.

social structure an organized pattern of behavior that governs people's relationships.

FIGURE 5.1 IS THIS YOUR STATUS SET?

Friend

Brother or Sister

Son or Daughter

Parent or Grandparent

Significant Other or Spouse

College Student

Employer or Employee

Registered Voter

Consumer

Member of Religious Group

Photodisc/Getty Images

Being a college student is only one of your current statuses. What other statuses comprise your status set?

Dionne, one of my students, is female, African American, 42 years old, divorced, mother of two, daughter, aunt, Baptist, voter, bank supervisor, volunteer at a soup kitchen, and country music fan. All of these socially defined positions (and others) make up Dionne's status set.

Status sets change throughout the life course. Because she'll graduate next year and is considering remarrying and starting an after-school program, Dionne will add at least three more statuses to her status set and will also lose the statuses of divorced and college student. As Dionne ages, she'll continue to gain new statuses (grandmother, retiree) and lose others (supervisor at a bank, or wife, if she is widowed).

Statuses are *relational*, or complementary, because they're connected to other statuses: A *husband* has a *wife*, a *real-estate agent* has *customers*, and a *teacher* has *students*. No matter how many statuses you occupy, every status is linked to those of other people. These connections between statuses influence our behavior and relationships.

For sociologists, **status** refers to a social position that a person occupies in a society (Linton, 1936). Thus, executive, secretary, physician, and nurse are all social statuses. Other examples of statuses are musician, voter, sister, parent, police officer, and friend.

Sociologists don't assume that one position is more important than another. A mother, for example, isn't more important than a father, and an adult isn't more important than a child. Instead, all statuses are significant because they determine social identity, or who we are.

5-2a Status Set

Every person has many statuses (see *Figure 5.1*). Together, they form her or his **status set**, a collection of social statuses that a person occupies at a given time (Merton, 1968).

status a social position that a person occupies in a society.

status set a collection of social statuses that a person occupies at a given time.

ascribed status a social position that a person is born into.

achieved status a social position that a person attains through personal effort or assumes voluntarily.

5-2b Ascribed and Achieved Status

Status sets include both ascribed and achieved statuses. An **ascribed status** is a social position that a person is born into. We can't control, change, or choose our ascribed statuses, which include sex (male or female), age, race, ethnicity, and family relationships. Your ascribed statuses, for example, might include *male*, *Latino*, and *brother*. Some argue that sex isn't really an ascribed status because people can have sex-change operations. However, a sex-change operation doesn't change the fact that someone was born a male or a female (except in a minority of cases, as you'll see in Chapter 9).

An **achieved status**, in contrast, is a social position that a person attains through personal effort or assumes voluntarily. Your achieved statuses might include college graduate, mother, and employee. Unlike our ascribed statuses, our achieved statuses can be controlled and changed. We have no choice about being a son or daughter (an ascribed status), but we have an option to become a parent (an achieved status).

Students sometimes think that religion and social class are ascribed rather than achieved statuses. It's true that someone may be born into a family that practices a certain religion or one that is poor, middle class, or wealthy. Because we can change these statuses through our own actions, however, neither religion nor social class is an ascribed status. A Catholic might convert to Judaism (or vice versa), and thousands of Americans born into poor or working-class families have become millionaires (see Chapter 8).

5-2c Master Status

A **master status** overrides other statuses and forms an important part of a person's social identity (Hughes, 1945; Becker, 1963). A master status is usually immediately apparent, makes the biggest impression, affects others' perceptions, and, consequently, often shapes a person's entire life.

Some master statuses—such as sex, age, and race—are ascribed. Whatever you do, for example, people will see you as male or female and react accordingly. Other master statuses are achieved. For many people, their occupation is their master status: Occupation tends to be the first thing we ask about when we meet someone; conveys much information about income, education, skills, and achievements; and lingers after we've left our jobs (e.g., someone is introduced as a former teacher

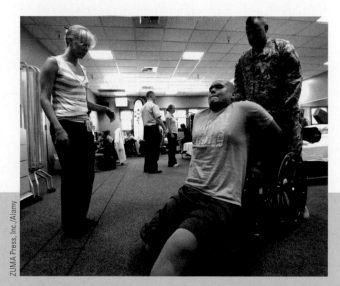

A physical disability can become a master status. One of the military's major tasks today is teaching severely injured soldiers to become as independent as possible.

or ex-military officer). Other achieved master statuses can be based on religion, community leadership, family wealth, or deviant behavior.

5-2d Status Inconsistency

Because we hold many statuses, some clash. **Status inconsistency** refers to the conflict that arises from occupying social positions that are ranked differently. Examples include a computer programmer who works as a bartender or a skilled welder who stocks shelves at Walmart because neither can find a better job in a weak economy. We'll cover status inconsistency in more detail in Chapter 7. For now, you should be aware that you occupy many statuses, and some of them may clash now and in the future.

5-3 ROLE

Each status is associated with one or more roles. A **role** is the behavior expected of a person who has a particular status. We *occupy* a status but *play* a role. In this sense, a role is the dynamic aspect of a status (Linton, 1936).

College student is a status, but the role of a college student requires many *formal* behaviors such as going to class, reading, thinking, completing weekly assignments, writing papers, and taking exams. *Informal* behaviors may include joining a student club, befriending classmates, attending football games, and even abusing alcohol on weekends.

Like statuses, roles are relational (complementary). Playing the role of professor requires teaching, advising students, being present during office hours, responding to email, and grading exams and assignments. Most professors are also expected to serve on committees, do research, publish articles and/or books, and perform services such as giving talks to community groups. Thus, the status of college student or professor involves numerous role requirements that govern who does what, where, when, and how.

Roles can be rigid or flexible. A person who occupies the status of secretary typically plays a role that's defined by rules about when to come to work, how to

master status a status that overrides other statuses and forms an important part of a person's social identity.

status inconsistency the conflict that arises from occupying social positions that are ranked differently.

role the behavior expected of a person who has a particular status.

answer the phone, and when to submit the necessary work and in what format. A boss, on the other hand, usually enjoys considerable flexibility: She or he has more freedom to come in late or leave early, to determine which projects should be completed first, and to decide when to hire or fire employees.

Because roles are based on mutual obligations, they ensure that social relations are fairly orderly. We know what we're supposed to do and what others expect of us. If professors fail to meet their role obligations by missing many classes or coming to classes unprepared, students may respond by studying very little, cutting classes, and submitting negative course evaluations. If students miss classes, don't turn in the required work, or cheat, professors can fail them or assign low grades.

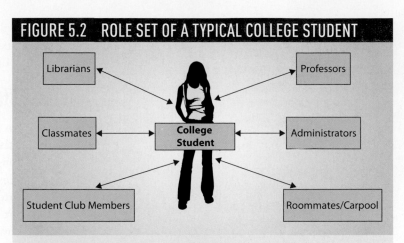

FIGURE 5.2 ROLE SET OF A TYPICAL COLLEGE STUDENT

Librarians — Professors — Classmates — College Student — Administrators — Student Club Members — Roommates/Carpool

How does interaction differ with these other role players in a college student's role set?

5-3a Role Performance

Roles define how we're *expected* to behave in a particular status, but people vary considerably in fulfilling the responsibilities associated with their roles. Many college students succeed, whereas others fail; some professors inspire their students, whereas others put them to sleep. These differences reflect **role performance**, the *actual* behavior of a person who occupies a status. For example, a professor may vary her or his role by demanding more of graduate students than of undergraduates. An instructor may also interact differently with male and female students and with a 19-year-old student than a 40-year-old student who's anxious about returning to college.

5-3b Role Set

We occupy many statuses and play many roles associated with each status. A **role set** refers to the different roles attached to a single status. Every role set includes rights and responsibilities associated with people with different statuses and role sets. *Figure 5.2* illustrates six roles of a typical college student. Because a different

set of norms governs each of these relationships, the student interacts differently with a classmate than with a reference librarian or a professor. All of these interactions, shaped by explicit or implicit rules, make up a student's role set.

These six roles reflect only one status—that of college student. If you think about other statuses that a college student may occupy (employee, son or daughter, parent, girlfriend or boyfriend, husband or wife), you can see that meeting the expectations of numerous role sets can create considerable role conflict and role strain.

5-3c Role Conflict and Role Strain

Playing many roles often leads to **role conflict**, the frustrations and uncertainties a person experiences when confronted with the requirements of two or more statuses. College students who have a job, especially if it's full-time, often experience role conflict because both employers and professors expect them to excel, but it's very difficult to meet these expectations. The role conflict increases if the student is a parent of young children or cares for an aging parent or grandparent who needs help on a daily basis.

Whereas role conflict arises from tensions *between* the roles of two or more statuses, **role strain** is the stress that arises from incompatible demands among roles *within* a single status. Students experience role strain when several exams are scheduled on the same day and they're also involved in time-consuming extracurricular activities. Faculty members experience role strain when the requirements of being a professor—teaching, research, and community service—sap their energy and time. Military chaplains report role strain in preaching about peace while blessing those about to go to war.

role performance the actual behavior of a person who occupies a status.

role set the different roles attached to a single status.

role conflict the frustrations and uncertainties a person experiences when confronted with the requirements of two or more statuses.

role strain the stress that arises from incompatible demands among roles within a single status.

TABLE 5.1 WHY DO WE EXPERIENCE ROLE CONFLICT AND ROLE STRAIN?

Reason	Example
Because many people are overextended, some roles are bound to conflict with others.	Students may study less than they want because employers demand that they work overtime or on weekends.
People have little or no training for many roles.	Parents are expected to turn out "perfect" kids even though they receive training for driving a car but none for parenting.
Some role expectations are unclear or contradictory.	Some employers pride themselves on having family-friendly policies but expect employees to work 12 hours a day, travel on weekends, and use vacation days to care for a sick child.
Highly demanding jobs often create difficulties at home.	Some jobs (being in the military, policing, and firefighting) require people to be away from their families for extended periods of time or during crises.

Like many other people, college students can reduce their role conflict and role strain by setting priorities and compartmentalizing their roles.

Almost all of us experience role strain because many inconsistencies are built into our roles. *Table 5.1* gives examples of some of the factors that create role conflict and role strain.

5-3d Coping with Role Conflict and Role Strain

Role conflict and role strain can produce tension, hostility, aggression, and stress-related physical problems that include insomnia, headaches, ulcers, eating disorders, anxiety attacks, chronic fatigue, nausea, weight loss or gain, and drug and alcohol abuse (Weber et al., 1997). Some role conflict and role strain may last only a few weeks, but others are long-lived (e.g., working in a stressful or low-paying job).

To deal with role conflict and role strain, some people *deny that there's a problem.* Employed mothers, especially those who are divorced, often become supermoms— providing home-cooked meals, attending their children's sports activities, and volunteering for a school's fundraising campaign even though they're exhausted, must do laundry at midnight, and neglect their own health and interests. Supermoms may succeed over a number of years in managing role conflict and role strain. Eventually,

however, they may become angry and resentful, or experience health or emotional problems (Douglas and Michaels, 2004).

There are five more effective ways to minimize role conflict and role strain. First, we can reduce role conflict through *compromise* or *negotiation.* To decrease the conflict between work and family roles, for example, many couples draw up schedules that require fathers to do more of the housework and child rearing (see Chapter 12).

Second, we can *set priorities.* If extracurricular activities interfere with studying, which is more important? Succeeding in college always requires making sacrifices, such as attending fewer parties and not seeing friends as often as we'd like.

Third, we can *compartmentalize* our roles. Many college students take courses in the morning, work part-time during the afternoon or evening, and devote part of the weekend to leisure activities. It's not always easy but usually possible to separate our various roles.

Fourth, we can decide *not to take on more roles.* One of the most effective ways to avoid role conflict is to just say "no" for requests to do volunteer work, pressure from family or friends to take on unwanted tasks (e.g., babysitting), or pleas to become involved in college or community activities.

Finally, we can *exit* a role or status. Withdrawing from community activities and club offices, for example, can decrease role conflict and role strain considerably. There's always pressure to remain in a role, but none of us is indispensable, and there are many people who are eager to replace us. Some role exits are painful and long-lived, however. Divorce, for instance, isn't a quick event but a process that's usually spread over many years (sometimes decades), during which two people (and their children) must redefine their expectations and adjust to a different household structure (see Chapter 12).

5-4 EXPLAINING SOCIAL INTERACTION

Statuses and roles are two critical components of social structure that shape our everyday relationships, but it is social interaction that provides the basis of these relationships: It affects who you are, how you behave, and what you say. Social interaction seems natural and simple, but why do we interact as we do? And why do people sometimes interpret the same words differently? Three micro-level perspectives—symbolic interaction, social exchange theory, and feminist theories—provide answers to these and other questions. These perspectives offer distinctive contributions, but they have one characteristic in common: Each explains how people interact in their daily lives. (*Table* 5.2 summarizes these approaches.)

5-4a Symbolic Interaction

For interactionists, the most significant characteristic of all human communication is that people take each other and the context into account (Blumer, 1969). When your professors ask, "How are you?" they expect a "Fine, thanks" and probably barely look at you. A doctor, on the other hand, usually looks you in the eye when asking "How are you?" and takes notes as soon as you start to reply. Thus, "How are you?" has different meanings in different social contexts and elicits different responses ("I'm doing okay" vs. "I've been having a lot of headaches lately"). In this sense, people construct reality.

Doublespeak

1. automotive internist		A. repairperson
2. internment excavation expert		B. bombing
3. service technician		C. spying
4. auto dismantler and recycler		D. gravedigger
5. air support		E. fake
6. previously owned		F. car mechanic
7. genuine imitation		G. junk dealer
8. surveillance		H. stock market crash
9. equity retreat		I. used
10. revenue enhancement		J. tax increase

Anwers: 1-F, 2-D, 3-A, 4-G, 5-B, 6-I, 7-E, 8-C, 9-H, 10-J

SOCIAL CONSTRUCTION OF REALITY

What people perceive and understand as reality is a creation of the social interaction of individuals and groups. "Human reality is socially constructed reality" because people impose their subjective meanings on interactions to make sense of the world around them (Berger and Luckmann, 1966: 172). That is, we produce, interpret, and share the reality of everyday life with others. This social construction of reality typically evolves through direct, face-to-face interaction, but the interaction can also be indirect, as in watching television or participating in online social networks.

Businesspeople, advertisers, politicians, educators, advocacy groups, and even social scientists use words

TABLE 5.2 SOCIOLOGICAL EXPLANATIONS OF SOCIAL INTERACTION

Perspective	Key Points
Symbolic Interactionist	• People create and define their reality through social interaction. • Our definitions of reality, which vary according to context, can lead to self-fulfilling prophecies.
Social Exchange	• Social interaction is based on a balancing of benefits and costs. • Relationships involve trading a variety of resources, such as money, youth, and good looks.
Feminist	• Females and males act similarly in many interactions but may differ in communication styles and speech patterns. • Men are more likely to use speech that's assertive (to achieve dominance and goals), whereas women are more likely to use language that connects with others.

deliberatively to shape or change our perceptions of reality. For example, doublespeak is "language that pretends to communicate but really doesn't. [It] makes the bad seem good, the negative appear positive, the unpleasant appear attractive or at least tolerable" (Lutz, 1989:1).

There are several kinds of doublespeak: *Euphemisms* are words or phrases that avoid a harsh, unpleasant, or distasteful reality; *gobbledygook* (or *bureaucratese*) overwhelms the listener with big words and long sentences; and *inflated language* makes simple everyday things seem complex (Lutz, 1989: 1–6). Take the quiz on the previous page to see how much doublespeak you understand or use.

SOCIAL INTERACTION AND SELF-FULFILLING PROPHECIES

Our perceptions of reality shape our behavior. In an oft-cited statement, known as the *Thomas Theorem,* sociologists W. I. Thomas and Dorothy Thomas (1928: 572) observed, "If men define situations as real, they are real in their consequences." That is, our behavior is a result of how we interpret a situation ("My mother-in-law hates me, so I go shopping when she drops in").

Carrying this idea further, sociologist Robert Merton (1948/1966) proposed that our definitions of reality can result in a **self-fulfilling prophecy**: If we define something as real and act on it, it can, in fact, become real. For example, a study of adults found that physical education teachers who publicly humiliated students often turned them off physical fitness for good. One man said, "To this day I feel totally inadequate in team-related activities and have a natural reflex to AVOID THEM AT ALL COSTS" (Strean, 2009: 217; capitalization in original). Thus, gym teachers' negative comments made students feel inadequate in sports and changed their behavior during adulthood, regardless of their ability.

Our perceptions of reality shape our behavior, but how do people define that reality? Interactionists use two research tools—*ethnomethodology* and *dramaturgical analysis*—to help answer this question.

ETHNOMETHODOLOGY

A term coined by sociologist Harold Garfinkel (1967), **ethnomethodology** is the study of how people construct and learn to share definitions of reality that make everyday interactions possible. That is, we base our interactions on common assumptions about what makes sense in specific situations (Schutz, 1967; Hilbert, 1992).

People make sense of their everyday lives in two ways. First, by observing conversations, people discover the general rules that we all use to interact. Second, people can understand interaction rules by breaking them. Over a number of years, Garfinkel instructed his students to purposely violate everyday interaction rules and then to analyze the results. In these exercises, some of his students went to a grocery store and insisted on paying more than was asked for a product. Others were instructed, in the course of an ordinary conversation and without indicating that anything unusual was happening, "to bring their faces up to the subject's until their noses were almost touching" (Garfinkel, 1967: 72).

When students violated such everyday interaction rules, they were sanctioned. Grocery clerks became hostile when the students insisted on paying more than the marked price for a product, and people backed off when their noses were almost touching: "Reports were filled with accounts of astonishment, bewilderment, shock, anxiety, embarrassment, and anger" (Garfinkel, 1967: 47).

Violating interaction rules, even unspoken ones, can trigger anger, hostility, and frustration. College students become upset if professors ignore teaching norms by using sarcasm and putdowns, coming to class unprepared or consistently late, constantly reading from the book, or discouraging students' comments and questions (Berkos et al., 2001).

DRAMATURGICAL ANALYSIS

Dramaturgical analysis is a research approach that examines social interaction as if occurring on a stage where people play different roles and act out scenes for the "audiences" with whom they interact. According to sociologist Erving Goffman (1959, 1967), life is similar to a play in which each of us is an actor, and our social interaction is much like theater because we're always on stage and always performing. In our everyday performances, we present different versions of ourselves to people in different settings (audiences).

> **self-fulfilling prophecy** a situation in which if we define something as real and act on it, it can, in fact, become real.
>
> **ethnomethodology** the study of how people construct and learn to share definitions of reality that make everyday interactions possible.
>
> **dramaturgical analysis** examines social interaction as if occurring on a stage where people play different roles and act out scenes for the audiences with whom they interact.

The chaotic and crowded back-stage conditions in many restaurant kitchens stand in stark contrast with the efficient and relaxed front-stage behavior of its servers. Can you think of a situation where your front- and back-stage behaviors differ dramatically?

Thinkstock Images/Stockbyte/Getty Images

Because most of us try to present a positive image of ourselves, much social interaction involves *impression management*, a process of suppressing unfavorable traits and stressing favorable ones (see Chapter 4). To control information about ourselves, we often rely on props to convey or reinforce a particular image. For example, physicians, lawyers, and college professors may line their office walls with framed diplomas, medical certificates, or community awards to give the impression that they're competent, respected, and successful.

According to Goffman, the presentation of a performance involves front- and back-stage behaviors. The *front stage* is an area where an actual performance takes place. In front-stage areas, such as living rooms or restaurants, the setting is clean and the servers or hosts are typically polite and deferential to guests. The *back stage* is an area concealed from the audience, where people can relax. Bedrooms and restaurant kitchens are examples of back stages. After guests have left, the host and hostess may kick off their shoes and gossip about their company. In restaurants, cooks and servers may criticize the guests, use vulgar language, and yell at each other. Thus, the civility and decorum of the front stage may change to rudeness in the back stage.

Another example of front- and back-stage behavior involves faculty and students. Professors want to create the impression that they're well prepared, knowledgeable, and hardworking. In back-stage areas such as their offices, however, faculty may complain to their colleagues that they're tired of teaching a particular course or grading terrible exams and papers, or they may confess that they don't always prepare for classes as well as they should. Students often plead with instructors for higher grades: "But I studied very hard" or "I'm an A student in my other courses." In back-stage areas such as dormitories, student lounges, or libraries, however, students may admit to their friends that they barely studied or that they have a low grade-point average.

5-4b Social Exchange Theory

The fundamental premise of **social exchange theory** is that social interaction is based on each person's trying to maximize rewards (or benefits) and minimize punishments (or costs). An interaction that elicits a reward, such as approval or a smile, is more likely to be repeated than an interaction that brings a cost, such as disapproval or criticism (Thibaut and Kelley, 1959; Homans, 1974; Blau, 1986). Interactions are most satisfying when there's a balance between giving and taking.

People bring various tangible and intangible resources to a relationship (money, status, intelligence, good looks, youth, power, affection). Any of a person's resources can be traded for more, better, or different resources that another person possesses. Marriages between older men and younger women often reflect an exchange of the man's power, money, and/or fame for the woman's youth, physical attractiveness, and ability to bear children.

Many of our cost–reward decisions are conscious and deliberate, but others are passive or based on long-term negative interactions. For example, much of the domestic violence research shows that women stay in abusive relationships because their self-esteem has eroded after years

social exchange theory posits that social interaction is based on each person's trying to maximize rewards (or benefits) and minimize punishments (or costs).

© Tyler Olson/Shutterstock.com

of criticism and ridicule from both their parents and their spouses or partners ("You'll be lucky if anyone marries you," "You're dumb," "You're ugly," and so on). In effect, the abused women (and sometimes men) believe that they have nothing to exchange in a relationship, that they don't have the right to expect benefits (especially when the abuser blames the victim for provoking the anger), or that enduring an abusive relationship is better than being alone (Walker, 2000; Bergen et al., 2005).

5-4c Feminist Theories

According to much feminist research, women and men are more similar than different in their interactions. An analysis of studies published since the 1960s concluded that men are generally more talkative than women, but talkativeness depends on the particular situation. During decision-making tasks, men are more talkative than women, but when talking about themselves or interacting with children, women are more talkative than men (Leaper and Ayres, 2007; Mehl et al., 2007).

Studies also show that cultural norms and gender role expectations shape our interaction patterns. Generally, women are socialized to be more comfortable talking about their feelings, whereas men are socialized to be dominant and take charge, especially in the workplace. Women often ask questions that probe for a greater understanding of feelings ("Were you glad it happened?"). Women are also more likely than men to do conversational "maintenance work," such as asking questions that encourage conversation ("What happened at the meeting?") (Lakoff, 1990; Farley et al., 2010).

Compared with female speech, men's speech often reflects *conversational dominance*, speaking more frequently and for longer periods. Men also show dominance by interrupting or ignoring others, reinterpreting the speaker's meaning, changing the topic, or giving unsolicited and unwanted advice (Tannen, 1990; Toth, 2011; Gurkoff and Ranieri, 2012).

In predominantly white organizations, many black professionals believe that they're held to different interaction standards than their white coworkers. Consequently, they tend to keep their emotions in check to avoid being labeled as "too sensitive," "angry," "irritating," or "unpleasant" (Wingfield, 2010).

Such interaction differences aren't innate because of one's sex, however. Instead, much depends on women's and men's domestic roles and their position and status in the job hierarchy. For example, fathers who do much of the parenting have communication styles that are similar to those of mothers. In the workplace, women and men who occupy high-level decision-making positions also have similar interaction styles with superiors and subordinates (Cameron, 2007).

You've seen, so far, that much of our interaction is verbal and face-to-face. Nonverbal communication is also important in shaping our everyday relationships.

5-5 NONVERBAL COMMUNICATION

Nonverbal communication refers to messages sent without using words. This silent language, a "language of behavior" that conveys our real feelings, can be more potent than our words (Hall, 1959: 15). For example, sobbing sends a much stronger message than saying "I feel very sad." Some of the most common nonverbal messages are silence, visual cues, touch, and personal space.

5-5a Silence

Silence expresses a variety of emotions: Agreement, apathy, awe, confusion, disagreement, embarrassment, regret, respect, sadness, thoughtfulness, and fear, to name just a few. In various contexts, and at particular points in a conversation, silence means different things to different people. Sometimes silence saves us from embarrassing ourselves. Think, for example, about the times you fired off an angry email or text message, regretted doing so an hour later, and then dreaded getting a fuming response.

Many of us have experienced the pain of getting "the silent treatment" from friends, family members, coworkers, or lovers. Not talking to people who are

nonverbal communication messages that are sent without using words.

"This concludes my lecture on non-verbal communication. Any comments or questions?"

Chris Wildt/CSL, CartoonStock Ltd

important to us builds up anger and hostility. Initially, the "offender" may work hard to make the close-mouthed person discuss a problem. Eventually, the target of the silent treatment gets fed up, gives up, or ends a relationship.

5-5b Visual Cues

Visual cues, another form of nonverbal communication, include gestures, facial expressions, and eye contact. Let's consider a few examples of such body language in our daily interactions.

GESTURES

Most of us think we know what certain gestures mean—folding your arms across your chest indicates a closed, defensive attitude; leaning forward often shows interest; shrugging your shoulders signals indifference; narrowing your eyes and setting your jaw shows defiance; and smiling and nodding shows agreement.

Finger-pointing is usually a gesture that directs attention outward, placing blame or responsibility on someone else. The common "talk to the hand" gesture sends a stronger message: "Go away!" or "I'm not listening to you." Few gestures, however, clearly convey a meaning by themselves. Instead, they must be interpreted in context. Because of habit or hearing problems, for instance, a coworker may always lean forward when listening, regardless of whether he or she is interested.

The same gesture may have different meanings in different countries:

▸ Tapping one's elbow several times with the palm of one's hand indicates that someone is sneaky (in Holland), stupid (in Germany and Austria), or mean or stingy (in South America).

▸ Twisting an index finger into one cheek means "She's beautiful" in Libya, and that something tastes good in Italy. The same gesture in southern Spain means that a man is effeminate, and "You're crazy!" in Germany.

▸ In many Middle Eastern countries, people view the shoes and the soles of one's feet as unclean. Thus, stretching out one's legs with the feet pointing at someone or crossing one's legs so that a sole faces another person is considered rude (Morris, 1994; Lynch and Hanson, 1999; Jandt, 2001; Donadio, 2013).

In a remarkable tarmac ritual, and regardless of weather, the ground crews at Japanese airports use many gestures as a jet pushes back from the gate. First, the crew members line up, snap to attention, and then, "in perfect unison," they bow. This gesture is directed toward the passengers on the plane ("Thank you for visiting Japan"), toward the plane's flight crew ("We respect your expertise and dedication"), and, ultimately, as recognition of the crew's commitment to service ("We have fulfilled our duties to the best of our abilities"). Then they straighten up, smile broadly, and wave goodbye to show the passengers that they had been welcome to Japan (Gottlieb, 2011: 44).

| UNITED STATED okay | SOUTH AMERICA not okay | JAPAN money | FRANCE zero | GERMANY vulgar gesture | OTHER COUNTRIES better check first |

Hemera Technologies/AbleStock.com/Getty Images/Jupiter Images

Departing jets get a wave from the ground crew in Tokyo.

FACIAL EXPRESSIONS

Facial expressions are visual cues of feelings, but they can be deceptive. First, our facial expressions don't always show our true emotions. Parents tell their children "Don't you roll your eyes at me!" or "Look happy when Aunt Minnie hugs you." Thus, children learn that displaying their real feelings—especially when they're negative—is often unacceptable.

Second, faces can lie about feelings. Parents often know when children are lying because the children avoid eye contact, cry, or blink and swallow frequently. Some doctors say that they can tell if their patients are lying because of signs such as avoiding eye contact, pausing, or voice inflections, but many adults monitor and control their facial expressions. They can deceive, successfully and over many years, because they've rehearsed the lies in their heads, are smooth talkers with a reputation for being trustworthy, or have gotten away with lying in the past. In nonverbal behavior, people can correctly distinguish truth from lies only about 54 percent of the time (Reddy, 2013; Shea, 2014).

Third, facial expressions can be misleading because of cultural variations. American businesspeople have

Is she being friendly? Or flirting?

grumbled that Germans are cool and aloof, whereas many German businesspeople have complained that Americans are excessively friendly and hide their true feelings with grins and smiles. The Japanese, who believe that it's rude to display negative feelings in public, smile more than Americans do to disguise embarrassment, anger, and other negative emotions (Jandt, 2001).

The only consistent research finding, across cultures, is that women smile more than men. This difference may be due to cultural norms that socialize women to hide negative feelings that might make people uncomfortable, and to show deference to men (Szarota, 2010). Regardless of the reason, men are more likely than women to mistake being friendly as being flirtatious and sexually seductive (La France et al., 2009; Rutter and Schwartz, 2012).

EYE CONTACT

Eye contact serves several social purposes (Eisenberg and Smith, 1971; Ekman and Friesen, 1984; Siegman and Feldstein, 1987). First, we get much information about other people by looking at their eyes. Eyes open wide show surprise, fear, or a flicker of interest. When we're angry, we stare in an unflinching manner. When we're sad or ashamed, our eyes may be cast down.

Second, appropriate eye contact depends on the social context. Especially during job interviews, eye contact conveys attentiveness, confidence, and respect. Managers expect eye contact from subordinates, but subordinates often complain that a manager's prolonged eye contact is domineering or intimidating. Looking at a coworker when speaking shows trust and interest, but an overly intense gaze can be disturbing (Chen et al., 2013). When two people like each other, they establish eye contact more often and for longer durations than when there's tension in the relationship.

Finally, cultural norms affect eye contact. In many Asian cultures, including Japan and some Caribbean cultures, meeting another's eyes can be rude. Asians are more likely than Westerners to regard a person who makes frequent eye contact as angry, unapproachable, and unpleasant (Akechi et al., 2013). In effect, then, cross-cultural communication breaks down when we don't understand cultural rules about acceptable body language.

5-5c Touch

Touching, another important form of nonverbal communication, sends powerful messages about our feelings and attitudes. Parents worldwide communicate with their infants through touch—stroking, holding, patting, rubbing, and cuddling.

Touching can be positive (hugging, embracing, kissing, and holding hands) or negative (hitting, shoving, pushing,

Rod Aydelotte-Pool/Getty Images

Pictured are former President Bush and Saudi Crown Prince Abdullah at Bush's Texas ranch in 2002. Many U.S. journalists raised questions about two men holding hands. In response, Arab Americans were quick to point out that Arab society sees the outward display of affection between male friends as an expression of respect and trust. Thus, government officials and military officers often hold hands as they walk together or converse with one another.

spanking). Other forms of touching are controlling. One of my students, for example, left a boyfriend because "He said he loved me and trusted me but he gripped my arm tightly every time I talked to another guy and pulled me away." Between intimate partners, a decline in the amount of touching may signal that feelings are cooling off.

GENDERED TOUCHING

Whether touching is viewed as positive or negative depends on the situation and one's gender. In higher education, even when the faculty member is well liked, male and female students perceive touching differently. When female professors touch male students on the arm while talking to them, the students view the gesture as friendly. When a male professor touches a female student on the arm, she may get nervous because she's afraid that the touching may escalate (Lannutti et al., 2001; Fogg, 2005). Generally, women are more likely than men to initiate hugs and touches that express support, affection, and comfort. In contrast, men more often use touching to assert power or show sexual interest (Wood, 2011).

CROSS-CULTURAL VARIATIONS IN TOUCHING

As with other forms of nonverbal communication, the interpretation of touching varies from culture to culture.

In some Middle Eastern countries, people don't offer anything to another with the left hand because it's used to clean oneself after using the toilet. And among many Chinese and other Asian groups, hugging, backslapping, and handshaking aren't as typical as they are in the United States because such touching is seen as too intimate (Lynch and Hanson, 1999).

According to one scholar, Americans—even strangers—seem to be hugging more. Nonetheless, she maintains, the United States is a "medium touch" culture: "More physically demonstrative than Japan, where a bow is the all-purpose hello and goodbye, but less demonstrative than Latin or Eastern European cultures, where hugs are robust and can include a kiss on both cheeks" (Drexler, 2013: C3).

5-5d Personal Space

In the example of airline seats at the beginning of this chapter, you saw that the distance that people establish between themselves is an important aspect of nonverbal communication. Americans are more annoyed by a passenger who invades their seat space (47 percent) than skimpy airline leg room (39 percent), long security lines (37 percent), or weather delays (20 percent) (Aguila, 2014). Personal space plays a significant role in our everyday nonverbal interactions, reflects power and status, and varies across societies.

WHEN IS OUR PERSONAL SPACE VIOLATED?

Our living space is public or private. In the public sphere, which is usually formal, we have clearly delineated spaces: "This is your locker," "That's her office," or "You're parking in my spot." We usually decorate our public spaces with businesslike artifacts such as awards or framed photos of an institution's accomplishments.

After a snowstorm, people in densely populated cities who shovel parking spots "mark" their personal spaces with lawn chairs, strollers, and even a table set for two, complete with a bottle of wine. Some of the most aggressive (and locally accepted) retaliation occurs in South Boston, where residents punish violators by slashing their tires or smashing their car windows (Goodnough, 2010).

Private spaces send a different message. They convey informality and a relaxed feeling. Private spaces—homes, apartments, and dorm rooms—are often mini-museums that reflect people's interests, hobbies, hygiene, and personalities.

People, particularly men, sometimes invade our personal space by standing too close to us in a line, leaning against us, using all of the shared armrest in tight airline seats, or plopping their feet on the chair next to us in a classroom, airport, or movie theatre. Not all space

IMAGEMORE Co., Ltd./Alamy

Why do many men use more public space than women? Because they're larger and have longer legs? Or because men, but not women, feel "totally empowered to take up a lot public space," even if it inconveniences someone else (see Badahur, 2013)?

intrusions are physical encroachments. Loud cell phone conversations are a good example of an auditory intrusion into our personal space (Rosenbloom, 2013).

In intimate relationships (as between family members), our personal space is often 2 feet or less because we're at ease with close physical proximity. In contrast, in public situations (someone speaking to a large audience or faculty in large classrooms), the personal space is often 12 feet or more because the speaker, who has a higher social status than the audience, has a formal relationship with the listeners and avoids close physical contact (Hall, 1966).

SPACE AND POWER

Space usage signifies who has privilege, status, and power (Chapman and Hockey, 1999; Falah and Nagel, 2005). Wealthy people can afford enormous apartments in the city or houses in the suburbs, whereas the poor are crowded into the most undesirable sections of a city or town or in trailer parks. Executives, including college presidents, usually have huge offices (even entire suites), whereas faculty members often share office space, even though many of their discussions with students are confidential. Generally, the higher the socioeconomic status, the greater the consumption of space, including large cars, reserved

parking spaces, private dining areas, first-class airline seats, and luxurious skyboxes at sports stadiums.

CROSS-CULTURAL VARIATIONS AND SPACE

Cultural norms and values determine how we use space. Americans not only stand in line but also have strict queuing rules: "Hey, the end of the line is back there" and "That guy is trying to cut into the line." In contrast, "along with Italians and Spaniards, the French are among the least queue-conscious in Europe" (Jandt, 2001: 109). In these countries and others, people routinely push into the front of a group waiting for taxis, food, and tickets, and nobody objects.

Americans maintain personal space in an elevator by moving to the corners or to the back. In contrast, an Arab male may stand right next to or touching another man even when no one else is in the elevator. Because most Arab men don't share American concepts of personal space in public places and in private conversations, they consider it offensive if the other man steps or leans away. In many Middle Eastern countries, however, there are stringent rules about women and men separating themselves spatially during religious ceremonies and about women avoiding any physical contact with men in public places (Jandt, 2001; Office of the Deputy Chief, 2006; see also Chapter 3).

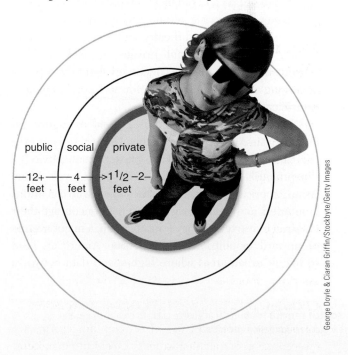

George Doyle & Ciaran Griffin/Stockbyte/Getty Images

ONLINE INTERACTION

Many of us now interact in *cyberspace*, an online world of computer networks. In cyberspace, **social media** are websites that enable users to create, share, and/or exchange information and ideas. Social media include social networking sites (e.g., Facebook and Twitter), gaming sites (e.g., *Second Life*), video sites (e.g., YouTube), and blogs (websites maintained by people who provide commentary and other material). This final section addresses two questions: Who's online and why? And is online interaction beneficial or harmful?

5-6a Who's Online and Why?

The percentage of Americans age 18 and older who are connected to the Internet increased from practically zero in 1994 to 87 percent in 2014. Fully 91 percent of U.S. adults have a cell phone. More than half (58 percent) of all cell phone users have smartphones, giving them Internet access all the time (Duggan, 2013; Fox and Rainie, 2014).

DEMOGRAPHIC VARIATIONS

Almost equal numbers of women (86 percent) and men (87 percent) use the Internet. Regarding age, only 57 percent of Americans aged 65 and older are online compared with 88 percent of those aged 50 to 64, 93 percent of those aged 30 to 49, and 97 percent of those aged 18 to 29 (Fox and Rainie, 2014). Of the 15 percent of Americans ages 18 and older who don't use the Internet, nearly half are 65 and older. The most common reasons (66 percent) for being offline include not wanting or needing the Internet, difficulty and frustration getting online, and worries about privacy or hackers. Only 19 percent of non-users are concerned about the expense of owning a computer or paying for an Internet connection (Zickuhr, 2013).

In terms of race and ethnicity, and as *Figure 5.3* shows, Asian Americans are the most connected group in the United States. This group's greater "connectivity" is due primarily to high education and income levels. Many Asian American parents are professionals who can afford computers and online service, and they encourage their children to use technology for education, a major avenue of upward mobility. Moreover, Asian Americans tend to live in urban areas where high-speed data networks

social media websites that enable users to create, share, and/or exchange information and ideas.

FIGURE 5.3 INTERNET AND SMARTPHONE USE, BY RACE AND ETHNICITY

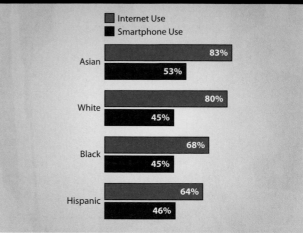

Note: The data are for people age 25 and older, 2012.

Source: Based on U.S. Census Bureau, "Computer and Internet Trends in America," Infographic, February 3, 2014. Accessed November 1, 2014 (census.gov/hhes/computer).

for both the Internet and smartphones are more readily available (File, 2013; see also Chapters 10 and 13).

The biggest digital divide is between social classes. The higher a family's income, the greater the likelihood that its members are Internet users (see *Figure 5.4*). Affluent families can buy home computers, pay for Internet service, and send their children to schools that provide computer-related instruction. Thus, the children from the poorest families are the most likely to lack technological skills.

FIGURE 5.4 SOCIAL CLASS AND INTERNET USE, 2014

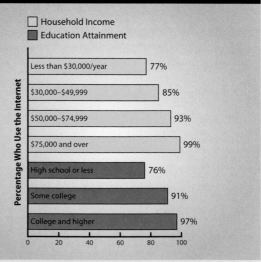

Source: Based on Fox and Rainie, 2014.

WHAT WE DO ONLINE

A **social networking site** is a website that connects people who share similar personal interests (e.g., Facebook) or professional interests (e.g., LinkedIn). In mid-2013, 72 percent of online adults used social networking sites, up from only 8 percent in 2005. Women (54 percent) are somewhat more likely than men to be on these sites, and a large majority (77 percent) of social networkers is white. The largest group of social networking users is young adults ages 18 to 29: Their participation rate jumped from 9 percent in 2005 to 89 percent in late 2012 (Brenner and Smith, 2013; Duggan and Brenner, 2013). Despite the growing popularity of social networking, Americans age 18 and older are more likely to use the Internet to find information on the Web (76 percent) and for email (92 percent) than for getting news, buying a product, or participating in social network sites (Purcell, 2011; see also Holcomb et al., 2013).

5-6b How Beneficial and Harmful Is Online Interaction?

You'll see in Chapter 16 that technological changes bring both benefits and costs. The same is true of online interaction, ranging from family ties to privacy issues.

FAMILY TIES

Social media has had a mixed effect on families. By age 3, many children have learned technological skills (creating a password, logging onto a computer, navigating some websites, using tablets). Video games such as *Minecraft* encourage school age children to build, explore, collaborate, and improve hand-eye coordination, problem-solving ability, and memory (Bilton, 2013; Dockterman, 2013; Lewin, 2013). On the other hand, as you saw in Chapter 4, pediatricians and other health professionals point out that excessive media use among young children is associated with obesity, poor school performance, aggression, and lack of sleep.

Almost 70 percent of parents with children age 8 or under have downloaded educational apps. Educational apps can foster creativity and learning, but hundreds of apps advertised as "educational" are rote, passive, and unproductive; stunt imaginative play; and contain advertisements, gender stereotypes, and violence. Even when apps are educational, there's a social class "app gap." For example, 57 percent of parents who earn less than $30,000 a year, compared with 80 percent of parents who earn $75,000 or more a year, have downloaded educational apps for their children (Hanes, 2013; Rideout, 2013).

Cell phones have increased the frequency of interaction between parents, especially if both are employed, to coordinate schedules and to chat with their children. Also, many parents report spending more time with their children by playing home video games. On the other hand, some children complain that they rarely receive their parents' full attention because a parent is often immersed in email, texting, or being online even when pushing a swing, driving, or during dinner (Young, 2011). In effect, some parents are paying more attention to their technology than their children.

About 89 percent of Americans say that mobile technology has increased their communication with friends and family members (Jones, 2014). On the other hand, an online survey at one university found that college students spent nearly nine hours a day on their smartphones—calling, texting, emailing, playing games, and using a number of applications (Facebook, Instagram, Twitter, Pinterest, iTunes, and so on). The researchers concluded that mobile technology allows us the freedom to gather information, communicate, and socialize. At the same time, excessive or obsessive usage of technology can cause conflict inside and outside the classroom with professors, employers, family, and friends (Roberts et al., 2014; see also Gellman, 2013).

Among adults who are married or in committed relationships, 74 percent say that the Internet, smartphones, and social media have had a positive impact primarily because email and texting facilitate communication and feeling closer to each other. On the other hand, 20 percent believe that the technology has had a negative effect on their relationships (e.g., a cell phone distracts a spouse or partner when the couple is together, or they argue about the amount of time one of them spends online) (Lenhart and Duggan, 2014).

ONLINE RELATIONSHIPS

The Internet has increased interaction, especially among people with similar interests. Millions of Americans use social media to plan religious activities, communicate with neighbors, petition politicians, and contact civic organizations. All of these activities have strengthened rather than diluted interpersonal and community ties, and have fostered a more diverse social network of people across social class and ethnic groups (Hampton et al., 2009; Smith, 2010).

An increasing number of U.S. adults use social networking to discuss important matters, get social support,

social networking site a website that connects people who share similar personal or professional interests.

and keep up or revive dormant relationships (Hampton et al., 2011). Also, some ethnic groups, especially Asian Americans, who are rarely featured on television, have become very successful on YouTube. They've found "millions of eager fans" who follow their comedy sketches and discuss subjects, particularly sex and race, that are taboo among older generations (Considine, 2011: ST6; see also Pipkin, 2013).

Millions of Americans are turning to the Internet to find romance. There are more than 1,500 online dating sites, many sites and dating apps are free, and geography-based technology allows people to arrange spur-of-the-moment dates. Subscribers can sift candidates based on height, age, mutual interests, and dozens of other traits. About 11 percent of Internet users (representing 9 percent of all American adults) have used online dating sites (e.g., Match.com, eHarmony.com, plentyoffish.com). Seventy percent of online daters believe that online dating helps people find a romantic match because they have access to a wide range of potential partners. On the other hand, only 5 percent of Americans who are currently married or in a long-term relationship met their partner online. Moreover, 54 percent of online daters, particularly women, say that people have lied about their age, height, weight, income, and occupation (Smith and Duggan, 2013; see also Benokraitis, 2014).

pearleye/iStockphoto.com

CONNECTIONS, COMPULSIONS, AND COMPLICATIONS

Online interaction develops networking skills, maintains ties with family members and friends—especially in faraway places—gives people an advantage in finding a job, connects people to their communities, and encourages political engagement (Peterson, 2013). On the other hand, there's a "constant compulsion to connect" (Suddath, 2013: 80). For example, 67 percent of cell phone owners check their phone, even when it's not ringing or vibrating, and 44 percent have slept with the phone next to their bed to make sure they didn't miss any calls or text messages during the night (Smith, 2012).

Some also maintain that social networking sites, especially Facebook, are superficial and give people a false sense of friendship and connection to others. The average Facebook-using teen has 300 "friends," 33 percent of whom she or he has never met in person (Madden et al., 2013). According to some critics, social networking sites are both inflating many people's egos and replacing the

time we might take to develop relationships with the few friends that we have (DiSalvo, 2010).

Among online adults, 67 percent are Facebook users, making Facebook the dominant U.S. social networking site. By comparison, only 20 percent use LinkedIn, and 16 percent use Twitter. Despite Facebook's popularity, 61 percent of Facebook users have taken a voluntary break from using the site at one time or another, and 27 percent plan to spend less time on the site in the future. The major reasons for a "Facebook vacation" included being too busy with other demands (21 percent), not liking the site or seeing it as a waste of time (20 percent), and too much drama, gossip, negativity, and conflict (9 percent) (Rainie et al., 2013).

Besides these reasons, several studies of young adults in their 20s have found that the more often they use Facebook, the unhappier they become, and regardless of gender, level of loneliness, or self-esteem. The root cause of the unhappiness is envy, even though Facebook users suspect that their "friends" Photoshop images and exaggerate their achievements, job success, vacations, and love life. Feelings of envy, anger, and loneliness are especially likely if Facebook users receive fewer positive comments, "likes," and general feedback than their friends do (Krasnova et al., 2013; Kross et al., 2013). Even adults admit that social media makes them feel bad about themselves when their friends describe awards, promotions, and "living a more fulfilling, fabulous life than I am" (Melton, 2013).

ONLINE HARASSMENT AND PUBLIC SHAMING

Nationally, about 24 percent of adolescents have experienced *cyberbullying*—deliberately using digital media to communicate false, embarrassing, or hostile information about someone. A greater percentage of females (79 percent) than males (60 percent) report being cyberbullied during the school year. Cyberbullying is less common than traditional bullying, but it has more profound negative outcomes that include depression, anxiety, severe isolation, and, most tragically, suicide (Patchin, 2013; Robers et al., 2013).

Cyberstalking, which is more common among adults than adolescents, is threatening behavior or unwanted advances using email, texting, social networking sites, and other electronic communication devices. About 80 percent of the victims are women, and 87 percent of the offenders are men. Cyberstalking may evolve into offline stalking, including abusive or harassing phone calls, vandalism, threatening or obscene mail, trespassing, and physical assault (National White Collar Crime Center, 2013).

Twitter users can be engaging, funny, and supportive. They can also be rude, hostile, aggressive, dismissive, and uncivil. Tweets, which are limited to 140 characters, are often written quickly and sent impulsively. Because tweets are short and easy to read, public accusations and shaming people with whom we disagree gets readers' attention. Blogs, by contrast, are much longer and produce "much more substantive, thoughtful critique": "Authors have time to sit with a post before publishing it, and commenters have time to read it and think about it before responding" (Fitzpatrick, 2013: B4).

The latest social media trend is "live-tweeting" a private conversation that occurs in a public place. For example, a man in Brooklyn overheard a couple arguing and breaking up on the rooftop of an apartment building, and live-tweeted the entire exchange through a sequence of tweets. Such material is then retweeted, described on media sites, laughed at, and mocked (Chittal, 2013).

PRIVACY

A major cost of online interaction is jeopardizing our privacy because email and text messages are neither anonymous nor confidential. Snapchat—an app that allows people to take a picture, send it to someone, and the photo is automatically deleted after 10 seconds—doesn't guarantee that the deletion always occurs (McLane, 2013). Many companies monitor and preserve their employees' email messages and can use them to discipline or fire people. It's also becoming increasingly common for college admissions officers and employers to search networking sites and Twitter before deciding whether to accept a student or to make a job offer (Belkin and Porter, 2012; Singer, 2013).

People often text and say things in emails they'd never say in person, especially when they're flirting, angry, frustrated, or tired. Emails and text messages don't disappear after being deleted but may last indefinitely in cyberspace. Divorce lawyers have successfully retrieved deleted emails, Facebook posts, tweets, Instagram photos, and the like to convince judges that a divorcing spouse shouldn't get financial support or custody of the children (Hillin, 2014).

© eteimaging/Shutterstock.com

Many Americans, especially teens, willingly give out personal information on social media sites—photos, phone number, address, age, education and work background, political and religious views, and so on (Madden et al., 2013). It takes a private investigator just a few clicks to get a composite picture of someone from public records, email messages, websites, blogs, and networking sites.

Also, smartphones know everything—where people go, what they search for, what they buy, what they do for fun, and when they go to bed. Tech companies such as Google and Facebook can track people on their cell phones and reach them with "individualized, hypertargeted ads." Thus, "if you research a Hawaiian vacation on your work desktop, you could see a Hawaii ad that night on your personal cell phone" (Miller and Sengupta, 2013: A1). Besides collecting data on its own users, Facebook has partnerships with four companies that collect behavioral data—from store transactions and customer email lists to divorce and Web browsing records—that it then sells to advertisers (Sengupta, 2013). Neither state nor federal laws prohibit the collection or sharing of such data by third parties.

6 | Social Groups, Organizations, and Social Institutions

LEARNING OBJECTIVES

6-1 Compare and illustrate the different types of social groups, explain how size and leadership affect groups, why people conform to group pressure, and the impact of social networks.

6-2 Describe and illustrate three types of formal organizations, the strengths and shortcomings of bureaucracies, and explain how informal groups affect organizations.

6-3 Compare the theoretical explanations of social groups and organizations, including their contributions and limitations.

6-4 Explain why social institutions are important and how they're interconnected.

After you finish this chapter go to **PAGE 117** for **STUDY TOOLS.**

Every year, about a dozen of our neighbors, all fervent Ravens football fans, get together to buy season tickets and to plan elaborate tailgating parties. They wear Ravens caps and purple jerseys, and most of the drivers attach pennants to their car antennas. These football enthusiasts are a social group.

6-1 SOCIAL GROUPS

A **social group** is two or more people who share some attribute and interact with one another. They have a sense of belonging or "we-ness." Friends, families, work groups, religious congregations, clubs, athletic teams, and World War II veterans are all examples of social groups. Each of us is a member of many social groups, but we identify more closely with some groups than others.

social group two or more people who share some attribute and interact with one another.

What do you think?

It's a good idea for companies to use video surveillance, including in restrooms.

(1   2   3   4   5   6   7)
strongly agree   strongly disagree

6-1a Types of Social Groups

Some groups are small and personal (families); others are large and impersonal (financial organizations). Some are highly organized and stable (political parties); others are fluid and temporary (high school classmates). The most basic types of social groups are primary and secondary groups.

PRIMARY AND SECONDARY GROUPS

A **primary group** is a a relatively small group of people who engage in intimate face-to-face interaction over an extended period (Cooley, 1909/1983). Primary groups, such as families and close friends, are our emotional glue. They have a powerful influence on our social identity because we interact with them on a regular basis over many years, usually throughout our lives. Because primary group members genuinely care about each other, they contribute to one another's personal development, security, and well-being. Our family and close friends,

Lew Robertson/Spirit/Corbis

particularly, stick with us through good and bad, and we feel comfortable being ourselves with them.

Compared with primary groups, a **secondary group** is a large, usually formal, impersonal, and temporary collection of people who pursue a specific goal or activity. Your sociology class is a good example of a secondary group. You might have a few friends in class, but students typically interact infrequently and relatively formally. When the semester (or quarter) is over and you've accomplished your goal of passing the course, you may not see each other again (especially if you're attending a large college or university). Other examples of secondary groups include sports teams, labor unions, and a company's employees.

> **primary group** a small group of people who engage in intimate face-to-face interaction over an extended period.
>
> **secondary group** a large, usually formal, impersonal, and temporary collection of people who pursue a specific goal or activity.

TABLE 6.1 CHARACTERISTICS OF PRIMARY AND SECONDARY GROUPS

	Characteristics of a Primary Group	Characteristics of a Secondary Group
Interaction	• Face-to-face • Usually small	• Face-to-face or indirect • Usually large
Communication	• Emotional, personal, and satisfying	• Emotionally neutral and impersonal
Relationships	• Intimate, warm, and informal • Usually long-term • Valued for their own sake (expressive)	• Typically remote, cool, and formal • Usually short-term • Goal-oriented (instrumental)
Individual Conformity	• Relatively free to stray from norms and rules	• Expected to adhere to rules and regulations
Membership	• Members aren't easily replaced	• Members are easily replaced
Examples	• Family, close friends, girlfriends and boyfriends, self-help groups, street gangs	• College classes, political parties, professional associations, religious organizations

Unlike primary groups, secondary groups are usually highly structured: There are many rules and regulations, people know (or care) little about each other personally, relationships are formal, and members are expected to fulfill particular functions. Whereas primary groups meet our *expressive* (emotional) needs, secondary groups fulfill *instrumental* (task-oriented) needs. Once a task or activity is completed—whether it's earning a grade, turning in a committee report, or building a bridge—secondary groups split up and become members of other secondary groups.

Table 6.1 summarizes the characteristics of primary and secondary groups. These characteristics are **ideal types**—general traits that describe a social phenomenon rather than every case. Ideal types provide composite pictures of how social phenomena differ rather than specific descriptions of reality. Because primary and secondary groups are ideal types, their characteristics can vary. Thus, primary group members may sometimes devote themselves to meeting instrumental needs (running a family-owned business), and secondary group members (military units and athletic teams) can develop lasting ties.

IN-GROUPS AND OUT-GROUPS

Rudyard Kipling's poem "We and They" captures the essence of in-groups and out-groups. Members of an **in-group** share a sense of identity and "we-ness" that typically excludes and devalues outsiders. **Out-group** members are people who are viewed and treated negatively because they're seen as having values, beliefs, and other characteristics different from those of an in-group. For example, "we" vegetarians are healthier than "you" meat eaters, "we" computer nerds are smarter than "you" fraternity and sorority "types," and so on.

CBS Photo Archive/Getty Images

Is the cast of characters on the television series *NCIS* an example of a primary group? A secondary group? Both? Neither?

ideal types general traits that describe a social phenomenon rather than every case.

in-group people who share a sense of identity and "we-ness" that typically excludes and devalues outsiders.

out-group people who are viewed and treated negatively because they're seen as having values, beliefs, and other characteristics different from those of an in-group.

All good people agree,
And all good people say,
All nice people, like Us, are We
And everyone else is They.
—from "We and They," by Rudyard Kipling

Based on ascribed or achieved statuses, almost everyone sees others as members of in-groups and out-groups. From ancient times to the present, people in various parts of the world have made "we" and "they" distinctions based on race, ethnicity, gender, sexual orientation, religion, age, social class, and other social and biological characteristics (Coser, 1956; Tajfel, 1982; Hinkle and Schopler, 1986). Such distinctions can promote in-group solidarity and cohesion. They can also create conflict and provoke inhumane actions, including wars, massacres of out-group members, and civil wars (see Chapter 3).

One person's in-group is another person's out-group. The general public and criminal justice system see gangs as dangerous out-groups. Many young people who join gangs, however, experience in-group benefits. Attractions include making money, although illegally; close relationships with family members and friends who are already involved in gang life; protection against violent family or community members and rival gangs; a sense of self-worth, support, and belonging; and the status of being an "outlaw" that the entertainment industry often romanticizes as sexy and exciting (Taylor and Smith, 2013).

REFERENCE GROUPS

An in-group or an out-group can become a **reference group**, people who shape our behavior, values, and attitudes (Merton and Rossi, 1950). Reference groups influence who we are, what we do, and who we'd like to be in the future. Unlike primary groups, however, reference groups rarely provide personal support or face-to-face interaction over time.

Reference groups might be people with whom we already associate (a college club or a recreational athletic team). They can also be groups that we admire and want to be part of (teachers or doctors). Each person has many reference groups. Your sociology professor, for example, may be a member of several professional associations, a golf enthusiast, a parent, and a homeowner. Identification with each of these groups influences her or his everyday attitudes and actions.

Like in-groups and primary groups, reference groups can have a strong impact on our self-identity, self-esteem, and sense of belonging because they shape our current and future attitudes and behavior. We typically add or drop reference groups throughout the life course. If you aspire to move up the occupational ladder, your reference group may change from entry-level employees to managers, vice presidents, and CEOs.

6-1b Group Size and Structure

Two important characteristics of a social group are its size and leadership. Both affect a group's interaction and dynamics (behavior over time).

GROUP SIZE: DYADS, TRIADS, AND BEYOND

German sociologist Georg Simmel (1858–1918) pioneered the study of **dyads**, groups with two members, and **triads**, groups with three members. A dyad (e.g., parent–child, lovers, husband–wife, close friends) is the most cohesive of all groups because its members tend to have a personal relationship and to interact more intensely than do people in larger groups. Dyads can be unstable, however, because both persons must cooperate. If either member doesn't fulfill her or his responsibilities, the dyad collapses (Simmel, 1902; Mills, 1958).

Adding only one member to a dyad has important consequences because it changes a group's dynamics. A triad is more stable than a dyad because the group can continue if a person drops out, and a member may patch up an argument that erupts between the other two. On the other hand, a third person who enters a dyad may intentionally create conflict to break up a relationship or to attain a dominant position. In any triad, two of the members can gang up against or leave out the third (Simmel, 1902; Mills, 1958). Triads also tend to interact less often and intensely than do dyads (e.g., think about a third roommate who moves in, another person who joins you and a coworker at lunch, or someone who tags along on a date).

As more members are added—a fourth, fifth, and so on—interaction changes even more and alliance possibilities increase. At about seven or eight members, the group may break down into dyads and triads because it's difficult to have a single conversation about the task at hand (Becker and Useem, 1942). As a result, larger groups often require some kind of leadership to keep the members focused.

GROUP LEADERSHIP

Several classic studies (Lewin et al., 1939; Bales, 1950) identified three basic types of group leaders. An **authoritarian leader** gives orders, assigns tasks, and

---

reference group people who shape our behavior, values, and attitudes.

dyad a group with two members.

triad a group with three members.

authoritarian leader gives orders, assigns tasks, and makes all major decisions.

makes all major decisions. There's a clear division between leaders and followers, and leaders focus on completing *instrumental* tasks (e.g., marketing a new smartphone). Authoritarian leadership is most effective when there's little time for group decision making and the leader is the most knowledgeable member of the group. On the other hand, authoritarian leaders tend to stifle creativity and may be viewed as controlling and bossy.

By contrast, a **democratic leader** encourages group discussion and includes everyone in the decision-making process. Democratic leaders have the final say, but tend to be concerned about meeting a group's *expressive* needs (e.g., maintaining harmony, high morale, and cohesiveness). Democratic leaders engage group members and encourage their creativity, but have also been blamed for not taking charge in times of crisis, and lower productivity than groups with authoritarian leaders.

A **laissez-faire** (pronounced les-ey-fair) **leader** offers little or no guidance to group members and allows them to make their own decisions. Laissez-faire leaders are similar to the permissive parents you met in Chapter 4: There are few rules, few demands, and people do pretty much what they want. Laissez-faire leadership is effective in situations where group members are highly qualified in an area of expertise. However, this leadership style often leads to poorly defined roles, low productivity, a lack of motivation, little cooperation between group members, and not meeting a group's goals.

It's difficult for a leader to meet both a group's instrumental and expressive needs because involving everyone in decision making may result in not accomplishing a group's objectives. Also, leadership styles are more effective in some settings than others. Imagine, for example, if your college professors were laissez-faire ("Take the exams when you think you're ready") or if your employers were always democratic ("Let's vote on who'll do what today").

6-1c Group Conformity

Most Americans see themselves as rugged individualists who have minds of their own (see the discussion of values in Chapter 3). A number of studies have shown, however, that many of us are profoundly influenced by group pressure. Four of the best known of these studies are by Solomon Asch, Stanley Milgram, Philip Zimbardo, and Irving Janis.

democratic leader encourages group discussion and includes everyone in the decision-making process.

laissez-faire leader offers little or no guidance to group members and allows them to make their own decisions.

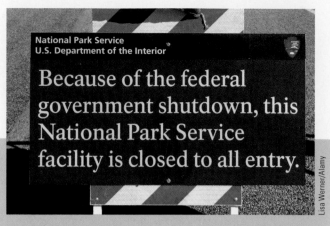

U.S. government shutdowns can result in a temporary halt of federal services because lawmakers, especially democratic group leaders, can't agree on consensus-based decisions.

ASCH'S RESEARCH

In a now classic study of group influence, social psychologist Solomon Asch (1952) told subjects that they were taking part in an experiment on visual judgment. After seating six to eight male undergraduates around a table, Asch showed them the line drawn on card 1 and asked them to match the line to one of three lines on card 2 (see *Figure 6.1*). The correct answer, clearly, is line C.

All but one of the subjects—who usually sat in the last chair—were Asch's confederates, or accomplices. In the first test, all the confederates selected the correct matching line. In the other tests, each of them, one by one, deliberately chose an incorrect line. Thus, the nonconfederates faced a situation in which seven other group members

FIGURE 6.1 CARDS IN ASCH'S EXPERIMENT

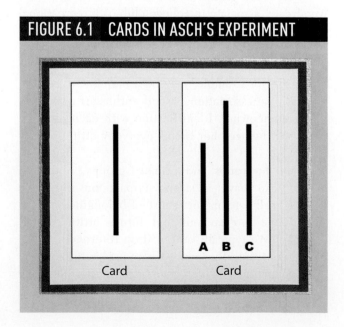

had unanimously agreed on a wrong answer. Averaged over all of the trials, 37 percent of the nonconfederate subjects ended up agreeing with the group's incorrect answers. When they were asked to judge the length of the lines alone, away from the influence of the group, they made errors only 1 percent of the time.

Asch's research demonstrated the power of groups over individuals. Even when we know that something is wrong, we may go along with the group to avoid ridicule or exclusion. Remember that these experiments were done in a laboratory and with people who didn't know each other. A group's influence on a person's attitudes and behavior can be even stronger when it's a real-life situation.

MILGRAM'S RESEARCH

In a well-known laboratory experiment on obedience, psychologist Stanley Milgram (1963, 1965) asked 40 volunteers to administer electric shocks to other study participants. In each experimental trial, one participant was a "teacher" and the other a "learner," one of Milgram's accomplices. The teachers were businessmen, professionals, and blue-collar workers.

The learner was strapped into a realistic-looking chair that supposedly regulated electric currents. The teacher read aloud pairs of words that the learner had to memorize. Whenever the learner gave a wrong answer, the teacher was told to apply an electric shock from a low of 15 volts to a high of 450 volts. The learner didn't actually receive a shock, but was told to fake pain and fear. When the learners shrieked in pain, the majority of the teachers, although distressed, obeyed the study supervisor and administered the shocks when told to do so.

Milgram's study was controversial. Ordering electric shocks raised numerous ethical questions about the participants' suffering extreme emotional stress (see Chapter 2; see also Perry, 2013, for a recent critique). However, the results showed that an astonishingly large proportion of ordinary people obeyed an authority figure's instructions to inflict pain on others. It's easy to sit back and say "I'd never do something like that," but we do. For example, many workers obey employers when told to ignore evidence that their product is unsafe (Tavris, 2013).

ZIMBARDO'S RESEARCH

The Stanford Prison Experiment conducted by social psychologist Philip Zimbardo also underscores the influence of groups on behavior (Haney et al., 1973; Zimbardo, 1975). Zimbardo recruited volunteers through a local California newspaper for an experiment on prison life. He then selected 24 young men, most of them college students.

On a Sunday morning, nine of the men were "arrested" at their homes as neighbors watched. The men were booked and transported to a mock prison that Zimbardo and his colleagues had constructed in the basement of the psychology building at Stanford University. The "prisoners" were searched, issued an identification number, and outfitted in a dresslike shirt and heavy ankle chains. Those assigned to be "guards" were given uniforms, billy clubs, whistles, and reflective sunglasses. The guards were told that their job was to maintain control of the prisoners but not to use violence.

All the young men quickly assumed the roles of either obedient and docile prisoners or autocratic and controlling guards. The guards became increasingly more cruel and demanding. The prisoners complied with dehumanizing demands (such as eating filthy sausages) to gain the guards' approval and bowed to their authority.

Zimbardo's study was scheduled to run for 2 weeks but was stopped after 6 days because the guards became increasingly aggressive. Among other things, they forced the prisoners to clean out toilet bowls with their bare hands, locked them in a closet, and made them stand at attention for hours. Instead of simply walking out or rebelling, the prisoners became withdrawn and depressed. Zimbardo ended the experiment because of the prisoners' stressed-out reactions.

The experiment raised numerous ethical questions about the harmful treatment of participants. It demonstrated, however, the powerful effect of group conformity: People exercise authority, even to the point of hurting others, or submit to authority if there's group pressure to conform (Zimbardo et al., 2000).

JANIS'S RESEARCH

Sometimes intelligent people, including those in highly responsible positions, make disastrous and irrational decisions. Why? Social psychologist Irving Janis (1972, 1982) cautioned presidents and other heads of state to be wary of **groupthink**—a situation in which in-group members make faulty decisions because of group pressures, rather than critically testing, analyzing, and evaluating ideas and evidence. To demonstrate group

groupthink a situation in which in-group members make faulty decisions because of group pressures, rather than critically testing, analyzing, and evaluating ideas and evidence.

Many social scientists have used Milgram's findings to explain the Abu Ghraib prison in Iraq, where U.S. soldiers humiliated prisoners because "I was just following orders."

SOMETIMES INTELLIGENT PEOPLE, EVEN THOSE IN HIGHLY RESPONSIBLE POSITIONS, MAKE DISASTROUS AND IRRATIONAL DECISIONS.

loyalty, Janis argues, individuals don't raise controversial issues, don't question weak arguments, or probe "soft-headed thinking." As a result, influential leaders often make decisions, based on their advisors' consensus, that turn out to be political and economic disasters.

Examples of groupthink "fiascoes" that Janis studied included U.S. failure to anticipate the attack on Pearl Harbor, the Bay of Pigs invasion, the escalation of the Vietnam War, and the ill-fated hostage rescue in Iran. More recently, a United States Senate study (2004) of intelligence agencies concluded that U.S. leaders' decision to invade Iraq in 2003 was based on a groupthink dynamic that relied on unproven and inaccurate assumptions, inadequate or misleading sources, and a dismissal of conflicting information showing that Iraq had no weapons of mass destruction. (It's still unclear whether or not groupthink affected the federal government's spying on U.S. citizens revealed in 2013.)

Janis and other researchers have focused on high-level decision making, but groupthink is common in all kinds of groups—student clubs, PTAs, search committees, juries, and community activists, for example (see Hansen, 2010). If group members are aware of groupthink, they can avoid some of the pitfalls by hammering out disagreements and seeking advice from informed and objective people outside the group.

6-1d Social Networks

A **social network** is a web of social ties that links individuals or groups to one another. It may involve as few as three people or millions.

social network a web of social ties that links individuals or groups to one another.

Network links between people or groups can be strong or weak. Some of our social networks, such as our primary and secondary groups, may be tightly knit, involve interactions on a daily basis, and have clear boundaries about who belongs and who doesn't. In other cases, our social networks connect us to large numbers of people whom we don't know personally, with whom we interact only rarely or indirectly, and the boundaries are fluid as people come and go. Examples of distant networks include members of the American Sociological Association and *Los Angeles Times* readers.

The Internet, in particular, has generated numerous social networks that unite people who have similar interests (see *Table 6.2*). Almost 75 percent of all adult Americans are active in some kind of voluntary group or organization. Internet users (80 percent) are more likely to participate in groups than non-Internet users (56 percent) (Rainie et al., 2011).

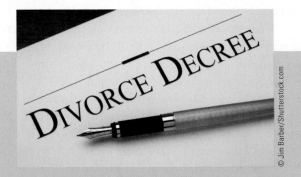

Do social networks affect divorce rates? A research team that analyzed three decades of divorce data found that people were 75 percent more likely to get divorced if a friend was divorced, and 33 percent more likely to end a marriage if a friend of a friend was divorced (McDermott et al., 2013; Chapter 12 discusses some of the micro- and macro-level reasons for divorce).

TABLE 6.2 HOW THE INTERNET AFFECTS SOCIAL NETWORKS

Percentage of U.S. adults who said the Internet has a "major impact" on their groups' ability to . . .

	Internet Users	Non-Internet Users
Communicate with members	75	44
Draw attention to an issue	68	43
Connect with other groups	67	40
Organize activities	65	41
Impact society at large	64	45
Recruit new members	55	38

Source: Based on Rainie et al., 2011: 31.

 ## 6-2 FORMAL ORGANIZATIONS

A **formal organization** is a complex and structured secondary group designed to achieve specific goals in an efficient manner. We depend on a variety of formal organizations to provide goods and services in a stable and predictable way, including city water departments, food producers and grocery chains that stock our favorite items, and garment industries and retailers that produce and sell the clothes we wear.

6-2a Characteristics of Formal Organizations

Formal organizations share some common characteristics:

▶ Social statuses and roles are organized around shared expectations and goals.

▶ Norms governing social relationships among members specify rights, duties, and sanctions.

▶ A formal hierarchy includes leaders or individuals who are in charge.

Modern, complex societies rely on formal organizations, including businesses and schools, to accomplish complex tasks. Organizations differ, however, in their goals, membership, and degree of hierarchy.

6-2b Types of Formal Organizations

Sociologist Amitai Etzioni (1975) identified three types of formal organizations based on their purpose and people's reasons for participating.

UTILITARIAN ORGANIZATIONS

Most people belong to a *utilitarian organization*, one that provides an income or other specific current or future material reward. Examples include government agencies, factories, corporations, and schools. Colleges are workplaces, but also offer students the opportunity to earn degrees.

NORMATIVE ORGANIZATIONS

People join *normative organizations* (also called *voluntary organizations* or *voluntary associations*) because of shared interests and to pursue goals that they consider personally worthwhile or rewarding. Examples include political parties, civic organizations (e.g., Habitat for Humanity), religious organizations, choirs, cultural groups, occupational groups (e.g., American Accounting Association), and groups that have a recreational focus (e.g., Beer Collectors of America) (see Rainie et al., 2011, for a detailed list of the most popular voluntary associations and some of their demographic characteristics).

Because helping others is a strong American value (see Chapter 3), every year, millions of Americans—many of whom have full-time jobs—are volunteers in a variety of local, regional, and national organizations, saving government agencies billions of dollars by providing needed services (see *Table 6.3*). In addition, 83 percent of Americans donate money to charitable organizations (Gallup Editors, 2013).

COERCIVE ORGANIZATIONS

Membership in *coercive organizations* is largely involuntary. People are pushed or forced to join these organizations because of punishment (prisons) or treatment (psychiatric hospitals, drug rehabilitation centers). Most coercive organizations are total institutions characterized by strict rules, the members' isolation, and a resocialization process (see Chapter 4).

TABLE 6.3 VOLUNTEERING IN THE UNITED STATES, 2012

Percentage of adults who volunteer	27
Total number of volunteers age 16 and older	65 million
Median annual hours per volunteer	50
Total dollar value of volunteer time	$171 billion

Sources: Based on Corporation for National & Community Service, 2012, and BLS News Release, 2013.

formal organization a complex and structured secondary group designed to achieve specific goals in an efficient manner.

Utilitarian, normative, and coercive organizations differ in their general characteristics and membership, but a single formal organization can fall into all three categories. For example, the military is a type of coercive organization because of its rigid rules, resocialization, and members' isolation from the rest of the population. It's a utilitarian organization that provides millions of jobs for both soldiers and civilians. The military is also a normative organization: Some people join during wartime because of civic responsibility; many volunteer their time and resources to support soldiers and their families during deployment; and others help veterans find jobs, housing, and health care for war-related injuries.

6-2c Bureaucracies

A **bureaucracy** is a formal organization designed to accomplish goals and tasks in an efficient and rational way. Bureaucracies aren't a modern invention but existed thousands of years ago in ancient Egypt, China, and Africa. Some bureaucracies function more smoothly than others, and some formal organizations are more bureaucratic than others. Whether they're relatively small (a 500-bed hospital) or huge (the Social Security Administration), bureaucracies have some common characteristics.

IDEAL CHARACTERISTICS OF BUREAUCRACIES

Max Weber (1925/1947) identified six key characteristics of the ideal type of bureaucracy. In Weber's model, the following characteristics describe what an efficient and productive bureaucracy *should* be like:

▸ **High degree of division of labor and specialization.** Individuals who work in a bureaucracy perform very specific tasks.

▸ **Hierarchy of authority.** Workers are arranged in a hierarchy in which each person is supervised by someone in a higher position. The resulting pyramids—often presented in organizational charts—show who has authority over whom and who is responsible to whom. Thus, there's a chain of command, stretching from top to bottom, that coordinates decision making.

▸ **Explicit written rules and regulations.** Detailed written rules and regulations cover almost every possible kind of situation and problem that

Dave Carpenter/CSL, CartoonStock Ltd

might arise, including hiring, firing, salary scales, rules for sick pay and absences, and everyday operations.

▸ **Impersonality.** There's no place for personal likes, dislikes, or tantrums. Workers are expected to behave professionally. An impersonal workplace in which all employees are treated equally minimizes conflict and favoritism and increases efficiency.

▸ **Qualifications-based employment.** People are hired based on objective criteria such as skills, education, experience, and standardized test scores. If workers perform well and have the necessary credentials and technical competence, they'll move up the career ladder.

▸ **Separation of work and ownership.** Neither managers nor employees own the offices they work in, the desks they sit at, the technology they use, or the products that they assemble, invent, or design.

A "rational matter-of-factness," Weber maintained, made bureaucracies more productive by "eliminating from official business love, hatred, and all purely personal, irrational, and emotional elements" (Weber, 1946: 216). Weber viewed bureaucracies as superior to other forms of organization because they're more efficient and predictable. He worried, however, that bureaucracies could become "iron cages" because people become trapped in them, "their basic humanity denied."

bureaucracy a formal organization designed to accomplish goals and tasks in an efficient and rational way.

© PzAxe/Shutterstock.com

SHORTCOMINGS OF BUREAUCRACIES

Weber described the ideal characteristics of a productive and efficient bureaucracy, but what's the reality? As you read through the following list of problems, think about the ones you've experienced while working in a bureaucracy or dealing with one.

▶ *Weak reward systems* reduce the motivation to do a good job and are thus a major source of inefficiency and lack of innovation (Barton, 1980). Besides low wages or salaries, weak reward systems include few or no health benefits, little recognition, unsafe equipment and work environments, and few incentives to be creative.

▶ *Rigid rules* discourage creativity and can lead to **goal displacement**, a preoccupation with rules and regulations rather than achieving the organization's objectives (Merton, 1968). Instead of questioning whether all the red tape is really necessary, bureaucrats are usually more concerned that people follow established procedures because "we've always done it this way."

▶ Rigid rules and goal displacement often lead to **alienation**, a feeling of isolation, meaninglessness, and powerlessness. Alienation—at all levels—may result in high turnover, tardiness, absenteeism, stealing, sabotage, stress, health problems, and in some cases, whistle-blowing (reporting organizational misconduct to legal authorities).

▶ *Communication problems* are common in bureaucracies. Because communication typically flows down rather than up the hierarchy, employees (and many managers below the highest echelons) rarely know what's going on. Supervisors and their subordinates may be reluctant to discuss problems or offer suggestions because they fear being criticized, demoted, or fired (Blau and Meyer, 1987; Fletcher and Taplin, 2002). Communication problems also waste time and resources. An employee prepared a monthly report the company no longer needed for nearly three years because no one ever mentioned it wasn't necessary anymore (Harnish, 2014).

▶ The **iron law of oligarchy** is the tendency of a bureaucracy to become increasingly dominated by a small group of people (Michels, 1911/1949). A handful of people can control and rule a bureaucracy

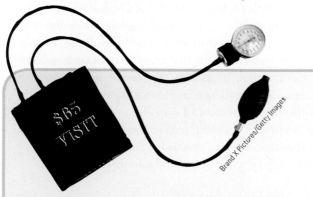

Health Care on Aisle 7!

To avoid medical and health insurance bureaucracies, more people are turning to retail clinics (also called "convenient care clinics"). The first one opened in 2000; there were about 1,400 in 2012; and they're expected to increase to at least 2,800 by 2015 (Accenture, 2013). The clinics—usually staffed by nurse practitioners—offer low-cost treatment for more than 25 common conditions (e.g., strep throat and ear infections), and provide health screening tests, immunizations, and physicals. The clinics are located in pharmacies, grocery stores, and "big box" stores such as Target and Walmart. There's no need to make an appointment and minimal (if any) waiting, the patients are in and out in about 15 minutes, and most of the clinics are open evenings and weekends (Spetz et al., 2013; *Consumer Reports*, 2014).

goal displacement a preoccupation with rules and regulations rather than achieving the organization's objectives.

alienation a feeling of isolation, meaninglessness, and powerlessness.

iron law of oligarchy the tendency of a bureaucracy to become increasingly dominated by a small group of people.

because the top officials and leaders monopolize information and resources. As a result, those at the top maintain their power and privilege.

▶ The cumulative effect of these and other bureaucratic dysfunctions can result in *dehumanization* because weak reward systems, rigid rules, and other problems limit organizational creativity and freedom. As a result, work becomes more automated and impersonal (see "The McDonaldization of Society").

Bureaucracies are plagued by other problems, such as favoritism and dishonest employee evaluations (Kilmann, 2011). They function, however, largely because of the internal development of informal groups and recent changes in performance reviews.

6-2d Getting Around Bureaucratic Deficiencies

Weber's model overlooked the many ways that workers get around rigid bureaucratic regulations and impersonality. Since around 2000, self-managing work teams have become more common, and many large corporations are rethinking age-old performance reviews that alienate employees. Let's begin with a brief look at some of the early studies' missteps and contributions regarding the importance of informal group networks.

HISTORIC STUDIES OF INFORMAL GROUP NETWORKS

For many years, and well into the 1930s, experts concentrated on organizational efficiency based on the principles of *scientific management* developed by Frederick Winslow Taylor, a mechanical engineer who wanted to improve industrial productivity. Taylor (1911/1967) considered workers—especially those in factories and on assembly lines—as mere adjuncts to machines, assumed that people don't like to work, maintained that employees need close supervision, and believed that management had to enforce cooperation because most workers aren't capable of handling even the simplest tasks.

A number of studies conducted by industrial psychologists and sociologists between 1927 and 1932 shook up Taylor's views. The research teams studied employees at the Western Electric Company's Hawthorne plant in Chicago. The *Hawthorne studies*, as they're often called, found that informal groups were critical to the organization's functioning (Roethlisberger and Dickson, 1939/1942; Mayo, 1945; Landsberger, 1958).

Informal social groups can promote an organization's goals if they collaborate, are cohesive, and motivate each other (Barnard, 1938). They can also resist an organization's goals and formal rules. In one of the Hawthorne studies, Roethlisberger and Dickson (1939/1942) spent 6 months observing a group of 14 men who wired telephone switchboards in what the company called the "bank-wiring room."

The bank-wiring room work group consisted of nine wiremen, three solderers, and two inspectors. Management offered financial incentives for higher productivity, but the work group pressured people to limit output. The wiremen developed the following norms that controlled the group's behavior:

1. **You shouldn't turn out too much work.** If you do, you're a "rate-buster" and a "speed king."

2. **You shouldn't turn out too little work.** If you do, you're a "chiseler."

3. **You shouldn't tell a supervisor anything that will be detrimental to a coworker.** If you do, you're a "squealer."

4. **You shouldn't "act officious."** If you're an inspector, for example, you shouldn't act like one.

If individuals violated any of these norms, the other workers used "binging" (striking a person on the shoulder) to punish them.

Why did the men control productivity rather than take advantage of the management's promise of higher wages? They feared that if they produced at a high level, they'd be required to produce at that level regularly. They also worried that high productivity would lead to layoffs because supervisors would conclude that fewer workers could achieve the same output. In addition, the bank-wiring room men experienced high morale and job satisfaction because they felt they had some control over their work. In contradiction to Taylor's scientific management perspective, then, the Hawthorne studies concluded that there's an important relationship between formal and informal organization.

MODERN WORK TEAMS

Since the Hawthorne studies, many companies have recognized that informal groups affect workers' productivity. As a result, numerous organizations have implemented alternative management strategies.

Today, *self-managing work teams* are the dominant model in most large organizations (Cloke and Goldsmith, 2002; Neider and Schriesheim, 2005). Contrary to Taylor's notion that workers do and managers think, self-managing work teams, sometimes referred to as *post-bureaucratic organizations*, involve groups of 10 to 15 people who take on the duties of their former supervisors. Instead of being told what to do by a boss, self-managing workers gather and interpret information, act on the information, and take collective responsibility for their actions—whether it's designing a new refrigerator or handling food service for a university. Because they're not held back by unresponsive managers, effective self-managing

The McDonaldization of Society

According to sociologist George Ritzer (1996: 1), the organizational principles that underlie McDonald's, the well-known fast-food chain, are beginning to dominate "more and more sectors of American society as well as of the rest of the world." McDonaldization has four components: efficiency, calculability, predictability, and control.

1. *Efficiency* means that consumers have a quick way of getting meals. In a society where people rush from one place to another and where both parents are likely to work or single parents are pressed for time, McDonald's (and similar franchises) offers an efficient way to satisfy hunger and avoid much "fuss and mess" (see also Jargon, 2013). Like their customers, McDonald's workers function efficiently: The menu is limited, the registers are automated, and employees perform their tasks rapidly and easily.

2. *Calculability* emphasizes the quantitative aspects of the products sold (portion size and cost) and the time it takes to get the products. Customers often feel that they're getting a lot of food for what appears to be a nominal sum of money and that a trip to a fast-food restaurant will take less time than eating at home. In reality, soft drinks are sold at a 600 percent markup because most of the space in the cup is taken up by ice. It costs at least twice as much and takes twice as long to get to the restaurant, stand in line, and pick up the food than to prepare a similar meal at home.

3. *Predictability* means that products and services will be the same over time and in all locales. "There's great comfort in knowing that McDonald's offers no surprises" (p. 10). The Quarter Pounder is the same regardless where customers order. Outside of the United States, "homesick American tourists in far-off countries can take comfort in the knowledge that they'll likely run into those familiar golden arches and the restaurant they have become so accustomed to" (p. 81). Workers, like customers, behave in predictable ways: "There are, for example, six steps to window service: greet the customer, take the order, assemble the order, present the order, receive payment, thank the customer and ask for repeat business" (p. 81).

4. *Control* means that technology shapes behavior. McDonald's controls customers subtly by offering limited menus and uncomfortable seats that encourage diners to eat quickly and leave. "Consumers know that they're supposed to line up, move to the counter, order their food, pay, carry the food to an available table, eat, gather up their debris, deposit it in the trash receptacle, and return to their cars. People are moved along in this system not by a conveyor belt, but by the unwritten, but universally known, norms for eating in a fast-food restaurant" (p. 105).

McDonald's controls employees more openly and directly. Workers are trained to do a limited number of things in precisely the same way, managers and inspectors make sure that subordinates toe the line, and McDonald's has steadily replaced human beings with technologies that include soft-drink dispensers that shut off when the cup is full and a machine that rings and lifts the French fries out of the oil when they're crisp. Such technology increases the corporation's control over workers because employees don't have to use their own judgment or need many skills to prepare and serve the food.

Efficiency, calculability, predictability, and control reflect a rational system (remember Weber?) that increases a bureaucracy's efficiency. On the other hand, Ritzer contends, McDonaldization reflects the "irrationality of rationality" because the results can be harmful. The huge farms that now produce "uniform potatoes to create those predictable French fries" of the same size rely on the extensive use of chemicals that then pollute water supplies. And the enormous amount of nonbiodegradable trash that McDonald's produces wastes our money because we—and not McDonald's—pay for landfills.

McDonald's is such a powerful model that many businesses have cloned its four dimensions of efficiency, calculability, predictability, and control. Examples include "McDentists" and "McDoctors," drive-in clinics designed to deal quickly with minor dental and medical problems; "McChild" care centers like KinderCare; and "McPaper" newspapers, such as *USA Today* (Ritzer, 2008).

EllenMoran/iStockphoto.com

TABLE 6.4 SOCIOLOGICAL PERSPECTIVES ON GROUPS AND ORGANIZATIONS

Theoretical Perspective	Level of Analysis	Main Points	Key Questions
Functionalist	Macro	Organizations are made up of interrelated parts and rules and regulations that produce cooperation in meeting a common goal.	• Why are some organizations more effective than others? • How do dysfunctions prevent organizations from being rational and effective?
Conflict	Macro	Organizations promote inequality that benefits elites, not workers.	• Who controls an organization's resources and decision making? • How do those with power protect their interests and privileges?
Feminist	Macro and micro	Organizations tend not to recognize or reward talented women and regularly exclude them from decision-making processes.	• Why do many women hit a glass ceiling? • How do gender stereotypes affect women in groups and organizations?
Symbolic Interactionist	Micro	People aren't puppets but can determine what goes on in a group or organization.	• Why do people ignore or change an organization's rules? • How do members of social groups influence workplace behavior?

teams are focused on their goals and are committed to the organization and its success (Colvin, 2012).

PERFORMANCE REVIEWS

Since the early 1980s, many companies have relied on "stack" or "forced" ranking to evaluate employees. These annual performance reviews grade workers against one another and rank them on a scale of one to five. The fours and fives are designated as underperformers and fired, demoted, or not promoted. Stack ranking creates a sense of failure and helplessness. The evaluations are done infrequently and become "management hammers" rather than reviews that provide workers frequent feedback on improving their performance (Suddath, 2013).

Some companies are adopting a more popular and effective practice, "calibration." This is a process in which managers across different divisions or departments ensure that their review processes are similar to each other regarding criteria and scoring. Calibration is based on facts and measurable tasks, not feelings, greatly reducing the likelihood of a leader's biases and inconsistent scoring (Culbert and Rout, 2010; Ovide and Feintzeig, 2013).

6-3 SOCIOLOGICAL PERSPECTIVES ON SOCIAL GROUPS AND ORGANIZATIONS

How do organizations operate? And how do social groups affect massive bureaucracies? In answering such questions, functionalism, conflict theory, feminist theories,

and symbolic interaction offer different insights. Taken together, these perspectives provide a multifaceted understanding of groups and organizations (*Table 6.4* summarizes these perspectives).

6-3a Functionalism: Social Groups and Organizations Benefit Society

Much of what you've read so far comes from functionalist perspectives, which emphasize that social groups and organizations are composed of interrelated, mutually dependent parts. As Weber pointed out, when a bureaucracy operates rationally and efficiently, workers and bosses cooperate to turn out a final product, whether it's an automobile or a can of soup.

Effective leadership is a primary factor in job satisfaction. Even in our sprawling government bureaucracies, and across all organizational levels, the happiest workers are those whose bosses reward motivation and commitment, encourage integrity, manage people fairly, and promote the employees' professional development and creativity (Partnership for Public Service, 2011).

A recent study of a large company with technology-based service jobs (e.g., retail clerks, airline gate agents, call-center workers) found that replacing a supervisor from the bottom 10 percent of the pool with one from the top 10 percent increased output almost as much as adding a tenth worker to a nine-worker team. The researchers concluded that productivity increases when the most promising workers are paired with top bosses who provide frequent feedback and teach better work

Since 1975, the number of employee-owned companies in the United States has grown from 1,600 to more than 11,500, and now represents about 12 percent of the private sector workforce. These firms are profitable because the workers are highly motivated: They have a stake in the business, control the decision making, contribute original ideas and cost-saving solutions, and depend on each other. Pictured here are people learning to bake bread at the King Arthur Flour Company's Baking Education Center in Norwich, Vermont, an employee-owned company that has expanded its product line (Case, 2010; Koba, 2013).

Courtesy King Arthur Flour Company

methods and skills (Lazear et al., 2013). Thus, good bosses don't threaten or punish employees, but provide concrete training to get the job done.

Some companies have boosted productivity by encouraging workers' interaction. Examples include providing group rather than individual breaks, and carving out meeting spaces that accommodate three or four people instead of 12 or more (Silverman, 2013).

Functionalists acknowledge that organizations (and particularly bureaucracies) can be dysfunctional (Fisman and Sullivan, 2013). Workers may be alienated because of weak reward systems, favoritism, and incompetent supervisors. Employers waste at least $5,000 a year per worker because employees who aren't challenged spend more than 2 hours a day pretending to work: They shop online, do email, and even surf online sites hoping to find leads for better jobs (Rothlin and Werder, 2008).

Many organizations are trying to correct such problems. Despite many people's complaints, employers have a legal right to use surveillance technology to monitor workers' use of company computers and cell phones to increase productivity (Barnes, 2010). Since 2008, business spending on employees has grown only 2 percent compared with 26 percent on equipment and software (Rampell, 2011). Many people contend that doing so increases unemployment rates. From a functionalist perspective, however, investing more in equipment than workers is a rational bureaucratic response to the rising costs of health care benefits and economic competition from other countries (see Chapter 11).

CRITICAL EVALUATION

Functionalism is useful in understanding how groups and organizations fulfill essential functions, including motivating people to get work done and achieving a common goal. For critics, especially conflict theorists, functionalists exaggerate cooperation and tend to gloss over dysfunctions such as worker dissatisfaction and alienation.

A related issue is whether informal social networks improve worker morale and control as much as functionalists claim. After all, a tedious and monotonous job is tedious and monotonous regardless of the degree of informal coworker interaction.

6-3b Conflict Theory: Some Benefit More Than Others

Conflict theorists contend that organizations promote inequality that benefits those at the top of the hierarchy, not workers. In many companies, those at higher levels are more comfortable hiring and promoting others similar to themselves (male, white, and middle class or higher). Phrases such as "fitting the mold" and "having common interests" are often code words for belonging to an in-group.

Inequality in income, status, and other rewards means that owners and managers can easily exploit workers. Those at the top dictate to those at the middle and the bottom. Because organizations serve elites, conflict theorists argue, they routinely ignore workers' needs and interests. What about performance reviews in rewarding the best employees? According to a management professor, "I've examined scores of empirical studies since the early 1980s and have not found convincing evidence that performance reviews are fair, accurate or consistent across managers, or that they improve organization effectiveness." Instead, subjective evaluations "are intimidating tools that make employees too scared to speak their minds" (Culbert, 2011: A25).

There's also considerable incompetence, waste, and corruption in the public sector because, in part, many government agencies and programs "operate in an informational stone age" (Schuck, 2013: A15). For example:

▶ In 2009 alone, Social Security made $6.5 billion in overpayments to people not entitled to receive them.

▶ In 2010, throughout the federal government, improper payments totaled $125 billion, but no one seems to know why.

▶ In 2012, the FBI's "spotty" database of state criminal and mental-health records allowed 3,000 unauthorized people with criminal and mental health histories to purchase firearms.

▶ Between 2008 and 2012, the Agriculture Department paid $22 million to farmers who had been dead for at least two years.

▶ The Social Security and Veterans' Administrations and the food stamp program are "rife with fraud" (Garcia-Diaz, 2013).

▶ In 2012, the Defense Department paid $13.4 million for helicopter parts that should have cost $4.4 million, but no one knows why (Ohlemacher, 2011; Schuck, 2013; Capaccio and Salant, 2014).

Federal workers' job satisfaction has been dropping since 2003. Over the years, the major complaints have included poor leadership, few opportunities for advancement, and meager rewards for good performance (Partnership for Public Service, 2013; see also Clowers, 2013). In the most recent survey of 82 federal agencies, 71 percent of the civilian employees gave their immediate supervisors high marks, including trust, respect, and support. In contrast, just 38 percent said that their senior leaders showed a high level of motivation and commitment to their jobs, and less than half trusted their senior leaders to maintain "high standards of honesty and integrity" (U.S. Office of Personnel Management, 2014).

According to many federal employees, besides wasting billions of taxpayer dollars, the agencies provide generous subsidies and tax concessions to the oil, coal, and agriculture industries as well as religious groups and foreign countries; don't fire incompetent workers; and have numerous (and unnecessary) levels of executives and managers rather than people working in the field. An employee in the Environmental Protection Agency complained, for instance, "The layers of management are insane. . . . It takes 13 steps and five layers to get a signature from our office director" (Rein, 2011: B4). Because of these and similar problems,

TABLE 6.5 MANY AMERICANS HAVE LITTLE CONFIDENCE IN ORGANIZATIONS

Percentage who said, in 2013, that they have "a great deal" or "quite a lot" of confidence in ...	
The church or organized religion	57
The presidency	36
The medical system	35
The U.S. Supreme Court	34
The public schools	32
The criminal justice system	28
Banks	26
Newspapers and television news	23 (each)
Big business	22
Organized labor	20
Health maintenance organizations (HMOs)	19
Congress	10

Note: A majority of Americans expressed high esteem only for the military (68 percent), small business (65 percent), and the police (57 percent).

Source: Based on Mendes and Wilke, 2013.

many Americans are disillusioned with a number of organizations (see *Table 6.5*).

CRITICAL EVALUATION

Among its contributions, conflict theory has underscored the inequality within groups and organizations that saps workers' motivation and limits their economic success. Critics, however, fault conflict theory for assuming that greater equality always leads to a more successful and productive organization. For example, even self-managing work teams can fail because of ineffective team leaders or management interference with teams (Barker, 1993; Lencioni, 2002).

In addition, do conflict theorists focus too much on organizational deficiencies? There are, after all, many organizations that are efficient and profitable as well as corporations that don't exploit their workers (Kaplan, 2010).

6-3c Feminist Theories: Men Benefit More Than Women

Feminist scholars emphasize that across all social classes, women (and especially minority women) consistently fare worse than men, especially in leadership roles. Women have enjoyed increased success in organizations since the 1980s, but rarely at the highest levels. For example, women hold more than half of all management and professional positions, but only 4.6 percent of *Fortune* 500 and *Fortune* 1000 CEO positions (Catalyst, 2014).

Large numbers of women still hit a **glass ceiling**—workplace attitudes or organizational biases that prevent them from advancing to leadership positions. Examples of obstacles include men's negative attitudes about women in the workplace (see Chapter 11), women's placement in staff positions that aren't on the career track to the top, a lack of mentoring, biased and damaging evaluations by male supervisors, and little or no access to highly visible committees or task forces (Barreto et al., 2009; Sandberg, 2013).

In contrast, many men who enter female-dominated occupations (e.g., nursing and teaching) receive higher wages and faster promotions, a phenomenon known as a **glass escalator**. In 2013, for example, men made up only 11 percent of all registered nurses. However, the average female nurse earned $56,472, almost 12 percent less than the $64,272 that the average male nurse earned. Men on glass escalators also move up to supervisory positions more quickly than do females (Goudreau, 2012; Hegewisch and Hudiburg, 2014).

CRITICAL EVALUATION

Feminist theories have expanded conflict explanations of groups and organizations by showing that many talented women are treated like outsiders. One weakness of feminist theories, however, is that much of the emphasis is still on white and black women in professional and managerial positions, even though Latinas and Asian American women comprise a large segment of these workers and those in the general labor force.

A second limitation is that even when feminist scholars say that both sexes suffer from organizational stereotypes, they offer little data on how such stereotypes affect men. Also, female supervisors don't always decrease an organization's gender inequality because they neither support nor promote women workers (see Huffman, 2013).

Nikada/Getty Images

6-3d Symbolic Interaction: People Define and Shape Their Situations

Whereas functionalist, conflict, and (some) feminist theorists examine groups and organizations on a macro level, symbolic interactionists focus on interpersonal relationships. Interactionists emphasize that an individual's perception and definition of a situation shape group dynamics and, consequently, organizations.

Group leaders or members can create or reinforce conformity (as shown in the Asch and Milgram studies). Informal groups can also determine what goes on in an organization by refusing to obey the rules and implementing their own (as in the Hawthorne studies). Thus, according to symbolic interactionists, individuals make choices, change rules, and mold their own identities instead of being manipulated (Kivisto and Pittman, 2001).

Symbolic interactionists also note that people's outcomes depend on how coworkers and bosses interpret the *same* behavior. For example, whether women are supervisors or clerical workers, if they lose their temper, they're overwhelmingly seen as too emotional, incompetent, out of control, weak, and worth less pay. Their angry male counterparts, in contrast, are often viewed as authoritative and in control (Brescoll and Uhlmann, 2008).

CRITICAL EVALUATION

Symbolic interaction has made important contributions in describing how members of groups and organizations interpret the world around them and, as a result, affect what goes on. Social networking sites, especially Facebook, have empowered employees to discuss everyday problems despite management's control (Kirkpatrick, 2010).

Despite these contributions, because symbolic interaction is a micro-level theory, it ignores macro-level factors that exploit workers and consumers. About 51 percent of Americans seek out "all natural" products, but much of the food contains genetically modified organisms, chemically processed vitamins, and artificial ingredients. The average consumer has little knowledge of and control over such false advertising, a macro-level factor (Esterl, 2013). In addition, and in contrast to interactionists' claims, most people can't shape or change their situations. Instead, formal organizations (the U.S. government and large companies)

glass ceiling workplace attitudes or organizational biases that prevent women from advancing to leadership positions.

glass escalator men who enter female-dominated occupations and receive higher wages and faster promotions than women.

often invade the privacy of citizens, workers, or consumers by collecting information and monitoring people's behavior.

We've looked at the general characteristics of social groups and formal organizations, both of which are part of larger structures that sociologists call *social institutions*. This final section introduces you to this important sociological concept and lays the foundation for analyses of social institutions in later chapters.

6-4 SOCIAL INSTITUTIONS

A **social institution** (or simply, *institution*) is an organized and established social system that meets one or more of a society's basic needs. Some social institutions are universal because they ensure a society's survival and affect practically all of a society's members. According to functionalists, there are five major social institutions worldwide:

▸ The *family* replaces the members of a society through procreation, socializes children, and legitimizes sexual activity between adults.

▸ The *economy* organizes a society's development, production, distribution, and consumption of goods and services.

▸ *Political institutions* maintain law and order, pass legislation, and form military groups for internal and external defense.

▸ *Education* helps to socialize children, transmits knowledge, and provides information and training for jobs and other work-related activities.

▸ *Religion* encompasses beliefs and practices that offer a sense of meaning and purpose in life.

Besides these core universal institutions (which we'll examine in Chapters 7 and 11–14), others include law, science, sports, and the military.

6-4a Why Social Institutions Are Important

Social institutions are abstractions, but they have an organized purpose, weave together norms and values, and, consequently, guide behavior. No two families are exactly alike, but the family institution reflects broadly shared cultural agreements about what a family should be and do. In the same vein, no two grocery stores are exactly alike, but

social institution an organized and established social system that meets one or more of a society's basic needs.

Five major social institutions

- family
- economy
- political system
- education
- religion

the economy as an institution creates and maintains a variety of rules that usually make food shopping predictable.

6-4b How Social Institutions Are Interconnected

Social institutions are linked to one another. Understanding institutions can tell us a lot about how a society functions and how we're connected to one another. Consider, for example, the linkages among six social institutions in addressing Americans' weight. *Medical researchers* tell us that obesity is the second top cause of preventable disease, disability, and death (the top cause is smoking). The most commonly used measure of weight status is body mass index, or BMI, which uses a simple calculation based on the ratio of a person's height and weight. For adults, a BMI between 25 and 29 is considered overweight, between 30 and 39 is obese, and extreme obesity is 40 or higher. Between 1971 and 2012, the prevalence of obesity among children ages 2 to 19 increased from 5 to 17 percent. In 1971, 15 percent of U.S. adults aged 20 and over were obese, and 1 percent were extremely obese. By 2012, 36 percent were obese, and 7 percent were extremely obese (National Center for Health Statistics, 2014; Ogden et al., 2014; see also Chapter 14).

Media organizations (television networks, magazines, and online sites) routinely broadcast such results and their negative health outcomes, including

Bill Pugliano/Getty Images

Sam Walton, the founder of Walmart, kept in touch with his employees, called them "associates," and promised them "limitless opportunities" in the company. As a result, the employees, despite their low wages, worked hard because they felt that they had a stake in the "Walmart family" (Frank, 2006).

On average, an 8-ounce soft drink contains at least 3 tablespoons of sugar. Over the course of a year, drinking one soda a day can make you 10 pounds fatter.

Caspar Benson/Getty Images

heart disease, strokes, asthma, diabetes, and early death (see, for example, Sharpe, 2013). On the other hand, the media also relies on ads from food companies that create and deliver unhealthy products because, in part, the *economy* depends on consumers to survive.

Many *educational institutions*, particularly K–12, offer healthier lunches that are lower in salt and fat, higher in whole grains, and offer more fresh fruit snacks and drinks (Jackson, 2010). However, school vending machines still have high fat snacks and sugary drinks. Children may not choose healthier food because, by the time they enter school, their *families* have already shaped their eating habits and preferences (see Chapter 4).

The *political system* has had mixed effects on weight issues. Congress hasn't passed laws to rid school vending machines of sugary snacks and beverages. A major reason is that almost all politicians rely on the food industry to help finance their reelection campaigns and to hire them as lobbyists after they leave public office

(see Chapter 11). First Lady Michelle Obama was silent when food and media companies banded together to kill a federal proposal aimed at banning advertising that uses popular cartoon characters to sell sugary products. On the other hand, she has convened medical professionals, business and government leaders, grassroots activists, and nutrition advocates to combat children's obesity. Also, New York City's Mayor Michael Bloomberg tried, unsuccessfully, to ban the sale of sugary drinks larger than 16 ounces (Hennessey, 2013; Saul, 2013).

As you've seen in this chapter, social groups, formal organizations, and social institutions shape our behavior. Nonetheless, we're not robots who succumb to bureaucracies. Instead, and despite numerous constraints, we can make better decisions if we understand how groups, organizations, and social institutions affect our everyday lives.

STUDY TOOLS 6

READY TO STUDY? IN THE BOOK, YOU CAN:

- ☐ Check your understanding of what you've read with the Test Your Learning Questions provided on the chapter review card at the back of the book.
- ☐ Rip out the chapter review card for a handy summary of the chapter and key terms.

ONLINE AT CENGAGEBRAIN.COM YOU CAN:

- ☐ Prepare for tests with quizzes.
- ☐ Review the key terms with Flash Cards.
- ☐ Play games to master concepts.

© Anelina/Shutterstock.com

7 | Deviance, Crime, and Social Control

© Anelina/Shutterstock.com

LEARNING OBJECTIVES

7-1 Differentiate between deviance and crime, and describe the key characteristics of deviance.

7-2 Evaluate the two major crime measures, and identify and illustrate the different types of crime.

7-3 Describe, illustrate, and evaluate functionalist perspectives on deviance.

7-4 Describe, illustrate, and evaluate conflict perspectives on deviance.

7-5 Describe, illustrate, and evaluate feminist perspectives on deviance.

7-6 Describe, illustrate, and evaluate symbolic interaction perspectives on crime.

7-7 Identify social control methods, including those of the criminal justice system.

After you finish this chapter go to **PAGE 137** for **STUDY TOOLS.**

Blair Seitz/Alamy

A rising number of brides in the United States and other countries are destroying their wedding dresses after the ceremony. In this trend, known as "Trash the Dress," the wedding photos include brides who climb trees, water ski, douse themselves with paint, roll around in the mud, and even set the dress on fire while they're still wearing it. Brides trash the dress because they want to do something different. Many find this practice reprehensible because there are alternatives such as selling the dress or donating it to someone who can't afford a wedding gown (Kirkova, 2013; Ruiz, 2013).

Is dress-trashing deviant? Why is it more acceptable to store a wedding gown that no one will ever wear? Or to donate rather than sell it? At one time or another, almost all of us are deviant, but what exactly is deviance? Before reading further, take the quiz on the next page to see how much you know about U.S. deviance and crime.

7-1 WHAT IS DEVIANCE?

Have you ever driven above the speed limit? Cheated on an exam? Engaged in underage drinking? All are examples of **deviance**—violations of social norms. Unlike the

What do you think?

There would be less crime if the punishments were more severe.

1	2	3	4	5	6	7

strongly agree   strongly disagree

TRUEORFALSE?

How Much Do You Know About U.S. Deviance and Crime?

1. All deviant behavior is criminal.

2. Deviance serves a useful purpose in society.

3. Most crime victims are women.

4. Crime rates have decreased since 1990.

5. People are more likely to be arrested for a drug violation than for driving while drunk.

6. Murder and assault are more common than illegal gambling and prostitution.

7. Suburbs are safer than cities.

8. Death penalties deter crime.

The answers for #2, #4, and #5 are true; the others are false. You'll see why as you read this chapter.

Crime, the violation of a society's formal laws, is an important category of deviance. Deviance becomes crime when it violates rules that have been written into law and enforced by a political authority. For example, trafficking children under age 18 for prostitution, pornography, or sex is deviant. In 2013, 47 states enacted 186 laws that punish sex traffickers with 25 or more years of imprisonment and as much as $50,000 in fines (Khadaroo, 2013).

7-1a Some Key Characteristics of Deviance

Deviance is universal because it exists in every society. Still, deviance can vary quite a bit over time, from situation to situation, from group to group, and from culture to culture (Sumner, 1906; Schur, 1968):

deviance a violation of social norms.

crime a violation of society's formal laws.

- **Deviance can be a trait, a belief, or a behavior.** People usually do something to be considered deviant. We can also be treated as outsiders simply because of our appearance, skin color, religious beliefs, or sexual orientation (Becker, 1963; see also Chapters 9, 10, and 13).

- **Deviance is accompanied by social stigmas.** A **stigma** is a negative label that devalues a person and changes her or his self-concept and social identity. Stigmatized individuals may react in many ways: They may alter their appearance (as through cosmetic surgery), associate with others like themselves who accept them (as in gangs), hide information about some aspect of their deviance (an ex-convict who doesn't reveal that status), or divert attention from a stigma by excelling in some area (as music or sports) (Goffman, 1963).

- **Deviance varies across and within societies.** What's appropriate or tolerated in one society may be deviant in another. In 2011, fewer than 2 percent of all births in South Korea were to unmarried women compared with 41 percent in the United States (Haub, 2013). As a 33-year-old single mother in South Korea explained, "Once you become an unwed mom… people treat you as if you had committed a crime. You fall to the bottom rung of society" (Sang-Hun, 2009: 6). In the United States, in contrast, there's little stigma in having a nonmarital child (see Chapters 9 and 12). Only 47 percent of French people believe that extramarital affairs are immoral compared with 84 percent of Americans and 94 percent of Palestinians and Turks ("Extramarital Affairs," 2014).

- **Deviance varies across situations.** What is normal in one context may be stigmatized in another. Almost 40 percent of Americans ages 30 to 39 have at least one tattoo, but many employers—especially those in businesses and federal agencies—don't hire people with visible "body art" (Maltby, 2013; "Tattoos in the Workplace…," 2014).

- **Deviance is formal or informal.** *Formal deviance* is behavior that violates laws. A major example is crime, a topic we'll examine shortly. In contrast, *informal deviance* is behavior that disregards accepted social norms, such as picking one's nose or teeth or scratching one's private parts in public, belching loudly, and not dressing appropriately (e.g., wearing jeans to a wedding reception or job interview).

- **Perceptions of deviance can change over time.** Many behaviors that were acceptable in the past are now seen as deviant. Only during the 1980s and 1990s did U.S. laws define date rape, marital rape, stalking, and child abuse as crimes. Smoking—widely accepted in the past—has been prohibited in many public places. In 2011, for the first time, a majority of Americans (59 percent) supported a smoking ban in *all* public places, a considerable increase from 31 percent in 2003 (Dugan, 2013). Also, increasing numbers of employers don't hire people who fail urine tests for nicotine usage.

On the other hand, most Americans now shrug off behaviors that were stigmatized in the past. Cohabitation, seen as sinful and immoral 40 years ago, is now widespread and considered normal. And Americans' support for legalizing marijuana for recreational use increased from only 12 percent in 1969 to 58 percent in 2013 (Swift, 2013).

More than 800 women in Ghana live in exile in witch camps. Most are poor widows or older women who are blamed for outbreaks of disease in their villages and the illness or death of relatives or neighbors (Küntzle and Blondé, 2013). Murders of women and children accused of sorcery have risen since 2009. Saudi Arabia, Tanzania, Nepal, India, Papua New Guinea, Tanzania, and Uganda still kill people suspected of witchcraft (Hoffman, 2014).

stigma a negative label that devalues a person and changes her or his self-concept and social identity.

7-1b Who Decides What's Deviant?

Because deviance is culturally relative and standards change over time, who decides what's right or wrong?

Are any of these people deviant? Why or why not?

BananaStock/Jupiter Images; MATTHIAS BEIN/Corbis Wire/Corbis; © Dmitry Naumov/Shutterstock.com

An important group is those who have authority or power. During our early years, parents and teachers specify acceptable and unacceptable behavior. As we get older, laws also define what's deviant (e.g., being forbidden to drive until age 16 or to purchase or consume alcohol until age 21). Members of Congress and state legislators pass numerous laws that people must obey, but often exempt themselves. Examples include having immunity from employment discrimination and sexual harassment lawsuits, and passing laws that allow guns in schools, bars, restaurants, public buildings, and churches, but not state legislative chambers (Skoning, 2013; Deprez and Selway, 2014).

Public attitudes and behavior also affect definitions of deviance. Being fat, accepted in the past, is now largely stigmatized; many people and researchers even use the word "overweight" instead of "fat." On the other hand, there has been considerable *normalization of deviance,* the gradual process through which unacceptable practices or standards become acceptable. Examples include the growing acceptance of gay and lesbian relationships, same-sex marriage, and legalized gambling in many states. So far, only Alaska, Colorado, Oregon, and Washington have legalized recreational marijuana, but other states are expected to do the same. Normalizing some types of deviance generates considerable income, particularly at the state level.

7-2 TYPES OF DEVIANCE AND CRIME

Sociologists are interested in two types of deviance. You'll see shortly that *noncriminal deviance*—such as suicide, alcoholism, lying, mental illness, and adult pornography—is common. Sociologists are especially interested in *criminal deviance,* behavior that violates laws, because most of the transgressions are visible, publicized by the media, and threaten people's lives or property. Measuring criminal deviance may seem straightforward, but isn't as simple as it appears.

7-2a Measuring Crime

Two of the most important sources of crime statistics are the FBI's *Uniform Crime Report* (UCR) and the U.S. Department of Justice's *National Crime Victimization Survey* (NCVS). Each has its strengths and weaknesses.

The UCR data are crimes reported to the police and arrests made each year. The statistics are useful in examining trends over time, but the UCR doesn't include offenses such as corporate crime, kidnapping, and Internet crimes; simple assaults (those not involving weapons or serious injury); nor the 56 percent of all violent crimes (e.g., rape, robbery, and aggravated assault) and 66 percent of property crimes (e.g., burglary and motor vehicle theft) that aren't reported to the police (Truman et al., 2013). Because of such limitations, many sociologists also use victimization surveys to measure the extent of crime.

A **victimization survey** interviews people about being crime victims. Because the response rates are at least 90 percent, the NCVS offers a more accurate picture of many offenses than does the UCR. The NCVS also includes both reported and unreported crime, and isn't affected by police discretion in deciding whether to arrest an offender. Still, many victims don't report a crime to either the police or interviewers for a variety of reasons, including fear of reprisal or of getting the offender in trouble, believing that the police can't or won't do anything to help, or the crime, particularly rape, is too painful to discuss. Some people don't want to admit having been victims—especially when the perpetrator is a family member or friend. Others may be too embarrassed to tell the interviewer that they were victimized while drunk or

victimization survey interviews people about being crime victims.

One of the biggest costs for retail stores is *wardrobing*, returning expensive clothes after wearing them. Such "borrowing" has also become prevalent in fine jewelry, high-priced cameras for weddings, men's clothes (such as business suits for special occasions), and even household tools. Wardrobing has increased 40 percent since 2009, and cost retailers almost $9 billion in 2013 (National Retail Federation, 2013; Timberlake, 2013). Who do you think pays for wardrobing fraud?

© ESLINE/Shutterstock.com

© Becky Stares/Shutterstock.com

engaged in drug sales (Hagan, 2008; Truman et al., 2013).

No one knows the extent of U.S. crime because even the best sources (the UCR and NCVS) provide only estimates, and data for higher-class lawbreakers are more elusive than for lower-class offenders. Let's begin with the latter, often called *street crimes*, offenses committed by ordinary people, usually in public places.

7-2b Street Crimes

A majority of Americans (55 percent) say crime is an "extremely" or "very serious" problem in the United States (Dugan, 2013). They're usually referring to *violent crimes* (e.g., murder, rape, robbery, aggravated assault) that use force or the threat of force against others. *Property crimes* (e.g., burglary, motor-vehicle theft, arson) involve the destruction or theft of property, but offenders don't use force or the threat of force.

INCIDENCE OF VIOLENT AND PROPERTY CRIMES

Of the almost 10.2 million crimes reported to the police in 2012, 88 percent were property crimes. There has been a much higher rate of property to violent crimes since at least 1993 (see *Figure 7.1*). The media are most likely to cover violent crimes because they inspire the greatest fear, but Americans are much more likely to be victimized by theft or burglary than to be murdered, raped, robbed, or assaulted with a deadly weapon. In fact, the average person is 433 times more likely to experience a theft than to be murdered (Glassner, 2010; Federal Bureau of Investigation, 2013).

Between 2003 and 2012, U.S. violent crimes decreased by 19 percent and property crimes by 14 percent (Federal Bureau of Investigation, 2013). The odds of being murdered or robbed are now less than half of what they were in the early 1990s, when violent crimes peaked (Oppel, 2011).

OFFENDERS AND VICTIMS

Most offenders are never caught, but street crimes aren't random. Of those arrested in 2012, for example:

▶ 25 percent were ages 15 to 21, and 38 percent were ages 22 to 34. In contrast, only 5 percent were 55 or older, and many of the crimes involved drug abuse violations and drunk driving rather than violent and property crimes.

▶ 74 percent were males. They accounted for 80 percent of people arrested for violent crime and 63 percent of those arrested for property crime.

▶ 69 percent were white, 28 percent were black, and the remaining 3 percent were American Indian or Asian American. (The FBI doesn't provide data on Latinos because it doesn't collect information on ethnicity.) Proportionately, African Americans make up about 13 percent of the general population but account for 28 percent of street crimes (Federal Bureau of Investigation, 2013).

Men are more likely than women to be crime victims, but the type of victimization differs. Males experience higher rates of homicide, robbery, and aggravated assault, whereas females are more likely to be victims of rape, sexual assault, and intimate partner violence. The victimization rates for violent crimes are higher among American Indians (49 per 1,000 persons age 12 and older) and blacks (34) than Latinos and whites (25 each) and Asian Americans (16). The rates are also much higher for people ages 12 to 17 and 18 to 24 (48 and 41, respectively) than people age

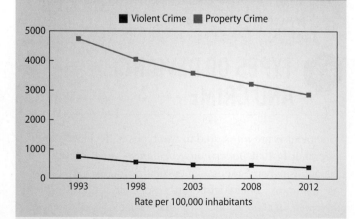

FIGURE 7.1 U.S. CRIME RATES HAVE DECREASED

■ Violent Crime ■ Property Crime

Rate per 100,000 inhabitants

Source: Based on Federal Bureau of Investigation, 2013, Table 1.

65 and older (6) (Truman et al., 2013). However, crimes against older people are especially traumatic because the victims are usually physically unable to fight back, and it takes them longer to recover from a crime both physically and emotionally (Klaus, 2005; Madigan, 2009).

Crime rates are higher in low-income areas, but does this mean that poor people are more deviant? Because police devote more resources to poor neighborhoods, those at the lower end of the socioeconomic ladder are more likely to be caught, arrested, prosecuted, and incarcerated. *Intellectual property theft*—which includes offenses such as software piracy, bootlegging musical recordings and movies, selling company trade secrets, and copyright violations—is a primarily middle-class crime, but the offenders are difficult to catch (Motivans, 2004).

7-2c Hate Crimes

A **hate crime** is a criminal act motivated by the offender's bias regarding race, religion, ethnicity, sexual orientation, gender, or disability. The number of hate crimes decreased from nearly 7,500 in 2002 to 5,790 in 2012, but about 65 percent aren't reported to the police. In 2012, 61 percent of hate crimes were against people and 39 percent against property. Among the former, 80 percent were directed at individuals; the remainder targeted businesses or financial institutions, the government, and religious organizations. Almost half of the incidents are racially motivated (see *Table 7.1*). A majority (55 percent)

TABLE 7.1 WHY PEOPLE COMMIT HATE CRIMES, 2012

Source of Bias	Percentage of All Cases	Largest Target Group
Race	48	Blacks (66%)
Religion	20	Jews (60%)
Sexual Orientation	19	Gay Men (55%)
Ethnicity/National Origin	12	Latinos (59%)
Disability	2	Mental (80%)
Total	101*	

*Percentage equals 101 due to rounding.

Source: Based on Federal Bureau of Investigation, 2013, Table 1 of hate crime statistics.

of the offenders were white, 23 percent were black, and 12 percent were of unknown races; 66 percent of the victims were black (Federal Bureau of Investigation, 2003, 2013; Sandholtz et al., 2013).

7-2d White-Collar Crime

White-collar crime refers to illegal activities committed by high-status people in the course of their occupations (Sutherland, 1949; Cressey, 1953). The activities don't involve violence because white-collar criminals use their status and powerful positions to enrich themselves or others. Typical white-collar crimes include embezzlement (stealing an employer's or client's money); identity theft; bribery; insider trading (illegal stock manipulation); and mass marketing, mortgage, and investment fraud (Federal Bureau of Investigation, 2012).

White-collar crimes cause significant public harm. The financial cost of white-collar crime is unknown, but estimated between $300 and $660 billion a year (cited in Huff et al., 2010). In 2012, Americans who experienced identity theft lost about $25 billion—almost twice as much as the $14 billion for all property crimes (Harrell and Langton, 2013).

White-collar crimes are lucrative because most offenders aren't caught. According to one estimate, there's only one FBI investigator per almost 4,000 financial services sales agents, personal financial advisers, and financial analysts. Thus, the odds of being arrested are very small (Rohrlich, 2010; see fbi.gov/wanted/wcc for annual updates of the FBI's most wanted white-collar criminals).

Some of the most affluent white-collar offenders may be caught, but until then enjoy lavish lifestyles, and for many years, that include yachts, private jets, and $2 million birthday parties (Lublin, 2013; Spitznagel, 2013). Even when white-collar criminals are arrested and convicted, few are incarcerated. According to a recent federal study, of the almost 3,800 suspected embezzlers, only 525 (14 percent) have been arrested and sentenced; of those arrested and sentenced, only half (7 percent) served any prison time (Motivans, 2013).

hate crime a criminal act motivated by the offender's bias regarding race, religion, ethnicity, sexual orientation, gender, or disability.

white-collar crime illegal activities committed by high-status people in the course of their occupations.

7-2e Corporate Crimes

Corporate crimes (also known as *organizational crimes*) are illegal acts committed by executives to benefit themselves and their companies. Corporate crimes include a vast array of illicit activities such as conspiracies to stifle free market competition, price-fixing, tax evasion, and false advertising. The target of the crime can be the general public, the environment, or even a company's own workers (see Chapters 11 and 15). Often, corporate offenders commit multiple crimes (fraud, insider trading, and perjury).

Corporate crimes are fairly common. In 2002, for example, Toyota Motor Corporation received but ignored numerous complaints about sudden acceleration that resulted in deaths. In 2010, Toyota finally admitted that the sticky gas pedals could cause drivers to lose control of their vehicles but delayed notifying the car owners and the U.S. Transportation Department. General Motors knew about a malfunctioning ignition switch as far back as 2001. The company started recalling cars only in 2014, however, after the ignition switch was linked to at least 16 deaths, a $35 million federal fine, and dozens of wrongful death and injury lawsuits (Vartabedian and Bensinger, 2010; Hirsch, 2014).

Corporate crimes are profitable. Fraudulent billings to health care programs are estimated between 3 and 10 percent of the nation's total health care expenditures of $2.4 trillion a year (Federal Bureau of Investigation, 2012; see also "Health-care Fraud…," 2014). The collapse of Enron, a major U.S. energy company, put more than 5,000 people out of work, wiped out more than $2 billion in employee pensions, and investors lost $60 billion in stock. Jeffrey Skilling, Enron's CEO, was convicted on 19 counts of conspiracy, securities fraud, insider trading, and other illegal business deals. His prison sentence was reduced from 24 to 14 years, and could be cut to 12 years for good behavior (McCoy, 2013; Fowler, 2013).

7-2f Cybercrime

Cybercrime (also called *computer crime*) refers to illegal activities that are conducted online. These high-tech crimes include defrauding consumers with bogus financial

In 2009, Bernard Madoff—a highly respected stockbroker and investment advisor—pleaded guilty to defrauding thousands of investors of almost $18 billion between 1991 and 2008. Madoff was sentenced to 150 years in a federal prison. So far, it's cost almost $1 billion to recover 59 percent of the lost money (Bloomberg News, 2014).

investments, embezzling, being paid to recommend stock on chat rooms, and stealing business data. Other offenses include sabotaging computer systems, hacking (gaining unauthorized access to computers), and stealing confidential information. Between 2005 and 2012, a small group of Russians resold 160 million credit card numbers that they had stolen from more than a dozen U.S. companies. In 2013, a small group of Eastern European hackers stole 110 million credit card records from Target, the nation's third-largest retail chain, but cybercriminals also include Americans (Popper and Sengupta, 2013; Harris et al., 2014).

In 2012 alone, the Internet Crime Complaint Center (2013) received almost 290,000 complaints that resulted in nearly $526 million in financial losses. The most common offenses were selling cars online for a full or partial payment and not delivering the vehicle; impersonating the FBI and other government agencies to get information about addresses, Social Security numbers, and other personal and financial data; and romance scams that manipulated victims who sent money or expensive gifts to their new "lovers."

Cybercrime and cyber espionage cost the United States about $100 billion a year. Preventing or dealing with cybercrime, particularly the theft of intellectual property, slows research and development (which results in less hiring), increases a business's or agency's operating costs, and results in considerable revenue loss (Lewis and Baker, 2013; Offutt, 2013; Ponemon Institute, 2013).

7-2g Organized Crime

Organized crime refers to activities of individuals and groups that supply illegal goods and services for profit.

---

corporate crimes (also known as *organizational crimes*) illegal acts committed by executives to benefit themselves and their companies.

cybercrime (also called *computer crime*) illegal activities that are conducted online.

organized crime activities of individuals and groups that supply illegal goods and services for profit.

Organized crime includes drug distribution, loan-sharking (lending money at illegal rates), illegal gambling, pornography, theft rings, hijacking cargo, and laundering money obtained from illegal activities (e.g., prostitution) by depositing the money in legitimate enterprises (e.g., banks) (Federal Bureau of Investigation, 2012).

One of the most profitable activities, *organized retail crime*, involves stealing billions of dollars in merchandise that is then sold online, a practice known as "e-fencing." Organized retail crime results in an estimated $15 to $30 billion in annual financial loss (Larence, 2011).

7-2h Victimless Crimes

Offenses that are the least likely to be reported—such as illicit drug use, prostitution, drunkenness, and illegal gambling—are called *public order crimes*, or *victimless crimes*. **Victimless crimes** are acts that violate laws but those involved don't consider themselves victims. For example, prostitutes argue that they're simply providing services to people who want them, and substance abusers claim that they aren't hurting anyone but themselves.

Some contend that this term is misleading because victimless crimes often lead to property and violent

crimes, as when addicts commit burglary, rob people at gunpoint, and engage in identity theft to get money for drugs. In a national survey, 70 percent of the respondents said that a family member's drug abuse had a negative effect on the emotional or mental health of at least one other family member, and 39 percent experienced financial problems because they went into debt to pay for an uninsured addict's treatment (Saad, 2006). In 2014, 36 percent of Americans—up from 15 percent in 1947—said that alcohol was a major cause of family problems. Drug use is also often a reason for divorce and parenting problems (White and Witkus, 2013; Jones, 2014).

The next sections examine four important sociological perspectives that help us understand why people are deviant. *Table 7.2* summarizes these theories.

7-3 FUNCTIONALIST PERSPECTIVES ON DEVIANCE

For functionalists, deviance is a normal part of society. Functionalists don't endorse undesirable behavior, but view deviance as both functional and dysfunctional.

7-3a Dysfunctions and Functions

Dysfunctions are the detrimental consequences of behavior, but what's dysfunctional for one group or individual may be functional for another. For example, gangs are functional because they provide members with a sense of belonging, identity, and protection. Gangs are also dysfunctional because they commit violent and property crimes (Egley and Howell, 2013).

TABLE 7.2 SOCIOLOGICAL EXPLANATIONS OF DEVIANCE

Theoretical Perspective	Level of Analysis	Key Points
Functionalist	Macro	• Deviance is both functional and dysfunctional. • Anomie increases the likelihood of deviance. • People are deviant when they experience blocked opportunities to achieve the culturally approved goal of economic success.
Conflict	Macro	• There's a strong association between capitalism, social inequality, power, and deviance. • The most powerful groups define what's deviant. • Laws rarely punish the illegal activities of the powerful.
Feminist	Macro and micro	• There's a large gender gap in deviant behavior. • Women's deviance reflects their general oppression due to social, economic, and political inequality. • Many women are offenders or victims because of patriarchal beliefs and practices and living in a rape culture.
Symbolic Interactionist	Micro	• Deviance is socially constructed. • People learn deviant behavior from significant others such as parents and friends. • If people are labeled or stigmatized as deviant, they're likely to develop negative self-concepts and engage in criminal behavior.

> **victimless crimes** acts that violate laws but those involved don't consider themselves victims.

Deviance is *dysfunctional* because it:

▶ **Creates tension and insecurity.** Any violation of norms—a babysitter who cancels at the last minute or the theft of your cell phone or tablet—makes life unpredictable and increases anxiety.

▶ **Erodes trust in personal and formal relationships.** Crimes such as date rape and stalking make many women suspicious of men, and victims of identity theft have had problems obtaining banking services or credit cards because financial organizations don't trust them (see Harrell and Langton, 2013).

▶ **Decreases confidence in institutions.** In 2001, there were numerous scandals involving Enron and other corporations. In the 2008 stock market crash, taxpayers had to pay for the financial industry's corporate fraud and mismanagement. Since then, millions of people, even those who didn't lose money, worry that their retirement funds may disappear in the future (see Chapter 12).

▶ **Is costly.** Besides personal costs to victims (e.g., fear, emotional trauma, physical injury), deviance is expensive. All of us pay higher prices for consumer goods and services (e.g., auto and property insurance), as well as higher taxes for prosecuting criminals and for building and maintaining prisons. And as you saw in the previous section, white-collar, corporate, cyber, and organized crime cost the U.S. economy hundreds of billions of dollars every year.

Deviance can also be *functional* because it provides a number of societal benefits (Durkheim, 1893/1964; Erikson, 1966; Sagarin, 1975). For example, deviance

▶ **Affirms cultural norms and values.** Negative reactions, such as expelling a college student who's caught cheating or incarcerating a robber, reinforce a society's norms and values about being honest and law-abiding.

▶ **Provides temporary safety valves.** Some deviance is accepted under certain conditions, enabling people to "blow off steam." Typically, a community and police will tolerate noise, underage drinking, loud music, and obnoxious behavior during college students' spring break because it's a short-lived nuisance.

▶ **Creates social unity.** Some deviant behavior unites a group, community, or society, especially when there's a common enemy. After the U.S. Navy Seals killed Osama bin Laden in 2011, most Americans—regardless of age, political views, social class, or race and ethnicity—applauded his death.

▶ **Bolsters the economy.** Deviance can benefit a community financially. Some towns welcome new prisons because they generate jobs and stimulate the economy (Semuels, 2010). Unlike a manufacturing plant that might close down or move to another country, prisons are rarely dismantled.

▶ **Triggers social change.** An estimated 450,000 people were killed or injured in 2009 because of "distracted driving" collisions. Consequently, by early 2014, 41 states and the District of Columbia banned text messaging for all drivers (Hanes, 2010; Insurance Institute for Highway Safety, 2010; Governors Highway Safety Association, 2014).

7-3b Anomie and Social Strain Theory

Functionalists offer many theories of deviance, but two of the most influential are anomie and strain theories. Both analyze why so many people engage in deviant behavior even though they share many of the same goals and values as those who conform to social norms.

DURKHEIM'S CONCEPT OF ANOMIE

Émile Durkheim (1893/1964, 1897/1951) introduced the term **anomie** to describe the condition in which people are unsure of how to behave because of absent, conflicting, or confusing social norms. During periods of rapid social change, such as industrialization in Durkheim's time, societal rules may break down. As many young people moved to the city to look for jobs in the nineteenth century, norms about proper behavior that existed in the countryside crumbled. Even today, as you'll see in Chapter 15, many urban newcomers experience anomie and miss the neighborliness that was common at home.

MERTON'S CONCEPT OF SOCIAL STRAIN

Robert Merton (1938) elaborated on the concept of anomie to explain how social structure helps to create deviance. According to Merton, Americans are socialized to believe that anyone can realize the American dream of accumulating wealth and being successful economically. To achieve the *cultural goal* of economic success, society emphasizes legitimate and *institutionalized means* such as education, hard work, saving, starting at the bottom and working one's way up, and making sacrifices instead of seeking quick gratification.

anomie the condition in which people are unsure of how to behave because of absent, conflicting, or confusing social norms.

John Lund/Blend Images/Getty Images

In many countries, including the United States, however, people don't always have access to institutionalized means for financial success because of poverty, low wages, or long-term unemployment. How do they respond? Merton's **strain theory** posits that people may engage in deviant behavior when they experience a conflict between goals and the means available to obtain the goals. Not all people turn to deviance to resolve social strain (see *Table 7.3*). Most of us *conform* by working harder and longer to become successful. The fact that you're reading this text shows that, using Merton's language, your mode of adaptation to strain is conformity—one of achieving success through an institutionalized means such as a college education.

Merton's other four modes of adaptation signify deviance. *Innovation* occurs when people endorse the cultural goal of economic success but turn to illegitimate means, especially crime, to achieve the goal. For many people living in inner-city ghettos, education, hard work, and deferred gratification are often unachievable. Thus, deviance is functional for those who turn to crime to make money.

In *ritualism*, people don't expect to be rich but get the necessary education and experience to obtain or keep a job. In many bureaucratic positions, employees don't make decisions and can't move up the ladder, but their jobs are relatively secure. Ritualists aren't criminals, but they're deviant because they've given up on becoming financially successful. Instead, they do what they're told and "go along to get along."

In *retreatism*, people reject both the goals and the means for success. Merton used examples such as vagrants, alcoholics, and drug addicts, some of whom give up because they believe it's impossible to succeed.

In *rebellion*, people feel so alienated that they want to change the social structure entirely by substituting new goals and means for the current ones. Contemporary examples are terrorists and U.S. militia groups that oppose the federal government.

7-3c Critical Evaluation

A major contribution of functionalism is showing how social structure, not just individual attitudes and behavior, produces or reinforces deviance and crime. Functionalism also helps us understand current and emerging forms of deviance. For instance, computer-related crimes have increased for a number of structural reasons such as Americans' greater use of the Internet and offenders' overcoming security technology.

Functionalist theories also have weaknesses: (1) the concepts of *anomie* and *strain theory* are limited because they overlook the fact that not everyone in the United States embraces financial success as a major goal in life; (2) the theories don't explain why women's crime rates are much lower than men's, especially since women generally have fewer legitimate opportunities for financial

strain theory posits that people may engage in deviant behavior when they experience a conflict between goals and the means available to obtain the goals.

TABLE 7.3   MERTON'S STRAIN THEORY OF DEVIANCE

Mode of Adaptation	Accept Cultural Goals?	Accept Institutionalized Means to Achieve Goals?	Examples
Conformity	Yes	Yes	College graduate
Innovation	Yes	No	Thief
Ritualism	No	Yes	Low-level bureaucrat
Retreatism	No	No	Alcoholic
Rebellion	No—seeks to replace goals	No—seeks to replace means for achieving goals	Antigovernment militia member

Source: Based on Merton, 1938. Reference Crediting: Merton, Robert K. 1938. "Social Structure and Anomie." American Sociological Review 3 (December): 672–682.

success than men; (3) crime rates have declined since 2000 despite chronic poverty and unemployment; and (4) functionalism doesn't explain why people commit some crimes (setting fires just for kicks or murdering an intimate partner) that have nothing to do with being successful (Anderson and Dyson, 2002; Williams and McShane, 2004).

The most consistent criticism is that functionalism typically focuses on lower-class deviance and crime. Conflict theorists have filled this gap by examining middle- and upper-class crime.

7-4 CONFLICT PERSPECTIVES ON DEVIANCE

Functionalists ask, "Why do some people commit crimes and others don't?" Most conflict theorists focus on capitalism, social inequality, power differences, who makes the laws, and ask, "Why are some acts defined as criminal while others aren't?" (Akers, 1997).

7-4a Capitalism, Social Inequality, and Deviance

A number of sociologists maintain that there's a strong association between capitalism, social inequality, and deviance. That is, people and groups with economic and political power view deviance as any behavior that threatens their own interests (Vold, 1958; Turk, 1969, 1976; Quinney, 1980; Kraska, 2004). As a result, newspapers and TV news stations (which are owned by the affluent) routinely focus on "street crimes" rather than "suite crimes."

The United States, like many other countries, is a capitalist economy based on the private ownership of property (see Chapter 11). Because capitalism is based on personal gain rather than collective well-being, it emphasizes the maximization of profits, an accumulation of wealth, and economic self-interest.

From a conflict perspective, capitalism is a major cause of crime for several interrelated reasons (Chambliss, 1969; Quinney, 1980; Snider, 1993). First, greed and self-interest perpetuate deviance. For example, a study of 36 countries found that 42 percent of board directors and senior managers—pressured to deliver growth and fearing personal pay cuts—"cooked the books" by understating

costs, overstating revenue, encouraging customers to buy unnecessary stock, and bribing powerful people to win or retain business (Ernst & Young, 2013).

Second, capitalism is a competitive system that encourages corporate crime. When only a few firms dominate an industry, they can easily eliminate competitors, conspire to fix prices, and falsify financial statements.

Third, because capitalism creates considerable social inequality, those at the top can exploit the unemployed and low-income workers. Among the most vulnerable are undocumented and migrant laborers: A number of the employers confiscate the workers' identification documents, forbid them to contact family members, and subject them to physical and sexual violence (Zhang, 2012; see also Chapter 10). People who are the most exploited, alienated, and powerless may turn to street crimes for survival.

7-4b Laws, Power, Social Control, and Deviance

For conflict theorists, criminal laws serve the interests of the capitalist ruling class. The most powerful groups in

CSI: Crime Scene Investigation premiered in 2000. Some of the main characters have changed over the years, but CSI is the most watched drama series in the world, airing in 70 countries from Argentina to Vietnam (Gorman, 2011). Why do you think this show is so popular?

society control the law, which defines what's deviant and who will be punished (Chambliss and Seidman, 1982; Reiman and Leighton, 2010).

Why do so many high-status people commit crimes? Because they can, according to conflict theorists. First, most white-collar and corporate crimes are *not criminalized* (Turk, 1969; Lilly et al., 1995). Generally, the wealthy and powerful make and enforce laws that protect their property. The laws against higher-status criminals are relatively lenient and seldom enforced, whereas laws against lower-status offenders, particularly those who commit property crimes, are harsher and enforced more often (Thio et al., 2012).

Second, the powerful have more *opportunities for deviance*. As you saw earlier, white-collar and corporate criminals have access to resources and the necessary occupational skills to defraud customers and employees. Compared with street criminals who rob banks, for example, bankers can steal huge amounts of money without leaving their desks.

Third, it's usually easy for powerful people to *hide their deviance*. Even though corporate fraud is a "widespread problem," it's very difficult to uncover and prosecute offenders in multi-layered corporate structures: The evidence may have been destroyed, and suspects often deny knowledge of the criminal activities or maintain that the "mistakes" were unintentional (Federal Bureau of Investigation, 2012).

Fourth, there are usually *few penalties* for high-status crimes because lawmakers and defendants share a common cultural background. In 2010, the U.S. Justice Department vowed to prosecute Wall Street crime. However, none of the senior executives who played leading roles in the 2008 financial crisis have stood trial. And, since 2009, the FBI has closed 747 mortgage-fraud cases

with little or no investigation. Fraud by the rich wipes out 40 percent of the world's wealth, but no one goes to jail (Smith, 2014; Taibbi, 2014).

When U.S. prosecutors bring civil charges, not a single individual has to pay even a dime of her or his own money in damages, and the fines often represent less than half of 1 percent of annual revenue. In 2009, top officials at Goldman Sachs agreed to pay a $550 million fine to settle accusations of securities fraud that made almost $52 billion in profits. The fine represented four days of the bank's revenues (Summers, 2010; see also Prins, 2011). If, using a comparable percentage, you made $100,000 a year illegally and had to pay a $1,000 fine, would there be a financial incentive for honesty?

7-4c Critical Evaluation

Conflict theories explain the linkages between capitalism, laws, power, and social class that benefit those at the top. Critics, however, point to several weaknesses. First, some contend that conflict theory exaggerates the importance of capitalism in explaining white-collar and corporate crime. Other capitalist societies—Japan, Finland, Norway, Sweden, and Switzerland—appear to have much lower business crime rates because of greater social solidarity and enforcement of laws that ban corruption and fraud (Leonardsen, 2004; Ernst & Young, 2013). In effect, some critics say, capitalism isn't a major reason for crime.

A second criticism is that conflict theory deemphasizes the crimes committed by low-income people. Street crimes are costly to society because of legal expenses, victim injuries, wage losses, incarceration, and crime prevention activities (Mokhiber, 2007). A related criticism is that conflict theories tend to ignore the fact that some affluent

In 2009, 36-year-old Ryan LeVin (pictured here), a super-rich heir to a family costume jewelry fortune, killed two men in Florida when he lost control of his Porsche while hurtling down a street at more than 100 miles an hour, then fled the scene. In 2011, he pleaded guilty to two counts of vehicular homicide and faced 30 years in prison. Because LeVin's attorney convinced the widows to accept a financial settlement (the amount is unknown), LeVin was sentenced to only 2 years of house arrest, which he served at one of his parents' two luxury seaside condos. LeVin never apologized to the victims' families. His major concern was getting back his Porsche (Shepherd, 2011; Mayo, 2012). In another recent case, a great grandson of the wealthy du Pont family was convicted of raping his three-year-old daughter. The felony requires a 10-year mandatory sentence, but a Superior Court judge gave him probation because the offender "will not fare well" in prison (Alman, 2014).

Sun Sentinel/Getty Images

people, including corporate executives, don't get away with their crimes (see Federal Bureau of Investigation, 2012).

Third, some critics maintain that conflict theory ignores the ways that crime is functional for society. As you saw earlier, deviance provides jobs and affirms law-abiding cultural norms and values. Finally, many contend, the most influential conflict theories focus almost entirely on men as both victims and offenders (Moyer, 2001; Belknap, 2007). Feminist theories have been filling this gap.

Siede Preis/Getty Images

a fraud. When it comes to lower-level corporate crime, however—which is often committed by individual bookkeepers, bank tellers, and accountants rather than groups—45 percent of the offenders are women (Association of Certified Fraud Examiners, 2012; Steffensmeier et al., 2013).

Men are much more likely than women to commit crimes, especially violent crimes, but girls' and women's arrest rates have increased for some offenses. Between 2003 and 2012, for example, men's arrests dropped across all offenses, whereas female arrests rose 20 percent for robbery, 30 percent for larceny-theft, and 21 percent for drunken driving (Federal Bureau of Investigation, 2013).

Why? Some posit that women are more likely to be employed in jobs that present opportunities for crimes such as fraud, property theft, and embezzlement. Because

7-5 FEMINIST PERSPECTIVES ON DEVIANCE

Until recently, most sociological research on deviance focused almost entirely on males. Nearly all sociologists were men who saw women as not worthy of much analytical attention or assumed that explanations of male behavior were equally applicable to females (Flavin, 2001; Belknap, 2007). Feminist scholars' analyses of female offenders and victims have greatly enriched our understanding of deviance.

7-5a Women as Offenders

Crime statistics show a large gender gap. Women commit proportionately fewer violent, property, and other crimes (see *Table 7.4*). In 2012, the gender gap narrowed only in three crimes—forgery, fraud, and embezzlement—and women outranked men only in prostitution and juvenile runaways.

Some believe that women's lower crime participation reflects their socialization to take fewer risks and "to an ethic of care" not to hurt others (England, 2005). Others say that women have fewer opportunities to commit crimes. For example, only 9 percent of people involved in high-level corporate crimes are women: They're generally excluded from conspiracies because of their subordinate position in companies, profit less because they play minor roles, and are rarely ringleaders who initiate or orchestrate

TABLE 7.4 ARRESTS IN THE UNITED STATES, SELECTED CRIMES, BY SEX, 2012

Offense	Percent Male	Percent Female
Violent crime (murder and nonnegligent manslaughter, forcible rape, robbery, and aggravated assault)	80	20
Property crime (burglary, larceny-theft, including thefts of bicycles and motor vehicle parts, shoplifting, pocket-picking, motor vehicle theft, arson)	63	37
Forgery and counterfeiting	63	37
Fraud (confidence games and bad checks, except forgeries and counterfeiting)	60	40
Embezzlement	52	48
Stolen property (buying, receiving, possessing)	80	20
Weapons (carrying, possessing, using)	92	8
Prostitution	32	67
Sex offenses (except forcible rape and prostitution)	92	8
Drug abuse violations	80	20
Gambling	88	12
Offenses against the family and children	73	27
Driving under the influence	75	25
Disorderly conduct	72	28
Total (all arrests)	74	26

Source: Based on Federal Bureau of Investigation, 2013, Table 42.

women on average are in low-income jobs, some resort to crime—especially shoplifting, petty theft, and prostitution—to survive financially or to support a family. Women's offenses are highest in cities, where their economic oppression and poverty are greatest (Heimer et al., 2006; Flower, 2010).

More women (64 percent) are drinking now than at any time in recent history, and driving while intoxicated. In the case of female adolescents, the higher arrest rates are partly a by-product of policy changes, including more aggressive policing and the greater likelihood that parents and school officials will call the police to deal with girls' unruly behavior (Chesney-Lind and Jones, 2010; Zahn et al., 2010; Glaser, 2013).

7-5b Women as Victims

Between 2002 and 2004, Ariel Castro, a Cleveland bus driver, kidnapped three women. The victims—ages 14, 16, and 20 at the time—were kept in Castro's house, chained in rooms and sometimes in the basement, and repeatedly sexually abused. One of the women gave birth to a child fathered by Castro. The women were rescued in 2013 when one of them broke through a door and called for help.

The Castro case is a recent example of the physical and sexual victimization of women and children. Many of the serious crimes (murder and robbery) are committed by men against men, but women and girls are almost always the victims of sexual assault, rape, intimate partner violence, stalking, sexual exploitation, sexual harassment, female infanticide, and other crimes that degrade and terrorize women. Even in neighborhoods where the household income is $75,000 or higher, 36 percent of women compared with only 13 percent of men, fear walking alone at night near their homes (Saad, 2010).

Feminist scholars offer several explanations for women's victimization. One is **patriarchy**, a hierarchical system in which cultural, political, and economic structures are controlled by men. In patriarchal societies, including the United States, because women have less access to power, they're at a disadvantage in creating and implementing laws that are sympathetic to female victims (Price and Sokoloff, 2004; DeKeseredy, 2011). Such inequity diminishes women's control over their lives and increases their invisibility as victims.

Another reason for female victimization is the effect of culture on gender roles. Many girls and women have been socialized to be victims of male violence because of societal images of women as weaker, less intelligent, and less valued than men: "Girls are rewarded for passivity and feminine behavior whereas boys are rewarded for

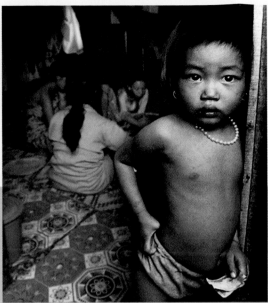

AP Images/David Longstreath

Holding money exchanged for sex, the child of a prostituted woman stands in the doorway of a brothel in Phnom Penh, Cambodia. Children, especially girls, who are raised in brothels are highly vulnerable to sexual exploitation. Many of the customers are Western men.

aggressiveness and masculine behavior" (Belknap, 2007: 243). In effect, then, both sexes internalize the belief that female victimization is normal. Because patriarchal societies rarely punish men who victimize women, or give them light sentences, many female victims feel trapped. For example, a Maryland judge sentenced a 20-year-old man to 3 years in prison for killing his girlfriend's puppy and 3 years in prison for beating the girlfriend over a 2-year period (Siegel, 2009).

A third reason for women's victimization is living in a **rape culture**, an environment in which sexual violence is prevalent, pervasive, and perpetuated by the media and popular culture. Men are victims, but 99.1 percent of those arrested for rape are males (Federal Bureau of Investigation, 2013). Here are a few examples of the pervasiveness of rape and sexual assaults in U.S. culture:

▶ 3 percent of men and 17 percent of women have been victims of an attempted or completed rape in their lifetimes (Rape Crisis Center, 2014);

patriarchy a hierarchical system in which cultural, political, and economic structures are controlled by men.

rape culture an environment in which sexual violence is prevalent, pervasive, and perpetuated by the media and popular culture.

- 25 percent of women nationwide have been raped by their spouses or ex-spouses (Tjaden and Thoennes, 2006);

- 27 percent of college women have been raped or have suffered attempted rape (Lauerman, 2013);

- 26,000 members of the military were sexually assaulted in 2012, 42 percent more than in 2010 (Department of Defense, 2013).

The actual number of violations is much higher. In 2013, for example, only 35 percent of victims reported rapes and sexual assaults to the police (Truman and Langton, 2014).

7-5c Critical Evaluation

Feminist sociologists have been at the forefront of studying female deviance and increasing awareness of crimes such as date rape, stalking, intimate partner violence, and the international sex trafficking of women and girls. Some critics contend, however, that because concepts such as patriarchy are difficult to measure, feminist research has yet to show, specifically, how patriarchy victimizes women. Others criticize feminist analyses for emphasizing male but not female violence, focusing on men's but not women's crimes, and not addressing the simultaneous effects of gender, social class, and race/ethnicity (Belknap, 2007; Friedrichs, 2009; Dao, 2013).

7-6 SYMBOLIC INTERACTION PERSPECTIVES ON DEVIANCE

For symbolic interactionists, deviance is socially constructed because it's in the eye of the beholder. They offer many theories to explain deviance, but two of the best known are differential association theory and labeling theory.

differential association theory asserts that people learn deviance through interaction, especially with significant others.

labeling theory holds that society's reaction to behavior is a major factor in defining oneself or others as deviant.

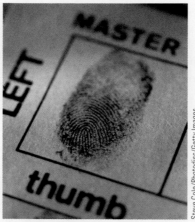

Steve Cole/Photodisc/Getty Images

7-6a Differential Association Theory

Differential association theory asserts that people learn deviance through interaction, especially with significant others such as family members and friends. Dale "Rooster" Bogle's family in Oregon is an example of this theory. Even though Rooster, the father, served time in prison for theft, he taught his sons and daughters to survive by stealing: "By the time the boys were 10 years old they were breaking into liquor stores for their dad or stealing tractor-trailer trucks, hundreds of them. The girls turned to petty crimes to support their drug addictions" (Butterfield, 2002: 1).

People become deviant, according to sociologist Edwin Sutherland, if they have more contact with significant others who violate laws than with those who are law-abiding. In effect, people learn techniques for committing criminal behavior, along with the values, motives, rationalizations, and attitudes that reinforce such behavior (Sutherland and Cressey, 1970). Thus, in Rooster's clan, all of the children wound up in prison because they grew up in an environment that taught deviance rather than conformity.

Sutherland emphasized that differential association doesn't occur overnight. Instead, people are most likely to engage in crime if they're exposed to deviant values (1) early in life, (2) frequently, (3) over a long period of time, and (4) from significant others and reference groups (e.g., parents, siblings, close friends, and business associates).

Considerable research supports differential association theory. Almost 47 percent of state prisoners have a parent or other close relative who has also been incarcerated. Even before age 13, children who associate with peers who smoke, use drugs, or commit crimes are more likely to do so themselves because they learn these behaviors from their friends (Adamson, 2010; Zahn et al., 2010; Roman et al., 2012; Huizinga et al., 2013). Thus, according to differential association theory, we're products of our socialization.

7-6b Labeling Theories

Have you ever been accused of something wrong that you didn't do? What about getting credit for something that you and others know you didn't deserve? In either case, did people start treating you differently? The reactions of others are the crux of **labeling theory**, which holds that society's reaction to behavior is a major factor in defining oneself or others as deviant. Some of the earliest sociological studies found that teenagers who misbehaved were tagged as delinquents. Such labeling changed the child's self-concept and resulted in more deviance and criminal behavior (Tannenbaum, 1938). During the 1950s and 1960s, two influential sociologists—Howard Becker and Edwin Lemert—extended labeling theory.

BECKER: DEVIANCE IS IN THE EYES OF THE BEHOLDER

According to Howard Becker (1963: 9), being a deviant or a criminal depends on how others react: "The deviant is one to whom that label has successfully been applied; *deviant behavior is behavior that people so label*" (emphasis in original). That is, it's not an act that determines deviance, but whether and how others react.

Some people are never caught or prosecuted for crimes they commit, and thus aren't labeled deviant. In other cases, people may be falsely accused (as cheating on taxes) and stigmatized. In effect, then, deviance is in the eye of the beholder because societal reaction, rather than an act, labels people as law-abiding or deviant. Moreover, labeling can lead to secondary deviance.

LEMERT: PRIMARY AND SECONDARY DEVIANCE

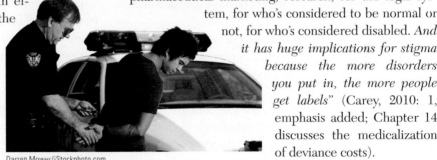
Darren Mower/iStockphoto.com

Edwin Lemert (1951, 1967) expanded on the effects of labeling by differentiating between primary and secondary deviance. **Primary deviance** is the initial act of breaking a rule. Primary deviance can range from relatively minor offenses, such as not attending a family member's funeral, to serious offenses, such as rape and murder.

Even if people aren't guilty of primary deviance, labeling can lead to **secondary deviance**, rule-breaking behavior that people adopt in response to others' reactions. A teenager who's caught trying marijuana may be labeled a "druggie." If the individual is rejected by others, he or she may accept the label, associate with drug users, and become involved in a drug-using subculture. According to Lemert, a single deviant act will rarely result in secondary deviance. The more times a person is labeled, however, the higher the probability that she or he will accept the label and engage in deviant behavior.

There's considerable evidence that labeling affects people's lives. For example, those who have been jobless for a year or longer report stigmas in job searches: Employers assume that they're lazy, don't really want to work, or there's something wrong with them. Some companies have explicitly barred the long-term unemployed from certain job openings, telling them outright in job ads that they need not apply (see Chapter 11).

THE MEDICALIZATION OF DEVIANCE

In 1952, the American Psychiatric Association published the first edition of the *Diagnostic and Statistical Manual of Mental Disorders (DSM)*, which describes mental-health disorders. The number of disorders increased from 106 in 1952 to nearly 400 in 2013, and increased in size from "a spiral-bound pamphlet" to "a tome of nearly 1,000 pages" (Tavris, 2013: C5). Some of the new disorders include "binge eating" and "bereavement" (American Psychiatric Association, 2013).

The *DSM*'s ever-changing diagnoses and labels is an example of the **medicalization of deviance**, diagnosing and treating a violation of social norms as a medical disorder. The DSM has far-reaching effects, according to one of the past editors: "Anything you put in that book … has huge implications not only for psychiatry but for pharmaceutical marketing, research, for the legal system, for who's considered to be normal or not, for who's considered disabled. *And it has huge implications for stigma because the more disorders you put in, the more people get labels*" (Carey, 2010: 1, emphasis added; Chapter 14 discusses the medicalization of deviance costs).

7-6c Critical Evaluation

Symbolic interaction theories have helped to explain deviance, but critics point to several weaknesses. Differential association theory doesn't explain impulsive crimes of rage (e.g., domestic violence), those committed by people who have grown up in law-abiding families, or why only a minority of children who grow up in poor, urban, disadvantaged communities join gangs and/or commit crimes. This theory also ignores the possibility that deviant values and behaviors can be unlearned, as when young children model teachers or other adults whom they respect (Williams and McShane, 2004; Graif and Sampson, 2009; Cuevas et al., 2013; Guerra et al., 2013).

Critics have faulted labeling theory on other points: The theory exaggerates the importance of judgments in

primary deviance the initial act of breaking a rule.

secondary deviance rule-breaking behavior that people adopt in response to others' reactions.

medicalization of deviance diagnosing and treating a violation of social norms as a medical disorder.

Trailer Trash? About 20 million low-income Americans live in mobile homes because they can't afford other housing. Most realize that they're stigmatized as "trailer trash," but maintain their dignity despite the labeling. They describe themselves as homeowners, many are employed, they differentiate themselves from those who commit crimes, and take care of their property (Kusenbach, 2009; Geoghegan, 2013).

changing a person's self-concept, doesn't explain why crime rates are higher in the South than in other parts of the United States or at particular times of the year (as before holidays), and doesn't tell us *why* people commit crimes. Some critics also contend that social reactions are the *result* rather than the "cause" of deviant behavior (Schurman-Kauflin, 2000; Benson, 2002). Conflict theorists, in particular, criticize symbolic interactionists for ignoring structural factors—such as poverty and low-paid jobs—that create or reinforce deviance and crime (Currie, 1985).

7-7 CONTROLLING DEVIANCE AND CRIME

Most of us conform to laws and cultural expectations. We usually conform, however, not because we're innately good, but because of **social control**, techniques and strategies that regulate people's behavior in society. The purpose of social control is to eliminate, or at least reduce, deviance.

social control the techniques and strategies that regulate people's behavior in society.

control theory proposes that deviant behavior decreases when people have strong social bonds with others.

sanctions rewards or punishments for obeying or violating a norm.

criminal justice system government agencies that are charged with enforcing laws, judging offenders, and changing criminal behavior.

7-7a Control Theory and Types of Social Control

Control theory proposes that deviant behavior decreases when people have strong social bonds with others. According to sociologist Travis Hirschi (1969), the bonds that strengthen our self-control are based on (1) *attachment* to other people (e.g., family, friends, and coworkers), (2) *commitment* to conformity (e.g., participating in community activities), (3) *involvement* in legitimate activities (e.g., getting an education, holding a job), and (4) *belief* in social norms and values (e.g., being moral, obeying laws).

Most conformity is due to the internalization of norms during the powerful socialization process. Because most of us care about the opinions of family members, friends, and teachers, we try to live up to their expectations (see Chapter 4). Thus, we conform because of *informal social controls* that we learn and internalize during childhood.

Formal social control also regulates social behavior. Unlike informal social control, formal social control exists outside of the individual. For example, a college dean might threaten a student with suspension or expulsion, a business might install security cameras, or a police officer might arrest someone.

7-7b Positive and Negative Sanctions

Most of us conform because of **sanctions**—rewards or punishments for obeying or violating a norm. *Positive sanctions* are rewards for desirable behavior. They include a variety of facial expressions (e.g., smiling), body language (e.g., hugging), comments (e.g., "Congratulations!"), and other forms of recognition (e.g., good grades, trophies, and promotions). Recently, a tourist's snapshot of a New York City police officer giving new boots to a barefoot homeless man on a frigid night "went viral" because of the officer's kind act.

Negative sanctions are punishments for violating a norm. They range from mild and informal expressions (e.g., frowns and gossip) to more severe and formal reactions (e.g., fines, arrests, and incarceration). The ultimate formal negative sanction in most societies, including the United States, is execution.

7-7c The Criminal Justice System and Social Control

Social institutions such as the family, education, and religion try to maintain *social* control over behavior; the criminal justice system has the *legal* power to control crime and punish offenders. The **criminal justice system** is made up of government agencies—including the police, courts, and prisons—that are charged with enforcing laws, judging offenders, and changing criminal behavior. The criminal justice system relies on three major strategies to control crime: Prevention and intervention, punishment, and rehabilitation. Punishment, as you'll see, is the least effective approach.

PREVENTION AND INTERVENTION

Crime prevention is much less expensive than punishment. For example, many crimes involve substance abuse, but the average annual cost per outpatient for alcohol or drug abuse treatment is $4,700 compared with $24,000 for incarceration (National Institute on Drug Abuse, 2012). Nearly a quarter-million inmates in state and federal prisons are age 50 and older, and many have serious health problems. An older offender's annual medical costs can exceed $100,000 (Johnson and Beiser, 2013).

Some crimes have plummeted because of technology. Car theft, for example, has decreased 40 percent since 2003 because most new automobiles are almost impossible to hot-wire: They come equipped with immobilizer systems that require an authorized key to start (Klein and White, 2011). Technology, however, hasn't stopped other crimes such as burglary because many offenders have become more sophisticated in avoiding home security systems.

Nationally, states with the most firearm laws have the lowest gun-related homicide and suicide rates (Fleegler et al., 2013). Stricter gun control laws are controversial, however. More than

© Andrey Burmakin/Shutterstock.com

half of Americans believe that firearm sale laws should stay the same or be more lenient (Swift, 2014). Some criminologists also question the effectiveness of passing new firearm laws because we live in a violent culture (Ross, 2013). A recent study of top-selling movies from 1950 to 2012 found that episodes of gun violence in PG-13 rated films now surpass the violence in R-rated films (Bushman et al., 2013). Thus, violence sells.

SOCIAL SERVICE AGENCIES AND COMMUNITY OUTREACH PROGRAMS

Many organizations—composed of law enforcement professionals, social workers, and nonprofit groups—try to prevent crime. Because of high prison costs, the U.S. Department of Justice, among other federal agencies, funds numerous initiatives that try to steer youth away from crime (Office of Juvenile Justice and Delinquency Prevention, 2010; Office of Justice Programs, 2011). "Scared straight" programs involve prison tours and face-to-face encounters, but an evaluation of nine scared-straight programs concluded that they don't deter

the teenage participants from deviant behavior. Still, the tours continue to be very popular (Petrosino et al., 2003; Robinson and Slowikowski, 2011).

POLICE

The primary role of the police is to enforce society's laws, but can the police prevent crimes? Police can head off some crimes by cruising high-risk areas in patrol cars or by having more officers on foot patrols. Concentrating on *hot spots,* areas of high criminal activity, reduces crime only temporarily, however, because criminals simply move to other parts of the city (U.S. Department of Justice, 1996; Braga, 2003). Increasing the number of foot patrols in Philadelphia— a city with one of the nation's highest homicide rates— decreased violent crime by 23 percent, but the patrols worked best if officers returned to the same street several times during a shift ("Policing Philadelphia…," 2013).

The sharpest increases in violent crimes have been in the suburbs surrounding large cities such as Houston, Atlanta, and Pittsburgh. As law enforcement resources shrink, criminals are focusing on suburbs because they aren't as well-policed as cities (McWhirter and Fields, 2012). Even if law enforcement budgets mushroomed, police couldn't prevent most crimes because they have no control over macro-level factors—poverty, unemployment, limited educational opportunities, and neighborhood deterioration—that are associated with high crime rates.

PUNISHMENT

Those who endorse a **crime control model** believe that crime rates increase when offenders don't fear apprehension or punishment. This perspective emphasizes protecting society and supports a tough approach toward criminals in sentencing, imprisonment, and capital punishment.

SENTENCING

After someone has been found guilty of a crime or has pleaded guilty, a judge (and sometimes a jury) imposes a *sentence,* or penalty. A sentence can be a fine, *probation* (supervision instead of serving time in jail), incarceration, or the death penalty. Those who receive *parole* are released from prison before the end of their full sentence on condition that they check in regularly with an officer and obey the law.

About 31 percent of those sentenced get probation; the others serve a term in a local jail or a state or

crime control model proposes that crime rates increase when offenders don't fear apprehension or punishment.

federal prison. Many people question the fairness of sentencing because those convicted of similar crimes can receive widely different sentences depending, among other factors, on race/ethnicity, social class, variations in state sentencing laws (including how many times a person has been arrested for a crime), and how juries and judges evaluate the seriousness of a crime (Durose and Langan, 2004; Whelan, 2007). Nationwide, for example, of those serving life-without-parole sentences for nonviolent offenses (e.g., possession, sale, or distribution of drugs; shoplifting; attempting to cash a stolen check), 65 percent are black and, across all racial-ethnic groups, the vast majority come from low-income families and don't have a high school diploma (Turner, 2013).

As state and local budgets shrink, even states with reputations for being tough on crime are embracing more lenient policies. Some states have begun to shorten the average number of years on probation and parole, which decreases supervision costs. Other states are reducing the number of people sent to prison: It costs an average of $79 a day to keep an inmate in prison but only about $3.50 a day to monitor the same person on probation (Richburg, 2009; see also Maruschak and Bonczar, 2013).

INCARCERATION

The number of U.S. prisoners has surged during the last few decades (see *Figure 7.2*). Since 2008, and for the first time in history, 1 in every 100 American adults is in prison, and, as a journalist notes, America has more prisoners than high school teachers (Knafo, 2013). The United States has less than 5 percent of the world's population, but almost 25 percent of the planet's prisoners. Besides the sheer number of inmates, the United States is also the global leader in per capita inmates (716 per 100,000 population), well ahead of nations such as Russia (475), Cuba (510), and the United Kingdom (148) (Walmsley, 2013).

FIGURE 7.2 THE NUMBER OF U.S. PRISONERS HAS NEARLY TRIPLED SINCE 1987

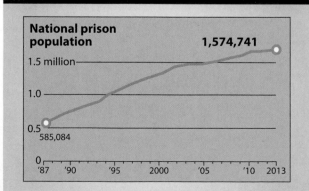

Sources: Based on Pew Center on the States 2010, and Carson, 2014.

TABLE 7.5 U.S. PRISONERS BY RACE/ETHNICITY AND SEX, 2013

Imprisonment Rate Per 100,000 U.S. Population		
Race/Ethnicity	Male	Female
White	466	51
African American	2,805	113
Latino	1,134	66
Other*	963	90

*Includes American Indians, Alaska Natives, Asians, Native Hawaiians, Pacific Islanders, and persons of two or more races.

Source: Based on Carson, 2014, Table 8.

Of all U.S. state and federal prisoners, 92 percent are men (Carson, 2014). However, imprisonment rates vary by race/ethnicity and gender. As *Table 7.5* shows, incarceration rates are highest for African Americans, both women and men. If current trends continue, 1 of every 3 black males born in 2013 can expect to go to prison in his lifetime, as can 1 of every 6 Latino males—compared with 1 of every 17 white males (The Sentencing Project, 2013).

The U.S. prison population decreased by more than 2 percent between 2010 and 2013 (Carson, 2014). Why the drop? Arrest rates have declined, as you saw earlier, but prison counts have dropped primarily because of economic reasons. Driven by budget crises, many states have pursued a variety of strategies such as imposing probation or lighter sentences for nonviolent crimes, releasing low-level offenders before they served their sentences, reducing the severity of sentences for property crimes, and accelerating the release of inmates who completed educational, vocational, and substance abuse treatment programs. Releasing prisoners early has been controversial, but state officials maintain that their budgets can't afford the high incarceration costs (Barrett, 2013; Goode, 2013; Kearney et al., 2014).

It's tempting to assume that the more people behind bars, the lower the crime rate. State spending on corrections surged from $15 billion in 1982 to $49 billion in 2010

IT'S TEMPTING TO ASSUME THAT THE MORE PEOPLE BEHIND BARS, THE LESS CRIME THERE WILL BE.

(Kyckelhahn, 2013). However, *recidivism* (being arrested for committing another offense after being released) has barely changed since 1999: Almost half of released prisoners are back behind bars within 3 years (Pew Center on the States, 2011).

CAPITAL PUNISHMENT

About 61 percent of Americans support *capital punishment* (the death penalty), down from 80 percent in the mid-1990s (Jones, 2014). By early 2013, 140 nations, including many developing countries, had abolished the death penalty or suspended executions. The United States executes more people any other country except China, Iran, Iraq, and Saudi Arabia. Thirty-two states have the death penalty, but the number of executions fell from 98 in 1999 to 39 in 2013, and only 2 percent of all counties—primarily in Texas, California, Arizona, Georgia, and Florida—account for more than half of all executions since 1976 (Dieter, 2013; Amnesty International, 2014; Death Penalty Information Center, 2014).

There are several reasons for fewer executions. First, and as you saw earlier, homicide rates have declined sharply since 1980. Second, public support for the death penalty has waned. Third, the nation's police chiefs believe that the death penalty doesn't deter crime. Fourth, there's considerable research which shows racial disparities in executions (e.g., black males, especially if the victims are white, are substantially more likely to be executed than white males who murder blacks). Fifth, most juries can now impose life sentences without the possibility of parole. Finally, some argue that the death penalty is a waste of money because inmates can spend 15 to 20 years appealing the sentence. In California, for example, it costs taxpayers $137 million per year for inmates who appeal a death sentence compared with $12 million per year for those serving life sentences (Death Penalty Information Center, 2009, 2014; Pilkington, 2013; The Sentencing Project, 2013; "Capital Punishment…," 2014).

REHABILITATION

Rehabilitation, a third approach to controlling deviance, maintains that appropriate treatment can change offenders into productive, law-abiding citizens. A record number of inmates is released from state prisons every year because of overcrowding, early parole, and lighter sentences. Social workers and other professionals believe that reentry programs can be successful, especially for those convicted of drug-related crimes (Miller, 2012).

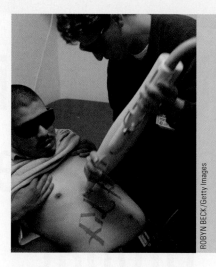

Among other services, Homeboy Industries, based in Los Angeles, provides previously incarcerated male and female gang members with job training and placement. Pictured here, Dr. Mark Benor uses a laser machine to remove a former gang member's tattoo at a free clinic funded by Homeboy Industries.

ROBYN BECK/Getty Images

Are rehabilitation programs successful? Only if they provide employment after release. Other effective rehabilitation efforts include learning a trade while in prison, earning a high school diploma or college degree while in prison or after release, and receiving services that address several needs (e.g., housing, employment, and medical services) rather than just one (e.g., drug abuse counseling) (Lowenkamp and Latessa, 2005; Cumberworth, 2010; Flower, 2010).

rehabilitation a social control approach which maintains that appropriate treatment can change offenders into productive, law-abiding citizens.

STUDY TOOLS 7

READY TO STUDY? IN THE BOOK, YOU CAN:

☐ Check your understanding of what you've read with the Test Your Learning Questions provided on the chapter review card at the back of the book.

☐ Rip out the chapter review card for a handy summary of the chapter and key terms.

ONLINE AT CENGAGEBRAIN.COM YOU CAN:

☐ Prepare for tests with quizzes.

☐ Review the key terms with Flash Cards.

☐ Play games to master concepts.

© Anelina/Shutterstock.com

8 | Social Stratification: United States and Global

LEARNING OBJECTIVES

8-1 Explain and illustrate social stratification systems and bases.

8-2 Describe the U.S. class structure and explain how and why social classes differ.

8-3 Describe poverty and how it varies and explain why people are poor.

8-4 Compare the different types of social mobility, describe recent trends, and explain what factors affect mobility.

8-5 Describe global stratification, its variations and consequences and the theoretical models that try to explain why inequality is universal.

8-6 Compare and evaluate the theoretical explanations of social stratification.

After you finish this chapter go to **PAGE 157** for **STUDY TOOLS.**

In 2012, American multi-billionaire Larry Ellison—founder of Oracle, the world's biggest database company—bought the Hawaiian island of Lanai for $500 million. The same year, 17 percent of Hawaii's residents were living in poverty (Hawaii Free Press, 2012; Kroll and Dolan, 2013). Many people in the United States and around the world are struggling to survive while others enjoy astonishing wealth. This chapter examines social stratification and why there are haves and have-nots.

8-1 SOCIAL STRATIFICATION SYSTEMS AND BASES

Social stratification is a society's ranking of people based on their access to valued resources such as wealth, power, and prestige. All societies are stratified, but some more than others. Sociologists distinguish among

social stratification a society's ranking of people based on their access to valued resources such as wealth, power, and prestige.

stratification systems based on the extent to which they're open or closed.

8-1a Closed and Open Stratification Systems

In a *closed stratification system*, movement from one social position to another is limited by ascribed statuses. An *open stratification system* allows movement up or down because mobility is influenced by people's achievements. Closed stratification systems are considerably more fixed than open ones, but neither is completely open or closed.

Two closed stratification systems exist today: slavery and castes. In a **slavery system**, people own others as property and have almost total control over their lives. The slaves have been kidnapped, inherited, or given as gifts to pay a debt. They're bought and sold as commodities, sometimes multiple times. The United Nations banned all forms of slavery worldwide in 1948, but

it persists in many countries in the Middle East, Africa, the Balkans, and Asia (U.S. Department of State, 2006).

In a **caste system**, a second type of closed stratification system, people's positions are ascribed at birth and largely fixed. That is, people's places in the hierarchy are primarily determined by inherited characteristics such as race, skin color, gender, family background, or nationality (see Chapter 5). People must marry within their caste; there are severe restrictions in one's choice of occupation, residence, and social relationships; and people rarely move from one caste to another.

A good example is India, where a caste system has existed for more than 3,000 years. At the top were the Brahmins. On the bottom rung were the Dalits (formerly

slavery system people own others as property and have almost total control over their lives.

caste system people's positions are ascribed at birth and largely fixed.

called "untouchables"), who were very poor and performed the most menial and unpleasant jobs, including collecting human waste and cleaning streets. Fearing being "polluted," persons of higher castes wouldn't interact with the Dalits (Mendelsohn and Vicziany, 1998).

India outlawed the caste system in 1949, but social distinctions are deeply entrenched, particularly in rural areas, and most people socialize with and marry within their own castes (Banerjee et al., 2009). Castes are so common in some parts of India that the national census now collects data on them (Haub, 2011). Dalits are becoming a potent political force, but caste discrimination is pervasive, including at many universities (Neelakantan, 2011).

In a **class system**—a relatively open stratification structure—people's positions are based on both birth and achievement. Because achieved characteristics (e.g., education, work skills, occupation) can change, people

class system people's positions are based on both birth and achievement.

social class people who have a similar standing or rank in a society based on wealth, education, power, prestige, and other valued resources.

wealth economic assets that a person or family owns.

income the money a person receives, usually through wages or salaries, but can also include other earnings.

in open stratification systems can move from one social class to another. A **social class** is a group of people who have a similar standing or rank in a society based on wealth, education, power, prestige, and other valued resources. A large majority of Americans (60 percent) believe that hard work leads to success (Pew Research Center, 2014), but is this really the case?

8-1b The Bases of Stratification

Conventional wisdom says that money can't buy happiness, but a study of 155 countries, including the United States, found that the richer people are, the happier they are (Stevenson and Wolfers, 2013). Income is an important stratification factor, but it's not the only one. Instead, sociologists use a multidimensional approach that includes wealth, prestige, and power.

WEALTH

Wealth is the abundance of economic assets that a person or family owns. It includes money, property (e.g., real estate), stocks and bonds, retirement and savings accounts, personal possessions (e.g., cars and jewelry), and income. **Income** is the money a person receives, usually through wages or salaries, but it can also include rents, interest on savings accounts, stock dividends, royalties, or business proceeds. Income and wealth differ in several important ways:

▶ Wealth is *cumulative*. It increases over time, especially through investment, whereas income is usually spent on everyday expenses.

▶ Because wealth is accumulated over time, much of it can be *passed on to the next generation*. With an estimated $82 billion net worth in 2013, Bill Gates, cofounder of Microsoft Corporation, is one of the wealthiest people in the world. If Gates's three

India's Dalits, such as those pictured here, still perform unpleasant tasks such as burning corpses and removing rubbish and human waste.

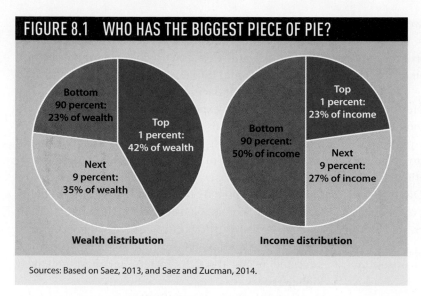

FIGURE 8.1 WHO HAS THE BIGGEST PIECE OF PIE?

Bottom 90 percent: 23% of wealth

Top 1 percent: 42% of wealth

Next 9 percent: 35% of wealth

Wealth distribution

Top 1 percent: 23% of income

Bottom 90 percent: 50% of income

Next 9 percent: 27% of income

Income distribution

Sources: Based on Saez, 2013, and Saez and Zucman, 2014.

children inherit only 1 percent of his fortune, each will get at least $820 million!

▶ Passing on wealth to the next generation *preserves privilege*. Wealthy people tend to have the greatest amount of *economic capital* (income and other monetary assets such as property), *cultural capital* (advanced degrees and assets such as style of speech, table manners, and physical appearance), and *social capital* (networks comprised of influential people). Differential access to all three types of capital reinforces and reproduces the existing class structure and inequality (Bourdieu, 1984).

U.S. wealth and income inequality is staggering. As *Figure 8.1* shows, the top 1 percent of Americans own 42 percent of all wealth and 23 percent of all income—the highest levels recorded since the government began collecting such data, and income inequality is the highest it's been since 1928 (Lowrey, 2013; Saez, 2013). We can get a better sense of income disparities by looking at the average household income. As *Figure 8.2* shows, there's a significant income gap between even the top 5 and 20 percent of households.

Both wealth and income gaps have *increased*. In 2013, the wealthiest 10 percent of households had 77 of the country's wealth, up from 67 percent in 1989. The wealth of the bottom 90 percent fell by more than 3 percent; in 2013, this group owned only 23 percent of all U.S. wealth, about as much as in 1940 (Levine, 2012; Saez and Zucman, 2014).

In 2003, the net wealth of the typical household, adjusted for inflation, was $87,992. Ten years later, it was down to $56,335. Thus,

between 2003 and 2013, ordinary Americans became 36 percent poorer; those in the top 5 percent got 14 percent richer (Pfeffer et al., 2014).

Five years after the Great Recession (between late 2007 and mid-2009) officially ended, few families were enjoying an economic recovery. The typical family's income and wealth was about 5 percent lower than in 2007, and middle-income families were poorer in 2014 than in 1989. In contrast, the median family income and wealth of the top 10 percent of earners rose by about 3 percent. The wealthiest 10 percent of households own 80 percent of all stock; the bottom 80 percent own only 8 percent. Thus, richer Americans benefit when the stock market recovers after a recession (Levine, 2012; DeNavas-Walt and Proctor, 2014).

Regardless of how the economy is doing, the rich keep getting richer. Between 1983 and 2010, the number of U.S. households that had $10 million or more increased by 426 percent. In 2014, the United States had 571 billionaires, up from 413 in 2011, who comprised nearly 25 percent of the world's billionaires. These super-rich Americans average $3.1 billion per person and control 37 percent of the nation's wealth (Wolff, 2012; Wealth-X and UBS, 2014).

PRESTIGE

A second basis of social stratification is **prestige**—respect, recognition, or regard attached to social positions.

prestige respect, recognition, or regard attached to social positions.

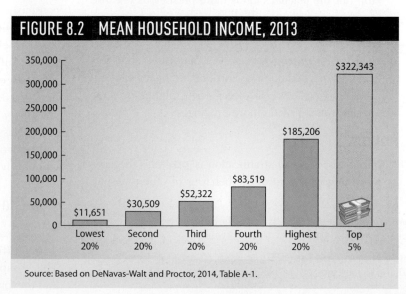

FIGURE 8.2 MEAN HOUSEHOLD INCOME, 2013

	Income
Lowest 20%	$11,651
Second 20%	$30,509
Third 20%	$52,322
Fourth 20%	$83,519
Highest 20%	$185,206
Top 5%	$322,343

Source: Based on DeNavas-Walt and Proctor, 2014, Table A-1.

Prestige is based on many criteria, including wealth, family background, power, and accomplishments. Regarding accomplishments, for example, every college convocation acknowledges students who graduate *cum laude*, *magna cum laude*, and *summa cum laude*.

We typically evaluate people according to the kind of work they do. Higher prestige occupations include physicians, lawyers, pharmacists, college professors, architects, dentists, and teachers. Legislators, police officers, actors, librarians, realtors, and firefighters are examples of jobs at a medium prestige level. Occupations at the lower prestige level consist of janitors, carpenters, bartenders, garbage collectors, truck drivers, and food servers (Smith et al., 2011).

The most prestigious occupations have several characteristics:

▶ They *require more formal education* (college or postgraduate degrees) and/or extensive training. Physicians, for example, must fulfill internship and residency requirements after receiving a medical degree.

▶ They're primarily *nonmanual and require more abstract thought*. All jobs have some physical activity, but an architect must use more imagination in designing a building than a carpenter, who usually performs very specific tasks.

▶ They *are paid more*, even though there are some exceptions. A realtor or a truck driver may earn more than a registered nurse, but registered nurses are likely to earn more over a lifetime because they have steady employment, health benefits, and retirement programs.

▶ They're *seen as socially more important*. An elementary school teacher may earn less than the school's janitor, but the teacher's job is more prestigious because teachers contribute more to a society's well-being.

▶ They *involve greater self-expression, autonomy, and freedom from supervision*. A dentist has considerably more freedom in performing her or his job than a dental hygienist. In effect, then, higher prestige occupations provide more privileges.

POWER

A third basis of social stratification is **power**—the ability to influence or control the behavior of others despite

---

power the ability to influence or control the behavior of others despite opposition.

socioeconomic status (SES) an overall ranking of a person's position in society based on income, education, and occupation.

---

opposition. Wealthy people generally have more power than others in dominating top government and economic positions. Even if they don't hold a political office or a corporate position, the most powerful people make the nation's important decisions.

People who experience *status consistency* are about equal in terms of wealth, prestige, and power. A Supreme Court judge, for instance, is usually affluent, enjoys a great deal of prestige, and wields considerable power. In many cases, however, there's *status inconsistency* (see Chapter 5) if a person ranks differently on stratification factors. Consider funeral directors. Their prestige is relatively low, but most have higher incomes than college professors, who are among the most educated people in U.S. society and have relatively high prestige.

We've looked at stratification systems and their bases, but how, specifically, do people in different social classes act? And how does social class affect our behavior?

8-2 SOCIAL CLASS IN AMERICA

To measure social class, sociologists typically use **socioeconomic status (SES)**—an overall ranking of a person's position in society based on income, education, and occupation. Some researchers ask people to identify the social classes in their communities (*reputational approach*), some ask people to place themselves in one of a number of classes (*subjective approach*), but most use SES indicators (*objective approach*).

Because there are different ways of measuring social class, sociologists don't always agree on the number of social classes in the United States. There's consensus, however, that there are four general classes: upper, middle, working, and lower. Except for the working class, many sociologists divide these classes further into more specific strata. Social class is a complex concept and there aren't rigid dividing lines, but *Figure 8.3* provides an overview of the U.S. social class structure. Besides income, education, and occupation, social classes also differ in values, power, prestige, social networks, and *lifestyles* (tastes, preferences, and ways of living).

8-2a The Upper Class

You saw earlier that very rich Americans control a vastly disproportionate amount of the total wealth and income. This group comprises two classes: the upper-upper class and the lower-upper class.

FIGURE 8.3 THE AMERICAN CLASS STRUCTURE

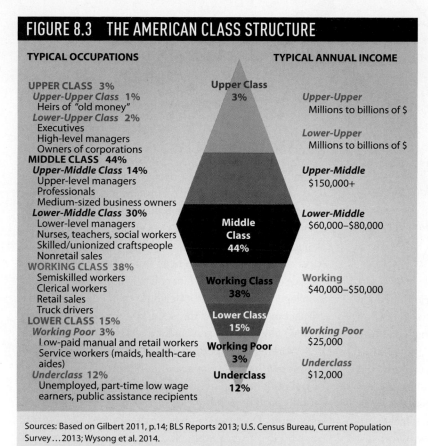

TYPICAL OCCUPATIONS

UPPER CLASS 3%
Upper-Upper Class 1%
  Heirs of "old money"
Lower-Upper Class 2%
  Executives
  High-level managers
  Owners of corporations
MIDDLE CLASS 44%
Upper-Middle Class 14%
  Upper-level managers
  Professionals
  Medium-sized business owners
Lower-Middle Class 30%
  Lower-level managers
  Nurses, teachers, social workers
  Skilled/unionized craftspeople
  Nonretail sales
WORKING CLASS 38%
  Semiskilled workers
  Clerical workers
  Retail sales
  Truck drivers
LOWER CLASS 15%
Working Poor 3%
  Low-paid manual and retail workers
  Service workers (maids, health-care aides)
Underclass 12%
  Unemployed, part-time low wage earners, public assistance recipients

Upper Class 3%

Middle Class 44%

Working Class 38%

Lower Class 15%

Working Poor 3%

Underclass 12%

TYPICAL ANNUAL INCOME

Upper-Upper
Millions to billions of $

Lower-Upper
Millions to billions of $

Upper-Middle
$150,000+

Lower-Middle
$60,000–$80,000

Working
$40,000–$50,000

Working Poor
$25,000

Underclass
$12,000

Sources: Based on Gilbert 2011, p.14; BLS Reports 2013; U.S. Census Bureau, Current Population Survey…2013; Wysong et al. 2014.

THE UPPER-UPPER CLASS

Upper-upper-class people are the *old rich* who have been wealthy for generations. Because they value their privacy, upper-upper-class members rarely appear on the lists of wealthiest individuals published by *Forbes* or other sources.

An inherited fortune brings power. Upper-upper-class white males, in particular, shape the economic and political climate through a variety of mechanisms: Dominating the upper levels of business and finance, holding top political positions in the federal government, underwriting thousands of think tanks and research institutes that formulate national policies, and shaping public opinion through the mass media (Zweigenhaft and Domhoff, 2006; Hacker and Pierson, 2010).

THE LOWER-UPPER CLASS

The lower-upper class—which is much more diverse than the upper-upper class—is the *nouveau riche*, those with "new money." Some, like the Kennedys, amassed fortunes several generations ago, but many of the new rich—Oprah Winfrey, President Bill Clinton, and Mark Zuckerberg—worked for their income rather than inherited it. Besides business entrepreneurs, the lower-upper class also includes high-level managers

of international corporations, those who earn at least a million dollars a year, and some highly paid athletes and actors, but the lifestyles vary considerably (Wood, 2012). Some lower-upper class members live modestly, but many flaunt their new wealth. Sociologist Thorstein Veblen (1899/1953) coined the term *conspicuous consumption*, a lavish spending on goods and services to display one's social status and enhance one's prestige.

Most of the *nouveau riche* are unbelievably wealthy. If Bill Gates cashed in all his wealth and spent $1 million every single day, it would take him 218 years to spend all his money (Oxfam, 2014). Because they lack the "right" ancestry and have usually made their money by working for it, however, lower-uppers aren't accepted into "old money" circles that have strong feelings of in-group solidarity. Still, lower-upper class members engage in lifestyles and rituals that parallel those of the upper-upper class, such as having personal chefs, taking exotic vacations (often in private planes), and joining country clubs (Sherwood, 2010).

8-2b The Middle Class

When given a closed-ended question (see Chapter 2), 13 percent of Americans identify themselves as upper-middle class, 44 percent as middle class, and 28 percent as lower-middle class (Pew Research Center, 2014). There's no firm answer about exactly where these classes begin and end, but there are some major

"And you, Al, what would you do if you had only a million?"

Conspicuous consumption isn't limited to the super-rich. Many upper-middle-class people, in particular, use status symbols to show that they've made it by buying "almost rich" cars (such as Jaguar X-type sedans that start at $30,000), upscale kitchen appliances, designer handbags, expensive jewelry, and by taking lavish vacations.

Hemera Technologies/Getty Images

differences between the *upper-middle class* and the *lower-middle class*.

THE UPPER-MIDDLE CLASS

Upper-middle-class members, although rich, live on earned income rather than accumulated or inherited wealth. The occupations of this group usually require a Ph.D. or advanced professional degree. People in this class include corporate executives and managers (but not those at the top), high government officials, owners of large businesses, physicians, and successful lawyers and stockbrokers. Most of these occupations have considerable on-the-job autonomy and freedom from supervision, but these people are three times more likely than those in the general population to work 50 or more hours per week (Dewan and Gebeloff, 2012).

THE LOWER-MIDDLE CLASS

The lower-middle class, more diverse than the upper-middle class, is composed of people in nonmanual occupations that require some training beyond high school, and professional occupations that require a college degree. Nonmanual jobs include office staff, low-level managers, owners of small businesses, medical and dental technicians, secretaries, police officers, and sales workers (e.g., insurance and real estate agents). Examples of professional occupations are nursing, social work, and teaching. Most families in the lower-middle class rely on two incomes to maintain a comfortable standard of living.

Unlike upper-middle-class jobs, those in the lower-middle class have less autonomy and freedom from supervision, and there's little chance for advancement. Except for some retirement funds, most have only modest

working poor people who work at least 27 weeks a year but whose wages fall below the official poverty level.

underclass people who are persistently poor and seldom employed, residentially segregated, and relatively isolated from the rest of the population.

savings to cover emergencies. Many buy used or inexpensive late-model cars, eat out at middle-income restaurants, and take occasional vacations, but they rarely have the income to buy luxury products without going deeply into debt.

8-2c The Working Class

About 31 percent of Americans identify themselves as working class (Dugan, 2012). This group consists of skilled and semiskilled laborers, including construction workers, assembly-line workers, truck drivers, auto mechanics, and skilled craft workers (e.g., carpenters and electricians). Most of the jobs are blue collar, but some—clerks and retail sales workers—are white collar. The jobs don't require a college education, and offer little or no opportunity for advancement. Most of the semiskilled jobs require little training, are mechanized, and are closely supervised.

Working-class people who purchase homes, including mobile homes, may experience foreclosure because of delinquent payments. Many use credit cards to pay off bills each month but then can barely pay the monthly minimum. Debts become overwhelming when borrowers who live from paycheck to paycheck suffer setbacks such as divorce, illness, or job loss.

8-2d The Lower Class

About 12 percent of Americans identify themselves as lower class (Pew Research Center, 2014). People in this group are at the bottom of the economic ladder because they have little education, few occupational skills, work in minimum wage jobs, or are often unemployed. Most of the lower class is poor, but sociologists often distinguish between the *working poor* and the *underclass*.

THE WORKING POOR

Almost 23 percent of America's poor are the **working poor**—people who work at least 27 weeks a year but whose wages fall below the official poverty level. Women are more likely than men to be among the working poor. Blacks and Latinos are more than twice as likely as Asians and whites to be in this group. Full-time workers are less likely than part-time workers to be working poor but, in 2012, more than 4 percent of those usually employed full time were among the working poor, primarily because of low earnings (BLS Reports, 2014).

THE UNDERCLASS

The **underclass**, which occupies the bottom rung of the economic ladder, consists of people who are persistently

Among America's working poor are hotel housekeepers. On average, those employed even at expensive hotels earn less than $19,000 a year working full time. The working poor do dirty and often demeaning work for next to nothing, barely scrape by, and live in constant fear of being fired (Wing and Schwartz, 2014).

poor, residentially segregated, and relatively isolated from the rest of the population. Most are chronically unemployed or drift in and out of jobs. They may work erratically or at part-time jobs, but their lack of skills, low education levels, and mental or physical disabilities make it difficult for them to find full-time jobs. Members of the underclass may have some earnings, but are often dependent on income from government programs, including Social Security, public assistance, and veterans' benefits. Some also have income from criminal activities (Gilbert, 2011; see also Chapter 7).

8-2e How Social Class Affects Us

Our social class, more than any other single variable, affects almost all aspects of our lives. Max Weber referred to the consequences of social stratification as **life chances**—the extent to which people have positive experiences and can secure the good things in life (food, housing, education, good health) because they have economic resources. Regarding health, for example:

▶ Poor children are less healthy than those at higher SES levels and are more likely to be unhealthy in adulthood. Unhealthy adults earn less, spend less

time in the labor force, and must often retire earlier (see Chapter 14).

▶ Living in a chronically disadvantaged neighborhood and experiencing violence increases anxiety and stress, accelerating the onset of diseases (Bird, et al., 2010; Jordan, 2013).

▶ Compared with the upper class, the lower class is four times more likely to report being in poor health, three times more likely to be unhappy, and twice as likely to be frequently stressed and depressed (Brown, 2012; Morin and Motel, 2012).

8-3 POVERTY

There are two ways to define poverty: absolute and relative. **Absolute poverty** is not having enough money to afford the basic necessities of life, such as food, clothing, and shelter ("what I need"). **Relative poverty** is not having enough money to maintain an average standard of living ("what I want").

8-3a The Poverty Line

The **poverty line** (also called the *poverty threshold*) is the minimal income level that the federal government considers necessary for basic subsistence. To determine the poverty line, the Department of Agriculture (DOA) estimates the annual cost of food that meets minimum nutritional guidelines and then multiplies this figure by three to cover the minimum cost of clothing, housing, health care, and other necessities. The poverty line, which in 2013 was $23,624 for a family of four (two adults and two children), is adjusted every year to include cost-of-living increases. If a family makes more than the poverty line, it's usually not eligible for public assistance (DeNavas-Walt and Proctor,

life chances the extent to which people have positive experiences and can secure the good things in life because they have economic resources.

absolute poverty not having enough money to afford the basic necessities of life.

relative poverty not having enough money to maintain an average standard of living.

poverty line the minimal income level that the federal government considers necessary for basic subsistence (also called the *poverty threshold*).

2014; see Schaefer and Edin, 2014, for an analysis of Americans living in "extreme poverty").

Is the poverty line realistic? Some critics contend that the poverty line is too high. They argue, for example, that many people aren't as poor as they seem because the poverty threshold doesn't include the value of noncash benefits such as food stamps, medical services (primarily Medicare and Medicaid), public housing subsidies, and unreported income (Eberstadt, 2009).

Others claim that the poverty line is too low because it doesn't include child care and job transportation costs or the cost-of-living expenses, particularly housing, that vary considerably across states, regions, and urban-rural areas. According to some economists, a family of four needs almost $60,000 a year to cover basic necessities, a considerably higher amount than the official poverty line of just under $24,000. Some also argue that poverty estimates exclude millions of Americans who live above the poverty line but rely on food banks, soup kitchens, and clothing thrift stores to survive (Fremstad, 2012; DeGraw, 2014).

8-3b Who Are the Poor?

In 2013, the official U.S. poverty rate was almost 15 percent (45.3 million Americans), and up from 11 percent in 2000, but poverty isn't random. Both historically and currently, the poor share some common characteristics that include age, gender, family structure, race, and ethnicity.

One in three poor Americans live in the suburbs. Compared with central cities, many suburbs offer safer streets, better schools, and a greater number of low-paying jobs (e.g., retail sales, landscaping, restaurants). Rents are rising, however, and many of the suburban poor worry about the costs of maintaining a car to get to work ("Broke in the 'Burbs," 2013; Kneebone and Berube, 2014).

AGE

Children under age 18 make up only 24 percent of the U.S. population, but 32 percent (almost 15 million) of all poor people, up from a low of 16 percent in 2001. Americans aged 65 and older make up about 14 percent of the total

Six Ways the Poor Pay More

Food: Not having a car to get to a supermarket chain or discount store means getting groceries at the corner store where staples are more expensive and produce is limited and often less fresh.

Transportation: About 45 percent of American households don't have access to public transportation. Old cars break down, can be expensive to fix, use a lot of gas, and can cost people a job.

Doing Business: Most neighborhood stores charge more for products because real estate and insurance are more expensive. Poor people also pay more for financial services. Banks are scarce and "check-cashing stores" charge up to 10 percent of a check's value to cash it.

Credit: Because poor people don't have credit, they must rely on "cash advance" stores if they can prove that they get a regular paycheck. If poor people qualify for such a loan, they pay an annual percentage rate of about 825 percent.

Everyday Hassles: The poor are hassled almost every day by bill collectors because of overdue payments, deal with Laundromat trips, carry groceries long distances, and contend with violence in high-crime areas.

Waiting: Poor people spend much of their time waiting—in food pantry lines, at blood banks to sell their blood, at bus stops to get to work, and for apartments in safer neighborhoods

Source: Brown, 2009; Schwartz, 2011; Bass and Campbell, 2013; Johnson, 2014.

population and almost 10 percent (4.2 million) of the poor. The poverty rate of older Americans is at an all-time low, and lower than that of any other age group, because government programs, particularly Medicare and Medicaid, have generally kept up with the rate of inflation. In contrast, many programs for poor children have been reduced or eliminated since 1980 (Addy et al., 2013; Korenman and Remler, 2013; Children's Defense Fund, 2014; DeNavas-Walt and Proctor, 2014).

GENDER AND FAMILY STRUCTURE

Women's poverty rates are higher than men's (16 and 13 percent, respectively), but family structure is an important factor. In 2013, 6 percent of married-couple families, 16 percent of male-headed families, and 31 percent of female-headed families lived in poverty (DeNavas-Walt and Proctor, 2014).

The term **feminization of poverty** refers to the disproportionate number of the poor who are women. Because of increases in divorce, nonmarital childbearing, and low-paying jobs, single-mother families are four to five times more likely to be poor than married-couple families that have two wage earners, and more likely to be extremely poor over many years (Gould, 2012; Mishel et al., 2012; Shriver, 2014).

RACE AND ETHNICITY

In absolute numbers, there are more poor whites (19 million) than poor Latinos (12.7 million), blacks (11 million), or Asians (1.7 million) (DeNavas-Walt and Proctor, 2014). Proportionately, however, and as *Figure 8.4* shows, whites and Asian Americans are less likely to be poor than other racial-ethnic groups.

8-3c Why Are People Poor?

Social scientists have a number of theories about poverty (see Cellini et al., 2008, for a nontechnical summary). Regarding two of the most common explanations, one blames the poor themselves, and the other emphasizes societal factors.

BLAMING THE POOR: INDIVIDUAL FAILINGS

About 60 percent of Americans believe that people can be successful if they're more industrious. In a recent national survey, for example, 38 percent said that people are rich because they work harder than others (Morin, 2012; Pew Research Center, 2014).

Such views reflect a still influential *culture of poverty* perspective, which contends that the poor don't succeed because they're "deficient": They have values, beliefs, and attitudes about life that differ from those of

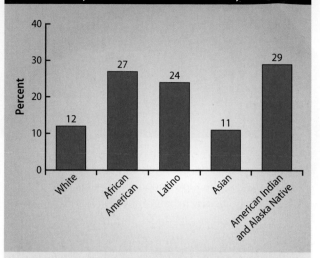

FIGURE 8.4 PERCENTAGE OF AMERICANS LIVING IN POVERTY, BY RACE AND ETHNICITY, 2013

Note: According to the most recent available data, 18 percent of Native Hawaiians and other Pacific Islanders were poor in 2011 (see Macartney et al., 2013).

Source: Based on DeNavas-Walt and Proctor, 2014, Table 3, and American FactFinder, 2013, TableS0201 (accessed November 10, 2014, factfinder2.census.gov.in).

people who aren't poor; they're more permissive in raising their children; and they're more likely to seek immediate gratification instead of planning for the future (Lewis, 1966; Banfield, 1974). The assertion that these values, beliefs, and attitudes are transmitted from generation to generation implies that the poor create their own problems through a self-perpetuating cycle of poverty ("like father, like son").

BLAMING SOCIETY: STRUCTURAL CHARACTERISTICS

In contrast to blaming the poor, most sociologists contend that macro-level structural factors create and sustain poverty. In a classic article, sociologist Herbert Gans (1971) maintained that inequality is functional because the poor:

▶ Ensure that society's dirty yet necessary work gets done (e.g., dishwashing and cleaning bedpans in hospitals).

▶ Subsidize the middle and upper classes by working for low wages.

▶ Buy goods and services that would otherwise be rejected (e.g., day-old bread, used cars, and the services of incompetent professionals).

feminization of poverty the disproportionate number of the poor who are women.

In 2013, almost 6.2 million Americans were homeless on a given night. Of all homeless people, 36 percent were families with children; 35 percent lived under bridges, in cars, or abandoned buildings; and 19 percent had jobs. The main causes of homelessness, in order or priority, are lack of affordable housing, poverty, and unemployment (Henry et al., 2013; United States Conference of Mayors, 2013).

▸ Absorb the costs of societal change and community growth (as when low-income people are pushed out of their homes by urban renewal and construction of expressways, parks, and stadiums).

Thus, according to Gans, poverty persists in the United States (and elsewhere) because many people benefit from the outcomes.

Which perspective is more accurate, blaming the poor or blaming society? Some people are poor because they're lazy and would rather get a handout than a job. However, much research shows that most people are poor because of economic factors (low wages, job loss, lack of affordable housing, inability to afford health insurance), which, in turn, can result in acute health and employment problems. Despite individual flaws, middle classes get government assistance through programs for the disabled, veterans, college students, older people, and the unemployed, and wealthy farmers receive generous federal subsidies (Paletta and Porter, 2013; Parramore, 2014; Rampell, 2014; "A Trillion in the Trough," 2014).

8-4 SOCIAL MOBILITY

Social mobility is movement up or down the social class hierarchy. The American dream is to move up the economic ladder, but that's not always the reality. In 2014, only 54 percent of Americans—down from 81 percent in 1998—believed that there's plenty of opportunity to get ahead, 59 percent said that it's almost impossible to achieve the American dream, and 26 percent expect to move down in social class over the next few years (Dugan and Newport, 2013; Morello and Clement, 2014). Are they right?

8-4a Types of Social Mobility

Intragenerational mobility is movement up or down the class hierarchy over one's lifetime. If Ashley begins as a nurse's assistant, becomes a registered nurse, and then a physician's assistant (PA), she experiences intragenerational mobility. **Intergenerational mobility** is movement up or down the class hierarchy relative to the position of one's parents. It's a change in social class across two or more generations. If Ashley's parents were blue-collar workers, her upward movement to the middle class is an example of intergenerational mobility. Intragenerational and intergenerational mobility can be downward or upward.

8-4b Recent Trends in Social Mobility

From the 1940s to the 1970s, U.S. upward mobility increased. Since then, many Americans have been less likely to rise to a higher social class. A child raised in a household in the poorest fifth of the income distribution has a 40 percent chance of being stuck at the bottom as an adult, and just a 9 percent chance of making it to the top 20 percent by adulthood. Those raised at the top of the 20 percent have a 37 percent of staying there, and only an 8 percent chance of tumbling to the bottom of the ladder (Winship, 2011; Reeves and Grannis, 2013; Chetty et al., 2014).

Between 1971 and 2011, there was a sharp decline in the share of middle-income adults (see *Figure 8.5*). Some have moved up the economic ladder, but many have been downwardly mobile. In 2014, for example, 40 percent of Americans said they were lower-middle or lower class, up from 25 percent in 2008 (Morin and Motel, 2012; Kochhar and Morin, 2014). Because of such changes, some researchers and analysts describe the middle class as "eroding," "sinking," "shrinking," and

social mobility movement up or down the social class hierarchy.

intragenerational mobility movement up or down the class hierarchy over one's lifetime.

intergenerational mobility movement up or down the class hierarchy relative to the position of one's parents.

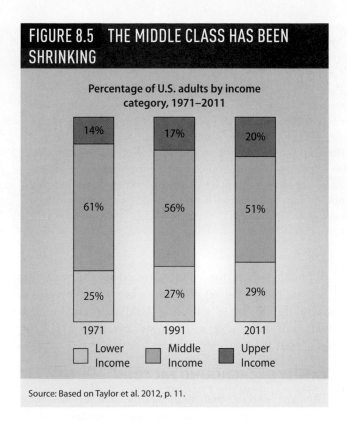

FIGURE 8.5 THE MIDDLE CLASS HAS BEEN SHRINKING

Percentage of U.S. adults by income category, 1971–2011

	1971	1991	2011
Upper Income	14%	17%	20%
Middle Income	61%	56%	51%
Lower Income	25%	27%	29%

Source: Based on Taylor et al. 2012, p. 11.

an "exclusive club" (see, for example, Galston, 2013; and Porter, 2013).

When social mobility occurs, most moves are short because social classes are pretty rigid (Winship, 2011; Wysong et al., 2014). A child from a working-class family, for example, is more likely to move up to the lower-middle class than to jump to the lower-upper class. The same is true of downward mobility: Someone from the middle class is more likely to slide into the working class than to drop to the underclass. Moreover, upward intergenerational mobility has become more limited because the rungs of the economic ladder have grown further apart (Chetty et al., 2014). As you saw earlier, income inequality has increased, making it more difficult for even upper-middle class people to move up.

A study of 22 high-income countries found that the United States ranked only 15th for upward intergenerational mobility (Corak, 2013). There's less upward mobility in the United States than in many other industrialized countries as well as some developing countries (e.g., Ecuador and Peru). It takes an average of six generations for American children to move up to the next fifth of the income distribution compared with three generations in Canada, Finland, Norway, and Denmark (Jäntti et al., 2006; Isaacs, 2008). In effect, countries with the greatest income inequality, such as the United States, are also the most likely to have limited upward intergenerational mobility (Greeley, 2013).

8-4c What Affects Social Mobility?

Upward intergenerational mobility doesn't always reflect people's talents, intelligence, or hard work. Instead, much social mobility depends on structural, demographic, and family background factors, all of which are interrelated.

STRUCTURAL FACTORS

Macro-level variables, over which people have little or no control, affect social mobility in several ways. First, *changes in the economy* spur upward or downward mobility. During an economic boom, the number of jobs increases, and many people have an opportunity to move up. During recessions, such as the Great Recession, long-term unemployment leads to downward mobility (Rose and Winship, 2009; see also Chapters 11 and 12).

Second, *government policies and programs* affect social mobility. Unlike the United States, European countries with the highest upward mobility rates have promoted greater equality. They have funded universal health care, which reduces the chance of people falling into poverty because of medical emergencies or poor health; have provided affordable housing; and have invested in technical schools where young people learn a high-paying trade (Foroohar, 2011; Deparle, 2012).

Third, *immigration* fuels upward mobility. Because many recent immigrants take low-paying jobs, groups that are already in a country move into higher-paying occupations (Haskins, 2008).

DEMOGRAPHIC FACTORS

Demographic factors also affect social mobility. Four of the most important are

J. K. Rowling worked as a teacher. After a divorce, she lived on public assistance, writing *Harry Potter and the Philosopher's Stone* during her daughter's naps. The book was published in 1997; by 2007, Rowling, a billionaire and one of the world's richest women, was considerably wealthier than the Queen of England.

AP Images/Kirsty Wigglesworth

Children raised in poor families can be upwardly mobile. Sonia Sotomayor, whose parents were born in Puerto Rico, is the first Latina to serve on the U.S. Supreme Court. Her family first lived in a South Bronx tenement and later a working-class housing project. Sotomayor's father, an alcoholic, had a third-grade education, didn't speak English, and was a tool and die worker. He died when Sotomayor was nine years old. Her mother—a telephone operator and later a practical nurse—stressed the value of education. Sotomayor excelled in school, received a full scholarship to Princeton University, and put in long hours to catch up with her peers' academic knowledge.

education, gender, race, and ethnicity, but geography also has an effect.

Education. Especially when the economy is slumping, people with college and graduate degrees fare better than those with a high school education or less. The latter often face long and frequent bouts of unemployment, must get by with temporary work, and may move down the socio-economic ladder because employment opportunities are greatest in high-skill and low-skill rather than middle-skill jobs (Steuerle et al., 2013; Kochhar and Morin, 2014).

Gender. Women's massive entry into the labor force since the 1980s has increased family income and many single women's upward mobility. In 1960, only 42 percent of women aged 25 to 54 were in the labor force compared with almost 75 percent in 2014 (U.S. Bureau of Labor Statistics, 2014). You'll see in later chapters that men's labor participation rates have been dropping, but men's mobility is less affected than women's by a divorce, nonmarital children, or widowhood.

Race and ethnicity. African Americans and Latinos, especially those from low-income families, usually experience little upward mobility. Black and Latino middle classes have grown since the 1970s, but both groups still lag significantly behind whites and Asian Americans in median family income (see Chapters 10 and 11).

Geography. Where one lives can stimulate or dampen upward mobility. For example, the probability of a child born into the poorest fifth of the population in San Jose, California, making it to the top fifth is 13 percent compared with only 4 percent for a child born in Charlotte, North Carolina. These disparities in upward mobility in different parts of the country are correlated with four factors: residential segregation (whether by income or race), the quality of schooling, how many children live with only one parent, and parents' social capital (Chetty et al., 2014).

FAMILY BACKGROUND FACTORS

You saw earlier that high-income households have more economic, cultural, and social capital than do low-income households. A study tracked 790 Baltimore children from the year they entered first grade until they turned 29 years old. The researchers found that the children's social class in adulthood was pretty much the same as the family into which they were born. For example, only 4 percent of the children from low-income families had a college degree by age 30 compared with 45 percent of those from higher-income backgrounds (Alexander et al., 2014).

Only 7 of the 44 U.S. presidents came from the lower-middle class or below. Abraham Lincoln, although born in a log cabin, had a father who was one of the wealthiest people in the community. And about a third of the students with low grades at Ivy League universities wouldn't be there if their parents weren't celebrities, well-known politicians, or others who donate millions to the school (see Chapter 11 and 13).

There are many advantages that high-income parents provide their children. Among young adults ages 19 to 22, 60 percent receive money from their parents. High- and low-income parents spend about 10 percent of their annual household income (almost $12,900 and $5,800, respectively) to help their adult children pay for bills, rent, tuition, and other expenses. However, the higher the parents' SES, the greater the financial assistance and children's opportunities to succeed (Wightman et al., 2012).

Our socialization affects what French sociologist Pierre Bourdieu (1984) called *habitus*—the habits of speech and lifestyle that determine where a person feels comfortable and knowledgeable. Upper-class parents cultivate flexibility, autonomy, and creativity because they expect their children to step into positions that require such characteristics. Poor and working-class parents stress obedience, honesty, and appearance—the marks of respectability that many employers expect. High-income parents have social connections to jobs and admission to particular schools or colleges. Because of such resources, children of top earners are likely to grow up to be top earners themselves (Khan, 2012; Corak, 2013).

8-5 GLOBAL STRATIFICATION

Global stratification refers to worldwide inequality patterns that result from differences in wealth, power, and prestige. All societies are stratified, but more than 75 percent of the world's population lives in countries where economic inequality has widened, increasing the possibility of social class tension and conflict (United Nations Development Program, 2013; World Economic Forum, 2013).

8-5a Wealth and Income Inequality

In 2014, global wealth reached a new all-time high of $263 trillion, but how is it distributed?

▶ The richest 1 percent of the world's adults own almost half of the world's wealth.

▶ The top 10 percent of the world population owns 86 percent of global wealth.

▶ The 85 richest people in the world have as much wealth as the 3.5 billion poorest (Fuentes-Nieva and Galasso, 2014; Shorrocks et al., 2014).

There are enormous wealth disparities across regions (see *Table 8.1*), and the inequalities have swelled. The incomes of the top 1 percent have increased 60 percent since 1990, and the income growth of the ultra-rich 0.01 percent has been even greater. Even in more economically egalitarian countries such as Norway and Sweden, the share of income going to the richest 1 percent increased by more than 50 percent between 1980 and 2012 (Oxfam, 2013; Fuentes-Nieva and Galasso, 2014).

Worldwide, income inequality is also pervasive. Across 131 countries, the richest 3 percent hold 20 percent of all income. The median per capita incomes in the wealthiest populations are more than 50 times those in the 10 poorest populations, all of which are in sub-Saharan Africa, but there's considerable economic stratification

TABLE 8.1 WEALTH PER ADULT, BY REGION AND SELECTED COUNTRIES, 2014

In 2013 U.S. Dollars	
World	**$56,016**
North America	$340,340
Europe	$145,977
Asia-Pacific	$44,715
Latin America	$22,997
China	$21,330
Africa	$5,080
India	$4,645

Source: Based on Shorrocks et al. 2014, Table 1.

within countries. In oil-rich Nigeria, for instance, the richest 6 percent hold 40 percent of the country's total income (Phelps and Crabtree, 2013b, 2014).

Moreover, living in a wealthy country doesn't mean that people enjoy income equality. Of 34 countries, the United States ranks 31st in income equality—far behind industrialized nations and just above some developing countries that include Turkey and Mexico (OECD, 2014a).

8-5b The Plight of Women and Children

Historically, currently, and across *all* nations, women and children experience the greatest inequality. An important measure of a country's health is its **infant mortality rate**, the number of babies under age 1 who die per 1,000 live births in a given year. The infant mortality rate ranges from a high of 109 in some low-income African nations to a low of 3 in affluent Western European countries (Haub and Kaneda, 2014).

Children's deaths have declined worldwide since 1990, but every year nearly 7 million poor children die before their fifth birthday; most of the deaths continue to be concentrated in sub-Saharan Africa and Southern Asia (e.g., Afghanistan, Bangladesh, Iran, Pakistan). Across and within countries, 87 percent of the under-five deaths are due to poverty (United Nations Development Program, 2013; You et al., 2013).

global stratification worldwide inequality patterns that result from differences in wealth, power, and prestige.

infant mortality rate the number of babies under age 1 who die per 1,000 live births in a given year.

There's a strong association between high infant and child mortality rates and women's low education levels. Investing in women's education reduces unwanted and unplanned births; lowers infant, child, and maternal mortality; increases women's labor force participation; and can help lift a family out of poverty (World Bank and Collins, 2013; World Economic Forum, 2013).

Education doesn't guarantee women's greater economic equality, however. In both rich and poor countries, women are less likely than men to be employed because of cultural attitudes about women's roles and child rearing responsibilities. When women work outside the home, the *gender pay gap*—the overall income difference between women and men in the workplace—ranges from a low of 3 percent in Slovenia to a high of 46 percent in Zambia (United Nations Development Program, 2013; Fuentes-Nieva and Galasso, 2014). And in all nations, women are paid less than men for similar work (see Chapters 9 and 12).

8-5c Other Consequences of Global Stratification

The consequences of global inequality are massive and profound. Worldwide, 22 percent of people (1.2 billion) live on $1.25 per day—the World Bank's international definition of *extreme poverty*—and 34 percent live on no more than $2.00 per day. The extreme poverty rate fell by 25 percent between 1981 and 2010—from 2.1 to 1.9 million. During the same period, however, the extreme poverty rate in low-income countries (where the annual per capita income is $1,035 or less) increased by 33 percent (from 249 to 352 million people). Across regions, 54 percent of the residents of sub-Saharan Africa and 38 percent of those in South Asia—compared with 1 percent in Canada, Europe, and the United States—live in extreme poverty (Olinto et al., 2013; Phelps and Crabtree, 2013a).

Because of severe destitution, almost one in eight people worldwide suffer from chronic hunger and undernutrition. In some impoverished regions, women and children must walk seven or more miles each day to fetch clean water. In Australia, Canada, and Germany—three of the most affluent countries—about 10 percent struggle to pay for food compared with 50 to 70 percent of those in poor African countries (Office of the United Nations . . . , 2010; FAO, IFAD, and WFP, 2013; Stokes, 2013a).

People in poor countries report high levels of unhappiness, physical and mental health problems, and have little chance to escape poverty. Economic inequality has other costs: The world's richest 1 percent use their wealth to get political favors, including bypassing environmental laws; have a global network of tax havens worth almost $19 trillion within their own countries and offshore; and harm democracy by reducing opportunities for upward mobility (Helliwell et al., 2013; Stevenson and Wolfers, 2013; Kirkland, 2014; Oxfam, 2014).

8-5d Why Is Inequality Universal?

Many theories try to explain why inequality is universal, but three of the most influential have been modernization theory, dependency theory, and world-system theory.

Modernization theory claims that low-income countries are poor because their leaders don't have the attitudes and values that lead to experimentation and using modern technology. Instead, policy makers adhere to traditional customs that isolate and prevent them from competing in a global economy. In effect, modernization theory blames poor nations for their poverty and other problems. After the key foundations of modernity and capitalism are in place, this perspective maintains, low-income countries will prosper.

Dependency theory contends that the main

We hear about the booming economies of India and China, but not everyone has benefited. A tent city in India houses many of the workers who earn about $1.30 a day building new office towers for the affluent (left). In China, a man in Shanghai begs as wealthier residents pass by (right).

© El Greco/Shutterstock.com

reason why low-income countries are poor is because they're pawns that high-income countries exploit and dominate. Rich nations wield an enormous amount of power by exporting jobs overseas, manipulating foreign aid, draining less powerful countries of their resources, penetrating other countries with multinational corporations, and coercing national governments to comply with their interests (e.g., by not passing environmental laws). In effect, according to dependency theorists, high-income countries benefit because the poor provide cheap labor and aren't powerful enough to protest even though they work in hazardous conditions and earn less than $1 a day.

More recently, *world-system theory*, similar to dependency theory, argues that "the economic realities of the world system help rich countries stay rich while poor countries stay poor" (Bradshaw and Wallace, 1996: 44). That is, countries such as the United States that dominate the world economy control the economies of low-income countries because their workers depend on external markets for jobs. High-income countries can extract raw materials (e.g., diamonds and oil) with little cost. They can also set prices, regardless of market values, for agricultural products that low-income countries export. Doing so forces many small farmers to abandon their fields because they can't pay for labor, fertilizer, and other costs (Carl, 2002; Alvaredo et al., 2013).

8-6 SOCIOLOGICAL EXPLANATIONS: WHY THERE ARE HAVES AND HAVE-NOTS

Why are societies stratified? *Table 8.2* summarizes the key points of the four theoretical perspectives. Let's begin by looking at a long-standing debate between functionalists and conflict theorists on why there are haves and have-nots.

8-6a Functionalist Perspectives: Stratification Benefits Society

Functionalists see stratification as both necessary and inevitable: Social class provides each individual a place in the social world, and motivates people to contribute to society. Without a system of unequal rewards, functionalists argue, many important jobs wouldn't be performed.

THE DAVIS–MOORE THESIS

Sociologists Kingsley Davis and Wilbert Moore (1945) developed one of the most influential functionalist perspectives on social stratification that persists today. The **Davis–Moore thesis**, as it's commonly called, asserts that social stratification benefits society. The key

Davis–Moore thesis the functionalist view that social stratification benefits a society.

TABLE 8.2 SOCIOLOGICAL EXPLANATIONS OF SOCIAL STRATIFICATION

Perspective	Level of Analysis	Key Points
Functionalist	Macro	• Fills social positions that are necessary for a society's survival • Motivates people to succeed and ensures that the most qualified people will fill the most important positions
Conflict	Macro	• Encourages workers' exploitation and promotes the interests of the rich and powerful • Ignores a wealth of talent among the poor
Feminist	Macro and micro	• Constructs numerous barriers in patriarchal societies that limit women's achieving wealth, status, and prestige • Requires most women, not men, to juggle domestic and employment responsibilities that impede upward mobility
Symbolic Interactionist	Micro	• Shapes stratification through socialization, everyday interaction, and group membership • Reflects social class identification through symbols, especially products that signify social status

arguments of the Davis–Moore thesis can be summarized as follows:

1. **Every society must fill a wide variety of positions and ensure that people accomplish important tasks.** Societies need teachers, doctors, farmers, trash collectors, plumbers, police officers, and so on.

2. **Some positions are more crucial than others for a society's survival.** Doctors, for example, provide more critical services to ensure a society's continuation than do lawyers, engineers, or bankers.

3. **The most qualified people must fill the most important positions.** Some jobs require more skill, training, or intelligence than others because they're more demanding, and it's more difficult to replace the workers. Pilots, for example, must have more years of training and aren't replaced as easily as flight attendants.

4. **Society must offer greater rewards to motivate the most qualified people to fill the most important positions.** People won't undergo many years of education or training unless they're rewarded by money, power, status, and/or prestige. If doctors and nurses earned the same salaries, there wouldn't be much incentive for people to spend so many years earning a medical degree.

According to the Davis–Moore thesis and other functionalist perspectives, then, stratification and inequality are necessary to motivate people to work hard and to succeed. For functionalists, social stratification is based on **meritocracy**—people's accomplishments. That is, people are rewarded for what they do and how well, rather than their ascribed status.

CRITICAL EVALUATION

Sociologist Melvin Tumin (1953) challenged the Davis–Moore thesis. First, he argued, societies don't always reward the positions that are the most important for the members' survival. If the multi-millionaire professional athletes, actors, and pop musicians went on strike, many of us would probably barely notice. If, on the other hand, those with much lower income—whether they're doctors, garbage collectors, teachers, truck drivers, or mail carriers—refused to work, society would grind to a halt. Thus, according to Tumin, there's little association between earnings and the jobs that keep a society going.

Singer Lady Gaga had an estimated net worth of $190 million in 2013. Do her earnings reflect her contribution to society, especially compared with teachers, physicians, trash collectors, computer scientists, and others?

© ChinellatoPhoto/Shutterstock.com

Second, Tumin claimed, Davis and Moore overlook the many ways that stratification limits upward mobility. Where wealth is differentially distributed, access to education, especially higher education, depends on family wealth. As a result, large segments of the population are likely to be deprived of the chance to discover what their talents are, and society loses.

Third, Tumin criticized Davis and Moore for ignoring the critical role of inheritance. In upper social classes, sons and daughters don't have to work because their inherited wealth guarantees a lifetime income and perpetuates privileges over generations.

Functionalist theories also don't explain why (1) intergenerational upward social mobility is more limited in the United States than in other industrialized (and even some developing) countries; (2) so many college graduates can find only low-paying jobs; and (3) racial/ethnic income and wealth gaps persist across all social classes (Kochhar et al., 2011; Legatum Prosperity Index, 2013; McKernan et al., 2013).

8-6b Conflict Perspectives: Stratification Harms Society

Like Tumin, conflict theorists maintain that social stratification is dysfunctional because it hurts individuals and societies. Karl Marx's (1934) analysis of social class and inequality has had a profound influence on modern sociology, especially conflict theory.

CAPITALISM BENEFITS THE RICH

Marx was aware that a diversity of classes can exist at any one time, but he predicted that capitalist societies would ultimately be reduced to two social classes: the

---

meritocracy a belief that social stratification is based on people's accomplishments.

capitalist class, or bourgeoisie; and the working class, or proletariat. The **bourgeoisie**, those who own and control capital and the means of production (e.g., factories, land, banks, and other sources of income), can amass wealth and power. The **proletariat**, workers who sell their labor for wages, earn barely enough to survive.

For conflict theorists, the economic struggles of the U.S. middle and working classes since the late 1970s weren't primarily the result of globalization and technological changes but, instead, a long series of government policies that overwhelmingly favored the rich (Hacker and Pierson, 2010; Buchheit, 2014a). This policy of **corporate welfare** consists of an array of subsidies, tax breaks, and assistance that the government has created for businesses. For example:

▶ Taxpayers are paying $12.2 trillion for the federal government's bailing out of mismanaged financial institutions. The executives of the companies still receive multimillion-dollar annual salaries and benefits ("Adding Up . . . ," 2011; Mider and Green, 2012; Sparshott, 2013).

▶ U.S. corporate income tax rates decreased from 53 percent in 1952 to 11 percent in 2010. In contrast, the tax burden of an average employee ranges from 33 to 41 percent. Companies benefit enormously from the taxpayer-supported highways they use, law enforcement and judicial systems that protect their intellectual and physical property, public education that provides a workforce, and military personnel who protect corporate assets abroad (Anderson et al., 2011; Buffet, 2011; see also Fitzgerald, 2013).

▶ Some companies pay no taxes at all. In 2010, General Electric—the nation's largest corporation—reported worldwide profits of $14.2 billion, received a federal tax benefit of $3.2 billion, paid its top CEO almost $12 million, and spent almost $42 million to lobby Congress for tax breaks (Lublin, 2010; Kocieniewski, 2011; see also Pinsky, 2014, for corporate welfare on the state level).

Considerable data support conflict theorists' contention that severe and growing inequality is dysfunctional for society. In the United States, for example:

▶ Since the end of the Great Recession in mid-2009, 95 percent of the economic gains have gone to the top 1 percent, and a mere 400 households have more money than do 60 percent of all households.

▶ In 2013, the total budget of major public assistance programs (e.g., Supplemental Security Income; Temporary Assistance for Needy Families; housing; food and nutrition services for women, infants, and children) was lower than the $300 billion earnings of the richest 400 Americans.

▶ The typical male earns less than he did 45 years ago (after adjusting for inflation) (Rank, 2013; Buchheit, 2014a; Edwards, 2014; Reich, 2014).

For conflict theorists, inequality undermines people's trust in political and economic institutions and, consequently, erodes national solidarity. Even powerful financiers and conservative economists have begun to worry that the increasing income and wealth gaps and low wages have made the United States less productive, have narrowed the consumer spending base, and have discouraged many people from entering or staying in the labor force (Lahart, 2013; Bruinius, 2014).

CRITICAL EVALUATION

Marxian and later conflict theories have their limitations. First, Marx predicted that as the numbers of oppressed workers increased, the proletariat would overthrow the bourgeoisie. Even though the concentration of corporate wealth has greatly increased during the past 100 years and many people have become poorer, there have been some protests but no revolutions in capitalist countries. Americans aren't demanding greater economic equality, and many go into debt to purchase luxury items such as TVs, jewelry, and pet supplies (Clifford and Martin, 2011). Thus, according to critics, conflict theorists are exaggerating the existence and effects of economic inequality.

Second, functionalists in particular contend that conflict theorists underrate people's ability to be upwardly mobile. If people really want to succeed, according to functionalists, they can do so by working hard and making sacrifices. Indeed, 81 percent of Americans are satisfied with the opportunities they have to succeed (Swift, 2013).

Third, some critics point out that conflict theorists ignore the fact that government programs have cut poverty by 40 percent since the 1960s. Moreover, 46 percent of American households pay no federal income taxes (and these tax breaks have nearly doubled since 1975) because they receive public assistance, get deductions for raising children under age 18 and/or paying for their education, don't report income, or have very low wages (Johnson et al., 2011; Grier, 2012; "A Smarter War on Poverty," 2014; Rosenberg, 2014).

bourgeoisie those who own and control capital and the means of production.

proletariat workers who sell their labor for wages.

corporate welfare subsidies, tax breaks, and assistance that the government has created for businesses.

8-6c Feminist Perspectives: Women Are Almost Always at the Bottom

For feminist scholars, functionalist and conflict theories are limited because they typically focus on men in describing and analyzing stratification and social class. As a result, women are largely invisible.

PATRIARCHY BENEFITS PRIMARILY MEN, NOT WOMEN

An estimated 70 percent of the world's poor are women. Often, in both rich and poor countries, gender inequality results in discrimination in the economy, politics, and access to medical services. Especially in low-income countries, almost 4 million girls and women are "missing" each year because of high death rates. About 20 percent are aborted because of a preference for sons, and more than a third die from complications of childbirth and pregnancy. Besides excessive deaths, most women are blocked from entering higher-paying occupations (e.g., finance and business), earn less than men in all jobs, have less access to land and credit, and have little decision-making power in households and politics (World Bank, 2012; Independent Television Service, 2014).

More than 100 countries ban women from certain jobs, accessing finance, owning businesses, or conducting legal affairs. And a "global pandemic" of violence against females affects one in three worldwide, including young girls who become sex slaves (UN Women, 2014).

Feminist theorists contend that, in a patriarchal system, men shape the stratification system because they control a disproportionate share of wealth, prestige, and power, and the feminization of poverty results in women's downward mobility. In addition, women often have to overcome economic inequities as well as juggle domestic and workplace responsibilities (see Chapters 11 and 12). The gender gaps in wealth, income, and household burdens make it harder for women to access resources and opportunities to build a strong future for themselves and their families.

CRITICAL EVALUATION

Some critics point out that feminist theorists often focus only on poor women in showing how patriarchy affects stratification and social class (see Kendall, 2002). Another criticism is that many feminist scholars don't explain why so many women succeed despite patriarchal barriers. Third, feminist theories don't account for some striking cross-cultural variations. In a study of 165 countries, for example, Canada, Australia, the Netherlands, and the United States ranked in the top ten regarding women's legal treatment, workplace participation, and access to education and health care, but well behind developing countries such as Cuba, Rwanda, and Burundi in women's representation in political offices (Ellison, 2011). Thus, feminist theories don't explain why women have greater political power in some patriarchal and developing countries than in many industrialized nations.

In 2013, more than 1,100 workers were killed and 2,000 were injured when an unsafe garment factory collapsed in Bangladesh. The country exports $20 billion in clothes annually to European and U.S. retailers, including Walmart. The $38-a-month minimum wage (for a six-day workweek) barely covers living expenses. More than 80 percent of the 4 million workers in Bangladesh's garment industry are women, mostly young and with little schooling (Al-Mahmood, 2013; Hammadi, 2013; Srivastava, 2013).

MUNIR UZ ZAMAN/Getty Images

© Gordana Sermek/Shutterstock.com

8-6d Symbolic Interaction Perspectives: People Create and Shape Stratification

Symbolic interactionists focus on how people reproduce social classes. They address micro-level issues such as how people learn their social positions in everyday life and how such learning affects their attitudes, behavior, and lifestyles.

SOCIAL CLASS AFFECTS PEOPLE'S ATTITUDES AND BEHAVIOR

People across social classes interact with and socialize their children differently. By age 2, children from higher-income homes know 30 percent more words than those from low-income homes. Early vocabularies increase reading comprehension and proficiency as early as kindergarten. Children as young as 4 years old are aware of and enact their social class. Among preschoolers, upper-middle-class children speak, interrupt, ask for help, and argue more often than do working-class children. Such behavior receives more adult attention and gives upper-middle-class children more opportunities to develop their language skills (Streib, 2011; Fernald et al., 2013).

Social contexts also affect mobility. The social classes into which children are born affect their aspirations, the skills they value and to which they have access, and the networks and resources on which they can draw. Doctors' children, for example, are more likely to become physicians themselves because they're exposed to medical discussions at home, are encouraged to pursue medicine as an occupation, and are embedded in social networks that provide them with information about how to become physicians (Jonsson et al., 2009).

Few people are demanding greater economic equality because social stratification is internalized. About 65 percent of Americans say that the U.S. economic system favors the wealthy, that the income gap between the rich and poor has widened since the early 2000s, and that not everyone has an equal chance to succeed. However, 43 percent believe that the rich are more hardworking and intelligent than the average person, only 47 percent think that the rich–poor gap is a big problem, and in both 1990 and 2012, 63 percent said that the country

benefits from having a rich class of people (Newport, 2012; Drake, 2013; Stokes, 2013b; DeSilver, 2014). Such attitudes may be reflecting many Americans' hopes that they, too, will be wealthy someday, or many Americans don't see the connection between structural economic inequality and their own financial situation.

CRITICAL EVALUATION

Symbolic interaction theories are important in understanding the everyday processes that underlie social stratification, but there are several weaknesses. First, the theories don't explain why—despite the same family background, resources, and socialization—some siblings are considerably more upwardly mobile than their brothers and sisters. Second, conflict theorists, especially, fault symbolic interactionists for ignoring structural factors—such as the economy, corporate welfare, and inherited wealth—that create and reinforce inequality. Between 2000 and 2011, for example, economic factors accounted for nearly half the increase in the number of people receiving food stamps (Ziliak, 2013).

STUDY TOOLS 8

READY TO STUDY? IN THE BOOK, YOU CAN:

☐ Check your understanding of what you've read with the Test Your Learning Questions provided on the chapter review card at the back of the book.

☐ Rip out the chapter review card for a handy summary of the chapter and key terms.

ONLINE AT CENGAGEBRAIN.COM YOU CAN:

☐ Prepare for tests with quizzes.

☐ Review the key terms with Flash Cards.

☐ Play games to master concepts.

© Anelina/Shutterstock.com

9 | Gender and Sexuality

After you finish this chapter go to **PAGE 177** for **STUDY TOOLS.**

LEARNING OBJECTIVES

9-1 Differentiate between sex and gender and describe societal reactions to LGBTs.

9-2 Explain how gender stratification and gendered institutions affect the family, education, workplace, and politics.

9-3 Describe contemporary sexual attitudes and practices, including sexual scripts and double standards.

9-4 Describe abortion and same-sex marriage trends and explain why both issues are controversial.

9-5 Describe and illustrate gender and sexual inequality across cultures.

9-6 Compare and evaluate the theoretical explanations of gender and sexuality.

For the average American male, compared with his female counterpart, three pairs of shoes are usually enough, he can shower and be ready in 20 minutes, his underwear is $10 for a three pack, and he rarely encounters a line in public restrooms. Do such descriptions stereotype women and men? Or contain a kernel of truth? This chapter examines how gender and sexuality shape our lives. First, however, take the short quiz on the next page to see how much you know about these topics.

9-1 SEX, GENDER, AND CULTURE

Some differences between women and men are biological; others are social creations. Both biology and social factors shape a person's identity, but sex and gender aren't synonymous.

9-1a How Sex and Gender Differ

Many people use *sex* and *gender* interchangeably, but they have different meanings. **Sex** refers to the biological characteristics with which we are born—chromosomes,

What do you think?

Today, women have more employment advantages than men.

1	2	3	4	5	6	7

strongly agree   strongly disagree

TRUEORFALSE?

How Much Do You Know About Gender and Sexuality?

1. About 10 percent of the U.S. population is gay.

2. If Americans could have only one child, a majority would prefer a boy.

3. More men than women use Twitter.

4. Americans would prefer to work for a male than a female boss.

5. U.S. abortion rates have declined since 1990.

6. Almost a third of Americans who are married or in a committed relationship met online.

The answers are on p. 161.

anatomy, hormones, and other physical and physiological attributes. These attributes influence our behavior (e.g., shaving beards, wearing bras), but *don't determine* how

we think or feel. Whether we see ourselves and others as feminine or masculine depends on gender, a more complex concept than sex.

Gender refers to learned attitudes and behaviors that characterize women and men. Gender is based on social and cultural expectations rather than on physical traits. Thus, most people are *born* either male or female, but we *learn* to be women or men because we internalize behavior patterns expected of each sex. In many societies, for example, women are expected to look young, thin, and attractive, and men are expected to amass as much wealth as possible.

9-1b Sex: Our Biological Component

Physical characteristics such as breasts and beards indicate whether someone is a male or female, but sex

sex the biological characteristics with which we are born.

gender learned attitudes and behaviors that characterize women and men.

isn't always clear-cut. Our cultural expectations dictate that we are female or male, but a number of people are "living on the boundaries of both sexes" (Lorber and Moore, 2007: 141). For example, **intersexuals** (this term has replaced *hermaphrodites*) are people whose medical classification at birth isn't clearly either male or female. About 1 in 2,000 to 4,000 children born each year are classified as intersex because they're born with both male and female external genitals or an incomplete development of internal reproductive organs. Australia was the first country, in 2011, to allow parents of intersex children and adults to choose "male," female," or "X" on birth certificates and other official documents such as passports. Germany followed suit by allowing parents to leave the sex category blank on birth certificates. Some parents seek surgery; others wait until a child is old enough to decide what to do (Bendavid, 2013; Bennett-Smith, 2013).

SEXUAL IDENTITY AND SEXUAL ORIENTATION

Our **sexual identity** is our awareness of ourselves as male or female and the ways that we express our sexual values, attitudes, feelings, and beliefs. Our sexual identity incorporates a **sexual orientation**—a preference for sexual partners of the same sex, of the opposite sex, of both sexes, or neither sex:

▶ **Homosexuals** (from the Greek root *homo*, meaning "same") are sexually attracted to people of the same sex. Male homosexuals prefer to be called *gay*, female homosexuals are called *lesbians*, and both gay men and lesbians are often referred to, collectively, as *gays*. *Coming out* is a person's public announcement of a gay or lesbian sexual orientation.

▶ **Heterosexuals**, often called *straight*, are attracted to people of the opposite sex.

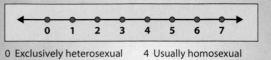

FIGURE 9.1 SEXUAL ORIENTATION CONTINUUM

0   1   2   3   4   5   6   7

0 Exclusively heterosexual
1 Predominantly heterosexual
2 Usually heterosexual
3 Bisexual
4 Usually homosexual
5 Predominantly homosexual
6 Exclusively homosexual
7 Asexual

Sources: Kinsey et al. 1948, p. 638; and Kinsey Institute 2011.

▶ **Bisexuals**, sometimes called *bis*, are attracted to both sexes.

▶ **Asexuals** lack any interest in or desire for sex.

Sexual orientation, like biological sex, isn't as clear-cut as many people believe. Alfred Kinsey (1948) and his associates' classic study found that most people weren't exclusively heterosexual or homosexual. Instead, they fell somewhere along a continuum in terms of sexual desire, attractions, feelings, fantasies, and experiences. Researchers have recently added *asexual* to Kinsey's classification (see *Figure 9.1*).

Most people's sexual identity corresponds with their biological sex, sexual attraction, and sexual behavior, but not always. Among Americans ages 15 to 44, for example, 99 percent report being *sexually attracted* only to the opposite sex, but 14 percent also *identify themselves* as gay or bisexual, and 18 percent have had *same-sex experiences* (Chandra et al., 2011).

PREVALENCE OF HOMOSEXUALITY

How many Americans are gay men, lesbians, and bisexuals? No one knows for sure because, among other reasons, researchers define and measure sexual orientation differently. For example, are people gay if they have ever engaged in same-sex behavior? What about heterosexuals who are attracted to people of the same sex? Moreover, respondents may not be willing to disclose this information, and people who engage in same-sex behavior often identify themselves as straight (Gates, 2011).

Researchers measure the prevalence of homosexuality by simply asking people to identify their sexual orientation. Americans estimate that 25 percent of the population is gay (Morales, 2011; Taylor et al., 2013). In fact, only about 3.5 percent of Americans identify themselves as lesbian, gay, and/or bisexual (Gates and Newport, 2012). This population is split nearly evenly between lesbians/gays and bisexuals, and women are more likely than men to be bisexual (see *Figure 9.2*).

intersexuals people whose medical classification at birth isn't clearly either male or female.

sexual identity our awareness of ourselves as male or female and the ways that we express our sexual values, attitudes, feelings, and beliefs.

sexual orientation a preference for sexual partners of the same sex, of the opposite sex, of both sexes, or neither sex.

homosexuals those who are sexually attracted to people of the same sex.

heterosexuals those who are sexually attracted to people of the opposite sex.

bisexuals those who are sexually attracted to both sexes.

asexuals those who lack any interest in or desire for sex.

FIGURE 9.2 ESTIMATED NUMBER AND PERCENT OF U.S. ADULTS WHO IDENTIFY AS LESBIAN, GAY, AND BISEXUAL

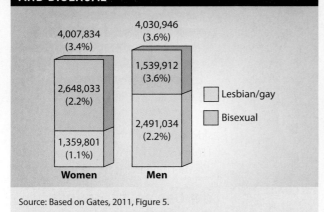

Source: Based on Gates, 2011, Figure 5.

WHAT DETERMINES OUR SEXUAL ORIENTATION?

A Hong Kong billionaire offered $65 million to any man who succeeded in marrying his daughter after she eloped with her female partner. The offer attracted 20,000 suitors, but all were unsuccessful (Nichols, 2014). Like this father, 36 percent of Americans (but down from 42 percent in 2003) believe that gays can become straight through prayer, therapy, and simply changing their lifestyles because sexual orientation is a "personal preference" (Kohut et al., 2012; Bennett-Smith, 2013).

Culture shapes people's sexual attitudes and behavior, but no one knows why we're straight, gay, bisexual, or asexual. Sexual orientation must have biological roots, according to some researchers, because homosexuality exists in all societies and, across cultures, the gay population is roughly the same—about 5 percent (Barash, 2012).

There's also growing scientific consensus that biological factors, particularly the early influence of sex hormones after conception and around childbirth, have a strong effect on sexual orientation. For example, pre-gay boys and pre-lesbian girls are gender-nonconforming at an early age compared with their heterosexual counterparts in terms of play and choice of clothes (see LeVay, 2011, for a comprehensive summary of these and other studies). Other researchers speculate that a combination of genetic and cultural factors influence our sexual orientation (Slater, 2013).

9-1c Gender: Our Cultural Component

Gender doesn't occur naturally, but is socially constructed. This means that gender aspects may differ across time,

TRUEORFALSE?

How Much Do You Know About Gender And Sexuality?

Answers to the quiz:

1. False. See p. 160.

2. False. In a recent Gallup poll, 40 percent of men compared with 28 percent of women said that if they could have only one child, they'd prefer a boy (Newport, 2011).

3. False. Among all online U.S. adults, 18 percent of women and 17 percent of men use Twitter (Duggan and Smith, 2013).

4. True. About 46 percent have no preference, but of the rest, 33 percent prefer a male boss to a female boss (20 percent) (Riffkin, 2014).

5. True. See p. 169.

6. False. Only 5 percent met online. The vast majority of relationships begin offline (Smith, 2014).

cultures, and even groups within a society. Let's begin with gender identity.

GENDER IDENTITY

People develop a **gender identity**, a perception of themselves as either masculine or feminine, early in life. Many Mexican baby girls but not boys have pierced ears, for example, and hairstyles and clothing for American toddlers differ by sex. Gender identity, which typically corresponds to a person's biological sex, is part of our self-concept and usually remains relatively fixed throughout life.

Transgender is an umbrella term for people whose gender identity and behavior differ from the sex to which they were assigned at birth. They comprise about 0.3 percent of the U.S. population (Gates, 2011).

Because transgender is independent of sexual orientation, people may identify as heterosexual, gay, bisexual, or asexual. Facebook users can now choose their gender identity from more than 50 possibilities, but here are

gender identity a perception of oneself as either masculine or feminine.

transgender an umbrella term for people whose gender identity and behavior differ from the sex to which they were assigned at birth.

some of the most common transgender categories (American Psychological Association, 2011):

▶ *Transsexuals* are people whose gender identity differs from their assigned sex. Some, but not all, undergo hormone treatment or surgery to change their physical sex to resemble their gender identity.

▶ *Cross-dressers* wear clothing that is traditionally or stereotypically worn by another gender in their culture. People who cross-dress are usually comfortable with their assigned sex and don't wish to change it.

▶ *Genderqueer* are people who identity their gender as falling outside the two-gender (male/female) system. They may define their gender as falling somewhere on a continuum between female and male, or a combination of gender identities and sexual orientations.

Lesbians, gay men, bisexuals, and transgender people (LGBT) are becoming increasingly more visible, but, as you'll see shortly, their acceptance is mixed.

Gender expression is the way a person communicates gender identity to others through behavior, clothing, hairstyles, voice, or body characteristics. Cross-dressing, girls' frilly dresses, and men's business suits are all examples of gender expression. Even if a person's gender identity is constant, gender expression can vary from situation to situation and change over time. For example, men's cosmetic surgery procedures increased over 273 percent between 1997 and 2013; half of American men over 18 now routinely use moisturizers, facial creams, or self-tanning lotions and sprays; and some of the National Basketball Association's "toughest players" have promoted products for Dove, La Mer, and other

Tibrina Hobson/Getty Images

After a sex-change operation in 2009, Chastity Bono—the daughter of singers Sonny and Cher—became Chaz, a male. His debut on *Dancing with the Stars*, a popular television program, was controversial because many people objected to including a transsexual contestant.

skin-care companies (Boyle, 2013; Holmes, 2013; American Society for Aesthetic Plastic Surgery, 2014).

GENDER ROLES, GENDER STEREOTYPES, AND SEXISM

Gender roles are the characteristics, attitudes, feelings, and behaviors that society expects of females and males. As you saw in Chapter 4, one of the purposes of socialization, which begins at birth, is to teach people appropriate gender roles. As a result, we learn to become male or female through interactions with family members, teachers, friends, and the larger society.

Americans are more likely now than in the past to pursue jobs and other activities based on their ability and interests rather than their sex. For the most part, however, our society still has fairly rigid gender roles and widespread **gender stereotypes**—expectations about how people will look, act, think, and feel because of their sex. We tend to associate stereotypically female characteristics with weakness and stereotypically male characteristics with strength. Consider, for example, how often we describe the same behavior differently for women and men:

▶ He's firm; she's stubborn.

▶ He's careful about details; she's picky.

▶ He's honest; she's opinionated.

▶ He's raising good points; she's "bitching."

▶ He's a man of the world; she's "been around."

Gender stereotypes fuel **sexism**, an attitude or behavior that discriminates against one sex, usually females, based on the assumed superiority of the other sex. In the late 1990s, and after receiving numerous rejections, a publisher finally accepted J. K. Rowling's manuscript of her Harry Potter book. Rowling followed the publisher's advice to sell the book under her initials, not her first name, Joanne. Even today, particularly for new science fiction and mystery authors, publishers instruct women to use

gender expression the way a person communicates gender identity to others through behavior, clothing, hairstyles, voice, or body characteristics.

gender roles the characteristics, attitudes, feelings, and behaviors that society expects of females and males.

gender stereotypes expectations about how people will look, act, think, and feel based on their sex.

sexism an attitude or behavior that discriminates against one sex, usually females, based on the assumed superiority of the other sex.

At an early age, sex-appropriate activities prepare girls and boys for future adult roles. As a result, there are few male ballet dancers and female auto mechanics.

Belinda Images/Belinda Images

male pseudonyms because "men prefer books written by men" (Cohen, 2012: D9).

Men also experience sexism. Here's what one of my students wrote during an online discussion of gender roles:

> Some parents live their dreams through their sons by forcing them to be in sports. I disagree with this but want my [9-year-old] to be "all boy." He's the worst player on the basketball team at school and wanted to take dance lessons, including ballet. I assured him that this was not going to happen. I'm going to enroll him in soccer and see if he does better.

Is this mother suppressing her son's natural dancing talent? We'll never know because she, like many parents, expects her son to fulfill sexist gender roles that meet with society's approval.

9-1d Societal Reactions to LGBTs

People's attitudes toward LGBTs are mixed. There's been greater acceptance in some countries, but considerable repression in others.

GREATER ACCEPTANCE

Australian passports and birth certificates now designate male, female, and transgender. In India, the 2011 national census for the first time offered three options: male, female, or a "third sex" that includes LGBTs (Cohn, 2011). In Thailand, which has the world's biggest transsexual population, an airline has recently recruited "third sex" flight attendants (Mutzabaugh, 2011).

In the United States, 92 percent of LGBT adults say society is more accepting of them than 10 years ago (Taylor et al., 2013). Numerous municipal jurisdictions, corporations, and small companies now extend more health care and other benefits to gay and lesbian employees and their partners than to unmarried heterosexuals who live together. A growing number of states and the U.S. Supreme Court have legalized same-sex marriages, federal courts are ruling that state laws limiting marriage to opposite-sex couples now violate the Constitution, and large numbers of Americans support policies that give lesbians and gays equal rights in the workplace and elsewhere (Cohen, 2013; Von Drehle, 2014).

There are many other indicators of greater LGBT acceptance. In 2012, the Army promoted the first openly gay female officer to brigadier general. In 2013, the Pentagon added benefits for same-sex partners, including services on U.S. military bases and pay if one is taken prisoner or missing in action (Wald, 2012; Cloud, 2013). Federal workers are now eligible for sex-change operations through some of the government's employee health-insurance plans (Mach and Cornell, 2013). And, since the mid-1990s, a number of LGBT characters have appeared in leading and supporting roles on prime-time TV in popular programs such as *The Simpsons*, *The Good Wife*, *Modern Family*, *Two and a Half Men*, and *American Dad* (GLAAD, 2013).

GREATER INTOLERANCE

According to one scholar, what makes gay people different from others is that "we are discriminated against, mistreated, [and] regarded as sick or perverted" (Halperin, 2012: B17). **Heterosexism**, a belief that heterosexuality is the only legitimate sexual orientation, pervades societal practices, laws, and institutions. Heterosexism can trigger **homophobia**, a fear and hatred of lesbians and gays.

Jonathan Daniel/Getty Images

In mid-2013, the Washington Wizards' Jason Collins appeared on the cover of *Sports Illustrated*. "I'm a 34-year-old NBA center, I'm black, and I'm gay," he announced (Collins and Lidz, 2013: 34). Why do you think his coming out received national coverage?

heterosexism belief that heterosexuality is the only legitimate sexual orientation.

homophobia a fear and hatred of lesbians and gays.

Homophobia often takes the form of *gay bashing:* threats, assaults, or acts of violence directed at LGBTs. Of the nearly 6,000 hate crimes reported to U.S. law enforcement agencies in 2012, almost 20 percent of the victims were LGBT, but much gay bashing isn't reported (Federal Bureau of Investigation, 2013; Wilson, 2014). Among transgender people, 41 percent have attempted suicide sometime in their lives—nearly nine times the national average—due primarily to rejection by family and friends; harassment, discrimination, and violence at school, work, and by law enforcement officers; and health care providers' refusal of treatment (Haas et al., 2014).

9-2 CONTEMPORARY GENDER STRATIFICATION AND INEQUALITY

A recent study concluded that "it will take until 2085 for women to reach parity with men in leadership roles in government/politics, business, entrepreneurship, and nonprofit organizations" (Klos, 2013). Why? From a sociological perspective, there's still widespread **gender stratification**—people's unequal access to wealth, power, status, prestige, opportunity, and other valued resources because of their sex. *Gendered institutions* are social structures that enable gender stratification. Let's look briefly at how gendered institutions perpetuate inequality in the family, education, the workplace, and politics, remembering that institutions intersect and affect each other (see Chapter 6).

9-2a Gender, Family Life, and Work Conflicts

Among parents with children under age 18 in the home, fathers spend more hours each week in paid work than do mothers, do less child care and housework, and have 3 hours per week more leisure time (see *Figure 9.3*). On average, fathers spend 4 hours per week versus mothers' 1 hour doing household repairs and maintenance (e.g., repairing cars, lawn care). Mothers, on the other hand, do much more cooking and cleaning than fathers (15 and 5 hours per week, respectively) (Wang, 2013).

gender stratification people's unequal access to wealth, power, status, prestige, and other valued resources because of their sex.

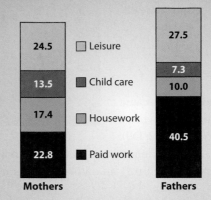

FIGURE 9.3 HOW EMPLOYED MOMS AND DADS SPEND THEIR TIME

Average number of hours parents with children under age 18 spend each week on...

Mothers: Leisure 24.5; Child care 13.5; Housework 17.4; Paid work 22.8

Fathers: Leisure 27.5; Child care 7.3; Housework 10.0; Paid work 40.5

Source: Wang, 2013.

Employed mothers spend from 33 to 55 percent more time than men do on child care and household tasks, but there's been a gradual shift in gender roles. Since 1965, fathers have spent less time in paid work and more time on child care and housework; mothers have spent more time in paid work and less time on housework but not child care (Parker and Wang, 2013; U.S. Bureau of Labor Statistics American Time Use Survey, 2013).

Family–work conflicts are also gendered. Among working parents with children under age 18, 56 percent of mothers and 50 percent of fathers say it's difficult to juggle work and family responsibilities. Fathers (46 percent) are much more likely than mothers (23 percent) to worry about not spending enough time with their children. Fathers' concerns aren't surprising because they devote half as much time as mothers do to child care (Parker and Wang, 2013; see also *Figure 9.3*).

9-2b Gender and Education

U.S. girls and women have made substantial educational progress in the last few decades. Despite the headway, there are gender differences at all educational levels. In public K–12 schools, as rank and pay increase, the number of women decreases. Among all full-time teachers, 89 percent of the teachers at the elementary level are women; the number falls to 58 percent in high school. Among principals, the number of women drops from 64 percent at elementary schools to 30 percent in high schools (Bitterman et al., 2013; Goldring et al., 2013).

TABLE 9.1 AS RANK INCREASES, THE NUMBER OF FEMALE FACULTY DECREASES

Rank	Percentage of Female Faculty Members
Professor	29
Associate Professor	42
Assistant Professor	49
Instructor	56

Note: Of the almost 762,000 full-time faculty in 2011, 44 percent were women.

Source: Based on Snyder and Dillow, 2013, Table 291.

Because women across all racial and ethnic groups are more likely than men to finish college, some observers have described this phenomenon as "the feminization of higher education." Women have sailed past men in earning associate's, bachelor's, and master's degrees, but earn only 37 percent of professional and doctoral degrees (see Chapter 13). Such data contradict the description of higher education as feminized.

Even when women earn doctoral degrees in male-dominated STEM (science, technology, engineering, and math) fields, they're less likely than men to be hired (Milan, 2012). Once hired, women faculty are less likely to be promoted. Since 2000, 45 percent of all Ph.D. degree recipients have been women (Snyder and Dillow, 2013), but as the academic rank increases, the number of female faculty decreases (see *Table 9.1*).

9-2c Gender and the Workplace

There has been progress toward greater workplace equality, but we still have a long way to go. In the United States (as around the world), many jobs are segregated by sex, there are ongoing gender pay gaps, and numerous women experience sexual harassment and bullying.

OCCUPATIONAL SEX SEGREGATION

Occupational sex segregation (sometimes called *occupational gender segregation*) is the process of channeling women and men into different types of jobs. As a result, a number of U.S. occupations are filled almost entirely by either women or men. For example, between 92 and 98 percent of all registered nurses, child care workers, secretaries, dental hygienists, and preschool and kindergarten teachers are women. From 96 to 99 percent of all pilots, mechanics, plumbers, and loading machine operators are men. Women have made progress in the higher-paying professional occupations that require at least a college degree, but 76 percent of doctors and dentists and 91 percent of engineers are men (BLS

Of the almost 2.9 million American teachers at elementary and middle schools, only 19 percent are men. A major reason is low salaries compared with other occupations, but gender stereotypes are also a factor (Bonner, 2007; BLS Reports, 2014).

Reports, 2014). The issue isn't women and men working in different spaces or locations, but that male-dominated occupations usually pay higher wages. And, as in education, women are much less likely than men to move up the occupational ladder (see Chapters 8 and 11).

THE GENDER PAY GAP

In 1980, women who worked full-time year-round earned 60 percent of what men earned. In late 2014, women earned 82 cents for every dollar men earned (BLS News Release, 2014), and it took 34 years to decrease the disparity by only 22 cents. To state this differently, *the average woman must work almost nine extra weeks every year to make the same wages as a man.*

This overall income difference between women and men is the **gender pay gap** (also called the *wage gap, pay gap,* and *gender wage gap*). The average woman who works full-time year-round for 47 years loses a significant amount of money because of the gender pay gap: $700,000 for high school graduates, $1.2 million for college graduates, and more than $2 million for women with a professional degree (e.g., medicine, law). Lower wages and salaries reduce women's savings, purchasing power, and quality of life, and bring them lower Social Security benefits after retirement (Soguel, 2009; Glynn and Powers, 2012).

occupational sex segregation (sometimes called *occupational gender segregation*) the process of channeling women and men into different types of jobs.

gender pay gap the overall income difference between women and men in the workplace (also called the *wage gap, pay gap,* and *gender wage gap*).

FIGURE 9.4 GENDER PAY GAP, BY EDUCATION

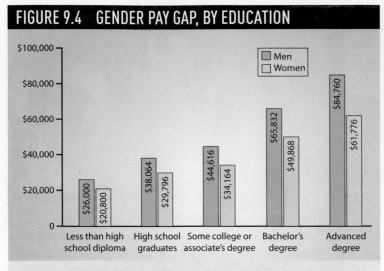

Note: These are the median annual earnings of year-round full-time workers age 25 and older in late 2013.

Source: Based on BLS News Release, "Usual Weekly Earnings…," 2014, Table 9.

Not only do women earn less than men at all educational levels, the gender pay gap widens as the level of educational attainment increases (see *Figure 9.4*). Female high school graduates earn 78 percent of what their male counterparts earn, but women's earnings drop to 73 percent of men's for those with advanced degrees. (Note that, as a group, women with advanced degrees earn less than men with a college degree.)

Why is there a gender pay gap? Women tend to choose fields with lower earnings (such as health care and education), whereas men are more likely to major in higher paying fields (such as engineering and computer science). Women on average also work fewer hours than men, usually to care for children or other family members. However, there's a 7 percent pay gap after accounting for a number of variables—including college major, occupation, hours worked, GPA, age, geographical region, and marital status—and the pay gap increases to 12 percent 10 years after college graduation (Corbett and Hill, 2012; AAUW, 2013; Tharenou, 2013; Frueh, 2014). Even when women choose high-paying occupations, they earn less than their male counterparts, and the wage gaps are greater in high-income than in low-income jobs (see Chapter 13).

SEXUAL HARASSMENT

Sexual harassment is any unwanted sexual advance, request for sexual favors, or other conduct of a sexual

sexual harassment any unwanted sexual advance, request for sexual favors, or other conduct of a sexual nature that makes a person uncomfortable and interferes with her or his work.

nature that makes a person uncomfortable and interferes with her or his work. It includes *verbal behavior* (e.g., pressure for dates, demands for sexual favors in return for hiring and promotion, the threat of rape), *nonverbal behavior* (e.g., indecent gestures, display of sexually explicit posters, photos, or drawings), and *physical contact* (e.g., pinching, touching, or rape). Between 1997 and 2013, the Equal Employment Opportunity Commission (EEOC) received almost 221,000 formal sexual harassment complaints, 83 percent of them from female employees (Equal Employment Opportunity Commission, 2011, 2014). Lawyers say the statistics would be much higher, but many companies now require new employees to agree to arbitrate complaints, including sexual harassment, as a condition of being hired (Green, 2011).

Nearly two-thirds of Americans say that sexual harassment is a serious problem in this country. About 25 percent of women have been sexually harassed at work. Only 40 percent of them reported the incident, however, because they believed that doing so wouldn't do any good or feared being fired or demoted (Clement, 2011).

9-2d Gender and Politics

Unlike a number of other countries (including Great Britain, Germany, India, Israel, Pakistan, Argentina, Chile, and Philippines), the United States has never had a woman serving as president or even vice president. In the U.S. Congress, 82 percent of the members are men. In several other important elective offices (governor, mayor, state legislator), only a handful of the decision makers are women (see *Table 9.2*). This number hasn't changed much since the early 1990s.

Women's voting rates in the United States have been higher than men's since 1984 (see Chapter 11). Why, in contrast, are there so few women in political office? There's a combination of reasons: (1) Women run for office at a far lower rate than men with similar credentials because women don't consider themselves qualified; (2) until very recently, female candidates have lagged behind their male counterparts in fund-raising, largely because donors—especially men, business groups, and local boards—believe that women can't win; (3) U.S. presidents appoint more men than women to important positions; and (4) there's a lingering sexism, among both men and women, that female politicians are both less feminine and compassionate than the average woman, and lack the leadership

TABLE 9.2 U.S. WOMEN IN ELECTIVE OFFICES, 2014

Political Office	Total Number of Office Holders	Percentage Who Are Women
Senate	100	20%
House of Representatives	435	18
Governor	50	10
State Legislator	7,383	24
Attorney General	50	16
Secretary of State	50	22
State Treasurer	50	14
State Comptroller	50	6
Mayor (100 largest cities)	100	13

Source: Based on material at the Center for American Women and Politics, 2014.

traits associated with male politicians (e.g., confident, assertive) (Luling, 2013; Steinhauer, 2013; Schneider and Bos, 2014).

9-3 SEXUALITY

In the movie *Annie Hall*, a therapist asks two lovers how often they have sex. The man rolls his eyes, and complains, "Hardly ever, maybe three times a week!" The woman exclaims, "Constantly, three times a week!" *Sexuality* is considerably more complex than just having sex, however, because it's a product of our sexual identity, sexual orientation and sexual scripts, and includes desire, expression, and behavior.

9-3a Contemporary Sexual Attitudes and Practices

Sex doesn't "just happen." It typically progresses through a series of stages such as approaching, flirting, touching, or asking directly for sex. Sexual attitudes and behavior can vary from situation to situation and change over time, including why we have sex.

WHY WE HAVE SEX

People have sex to reproduce and to experience physical pleasure, but there are other reasons. The majority of teenagers ages 15 to 19 have their first sexual intercourse with someone that they describe as "going steady," but the percentage is higher for females (70 percent) than males (56 percent). Nationwide, 5 percent of male and 12 percent of female high school students have been physically forced to have unwanted sexual intercourse. Teenagers are also more likely to engage in sex at any early age if they use alcohol or other drugs or experience domestic violence (Martinez et al., 2011; Eaton et al., 2012). A study of nearly 2,000 college students identified 237 reasons for having sex that ranged from the physical (stress reduction) to the spiritual (to get closer to God) and from the altruistic (to make the other person feel good) to the spiteful (to retaliate against a partner who had cheated) (Meston and Buss, 2007).

SEXUALITY THROUGHOUT THE LIFE COURSE

Contrary to some stereotypes, adolescents aren't sexually promiscuous and older people aren't asexual. Among teenagers aged 15 to 19, the percentage who ever had sexual intercourse declined from 51 percent in 1988 to 43 percent in 2010. By 2008, however, almost half of teens in this age group had had oral but not vaginal sex (Chandra et al., 2011). Adolescents who have oral sex prior to vaginal intercourse may be doing so because they view oral sex as "not really sex." Rather, they see it as a way to delay vaginal intercourse, to maintain one's virginity (especially among those who are religious), and to avoid the risk of pregnancy and STDs (Regnerus and Uecker, 2011; Copen et al., 2012).

By age 44, 98 percent of Americans have had vaginal intercourse; by the same age, 6 percent of men and 12 percent of women have had oral, anal, or vaginal contact with a same-sex partner. Fewer than 4 percent of Americans identify themselves as LGBT, but 15 percent of women and 5 percent of men have been sexually attracted to the same sex. Among those aged 15 to 21 who identify themselves as straight, 11 percent of females and 4 percent of males have had same-sex sexual contact (Chandra et al.,

Alex Gregory/The New Yorker Collection/CartoonBank.Com

"*Over all, I liked it, but I have a couple of notes.*"

2011; McCabe et al., 2011). Thus, as noted earlier, sexual identity, attraction, and behavior overlap.

A majority of adults ages 45 and older agree that a satisfying sexual relationship is important, but it's not their top priority. Marital sexual frequency may decrease because concerns about earning a living, making a home, and raising a family become more pressing than lovemaking. Others may be going through a divorce, dealing with unemployment, helping to raise grandchildren, or caring for aging parents—all of which sap people's interest in sex (Jacoby, 2005; ConsumerReports .org, 2009).

As people age, they experience lower levels of sexual desire and some sexual activities, but fairly slowly. About a third of married couples age 70 and older have vaginal intercourse, and among 75- to 85-year-olds, nearly 23 percent have sex four or more times a month (Lindau et al., 2007; Herbenick et al., 2010; Reece et al., 2010). In other cases, couples in their seventies and eighties emphasize the importance of emotional intimacy and companionship, and are satisfied with kissing, cuddling, and caressing (Heiman et al., 2011; Lodge and Umberson, 2012).

9-3b Sexual Scripts and Double Standards

We like to think that our sexual behavior is spontaneous, but all of us have internalized sexual scripts. A **sexual script** specifies the formal and informal norms for acceptable or unacceptable sexual behavior. Social scripts can change over time and across groups, but are highly gendered in two ways—women's increasing hypersexualization and a persistent sexual double standard.

THE "SEXY BABES" TREND

Sexualized social messages are reaching ever younger audiences, teaching or reinforcing the idea that girls and women should be valued for how they look rather than their personalities and abilities. For example, there are "bikini onesies" for infant girls, sexy lingerie for girls 3 months and older, and padded bras for 7- and 8-year olds (that's right, for 7- and 8-year-olds!).

Many girls are obsessed about their looks, and from an early age, for a variety of reasons, including their mothers' role modeling. By age 9, girls start imitating the clothes, makeup, and behavior of mothers who dress and act in highly sexualized ways (Starr and Ferguson, 2012).

Media images also play a large role in girls' hypersexualization. Girls and boys see cheerleaders (with increasingly sexualized routines) on TV far more than they see female basketball players or other athletes. A study of the 100 top-grossing films in 2008 found that 13- to 20-year-old females were far more likely than their male counterparts to be depicted in revealing clothes (40 percent vs. 7 percent), partially naked (30 percent vs. 10 percent), and physically attractive (29 percent vs. 11 percent). In television shows, women are now represented in more diverse roles—as doctors, lawyers, and criminal investigators—but they're often sexy ("hot"). Also, top female athletes regularly pose naked or semi-naked for men's magazines (Hanes, 2011; Smith and Choueiti, 2011).

Who benefits from girls' and women's hypersexualization? Marketers who convince girls (and their parents) that being popular and "sexy" requires the right clothes, makeup, hair style, and accessories, create a young generation of shoppers and consumers who will increase business profits more than ever before (Lamb and Brown, 2007; Levin and Kilbourne, 2009).

THE SEXUAL DOUBLE STANDARD

Some believe that the **sexual double standard**—a code that permits greater sexual freedom for men than women—has eroded. Others argue that it persists. Among U.S. adolescents, the higher the number of sexual partners, the greater the boy's popularity. In contrast, girls who have more than eight partners are far less popular than their less-experienced female peers. By age 44, many more men (21 percent) than women (8 percent) report having had at least 15 sex partners; and, over a lifetime, men are more likely than women to have sex outside of marriage (19 percent and 14 percent,

Galia Slayen built a life-sized Barbie to show what she would look like if she were a real woman. (Slayen used a toy for the head because she wasn't able to create a proportional head) ("Life Size Barbie...," 2011).

CB2/ZOB/WENN.com/Newscom

sexual script specifies the formal and informal norms for acceptable or unacceptable sexual behavior.

sexual double standard a code that permits greater sexual freedom for men than women.

respectively) (Kreager and Staff, 2009; Chandra et al., 2011; Drexler, 2012).

Another example of the sexual double standard is *hooking up*—which can mean anything from kissing to sexual intercourse. The prevalence of hooking up has increased only slightly since the late 1980s, but is now more common than dating at many high schools and colleges (Kalish and Kimmel, 2011; Monto and Carey, 2014).

Hooking up has its advantages. Many men prefer hookups because they're inexpensive compared with dating. For many college women, hookups offer sex without becoming involved in time-consuming relationships that compete with schoolwork, dealing with boyfriends who become demanding or controlling, and experiencing breakups (Bogle, 2008; Rosin, 2012).

Hooking up also has disadvantages, especially for women, because it reinforces a sexual double standard. For example, men are more likely than women to perform sexual acts that a partner doesn't like; more than twice as many men as women experience an orgasm because the men typically show no affection and aren't concerned about what pleases women sexually; and women who hook up generally get a reputation as "sluts" (England and Thomas, 2009; Armstrong et al., 2010, 2012). About half of women, compared with only 25 percent of men, have regretted having casual sex (Galperin et al., 2013).

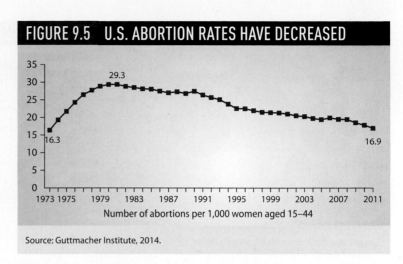

FIGURE 9.5   U.S. ABORTION RATES HAVE DECREASED

29.3

16.3

16.9

Number of abortions per 1,000 women aged 15–44

Source: Guttmacher Institute, 2014.

9-4   SOME CURRENT SOCIAL ISSUES ABOUT SEXUALITY

Most Americans see sex as a private act, but others believe that that there should be government control of some sexual expressions and decisions. People disagree about social policies on sex-related topics such as teenagers' birth control, prostitution, reproductive technologies that include genetic engineering (see Chapter 16), and teen pregnancy (see Chapter 12). Two of the most controversial and politically contested issues continue to be abortion and same-sex marriage.

9-4a Abortion

Abortion is the expulsion of an embryo or fetus from the uterus. It can occur naturally—in *spontaneous abortion* (miscarriage)—or induced medically. Abortion was outlawed in the nineteenth century, but has been legal since the U.S. Supreme Court's *Roe v. Wade* ruling in 1973.

TRENDS

Every year, 40 percent of unintended pregnancies end in abortion. Over a lifetime, 33 percent of women have an abortion by age 45 (Guttmacher Institute, 2014). The *abortion rate*, or the number of abortions per 1,000 women ages 15 to 44, increased during the 1970s, then decreased, and has dropped to its lowest point since 1973 (see *Figure 9.5*).

Why have abortion rates decreased? As of early 2014, at least half of the states had restricted access to abortion, but national data show that most of the decline in abortions is due to an overall drop in pregnancy rates, delaying childbearing until the economy improves, a more effective usage of contraceptives, and greater access to emergency contraception that prevents pregnancy (Pazol et al., 2013; Jones and Jerman, 2014).

Abortion is most common among women who are young (in their twenties), white, and never married (see *Figure 9.6*). Almost 40 percent have some college or an associate's degree, and another 20 percent have at least a college degree. Women who get abortions share several characteristics: 61 percent already have one or more children, 69 percent have incomes below or near the poverty level, 74 percent are financially unable to support a baby, and low-income women are much more likely than higher-income women to have experienced intimate partner violence that included being impregnated against their will. Abortion cuts across all religious groups, but 28 percent of women are Catholic and 37 percent are Protestant (including born-again/evangelical Christians) (Jones et al., 2010, 2013; Guttmacher Institute, 2014).

abortion   the expulsion of an embryo or fetus from the uterus.

FIGURE 9.6 WHO HAS ABORTIONS?

Age

- 20 to 24: 33%
- 25 to 29: 24%
- 19 and Younger: 18%
- 30 to 39: 22%
- 40 and Older: 3%

Race/Ethnicity

- Other*: 9%
- White: 36%
- Black: 30%
- Latinas: 25%

Marital Status

- Never Married: 45%
- Married: 15%
- Separated, Divorced, or Widowed: 11%
- Living Together: 29%

*Other refers to Asian/Pacific Islanders, American Indians, and Alaska Natives.

Sources: Based on Guttmacher Institute, 2013, 2014.

WHY IS ABORTION CONTROVERSIAL?

Since its legalization, abortion has been one of the most persistently contentious issues in U.S. politics and culture. More Americans describe themselves as "pro-life" (48 percent) than "pro-choice" (45 percent), and 7 percent aren't sure. Both anti- and pro-abortion groups agree on some issues, such as requiring a patient's informed consent, but 26 percent of Americans want abortion legal in all cases and 20 percent want it illegal in all cases (Saad, 2013).

Antiabortion groups believe that the embryo or fetus isn't just a mass of cells but a human being from the time of conception and, therefore, has a right to life. In contrast, abortion rights advocates point out that, at the moment of conception, the organism lacks a brain and other specifically and uniquely human attributes, such as consciousness and reasoning, and that a pregnant woman—not legislators—should decide whether or not to bear children.

Antiabortion groups maintain that abortion is immoral and endangers a woman's physical, mental, and emotional health. Whether abortion is immoral is a religious and philosophical question. On a physical level, a legal abortion in the first trimester (up to 12 weeks) is safer than driving a car, playing football, motorcycling, getting a penicillin shot, or continuing a pregnancy. There's also no evidence, despite the claims of abortion opponents, that having an abortion increases the risk of breast cancer or causes infertility (Sheppard, 2013; Guttmacher Institute, 2014).

What about mental and emotional health? Antiabortion activists argue that abortion leads to postabortion stress disorders, depression, and even suicide. Several high quality national studies conducted since the 1980s

have consistently found that abortion poses no hazard to an adolescent or adult woman's mental health, doesn't increase emotional problems such as depression or low self-esteem, and doesn't lead to drug or alcohol abuse or suicide. An unwanted pregnancy, not abortion, increases the risk of mental health problems (Major et al., 2008; Academy of Medical Royal Colleges, 2011; Steinberg and Finer, 2011).

A strong majority (63 percent) of Americans want to keep abortion legal, but it's virtually impossible for many women to get legal abortions in 89 percent of all U.S. counties. Between 2001 and 2010, states passed 189 laws restricting access to abortions. During 2013 alone, 22 states enacted 70 abortion restrictions. Thus, antiabortion legislation is increasing. Some of the restrictions have included cutting public funding for abortions for low-income women, passing licensing requirements that make it impossible for abortion clinics to remain open, and requiring women to view an ultrasound image of the fetus at least 24 hours before an abortion (Lipka, 2014; Nash et al., 2014).

9-4b Same-Sex Marriage

Same-sex marriage (also called *gay marriage*) is a legally recognized marriage between two people of the same biological sex and/or gender identity. Although still controversial, same-sex marriage is becoming more acceptable in the United States and some other countries.

TRENDS

As this book goes to press, court decisions, state legislatures, or voters have legalized gay marriage in 35 states and the District of Columbia. Worldwide, since 2001, 19 countries, mostly in Europe, have legalized same-sex marriage so far ("Gay Marriage," 2014; Masci et al., 2014).

The Defense of Marriage Act of 1996, known as DOMA, defined marriage as a "legal union between one man and one woman." In 2013, the U.S. Supreme Court struck down part of DOMA as unconstitutional. The court didn't rule on the question of whether the Constitution guarantees a right to same-sex marriage, but said that same-sex couples in states that recognize their marriages will have equal access to more than 1,100 federal benefits that apply to other married couples. Some of the benefits include filing together for federal taxes, refusing to testify against spouses in federal criminal and civil cases even in states that don't recognize gay marriage, and getting up

same-sex marriage (also called *gay marriage*) a legally recognized marriage between two people of the same biological sex and/or gender identity.

TABLE 9.3 WHY DO AMERICANS FAVOR OR OPPOSE SAME-SEX MARRIAGES?

What do *you* think? What other reasons can you add for each side of the debate?

Same-sex marriage should be legal because . . .	Same-sex marriage should not be legal because . . .
• Gay marriages strengthen families and long-term unions that already exist. Children are better off with parents who are legally married.	• Children need a mom and a dad, not two dads or two moms.
• There are no scientific studies showing that children raised by gay and lesbian parents are worse off than those raised by heterosexual parents.	• There are no scientific studies showing that children raised by gay and lesbian parents are better off than those raised by heterosexual parents.
• Every person should be able to marry someone that she or he loves.	• People can love each other without getting married.
• Gay marriages are good for the economy because they boost businesses such as restaurants, bakeries, hotels, airlines, and florists.	• What's good for the economy isn't necessarily good for society, especially its moral values and religious beliefs.

Sources: Bennett and Ellison, 2010; Olson, 2010; Sullivan, 2011; Sprigg, 2011; Whitehead, 2011.

to seven days of extra military leave to marry (Koseff and Fritze, 2013; Phelps, 2014). We'll examine same-sex marriages and families in Chapter 12, but why is gay marriage such a contentious issue?

WHY IS SAME-SEX MARRIAGE CONTROVERSIAL?

More than half (55 percent) of Americans support same-sex marriage (up from 37 percent in 2006). Opposition is greatest among those who regularly attend religious services, live in the South, are 55 and older, and hold conservative views on family issues (Gallup Historical Trends, 2014; McCarthy, 2014).

Those who favor same-sex marriage argue that people should have the same legal rights regardless of sexual orientation. Those who oppose same-sex marriage contend that such unions are immoral, weaken traditional notions of marriage, and are contrary to religious beliefs. *Table 9.3* summarizes some of the major pro and con arguments in this ongoing debate.

© CREATISTA/Shutterstock.com

9-5 GENDER AND SEXUALITY ACROSS CULTURES

There's considerable variation worldwide regarding gender inequality and sexual oppression. Such variations show that our behavior is learned, not innate.

9-5a Gender Inequality

The Global Gender Gap Index (GGGI) measures women's status and quality of life in 142 countries, representing more than 90 percent of the world's population. The GGGI is based on key indicators in four fundamental categories: economic opportunity and participation, educational attainment, political empowerment (women and men in the highest political positions), and health and survival. The score ranges from 0 (inequality) to 1 (equality) (Hausmann et al., 2014).

The GGGI isn't an overall measure of a country's development or wealth; rather, it gauges the relative equality between men and women on an indicator. For example, Saudi Arabia, one of the wealthiest countries in the world, and which has some of the most educated women in the world (including STEM college and advanced degrees), ranks near the bottom in women's economic and political participation (0.38 and 0.07, respectively) (Hausmann et al., 2014).

Of the 142 countries that the World Economic Forum analyzed in 2014, Iceland, Norway, Finland, Sweden, and New Zealand—as in the past—had the highest overall GGGI score. The United States ranked only 20th, and behind some poor countries such as Burundi, Lesotho, Nicaragua, and Rwanda. The greatest gender inequality was in the Middle East, North Africa, and sub-Saharan Africa; the five bottom countries in gender equality were Chad, Mali, Pakistan, Syria, and Yemen. In *all countries and regions*, the greatest gender gaps are in economic opportunity and participation and political leadership (Hausmann et al., 2014).

ECONOMIC PARTICIPATION

Women's economic participation depends on cultural attitudes, values, customs, and laws. Regarding attitudes, the biggest gender gaps on whether women and men should have the same rights, including economic participation, are in Egypt, Nigeria, Kenya, Indonesia, Jordan, and Pakistan. In Egypt, only 11 percent of men and 36 percent of women believe that women should be able to work outside the home (Horowitz et al., 2010). Thus, not all women endorse equal economic rights.

Countries that have closed education gaps and have high levels of women's economic participation—the Scandinavian countries, United States, Canada, New Zealand, and Australia—demonstrate strong economic growth (Worley, 2014). However, gender gaps still persist in senior positions, wages, and leadership. For example, Germany is Europe's No. 1 economy, but also has the largest pay gap between men and women in the European Union. Of 191 executives on the management boards of Germany's 30 biggest companies, only 12 are women, a 20 percent decrease from a year before (de Pommereau, 2013; Webb, 2013).

POLITICAL LEADERSHIP

Worldwide, only 19 percent of those holding seats in national legislatures are women. Rwanda has 56 percent, followed by seven countries in which women comprise 40 to 46 percent of those in high-level political positions. In contrast, women make up only 18 percent of the U.S. Congress (Clifton and Frost, 2011; Center for American Women and Politics, 2014).

Of the 197 world leaders who are presidents or prime ministers, only 11 percent are women (Institute for Women's Leadership, 2011). In 188 countries, women occupy only 21 percent of the positions in decision-making bodies that are comparable with our Congress. The United States ranks 80th in women's political leadership, well below many African, European, and Asian countries, and even below most of the Arab countries that many Westerners view as repressing women (Inter-Parliamentary Union, 2013).

9-5b Sexual Inequality

Globally, women have fewer rights and opportunities than men. There's been more acceptance of homosexuality in some countries, but heterosexism prevails.

VIOLENCE AGAINST FEMALES

In some of the most patriarchal countries around the world, women experience sexual violence and male dominance because of custom, laws, or minimal government protection. Worldwide, 35 percent of women have experienced physical and/or sexual violence by an intimate partner or another male (World Health Organization, 2013; UN Women, 2014). The rates are much higher in many countries. For example,

▶ In Afghanistan and some African countries, about 80 percent of girls—some as young as 8 years old—are forced into marriages; 87 percent of Afghan women report experiencing physical, psychological, or sexual abuse (Peter, 2012; Schnall, 2012; "Child Brides...," 2014).

▶ In Pakistan, 90 percent of women experience domestic violence in their lifetimes. Over 1,000 females are victims of "honor killings" every year. An *honor killing* is the murder of a family member, usually a female by a male, who is believed to have shamed the family by engaging in premarital and extramarital sex, and even wearing inappropriate clothes (Human Rights Now, 2011; Sahgal and Townsend, 2014).

▶ In the Democratic Republic of Congo, almost 1,160 women are raped every day by soldiers, strangers, and intimate partners (Peterman et al., 2011).

▶ In India, rape and gang rape are epidemic, but less than a quarter of reported crimes end in conviction. As many as 100,000 women a year are killed over dowry disputes (the money or goods that a wife brings to her husband at marriage) (Harris, 2013; "Ending the Shame...," 2013).

Female genital mutilation/cutting (FGM/C) is a partial or total removal of the female external genitalia. Most of the 140 million girls and women who have undergone FGM/C live in 29 African countries, Indonesia,

Little girls like this one scream and writhe in pain during FGM/C (see text). Complications include hemorrhaging to death, a rupture that causes continual dribbling of urine or feces for the rest of the woman's life, severe pain during sexual intercourse, and death during childbirth if the baby can't emerge through the mutilated organs.

and some Middle Eastern countries. Most of the girls are between 3 and 12 years old. The operator is typically an elderly village woman who uses a knife or other sharp object and doesn't administer an anesthetic. Cultures justify FGM/C on the grounds that it controls a girl's sexual desires and preserves her virginity, a prerequisite for marriage (UNICEF, 2013; Feldman-Jacobs and Clifton, 2014; Jones, 2014).

VIOLENCE AGAINST MALES

Male violence is common in some total institutions (see Chapter 4). Between 2009 and 2011, 69 percent of male inmates experienced sexual abuse by other inmates or guards (Beck et al., 2014). In 2012, 53 percent of male service members were sexually assaulted, but only 13 percent filed a report: The perpetrator often held a higher rank; a male who's sexually assaulted is stigmatized as gay, weak, and/or not a real "warrior," and is subsequently treated like an outcast (Brown, 2013; Department of Defense Sexual Assault..., 2013).

You saw earlier that Americans are more accepting of homosexuality, and that more nations are legalizing gay marriage. In contrast, many countries in Asia, Africa, and the Middle East don't tolerate LGBTs. Gay sex is illegal in 78 nations, including 38 of Africa's 54 countries. Gay men, particularly, may be legally tortured, stoned, imprisoned, or killed ("Deadly Intolerance," 2014; "The Gay Divide," 2014; Pflanz, 2014).

Russia's parliament recently, and unanimously, passed a law that bans LGBT relationships and forbids distributing material on gay rights. Russians are more accepting of extramarital affairs, gambling, and drinking alcohol (a major cause of men's death before age 55) than homosexuality (Council for Global Equality, 2014; Poushter, 2014).

9-6 SOCIOLOGICAL EXPLANATIONS OF GENDER AND SEXUALITY

Gender and sexuality affect all people's lives, but why is there so much variation over time and across cultural groups? The four sociological perspectives answer this and other questions about gender and sexuality somewhat differently (*Table 9.4* summarizes these theories).

9-6a Functionalism

Functionalists view women and men as having distinct roles that ensure a family's and society's survival. These roles help society operate smoothly, and have an impact on the types of work that people do.

DIVISION OF GENDER ROLES AND HUMAN CAPITAL

Some of the most influential functionalist theories, developed during the 1950s, proposed that gender roles differ because women and men have distinct roles and responsibilities. A man (typically a husband and father) plays an *instrumental role* of economic provider; he's competitive and works hard. A woman (typically a wife and mother) plays an *expressive role*; she provides the emotional nurturance that sustains the family unit and supports the father/husband. Instrumental and expressive roles are complementary, and each person knows what's expected: If the house is clean, she's a "good wife"; if the bills are paid, he's a "good husband." The duties are specialized, but both roles are equally important in meeting a family's needs and ensuring a society's survival (Parsons and Bales, 1955; Betcher and Pollack, 1993).

Such traditional gender roles help explain occupational sex segregation because people differ in the

TABLE 9.4 SOCIOLOGICAL EXPLANATIONS OF GENDER AND SEXUALITY

Theoretical Perspective	Level of Analysis	Key Points
Functionalist	Macro	• Gender roles are complementary, equally important for a society's survival, and affect human capital. • Agreed-on sexual norms contribute to a society's order and stability.
Conflict	Macro	• Gender roles give men power to control women's lives. • Most societies regulate women's, but not men's, sexual behavior.
Feminist	Macro and micro	• Women's inequality reflects their historical and current domination by men, especially in the workplace. • Many men use violence—including sexual harassment, rape, and global sex trafficking—to control women's sexuality.
Symbolic Interactionist	Micro	• Gender is a social construction that emerges and is reinforced through everyday interactions. • The social construction of sexuality varies across cultures because of societal norms and values.

amount of human capital that they bring to the labor market. *Human capital* is the array of competencies—including education, job training, skills, and experience—that have economic value and increase productivity.

From a functionalist perspective, what individuals earn is the result of the choices they make and, consequently, the human capital that they accumulate to meet labor market demands. Women diminish their human capital because they choose lower paying occupations (social work rather than computer science), as well as postpone or leave the workforce for childbearing and child care. When they return to work, women have lower earnings than men because, even in higher paying occupations, their human capital has deteriorated or become obsolete (Kemp, 1994).

WHY IS SEXUALITY IMPORTANT?

For functionalists, sexuality is critical for reproduction, but people should limit sex to marriage and forming families. Functionalists view sex outside of marriage as dysfunctional because most unmarried fathers don't support their children. The offspring often experience poverty and a variety of emotional, behavioral, and academic problems (Avellar and Smock, 2005).

You might be tempted to dismiss the functionalist view of limiting sex to marriage as outdated. Worldwide, however, sex outside of marriage is prohibited and *arranged marriages*—in which parents or relatives choose their children's future mates—are the norm. Most children agree to arranged marriages because of social custom and out of respect for their parents' wishes. The matches solidify relationships with other families and ensure that the woman's sexual behavior will be confined to her husband, avoiding any doubt about the offspring's parentage (see Benokraitis, 2014).

CRITICAL EVALUATION

Critics fault functionalist perspectives on three counts. First, even during the 1950s, white middle-class male sociologists ignored almost a third of the labor force that was composed of working-class, immigrant, and minority women who played *both* instrumental and expressive roles. Second, functionalists tend to overlook the fact that many people don't have a choice of playing only instrumental or expressive roles because most families rely on two incomes for economic survival. Third, the human capital model assumes that women have lower earnings than men because they "choose" lower paying occupations. As you saw earlier, however, there's a gender pay gap across all occupations, even those that require advanced degrees.

Functionalism frowns on sexual relationships outside of marriage. Compared with married couples, for example, those who cohabit have poorer quality relationships and lower happiness levels (Wilcox and Marquardt, 2011; Sassler et al., 2012; Wiik et al., 2012). As you'll see in Chapter 12, however, marriage doesn't guarantee long or happy relationships.

9-6b Conflict Theory

Conflict theorists see gender inequality as built into the social structure, both within and outside the home. In both developing and industrialized countries, men control most of a society's resources and dominate women. Like functionalists, conflict theorists see sexuality as a key component of a society's organization, but they view sexuality as reflecting and perpetuating sexism and discrimination.

CAPITALISM AND GENDER INEQUALITY

Conflict theorists maintain that capitalism, not complementary roles, explain gender roles and men's social and economic advantages. Influenced by the work of Karl Marx, conflict theorists argue that women's inequality is largely due to economic exploitation—both as underpaid workers in the labor force and doing most of the unpaid domestic work, especially housework and caring for aging family members. In effect, gender roles are profitable for business. Companies can require their male employees to work long hours or make numerous overnight business trips and don't have to worry about workers demanding payment for child care services that cost almost $63,000 in 2014 (Carey and Trap, 2014).

Conflict theorists agree that all men aren't equally privileged, and that women in upper classes have more economic power, status, and prestige than men in lower classes. However, within social classes, men typically enjoy more power and control than women.

IS GENDER INEQUALITY LINKED TO SEXUAL INEQUALITY?

From a conflict perspective, gender inequality gives men economic, political, and/or interpersonal power to control or dominate women's sexual lives. Most of the victims of domestic violence and rape, in the United States and around the world, are women and girls. In workplace sexual harassment cases, the offender is typically a male supervisor. For example, 66 percent of American female restaurant workers experience sexual harassment by a manager, owner, or supervisor on at least a monthly basis. In prostitution and sex trafficking, almost all of the victims worldwide are poor women and girls (U.S. Department of State, 2013; Basu, 2014; Restaurant Opportunities Center United et al., 2014).

Constantly at Risk: Afghan Policewomen

A United Nations report found that about 90 percent of the female police officers in Afghanistan have experienced or witnessed sexual violence and harassment by their male colleagues. Officials dismissed the findings because "If an Afghan policewoman is being raped or sexually harassed, [she] would report it." The women don't report the crimes because they fear being fired: They're often the only person earning money in the family and desperately need their salaries—usually around $240 per month. The women believe the sexual assaults would increase when the offender is a police commander or one of his close friends. Many Afghans already think that policewomen have loose morals because they work in public with men who aren't relatives. Reporting sexual harassment and violence could increase the chances of a woman's being forced to quit or even killed by a male relative for dishonoring the family (Rubin, 2013: A4).

Aref Karimi/AFP/Getty Images

In some societies, particularly in the Middle East and some African countries, men dictate how women should dress and whether they can travel, work, receive health care, attend school, or start a business; dismiss women's charges of sexual assaults (including gang rapes); punish real or imagined instances of women's, but not men's, marital infidelity; and blame girls for child rape because they're "seducing" older men (Neelakantan, 2006). Many women have internalized such gender inequality. For example, between 31 to 40 percent of the women in Uganda believe that wife beating is acceptable if the wife argues with her husband or refuses to have sex with him (Clifton and Frost, 2011).

CRITICAL EVALUATION

Conflict theory is useful in showing how social structures reinforce men's domination of women, but critics point out several limitations. First, conflict theorists focus on male–female competition rather than cooperation. You saw earlier, for instance, that among employed parents, fathers' child care contributions have increased since the 1960s. Second, women aren't as submissive as some conflict theorists claim. Like men, women often barter with their employers to get what they want ("I'll work late all this week if I can have Friday off"). Third, conflict theory emphasizes the differences between women and men rather than their common goals, similar attitudes, and comparable power plays to increase their economic and political influence.

A fourth criticism is that conflict theory often ignores women's exploitation of other women. In the United States, women comprise 19 percent of those involved in the sex trafficking industry. The thriving underground sex economy relies heavily on nannies, secretaries, and escort services and brothels owned or operated by females (Banks and Kyckelhahn, 2011; Dank et al., 2014).

9-6c Feminist Theories

Feminist scholars agree with conflict theorists that gender stratification benefits men and capitalism, but emphasize that women's subordination also includes their daily vulnerability to male violence (Katz, 2006). Feminist scholars, more than any other group of theorists, are especially concerned about men's controlling women's sexual lives.

LIVING IN A GENDERED WORLD

All feminists (female and male) agree on three general points: (1) men and women should be valued equally; (2) women should have more control over their lives; and (3) political, economic, family, and other institutions can reduce gender inequality. Let's look briefly at three prominent feminist explanations of gender roles.

Liberal feminism maintains that gender equality can be achieved through equal rights and opportunities. This approach emphasizes that gendered socialization creates inequality by teaching children culturally accepted masculine and feminine attitudes and behavior that limit their options in the family, education, and the workplace (Jagger and Rothenberg, 1984).

Radical feminism contends that patriarchy is the major reason for men's dominance over women, including sexual exploitation. The control is deliberate, brings men many benefits, and is supported by gender inequity in medicine, religion, science, law, and other institutions. For radical feminists, all women are potential victims of violence and all men are capable of violence. In this sense, patriarchy is more critical than social class in giving men power over women's lives (Bart and Moran, 1993; Lorber, 2005).

Multicultural feminism (sometimes referred to as *racial ethnic feminism* or *multiracial feminism*) asserts that gender, race, and social class intersect to form a hierarchical stratification system that shapes women's and men's attitudes, experiences, and behavior. Privileged women have less status than privileged men, but upper-class white men *and* women subordinate lower-class women *and* minority men (Andersen and Collins, 2010).

Understanding the interconnections between gender, race/ethnicity, and social class provides a more comprehensive picture of living in a gendered world than does any single variable. For example, white, college-educated, employed, divorced mothers are more likely to receive the full amount of court-ordered child-support payments than are never married black mothers who don't have a high school diploma and depend on public assistance (Grall, 2013).

SEXUALITY, SOCIAL CONTROL, AND COMMERCIALIZING SEX

Feminist theorists often point to sexuality as the root of inequality between women and men, both interpersonally and within institutions. Men assert their power and control—across cultures and over time—through rape, intimate partner violence, sexual harassment, exploiting women through prostitution and pornography, forcing women to marry against their will, honor killings, FGM/C, and punishing women, but not men, for seeking a divorce or committing adultery (Magnier, 2009; World Health Organization, 2013).

Sex is big business. From 2001 to 2007, Internet pornography in the United States alone increased from a $1 billion to a $3-billion-a-year industry (Covenant Eyes, Inc., 2013). Men's average testosterone levels decline naturally with aging, but are also due to obesity and alcohol. Pharmaceutical companies "have seized on the decline in testosterone levels as pathological and applicable to every man." As a result, the number of testosterone prescriptions given to American men has tripled since 2001, and sales of all testosterone-boosting drugs are estimated to soar from $2 billion in 2012 to $5 billion by 2017. Prescriptions are surging even though national studies have shown that testosterone drugs increase the risk of heart attacks (La Puma, 2014: A21; O'Connor, 2014).

A recent study examined more than 3,200 full-page ads in *Cosmopolitan*, *Redbook*, *Esquire*, *Playboy*, *Newsweek*, and *Time*. Ads using sex increased from 15 percent in 1983 to 27 percent in 2003, used sex to sell everything

Many sexually healthy men in their 20s, 30s, and 40s now use impotence drugs—such as Viagra and Levitra—because they believe the myth that a "real man" is always ready to be a sexual superman. Why is this an example of commercializing sex?

from alcohol to shoes, and most of the models were females (Reichert et al., 2012).

Pornography, testosterone drugs, ads, and the earlier discussion of the "sexy babes" trend are just a few examples of the increasing *commercialization of sex*, making sexuality a commodity that can be sold for financial gain. Commercializing sex demeans both women and men, but the "products" are usually women.

CRITICAL EVALUATION

Feminist perspectives provide insightful analyses of gender inequality, but have limitations. Liberal feminism typically emphasizes women's, but not men's, oppression in capitalistic societies. Radical feminism has been criticized for being too narrow; for example, men aren't universally violent or sexually exploitative. Multicultural feminism is inclusive, but this strength can also be a weakness. Regarding the gender pay gap, for example, which variable is the most important—race, ethnicity, or social class? Because gender inequality has many causes, it's not clear which factors should be given priority in implementing change.

Some critics contend that feminist theorists focus too much on men's sexual domination and power and minimize the importance of sexuality based on love and affection. And, like conflict theorists, feminist scholars are sometimes accused of glossing over women's exploitation of others. In Iraq, for example, sex traffickers are often women who target the youngest girls because virgins bring the highest prices (Naili, 2011).

9-6d Symbolic Interaction

Whereas functionalist, conflict, and some feminist theories are macro level, symbolic interactionists focus on the everyday processes that produce and reinforce gender roles. We "do" gender, sometimes consciously and sometimes unconsciously, by adjusting our behavior and our perceptions depending on the sex of the person with whom we're interacting (West and Zimmerman, 2009). Our expression of our sexuality, similarly, isn't inborn but a product of socialization, and what families and other societal groups deem as appropriate and inappropriate behavior (Hubbard, 1990).

GENDER IS A SOCIAL CONSTRUCTION

For symbolic interactionists, gender is a social creation, and gender inequality is shaped by learning gender roles through daily social interaction. For example, negative stereotypes about girls' abilities in math and science can significantly lower girls' test performance and lower their aspirations in STEM careers. When teachers tell girls and boys that both are equally capable in math and science, "the difference in performance essentially disappears" (Hill et al., 2010: 2). Thus, social interaction is important in cultivating or stifling abilities and interests.

In effect, believing in gender differences can actually *produce* differences. Women who begin college intending to become engineers are more likely than men to change their major and choose another career because they lack confidence, not competence. As one of the most sex-segregated professions, engineering carries ingrained cultural stereotypes that men are more naturally suited to the field. As a result, women's professional self-confidence falters and they change majors (Cech et al., 2011; see also Chapter 13).

HOW DO WE CONSTRUCT SEXUALITY?

For symbolic interactionists, sexuality is also socially constructed. As with gender roles, we *learn* to be sexual and to express our sexuality differently over time and across cultures because the people around us affect our sexual behavior. For example, students who attend conservative religious high schools are less likely than their public school counterparts to report same-sex attraction or an LGBT self-identity in young adulthood (Wilkinson and Pearson, 2013).

In many Middle East countries, men have premarital sex but don't marry women who aren't virgins (Fleishman and Hassan, 2009). In the United States, as you saw earlier, men who have casual sex are "studs" whereas women are "sluts." Thus, sexual double standards are socially constructed.

Because sexuality is socially constructed, attitudes and behavior can change. Many Americans have become more comfortable with bisexuals, lesbians, and gay men. In China, a university that planned to publicly shame students who engaged in "uncivilized behavior" (e.g., hugging or kissing in public) withdrew the policy after widespread protests that the surveillance violated the students' privacy (Meng, 2011).

Photodisc/Thinkstock

CRITICAL EVALUATION

Despite their contributions, interactionists have been criticized for ignoring the social structures that create and maintain gender inequality. Female soldiers in the Iraq and Afghanistan wars "have done nearly as much in battle as their male counterparts: Patrolled streets with machine guns, served as gunners on vehicles, disposed of explosives, and driven trucks down bomb-ridden roads" (Alvarez, 2009: A1). It was only in 2013, however, that the Pentagon lifted a ban on women serving in combat roles, an important criterion in promotions.

Symbolic interactionists show that behavior is socially constructed and, consequently, sexual attitudes and practices vary across cultures. However, the theories don't explain why siblings, even identical twins—who are socialized similarly—may have different sexual orientations. A related weakness is that symbolic interactionists don't explain why, historically and currently, women around the world are considerably more likely than men to be subjected to sexual control and exploitation. Such analyses require macro-level analyses that examine family, religious, political, and economic institutions.

STUDY TOOLS 9

READY TO STUDY? IN THE BOOK, YOU CAN:

☐ Check your understanding of what you've read with the Test Your Learning Questions provided on the chapter review card at the back of the book.

☐ Rip out the chapter review card for a handy summary of the chapter and key terms.

ONLINE AT CENGAGEBRAIN.COM YOU CAN:

☐ Prepare for tests with quizzes.

☐ Review the key terms with Flash Cards.

☐ Play games to master concepts.

© Anelina/Shutterstock.com

10 | Race and Ethnicity

After you finish this chapter go to **PAGE 197** for **STUDY TOOLS.**

In 2014, Coca-Cola aired an ad during the Super Bowl that portrayed U.S. ethnic diversity by featuring "America the Beautiful" sung in several languages. After the ad, Twitter lit up with criticism. Some of the commenters said that it's disrespectful to sing "America the Beautiful" in any language other than English, and many disparaged immigrants who don't learn English. Others pointed out that the song isn't our national anthem, perhaps it should've been sung in one of the American Indian tongues that were here before Europeans arrived, and that we don't have an official national language (Sahgal, 2014).

Such discourse illustrates our diversity and divisions. This chapter examines the impact of race and ethnicity on our lives, why racial-ethnic inequality is still widespread, and the growth of interracial and interethnic relationships. First, however, take the short quiz on the next page to see how much you know about these topics.

What do you think?

The United States is a melting pot.

1	2	3	4	5	6	7

strongly agree   strongly disagree

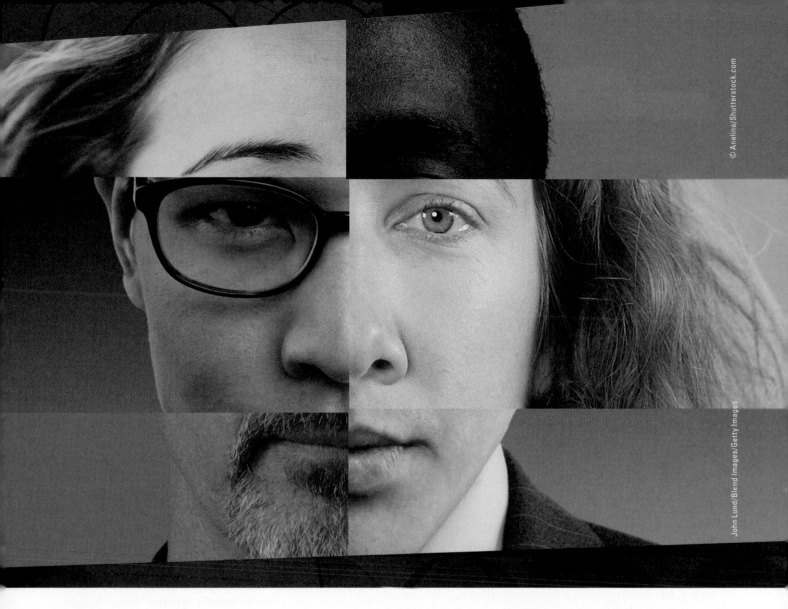

10-1 U.S. RACIAL AND ETHNIC DIVERSITY

The United States is the most multicultural country in the world, a magnet that draws people from hundreds of nations and is home to millions of Americans who are bilingual or multilingual. Of the almost 320 million U.S. population, 13 percent are foreign born, and this population is expected to grow to 23 percent by 2050. As a result, America's multicultural umbrella includes at least 150 distinct ethnic or racial groups (Passel and Cohn, 2008; Acosta et al., 2014).

By 2025, only 58 percent of the U.S. population is projected to be white—down from 86 percent in 1950 (see *Figure 10.1*). By 2050, whites may make up only 47 percent of the total population because Latino and Asian populations are expected to triple in size (Passel et al., 2012).

TRUE OR FALSE?

How Much Do You Know About U.S. Racial and Ethnic Groups?

1. Race is determined biologically.

2. The United States has the most foreign-born residents of any country in the world.

3. People who are prejudiced also discriminate.

4. Among Asian Americans, the most economically successful are Asian Indians.

5. About 85 percent of Americans report being only one race.

6. Minority groups are small in number.

The answer to #4 is true; the others are false. You'll see why as you read this chapter.

FIGURE 10.1 RACIAL AND ETHNIC COMPOSITION OF THE U.S. POPULATION, 1950–2025

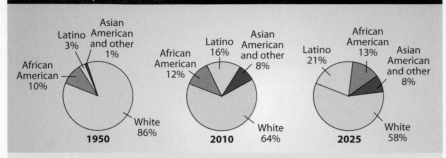

1950
- Latino 3%
- Asian American and other 1%
- African American 10%
- White 86%

2010
- Latino 16%
- African American 12%
- Asian American and other 8%
- White 64%

2025
- Latino 21%
- African American 13%
- Asian American and other 8%
- White 58%

Note: "Asian American and other" includes American/Indian/Alaskan Native, Native Hawaiians and Pacific Islanders, some other race, and those who identify themselves with with two or more races.

Sources: U.S. Census and Population Division, U.S. Census Bureau, 2008; Passel et al., 2011.

look different are miniscule compared with the genes that make us similar (Graves, 2001; Pittz, 2005).

If our DNA is practically identical, why are we so obsessed with race? People react to the physical characteristics of others, and those reactions have consequences. Skin color, hair texture, and eye shape, for example, are easily observed and mark groups for unequal treatment. As long as we act on the basis of these characteristics, our life experiences will differ in access to jobs and other resources, how we treat people, and how they treat us (Duster, 2005).

The multiracial population is also increasing. The number of Americans who identify themselves as being of two or more races is projected to more than triple—from almost 6 million (2 percent) in 2010 to more than 16 million (4 percent) in 2050 (U.S. Census Bureau News, 2008).

10-2 THE SOCIAL SIGNIFICANCE OF RACE AND ETHNICITY

All of us identify with some groups and not others in terms of sex, age, social class, and other factors. Two of the most common and important sources of self-identification, as well as labeling by others, are race and ethnicity.

10-2a Race

A **racial group** refers to people who share visible physical characteristics, such as skin color and facial features, that members of a society consider socially important. Contrary to the popular belief that race is determined biologically, it's a *social construction*, a societal invention that labels people based on physical appearance, social class, or other characteristics (see Daniel, 2014).

Possibly only six of the human body's estimated 35,000 genes determine the color of a person's skin. Because all human beings carry 99.9 percent of the same genetic material (DNA), the "racial" genes that makes us

10-2b Ethnicity

An **ethnic group** (from the Greek word *ethnos*, meaning "nation") refers to people who identify with a common national origin or cultural heritage. Cultural heritage includes language, geographic roots, food, customs, traditions, and religion. Ethnic groups in the United States include Puerto Ricans, Chinese, Serbs, Arabs, Swedes, Hungarians, Jews, and many others. Like race, ethnicity can be a basis for unequal treatment, as you'll see shortly.

10-2c Racial-Ethnic Group

People who have distinctive physical and cultural characteristics are a **racial-ethnic group**. Some people use the terms *racial* and *ethnic* interchangeably, but

There's much variation in skin color across and within groups. People of African descent have at least 35 different shades of skin tone (Taylor, 2003). So, can you determine someone's race simply by looking at her or him?

---

racial group people who share visible physical characteristics that members of a society consider socially important.

ethnic group people who identify with a common national origin or cultural heritage.

racial-ethnic group people who have distinctive physical and cultural characteristics.

remember that *race* refers to physical characteristics with which we are born, whereas *ethnicity* describes cultural characteristics that we learn. The term *racial-ethnic* includes both physical and cultural traits.

Describing racial-ethnic groups has become more complex because the U.S. government allows people to identify themselves in terms of both race and ethnicity. In the 2000 and 2010 U.S. Census, for example, Latinos could check off "black" or "white" for race and "Cuban" for ethnic origin. Such choices generate dozens of racial-ethnic categories. People prefer some "labels" to others, and racial and ethnic self-identification varies.

10-2d What We Call Ourselves

In 1976, the U.S. government began to use the word *Hispanic* or *Latino* to categorize Americans who trace their roots to Spanish-speaking countries. About 70 percent say that it doesn't matter if they're referred to as *Latino* or *Hispanic*. If given a choice, however, 51 percent prefer to identify themselves by the family's country of origin (e.g., Mexican, Cuban, Salvadoran) (Pew Hispanic Center, 2012; Jones, 2013).

Since 1900, the Census Bureau has used *Negro, black*, and *African American*, but plans to drop *Negro* from its surveys because few older Americans still use the term (Yen, 2013). Currently, 65 percent of this group says it doesn't matter to them whether they're called *black* or *African American*, and this trend hasn't changed much since 1991 (Jones, 2013). Many people, including African American scholars, use *black* and *African American* interchangeably.

We see similar variations in the usage of *Native American* and *American Indian*. These groups prefer their tribal identities (such as Cherokee, Apache, and Lumbi) to being lumped together under a single term. Among white Americans, even some third and fourth generations identify themselves as Irish, German, Ukrainian, and so on.

JStaley401/Getty Images

10-3 OUR CHANGING IMMIGRATION MOSAIC

The United States, a nation of immigrants, has historically and currently both welcomed immigration and feared its consequences. What fuels public debate is the rising number of foreign-born people in the United States.

10-3a The Foreign-Born Population

The current proportion of foreign-born U.S. residents is smaller than in the past. In 1900, 15 percent of the total U.S. population was foreign born compared with 13 percent in 2012 (Grieco et al., 2012).

The United States has one of the highest foreign-born rates in the world—1 in 8 people, but there's been a significant shift in the immigrants' country of origin. In 1900, almost 85 percent of immigrants came from Europe compared with only 12 percent in 2012. Today, the foreign born come primarily from Asia (mainly China and the Philippines) and Latin America (mainly Mexico) (*Figure 10.2*; see also Grieco et al., 2012).

The United States admits more than 1 million immigrants every year—more than any other nation. A major change has been the rise of *undocumented* (also called *unauthorized*) *immigrants*—from 180,000 in the early 1980s to 11.7 million in 2012. Undocumented immigrants make up 28 percent of all foreign-born U.S. residents, and 4 percent of the nation's population (Passel et al., 2013; Pew Hispanic Center, 2013). An estimated 59 percent are from Mexico, 16 percent are from Central and Latin America, 10 percent are from Asia, and 15 percent are from other countries, including Canada and Europe. Most immigrants—legal and undocumented—come to the United States for economic opportunities (Hoefer et al., 2012).

10-3b U.S. Reactions to Immigration

Of 21 top policy issues, dealing with unauthorized immigration is only 17th in priority (Pew Research Center for the People & the Press, 2014). Between 45 to 78 percent of Americans favor giving undocumented immigrants a chance to become U.S. citizens if they meet certain requirements including having jobs, paying fines and back taxes, and passing background checks (Mendes, 2013; Steinhauser, 2014).

Because federal immigration reform has languished in Congress, in 2010 and 2011 Arizona and five other states passed their own laws. The legislation barred undocumented immigrants from receiving state or local public benefits, from getting a driver's license, and allowed police to check the immigration status of people lawfully stopped for other reasons (Hing, 2011).

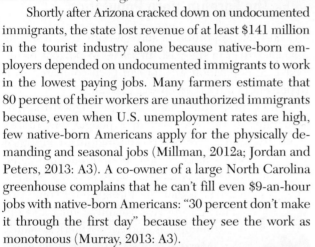
Tim Boyle/Bloomberg/Getty Images

Shortly after Arizona cracked down on undocumented immigrants, the state lost revenue of at least $141 million in the tourist industry alone because native-born employers depended on undocumented immigrants to work in the lowest paying jobs. Many farmers estimate that 80 percent of their workers are unauthorized immigrants because, even when U.S. unemployment rates are high, few native-born Americans apply for the physically demanding and seasonal jobs (Millman, 2012a; Jordan and Peters, 2013: A3). A co-owner of a large North Carolina greenhouse complains that he can't fill even $9-an-hour jobs with native-born Americans: "30 percent don't make it through the first day" because they see the work as monotonous (Murray, 2013: A3).

Some cities, towns, and counties—including those in states with hostile laws—have passed ordinances and resolutions that welcome undocumented immigrants. The residents believe that newcomers, both legal and illegal, are vital for economic growth and breathing new life into small rural towns and cities that have been "staggering toward the grave" (Sulzberger, 2011; Medrano, 2013).

Many researchers argue that, in the long run, easing undocumented immigrants' path to citizenship brings more benefits than costs. Few Americans realize that undocumented immigrants contribute significantly to state and local taxes—nearly $11 billion in 2010 alone—and that their average tax rate is almost 1 percent higher than that of the nation's wealthiest people. Allowing undocumented immigrants to work legally would increase state and local taxes by an additional $2 billion a year because about half of undocumented workers are "off the books" and don't pay income taxes (Citizens for Tax Justice, 2013; Institute on Taxation and Economic Policy, 2013).

On a number of measures, immigrants and their children—both legal and undocumented—outperform native-born Americans. Children of immigrants are more likely than their U.S.-born peers to live in a two-parent household; to have a parent with a secure job; to earn a college degree; to attend religious services regularly; and to be "prodigious job creators." They're also less likely than their native-born counterparts to be born underweight or to have a physical disability; to break laws once

Glenn Frank/iStockphoto.com

© Alexey Stiop/Shutterstock.com

Guest workers are permitted to work in the United States on a temporary basis because of labor shortages, especially in agriculture. After millions of undocumented immigrants fled to other states from Arizona, Alabama, and Georgia because of their harsh laws, many employers complained that couldn't find enough laborers—in construction and landscaping—to clean up and rebuild communities in the aftermath of devastating tornados (Shepherd, 2011).

in the United States; and to have nonmarital children (Allott, 2013; Hernandez and Napierala, 2013; "Second-Generation Americans," 2013; Alsever, 2014).

10-4 DOMINANT AND MINORITY GROUPS

Race and ethnicity often make people feel like outsiders, or "the other." Otherness is being different or having characteristics that set you apart from the dominant group (Thorpe-Moscon and Pollack, 2014).

10-4a What Is a Dominant Group?

A **dominant group** is any physically or culturally distinctive group that has the most economic and political power, the greatest privileges, and the highest social status. As a result, it can treat other groups as subordinate. In most societies, for instance, men are a dominant group because they have more status, resources, and power than women (see Chapters 8 and 9).

Dominant groups aren't necessarily the largest in number. From the seventeenth century until 1994, about 10 percent of the population in South Africa was white and had almost complete control of the black population. Because of *apartheid*, a formal system of racial segregation, the black residents couldn't vote, lost their property, and had minimal access to education and politics. Apartheid ended in 1994, but most black South Africans are still a minority because whites "hold the best jobs, live in the most expensive homes, and control the bulk of the country's capital" (Murphy, 2004: A4).

10-4b What Is a Minority?

Sociologists describe Latinos, African Americans, Asian Americans, Middle Eastern Americans, and American Indians as minorities. A **minority** is any group that may be treated differently and unequally because of its physical, cultural, or other characteristics. The characteristics include gender, age, sexual orientation, religion, ethnicity, or skin color. Minorities may be larger in number than a dominant group, but they have less power, privilege, and social status. For example, middle-class African Americans are much more likely than their white counterparts to be called by debt collectors even though both groups have similar debt levels and repayment rates (Ruetschlin and Asante-Muhammad, 2013). *Table 10.1* offers other examples of everyday privileges associated with skin color.

TABLE 10.1 AM I PRIVILEGED?

Most white people don't feel privileged because they aren't wealthy. Nonetheless, they enjoy everyday benefits, and take them for granted, simply because they're members of the dominant group. Would you add other advantages of being white?

1. I can go shopping and feel fairly sure that I won't be followed or harassed by store detectives.

2. If a traffic cop pulls me over, I can be sure that I haven't been singled out because of my race.

3. I can be late to a meeting without having the lateness reflect on my race.

4. I can turn on the television and see people who look like me represented in positive ways and in a wide range of roles.

5. I live in a safe neighborhood with good schools.

6. I never think twice about calling the police when trouble occurs.

7. I can wear a hoodie and not worry about others thinking that I'm up to no good.

8. I don't get dirty looks if I listen to loud music at a gas station.

Sources: McIntosh, 1995; Independent Television Service, 2003; Williams, 2014.

10-4c Some Patterns of Dominant-Minority Group Relations

To understand some of the complexity of dominant-minority group relations, think of a continuum. At one end of the continuum is genocide; at the other end is pluralism (see *Figure 10.3*).

GENOCIDE

Genocide is the systematic effort to kill all members of a particular ethnic, religious, political, racial, or national group. By 1710, the colonists in America had killed thousands of Indians in skirmishes, poisoned others, and promoted scalp bounties. In 1851, the governor of California officially called for the extermination of all Indians in the state (de las Casas, 1992; Churchill, 1997). Worldwide, and between 1915 to 1995 alone, well over 74 million people have been victims of genocide in Africa, Cambodia, China, Eastern Europe, and Turkey

dominant group any physically or culturally distinctive group that has the most economic and political power, the greatest privileges, and the highest social status.

minority a group of people who may be treated differently and unequally because of their physical, cultural, or other characteristics.

genocide the systematic effort to kill all members of a particular ethnic, religious, political, racial, or national group.

FIGURE 10.3 CONTINUUM OF SOME DOMINANT-MINORITY GROUP RELATIONS

INTOLERANCE
INEQUALITY

ACCEPTANCE
EQUALITY

Genocide
Systematic efforts to destroy minorities (e.g., American Indians)

Segregation
Physical and social separation of dominant and minority groups (e.g., housing segregation)

Acculturation
A minority group adopts the language, values, and other characteristics of the dominant group (e.g., learning and speaking English)

Pluralism
There is no dominant group because all groups share power and other resources fairly equally (e.g., possibly Switzerland)

(Niewyk and Nicosia, 2000; United Human Rights Council, 2004).

SEGREGATION

Segregation is the physical and social separation of dominant and minority groups. In 1954, the Supreme Court ruling in *Brown v. Board of Education* declared *de jure,* or legal, segregation unconstitutional. It was followed by the passage of a variety of federal laws that prohibited racial segregation in public schools, as well as discrimination in employment, voting, and housing.

De facto, or informal, segregation has replaced *de jure* segregation in the United States and many other countries. Some *de facto* segregation may be voluntary, as when minorities prefer to live among their own racial or ethnic group. In most cases, however, *de facto* segregation is due to discrimination. In the case of equally qualified homeseekers, for instance, realtors show minorities fewer homes and apartments than they do whites. Because minorities have fewer choices, the time and cost of their housing searches rise (Turner et al., 2013; see also Rugh and Massey, 2010).

ACCULTURATION AND ASSIMILATION

Many minority group members blend into U.S. society through **acculturation**, the process of adopting the language, values, beliefs, and other characteristics of the host culture (e.g., learning English, celebrating

Thanksgiving). Acculturation doesn't include intermarriage, but the newcomers merge into the host culture in most other ways. **Assimilation** involves conforming to the dominant group's culture, adopting its language and values, and intermarrying with that group. Journalists often use *assimilation* and *acculturation* interchangeably, but the former absorbs, rather than just changes, a group culturally.

PLURALISM

In **pluralism**, sometimes called *multiculturalism*, minority groups maintain many aspects of their original culture—including using their own language and marrying within their own racial or ethnic group—while living peacefully with the host culture. Pluralism is especially evident in urban areas with large racial and ethnic communities (e.g., "Little Italy," "Greek Town," "Little Korea," "Spanish Harlem"). U.S. minorities have numerous ethnic newspapers and radio stations, and the same constitutional rights (such as freedom of speech) as the dominant group. Nonetheless, people of various skin colors and cultures don't always experience the same social standing, and there can be considerable racial-ethnic friction.

10-5 SOME SOURCES OF RACIAL-ETHNIC FRICTION

During a recent trip to Switzerland, Oprah Winfrey asked a clerk at an expensive shop to show her a $37,000 purse. The clerk refused: "That one will cost too much, you won't be able to afford that." Winfrey didn't know for sure whether or not the snub was motivated by racism, but said that the incident was what "people with black or brown skin experience every day" (Goyette, 2013). In the United States, a Montana federal judge circulated a racist email that compared President Obama's mother

segregation the physical and social separation of dominant and minority groups.

acculturation the process of adopting the language, values, beliefs, and other characteristics of the host culture.

assimilation conforming to the dominant group's culture, adopting its language and values, and intermarrying with that group.

pluralism minority groups maintain many aspects of their original culture while living peacefully with the host culture.

AP Images

The Washington Post/Getty Images

to an animal because she gave birth to a biracial son (Hamedy, 2014). The judge later apologized, but both of these incidents show that racism persists in the United States and other countries.

10-5a Racism

Racism is a set of beliefs that one's own racial group is inherently superior to other groups. Using this definition, anyone can be racist if she or he believes that another group is inferior. It's a way of thinking about racial and ethnic differences that justifies and preserves the social, economic, and political interests of dominant groups.

Whereas blacks see racism as continuing, many whites view it as a problem that has been pretty much "solved." After all, we've elected an African American president twice, have a Latina Supreme Court justice, many law firms advertise summer positions that are limited to minority candidates, and minorities now own many small businesses. One outcome of such changes is that many whites now believe that antiwhite bias is a bigger social problem than antiblack bias (Bernstein, 2011; Norton and Sommers, 2011). You'll see that such beliefs aren't supported by data, but racism fuels both prejudice and discrimination.

10-5b Prejudice

Prejudice is an *attitude* that prejudges people, usually in a negative way, who are different from "us" in race, ethnicity, or religion. If an employer assumes, for example, that white workers will be more productive than blacks or Latinos, she or he is prejudiced. Prejudice isn't one-sided because *anyone* can be prejudiced ("White people can't be trusted" or "Black women can't afford expensive purses"). Prejudice is most evident in stereotypes and scapegoating.

A **stereotype** is an oversimplified or exaggerated generalization about a group of people (see, for example, www.stuffwhitepeoplelike.com). Stereotypes can be positive ("African Americans are great athletes") or negative ("African Americans are violent"). Whether positive or negative, stereotypes distort reality. Some blacks are great athletes and some are violent, just like people in other groups. Whatever the intention, stereotypes lump people together and reinforce the belief that many traits are biological and fixed. Once established, stereotypes are difficult to change because people often dismiss any evidence to the contrary as an exception ("For an Asian, Jeremy Lin is a terrific basketball player").

Stereotypes can lead to a displacement of anger and aggression on **scapegoats**, individuals or groups whom people blame for their own problems or shortcomings ("They didn't hire me because the company wants blacks" or "I didn't get into that college because Asians Americans are at the top of the list"). Minorities are easy targets because they typically differ in physical appearance and are usually too powerless to strike back (Allport, 1954; Feagin and Feagin, 2008).

At one time or another, almost all newcomers to the United States have been scapegoats. Especially in times of economic hardship, the most recent immigrants often become scapegoats

© pedalist/Shutterstock.com

racism a set of beliefs that one's own racial group is inherently superior to other groups.

prejudice an attitude that prejudges people, usually in a negative way.

stereotype an oversimplified or exaggerated generalization about a group of people.

scapegoats individuals or groups whom people blame for their own problems or shortcomings.

("Latinos are replacing Americans in construction jobs"). Prejudice, stereotypes, and scapegoating are attitudes, but they often lead to discrimination.

10-5c Discrimination

Discrimination is *behavior* that treats people unequally or unfairly because of their group membership. It encompasses all sorts of actions, ranging from social slights (e.g., not inviting minority coworkers to lunch) to rejection of job applications and racially motivated hate crimes. Discrimination can be subtle (e.g., not sitting next to someone) or blatant (e.g., racial slurs), and it occurs at individual and institutional levels.

Individual discrimination is harmful action on a one-to-one basis by a dominant group member against someone in a minority group. Nationally, 74 percent of blacks, compared with only 31 percent of whites, believe that they have been discriminated against because of their race (Romano and Samuels, 2012).

In **institutional discrimination** (also called *institutionalized discrimination, systemic discrimination*, and *structural discrimination*), minority group members experience unequal treatment and fewer opportunities because of the everyday operations of a society's laws, rules, policies, practices, and customs. Institutional discrimination is widespread. Recently, Wells Fargo Home Mortgage, one of the largest and most powerful U.S. banks, settled a $175 million lawsuit which alleged that minorities, particularly blacks and Latinos, experienced thousands of home foreclosures because of discriminatory lending practices (Broadwater, 2012).

Discrimination, both individual and institutional, also occurs *within* racial-ethnic groups. For example, immigrants from both Mexico and Central American countries speak the same language (Spanish) and migrate to Los Angeles in large numbers. El Salvadorans, however, believe that many Mexican American business owners discriminate against Central and South Americans, even those in low-paying jobs, because of the cultural differences in accents, food preferences, and music (Bermudez, 2008).

---

discrimination behavior that treats people unequally or unfairly because of their group membership.

individual discrimination harmful action on a one-to-one basis by a dominant group member against someone in a minority group.

institutional discrimination unequal treatment and fewer opportunities that minority groups members experience because of the everyday operations of a society's laws, rules, policies, practices, and customs.

TABLE 10.2 RELATIONSHIP BETWEEN PREJUDICE AND DISCRIMINATION

		Does the Person Discriminate?	
		Yes	No
Is the Person Prejudiced?	Yes	Prejudiced discriminator (e.g., a prejudiced person who attacks minority group members verbally or physically)	Prejudiced nondiscriminator (e.g., a prejudiced person who goes along with equal employment opportunity policies)
	No	Unprejudiced discriminator (e.g., an unprejudiced person who joins a club that excludes minorities)	Unprejudiced nondiscriminator (e.g., an unprejudiced employer who hires minorities)

Source: Based on Merton 1949. Reference crediting: Merton, Robert K 1949. "Discrimination and the American Creed." Pp. 99–126 in Discrimination and National Welfare, edited by Robert M. MacIver. New York: Harper.

10-5d Relationship Between Prejudice and Discrimination

Sociologist Robert Merton (1949) created a model showing how the relationship between prejudice and discrimination can vary. His model includes four types of people and their possible response patterns (see *Table 10.2*).

Unprejudiced nondiscriminators aren't prejudiced and don't discriminate. They believe in the American creed of freedom and equality for all and cherish egalitarian values. They may not do much, however, individually or collectively, to change discrimination. In contrast, but equally consistent in attitude and action, are *prejudiced discriminators* who are both prejudiced and discriminate. They're willing to defy laws, such as not renting to minorities, to express their beliefs.

Unprejudiced discriminators aren't prejudiced, but they discriminate because it's expedient or in their own self-interest to do so. If, for example, an insurance company charges higher automobile insurance premiums to people with low occupational and educational levels

"You look like this sketch of someone who's thinking about committing a crime."

David Sipress/The New Yorker Collection/Cartoon Bank.Com

(often minorities), agents will implement these policies even though they themselves aren't prejudiced.

Prejudiced nondiscriminators are prejudiced, but don't discriminate. Despite their negative attitudes, they hire minorities and are civil in everyday interactions because they believe they must conform to antidiscrimination laws or situational norms. If, for example, most of their neighbors or coworkers don't discriminate, prejudiced nondiscriminators will go along with them.

10-6 MAJOR RACIAL AND ETHNIC GROUPS IN THE UNITED STATES

Four states (California, Hawaii, New Mexico, and Texas) and the District of Columbia are now "majority-minority," meaning that minorities make up more than half of the population ("Asians Fastest-Growing...," 2013). Of the major racial-ethnic groups, some encounter more barriers than others, but all have numerous strengths that enhance U.S. society. Let's begin with white ethnic groups whose ancestors are from Europe.

10-6a European Americans: A Declining Majority

During the seventeenth century, English immigrants settled the first colonies in Massachusetts and Virginia. Other white Anglo-Saxon Protestants (WASPs), who included people from Wales and Scotland, quickly followed. Most of these groups spoke English. Some of the immigrants were affluent, but many were poor or had criminal backgrounds.

DIVERSITY

About 58 percent of the U.S. population has a European background. The largest groups have ancestors from Germany, Ireland, England, Italy, Poland, France, and the Scandinavian countries (see *Table 10.3*).

CHARACTERISTICS AND CHANGES

WASPs generally looked down on later waves of immigrants from southern and eastern Europe. They viewed the newcomers as inferior, dirty, lazy, and uncivilized because they differed in language, religion, and customs. New England, which was 90 percent Protestant, was particularly hostile to Irish Catholics, characterizing them as irresponsible and shiftless (Feagin and Feagin, 2008).

TABLE 10.3 AMERICANS OF EUROPEAN DESCENT

Of the many European ancestries that Americans report, the following comprise at least 1.5 percent of the U.S. population.

Ancestry	Number (In Millions)	Percentage of Total U.S. Population
German	50.7	16.5
Irish	36.9	12.0
English	27.7	9.0
Italian	18.1	5.8
Polish	10.1	3.3
French	9.4	3.0
Scottish	5.8	1.9
Dutch	5.0	1.6
Norwegian	4.6	1.5

Note: This survey was based on a total estimated population of almost 307.1 million Americans in 2009.

Source: Based on U.S. Census Bureau, 2012, Table 52.

All the later waves of European immigrants faced varying degrees of hardship in adjusting to the new land because the first English settlers had a great deal of power in shaping economic and educational institutions. In response to prejudice and discrimination, many of the immigrants founded churches, schools, and recreational activities that maintained their language and traditions (Myers, 2007).

Despite stereotypes, prejudice, and discrimination, European immigrants began to prosper within a few generations. They surmounted numerous obstacles and became influential in all sectors of U.S. life. Overall, they now fare much better financially than most other groups in the United States. This doesn't mean that all are rich. In fact, in absolute numbers, poor whites outnumber those of other racial-ethnic groups (see Chapter 8).

10-6b Latinos: The Largest Minority

About 1 in 3 Americans is a member of a racial or ethnic minority, but Latinos are the largest group, constituting 17 percent of the nation's population. The size and growth of the Latino population this century is due mainly to births in the United States, not recent immigration, because, on average, Latinas have three births each, one more than white, black, and Asian or Pacific Islander women (U.S. Census Bureau Population Division, 2012; Krogstad and Lopez, 2014).

Dominican-born Alfredo Rodriguez is one of numerous successful Latino businessmen who are rebuilding neglected inner-city neighborhoods. In 1985, Rodriguez bought his first grocery store in Queens, New York, with the $25 a week his mother had been setting aside for him for a decade. In 2002, he purchased a supermarket in Newark, New Jersey, to meet the needs of local Latino shoppers. Five years later, his Xtra Supermarket had annual sales of $9 million (Rayasam, 2007).

Jeffrey Macmillan for U.S News & World Report

DIVERSITY

Worldwide, only Mexico has a larger Latino population (120 million) than the United States (54 million). More persons of Puerto Rican origin now live in the 50 states and the District of Columbia than in Puerto Rico ("Hispanic Heritage Month…," 2014; Cohn et al., 2014).

Some Latinos trace their roots to the Spanish and Mexican settlers who established homes and founded cities in the Southwest before the arrival of the first English settlers on the East Coast. Others are recent immigrants or children of the immigrants who arrived in large numbers at the beginning of the twentieth century. Of the Latinos living in the United States, most are from Mexico (see *Figure 10.4*), but Spanish-speaking people from different countries vary widely in their customs, cuisines, and cultural practices.

CHARACTERISTICS AND CHANGES

About 21 percent of Americans age 5 and older speak a language other than English at home. In this group,

74 percent speak Spanish. The number of U.S. residents who speak Spanish at home increased by 121 percent between 1990 and 2012, whereas the number who spoke Italian, French, German and other Indo-European languages declined during this period (Ryan, 2013; "Hispanic Heritage Month…," 2014).

The median household income of Latinos is lower than the national median of $51,939, but higher than that of American Indians and African Americans (see *Figure 10.5*). Almost 41 percent of Latino families earn $50,000 a year or more, up considerably from only 7 percent in 1972, but almost 25 percent of Mexican Americans and Puerto Ricans and 17 percent of Cuban Americans live below the poverty line (U.S. Census Bureau, 2012; DeNavas-Walt and Proctor, 2014).

FIGURE 10.4 LATINOS BY ORIGIN, 2012

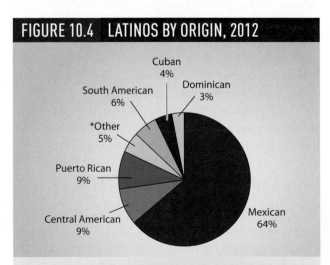

Cuban 4%
Dominican 3%
South American 6%
*Other 5%
Puerto Rican 9%
Central American 9%
Mexican 64%

Note: Central American includes countries such as El Salvador, Honduras, and Guatemala; South American includes countries such as Argentina, Bolivia, and Venezuela.

*Includes people from Spain and those who didn't specify a country of origin.

Source: Based on U.S. Census Bureau, 2012 American Community Survey, Table B03001, accessed March 20, 2014 (factfinder2.census.gov).

FIGURE 10.5 U.S. MEDIAN HOUSEHOLD INCOME, BY RACE AND ETHNICITY, 2013

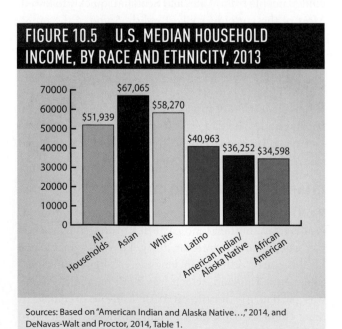

All Households	$51,939
Asian	$67,065
White	$58,270
Latino	$40,963
American Indian/Alaska Native	$36,252
African American	$34,598

Sources: Based on "American Indian and Alaska Native…," 2014, and DeNavas-Walt and Proctor, 2014, Table 1.

Latinos have among the lowest education levels (see *Figure 10.6* on p. 190), but there's considerable variation across subgroups. Almost half of Venezuelans have a bachelor's degree or higher compared with only 8 to 9 percent of those from Guatemala, Mexico, and Salvador (Ogunwole et al., 2012). As with other groups, the socioeconomic status of Latinos reflects a number of interrelated factors, particularly education, occupation, English language proficiency, and recency of immigration. For example, 21 percent of second-generation Latinos have at least a college degree compared with 11 percent of foreign-born Latinos, and higher median household incomes—$48,400 and $34,600, respectively ("Second-Generation Americans," 2013).

Almost 78 percent of second-generation Latinos, compared with 58 percent of all U.S. adults, believe that most people can get ahead if they're willing to work hard ("Second-Generation Americans," 2013). Such attitudes help explain why many Latinos are successful. They own almost 2.3 million businesses, up almost 44 percent from 2002. Between 1995 and 2005, many Latino immigrants earned better hourly wages than their predecessors because they tended to be older, better educated, and more likely to be employed in construction than in agriculture (Kochhar, 2007; "Hispanic Heritage Month…," 2014).

Few people know that Madame C. J. Walker (1867–1919) manufactured hair care products for black women and was one of the first American women to become a millionaire. Or that Dr. Charles R. Drew (1904–1950) was a renowned surgeon, teacher, and researcher. He founded two of the world's largest blood banks, saving untold lives during and since World War II.

10-6c African Americans: A Changing Minority

The 45 million African Americans are the second largest minority group in the United States, making up 13 percent of the population ("Black (African-American)…," 2014). This percentage includes those of more than one race.

DIVERSITY

Most African Americans share a common characteristic: They're members of the only group ever brought to the United States involuntarily and legally enslaved. The term *African American* encompasses tremendous diversity, including native-born Americans with black, white, American Indian, and/or Latino ancestors, as well as recent immigrants from Africa and elsewhere. Of the 3.3 million foreign-born blacks, 52 percent are from the Caribbean, 33 percent from Africa, and 11 percent from a Central or South American country. The largest African-born populations are from Nigeria and Ghana in Western Africa ("Who is 'Black' in America?" 2011; Gambino et al., 2014).

CHARACTERISTICS AND CHANGES

African Americans make up 13 percent of the U.S. population but 27 percent of those living in poverty (DeNavas-Walt and Proctor, 2014). The median family income of African Americans is the lowest of all racial-ethnic groups (see *Figure 10.5* on p. 188). Especially in the country's inner cities, many young black men are high school dropouts, jobless, or incarcerated (Mincy, 2006; U.S. Census Bureau, 2012). Some cities have implemented mentoring programs that are increasing high school graduation and college enrollment rates (Hodge-Williams, 2014).

Blacks have lower education levels than Asians or whites (see *Figure 10.6*). The proportion of African Americans who earned a college degree or higher increased from 17 percent in 2002 to 22 percent in 2013. During this period, however, the number of black households with annual incomes of $50,000 or more dropped from 36 to 35 percent (DeNavas-Walt and Proctor, 2014).

Household income depends on many factors, including type of occupation, the number of earners, and financial support from family and friends (McKernan et al., 2012; see also Chapters 8 and 11). Nonetheless, some economists estimate that at least 30 percent of the black–white wage gap is due to differential treatment in hiring, firing, and pay rather than differences in formal schooling or specific job skills (Fryer et al., 2011).

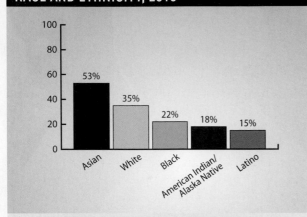

FIGURE 10.6   EDUCATIONAL ATTAINMENT, BY RACE AND ETHNICITY, 2013

Sources: Based on U.S. Census Bureau, "Educational Attainment in the United States: 2013 Detailed Tables," Table 1, accessed March 22, 2014 (census.gov), and "American Indian and Alaska Native...," 2014.

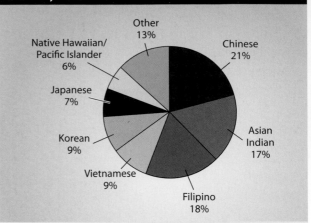

FIGURE 10.7   ASIAN AMERICANS BY ORIGIN, 2012

Note: "Other" includes people from at least 15 countries, including Laos, Cambodia, and Sri Lanka.

Source: Based on U.S. Census Bureau, 2012, American Community Survey, Tables B02018 and B02019. Accessed March 22, 2014 (factfinder2.census. gov).

Despite discrimination, many African Americans are successful. They own almost 6 percent of U.S. businesses. Between 2002 and 2007, the number of black-owned businesses increased by 61 percent to almost 2 million, more than triple the national rate of 18 percent, and generated almost $137 billion in sales in 2007 (*Black-Owned Firms...*, 2006; "Census Bureau Reports...," 2010; U.S. Census Bureau, 2012).

10-6d Asian Americans: A Model Minority?

The more than 20 million Asian Americans comprise almost 6 percent of the U.S. population. In 2012, Asians became the fastest-growing racial or ethnic group in the country; over 60 percent of the growth came from international migration. Hawaii is the only state with an Asian majority of 57 percent ("Asians Fastest-Growing...," 2013; "Asian/Pacific American...," 2014).

DIVERSITY

Asian Americans encompass a broad swath of cultural groups. They come from at least 26 countries in East and Southeast Asia (e.g., China, Korea, Vietnam, Cambodia, the Philippines) and South Asia (especially India, Pakistan, and Sri Lanka), and speak at least 19 languages in the United States (Ryan, 2013; "Asian/Pacific American...," 2014).

These diverse origins mean that there are huge differences in languages and dialects (and even alphabets), religions, cuisines, and customs. Chinese are the largest Asian American group, followed by Filipinos and Asian Indians (see *Figure 10.7*). At least 83 percent of Asian Americans trace their roots to only six countries—China, India, Japan, Korea, the Philippines, and Vietnam (Taylor et al., 2012).

CHARACTERISTICS AND CHANGES

Compared with whites and other racial-ethnic groups, many Asian Americans are doing well. They have the highest education levels (see *Figure 10.6*), and almost 21 percent, compared with 11 percent of the general population, have a graduate or professional degree. Because of their high education levels, almost half of Asian Americans are in highly skilled and high-paying occupations such as information technology, science, engineering, and medicine ("Asian/Pacific American...," 2014). Consequently, they have the highest median household income (see *Figure 10.5* on page 188).

In 2007, Asian Americans owned 1.5 million businesses, an increase of 40 percent from 2002. They own about 43 percent of the 47,000 hotels and motels in the United States. In many cases, the owners bought run-down lodgings and converted them to upscale Sheraton and Hilton hotels (Yu, 2007; "Asian/Pacific American...," 2014).

Because of their educational and economic success, Asian Americans are often hailed as a "model minority." Such labels are misleading, however, because there's considerable variation across subgroups. For example, more than 70 percent of those from India and Taiwan have a bachelor's degree or higher compared with only 12 to 14 percent of Hmong, Laotians, and Cambodians. More than 63 percent of Hmong, Laotians, and Cambodians and 50 percent of Native Hawaiians and Vietnamese haven't attended college compared with only 23 percent of Filipinos and Asian Indians (Teranishi, 2011; Ogunwole et al., 2012).

Fully 93 percent of Asian Americans describe members of their country of origin group as "very hardworking";

Many U.S. malls are struggling to survive, but those catering to Asian Americans are thriving. The malls are successful because they sell familiar foods (rather than non-necessities such as diamond necklaces) and provide customers a sense of community (Haq, 2009).

just 57 percent say the same about Americans as a whole (Taylor et al., 2012). You'll see in Chapter 13 that many Asian American college students succeed, and despite discrimination, because they work harder than others. However, the model minority concept stereotypes Asian Americans and has negative effects. For example, it pigeonholes group members into specific careers and doesn't offer the chance "to see themselves as something other than doctors, engineers, or accountants"; assumes that all minorities can overcome institutionalized discrimination; and holds all Asian Americans to unrealistically high expectations (Kay et al., 2013; Leung, 2013).

10-6e American Indians: A Growing Nation

American Indians used to be called the "vanishing Americans," but they've "staged a surprising comeback" due to higher birth rates, a longer life expectancy, and better health services (Snipp, 1996: 4). The 5.2 million American Indians and Alaska Natives (AIANs) make up almost 2 percent of the U.S. population, and are expected to increase to almost 3 percent by 2060 ("American Indian and Alaska Native...," 2014).

DIVERSITY

Like Asian Americans, American Indians and Alaska Natives are a heterogeneous group. Of the 566 federally recognized tribes, 8 have more than 100,000 members. The Cherokee, with almost 820,000 members, is the largest, followed by the Navajo and Choctaw (Norris et al., 2012). Tribes speak 150 native languages, although many are quickly vanishing. For example, only about 25 people nationwide speak Comanche. Tribes also vary widely in their

religious beliefs and cultural practices. Thus, a Comanche-Kiowa educator cautions, "Lumping all Indians together is a mistake. Tribes ... are sovereign nations and are as different from another tribe as Italians are from Swedes" (Pewewardy, 1998: 71; Ashburn, 2007; Mangan, 2013).

CHARACTERISTICS AND CHANGES

AIANs are a unique minority group because they're not immigrants and have been in what is now the United States longer than any other group. They have experienced centuries of subjugation, exploitation, and political exclusion (e.g., not having the right to vote until 1924). Some tribes are still trying to reclaim billions of dollars that the federal government has squandered or mismanaged (Wilkinson, 2006; Volz, 2012; Kindy, 2013).

Many AIANs are better off today than they were a decade ago, but long-term institutional discrimination has been difficult to shake. For example, 29 percent live below the poverty line compared with 15 percent of the general population. The median household income of AIANs is slightly higher than that of African Americans, but lower than that of other racial-ethnic groups (see *Figure 10.5* on p. 188).

Their educational levels have increased, but only 18 percent of AIANs have a bachelor's degree or higher compared with 29 percent of the general population. One of four civilian-employed AIANs works in a management or professional occupation. Even when AIANs are similar to whites in age, gender, education level, marital status, state of residence, and other factors, their odds of being employed are 31 percent lower than those of whites. This suggests, according to some researchers, that AIANs are experiencing discrimination in the labor market (Austin, 2013; "American Indian and Alaska Native...," 2014).

Despite numerous obstacles, AIANs have made considerable economic progress by insisting on self-determination and the rights of tribes to run their own affairs. American Indian tribes now own 37 percent of the U.S. gambling industry. Few tribes benefit, however, because many casinos are in remote areas that don't attract tourists. In other cases, powerful tribes have cast out some clans to increase their own share of the profits. Outside of gaming, the number of AIAN-owned businesses grew from fewer than 5 in 1969 to nearly 237,000 in 2007, most of them in construction and repair, maintenance, and personal and laundry services (Dao, 2011; Millman, 2012b; "American Indian and Alaska Native ...," 2011).

Recently, the Justice Department announced that it wouldn't block tribes from growing or selling marijuana on their lands, even in states that ban the practice. Some tribes see marijuana sales as a potential source of revenue, similar to cigarette sales and casino gambling, "which

Mohegan Sun in southern Connecticut is one of the largest casinos in the United States. It has spent some of its profits on college scholarships, a $15 million senior center, and health insurance for tribal members. In contrast, some of the poorest tribes, such as the Navajo and Hopi, who have rejected gaming for religious reasons, have many members who live in poverty without kitchen facilities (like stoves or refrigerators) or indoor plumbing.

have brought a financial boon to reservations across the country." Others, "mindful of the painful legacy of alcohol abuse in their communities," are opposed to selling or using marijuana on their territory (Phelps, 2014: 6).

10-6f Middle Eastern Americans: An Emerging Minority

The Middle East is "one of the most diverse and complex combinations of geographic, historical, religious, linguistic, and even racial places on Earth" (Sharifzadeh, 1997: 442). It encompasses about 30 countries, including Armenia, Turkey, Israel, Iran, Afghanistan, Pakistan, and 22 Arab nations (e.g., Algeria, Iraq, Kuwait, Saudi Arabia, and the United Arab Emirates).

DIVERSITY

Of the 57 million people in the United States who speak a language other than English at home, more than 3 percent speak Middle Eastern languages such as Armenian, Arabic, Hebrew, Persian, or Urdu (U.S. Census Bureau, 2012). There are more than 1.5 million Americans of Arab descent, accounting for less than 0.5 percent of the total population (Asi and Beaulieu, 2013). Those who identify themselves as Arab Americans come from many different countries (see *Figure 10.8*). As in the case of Asian American families, Middle Eastern families make up a heterogeneous population that is a "multicultural, multiracial, and multiethnic mosaic" (Abudabbeh, 1996: 333).

CHARACTERISTICS AND CHANGES

Middle Eastern Americans tend to be better educated and wealthier than other Americans. About 46 percent have a college degree or higher compared with 29 percent of the general population; 20 percent have a post-graduate degree, which is nearly twice the U.S. average of 10 percent. In 2010, the median income of Arab American households was almost $57,000 compared with $52,000 for all U.S. households. As in other groups, however,

Steve Jobs—the late co-founder, chairman, and CEO of Apple Inc.—was born to unmarried university students. His father was a Syrian-born Muslim, and his mother was a U.S.-born Catholic of Swiss descent. The father put the baby up for adoption because his girlfriend's family wouldn't allow her to marry an Arab ("Steve Jobs...," 2011).

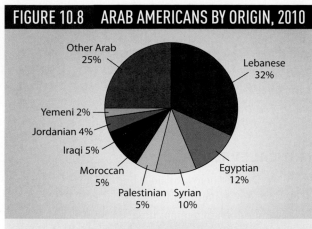

FIGURE 10.8 ARAB AMERICANS BY ORIGIN, 2010

Other Arab 25%
Lebanese 32%
Yemeni 2%
Jordanian 4%
Iraqi 5%
Moroccan 5%
Palestinian 5%
Syrian 10%
Egyptian 12%

Note: "Other Arab" includes those from the Middle East and North Africa.

Source: Based on Asi and Beaulieu, 2013, Table 1.

there are wide variations. About Lebanese have higher median family incomes ($67,300) than Iraqis ($32,000) (Asi and Beaulieu, 2013; Motel and Patten, 2013).

Not all Middle Eastern Americans are successful, of course. Lebanese and Syrians have the lowest poverty rates, 11 percent, compared with more than 26 percent for those from Iraq, and 23 percent of all foreign-born Middle Easterners (Arab American Institute Foundation, 2012; Motel and Patten, 2013).

10-7 SOCIOLOGICAL EXPLANATIONS OF RACIAL-ETHNIC INEQUALITY

As with other topics, the four major sociological theories help us understand racial-ethnic relations. (*Table 10.4* summarizes the key points of each perspective.)

10-7a Functionalism

Those who criticize immigrants for not becoming Americanized quickly enough reflect a functionalist view of racial-ethnic relations. That is, if a society is to work harmoniously, newcomers must adopt the dominant group's values, goals, and particularly, language. Doing so increases a society's cultural solidarity.

STABILITY AND COHESION

Immigration is functional for the host nation if it gains needed workers. Highly educated and skilled immigrants fill important positions in medicine, science, and business. Many employers also rely on immigrants to work in fields, orchards, and vineyards at low wages; others are actively recruiting immigrants for decent-paying but "dirty" jobs at meat and fish factories that native-born Americans avoid (Newkirk and Douban, 2012; Davey, 2014).

Racial-ethnic inequality is dysfunctional, but maintains or increases many dominant group members' solidarity. Institutional discrimination is particularly effective in relegating many minorities to low-paying jobs, low quality schools, and inferior housing, making more of these scarce resources available to the dominant group. In inner-city neighborhoods, as you saw in Chapter 8, fringe bankers (check advance stores and pawnshops) profit by offering financial services to low-income residents who can't get bank loans or credit cards. Thus, racial-ethnic inequality persists because it benefits dominant groups.

CRITICAL EVALUATION

Functionalism helps us understand why acculturation and assimilation increase social solidarity, but overlooks their negative outcomes. For example, second-generation immigrants are more likely than their foreign-born peers to join gangs, commit crimes, and experience

TABLE 10.4 SOCIOLOGICAL EXPLANATIONS OF RACIAL-ETHNIC INEQUALITY

Theoretical Perspective	Level of Analysis	Key Points
Functionalist	Macro	Immigration provides needed workers; acculturation and assimilation increase social solidarity; racial-ethnic inequality can be dysfunctional, but benefits dominant groups.
Conflict	Macro	There's ongoing strife between dominant and minority groups; powerful groups maintain their advantages primarily through economic exploitation; race is a more important factor than social class in perpetuating racial-ethnic inequality.
Feminist	Macro and micro	Minority women suffer from the combined effects of racism and sexism; gendered racism occurs within and across racial-ethnic groups.
Symbolic Interactionist	Micro	Because race and ethnicity are socially constructed, social interaction can increase or reduce racial and ethnic hostility; antagonistic attitudes toward minorities, which are learned, can be lessened through cooperative interracial and interethnic contacts.

obesity and other health problems (Bersani, 2014; see also Chapter 14).

Another weakness is that functionalists, by focusing on order and stability, ignore racial-ethnic inequalities that often spawn tension and discord. Although, as you saw earlier, immigrants have revitalized many communities in the Midwest, their presence has been controversial and divisive (Martinez, 2011). Functionalists acknowledge that inequality is dysfunctional, but this isn't their major focus. Thus, functionalists seem to accept discrimination as inevitable (Chasin, 2004).

10-7b Conflict Theory

Conflict theorists see ongoing strife between dominant and minority groups. Dominant groups try to protect their power and privilege, whereas subordinate groups struggle to gain a larger share of societal resources.

ECONOMIC AND SOCIAL CLASS INEQUALITY

For most conflict theorists, economic inequality generates racial-ethnic inequality. According to a classic explanation, there's a "split labor market." Jobs in the *primary labor market*, held primarily by white workers, provide better wages, health and pension benefits, and some measure of job security. In contrast, workers in the *secondary labor market* (e.g., fast-food employees) are largely minorities and easily replaced. Their wages are low, there are few fringe benefits, and working conditions are generally poor (Doeringer and Piore, 1971; Bonacich, 1972).

Such economic stratification pits minorities against each other and low-income whites. Because these groups compete with each other instead of uniting against exploitation, capitalists don't have to worry about increasing

© 2001 Brian Fairrington and PoliticalCartoons.com/Cagle Cartoons, Inc.

gendered racism the overlapping and cumulative effects of inequality due to racism *and* sexism.

wages or providing safer work environments. Some conflict theorists maintain that race is a more important factor than social class in explaining and perpetuating racial-ethnic inequality. They note, for example, that even middle-class African Americans (and their counterparts in other minority groups) experience discrimination on a daily basis that reminds them of their subordinate position in U.S. society (Feagin and Sikes, 1994).

CRITICAL EVALUATION

Conflict theories help explain why discrimination occurs and persists, as when members of privileged groups benefit by subordinating minorities economically. However, conflict theorists often assume that racial inequality is conscious, deliberate, widespread, and inescapable. In contrast, large majorities of both blacks and Latinos say that immigration and income, not race or ethnicity, are the primary sources of social conflict in the United States (Morin, 2009). Such data suggest a substantial consensus among some minorities that racial inequality isn't due entirely to racism.

There's much evidence that racial inequities persist in employment, but the discrimination isn't always as conscious as some conflict theorists maintain. A study of 700 retail stores found that store managers tended to hire members of their own racial or ethnic group. The researchers attributed the outcomes to factors such as segregated neighborhoods and hiring networks (i.e., institutional discrimination), rather than to prejudice and deliberate discrimination (Giuliano et al., 2009).

10-7c Feminist Theories

Walk through almost any hotel, large discount store, nursing home, or fast-food restaurant in the United States. You'll notice two things: Most of the low-paid employees are women, and predominantly minority women. For feminist scholars, such segregation of minority women is due to gendered racism.

GENDERED RACISM

Gendered racism refers to the overlapping and cumulative effects of inequality due to racism *and* sexism. Many white women encounter discrimination on a daily basis (see Chapter 9). Minority women, however, are also members of a racial-ethnic group, bringing them a double dose of inequality. If social class is included, some minority women experience *triple oppression*. Many affluent women, in particular, have no qualms about exploiting recent immigrants, especially Latinas, who perform demanding housework at very low wages (Hondagneu-Sotelo, 2001).

Some viewers and movie analysts have praised *Dear White People* for addressing everyday racism. Others have criticized the film for featuring light-skinned African Americans in some of the leading roles, and for stereotyping both blacks and whites. What do you think?

Gendered racism also occurs *within* racial-ethnic groups. According to a black male sociologist, scholars rarely discuss black male privilege, which is characterized by having advantages over black women. Examples include being promoted more often and getting higher pay than their black female counterparts who are equally skilled and educated (National Public Radio, 2010).

CRITICAL EVALUATION

Feminist perspectives have sharpened our understanding of the effects of gendered racism and minority women's subordinate status in U.S. society. Because all of us have internalized institutional discrimination, however, minority group members may also be guilty of reinforcing gendered racism in schools, workplaces, and other situations. African American girls as young as 15 have to cope with black men's sexual harassment, stereotypes that black girls and women are sexually promiscuous, and many black men's preference for women with lighter complexions, facial features, and hair that are closer to European standards of beauty (Friedman, 2011; Thomas et al., 2011). A related weakness is that feminist scholars seldom explore women's current participation in gendered racism (e.g., affluent black women and Latinas who exploit domestic and farm workers).

10-7d Symbolic Interaction

According to symbolic interactionists, we learn attitudes, norms, and values throughout the life course. Because, as you saw earlier, race and ethnicity are constructed socially, labeling, selective perception, and social contact can have powerful effects on everyday intergroup relations.

LABELING, SELECTIVE PERCEPTION, AND THE CONTACT HYPOTHESIS

We learn attitudes toward dominant and minority groups through labeling and selective perception, both of which can increase prejudice and discrimination. For example, a comprehensive study of major U.S. news magazines (e.g., *Time*, *U.S. News & World Report*) concluded that labeling immigrants as a "problem" and a "menace" ignored "the broader array of Latino roles and contributions to American communities" (Gavrilos, 2006: 4; see also "Media Coverage of Hispanics," 2009).

Some media critics have lambasted black sitcoms (e.g., *House of Payne*, *Let's Stay Together*) as stereotypical and artificial portrayals of African American life (Abrams, 2011; Crockett, 2011). Others have similarly criticized popular television shows such as *The Big Bang Theory* and *2 Broke Girls* for stereotyping South Asian men as nerdy

After Hurricane Katrina in 2005, the caption on a photo of a young black man wading through water described him as "looting."

A caption described a white man and a light-skinned woman as "finding" goods at a flooded local grocery store.

For symbolic interactionists, images shape our perceptions of racial and ethnic groups.

scientists, asexual, speaking broken English, and being so-cially awkward (Deggans, 2012; Dholia, 2014).

Negative images create and reinforce racial and ethnic stereotypes, but people can decrease labeling and selective perception. The **contact hypothesis** posits that the more people get to know members of a minority group personally, the less likely they are to be prejudiced against that group. Such contacts are most effective when dominant and minority group members have approximately the same status (e.g., both are coworkers or bosses), when they share common goals (e.g., work on a project), when they cooperate rather than compete, and if an authority figure supports intergroup interaction (e.g., an employer requires white supervisors to mentor minority workers) (Allport, 1954; Kalev et al., 2006).

CRITICAL EVALUATION

Symbolic interaction is valuable in helping us understand how race and ethnicity shape our everyday lives. If we're aware of the negative effects of labeling and selective perception, for example, we can change our attitudes and behavior. It's not clear, however, why labeling, selective perception, and racial bias are more common among some people than others, especially when they're similar on a number of variables such as social class, gender, age, religion, race, and ethnicity (Dovidio, 2009).

Another weakness is that symbolic interaction tells us little about the social structures that create and maintain racial-ethnic inequality. For instance, people who aren't prejudiced can foster discrimination by simply going along with the inequitable policies that have been institutionalized in education, the workplace, and other settings (i.e., the unprejudiced discriminators described earlier).

10-8 INTERRACIAL AND INTERETHNIC RELATIONSHIPS

President Obama, the son of a black father from Kenya and a white mother from Kansas, identified himself as "black" on the 2010 census questionnaire, although he could have checked off both black and white or "other."

---

contact hypothesis posits that the more people get to know members of a minority group personally, the less likely they are to be prejudiced against that group.

miscegenation marriage or sexual relations between a man and a woman of different races.

White supremacist Craig Cobb made headlines worldwide when he tried, unsuccessfully, to establish a white-only town in North Dakota. Shortly after that, Cobb's DNA test showed that he was 14 percent black. How awkward, a journalist observed ("DNA Reveals...," 2013: 8).

The president is just one of the growing number of Americans who are biracial or multiracial.

10-8a Growing Multiracial Diversity

The 2000 U.S. Census for the first time allowed people to mark more than one race, which generated about 126 categories and combinations. In 2010, 97 percent of Americans reported being only one race, but almost 3 percent (9 million people) self-identified as being two or more races, up from 2.4 percent in 2000. Every state saw its multiple-race population jump by at least 8 percent, but Native Hawaiians/Pacific Islanders were the most likely to identify themselves as belonging to two or more races (see *Figure 10.9*).

10-8b Interracial Dating and Marriage

Laws against **miscegenation**, marriage or sexual relations between a man and a woman of different races, existed in America as early as 1661. It wasn't until 1967, in the U.S. Supreme Court's *Loving v. Virginia* decision, that antimiscegenation laws were overturned nationally, but not everyone is happy about interracial relationships. Recently, for example, a Baptist congregation in Kentucky voted to ban interracial couples from becoming members or participating in worship activities (Waldron, 2011).

In 2013, 87 percent of Americans approved of black–white marriages, up from only 4 percent in 1958 (Newport, 2013). A large majority (63 percent) of

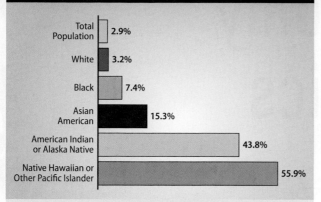

FIGURE 10.9 PERCENTAGE OF AMERICANS WHO IDENTIFIED THEMSELVES AS BEING TWO OR MORE RACES, 2010

Total Population	2.9%
White	3.2%
Black	7.4%
Asian American	15.3%
American Indian or Alaska Native	43.8%
Native Hawaiian or Other Pacific Islander	55.9%

Note: Of the almost 309 million U.S. population in 2010, only 0.3 percent said they were three or more races.

Source: Based on Jones and Bullock, 2012, Figure 10.

greater the potential for intermarriage, because educated minority group members often attend integrated colleges, and their workplaces and neighborhoods are more racially mixed than in the past (Chen and Takeuchi, 2011; Wang, 2012).

People often marry outside of their racial-ethnic groups because of a shortage of potential spouses within their own group. Because the Arab American population is so small, 80 percent of U.S.-born Arabs have non-Arab spouses. Acculturation also affects intermarriage rates. Intermarriage is more common among second-generation immigrants (15 percent) than those in the first generation (8 percent). By the third generation, 31 percent of Latinos and Asian Americans have a spouse of a different race or ethnicity from themselves (Kulczycki and Lobo, 2002; "Second-Generation Americans," 2013).

Americans say that they'd be fine with a family member's marriage to someone of a different racial or ethnic group (Wang, 2012).

Attitudes about interracial relationships are changing, but what about behavior? Racial-ethnic intermarriages have increased slowly—from only 0.7 percent of all marriages in 1970 to nearly 10 percent in 2010. In 2010, also, 18 percent of heterosexual unmarried couples were of different races and 21 percent of same-sex couples were interracial. A record 15 percent (1 in 7) of new U.S. marriages were between people of different races or ethnicities. Asian Americans have the highest intermarriage rates and whites and blacks the lowest (Lofquist et al., 2012; Wang, 2012).

The increase in intermarriage is due to many interrelated factors—both micro and macro level—that include everyday contact and changing attitudes. For example, we tend to date and marry people we see on a regular basis. The higher the educational level, the

STUDY TOOLS 10

READY TO STUDY? IN THE BOOK, YOU CAN:

☐ Check your understanding of what you've read with the Test Your Learning Questions provided on the chapter review card at the back of the book.

☐ Rip out the chapter review card for a handy summary of the chapter and key terms.

ONLINE AT CENGAGEBRAIN.COM YOU CAN:

☐ Prepare for tests with quizzes.

☐ Review the key terms with Flash Cards.

☐ Play games to master concepts.

© Anelina/Shutterstock.com

11 | The Economy and Politics

LEARNING OBJECTIVES

11-1 Compare and illustrate the different types of global economic systems.

11-2 Describe corporations and explain how corporate and political power are interwoven.

11-3 Explain how and why macro-level variables have changed the U.S. economy and jobs.

11-4 Compare and illustrate the different types of global political systems.

11-5 Explain the relationship between power, authority, and politics.

11-6 Describe the U.S. political system, and explain who votes, who doesn't, and why.

11-7 Compare and evaluate the theoretical explanations of the economy and politics.

After you finish this chapter go to **PAGE 221** for **STUDY TOOLS.**

In 2014, 42 percent of Americans said that they were worse off financially than in 2013, and 89 percent wanted the economy to be a top priority for the president and Congress (Dugan, 2014; Newport and Wilke, 2014). Such national data show that many Americans see a close linkage between their personal lives, the economy, and politics.

The **economy** determines how a society produces, distributes, and consumes goods and services. In **politics** individuals and groups acquire and exercise power and authority and make decisions. Societies worldwide differ in the kinds of economic and political systems they develop because of many factors such as globalization and technology. Let's begin by looking at global economic systems.

economy determines how a society produces, distributes, and consumes goods and services.

politics individuals and groups acquire and exercise power and authority and make decisions.

What do you think?

The country would be better off if there were more women in political office.

| 1 | 2 | 3 | 4 | 5 | 6 | 7 |

strongly agree · · · · · strongly disagree

11-1 GLOBAL ECONOMIC SYSTEMS

The two major economic systems around the world are capitalism and socialism. A handful of countries endorse communism, but in actual practice, economies are usually some mixture of both capitalism and socialism.

11-1a Capitalism

Capitalism is an economic system based on private ownership of property and competition in producing and selling goods and services. Ideally, capitalism has four essential characteristics (Smith, 1776/1937; Heilbroner and Thurow, 1998):

▶ **Private ownership of the means of production.** Property (e.g., real estate, banks, utilities) belongs to individuals or organizations rather than the state or the community.

▶ **Market competition.** Owners of production compete in deciding what goods and services to produce and setting prices that offer consumers the greatest value in price and quality.

▶ **Profit.** Selling something for more than it costs to produce generates profits and an accumulation of wealth for individuals and companies.

▶ **Investment.** By investing profits, capitalists can increase their own wealth. Workers, too, can save and invest their money.

In reality, capitalism doesn't function ideally because of abuses, greed, and worker exploitation. It usually results in monopolies and oligopolies rather than a free market that encourages competition. Federal laws prohibit the establishment of a **monopoly**, the domination of a particular market or industry by one person or company. With little or no competition, a company would be able to dictate prices and lower the quality of goods and services.

capitalism an economic system based on private ownership of property and competition in producing and selling goods and services.

monopoly domination of a particular market or industry by one person or company.

An **oligopoly**, which is legal, is the domination of a market by a few large producers or suppliers. In the mid-1980s, 50 companies controlled 90 percent of all U.S. media. Today, it's six companies—including CBS, Time Warner, and the Walt Disney Corporation. There were hundreds of airlines in the mid-1980s, but four giant airlines (United, Delta, American, and Southwest) now control 85 percent of the U.S. market. CEOs maintain that merging companies increases efficiency and safety. Critics say, however, that oligopolies raise prices, stifle competition, and that consumers have fewer choices in buying goods and services (Fox, 2013; Yglesias, 2013; Morris, 2014). Moreover, entrepreneurs have little chance of breaking into an industry dominated by an oligopoly.

11-1b Socialism

Socialism is an economic system based on the public ownership of the production of goods and services. Ideally, socialism is the opposite of capitalism and has the following characteristics:

▶ **Collective ownership of property.** The community, rather than the individual, owns property. The state owns utilities, factories, land, and equipment, but distributes them equally among all members of a society.

▶ **Cooperation.** Working together and providing social services to all people are more important than competition.

▶ **No profit motive.** Private profits that are fueled by greed and exploitation of workers are forbidden.

▶ **Collective goals.** The state is responsible for all economic planning and programs. People are discouraged from accumulating individual profits and investments, and are expected to work for the greater good.

There have been many socialist governments during the last 150 years, but none has reflected pure socialism. Competition and individual profits are officially forbidden, but government officials, top athletes, and high-ranking party members enjoy more freedom, larger

Walt Disney Company is an example of an oligopoly. Among other holdings, it owns publishing companies, web portals, 19 major television and cable stations, at least 60 radio stations, numerous magazines, and 14 theme parks and resorts around the world. All of these outlets promote and sell Disney products, increasing the corporation's profits.

apartments, higher incomes, and greater access to education and other resources than the rank and file.

11-1c Communism

Communism is a political and economic system in which property is communally owned and all people are considered equal. Whereas socialism allows some free market economy, and people receive resources according to how hard they work, communism demands that all production be owned by the public, and that people receive resources according to their need.

Karl Marx (1867/1967) believed that socialist states would evolve into communist societies. This hasn't happened and most formerly communist countries have collapsed. Today, only China, Cuba, Laos, North Korea, and Vietnam practice some form of communism, but have also incorporated capitalism. In contrast to its principles, communism has failed, according to historians, because of widespread corruption and mismanagement, oppression, imprisonment or execution of people who questioned coercive policies, economic inefficiency that resulted in shortages of food and other goods, and few rewards for working hard (Brown, 2010; Johnson, 2010; Pollick, 2014).

11-1d Mixed Economies

Welfare capitalism (also called *state capitalism*) is an economic system that combines private ownership of property, market competition, and a government's

oligopoly domination of a market by a few large producers or suppliers.

socialism an economic system based on the public ownership of the production of goods and services.

communism a political and economic system in which property is communally owned and all people are considered equal.

welfare capitalism (also called *state capitalism*) an economic system that combines private ownership of property, market competition, and a government's regulation of many programs and services.

Reuters/Landov

China is an example of a country with a mixed economy: It espouses communism, practices socialism, and has endorsed many aspects of capitalism. China's economic boom hasn't benefited all of its citizens. Many urban centers are thriving—offering those in upper and middle classes high-rise apartments, stores, and restaurants. In contrast, millions of Chinese, including many older people, like the one pictured here, survive by scouring trash bins for plastic bottles to recycle.

regulation of many programs and services. Many countries, including the United States, have some form of welfare capitalism. Most industry is private, but the government owns or operates some of the largest industries and services (e.g., education, transportation, postal services) that benefit the entire population (Jacoby, 1997; Esping-Andersen, 1990; Hacker, 2002).

Crony capitalism is an economy based on close relationships between business people and the government. In this system, the government gives wealthy people and large corporations preferential treatment through special tax breaks (called *tax credits* and *deductions*), direct payment or loans (called *subsidies*, not welfare), and provides grants that rarely require competition in an open market. From 2000 to 2012, for example, all but one of the Fortune 100 companies received almost $1.3 trillion from the federal government in tax credits, contracts, grants, loans, and various subsidies (Andrzejewski, 2014; see also the discussion of corporate welfare in Chapter 8).

Large companies drive both capitalism and mixed economies. Corporations, in particular, wield enormous influence both at home and abroad.

11-2 CORPORATIONS AND THE ECONOMY

When Bob Thompson sold his construction company for $442 million, he distributed almost a third of the profits to his 550 workers, as well as to some retirees and widows. About 80 became instant millionaires ("Boss Sells His Company...," 1999). Unlike Thompson, most corporate heads have luxurious lifestyles, eliminate employees' pensions, retire as billionaires, and pass on their massive wealth to their heirs (see Chapter 8). Corporate and political power are interwoven, but let's begin by looking at some of the characteristics of corporations.

11-2a Corporations

A **corporation** is an organization that has legal rights, privileges, and liabilities apart from those of its members. Until the 1890s, there were only a few U.S. corporations—in textiles, railroads, and the oil and steel industries. Today, there are almost 6 million, most created for profit, but a mere 0.5 percent bring in 90 percent of all corporate income (U.S. Census Bureau, 2012).

Whereas many industrialized nations have closed corporate tax loopholes, the United States has expanded them. As a result, many corporations exercise more power than do governments, and have amassed enormous wealth. A number of U.S. corporations use legal tax loopholes to avoid paying tens of billions of dollars every year on overseas income; 29 companies alone have more cash than the U.S. Treasury Department; and the profits of the Fortune 500 corporations have soared (Durden, 2011; Jilani, 2011; Niquette and Rubin, 2014).

11-2b Conglomerates

A **conglomerate** is a corporation that owns a collection of companies in different industries. Conglomerates emerged during the 1960s and grow by acquiring companies through mergers. Mergers might increase the value of shareholders' stock, but typically make chief executives "truly, titanically, stupefyingly rich" (Morgenson, 2004: C1). An example of a conglomerate is Kraft Foods, which owns companies that produce snacks, beverages, pet foods, a variety of groceries and convenience foods, and has ties with other corporations such as Starbucks (see www.kraft.com/brands).

Conglomerates diversify business risk by participating in a number of different markets, and can ward off

corporation an organization that has legal rights, privileges, and liabilities apart from those of its members.

conglomerate a corporation that owns a collection of companies in different industries.

emerging competition. However, their huge bureaucracies are costly to maintain and can bog down decision making as markets and technology quickly change. Conglomerates were popular in the United States and Europe until the 1960s, but only a few dozen survive in these countries today (Ray, 2013).

11-2c Interlocking Directorates

In an **interlocking directorate**, the same people serve on the boards of directors of several companies or corporations. An estimated 15 to 20 percent of corporate directors sit on two or more corporate boards. Some interlocking directorates are especially powerful because they include past U.S. presidents and past members of Congress who provide connections in Washington, D.C. Interlocking directors develop similar economic perspectives; those who sit on two or more corporate boards are especially likely to receive appointments to influential government advisory committees (Domhoff, 2013). Thus, a very small group of people has the power to shape the national agenda.

11-2d Transnational Corporations and Conglomerates

Interlocking directorates have become more influential than ever because of the proliferation of transnational corporations. A **transnational corporation** (also called a *multinational corporation* or an *international corporation*) is a large company that's based in one country but operates across international boundaries. By moving production plants abroad, large U.S. corporations can avoid trade tariffs, bypass environmental regulations, and pay low wages. The political leaders of many countries welcome transnational corporations to stimulate their economies, create jobs, and enrich their personal bank accounts (Caston, 1998).

The most powerful are **transnational conglomerates** (also called *multinational conglomerates*), corporations that own a collection of different companies in various industries in a number of countries. General Electric is the world's biggest transnational conglomerate with 70 percent of its businesses and more than half of its employees based abroad. It owns hundreds of companies in the United States, and has subsidiaries in at least 27 countries and on every continent ("Biggest Transnational Companies," 2012). Of the 25 most profitable conglomerates worldwide, 17 are based in the United States. In effect, then, a very small group of corporations holds considerable global economic power (McGregor and Hamm, 2008).

11-3 WORK IN U.S. SOCIETY TODAY

Poll after poll shows that financial hardship is a top issue for many Americans. Many worry about the economy in general, losing their job, and paying off debt (Dugan, 2014; Wilke and Newport, 2014). Economic problems vary by social class, but much of the hardship has to do with changes in **work**, a physical or mental activity that produces goods or services. Even those who work hard

Many U.S. factories today need only a few skilled workers to manage the high-tech machines that have replaced people.

---

interlocking directorate the same people serve on the boards of directors of several companies or corporations.

transnational corporation (also called a *multinational corporation* or an *international corporation*) a large company that's based in one country but operates across international boundaries.

transnational conglomerate (also called a *multinational conglomerate*) a corporation that owns a collection of different companies in various industries in a number of countries.

work a physical or mental activity that produces goods or services.

have lost economic ground because of macro-level variables such as deindustrialization, globalization, offshoring, and weakened labor unions.

11-3a Deindustrialization and Globalization

Many Americans have been casualties of **deindustrialization**, a process of social and economic change resulting from the reduction of industrial activity, especially manufacturing. Since 2000, 32 percent of U.S. manufacturing jobs have disappeared (Hargrove, 2011). In mid-2012, 304,000 manufacturing plants were operating in the United States, 27,000 fewer than at the end of 2007 (Philips, 2012). A major reason for deindustrialization is that, beginning in the early 1960s, employers easily replaced workers with the lowest skill levels, usually those on assembly lines, with automation. Machines and robots have also replaced many blue collar jobs that provided a comfortable standard of living in the past.

The number of manufacturing jobs plummeted from a peak of almost 20 million in 1979 to less than 12 million by the end of the 2007–2009 recession (Barker, 2011). Since then, manufacturing has been rebounding. It was the fourth largest U.S. employer in 2011, behind only health care, retail trade, and food services, and the average wage is almost $30.00 an hour ("Manufacturing Day...," 2013).

The work has changed, however: "Today's U.S. factories aren't the noisy places where your grandfather knocked in four bolts a minute for eight hours a day. Dungarees and lunch pails are out; computer skills and specialized training are in, [and] workers must be able to master the machines" (Foroohar and Saporito, 2013: 24). Manufacturers complain about a shortage of skilled candidates for many jobs, particularly operating or programming computer-controlled cutting tools, repairing sophisticated machinery, and designing software that runs a factory's operations (Hagerty, 2013; Philips, 2014).

Deindustrialization was also hastened by **globalization**, the growth and spread of investment, trade, production, communication, and new technology around the world. One example of globalization is a car that's assembled in the United States with practically all of its parts manufactured and produced in Germany, Japan, South Korea, or developing countries.

Proponents argue that everyone benefits from globalization because it creates millions of jobs and brings affordable goods and services (e.g., cell phones) to millions of people around the world. Critics contend that while globalization has improved the living standards of millions of workers in developing countries such as China and India, it has reduced the number of workers in Western economies. They also maintain that, by and large, globalization benefits only the world's most powerful transnational corporations: It increases their profits by expanding their worldwide base of consumers, exploits poor people in developing countries, creates greater income inequality between workers and the global elite, and gives corporations unprecedented political power (Schaeffer, 2003; Porter, 2014).

11-3b Offshoring and Labor Unions

Offshoring refers to sending work or jobs to another country to cut a company's costs at home. Sometimes called *international outsourcing* or *offshore outsourcing*, the transfer of manufacturing jobs overseas has been going on since at least the 1970s.

Between 2001 and 2011, U.S. companies moved more than 2.7 million jobs to China, 77 percent of them in manufacturing and primarily blue-collar (Scott, 2012). During the same period, U.S. firms also offshored 28 percent of high-level, well-paid information technology (IT) jobs, including those in accounting, computer science, and engineering (National Science Board, 2012). Most of the offshored jobs go to India and China, but many have also moved to Canada, Hungary, Mexico, Poland, Russia, Egypt, Venezuela, Vietnam, and South Africa.

Companies can get accounting services in India that cost much less than in the United States (see *Figure 11.1*). In Mexico, the average General Motors worker earns wages and benefits that cost less than $4 an hour compared with $55 an hour in the United States. Because of such large wage differences, American consumers can purchase goods and services at low prices, and the majority doesn't care where products are made (Black, 2010; Guarino, 2013; Newport et al., 2014).

P_Wei/iStockphoto.com

deindustrialization a process of social and economic change resulting from the reduction of industrial activity, especially manufacturing.

globalization the growth and spread of investment, trade, production, communication, and new technology around the world.

offshoring sending work or jobs to another country to cut a company's costs at home.

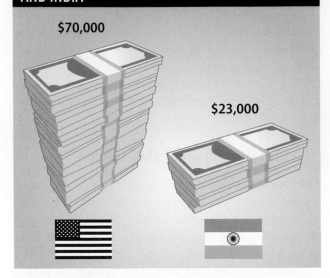

FIGURE 11.1 AVERAGE ANNUAL EARNINGS OF AN ACCOUNTANT IN THE UNITED STATES AND INDIA

$70,000

$23,000

Deindustrialization, globalization, and offshoring have weakened *labor unions*, organized groups that seek to improve wages, benefits, and working conditions. Union membership has dropped sharply—from 35 percent of the workforce in the mid-1950s to 11 percent in 2013 ("Union Members—2013," 2014). In the 1950s, 75 percent of Americans approved of labor unions compared with only 54 percent in 2013 (Dugan, 2013). Five states prohibit unionization; in 2012, 24 states significantly restricted collective bargaining rights; and anti-union groups have helped defeat efforts to unionize autoworkers in Tennessee and other states (Trottman and Maher, 2013; Whitesides, 2014).

Do we still need unions? Opponents argue that unions have too much influence, that their members are overpaid, and that unions drain state resources because of high pensions and salaries of public sector employees (e.g., teachers, nurses, sanitation workers, police). Critics also contend that unions have limited employers' flexibility in hiring and firing decisions, and that ever-increasing labor and health care costs have forced employers to move their operations overseas to remain competitive (McKinnon, 2011; Schlesinger, 2011).

Proponents argue that states have experienced budget deficits because of the housing crisis and a recession that was due to Wall Street greed and not overpaid union members. Others point out that union

members have made numerous concessions like decreasing their wages and benefits. Most important, historically, unions have benefited almost all workers by insisting on paid holidays and vacations, greater workplace safety, overtime pay, and by challenging corporations that are becoming more dominant and powerful in both the economy and politics (Klein, 2011; Welch, 2011; Gould and Shierholz, 2012).

11-3c How Americans' Work Has Changed

Only 29 percent of U.S. workers are "completely satisfied" with their pay (Saad, 2013). The U.S. economy started adding more jobs in 2012, but millions of Americans still have only low-paying and part-time jobs. If these strategies fail, they find themselves among the unemployed.

LOW-WAGE JOBS

A recent study of 702 U.S. occupations found that automation and globalization have wiped out millions of middle-income jobs (e.g., sales, accounting, office and administrative support). By 2025, 230 million white collar jobs and 47 percent of all U.S. employees are expected to be replaced by computers (Frey and Osborne, 2013; Manyika et al., 2013).

Except for registered nurses, whose median annual salary is about $65,600, seven of the occupations expected to add the most jobs through 2020 are low paying, up to a maximum of about $24,000 a year. These occupations include home health aides and personal care aides who assist older and disabled people. Other fast-growing but low-paying jobs—from about $22,000 to $31,000 a year—include retail salespersons, general office clerks, and nurses' aides (Sommers and Franklin, 2012; Maher, 2013; see also "Labor Day 2014...," 2014).

The federal minimum wage, which rose from $6.55 to $7.25 an hour in 2009, increased the wages of less than 4 percent of the workforce. A worker earning a minimum wage today is worse off than one who made the base hourly wage of $1.60 in 1968. Taking inflation into account, according to some economists, the minimum wage should be about $15.00 an hour. The current federal minimum wage is lower than that of 19 states and the District of Columbia, but only three states pay $9.00 an hour or more (Klein and Leiber, 2013; Trumbull, 2014).

Many small businesses are starting their workers at $10 an hour, but some of the wealthiest corporations contend that increasing the federal minimum wage would result in much higher consumer costs. If, however, the nation's largest low-wage employer, Walmart, paid its 1.4 million U.S. workers $12 per hour and passed every penny of the costs to consumers, the average Walmart customer would pay just 46 cents more per shopping trip, or about $12 a year. It would cost McDonald's only 4 cents more per meal to provide all its workers a $10.10 minimum wage that President Obama proposed to Congress (Jacobs et al., 2011; Velasco, 2013). In contrast, Costco, a discount retailer, has huge profits yet pays employees an average of $21 per hour, not including overtime ("Wages Maximus," 2014: 33).

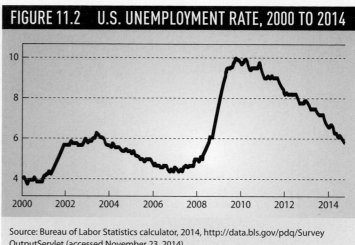

FIGURE 11.2 U.S. UNEMPLOYMENT RATE, 2000 TO 2014

Source: Bureau of Labor Statistics calculator, 2014, http://data.bls.gov/pdq/Survey OutputServlet (accessed November 23, 2014).

PART-TIME WORK

Of the almost 27 million part-timers (those who work less than 35 hours a week), 31 percent are involuntary because people can't find suitable full-time employment or employers have reduced their hours (BLS News Release, 2014). Part-time work—traditionally parceled out to lower-level hourly employees like cashiers and administrative assistants—has spread to white-collar and professional sectors and now includes marketing directors, engineers, and financial officers. Employers save on costs—including vacations and health and unemployment insurance—by hiring more (or all) part-time workers. They admit that the downside includes high turnover and employees who are less committed to turning out a better product or providing a better service (Davidson, 2011).

UNEMPLOYMENT AND DISCOURAGED WORKERS

The U.S. unemployment rate surged between 2008 and 2010, but has dropped since then (see *Figure 11.2*). Despite the drop, only 63 percent of Americans age 16 and older have a job or are looking for one—the lowest share of the population participating in the labor force since 1978. Before and since the recession, unemployment rates have been highest among teenagers, single female heads of households, Latinos, blacks, those with less than a college degree, and people employed in agriculture, construction, and food and hotel industries (Gable, 2013; Bureau of Labor Statistics, 2014; BLS News Release, 2014).

About 36 percent of the jobless are *long-term unemployed*, people who have searched for but not found a job for 27 weeks or longer (Bureau of Labor Statistics, 2014). The longer that people are unemployed, the harder it is to find a job: Only 11 percent find steady, full-time employment a year later. Many of the long-term unemployed become **discouraged workers**,

people who stop searching for employment because they believe that job hunting is futile. About 35 percent of Americans are discouraged workers who aren't included in the Census Bureau's unemployment rate (Krueger et al., 2014).

UNDEREMPLOYED WORKERS

Besides not counting discouraged workers, unemployment rates exclude the **underemployed**—people who have part-time jobs but want full-time work or whose jobs are below their experience, skill, and education levels. In 2014, an estimated 17 percent of Americans were underemployed compared with only 8 percent in 2007 (Mishel et al., 2012; Gallup, 2014).

Among the underemployed, almost 33 percent of those who graduated from college between 2006 and 2010 said that their postcollege job was below their education and skill level. Underemployed Americans are better off financially than the unemployed, but both groups experience similar levels of sadness, stress, anger, worry, and depression (Godofsky et al., 2011; Mishel et al., 2012).

JOB SATISFACTION AND STRESS

A whopping 70 percent of Americans report being "not engaged" or "actively disengaged" in their jobs, meaning that they're not emotionally involved in or enthusiastic

discouraged workers people who stop searching for employment because they believe that job hunting is futile.

underemployed people who have part-time jobs but want full-time work or whose jobs are below their experience, skill, and education levels.

In good times and in bad, African American men have the highest unemployment rates. And compared with their white counterparts, black workers are unemployed longer after their unemployment insurance benefits end. Pictured here are laid-off workers examining listings at a job fair.

TABLE 11.1 VACATION DAYS, ON AVERAGE, THAT WORKERS RECEIVE AND USE

	Receives	Uses
France	30	30
Denmark	30	29
Spain	30	26
Italy	28	20
Japan	18	7
United States	14	10
Malaysia	12	10
Thailand	11	8
South Korea	10	7

Source: Based on "Vacation Habits...," 2013).

about their work (Gallup, 2013). People with a high school diploma or less are slightly more likely to be engaged in their work than those with a college degree (34 percent and 28 percent, respectively). Perhaps workers with only a high school diploma feel lucky to have a job at all when unemployment rates are high. In contrast, many college graduates have invested time and money in a degree, but believe that they're overqualified for their jobs, and have high expectations that their managers don't meet (Sorenson and Garman, 2013).

In most European countries, the typical work week is 35 hours—comparable with the U.S. definition of part-time work. Americans work more hours (1,804 per year) and more weeks per year (48 weeks) than people in any other industrialized country except Japan, and they have the highest productivity rate in the world. In 2014, full-time employees worked an average of 47 hours a week, and nearly 40 percent worked at least 50 hours a week (Begala, 2011; Saad, 2014).

The United States is the only industrialized nation in the world that doesn't legally guarantee workers a paid vacation; 23 percent of Americans have no paid vacation *and* no paid holidays. In contrast, the typical worker, especially in Western Europe, has at least 6 weeks of paid vacation, regardless of job seniority or the number of years worked (Ray et al., 2013). U.S. workers have

the least vacation days and don't use all of them (see *Table 11.1*). Workers in France receive and take 30 days, but 90 percent complain that they're "vacation deprived" ("Vacation Habits...," 2013).

About 60 percent of Americans shrink their vacations to a few days or long weekends because they can't afford a vacation, have too much to do on the job, fear being laid off, or worry about not being promoted when their company keeps pushing employees to put in more hours. Employees who forfeit paid time off don't get more raises or bonuses and report higher stress levels than those who take all of their vacation time, but employers save almost $53 billion a year when people work for free ("Vacation Habits...," 2013; U.S. Travel Association, 2014).

11-3d Women and Minorities in the Workplace

One of the most dramatic changes in the United States during the twentieth century was the increase of women in the labor force (see *Table 11.2*). Many factors have contributed to the surge in women's employment, especially since the 1970s, including the growth in the number of college-educated women (who, consequently, had more job opportunities), an increase in the number of

TABLE 11.2 WOMEN AND MEN IN THE LABOR FORCE, 1890–2013

	Percentage of Men and Women in the Labor Force		Women as a Percentage of All Workers
Year	Men	Women	
1890	84	18	17
1900	86	20	18
1920	85	23	20
1940	83	28	25
1960	84	38	33
1980	78	52	42
1990	76	58	45
2013	70	57	47

Sources: Based on BLS Reports, 2014, and Labor Force Statistics from the Current Population Survey, Table 2, "Employment Status," 2014. Accessed April 15, 2014 (bls.gov/cps/tables.htm).

working single mothers, and the higher costs of home-ownership that require two incomes.

Largely because of their higher educational attainment, 29 percent of women in two-income marriages bring home the bigger paycheck, up from only 4 percent in 1970 ("Wives Who Earn More . . . ," 2014). As a group, however, women have lower earnings than men in both the highest and lowest paying occupations, and the wage gaps are greater in high-income jobs (see *Figure 11.3*).

There are earnings disparities across racial-ethnic groups, but the differences are especially striking by sex. As you examine *Figure 11.4*, note two general characteristics. First, earnings increase—across all racial-ethnic groups and for both sexes—as people go up the occupational ladder. But across all occupations, men have higher earnings than women of the same racial-ethnic group. At the bottom of the occupational ladder are African American women and Latinas, with the latter faring worse than any of the other groups. Thus, both gender *and* race-ethnicity affect earnings.

FIGURE 11.3 WOMEN EARN LESS THAN MEN WHETHER THEY'RE CEOs OR COOKS

These are five of the highest and lowest paid occupations of full-time, year-round U.S. workers in 2013. How might you explain why the earnings differ by sex, especially in the highest paid jobs?

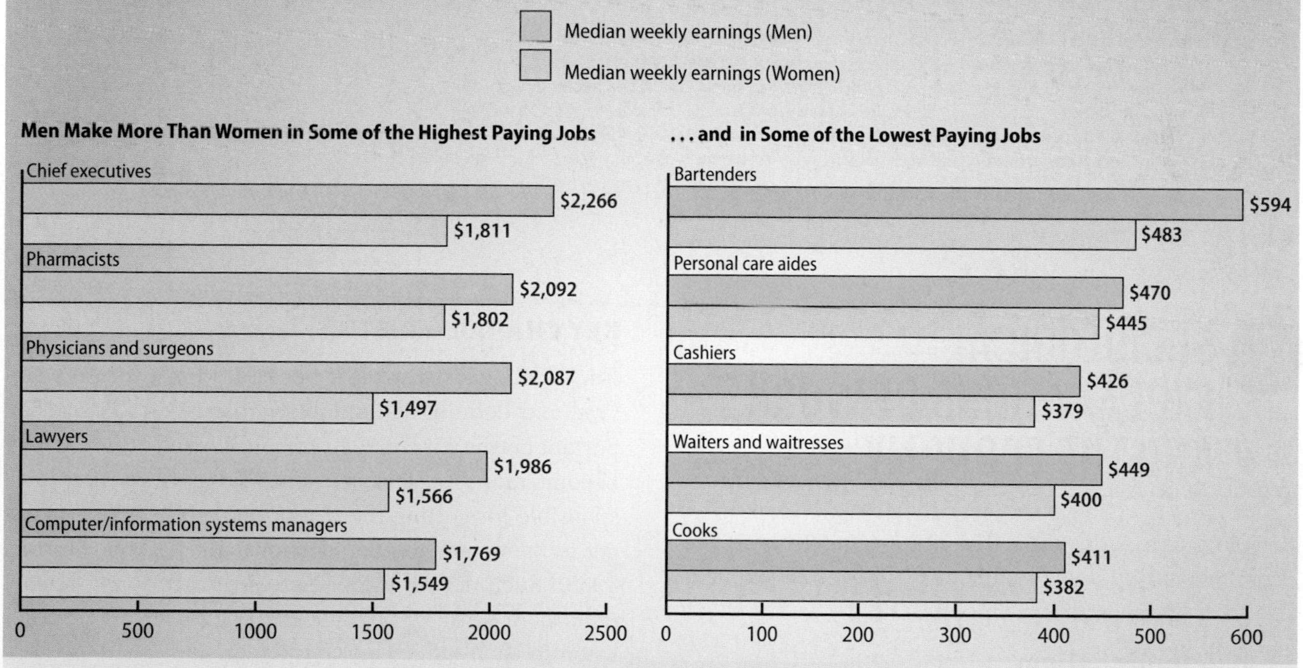

Median weekly earnings (Men)
Median weekly earnings (Women)

Men Make More Than Women in Some of the Highest Paying Jobs

Chief executives
Men: $2,266
Women: $1,811

Pharmacists
Men: $2,092
Women: $1,802

Physicians and surgeons
Men: $2,087
Women: $1,497

Lawyers
Men: $1,986
Women: $1,566

Computer/information systems managers
Men: $1,769
Women: $1,549

. . . and in Some of the Lowest Paying Jobs

Bartenders
Men: $594
Women: $483

Personal care aides
Men: $470
Women: $445

Cashiers
Men: $426
Women: $379

Waiters and waitresses
Men: $449
Women: $400

Cooks
Men: $411
Women: $382

Note: Some of the differences between men's and women's median weekly earnings may seem small, but multiply each figure by 52 weeks. Thus, in annual earnings, male physicians and surgeons average almost $109,000 compared with $78,000 for females.

Source: Based on Labor Force Statistics from the Current Population Survey, 2014, Table 39, "Median Weekly Earning of Full-time Wage and Salary Workers by Detailed Occupation and Sex." Accessed April 15, 2014 (bls.gov/cps/cpsaat39.htm).

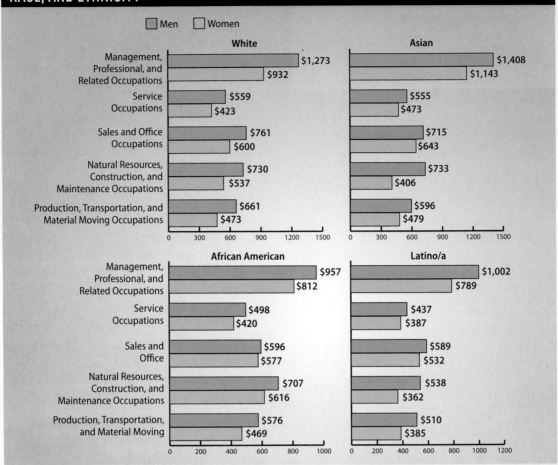

FIGURE 11.4 MEDIAN WEEKLY EARNINGS OF FULL-TIME WORKERS BY OCCUPATION, SEX, RACE, AND ETHNICITY

☐ Men ☐ Women

White

Management, Professional, and Related Occupations	Men $1,273 / Women $932
Service Occupations	Men $559 / Women $423
Sales and Office Occupations	Men $761 / Women $600
Natural Resources, Construction, and Maintenance Occupations	Men $730 / Women $537
Production, Transportation, and Material Moving Occupations	Men $661 / Women $473

Asian

Management, Professional, and Related Occupations	Men $1,408 / Women $1,143
Service Occupations	Men $555 / Women $473
Sales and Office Occupations	Men $715 / Women $643
Natural Resources, Construction, and Maintenance Occupations	Men $733 / Women $406
Production, Transportation, and Material Moving Occupations	Men $596 / Women $479

African American

Management, Professional, and Related Occupations	Men $957 / Women $812
Service Occupations	Men $498 / Women $420
Sales and Office	Men $596 / Women $577
Natural Resources, Construction, and Maintenance Occupations	Men $707 / Women $616
Production, Transportation, and Material Moving	Men $576 / Women $469

Latino/a

Management, Professional, and Related Occupations	Men $1,002 / Women $789
Service Occupations	Men $437 / Women $387
Sales and Office	Men $589 / Women $532
Natural Resources, Construction, and Maintenance Occupations	Men $538 / Women $362
Production, Transportation, and Material Moving	Men $510 / Women $385

Note: Examples of the occupations include the following:
Managerial and Professional—Executives, managers, public administrators
Service—Private household workers, police, firefighters, people in food and health services, janitors
Sales and Office—Supervisors, sales representatives, office administrators
Natural Resources, Construction, and Maintenance—Farmers, fishing industries, construction workers, auto mechanics, machine repairers
Production, Transportation, and Material Moving—Truck drivers, assembly-line workers, equipment cleaners

Source: Based on Bureau of Labor Statistics, *The Editor's Desk*, 2011. Accessed April 15, 2014 (bls.gov/opub/ted/2011/ted_20110914.htm).

11-4 SOCIOLOGICAL EXPLANATIONS OF WORK AND THE ECONOMY

Sociological theories offer differing perspectives on the economy and its impact on society. *Table 11.3* summarizes each perspective's key points.

11-4a Functionalism

Functionalists typically emphasize the benefits of work and the economy. They also see capitalism as bringing prosperity to society as a whole.

KEY CHARACTERISTICS

For functionalists, work is necessary for a society's survival and defining its members' roles. Work is also important because it connects people to each other: As jobs become more specialized, a small group of workers is responsible for getting the work done, and coworkers can get to know one another (Parsons, 1954, 1960; Merton, 1968). Such networks increase workplace solidarity and enhance a sense of belonging to a group where people listen to each other's ideas, interact, and "share a vision for the work [they] do together" (Gardner, 2008: 16).

Besides providing income, work has social meaning. For many Americans, jobs offer a sense of accomplishment and feeling valued; stability, order, and a daily

TABLE 11.3 SOCIOLOGICAL EXPLANATIONS OF WORK AND THE ECONOMY

Theoretical Perspective	Level of Analysis	Key Points
Functionalist	Macro	Capitalism benefits society; work provides an income, structures people's lives, and gives them a sense of accomplishment.
Conflict	Macro	Capitalism enables the rich to exploit other groups; most jobs are low-paying, monotonous, and alienating; productivity isn't always rewarded.
Feminist	Macro and micro	Gender roles structure women's and men's work experiences differently and inequitably.
Symbolic Interactionist	Micro	How people define and experience work in their everyday lives affects their workplace behavior and relationships with coworkers and employers.

rhythm; and a network of interesting social and professional contacts (Katzenbach, 2003; Brooks, 2007). Functionalists also maintain that wage inequities spur people to persevere and to set higher goals. Low-paying jobs, for example, can motivate people to work harder or obtain further education to move up the economic ladder (see Chapter 8).

CRITICAL EVALUATION

Critics point out that work is a significant source of stress for 69 percent of Americans and often leads to myriad health problems (American Psychological Association, 2014). Many people have unstimulating and low-paid jobs that require repetitive and routine tasks, instead of providing a connection to a product or a group (see the discussion of McDonaldization in Chapter 6). Even in higher paying jobs, only a third of physicians and nurses report being engaged in their work (Gallup, 2013).

Functionalists also gloss over many U.S. corporations' disinterest in workers' well-being. In the private sector, 30 percent of full-timers and 74 percent of part-timers don't have paid sick days (Williams and Gault, 2014).

AP Images/Connecticut Post, Ned Gerard

Even when business improves, many employers convert full-time jobs to part-time or temporary positions, and have been cutting health benefits (Schultz, 2011).

11-4b Conflict Theory

Whereas functionalists emphasize the economy's benefits, conflict theorists argue that capitalism creates social problems. They contend, for example, that globalization has led to job insecurity, and that a handful of transnational conglomerates have enormous power.

KEY CHARACTERISTICS

For conflict theorists, low wages alienate employees rather than motivate them to work harder, as functionalists claim, because many workers realize that their labor benefits the wealthy and not themselves. Many people are stuck in low-paying, monotonous jobs and don't always get promotions regardless of their skills, attitudes, perseverance, or productivity (Clements, 2012).

Conflict theorists also argue that capitalism enables the rich and powerful to exploit other groups, resulting in huge wealth disparities. In most industrialized countries, CEOs earn 10 to 25 times more than the average worker. In the United States, CEOs—including those responsible for the Great Recession's financial crisis—earn 273 times more than the typical worker. In the 100 largest companies, CEOs earn 495 times more than the rank-and-file employees. From 1978 to 2013, CEO pay increased by 937 percent compared with only 10 percent for workers (Smith and Kuntz, 2013; Mishel and Davis, 2014).

For functionalists, the economy provides order and stability. Many unemployed men—including recent Asian and Latino immigrants—are joining the military because it offers steady work, college education benefits, and an opportunity for career advancement.

ARE CONFLICT THEORISTS TOO QUICK TO BLAME CAPITALISM FOR ECONOMIC PROBLEMS?

CRITICAL EVALUATION

Critics fault conflict theorists for emphasizing economic constraints rather than choices. For example, middle-class Americans are accumulating debt faster than they're saving for retirement. Some are using credit cards to keep up with bills, but many are getting deeper in debt because they're living beyond their means (Fellowes and Spiegel, 2013).

Also, according to functionalists, conflict theorists underestimate people's ability to get ahead. Many young people are thronging schools that teach high-tech skills in welding and other trades, fast-food workers are unionizing to raise the industry's minimum wage to $15.00 an hour, and 63 percent of small-business owners find qualified new employees through word-of-mouth referrals (Berfield, 2013; Jacobe, 2013; Phillips, 2014). Thus,

In 2007, both General Motors and federal agencies ignored internal reports that a faulty ignition switch was causing injuries and fatal crashes. The part costs less than $10.00 wholesale and takes less than an hour to fix. It was only in 2014, after lawsuits, that General Motors started recalling 2.6 million Chevrolet Cobalts and other small cars to replace the ignition switches (Fletcher and Mufson, 2014; Wald, 2014). Why does this issue illustrate conflict theory?

motherhood penalty (also called *motherhood wage penalty* or *mommy penalty*) a pay gap between women who are and aren't mothers.

many Americans are using several strategies to succeed economically.

11-4c Feminist Theories

Feminist scholars agree with conflict theorists that there's widespread workplace inequality. They emphasize, however, that gender is a critical factor in explaining the inequity.

KEY CHARACTERISTICS

Nationally, women and men are equally satisfied with job security and relations with coworkers. Pay is the primary exception, and women are much more dissatisfied than men (Saad, 2014). Such data aren't surprising because at all income levels and in all occupations, women—especially Latinas and black women—earn less than men (see *Figure 11.4* on p. 208).

Employment rates increased after the recession, but the gains for women have been largely in low-paying jobs, particularly waitressing, in-home health care, food preparation, and housekeeping. From 2009 to 2012, 60 percent of the increase in women's employment, versus 20 percent for men, was in jobs that pay less than $10.10 an hour (National Women's Law Center, 2014). Both historically and currently, as you saw in Chapters 9 and 10, women have lower earnings than men because of occupational sex segregation, gender pay gaps, glass ceilings, and glass escalators.

Many women's earnings suffer from a **motherhood penalty** (also called *motherhood wage penalty* or *mommy penalty*), a pay gap between women who are and aren't mothers. On average, women without children make 90 cents to a man's dollar, mothers make 73 cents to a man's dollar, and single mothers make only 60 cents to a man's dollar (Rowe-Finkbeiner, 2012; Entmacher et al., 2014).

After declining for several decades, the share of stay-at-home mothers rose from 23 percent in 1999 to 29 percent in 2012. There are many reasons for this increase, but low wages and motherhood penalties are at the top of the list. Many mothers in two-income households have dropped out of the labor force because child care costs rose more than 70 percent from 1985 to 2011. Nationally, families whose youngest child is under age 5 pay, on average, $180 a week for child care. If the jobs available to mothers paid more, many would go to work and use child care (Katz and Tanzi, 2013; Cohn et al., 2014).

CRITICAL EVALUATION

Critics note that men don't always dominate women in the workplace. There are many situations where female supervisors have power and authority over women and men, and top female managers aren't always concerned about gender wage gaps or support qualified women's promotions (Huffman, 2013). Second, many feminist scholars maintain that capitalism exploits women by crowding them into lower paying occupations. According to critics, however, there's considerable economic gender inequality in socialist, communist, and mixed economies in both industrialized and developing nations (Cudd and Holstrom, 2010).

Third, some people question whether the economy reinforces sex discrimination or simply reflects cultural sexism. For example, 51 percent of Americans, including women, say that children are better off if the mother is at home and doesn't work; only 8 percent say the same about stay-at-home dads (Cohn et al., 2014). Thus, patriarchal values may be a more critical factor than discrimination in explaining many women's economic inequality.

11-4d Symbolic Interaction

Symbolic interactionists rely on micro-level approaches to explain the day-to-day meaning of work. They're especially interested in how work shapes people's self-identity.

KEY CHARACTERISTICS

Symbolic interaction has provided numerous insights on how people define and experience work. Regardless of income level, 63 percent of U.S. workers who are engaged in their jobs, and even 42 percent of those who aren't engaged, say they would continue to work in their current job if they won a $10 million lottery. All of them might quit, of course, but such responses show that work provides many people with "a source of identity, purpose, and satisfaction that money alone may not replace" (Harter and Agrawal, 2013).

Interactionists have also studied how people are socialized into their jobs and the informal rules that shape behavior. In medicine, surgeons teach their interns and residents that it's normal to make some mistakes, but that carelessness and continued errors are unprofessional (Bosk 1979). Assembly-line workers also control coworkers who overproduce or underproduce by punishing or rewarding them (see Chapter 6).

CRITICAL EVALUATION

The most common criticism is that symbolic interaction, although providing in-depth analyses, sacrifices scope. Studies of small work groups can tell us much about people's interaction. We don't know, however, if the findings are applicable to large companies or different parts of the country because the research is based on small and nonrepresentative samples.

Another weakness is that interactionism neglects macro-level social forces that affect people's work and

choices. In Baltimore, for example, manufacturing employment dropped by 38 percent between 2000 and 2012 when large companies like General Motors and Westinghouse shut down plants. A community college was eager to help laid-off steelworkers retool their skills, but "any workers who tried to enroll met a tangle of overlapping and sometimes conflicting requirements from the college and the county, state, and federal governments." Others didn't sign up for the programs after learning that, even with new credentials, they would probably get only $14 an hour jobs, quite a drop from $28 an hour jobs in the past (Mangan, 2013: A24). Thus, organizational obstacles can thwart individual choices and efforts.

11-5 GLOBAL POLITICAL SYSTEMS

Every society has a **government**, a formal organization that has the authority to make and enforce laws. Governments are expected to maintain order, provide social services, regulate the economy, establish educational systems, create armed forces to discourage (real or imagined) attacks by other countries, and ensure their residents' safety.

Worldwide, governments vary from democratic to totalitarian, but there are also authoritarian governments and monarchies. Let's begin by looking at democracy.

11-5a Democracy

A **democracy** is a political system in which, ideally, citizens have control over the state and its actions. Democracies are based on several principles:

▶ Individuals participate in governmental decisions, and select leaders who are responsive to the wishes of the majority of the people.

▶ Suffrage (the right to vote) is universal, and elections are frequent, free, fair, and secret.

▶ The government recognizes individual rights, such as freedom of speech (including dissent), press, and assembly, and the right to organize political parties whose members compete for public office.

▶ The "rule of law" requires everyone to obey the law and to be held accountable if they violate it.

government a formal organization that has the authority to make and enforce laws.

democracy a political system in which, ideally, citizens have control over the state and its actions.

FIGURE 11.5 POLITICAL FREEDOM AROUND THE WORLD, 2014

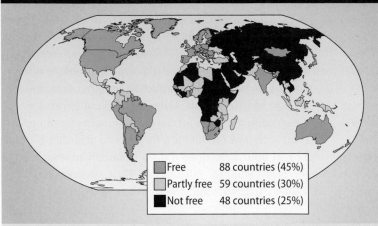

Free — 88 countries (45%)
Partly free — 59 countries (30%)
Not free — 48 countries (25%)

This map shows the degree of political freedom, measured by political rights and civil liberties, in 195 countries around the world. *Political rights* **include free and fair elections, competitive parties, and no discrimination against minorities.** *Civil liberties* **include freedom of speech and the press; freedom to practice one's religion; and the freedom to discuss political issues without fear of physical violence or intimidation by the government.**

Source: Freedom House, 2014.

▶ A system of terror that relies on secret police and the military to intimidate people into conformity and to punish dissenters.

▶ Total control by the government over other institutions, including the military, education, family, religion, economy, media, and all cultural activities, including the arts and sports.

Contemporary examples of the most repressive totalitarian governments include Central African Republic, North Korea, Saudi Arabia, Somalia, Sudan, Syria, and Uzbekistan. Within these countries and territories, state control over daily life is pervasive and wide-ranging, independent organizations and political opposition are banned or suppressed, and those who criticize the government are imprisoned, tortured, or killed (Makinen, 2014; Puddington, 2014).

Worldwide, 35 percent of the world's population (almost 2.5 billion people) lives in countries with repressive governments that deny basic political and civil rights. People in Sub-Saharan and North Africa, the Middle East, China, and Russia are the least likely to have political freedom (see *Figure 11.5*). Despite some countries' attempts to establish democracies, democracy declined around the world in 2014 for the eighth consecutive year (Kurlantzick, 2014; Puddington, 2014).

11-5b Totalitarianism and Dictatorships

At the opposite end of the continuum from democracy is **totalitarianism**, a political system in which the government controls almost every aspect of people's lives. Totalitarianism has several distinctive characteristics (Taylor, 1993; Tormey, 1995; Arendt, 2004):

▶ A pervasive ideology that legitimizes state control and instructs people how to act in their public and private lives.

▶ A single political party controlled by one person, a *dictator*—a supreme, sometimes idolized leader—who stays in office indefinitely.

totalitarianism the government controls almost every aspect of people's lives.

authoritarianism the state controls the lives of citizens but permits some degree of individual freedom.

11-5c Authoritarianism and Monarchies

Many nations have some version of democracy or totalitarianism. However, a number of countries are characterized by **authoritarianism**, a political system in which the state controls the lives of citizens but permits some degree of individual freedom. Authoritarian governments,

Vladimir Putin has served as Russia's Prime Minister and/or President since 1999. In 2014, he occupied and annexed the Ukraine's Crimean peninsula, an important military base and port on the Black Sea. Because Putin once described the Soviet Union's collapse as the twentieth century's worst disaster, some political analysts believe that seizing Crimea is just a first step toward usurping other countries that became independent in the early 2000s ("The End of the Beginning?," 2014; Wynnyckj, 2014).

ALEXEY DRUZHININ/AFP/Getty Images

In 1995, at age 3, Oyo Nyimba Kabamba Iguru became the world's youngest monarch when he was crowned as king of the Ugandan Kingdom of Toro. At age 18, he officially took control of the kingdom.

PETER BUSOMOKE/Getty Images

like Russia, may permit elections, often controlled, but prohibit free speech, punish those who publicly question state policies, and imprison people without giving them a trial (Whitson, 2013; "Don't Think, Just Teach," 2014; Ford, 2014; Kirişci, 2014; Ruth, 2014).

A **monarchy**, the oldest type of authoritarian regime, is a political system in which power is allocated solely on the basis of heredity and passes from generation to generation. A member of a royal family, usually a king or queen, reigns over a kingdom. A monarch's power and authority are legitimized by tradition and religion. There are more than 40 monarchies around the world. In some countries—especially in the Middle East and parts of Africa—monarchs have absolute control. Those in Norway, Denmark, Belgium, Britain, Japan, and many other countries have little political power because they're limited by democratic constitutions and serve primarily ceremonial roles (Glauber, 2011).

We see, then, that political systems vary worldwide. Despite the variation, political leaders wield considerable power and authority.

11-6 POLITICS, POWER, AND AUTHORITY

Power and authority are important concepts in sociological analyses of politics. The use and misuse of power and authority affect the quality of our everyday lives.

11-6a Power

Power is the ability of a person or group to influence others, even if they resist. For example, a government can quash a peaceful protest or, conversely, protect people's right to protest. *Legitimate power* comes from having a role, position, or title that people accept as legal and appropriate (e.g., Congress has the right to pass laws, the police have the right to enforce laws). *Coercive power* relies

on force or the threat of force to impose one's will on others (e.g., a dictator who imprisons or executes protestors).

Because power is relational, it's expressed downward and upward. With downward power, state senators and representatives affect their constituents' lives because of the laws they pass on taxes, the environment, and other issues. In upward power, voters can protest, refuse to follow laws, and vote politicians out of office.

Whether legitimate or coercive, power—especially political power—is about controlling others. Because people may revolt against sheer force, many governments depend on authority to establish order, shape people's attitudes, and control their behavior.

11-6b Authority

Authority, the legitimate use of power, has three characteristics. First, people *consent* to authority because they believe that their obedience is for the greater good (e.g., following traffic rules). Second, people see the authority as *legitimate*—valid, justifiable, and necessary (e.g., paying taxes to support public education, increase national security, and remove trash and snow). Third, people accept authority because it's *institutionalized*, accepted by a large number of people, in organizations (e.g., police departments and government agencies).

Max Weber (1925/1978) described three ideal types of legitimate authority: traditional, charismatic, and rational-legal (see *Table 11.4*). Remember that ideal types are models that describe the basic characteristics of any phenomenon (see Chapters 1 and 5). In reality, these types of authority often overlap.

TRADITIONAL AUTHORITY

Traditional authority is power based on customs that justify the ruler's position. The source of power is personal because the ruler inherits authority on the basis of long-standing customs, traditions, or religious beliefs. Traditional authority is most common in nonindustrialized societies where power resides in kinship groups, tribes, and clans. In many African, Middle Eastern, and Asian countries today, power is passed down among men within a family line (Tétreault, 2001; see also Chapter 8).

monarchy power is allocated solely on the basis of heredity and passes from generation to generation.

power the ability of a person or group to influence others, even if they resist.

authority the legitimate use of power.

traditional authority power based on customs that justify the ruler's position.

TABLE 11.4 WEBER'S THREE TYPES OF AUTHORITY

Type of Authority	Description	Source of Power	Examples
Traditional	Power is based on customs, traditions, and/or religious beliefs.	Personal	Medieval kings and queens, emperors, tribal chiefs
Charismatic	Power is based on exceptional personal abilities or a calling.	Personal	Adolf Hitler, Gandhi, Martin Luther King, Jr.
Rational-legal	Power is based on the rules and laws that are inherent in an elected or appointed office.	Formal	U.S. presidents, members of Congress, state officials, police, judges

CHARISMATIC AUTHORITY

Charismatic authority is power based on exceptional individual abilities and characteristics that inspire devotion, trust, and obedience. Like traditional authority, charismatic authority is personal and reflects extraordinary deeds or even a belief that a leader has been chosen by God, but the leaders don't pass their power down to their offspring.

Charismatic leaders can inspire loyalty and passion whether they're heroes or tyrants. Examples of the latter include historical figures Adolf Hitler, Napoleon, Ayatollah Khomeini, and Fidel Castro. These and other dictators have been spellbinding orators who radiated magnetism, dynamism, and tremendous self-confidence, and promised to improve a nation's future (Taylor, 1993).

RATIONAL-LEGAL AUTHORITY

Rational-legal authority is power based on the belief that laws and appointed or elected political leaders are legitimate. Unlike traditional and charismatic authority, rational-legal authority comes from rules and regulations that pertain to an office rather than to a person. For example, anyone running for mayor must have specific qualifications (e.g., U.S. citizenship). When a new mayor (or governor or other politician) is elected, the rules don't change because power is vested in the office rather than the person currently holding the office.

MIXED AUTHORITY FORMS

People may enjoy more than one type of authority. Some historians believe that presidents Abraham Lincoln, Theodore Roosevelt, John F. Kennedy, and Ronald

charismatic authority power based on exceptional individual abilities and characteristics that inspire devotion, trust, and obedience.

rational-legal authority power based on the belief that laws and appointed or elected political leaders are legitimate.

political party an organization that tries to influence and control government by recruiting, nominating, and electing its members to public office.

ChinaFotoPress/Getty Images

First Lady Michelle Obama is an example of a charismatic political leader. She works to end childhood obesity, helps military families, and visits other countries as a good will ambassador.

Reagan enjoyed charismatic appeal beyond their rational-legal authority. In some countries, including Japan and England, emperors, kings, and queens have traditional authority and perform symbolic state functions (e.g., attending ceremonies). Both of these countries also have parliaments that exercise legal-rational authority in determining laws and policies.

11-7 POLITICS AND POWER IN U.S. SOCIETY

Politics plays a critical role in our everyday lives. Three important components of the political system are political parties, special-interest groups, and voters.

11-7a Political Parties

A **political party** is an organization that tries to influence and control government by recruiting, nominating, and electing its members to public office. Political

TABLE 11.5   HOW DO DEMOCRATS AND REPUBLICANS DIFFER?

Issue	Many Democrats Believe That...	Many Republicans Believe That...
Economy	The government should raise taxes on the wealthy and corporations, and expand programs for the poor.	The government should lower taxes on the wealthy and corporations to encourage investment and economic growth; poor people are too dependent on the government.
Abortion	Women should have the right to choose abortion in practically all cases; the government should pay for abortions for low-income women.	The unborn should be protected in all cases; there should be no public money for abortions.
Gay Rights	Gay men and lesbians should have the right to marry and to adopt children. They should be fully accepted in the military and other institutions.	Marriage and adoption should be limited to heterosexuals. Inclusion of gay men and lesbians in the armed forces creates interpersonal problems.
Education	The government should strengthen public schools by raising salaries for teachers and decreasing classroom sizes. It shouldn't use tax dollars to help students attend private schools.	The government should increase state and local control of schools. It should use tax dollars to fund students to attend private schools of their choice if their local school is underperforming.

Sources: Doherty et al., 2014; Johnson, 2014; Wilke and Newport, 2014.

parties provide a way for citizens to shape public policy at the local, state, and national levels.

FUNCTIONS OF U.S. POLITICAL PARTIES

Political parties engage in a wide variety of activities— everything from mailing flyers and calling voters to drafting laws if a candidate is elected. Parties perform a number of vital political functions, especially recruiting candidates for public office, organizing elections, and if elected, running the government.

Ideally, that's how political parties *should* work. In reality, parties perform some of these functions more effectively than others. For example, parties typically concentrate on winning elections rather than making promised changes once candidates are in office.

THE TWO-PARTY SYSTEM

Many democracies around the world have a number of major political parties: 3 in Canada, 5 in Germany, 9 in Italy, more than 20 in Israel, and about 1,050 in India. Thus, compared with many other countries, the two-party system of Democrats and Republicans in the United States is uncommon. Political parties base their activities on an *ideology*, a set of ideas that constitute a person's or group's beliefs, goals, expectations, and actions. *Table 11.5* offers some examples of major ideological differences between Democrats and Republicans.

Until the late 1980s, about equal percentages of Americans identified themselves as either Republicans or Democrats. Because of dissatisfaction with both parties, however, in late 2013, 22 percent identified themselves as Republicans, 29 percent as Democrats, and a record 46 percent as independent, up from 30 percent in 1991. A large majority (58 percent) of Americans say that we need a third major political party that would

represent the people (Jones, 2014). Because there's no national Independent Party, however, independents usually vote for Democrats, Republicans, or don't vote at all.

11-7b Special-Interest Groups

Some people are active in political parties. Others join a **special-interest group** (also called an *interest group*), people who seek or receive benefits or special treatment. Whereas political parties include diverse individuals, special-interest groups are usually made up of people who are similar in social class and political objectives. There are thousands of U.S. special-interest groups that include everything from agribusiness to unions. Special interest groups use many tactics to influence the government, but the most effective are campaign contributions and lobbying.

CAMPAIGN CONTRIBUTIONS

Money is usually the most important factor in winning an election. Money doesn't guarantee a victory, but candidates with little financing usually have little public visibility. Among all presidential and congressional candidates, the spending doubled from $3.1 billion in the 2000 election to $6.3 billion in the 2014 election (Center for Responsive Politics, 2014a; Montanaro et al., 2014).

Contributions come from individuals and national, state and local party committees, but the most powerful are **political action committees (PACs)**, special-interest groups that raise money to elect one or more

special-interest group (also called an *interest group*) a group of people that seeks or receives benefits or special treatment.

political action committee (PAC) a special-interest group that raises money to elect one or more candidates to public office.

In 5 to 4 decisions in 2010 and 2014, the Supreme Court overturned previous limits on federal campaign donations. Now, individuals, corporations, unions, political parties, and PACs can contribute almost as much as they want (Center for Responsive Politics, 2014c; Common Cause, 2014).

candidates to public office. PACs have mushroomed—from 608 in 1974 to almost 15,700 in 2014 (35 percent of them representing corporate interests), and have received billions of dollars. Some PACs, like the International Brotherhood of Electrical Workers, support primarily Democrats, whereas the National Rifle Association contributes only to Republicans. A number of PACs, however—including the Credit Union National Association and the National Association of Realtors—contribute to both parties to cover their bases regardless of who's elected (Center for Responsive Politics, 2014b; Federal Election Commission, 2014). PACs provide access to politicians and obligate lawmakers to pass legislation that a PAC wants. In contrast with small business firms with few resources, national and transnational corporations can contribute huge amounts of money to PACs, increasing the corporations' influence.

LOBBYISTS

A **lobbyist** is someone hired by a special-interest group to influence legislation on the group's behalf. At the federal level, the number of lobbyists rose from 10,400 in 1998 to almost 12,300 in 2013, an average of 23 lobbyists per congressional member. Some of the best-paid lobbyists include former members of Congress, former congressional aides, and lawmakers' relatives (Drutman and Furnas, 2014; "Lobbying Database," 2014).

In 2013, lobbyists spent $3.2 billion to send members of Congress and their family members on expensive vacations, and, most importantly, made generous campaign contributions. In return, corporations and

lobbyist someone hired by a special-interest group to influence legislation on the group's behalf.

other organizations received favorable legislation, such as Congress passing lax gun and banking laws, approving corporate mergers, and not investigating monopolies (Draper, 2013; Protess, 2013; Schweizer, 2013; Hamburger and Gold, 2014).

Lobbying also generates *earmarks* (known commonly as "pork")—funding requests by Congress to provide federal money to companies, projects, groups, and organizations in their district—that don't require external competitive bidding. Despite calls to ban earmarks, in 2012 taxpayers paid $3.3 billion for pork spending. Some examples of recent pork barrel projects include $200 million for corporations to promote their products overseas, $4.9 million for 11 states to determine how to use wood, and a $350 million tower in Mississippi to test rocket engines that NASA didn't want, didn't need, and hasn't used but costs $840,000 annually to maintain (Kennedy and Gelber, 2013; Salant, 2014).

11-7c Who Votes, Who Doesn't, and Why

In 2012, 62 percent of Americans voted in the presidential election, down slightly from almost 64 percent in 2008. These were larger than usual percentages, but lower than the numbers of those who voted in presidential elections during the 1950s and 1960s (File and Crissey, 2010; File, 2013).

Of 110 countries around the world that hold democratic presidential elections, 69 have higher voter turnout rates than the United States; in 17 countries, at least 80 percent of the eligible population votes (Friedman, 2012; International Institute for Democracy and Electoral Assistance, 2014).

Why are U.S. voting rates so low? Who votes, who doesn't, and why reflect demographic characteristics, attitudes about politics, and situational and structural factors.

DEMOGRAPHIC FACTORS

Many demographic factors (e.g., marital status, geographic region, religion) affect registration and voting. The most important, however, are sex, age, social class, and race and ethnicity.

Sex. In every presidential election since 1996, women—and across all racial/ethnic groups—have voted at higher rates than men. In the 2012 election, 64 percent of women and 60 percent of men voted. The biggest gender voting gap was among African Americans—black women voted at higher rates than black men by almost 9 percentage points (File, 2013). The reasons for these differences are unclear, but may reflect higher turnouts by women who believe that social services, abortion rights, and health care coverage are threatened (Jones, 2012).

TABLE 11.6 VOTER RATES, BY SELECTED CHARACTERISTICS, 2012

A. Age	
18–24	41%
25–44	57
45–64	68
65–74	74
75 and older	70

B. Education	
High school, no diploma	38%
High school graduate	53%
Some college or associate degree	64%
Bachelor's degree	75%
Advanced degree	81%

C. Annual Family Income	
Under $20,000	50%
$20,000 to $29,999	56%
$30,000 to $39,999	58%
$40,000 to $49,999	63%
$50,000 to $74,999	68%
$75,000 to $99,999	74%
$100,000 to $149,999	77%
$150,000 and over	90%

Source: Based on U.S. Census Bureau, Voting and Registration, 2013, Tables 5 and 7.

Age. Historically, and across all elections, voting rates increase with age. In 2012, voting rates were much higher among older than younger people (see *Table 11.6a*). Why the age difference? A key factor is registration. Only 55 percent of young people (ages 18–24) are registered to vote compared with 79 percent of those age 65 and older (U.S. Census Bureau, Voting and Registration, 2013).

Young people have lower registration rates because they're more geographically mobile than older people and are less likely to reregister after a move. Moreover, they often feel uninformed about politics, believe that elections aren't relevant in their lives, and may also be preoccupied with major life events such as going to college and finding jobs (Schachter, 2004; Child Trends Data Bank, 2013). In effect, young people's low voting rates means that they have little effect on political processes.

Social Class. Social class has a significant impact on voting behavior. At each successive level of educational attainment, the voting rate increases, and those with advanced degrees are the most likely to vote (see *Table 11.6b*). People with higher educational levels are usually more informed about and interested in the political process and more

likely to believe that their vote counts ("Who Votes...," 2006; File and Crissey, 2010).

Voting rates also increase with income levels (see *Table 11.6c*). People with higher incomes are more likely to be employed and to have assets (e.g., houses, stock) and, therefore, are more likely to vote to protect or increase their resources. In contrast, low-income people usually have few assets and may be too disillusioned with the political system to vote ("Who Votes...," 2006; File and Crissey, 2010).

Race and Ethnicity. The 2012 presidential election was the first time since at least 1996, the earliest year with comparable data, that African Americans voted at a higher rate than whites (see *Figure 11.6*). Here again, a key to voter turnout is registration. Across all racial-ethnic groups—including both native-born and naturalized citizens—the higher the voter registration, the higher the voting rate (File and Crissey, 2010).

In 2012, blacks and whites had a higher voter registration rate (74 percent) than Latinos (59 percent) and Asian Americans (56 percent), but why did black voters outpace whites? The increase in black voting, especially since 2008, stems from enthusiasm for Barack Obama's presidential candidacy, high levels of voter mobilization, and a decline of the white population. In both 2008 and 2012, however, much of the surge in turnout rates was driven by blacks ages 45 and older, particularly women, who voted in record numbers (File, 2013).

In contrast to blacks, Latinos and Asian Americans who are eligible to vote don't do so, and their turnout fell between 2008 and 2012 (see *Figure 11.6*). About 75 percent of both groups were born in the United States, but have lower voter registration rates than blacks and whites. Latinos have lower education and income levels

FIGURE 11.6 VOTING RATES IN PRESIDENTIAL ELECTIONS, BY RACE AND ETHNICITY: 1996–2012

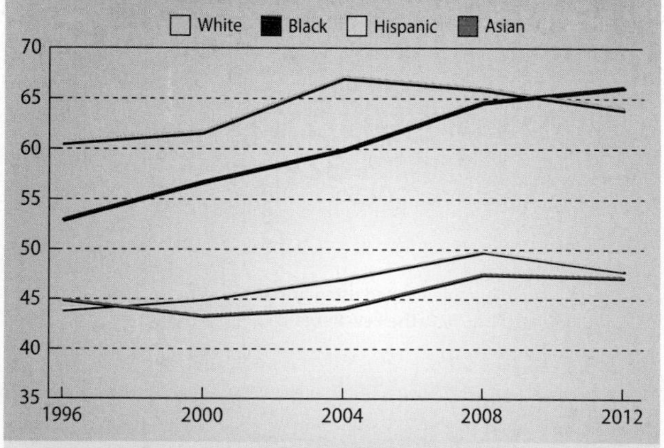

Note: Rates based on total citizen population age 18 and older.

Source: Based on File, 2013, Figure 1.

than most other groups, both of which are associated with lower voter turnout rates. Low Asian American voter turnout is puzzling: They have the highest education and income levels (see Chapters 8 and 10), but even among the college-educated, voter turnout has lagged behind whites, blacks, and Latinos (Jones, 2013; Krogstad, 2014).

ATTITUDES

Millions of Americans have a low opinion of the government and Congress. When Gallup asked Americans to rate 22 occupations according to honesty and ethics, those at the bottom of the list were state officeholders (14 percent), car salespeople (9 percent), members of Congress (8 percent), and lobbyists (6 percent) (Swift, 2013; McCarthy, 2014).

Such distrust may help explain why many people don't vote: They believe that their lives won't improve regardless of who's elected. On the other hand, few political incumbents are voted out of office election after election. Why? Only 17 percent of registered voters say that members of Congress should be re-elected, but almost half believe that *their* congressional representatives are doing a good job (Dugan and Hoffman, 2014). Perhaps, then, "the government we have is the government we deserve" (Cillizza, 2014).

How Would *You* Describe Congress?

In 2010, the Pew Research Center asked a nationally representative sample of Americans to provide one word that described Congress. The responses were put into a "word cloud." The larger the word, the more times it was used (Auxier, 2010). What one word would *you* use to describe Congress?

Source: Word cloud graphic, created using http://wordle.net, from "Congress in a Wordle," March 22, 2010, The Pew Research Center for the People and the Press.

---

pluralism   a political system in which power is distributed among a variety of competing groups in a society.

SITUATIONAL AND STRUCTURAL FACTORS

Situational and structural factors also encourage or discourage voting. Some of the high turnout rates in other countries are partly due to *compulsory voting*: The government imposes fines for not voting, may make it difficult for nonvoters to get benefits such as public assistance, or requires evidence of voting to get a passport or a driver's license. Generally, these kinds of sanctions send the message that voting isn't a privilege but a civic responsibility. In many European and other countries, voters are registered automatically when they pay taxes or receive public services, elections are held on weekends, and people vote electronically (Seward, 2012; Friedman, 2012; International Institute for Democracy and Electoral Assistance, 2014).

Instead of making voting easier, 34 states (particularly those controlled by Republican legislatures) have passed or are considering laws that will make voter registration more difficult. Some of the new rules require government-issued photo identification (e.g., a driver's license or passport) and a birth certificate or proof of citizenship. Some states have reduced early voting, limited the time polls are open, done away with same-day voter registration, and moved polling locations outside of areas with large racial-ethnic populations. Some state legislators contend that the new laws will crack down on voter fraud in future elections, but opponents argue that the new regulations will stifle voter turnout, especially among students, the poor, and minorities, who tend to vote for Democrats (Barnes, 2014; Whiteaker et al., 2014; Yaccino and Alvarez, 2014).

Because many Americans don't vote, who has the most power? And do government leaders represent the average citizen? Such questions have generated considerable debate among sociologists and other social scientists.

11-8 SOCIOLOGICAL PERSPECTIVES ON POLITICS AND POWER

According to symbolic interaction, power is socially constructed through interactions with others; people learn to be loyal to a political system—whether it's a democracy or a monarchy—and to show respect for its symbols and leaders. For the most part, however, sociologists rely on functionalist, conflict, and feminist theories to explain political power. *Table 11.7* summarizes these perspectives.

11-8a Functionalism: A Pluralist Model

For functionalists, the people rule through **pluralism**, a political system in which power is distributed among a variety of competing groups in a society (Riesman, 1953; Polsby, 1959; Dahl, 1961).

TABLE 11.7 SOCIOLOGICAL EXPLANATIONS OF POLITICAL POWER

	Functionalism: A Pluralist Model	Conflict Theory: A Power Elite Model	Feminist Theories: A Patriarchal Model
Who has political power?	The people	Rich upper-class people—especially those at top levels in business, government, and the military	White men in Western countries; most men in traditional societies
How is power distributed?	Very broadly	Very narrowly	Very narrowly
What is the source of political power?	Citizens' participation	Wealthy people in government, business corporations, the military, and the media	Being white, male, and very rich
Does one group dominate politics?	No	Yes	Yes
Do political leaders represent the average person?	Yes, the leaders speak for a majority of the people.	No, the leaders are most concerned with keeping or increasing their personal wealth and power.	No, the leaders are rarely women who have decision-making power.

KEY CHARACTERISTICS

In the pluralist model, individuals have little direct power over political decision making but can influence government policies through special interest groups—unions, professional organizations, and so on (see *Figure 11.7*). The various groups rarely join ranks because they concentrate on single issues such as health care, pollution, or education. This focus on different issues fragments groups, but also results in a broad representation of interests and a distribution of power.

Because there are a number of single-issue groups, functionalists maintain, there are multiple leaderships (Dye and Ziegler, 2003). As a result, many leaders, not just a few, can shape decisions that represent numerous groups and issues. Pluralists note that people also have power outside of interest groups: They can vote, run for office, contact lawmakers, and put specific issues on a ballot. Therefore, there are continuous checks and balances as individuals and groups vie for power and try to influence laws and policies.

CRITICAL EVALUATION

Does pluralism work as democratically as functionalists maintain? Critics argue that interest groups have unequal resources. The poor and disadvantaged rarely have the skills and educational backgrounds to organize or promote their interests. In contrast, wealthy individuals and organizations can influence government through political contributions and personal connections.

FIGURE 11.7 PLURALIST AND POWER ELITE PERSPECTIVES OF POLITICAL POWER

PLURALIST MODEL

Power is dispersed among multiple groups that influence the government. For example,

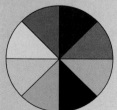

- government employees
- victims' rights groups
- labor unions
- banks and other financial institutions
- realtors and home builders
- teachers
- environmental groups
- women's rights groups

POWER ELITE MODEL

Power is concentrated in a very small group of people who make all the key decisions. For example,

Top level (1 percent)—CEOs of large corporations, high-ranking lawmakers in the executive branch, and top military leaders

Middle level (8 percent)—most members of Congress, lobbyists, entrepreneurs of small businesses, leaders of labor unions and other interest groups (in law, education, medicine, etc.), and influential media commentators

The masses (91 percent)—people who are unorganized and exploited and either don't know or don't care about what's going on in government

Critics also maintain that pluralists aren't realistic about the emergence of multiple leaderships because the affluent control the government. For example:

▶ In 2013, a first-year House of Representatives member had a median net worth of almost $808,000 compared with only $45,000 of the average American.

▶ In 2014, only 5 percent of Americans had a net worth of $1 million or more compared with 50 percent of those in Congress, most of whom are multimillionaires.

▶ All of the Supreme Court Justices and top executive branch members are also multimillionaires ("Millionaires' Club...," 2014; Rodrigue and Reeves, 2014).

Because of such wealth inequality, according to an owner of a one-man glass business, most politicians are totally disconnected from the lives of ordinary Americans. What would he do if he were president? "I'd fire everyone in the House and the Senate, and put working class people in who actually know what it's like to be out here" (Geller, 2014).

11-8b Conflict Theory: A Power Elite Model

Conflict theorists contend that the United States is ruled not by pluralism but a **power elite**, a small group of influential people who make the nation's major political decisions. Sociologist C. Wright Mills (1956) coined the term *power elite* to describe a pyramid of power that he believed characterized American democracy.

KEY CHARACTERISTICS

According to Mills, the power elite is made up of three small but dominant groups of people at the top level who run the country: Political leaders, corporate heads, and military chiefs (see *Figure 11.7*). Practically all of the members of these groups are white, Anglo-Saxon, Protestant men who form an inner circle of power. Those at the middle power level include members of Congress, lobbyists, influential media commentators, and others. The bottom level, and the largest and least powerful group, the masses, is composed of everyone else.

The power elite tolerates the masses—including their elections and laws—but in the end simply does what it wants. In 2013, almost 80 percent of Americans said there should be limits on the amount of money that congressional candidates could spend on their campaigns (Saad, 2013). The following year, the Supreme Court did just the opposite by striking down laws that

power elite a small group of influential people who make the nation's major political decisions.

limited how much a donor may give overall to political candidates, parties, and PACs.

Contemporary conflict theorists maintain, like Mills, that the United States isn't a democracy because of the close ties and "revolving door" between the power elite in politics, business, and the military. For example, fewer than 31,390 people—less than 0.01 percent of the nation's population—contributed more than $1.6 billion to political campaigns in 2012, and nobody was elected to Congress without the economic elite's money. In return, politicians pass favorable legislation such as $83 billion taxpayer-financed subsidies to the country's banks, award earmarks and lucrative military and other contracts, and may later serve as corporate lobbyists (Drutman et al., 2013; Gilson, 2014; Kristof, 2014).

CRITICAL EVALUATION

Functionalists accuse conflict theorists of exaggerating the power elite's importance. Functionalists point out, for example, that "the masses" aren't just puppets but are aware of the power elite's influence. As a result, they vote, mobilize, support particular interest groups, and protest current political and corporate policies.

Also, according to critics, the power elite perspective assumes, incorrectly, that the wealthy people at the top have the same interests and goals. Because the Democratic and Republican parties and their top officials endorse very different agendas, the power elite are rarely unified.

Moreover, critics maintain, the wealthy can't always buy political offices. In 2010, for example, Meg Whitman (the founder of eBay) spent almost $142 million of her personal fortune in her bid for California governor. She lost to Jerry Brown, who contributed only $24,000 of his own money to his $4 million campaign funds (Crowley, 2010; Palmeri and Green, 2010).

11-8c Feminist Theories: A Patriarchal Model

As you saw in Chapter 9, relatively few women, especially minority women, serve in Congress and at the highest government levels. In a recent national poll, 62 percent of men and 69 percent of women said that the country would be better off if there were more women in political office (Nelson, 2013). Why, then, are there so few female officeholders?

KEY CHARACTERISTICS

Feminist theorists maintain that women are generally shut out of the most important political positions because, in a patriarchal society, men rule. A "gender gap in political ambition" begins early in the socialization process. Parents often encourage their sons, not their daughters, to

In 2012, at age 63, Elizabeth Warren (left), a Massachusetts Democrat, ran for the U.S. Senate. The *Boston Herald* repeatedly referred to her as "granny," "Liz," and "Lizzy." She's now called Madame Senator. Some of the *Huffington Post*'s feminist writers said that Michele Bachmann (right)—a candidate for the Republican nomination in the 2012 presidential election—might have won "if she had let down her necklines" (Seltzer, 2012). Do male candidates ever experience similar sexism?

think about politics as a career. As a result, men are more likely than women to take government or political science classes and to join political clubs in college. In adulthood, men are more likely than women to be encouraged to run for a political office. Thus, despite comparable educational and occupational backgrounds, men are almost 60 percent more likely than women to view themselves as "very qualified" to run for office (Lawless and Fox, 2013).

Feminist research shows that female lawmakers significantly reshape policies only when they have parity with men. Regardless of political ideology, when women make up 60 percent or more of a political decision-making group, they speak up as much as men, advocate for women's and children's rights, and don't let men interrupt them (Mendelberg and Karpowitz, 2012).

A larger number of female officeholders alone doesn't guarantee greater gender equality. Some Republican women in Congress have opposed equal pay for equal work laws because, they maintain, well-qualified women are already paid the same as men, that such legislation would increase the number of civil lawsuits, and would hurt businesses (Mukherjee, 2013; Lowery, 2014; McDonough, 2014).

Unlike their male counterparts, female political candidates encounter sexist insults from hosts of TV shows (particularly Fox News), sexist debate questions from both male and female moderators, and reporters' obsession with the women's fashion choices—all of which taint the women's credibility. One outcome is that women are less likely than men to consider running for political office (Lawless and Fox, 2012; Seltzer, 2012).

CRITICAL EVALUATION

Critics fault feminist explanations for three reasons. First, many women's organizations tend to focus on single issues like abortion or sexual violence. Such splintering dilutes women's mobilization within a political party and decreases their chances of winning office.

Second, there are powerful women in the most influential congressional committees (O'Keefe, 2014), and some of the world's political elite are women. A few examples are Angela Merkel, Chancellor of Germany; Dilma Rousseff, President of Brazil; Geun-hye Park, President of South Korea; Christine Lagarde, Managing Director of the International Monetary Fund; Janet Yellen, Chair of the U.S. Federal Reserve; and Hillary Clinton, former U.S. Secretary of State.

One might also question many feminists' contention that patriarchy is the root of political power differences between women and men. Some of the most patriarchal societies—those in parts of Africa, Latin America, and Arab countries, for example—have more women in high-ranking political positions than does the United States, which professes equality between the sexes. Women have the fewest rights in societies where male political leaders are religious zealots who don't believe in gender equality (see Jones, 2014). Thus, religion may be more important than patriarchy in quashing women's political leadership.

STUDY TOOLS 11

READY TO STUDY? IN THE BOOK, YOU CAN:

☐ Check your understanding of what you've read with the Test Your Learning Questions provided on the chapter review card at the back of the book.

☐ Rip out the chapter review card for a handy summary of the chapter and key terms.

ONLINE AT CENGAGEBRAIN.COM YOU CAN:

☐ Prepare for tests with quizzes.

☐ Review the key terms with Flash Cards.

☐ Play games to master concepts.

12 | Families and Aging

After you finish
this chapter go to
PAGE 242 for
STUDY TOOLS.

LEARNING OBJECTIVES

12-1 Describe how families are similar and different in the United States and worldwide.

12-2 Describe how and explain why U.S. families are changing.

12-3 Describe, illustrate, and explain why intimate partner violence, child maltreatment, and elder abuse occur.

12-4 Describe, illustrate, and explain how the U.S. older population is changing, and its impact on our society.

12-5 Compare and evaluate the theoretical explanations of families and aging.

Cohabitation, marriage, divorce, and grandparents raising grandchildren—these are just some of the family characteristics that continue to change. This chapter examines the connections between families and our aging society. Before considering what sociologists mean by *family* and *aging*, take the short quiz on page 224 to see how much you already know about these topics.

12-1 WHAT IS A FAMILY?

Ask five of your friends to define *family*. Their definitions will probably differ not only from each other's but also from yours. For our purposes, a **family** is an intimate group consisting of two or more people who (1) have a committed relationship, (2) care for one another and any children, and (3) share close emotional ties and functions. This definition includes households (e.g., foster families, same-sex couples) whose members aren't related by birth, marriage, or adoption.

family an intimate group consisting of two or more people who (1) have a committed relationship, (2) care for one another and any children, and (3) share close emotional ties and functions.

What do you think?

I would undergo medical treatments to live to age 120 or longer.

1	2	3	4	5	6	7

strongly agree   strongly disagree

Contemporary households are complex; family structures vary across cultures and have changed over time. In some societies, a family includes uncles, aunts, and other relatives. In others, only parents and their children are viewed as a family.

12-1a How Families Are Similar Worldwide

The family, a social institution, exists in some form in all societies. Worldwide, families are similar in fulfilling some functions that have persisted over time.

FAMILY FUNCTIONS

Families vary considerably in the United States and globally but fulfill five important functions that ensure a society's survival (Parsons and Bales, 1955):

▶ **Sexual activity.** Every society has norms regarding who may engage in sexual relations, with whom, and under what circumstances. In the United States sexual intercourse is banned with someone younger than 18 (or 16 in some states), but some societies permit marriage with girls as young as 8 (see Chapter 9).

One of the oldest rules that regulates sexual behavior is the **incest taboo**, cultural norms and laws that forbid sexual intercourse between close blood relatives (e.g., brother and sister, father and daughter, uncle and niece).

▶ **Procreation and socialization.** Procreation is an essential family function because it replenishes a country's population. Through socialization, children acquire language; absorb the accumulated knowledge, attitudes, beliefs, and values of their culture; and learn the social and interpersonal skills they need to function effectively in society (see Chapters 3 and 4).

▶ **Economic security.** Families provide food, shelter, clothing, and other material resources for their members.

▶ **Emotional support.** Families supply the nurturance, love, and emotional sustenance that people need to be happy, healthy, and secure. Our friends may come and go, but our family is usually our emotional anchor.

▶ **Social class placement.** Social class affects all aspects of family life. Initially, our social position is based on our parents' social class, but we can move up or down the social hierarchy in adulthood (see Chapter 8).

Some sociologists include *recreation* as a basic U.S. family function. This function isn't critical for survival, but since the 1950s many parents have spent much more time with their children on leisure activities (e.g., visiting amusements parks, playing video games together) that bolster interpersonal bonds.

incest taboo cultural norms and laws that forbid sexual intercourse between close blood relatives.

MARRIAGE

Marriage, a socially approved mating relationship that people expect to be stable and enduring, is also universal. Countries vary in their specific norms and laws dictating who can marry whom and at what age, but marriage everywhere is an important rite of passage that marks adulthood and its related responsibilities, especially providing for a family.

ENDOGAMY AND EXOGAMY

All societies have rules, formal or informal, about the "right" marriage partner. **Endogamy** (often used interchangeably with *homogamy*) is a cultural practice of marrying within one's group. The partners are similar in religion (e.g., Catholics marrying Catholics), race or ethnicity (e.g., blacks marrying blacks), social class (e.g.,

marriage a socially approved mating relationship that people expect to be stable and enduring.

endogamy (often used interchangeably with *homogamy*) cultural practice of marrying within one's group.

exogamy (often used interchangeably with *heterogamy*) cultural practice of marrying outside one's group.

nuclear family a family form composed of married parents and their biological or adopted children.

college-educated marrying college-educated), and/or age (e.g., young people marrying young people). Across the Arab world and in some African nations, marrying first or second cousins is not only commonplace but desirable because such unions reinforce kinship ties and increase a family's resources (Bobroff-Hajal, 2006).

Exogamy (often used interchangeably with *heterogamy*) is a cultural practice of marrying outside one's group, such as not marrying one's relatives. In the United States, 25 states prohibit marriage between first cousins, but violations are rarely prosecuted (National Conference of State Legislatures, 2014). U.S. exogamy is growing. For example, interracial marriages rose from 3 percent of all marriages in 1980 to 10 percent by 2010, but the spouses are usually homogamous (similar) in social class and age (Wang, 2012). Thus, whom we marry isn't entirely a matter of choice but is limited by cultural mate selection norms and values.

12-1b How Families Differ Worldwide

Families also differ around the world. Some variations affect the family's structure whereas, others regulate where people reside and who has the most household power and authority.

NUCLEAR AND EXTENDED FAMILIES

In Western societies, the typical family form is the **nuclear family** composed of married parents and their biological or adopted children. In much of the world, however,

© Mike Baldwin / Cornered

"Empty-nesters. They're hoping to sell before the flock tries to move back in."

Mike Baldwin/CSL, CartoonStock Ltd

Michael S Yamashita/Documentary Value/Corbis

The Mosuo in China's Himalayas is an example of a matriarchal society. For the majority of Mosuo, a family is a household consisting of a woman, her children, and the daughters' offspring. A man will join a lover for the night and then return to his mother's or grandmother's house in the morning. Any offspring resulting from these unions belong to the female; she and her relatives raise the children (Barnes, 2006).

the most common family form is the **extended family**, composed of parents, children, and other kin (e.g., uncles and aunts, nieces and nephews, cousins, grandparents).

As the number of single-parent families increases in industrialized countries, extended families are more common. By helping out with household tasks and child-care, other adult members make it easier for a single parent to work outside the home. Many Americans assume that the nuclear family is the most common arrangement, but such families have declined—from 40 percent of all U.S. households in 1970 to 20 percent in 2012 (Vespa et al., 2013).

RESIDENCE PATTERNS

In a **patrilocal residence pattern**, newly married couples live with the husband's family. In a **matrilocal residence pattern**, they live with the wife's family. In a **neolocal residence pattern**, the couple sets up their own residence.

Around the world, the most common residence pattern is patrilocal. In industrialized societies, married couples are typically neolocal. Since the early 1990s, however, the tendency for young married adults to live with the parents of either the wife or the husband—or sometimes with the grandparents of one of the partners—has increased. Such "doubled-up" U.S. households have always existed, but escalated during the 2007–2009 recession for economic reasons (Parker, 2012).

As a result, a recent phenomenon is the **boomerang generation**, young adults who move back into their parents' home after living independently for a while or never leave home in the first place. In 2014, for example, and five years after the 2007–2009 recession ended, 14 percent of adults between the ages of 24 and 34 were living at home with their parents (Jones, 2014). Parents try to launch their children into the adult world, but like boomerangs, some keep coming back.

AUTHORITY AND POWER

Residence patterns often reflect who has authority and power in the family. In a **matriarchal family system**, the oldest females (usually grandmothers and mothers) control cultural, political, and economic resources and, consequently, have power over males. Some American Indian tribes were matriarchal, and in some African countries, the oldest women have considerable authority and influence. For the most part, however, matriarchal societies are rare.

Worldwide, a more typical pattern is a **patriarchal family system**, in which the oldest males (grandfathers, fathers, and uncles) control cultural, political, and economic resources and, consequently, have power over females. In some patriarchal societies, women have few rights within or outside the family; they may not be permitted to work outside the home or attend college. In other patriarchal societies, women may have considerable decision-making power in the home but few legal or political rights, such as getting a divorce or running for political office (see Chapter 9).

In an **egalitarian family system**, both partners share power and authority fairly equally. Many Americans

extended family a family form composed of parents, children, and other kin.

patrilocal residence pattern newly married couples live with the husband's family.

matrilocal residence pattern newly married couples live with the wife's family.

neolocal residence pattern a newly married couple sets up its own residence.

boomerang generation young adults who move back into their parents' home or never leave it in the first place.

matriarchal family system the oldest females control cultural, political, and economic resources and, consequently, have power over males.

patriarchal family system the oldest males control cultural, political, and economic resources and, consequently, have power over females.

egalitarian family both partners share power and authority fairly equally.

think they have egalitarian families, but our families tend to be patriarchal. For example, employed women shoulder almost twice as much housework and childcare as men, and are more likely than men to provide caregiving to aging family members (Cynkar and Mendes, 2011; see also Chapter 9).

COURTSHIP AND MATE SELECTION

Sociologists often describe U.S. dating as a **marriage market**, a courtship process in which prospective spouses compare the assets and liabilities of eligible partners, and choose the best available mate. Marriage markets don't sound very romantic, but open dating fulfills several important functions: recreation and companionship; a socially acceptable way of pursuing love; opportunities for sexual intimacy and experimentation; and finding a spouse (Benokraitis, 2014).

Many societies, in contrast, discourage open dating and have **arranged marriages** in which parents or relatives choose the children's spouses. An arranged marriage is a family rather than an individual decision that increases solidarity between families and preserves endogamy. Children may have veto power, but they believe that if partners are compatible, love will result. In some of India's urban areas, arranged marriages also rely on nontraditional methods (e.g., online dating services) to find prospective spouses (Cullen and Masters, 2008; see also Chapter 9).

MONOGAMY AND POLYGAMY

In **monogamy**, one person is married exclusively to another person. Where divorce and remarriage rates are high, as in the United States, people engage in **serial monogamy**. That is, they marry several people but one at a time—they marry, divorce, remarry, divorce, and so on.

Polygamy, in which a man or woman has two or more spouses, is subdivided into *polygyny*—one man married to two or more women—and *polyandry*—one woman is married to two or more men. Nearly 1,000 cultures around the world allow some form of polygamy, either officially or unofficially (Epstein, 2008).

Although rare, there are pockets of polyandry in some remote and isolated parts of India, and among the Pimbwe in western Tanzania, Africa. Polyandry serves several functions: The family is more likely to survive in harsh environments if there's more than one husband to provide food, and if one husband dies, the others care for the widow (Borgerhoff Mulder, 2009; Polgreen, 2010).

In contrast to polyandry, polygyny is common in many societies, especially in some regions of Africa, South America, and the Middle East. Men benefit from polygyny in several ways: They can have many legal sexual partners, more chances to have male heirs, a bigger income if some of the wives are employed, and high social status because they can support multiple wives and children (Al-Jassem, 2011; Nossiter, 2011).

Western and industrialized societies forbid polygamy, but there are pockets of isolated polygynous groups in the United States, Europe, and Canada. The Church of Jesus Christ of Latter-Day Saints (Mormons) banned polygamy in 1890 and excommunicates members who follow such beliefs. Still, males of the Fundamentalist Church of Jesus Christ of Latter Day Saints (FLDS), a polygynous sect that broke away off from the mainstream Mormon Church more than a century ago, head an estimated 300,000 families in Texas, Arizona, Utah, and Canada. Wives who have escaped from these groups have reported forced marriage between men in their 60s and girls as young as 10 years old, sexual abuse, and incest (Janofsky, 2003).

India's Ziona Chana has the world's largest polygynous family—39 wives, 94 children, and 33 grandchildren (so far). They all live together in a 100-room mansion ("After 39 Wives...," 2013).

marriage market prospective spouses compare the assets and liabilities of eligible partners and choose the best available mate.

arranged marriage parents or relatives choose the children's spouses.

monogamy one person is married exclusively to another person.

serial monogamy individuals marry several people, but one at a time.

polygamy a man or woman has two or more spouses.

12-2 HOW U.S. FAMILIES ARE CHANGING

American families have changed considerably since the 1950s. Some of the most important changes are related to marriage, divorce, cohabitation, nonmarital childbearing, and single-parent families.

12-2a Marriage and Divorce

The United States has one of the highest marriage and divorce rates in the world. By age 65, 95 percent of Americans have been married at least once; over a lifetime, about half of first marriages end in divorce (Copen et al., 2012; Vespa et al., 2013). Despite the high divorce rate, most people aren't disillusioned about marriage. Indeed, nearly 85 percent of Americans who divorce remarry, half of them within 4 years, and 8 percent have been married three times or more (Kreider and Ellis, 2011; Livingston, 2014).

TRENDS

U.S. marriage and divorce rates rose steadily during the twentieth century, but have declined since 1990 (see *Figure 12.1*). In 1960, only 9 percent of U.S. adults age 25 and older had never been married. By 2012, 20 percent had never been married—an historic high. Marriage rates are expected to decrease further because the number of newlyweds has been falling since 2012 (Cohn et al., 2011; Wang and Parker, 2014).

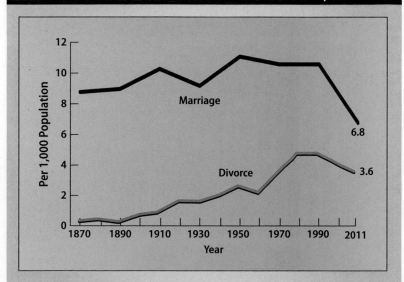

FIGURE 12.1 U.S. MARRIAGE AND DIVORCE RATES, 1870–2011

Sources: Based on Plateris 1973, Table 1; U.S. Census Bureau 2012, Table 78; and CDC/NCHS National Vital Statistics System 2013.

Divorce rates are high, but *lower* today than they were between 1980 and 2009 (see *Figure 12.1*). On average, first marriages last about 9 years, but many marriages are lasting longer than they did a decade ago. In 2009, 55 percent of couples had been married for at least 15 years, 35 percent had reached their twenty-fifth anniversary, and 6 percent had been married 50 years or longer. These percentages are several points higher than they were in 1996 (Kreider and Ellis, 2011; Aughinbaugh et al., 2013).

WHY ARE MARRIAGE AND DIVORCE RATES FALLING?

There are several *macro-level reasons* for declining marriage and divorce rates. First, U.S. values have been changing. In 2010, 39 percent of Americans said that marriage is becoming obsolete, up from 28 percent in 1978 (Cohn et al., 2011; Fry, 2012). In 2013, 64 percent of Americans, down from 76 percent in 2006, said that people should marry, and only 44 percent believed that having children is a "very important" reason to marry. Only 30 percent of Americans think that premarital sex is immoral, open marriage markets provide many opportunities for nonmarital sex, and nonmarital births are now socially acceptable (Cohn, 2013; Poushter, 2014). Thus, the traditional reasons for marriage have waned.

Second, the economy affects marriage and divorce rates. Economic depressions, recessions, and unemployment tend to delay marriage, especially for men. Because of the recent recession, 20 percent of 18- to 34-year-olds have postponed marriage. Moreover, when incomes plummet and people are insecure about their jobs, unhappy married couples tend to stay together: They can't afford to divorce and risk the possibility of not being able to maintain separate households (Cohn, 2012; Martin et al., 2014).

Demographic variables also affect marriage and divorce rates. The median age at first marriage in 2013 was 29 for men and almost 27 for women, up from 23 for men and 21 for women in 1970 (Vespa et al., 2013). People with college degrees have higher marriage rates than those with less education. In 2010, 64 percent of Americans with college degrees were married, compared with 47 percent of those with a high school diploma or less, who are more likely to cohabit than marry. And, increasingly, the highly educated are marrying each other. In effect, there's a growing "marriage divide" between those with and without college degrees (Cohn et al., 2011; Fry, 2012; Greenwood et al., 2014).

Americans with a bachelor's degree or higher also have lower divorce rates than those without college degrees. By age 46, 30 percent of those with a college degree or higher are divorced compared with 59 percent with less than a high school diploma (Aughinbaugh et al., 2013). College graduates have lower divorce rates not because they're smarter but because going to college postpones marriage. As a result, better-educated couples are often more mature and capable of dealing with personal crises. They also have higher incomes and health care benefits, both of which lessen marital stress over financial problems (Kreider and Ellis, 2011).

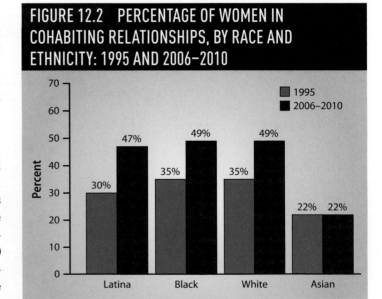

Rubberball/Mike Kemp/Jupiter Images

There are also *micro-level (individual) reasons* for falling marriage and divorce rates. About 84 percent of unmarried Americans say that love is a "very important" reason to marry. As a result, a third are still waiting for their "ideal mate" or "one true love" (Wang and Parker, 2014).

The most common micro-level reasons for divorce include infidelity, communication and financial problems, substance and spousal abuse, premarital doubts, continuous disagreements about how to raise and discipline children, and expecting to change a partner after marriage (see Benokraitis, 2014, for a discussion of these studies). Because many people are delaying marriage, they're usually more mature in handling the challenges of married life, which decreases the likelihood of divorce (Glenn et al., 2010).

12-2b Cohabitation

Cohabitation is an arrangement in which two unrelated people aren't married but live together and are in a sexual relationship (shacking up, in plain English). Because it's based on emotional rather than legal ties, "cohabitation is a distinct family form, neither singlehood nor marriage" (Brown, 2005: 33).

TRENDS

In 2013, married couples comprised 49 percent of all households, a sharp decline from 78 percent in 1950. Some of the decline was due to falling marriage rates and those who haven't remarried yet, but most of the decrease was due to cohabitation. The number of unmarried heterosexual couples surged from 430,000 in 1960 to almost 8.1 million in 2013. This number rises by at least another 605,000 if we include same-sex unmarried partners. However, only about 7 to 9 percent of the population is cohabiting at any given time, as in 2013 (Vespa et al., 2013).

By age 30, 74 percent of U.S. women aged 15–44 have cohabited compared with 62 percent in 1995 (Copen et al., 2013). Between 2005 and 2010, Asian women were the least likely to cohabit, and the only group that didn't experience an increase in cohabitation (see *Figure 12.2*). Asian women have low cohabitation rates because marriage is the norm and premarital sex and cohabitation are stigmatized (see Chapters 9 and 10). In contrast, many Hispanics come from countries in Latin and Central America that have much higher cohabitation rates than those in the United States (Oropesa and Landale, 2004).

Women's cohabitation rates have increased, but the likelihood of cohabiting decreases as their educational levels increase. On average, college-educated women cohabit for the shortest periods (17 months), and are more likely than those without college degrees to transition to marriage (Copen et al., 2012, 2013).

WHY HAS COHABITATION INCREASED?

Cohabitation is the only option for same-sex partners who live in states that prohibit same-sex marriages (see Chapter 9). For those who live in states that have legalized same-sex marriage and for opposite-sex partners, cohabitation serves many purposes.

FIGURE 12.2 PERCENTAGE OF WOMEN IN COHABITING RELATIONSHIPS, BY RACE AND ETHNICITY: 1995 AND 2006–2010

Percent (y-axis, 0–70)

1995:
- Latina: 30%
- Black: 35%
- White: 35%
- Asian: 22%

2006–2010:
- Latina: 47%
- Black: 49%
- White: 49%
- Asian: 22%

Note: These data are for never-married women aged 22–44.

Sources: Based on Cohn et al. 2011; "Marital Status" 2013; U.S. Census Bureau, Current Population Survey...2013.

cohabitation two unrelated people aren't married but live together and are in a sexual relationship.

Do you think that women or men benefit more from cohabitation? Why?

On a *micro level*, some people drift gradually into **dating cohabitation**, a living arrangement in which a couple that spends a great deal of time together decides to move in together. Dating cohabitation is essentially an alternative to singlehood because the decision may be based on a combination of reasons (e.g., convenience, finances, companionship, sexual accessibility), but there's no long-term commitment. In this type of cohabitation, and especially among young adults, there's considerable *serial cohabitation*, living with different sexual partners over time. Even if there's an unplanned pregnancy, the man, especially, may decide to move on to another cohabiting arrangement (Manning and Smock, 2005; Wartik, 2005).

For many people, premarital cohabitation is a step between dating and marriage. In **premarital cohabitation**, the couple lives together before getting married. They may or may not be engaged but plan to marry. Such "almost-married" cohabitation may be especially attractive to partners who wonder if they can deal successfully with problems that arise from differences in personalities, interests, finances, ethnicity, religion, or other issues.

On a *macro level*, and across all ethnic groups, 3 out of 4 people who cohabit say that they're delaying marriage because "Everything's there except money." Many low-income black women and Latinas don't want to marry because they believe that their live-in partners will be poor providers, unemployed, unfaithful, irresponsible fathers, or immature even though "he's the love of my life" (Xie et al., 2003; Lichter et al., 2006).

Those with high income and education levels have little to gain from cohabitation. Unlike people in lower socioeconomic groups, they're more likely to have jobs, to afford their own housing, and, consequently, to have more options in living independently even though they're involved romantically (Fry and Cohn, 2011; Sassler and Miller, 2011).

Does cohabitation lead to better marriages? Recent studies show that women who cohabit are no more likely to divorce than those who didn't cohabit *if* there's a commitment (definite plans to marry) and the cohabitation is short. However, marital success also depends on the cohabitors' age, socioeconomic status, commitment, and attitudes towards marriage (Reinhold, 2010; Manning and Cohen, 2012). Cohabitation, like any other relationship, has benefits and costs (see *Table 12.1*).

dating cohabitation a couple that spends a great deal of time together decides to move in together.

premarital cohabitation a couple lives together before getting married.

TABLE 12.1 SOME BENEFITS AND COSTS OF COHABITATION

Benefits	Costs
• Couples can pool their resources instead of paying for separate housing, utilities, and so on. They can also have the emotional security of an intimate relationship but maintain their independence by spending time with their friends and family members separately (McRae, 1999; Fry and Cohn, 2011).	• U.S. laws don't specify a cohabitant's rights and responsibilities. For example, there's no automatic inheritance if a partner dies without a will, and it's more difficult to collect child support from a cohabiting partner than an ex-spouse (Silverman, 2003; Grall, 2013).
• Couples who postpone marriage have a lower likelihood of divorce because being older is one of the best predictors of a stable marriage (Copen et al., 2012).	• Compared with married couples, cohabitants have a poorer quality of relationship and lower levels of happiness and satisfaction (Sassler et al., 2012; Wiik et al., 2012).
• Couples find out how much they really care about each other when they have to cope with unpleasant realities (e.g., a partner who doesn't pay bills or rarely showers).	• Cohabitation dilutes intergenerational ties. Compared with their married peers, the longer people live together, the less likely they are to give or receive help from their parents, and to be involved in extended family activities (Eggebeen, 2005).
• Children in cohabiting households reap economic advantages by living with two adult earners instead of a single mother (Cabrera et al., 2008).	• Because cohabiting parents are more than twice as likely as married parents to break up, children's academic, emotional, behavioral, and financial problems often increase (Fomby and Estacion, 2011; Rackin and Gibson-Davis, 2012).

12-2c Nonmarital Childbearing

More than 9 in 10 Americans aged 45 or older either have children (86 percent) or wish they had (7 percent). Such positive attitudes toward childbearing are virtually the same as in 1990, but since then U.S. birth rates have dropped by 11 percent (Martin et al., 2013; Newport and Wilke, 2013). A notable exception is nonmarital childbearing, which has increased significantly since 1970. Nearly half of American men ages 15 to 44 report that at least one of their children was born outside of marriage, and 31 percent say that all of their children were born outside of marriage (Livingston and Parker, 2011).

TRENDS

In 1950, only 3 percent of all U.S. births were to unmarried women. In 2013, there were more than 1.6 million, accounting for 41 percent of all U.S. births. Thus, 4 in 10 American babies are now born outside of marriage, a new record (Curtin et al., 2014).

Births to unmarried women vary widely across racial-ethnic groups. White women have more nonmarital babies than do other groups. Proportionately, however, nonmarital birthrates are highest for black women and lowest for Asian American women (see *Figure 12.3*).

As in marriage and cohabitation, nonmarital births vary by social class. Among college graduates, only 9 percent of recent births were outside of marriage compared with 57 percent for high-school dropouts (see *Table 12.2a*). And the percentage of nonmarital births decreases as household income increases (see *Table 12.2b*). Consequently, the poorest and least-educated women are the

most likely to have babies outside of marriage, decreasing the children's chances of moving up the social class ladder (Sawhill and Venator, 2014).

Unmarried teenage births peaked in the early 1990s, have declined steadily since then, and are now at their lowest level since 1968. Teenagers account for 15 percent of all nonmarital births compared with 74 percent for women ages 20 to 34. Thus, the number of births to unmarried women ages 20 to 34 is almost five times greater than for teenagers. Still, 86 percent of teenage births are nonmarital, compared with 60 percent for women in their 20s, and 14 percent for those ages 30 to 34. Moreover, 20 percent of unmarried teenagers ages 15 to 19 have a repeat birth: 86 percent have a second child, and 15 percent have 3 to 6 children (Gavin et al., 2013; Shattuck and Kreider, 2013).

WHY HAS NONMARITAL CHILDBEARING INCREASED?

On a *micro (individual) level,* many teens and young adults use contraception inconsistently or incorrectly because they think they or their partners are sterile, they think not taking the pill a few times "doesn't really matter," or the males don't want to use

TABLE 12.2   PERCENTAGE OF NONMARITAL BIRTHS TO U.S. WOMEN, BY EDUCATION AND HOUSEHOLD INCOME, 2011

A. Educational Attainment	
Less than high school	57%
High school graduate	49%
Some college	40%
Bachelor's degree or more	9%

B. Annual Household Income	
Less than $10,000	69%
$10,000 to $24,999	55%
$25,000 to $49,999	42%
$50,000 to $99,999	26%
$100,000 to $199,999	17%
$200,000 and above	9%

Source: Based on Shattuck and Kreider 2013, Table 2.

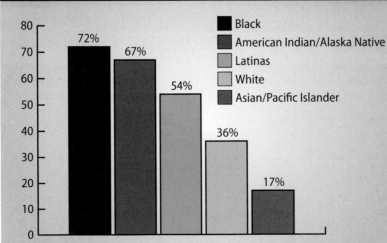

FIGURE 12.3   PERCENTAGE OF U.S. BIRTHS TO UNMARRIED WOMEN, BY RACE AND ETHNICITY, 2012

Legend:
- Black
- American Indian/Alaska Native
- Latinas
- White
- Asian/Pacific Islander

Black: 72%
American Indian/Alaska Native: 67%
Latinas: 54%
White: 36%
Asian/Pacific Islander: 17%

Source: Gibbs and Payne 2011, Figure 2. Reference crediting: Payne, K. K., and L. Gibbs. 2011. "Marital Duration at Divorce, 2010." NCFMR, FP-11-13. Accessed March 6, 2012 (ncfmr.bgsu.edu).

THE NUMBER OF BIRTHS TO UNMARRIED WOMEN AGES 20 TO 34 IS ALMOST FIVE TIMES GREATER THAN FOR TEENAGERS.

A recent study found that the reality TV show *16 and Pregnant* reduced U.S. teen births by almost 6 percent in the 18 months following its release (Kearney and Levine, 2014a). Should middle schools require adolescents to watch such programs?

condoms (Frohwirth et al., 2013). About 67 percent of all unmarried or cohabiting women report having unintended or unwanted pregnancies, but may not get abortions because of personal or religious values (Wildsmith et al., 2011; Curtin et al., 2014).

Demographic variables also affect nonmarital child-bearing. Social class is a key factor in nonmarital births (see *Table 12.2*). Educated women are more likely than less-educated women to postpone parenthood until marriage, to be more aware of family planning, and to have more decision-making power in their relationships, such as insisting that men use condoms (Guttmacher Institute, 2009; Livingston and Cohn, 2013).

Gender, race, and ethnicity intersect, however. Among college-educated women, 41 percent of black women, compared with 22 percent of Latinas and only 9 percent of white women, choose single motherhood. Nonmarital childbearing is a "rational choice" for black women who don't want to marry less educated black men or across racial lines, believe that they may never marry, or that marriage will come too late for them to bear children (Keels, 2014).

Macro-level variables increase nonmarital birthrates in several ways. Until the 1970s, births outside of marriage were virtually unheard of and often kept as secret as possible. Because U.S. attitudes have changed, nonmarital childbearing has become normal. In South Korea and several other industrialized countries, in contrast, nonmarital birth rates are much lower because unmarried mothers are stigmatized (Borowiec, 2014; Livingston and Brown, 2014).

About 40 percent of U.S. female teens ages 15 to 17 have had sex, but 80 percent of them hadn't received any formal sex education before the first time they had sex. Publicly funded family planning services that provide counseling and contraceptives, particularly for teenagers and young adults, reduce unintended nonmarital pregnancies and births. Many states, however, have decreased funding for sex education and family planning organizations (Kavanaugh et al., 2013; Cox et al., 2014; see also Chapter 9).

In what some researchers describe as a pattern of "negative assimilation," 41 percent of second-generation Latinas and Asian mothers are unmarried, compared with 23 percent of recent immigrant women ("Second-Generation Americans," 2013). Thus, many second-generation women are espousing U.S. values that births outside of marriage are acceptable.

12-2d Single-Parent Households

Most children spend the majority of their childhood living with two parents, but single-parent families are increasingly common. Women head most one-parent families.

TRENDS

In 2013, 64 percent of children under age 18 lived with two married parents, down from 77 percent in 1980 (Vespa et al., 2013; Federal Interagency Forum on Child and Family Statistics, 2013). Asian American children are the most likely to grow up in two-parent homes (see *Figure 12.4*). Until 1980, married couple families were the norm in African American families. Since then, black children have been more likely than children in other racial-ethnic groups to grow up with only one parent, usually the mother. About 65 percent of Latino children now live in two-parent families, down from 78 percent in 1970 (Lugaila, 1998).

Nationally, of the nearly 650,000 same-sex households, 19 percent are raising children. Among those living alone or with a partner or spouse, 48 percent of LGBT women and 20 percent of LGBT men are raising a child under age 18. Between 2000 and 2010, the number of

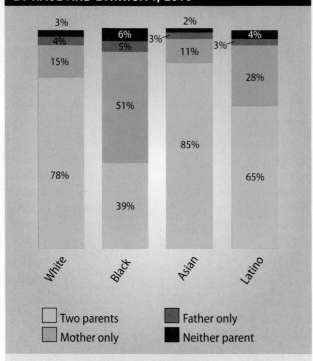

FIGURE 12.4 WHERE U.S. CHILDREN LIVE, BY RACE AND ETHNICITY, 2013

White: Two parents 78%, Mother only 15%, Father only 4%, Neither parent 3%

Black: Two parents 39%, Mother only 51%, Father only 5%, Neither parent 6%, 3%

Asian: Two parents 85%, Mother only 11%, Father only 3%, Neither parent 2%

Latino: Two parents 65%, Mother only 28%, Father only 3%, Neither parent 4%

☐ Two parents   ☐ Father only
☐ Mother only   ☐ Neither parent

Notes: The numbers may not sum to 100% due to rounding. "Two parents" includes those married (65 percent) and unmarried (4 percent).

Source: Based on U.S. Census Bureau, Current Population Survey, 2013 Annual Social and Economic Supplement, Table C9. Accessed May 10, 2014 (census.gov).

same-sex couples with adopted children increased from 9 to 20 percent (Gates, 2013). Thus, large numbers of lesbians and gay men—single, partnered, or married—are parents.

WHY HAVE FEMALE-HEADED HOUSEHOLDS INCREASED?

The reasons for the growing number of female-headed households are similar to those for the increase in nonmarital childbearing. On a *micro level*, having sex at an early age, and not using contraception at all, incorrectly, or sporadically results in unintended or unwanted pregnancies, especially among teenagers. Among unmarried parents, 63 percent have multiple partners. In these relationships, most of the fathers invest little time and money in their children because the fathers are young, have low education levels, or are incarcerated. The fathers don't live with their children, know little about them, and move from one sexual relationship to another (Scommegna, 2011; Dodson and Luttrell, 2011).

Demographic variables, especially social class, also affect the rise of female-headed households. College-

fictive kin nonrelatives who are accepted as part of a family.

educated people, as you've seen, tend to postpone marriage, marry rather than cohabit, and delay parenthood—factors that decrease the likelihood of divorce, single-parent households, and, consequently, children's experiencing poverty (Redd et al., 2011).

Not all female-headed households are poor, of course. Among mother-only families, 17 percent of the women have at least a college degree, and 38 percent own their own homes (Vespa et al., 2013). Instead of waiting for "Mr. Right," women with economic resources are deciding to raise children on their own.

In many African American and Latino families, adult males (e.g., sons, brothers, uncles, grandfathers) often provide emotional and financial support to their female kin who head households (Sudarkasa, 2007). Many black families also welcome **fictive kin**, nonrelatives who they accept as part of the family. Fictive kin have strong bonds with biological family members and provide important services (e.g., caring for children when young mothers are employed or negligent) (Billingsley, 1992; Dilworth-Anderson et al., 1993). A recent variation of fictive kin includes single mothers—many of whom are white, unmarried, college-educated women—who turn to one another for companionship and childcare help (Bazelon, 2009).

At the *macro level*, values and views about single parenthood have shifted. The share of Americans who view the growing trend of single mothers as a "big problem" decreased from 71 percent in 2007 to 64 percent in 2013 (Wang et al., 2013). The economy also affects the

Is the absent black father a stereotype? A recent national study of white, black, and Latino men found that, among fathers who don't live with their children, black fathers were the *most* likely to parent their children from birth to age 18. The parenting included, and on a daily basis, feeding, bathing, dressing, playing with and reading to children, taking them to and from activities, and helping them with or checking homework (Jones and Mosher, 2013).

number of female-headed households. As in the case of cohabitation and nonmarital childbearing, low-income women often drift into parenthood with low-income or unemployed men. Seeking financial security and stability, women gain little or nothing by marrying the babies' fathers or other low-income men.

12-3 FAMILY CONFLICT AND VIOLENCE

Conflict is a normal part of family life, but violence is *not* normal. Over a lifetime, we're much more likely to be assaulted or killed by a family member or current/former spouse, boyfriend, or girlfriend than by a stranger (Truman and Langton, 2014).

12-3a Intimate Partner Violence

Intimate partner violence (IPV) is abuse that occurs between people in a close relationship. The term *intimate partner* refers to current and former spouses, couples who live together, and current and former boyfriends or girlfriends.

TRENDS

IPV is pervasive in U.S. society. Nationwide, 36 percent of females and 29 percent of males have been victims of IPV at some time in their lives (Black et al., 2011). Nearly half of all adults have experienced psychological aggression by an intimate partner, but female victims outnumber males in other forms of violence (see *Figure 12.5*).

IPV resulted in 1,517 homicides in 2010; 89 percent of the victims were females. When victims survive IPV assaults, 40 percent of females compared with 5 percent of males suffer a serious physical injury (e.g., gunshot or knife wounds, internal injuries, broken bones) (Catalano, 2013). Women are also much more likely than men to experience IPV over a lifetime regardless of age, race or ethnicity, or social class (Black et al., 2011; Truman and Morgan, 2014).

From 1994 to 2012, the IPV rate declined 67 percent—from 9.8 per 1,000 persons age 12 or older in 1994 to 3.2 in 2012. During this period, 82 percent of the victims were females. In 2012, there were nearly 1 million intimate partner victimizations, but this number is conservative. An estimated 56 percent of victims don't report IPV to the police because they're ashamed, believe that no one can help, or fear reprisal (Truman and Morgan, 2014).

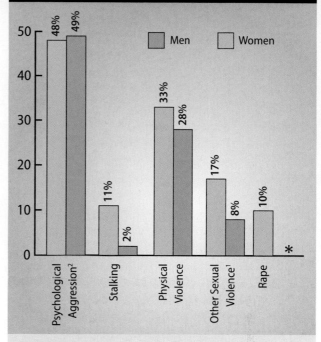

FIGURE 12.5   LIFETIME PREVALENCE OF EXPERIENCING INTIMATE PARTNER VIOLENCE, BY SEX, 2010

* Too few cases reported for reliable estimate.
[1] Includes any unwanted sexual contact.
[2] Includes name-calling, humiliation, insults, threats of physical harm, and controlling behavior (e.g., making all decisions, keeping someone from seeing her or his family or friends).

Source: Copen et al. 2013, Figure 2.

WHY DOES INTIMATE PARTNER VIOLENCE OCCUR?

The reasons for IPV are complex and often the product of multiple individual, demographic, and societal factors. On a *micro level*, violence escalates if one or both partners engage in heavy alcohol or drug use, if they have more children than they can afford (which intensifies financial problems), or if either partner has been raised in a violent household, has low self-esteem, and has a controlling personality. The most common disagreements that can escalate into abuse and violence have four sources: gender role expectations (who does what housework); money (saving and spending); children (especially discipline); and infidelity, both personally and online (Benokraitis, 2014; Spivak et al., 2014).

Regarding *demographic variables*, IPV often begins at a young age. Among those who experienced rape,

intimate partner violence (IPV) abuse that occurs between people in a close relationship.

physical violence, and/or stalking by an intimate partner the first time, 22 percent of females and 15 percent of males were between 11 and 17 years old; 47 percent of females and 39 percent of males were between 18 and 24 years old (Black et al., 2011). IPV cuts across all social classes, but women living in households with an annual income of less than $7,500 are nearly seven times more likely to be abused than those living in households with an annual income of $75,000 or more (Thompson et al., 2006).

Macro-level variables, particularly unemployment and poverty, increase the likelihood of financial stress and violence. The absence of legal or social sanctions against IPV and a scarcity of shelters discourage victims from trying to escape abuse (Matjasko et al., 2013). A study of girls aged 11 to 16 who had experienced sexual assault concluded that girls (and later women) often didn't report the incidents because such violence had been "normalized in their communities." The girls were ashamed and feared retribution, but also believed that men "can't help it," perceived everyday harassment and abuse as "normal male behavior," and assumed that male authority figures (e.g., police officers, judges) would accuse the girls of overreacting (Hlavka, 2014).

12-3b Child Maltreatment

Child maltreatment (also called *child abuse*) includes a broad range of behaviors that place a child at serious risk or result in serious harm, including physical and sexual abuse, neglect, and emotional mistreatment. The victims often experience several types of maltreatment.

TRENDS

U.S. child maltreatment rates have decreased since 2007. Because only a fraction of the cases is reported, however, federal agencies assume that millions of children experience abuse and neglect on a daily basis. In 2012, an estimated 686,000 children were victims of child maltreatment, representing almost 10 per 1,000 U.S. children, and 1,640 died of abuse and neglect (U.S. Department of Health and Human Services, 2013).

Neglect is the most common type of abuse (see *Figure 12.6*). From birth to age 17, girls (51 percent) are slightly more likely than boys (49 percent) to be maltreated. The most vulnerable children, those younger than 3 years old, account for 27 percent of all child victims; another 20 percent are ages 3 to 5. Victimization rates per

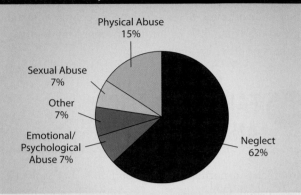

FIGURE 12.6   TYPES OF CHILD MALTREATMENT, 2012

Note: "Neglect" includes medical neglect (almost 2 percent of these cases). "Other" includes categories that some states report, such as a parent's drug/alcohol abuse.

Source: Based on U.S. Department of Health and Human Services 2013, Table 3-8.

1,000 children vary considerably by race and ethnicity: 14.2 for African Americans, 12.4 for American Indian/Alaska Natives, 10.3 for multiracial, 8.7 for Pacific Islanders, 8.4 for Latinos, 8.0 for whites, and 1.7 for Asians (U.S. Department of Health and Human Services, 2013).

Almost 81 percent of the perpetrators are one or both parents; in 37 percent of the cases, the abusers are mothers alone. An additional 7 percent of assailants are relatives, and 4 percent are the parents' intimate partners, usually boyfriends (U.S. Department of Health and Human Services, 2013).

WHY DOES CHILD MALTREATMENT OCCUR?

On a *micro level*, child mistreatment is most common in single-parent and stepparent households, those with parental substance abuse and mental illness, if there's adult or sibling violence, and if the child has emotional, developmental, or physical disabilities (Finkelhor et al., 2011). *Demographic variables* that increase the likelihood of child abuse include low socioeconomic status, a parent's young age, and parental conflict before, during, and after a hostile divorce (Truman and Smith, 2012; White and Lauritsen, 2012). On a *macro level*, economic hardship, unemployment, and poverty increase the likelihood of stress that leads to child abuse, including infant deaths due to severe head traumas (Lindo et al., 2013; Eckenrode et al., 2014; National Academy of Sciences, 2014).

Almost 26 percent of all children live in homes where parents or other adults engage in violence (Hamby et al., 2011). Whether children are targets of abuse or see it, a growing body of research has linked childhood experience

child maltreatment (also called *child abuse*) a broad range of behaviors that place a child at serious risk, including physical and sexual abuse, neglect, and emotional mistreatment

of violence with lifelong developmental problems, including depression, delinquency, suicide, alcoholism, low academic achievement, unemployment, and medical problems in adulthood (Zimmerman and Pogarsky, 2011; Hibbard et al., 2012; Sacks et al., 2014).

12-3c Elder Abuse and Neglect

Elder abuse (sometimes called *elder mistreatment*) is any knowing, intentional, or negligent act by a caregiver or other person that causes harm to people age 65 or older. This term includes physical, psychological, and sexual abuse; neglect; isolation from family and friends; deprivation of basic necessities such as food and heat; not providing needed medications; and financial exploitation.

TRENDS

A national study of people ages 60 and older found that almost 12 percent had experienced at least one of the following types of mistreatment: emotional (4.6 percent), physical (1.6 percent), sexual (0.6 percent), financial (5.2 percent), and neglect (6 percent) (Acierno et al., 2010; see also Laumann et al., 2008). The actual rates are probably much higher because only an estimated 20 percent of all elder abuse and neglect is reported (Administration on Aging, 2013).

Who are the victims? About 83 percent are white; the average age is 76; 76 percent are women; 84 percent live in their own homes; 86 percent have a chronic disease or other health condition; 57 percent are married or cohabiting; 53 percent haven't graduated from high school; 50 percent suffer from dementia, Alzheimer's, or other mental illness; 46 percent feel socially isolated; and the average combined household income is less than $35,000 a year (Acierno et al., 2009; Jackson and Hafemeister, 2011). These data suggest that the most likely

The movie *Precious* (2009) received many awards, but some critics contended that the film stereotyped black female teenagers—especially those in low-income households—as obese, illiterate, and living with abusive mothers (Amusa, 2010). If you've seen the movie, do you agree with such criticism or not? Why?

victims of elder abuse are those who are the most vulnerable physically, mentally, socially, and financially.

Who are the perpetrators? Most are adult children, spouses or cohabiting partners, or other family members. Less than a third are acquaintances, neighbors, or nonfamily service providers. The average age of the abuser is 45; 77 percent are white; 61 percent are males; 82 percent have a high school diploma or less; 50 percent abuse alcohol and/or other drugs; 46 percent have a criminal record; 42 percent are

elder abuse (sometimes called *elder mistreatment*) any knowing, intentional, or negligent act by a caregiver or other person that causes harm to people age 65 or older.

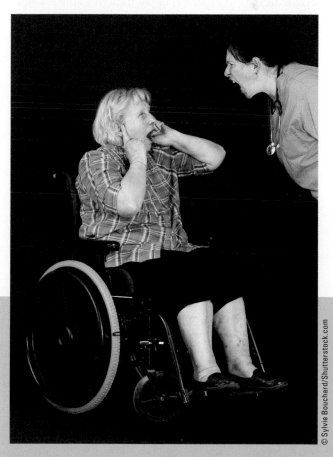

Adult children, spouses, and cohabiting partners are the most likely to mistreat older people.

financially dependent on the elder; 37 percent live with the elder; 29 percent are chronically unemployed; and 25 percent have mental health problems (Jackson and Hafemeister, 2011).

WHY DO ELDER ABUSE AND NEGLECT OCCUR?

On a *micro level*, abuse of alcohol and other drugs is more than twice as likely among family caregivers who abuse elders as among those who don't. Both victims and offenders often report a childhood history of witnessing or experiencing family violence, poor family relationships in the past and currently, and communication problems. In addition, older people with cognitive impairment—due to dementia (deteriorated mental condition) after a stroke, or the onset of Alzheimer's disease—are abused at higher rates than those without such disabilities (Heisler, 2012; National Center on Elder Abuse, 2012).

On a *macro level*, a shared residence is a major risk factor for elder mistreatment because the caregiver(s) may depend on the older person for housing, whereas the elder is dependent on the caregiver(s) for physical help. These situations compound the likelihood of everyday tensions and conflict. Financial stress may also increase the risk of experiencing or inflicting abuse. Unlike low-income families, those in the middle class aren't eligible for admission to public facilities, yet few can pay for the in-home nursing care, high-quality nursing homes, and other services that upper-class families can afford (Acierno et al., 2009; Jackson and Hafemeister, 2011).

12-4 OUR AGING SOCIETY

One of our friends, 55, was shocked and insulted when he ordered a meal at a fast-food restaurant, and the young worker called out "one senior meal!" The *Baltimore Sun* (2012) recently reported on the death of an "elderly" man in a house fire. He was 62.

12-4a When Is "Old"?

What images come to mind when you hear the word *old*? In a recent national survey, people in their 40s said that a person is old at age 63, those in their 70s said age 75, and a 90-year-old woman said that a woman isn't old "until she hits 95" (*AARP Magazine*, 2014: 40). So, how old is "old"?

It depends on whom you ask because "old" is a social construction. In some African nations, where

life expectancy the average expected number of years of life remaining at a given age.

AP Images/Itsuo Inouye

Between the ages of 65 and 91, avid mountaineer Hulda Crooks scaled Mount Whitney (the highest mountain in the continental United States) 23 times. She died at the age of 101 in 1997.

people rarely live past 50 because of diseases and civil wars, 40 is old. In industrialized societies, where the average person lives to age 78, 40 is considered young. Still, regardless of how we feel physically, society usually defines old in chronological age. In the United States, for instance, people are deemed old at age 65, 66, or 67 because they can retire and become eligible for Medicare and Social Security benefits.

Many older Americans are vigorous and productive, but others experience physical and mental limitations as they age. There are significant differences between the *young-old* (65–74 years old), the *old-old* (75–84 years old), and the *oldest-old* (85 years and older) in their ability to live independently, to work, and in their health. Generally, a 65-year-old woman is much less likely than an 85-year-old woman to need caregiving from family and friends.

12-4b Life Expectancy

Life expectancy is the average expected number of years of life remaining at a given age. American children born in 2010 have a life expectancy of almost 79 years (compared with 71 in 1970 and 47 in 1900).

Life expectancy varies by sex, social class, and race-ethnicity. Historically, currently, and across all racial-ethnic

TABLE 12.3 U.S. LIFE EXPECTANCY AT BIRTH, BY SEX AND RACE-ETHNICITY, 2012		
	Males	**Females**
Black and American Indian/ Alaska Native	72	78
White and Asian/Pacific Islander	77	82
Latinos	79	84

Source: Based on Ortman et al. 2014, Table 1.

groups, women live longer than men, and Latinas have the longest lifespans (see *Table 12.3*).

The higher rates of cigarette smoking, heavy drinking, gun use, employment in hazardous occupations, and risk-taking in recreation and driving are responsible for many males' shorter lifespans. The life expectancy gender gap has narrowed since the 1990s, however, because there's been an increase in women's smoking, use of alcohol and other drugs, obesity (which increases the risk of hypertension and heart disease), and stresses due to multiple roles (e.g., juggling employment while caring for children and older family members) ("Catching Up," 2013; see also Chapter 14).

Generally, higher socioeconomic levels increase life expectancy. People with a college degree and higher are more likely than less educated adults to be employed, to be financially secure, and to have employment-related health insurance. They're also more likely to exercise, to not smoke, to drink alcohol in moderation, and to maintain a healthy body weight. An exception is Latinos, who have among the lowest education levels and household incomes, but the highest life expectancy rates.

Researchers hypothesize that Latinos have longer life spans both because only the healthiest migrate to the United States and because many foreign-born Latinos return to their country of origin when their health deteriorates. Also, recent immigrants, compared with whites, are less likely to smoke and more likely to have strong support networks during illness (Pollard and Scommegna, 2013; Scommegna, 2013b, 2013c).

12-4c How Our Graying Nation Is Changing

The number of Americans age 65 and older is booming, increasing, and becoming more ethnically and racially diverse. For example,

▶ More than 43 million (14 percent of the total population) are 65 or older, a dramatic increase from just 4 percent in 1900, and will comprise more than 20 percent of U.S. residents by 2030.

▶ One of the fastest growing groups is the oldest-old, whose numbers increased from 100,000 in 1900 to almost 6 million in 2012. By 2030, this group will comprise almost 3 percent of the population.

▶ By 2030, 28 percent of the 65-and-older population will be minority, up from 20 percent in 2010; 23 percent of the 85-and-older population will be minority, up from 15 percent in 2010 (Federal Interagency Forum on Aging-Related Statistics, 2012; "U.S. Census Bureau Projections…," 2012; Ortman et al., 2014).

Our aging population has three important (and interrelated) implications. First, as you'll see shortly, health care and caregiving needs, services, and costs will surge. Second, the number of older Americans is increasing, whereas the proportion of young people is decreasing (see *Figure 12.7*). As a result, many adult children—even in their sixties—will be caring for aging parents, grandparents, and other older relatives for more years than in the past. Disability rates among older Americans declined during the 1980s and 1990s, but have recently increased, particularly among **baby boomers** (people born between 1946 and 1964) and those 10 years younger than baby boomers. The higher disability rates might be reflecting better medical diagnoses, but baby boomers are

baby boomers people born between 1946 and 1964.

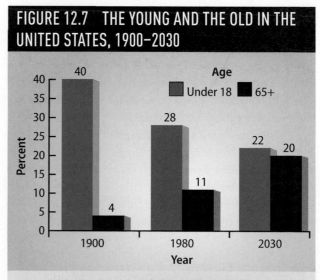

FIGURE 12.7   THE YOUNG AND THE OLD IN THE UNITED STATES, 1900–2030

Sources: Based on U.S. Senate Special Committee on Aging et al. 1991: 9, and Ortman et al. 2014, Table 2.

more likely than the previous generation to be obese, have diabetes, or high blood pressure (Scommegna, 2013a). Thus, the boomer generation might be sicker and for more years than their predecessors who had lower life expectancies.

Third, as the number of older people increases, so does the **old-age dependency ratio**, the number of people age 65 and older who aren't in the labor force relative to the number of working-age adults ages 18 to 64. In effect, this ratio shows the burden on the working-age population to support those 65 and older who aren't employed. The old-age dependency ratio has risen. By 2020, only 3 working-age people will be supporting each older person, compared with 14 workers per older person in 1900 (Jacobsen et al., 2011; Ortman et al., 2014). Thus, most people will have to pay much higher federal and state taxes to sustain our ever-growing older population.

12-4d Some Current Aging Issues

The *sandwich generation* is composed of midlife women and men who care for aging parents, are raising a child under age 18, or supporting a grown child. About 71 percent who do so are ages 40 to 59 (Parker and Patten, 2013). The sandwich generation experiences considerable stress because older people usually need assistance at a time when their adult children's and grandchildren's lives are complicated and demanding. Three specific outcomes that the sandwich generation is most likely to experience are the rise of multigenerational households, right-to-die issues, and competition for scarce resources.

MULTIGENERATIONAL HOUSEHOLDS

The share of the U.S. population living in multigenerational households rose from 12 to 16 percent between 1980 and 2010. Asians (28 percent) are more likely to live multigenerational households than blacks (26 percent), Latinos (25 percent), and whites (14 percent) (Taylor et al., 2013; Vespa et al., 2013; see also Chapter 4 on multigenerational households).

There are many reasons for several generations living under one roof. Graying boomers with small retirement savings may move in with their adult children and grandchildren to avoid poverty. Adult children may move back home with their parents (often with children and girlfriends/boyfriends in tow) because of unemployment,

low wages, or home foreclosures. In 2012, 10 percent (7.1 million) children lived with a grandparent—up from 5 percent in 1990—and the majority lived in the grandparent's home. Extended family living provides intergenerational support, including childcare for employed parents. Cultural values, particularly among recent immigrants, reflect long-standing practices of caring for aging parents and grandparents (Ellis, 2013; Fry and Passel, 2014; "National Grandparents Day...," 2014).

The effects of multigenerational households are mixed. Grandparents who live with their adult children and grandchildren may provide crucial economic and childcare support. On the other hand, if the multigenerational household is struggling financially, if the grandparents are also raising their own children, and if there's mother–grandmother conflict, grandparents experience emotional problems, including unhappiness, stress, worry, and anger (Barnett et al., 2012; Deaton and Stone, 2013).

THE RIGHT TO DIE

Living longer and with more years of chronic diseases and disabilities raises questions about later-life choices. Both historically and currently, one of the most controversial issues is regarding end-of-life decisions. Oregon's Death with Dignity Act, which took effect in 1997, authorized prescriptions for lethal doses when two doctors agreed that a patient would die within six months and freely

Medical care for patients in the last year of life accounts for more than 25 percent of annual Medicare costs (Carr, 2012). Is this a good investment of our resources?

old-age dependency ratio the number of people age 65 and older who aren't in the labor force relative to the number of working-age adults ages 18 to 64.

chose *physician-assisted suicide* (also called *physician-aid-in-dying*). Modeling Oregon's law, four other states (Montana, New Mexico, Vermont, and Washington) have passed similar legislation.

In a recent Gallup poll, 70 percent of Americans said that doctors should be allowed to "end the patient's life by some painless means" (McCarthy, 2014), but doing so is illegal in most of the country. Advocates argue that terminally ill people who face a long and painful death and huge medical costs should have the legal right to die with dignity on their own terms instead of lingering in a nursing home, hospital, or hospice. Opponents maintain that actively ending a life, regardless of a person's frailty or suffering, is a moral violation, and that patients might be pushed to die early for the caregiver's convenience (Eckholm, 2014).

COMPETITION FOR SCARCE RESOURCES

When Congress passed the Social Security Act in 1935, life expectancy was just below 62 years compared with almost 79 today. In the years ahead, the increasing numbers of older Americans will put a significant strain on the nation's health care services and retirement income programs.

The size of the U.S. older population (13 percent) is about half that of the population age 18 and younger (26 percent). Nonetheless, the federal government spends 31 percent on the elderly, primarily through Social Security and Medicare, compared with 10 percent on children (Isaacs et al., 2012; Matthews and Schmit, 2014). The older people get, the higher the medical costs. For example,

▶ A study of 1.8 million Medicare recipients who died in 2008 found that a third had surgery in the last year

of life. Nearly 1 in 5 had surgery in the last month of life, and nearly 1 in 10 had surgery in the last week of life. Among those undergoing end-of-life surgery, almost 60 percent were 80 and older (Kwok et al., 2011).

▶ About 60 percent of prostate cancer is diagnosed in men aged 65 or older. It costs $93,000 per patient to treat advanced prostate cancer; the treatment prolongs life by an average of four months (Beil, 2012).

Younger generations can't count on federally financed health care and retirement benefits in the future. Medicare funds are supposed to last until 2026, and Social Security through 2033 (The Board of Trustees…, 2013), but there are no guarantees. As baby boomers age, the costs of these programs will mushroom. Because boomers have higher educational levels than previous generations, they're likely to demand more and more expensive health care services (see Chapter 14).

Despite rising health care costs, 20 percent of Americans want to live to age 100—about 21 years longer than the current average U.S. life expectancy. Moreover, 38 percent support medical treatments that would allow them to live to at least age 120 (Lipka and Stencel, 2013; Lugo et al., 2013). Who'll pay for such radical life extensions?

12-5 SOCIOLOGICAL EXPLANATIONS OF FAMILY AND AGING

The four sociological perspectives are useful in understanding families and aging. *Table 12.4* summarizes the key points of these theories.

TABLE 12.4 SOCIOLOGICAL PERSPECTIVES ON FAMILIES AND AGING		
Theoretical Perspective	**Level of Analysis**	**Key Points**
Functionalist	Macro	Families are important in maintaining societal stability and meeting family members' needs.
		Older people who are active and engaged are more satisfied with life.
Conflict	Macro	Families promote social inequality because of social class differences.
		Many corporations view older workers as disposable.
Feminist	Macro and micro	Families both mirror and perpetuate patriarchy and gender inequality.
		Women have an unequal burden in caring for children as well as older family members and relatives.
Symbolic Interactionist	Micro	Families construct their everyday lives through interaction and subjective interpretations of family roles.
		Many older family members adapt to aging and often maintain previous activities.

12-5a Functionalism

We began this chapter by looking at the vital functions that families perform—such as procreation and economic security—that promote societal stability and individual well-being. Functionalists recognize that families differ in structure (e.g., nuclear vs. extended), but believe that their similar functions ensure a society's continuity.

STABILITY AND ACTIVITY

For many functionalists, marriage, followed by procreation, is critical in promoting social order and cohesion. Parenthood is a crucial social role. One of its tasks, socialization, is essential to maintaining any culture. Problems arise, for instance, when parents can't or don't provide their children with the necessary financial and emotional support or when they divorce.

Retirement benefits society because it provides younger people jobs. Even in retirement, however, **activity theory** proposes that many older people remain engaged in numerous roles and activities, including work. Moreover, those who are active adjust better to aging and are more satisfied with their lives (see Atchley and Barusch, 2004, for a summary of some of this research).

CRITICAL EVALUATION

Functionalism has several weaknesses. First, it's questionable whether some family functions are as universal or necessary as functionalists claim. Procreation often occurs outside of marriage, and the government has assumed some of the family's functions, including caring for some children (as in foster homes) and the aged (through Medicare, for example) (Lundberg and Pollak, 2007).

Second, is activity theory as representative of older people as some functionalists claim? Many people continue to work even in their 80s, but usually not by choice. They can't afford to retire, even though they're in poor health and unhappy in their low-income jobs. As health deteriorates, many older people become less active and more isolated (Lee, 2009).

Third, 10 percent of the huge wave of baby boomers says that they'll never retire, primarily because they've saved very little and carry too much debt (Harter and Agrawal, 2014). This means that there isn't always an orderly progression to retirement and opening up job slots to the younger generation.

activity theory proposes that many older people remain engaged in numerous roles and activities, including work.

12-5b Conflict Theory

Conflict theorists agree that families serve important functions, but point out that some groups benefit more than others. Families are sources of social inequality that mirror the larger society, and the inequities persist from birth to old age (Cruikshank, 2009).

INEQUALITY, SOCIAL CLASS, AND POWER

For conflict theorists, families in high-income brackets have the greatest share of capital, including wealth that they can pass down to the next generation. An inheritance reduces the likelihood that all families can compete for resources that include education, decent housing, and health care (see Chapter 8).

Economic inequality affects almost all aspects of family life. Child abuse and neglect are highest in U.S. counties with the greatest gaps between rich and poor. Moreover, the more distance between these social classes, the less likely that the poor will receive needed services and support that reduce child maltreatment rates (Eckenrode et al., 2014).

Unequal access to resources continues into old age. Most employers insist that they value older workers' loyalty, work ethic, reliability, and experience, but many are less likely to hire or retain older people because they're usually more expensive than younger workers. Because many large companies have cut their pension plans, numerous older workers must work long after they expected to retire (Kromer and Howard, 2013; Winerip, 2013).

CRITICAL EVALUATION

Conflict theory is limited for several reasons. First, conflict theorists tend to overlook the fact that many families, especially those in the middle and upper classes, fund public assistance programs such as Medicaid (a social health care program for low-income individuals and families). They also pay billions each year to cover the costs of unintended pregnancies and the resulting infant care expenses among poor women (Sonfield and Kost, 2013).

Second, conflict theory links family inequality to capitalism and social class, but there's also considerable family inequality in countries that aren't capitalist (see Chapters 8 and 11). Thus, social class affects but doesn't determine whether a child will succeed economically.

Third, many older people don't have to struggle for resources. As you saw earlier, the U.S. government's spending on children has declined but increased for people age 65 and older. Older people, regardless of social class, have generous health care coverage through Medicare or Medicaid. They vote in large numbers,

In most Latino communities, the *quinceañera* (pronounced "keen-say-ah-NYAIR-ah") is an important coming-of-age ritual that celebrates a girl's entrance into adulthood on her fifteenth birthday. It's an elaborate and dignified religious and social event that reaffirms strong ties with family, relatives, and friends. According to some critics, however, the budget—which often starts at $10,000—should be saved for college instead of "blown on a party" ("The Power of a Party," 2013: 29).

have influential lobbyists in Congress, and flood lawmakers with hundreds of thousands of letters and email messages when there's a threat of cutting Medicare or Social Security payments (Johnson, 2011; see also Chapter 11).

12-5c Feminist Theories

Feminist scholars agree with conflict theorists that there's considerable inequality between low-income and wealthy families in accessing necessary resources. However, feminist theorists emphasize the inequality of gender roles in families, especially in patriarchal societies (including the United States).

GENDER ROLES AND PATRIARCHY

For feminist scholars, families both mirror and perpetuate patriarchy and gender inequality. In most countries, males pass laws about property and inheritance rights, marriage and divorce, and many other regulations that give men authority over women. The United States is similar to most other nations in that men (particularly lawmakers and Catholic Church officials) control women's decisions about reproduction and access to abortion that, in turn, increases unwanted pregnancies and nonmarital birth rates.

Male aggression against women and children is common in patriarchal societies in which men hold most of the power, status, and privilege. Females, on the other hand, are marginalized and socialized to accept male domination. In many cases, legal, political, and religious institutions don't take violence against women and children seriously (Lindsey, 2005). Recently, for instance, a judge in Florida "sentenced" a man charged with domestic battery to take his wife, the victim, to Red Lobster for dinner and then bowling (McEwan, 2012). Thus, some judges still view domestic violence charges as frivolous.

Feminist sociologists have documented the ways that sexism, racism, and classism intersect to oppress women and children. The poorest older adults are most likely to be minority women. Out of 190 nations, the United States, Papua New Guinea, and Swaziland are the only countries that have no national paid parental leave policy. As a result, caregivers, who are predominantly women, must often leave their jobs or work only part-time to care for children and aging parents (Walsh, 2011; Fox et al., 2013).

CRITICAL EVALUATION

Feminist explanations of families and aging have several limitations. First, according to critics, feminist scholars overstate women's oppression. For example, a record 40 percent of all households with children under the age of 18 include mothers who are either the sole or primary source of family income, up from just 11 percent in 1960 (Wang et al., 2013).

A second criticism is that feminist theories tend to gloss over data which show that intimate partner violence is often mutual. Several dozen studies have found that women initiate from 30 to 73 percent of violent incidents (Straus, 2011, 2013). You'll recall that mothers mistreat their children, and almost 40 percent of the perpetrators of elder abuse are women (Jackson and Hafemeister, 2011).

Third, the poorest older adults are most likely to be minority women, but social class and marital status are important factors. Older married women and those in higher socioeconomic levels are less likely to experience poverty in later life than those who are single (never married, divorced, or widowed) and from lower socioeconomic levels (Hokayem and Heggeness, 2014; Hunter, 2014). Thus, patriarchy alone doesn't determine family gender roles, even in later life.

12-5d Symbolic Interaction

For symbolic interactionists, people create subjective meanings of what a family is and does. Thus, people learn, through interaction with others, how to act as a parent, a grandparent, a teenager, a stepchild, and so on throughout the life course.

LEARNING FAMILY AND AGING ROLES

Throughout the socialization process, family members establish trust and build emotional bonds. Between 90 and 97 percent of Americans believe that it's important for *both* mothers and fathers to provide discipline and emotional support, and to teach values ("The New American Father," 2013).

Interactionists often use exchange theory to explain mate selection and family roles. The fundamental premise of **exchange theory** is that people seek through their social interactions to maximize their rewards and minimize their costs. In mate selection, people trade their resources (e.g., wealth, good looks, youth, and/or status) for more, better, or different assets. People may stay in unhappy marriages and other intimate relationships because the rewards seem equal to the costs. Many women tolerate abuse because they fear loneliness or losing the economic benefits that a man provides (Choice and Lamke, 1997; Khaw and Hardesty, 2009; see also *Table 12.1* on page 229 on the benefits and costs of cohabitation).

A well-known psychologist who interviewed more than 200 couples over a 20-year period found that the difference between lasting marriages and those that split up was a "magic ratio" of 5 to 1. That is, if there are five positive interactions between partners for every negative one, the marriage is likely to be stable over time (Gottman, 1994). Thus, we can improve our family relationships by learning to interact in positive ways.

Stereotypes about older people are deeply rooted in U.S. society. About 84 percent of Americans age 60 and older report incidents of **ageism**, or discrimination against older people, including insulting jokes, disrespect, patronizing behavior, and assumptions about being frail or unhealthy (Roscigno, 2010). **Continuity theory** posits that older adults can substitute satisfying new roles for those they've lost (Atchley and Barusch, 2004). For

example, a retired music teacher can join a local chorus or orchestra. Thus, developing new roles may lessen some of the emotional distress due to ageism.

CRITICAL EVALUATION

A common criticism is that symbolic interaction, a micro-level perspective, doesn't address macro-level constraints. For example, U.S. nonmarital teen births have fallen primarily because of structural factors (e.g., greater access to family planning services, expanded educational opportunities for disadvantaged young women) rather than teen attitudes about childbearing (Kearney and Kost, 2014b).

Second, exchange theory is limited because people don't always calculate the potential costs and rewards of every decision. In the case of women who care for older family members, genuine love, concern, and a sense of obligation override cost–benefit decisions, even when the person receiving care is abusive (Jackson and Hafemeister, 2013).

Third, continuity theory ignores societal obstacles that discourage older people from engaging in satisfying new roles. After retirement, many must get by on low, fixed incomes and can't afford recreational activities that they enjoyed in the past. Also, older people who experience illnesses because of low-quality health care over a lifetime may become too sick to participate in family and community activities or pursue hobbies.

---

exchange theory posits that people seek through their social interactions to maximize their rewards and minimize their costs.

ageism discrimination against older people.

continuity theory posits that older adults can substitute satisfying new roles for those they've lost.

STUDY TOOLS 12

READY TO STUDY? IN THE BOOK, YOU CAN:

☐ Check your understanding of what you've read with the Test Your Learning Questions provided on the chapter review card at the back of the book.

☐ Rip out the chapter review card for a handy summary of the chapter and key terms.

ONLINE AT CENGAGEBRAIN.COM YOU CAN:

☐ Prepare for tests with quizzes.

☐ Review the key terms with Flash Cards.

☐ Play games to master concepts.

© Anelina/Shutterstock.com

USE THE TOOLS.

- Rip out the Review Cards in the back of your book to study.

Or Visit CourseMate to:

- Read, search, highlight, and take notes in the Interactive eBook
- Review Flashcards (Print or Online) to master key terms
- Test yourself with Auto-Graded Quizzes
- Bring concepts to life with Games, Videos, and Animations!

Go to CourseMate for **SOC4** to begin using these tools.
Access at **www.cengagebrain.com**

Complete the Speak Up
survey in CourseMate at
www.cengagebrain.com

f Follow us at
www.facebook.com/4ltrpress

13 | Education and Religion

After you finish this chapter go to **PAGE 268** for **STUDY TOOLS.**

Education and religion, two important social institutions, teach values, shape attitudes, maintain traditions, bring people together, control behavior, and grapple with social change. The second half of this chapter examines U.S. and global religions. Let's begin with education.

13-1 EDUCATION AND SOCIETY

Education transmits attitudes, knowledge, beliefs, values, norms, and skills to a society's members. Education can be formal or informal and can occur in a variety of settings. **Schooling**, a narrower term, is formal training and instruction provided in a classroom setting.

U.S. education and schooling have undergone four significant changes since the beginning of the twentieth century: Universal education has expanded, community

---

education a social institution that transmits attitudes, knowledge, beliefs, values, norms, and skills.

schooling formal training and instruction provided in a classroom setting.

What do you think?

I'm optimistic about getting a good job after graduating from college.

1	2	3	4	5	6	7

strongly agree   strongly disagree

colleges have flourished, public higher education has burgeoned, and student diversity has increased as more women and racial-ethnic groups enrolled at colleges and universities. As a result of these and other changes,

a record number of Americans have completed high school (90 percent) and obtained a bachelor's or higher degree (34 percent) (Kena et al., 2014).

13-2 SOCIOLOGICAL PERSPECTIVES ON EDUCATION

Sociologists agree that education and schooling are important, but offer different explanations of their purpose and outcomes. *Table 13.1* summarizes the four major perspectives.

13-2a Functionalism: What Are the Benefits of Education?

For functionalists, education contributes to society's stability, solidarity, and well-being, and provides people with an opportunity for upward mobility. In their analyses, functionalists distinguish between manifest and latent functions.

MANIFEST FUNCTIONS OF EDUCATION

Some of the functions of education are *manifest*; that is, they're open, intended, and visible. For example:

▶ Schools are *socialization agencies* that teach children how to get along with others and prepare them for adult economic roles (Durkheim, 1898/1956; Parsons, 1959).

▶ Education *transmits knowledge and culture*. Schools teach skills like reading, writing, and counting; they also instill cultural values that encourage competition, achievement, and democracy (see Chapter 3).

▶ Similar values increase *cultural integration*, the social bonds that people have with each other and with the community at large, and *societal cohesion*.

▶ Education promotes *cultural innovation*. Faculty at research universities receive billions of dollars every year to develop computer technology, treatments for diseases, and programs to address social problems.

▶ Education *benefits taxpayers* because more highly educated people tend to pay higher taxes, are less likely to rely on public assistance programs, and lead

TABLE 13.1 SOCIOLOGICAL EXPLANATIONS OF EDUCATION

Theoretical Perspective	Level of Analysis	Key Points
Functionalist	Macro	Contributes to society's stability, solidarity, and cohesion and provides opportunities for upward mobility
Conflict	Macro	Reproduces and reinforces inequality and maintains a rigid social class structure
Feminist	Macro and micro	Produces inequality based on gender
Symbolic Interactionist	Micro	Teaches roles and values through everyday face-to-face interaction and behavior

healthier lifestyles, reducing state and federal health care costs (Baum et al., 2013).

Many Americans agree with functionalists that a college degree is "very important": 70 percent (up from only 36 percent in 1979) believe that a college education increases job prospects; 74 percent say that a college degree leads to a better quality of life; and in a survey of incoming first-year college students at four-year colleges and universities, 88 percent said that the most important reason for going to college is to get a good job (Newport and Busteed, 2013; Schneider, 2013; Calderon and Sorenson, 2014). Among college graduates, 74 percent said their education helped them grow intellectually (Parker et al., 2011).

Besides expanding a person's intellectual horizons, education increases earnings. On average, college graduates earn 60 to 80 percent more per year than high school graduates. Even academically marginal students who manage to get a four-year degree have earnings that are as high, on average, as those of their peers with higher grades (Oreopoulos and Petronijevic, 2013).

Over a lifetime, education affects earnings five times more than other demographic factors like sex, race, ethnicity, and age (Schramm et al., 2013; Daly and Bengali, 2014). By age 64, people with a bachelor's degree earn six times more than those with only a high school diploma (Selingo, 2013). Not surprisingly, lifetime earnings are even higher for those with advanced degrees (see *Figure 13.1*).

Regardless of race, ethnicity, family background, and marital status, people with a bachelor's degree live longer, report better physical and emotional health, and are better able to handle stress because of their economic and social position. They're also more likely than non-college graduates to hold jobs that offer a greater sense of accomplishment, more autonomy, more opportunities for creativity, and more social interaction with coworkers (Rheault and McGeeney, 2011; Oreopoulos and Petronijevic, 2013).

LATENT FUNCTIONS OF EDUCATION

Education also has *latent functions*— hidden, unstated, and sometimes unintended consequences. For example:

▶ Schools *provide childcare*, particularly after-school programs, for the growing number of single-parent and two-income families.

FIGURE 13.1 EDUCATION PAYS OFF

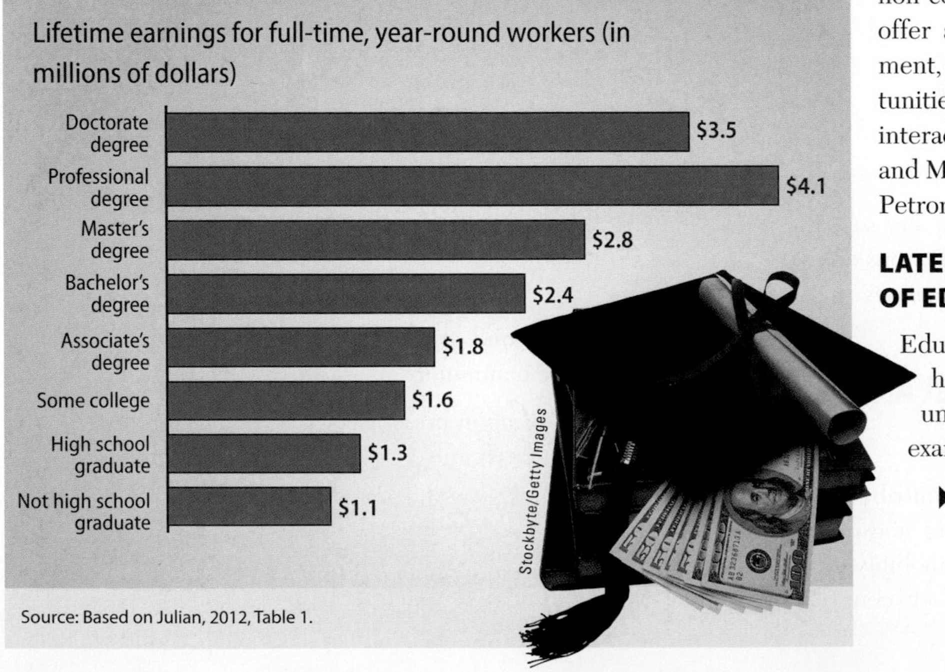

Lifetime earnings for full-time, year-round workers (in millions of dollars)

Doctorate degree	$3.5
Professional degree	$4.1
Master's degree	$2.8
Bachelor's degree	$2.4
Associate's degree	$1.8
Some college	$1.6
High school graduate	$1.3
Not high school graduate	$1.1

Stockbyte/Getty Images

Source: Based on Julian, 2012, Table 1.

- High schools and colleges are *matchmaking institutions* that bring together unmarried people.

- Education *decreases job competition*; the more time that young adults spend in school, the longer the jobs of older workers are safe.

- Educational institutions *create social networks* that can lead to jobs or business opportunities.

- Education is *good for business*. Thousands of companies offer services that tutor and test students and produce textbooks and related materials.

CRITICAL EVALUATION

Do functionalists exaggerate education's benefits? About 35 percent of the world's billionaires don't have a bachelor's degree, and some even dropped out of high school; 22 members of the 2013–2014 Congress have only a high school diploma (Manning, 2013; Wealth-X and UBS, 2014). Moreover, some of the highest paying jobs (e.g., radiation therapists, dental hygienists, commercial pilots, registered nurses) that will be in high demand in the coming decades require only a two-year degree or vocational training (Ogunro, 2012).

Among those with a bachelor's degree, 41 percent say that their work doesn't require a college degree (Saad, 2013). In addition, what people study affects their financial payoff. The lifetime earnings of people with bachelor's degrees vary dramatically. Earnings vary by occupation even with the same major field of study. Someone with a degree in the social sciences who works in management can earn $3.4 million versus $1.9 million with a job in education (Julian, 2012; Carnevale and Cheah, 2013). Thus, the combination of what people study in college *and* the careers they pursue after graduation can make a big difference in lifetime earnings.

Also, according to some critics, functionalists tend to gloss over education's dysfunctions, including rising college costs and student loan debt, topics we'll consider shortly. Conflict theorists, especially, argue that educational institutions produce and reproduce inequality.

13-2b Conflict Theory: Does Education Perpetuate Social Inequality?

From preschool to graduate school, conflict theorists maintain, education creates and perpetuates social inequality based on social class, race, and ethnicity. Schools are also gatekeepers that control and maintain the status quo.

SOCIAL CLASS, RACE/ETHNICITY, AND ACHIEVEMENT GAPs

An **achievement gap** is the difference in academic performance that shows up in grades, standardized test scores, and college completion rates. College enrollments have surged since 1990, especially among minority groups, but who graduates? Asian Americans have made the largest gains (see *Figure 13.2*), but there's considerable variation across subgroups. Of the 48 groups in the Asian population, 87 percent of Asian Indians have at least a bachelor's degree compared with only 21 percent of Samoans (Teranishi, 2011). Such variations, as with other minority subgroups, are due to many factors, including English language proficiency, but the best predictor of educational attainment is social class.

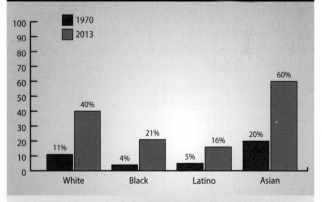

FIGURE 13.2 ATTAINMENT OF BACHELOR'S DEGREE OR HIGHER, BY RACE AND ETHNICITY, 1970 AND 2013

- 1970
- 2013

White: 11% (1970), 40% (2013)
Black: 4% (1970), 21% (2013)
Latino: 5% (1970), 16% (2013)
Asian: 20% (1970), 60% (2013)

Note: 1970 data aren't available for American Indians/Alaska Natives and Pacific Islanders. In 2013, 17 percent and 25 percent, respectively, had a bachelor's or higher degree.

Sources: Based on U.S. Census Bureau, 2012, Table 229; Kena et al., 2014, Figure 3.

achievement gap the difference in academic performance that shows up in grades, standardized test scores, and college completion rates.

BananaStock/Jupiter Images

Blend Images/Ariel Skelley/Getty Images

Fifty years ago, the largest achievement gap was between blacks and whites; today it's between social classes. The gap between children from high- and low-income families has grown by about 40 percent since the 1960s (Reardon, 2011).

CUMULATIVE REASONS FOR ACHIEVEMENT GAPS

When the 2007–2009 recession began, a number of schools experienced federal, state, and local budget cuts. Five years after the recession, class sizes at many public schools increased, but staff positions—including teachers, reading and math specialists, and guidance counselors—continued to decrease. Wealthier communities, in contrast, increased property taxes to accommodate higher enrollments (Rich, 2013).

Many of the nation's schools now have a high level of "double segregation" because students are increasingly separated not only by race but also by income. Nationally, 40 percent of black students attend a high-poverty public school compared with only 6 percent of white students. Educational problems linked to poverty *and* racial segregation include less-experienced and less-qualified teachers, high teacher turnover, inadequate facilities (e.g., computers, science labs), outdated textbooks, high dropout rates, and school buildings that are falling apart (Jordan, 2014; Reich, 2014).

Poverty and inferior schooling are significant academic roadblocks for low-income students. They're less likely to attend high schools that offer high-level courses in mathematics and science. This can mean the difference between passing or failing required courses during the first few years of college. Even when top students from low-income public high schools enter selective colleges, they often struggle to maintain passing grades. ("Selective" schools are those that admit a relatively low percentage of applicants.) Many college youths from low- and middle-income families must also work part-time or full-time and, consequently, have more stress and less time for academic work (Ross et al., 2012; Black et al., 2014).

EDUCATION AND SOCIAL CONTROL

Functionalists see education as an avenue for upward mobility. Conflict theorists maintain that education restricts mobility because of a hidden curriculum, credentialism, and privilege.

Hidden Curriculum. Every school has a formal curriculum that includes reading, writing, and learning other skills. Schools also have a **hidden curriculum**, practices that transmit nonacademic knowledge, values, attitudes, norms, and beliefs that legitimate "economic inequality and the staffing of unequal work roles" (Bowles and Gintis, 1977: 108).

Schools in low-income and working-class neighborhoods tend to stress obedience, following directions, and punctuality so that students can fill low-paid jobs (e.g., restaurants, nursing homes, hospitals) that require these characteristics (Kozol, 2005). Schools in middle-class neighborhoods emphasize proper behavior and appearance, cooperation, conforming to rules, and deference to authority because many of these students will go to college and work in bureaucracies that require such attributes (Hedges, 2011). In contrast, selective schools encourage leadership, creativity, independence, and people skills—all prized characteristics in elite circles (Persell and Cookson, 1985; see also Chapter 8). In effect, then, the hidden curriculum reproduces the existing class structure and provides workers for jobs and occupations in the stratification hierarchy.

Credentialism. Have you noticed that many faculty, doctors', lawyers', and dentists' offices are usually wallpapered with framed degrees? Such tangible symbols of people's achievements reflect **credentialism**, an emphasis on certificates or degrees to show that people have certain skills, educational attainment levels, or job qualifications.

hidden curriculum school practices that transmit nonacademic knowledge, values, attitudes, norms, and beliefs.

credentialism an emphasis on certificates or degrees to show that people have certain skills, educational attainment levels, or job qualifications.

Functionalists maintain that credentialism rewards people for their accomplishments, sorts out those who are the most qualified for jobs, and stimulates upward social mobility. Conflict theorists contend, however, that for many positions, people can gain skills on the job with a few weeks of training or succeed because of ability or other factors.

Because of a large supply of high school graduates, employers can demand higher levels of education even though some jobs (e.g., retail sales, law enforcement) don't require a college degree for competent performance, a process called *credential inflation*. As more people obtain a college degree, its value diminishes, and students from low-income families, who are the least likely to have access to a college education, fall further behind (Bollag, 2007).

Privilege. According to one observer, "We're the only rich nation to spend less educating poor kids than we do educating kids from wealthy families" (Reich, 2014). Many selective colleges claim that they're committed to admitting talented low-income and minority students. In fact, in 2010–2011, colleges and universities awarded $5.3 billion in financial aid to students from high-income families. Among students with a GPA of 3.5 or higher on a 4.0 scale, white students received more than three times as much in merit-based scholarships as minority students

Universal Pictures/Fotos International/Getty Images

How important is graduating from a prestigious college or university? Nationally, 84 percent of business leaders say that the amount of knowledge a job candidate has in a particular field is "very important." Only 9 percent say that where a person attended school is very important (Calderon and Sidhu, 2014). Steven Spielberg, the well-known movie producer and director, graduated from California State University at Long Beach after being rejected by the more prestigious University of Southern California and University of California at Los Angeles film schools.

(Baum and Payea, 2011; Kantrowitz, 2011). Among students with similar high school GPAs and SAT scores, since 1995 the proportion of high-income students receiving scholarships from colleges, the federal government, or the states has increased but has fallen for low-income students (Burd, 2013; Carey, 2013; Mettler, 2014).

Another privileged group is *legacies*, the children of alumni who have "reserved seats" regardless of their accomplishments or ability. For example, President George W. Bush—who had mediocre high school grades and standardized test scores—was admitted as a legacy at Yale University, which his father and grandfather had attended (Golden, 2006). Legacies also include students with inferior academic records whose parents, including celebrities, make million-dollar donations. Among selective colleges and universities, almost 75 percent use legacies, which account for up to 30 percent of some universities' student body (Massey, 2007; Hurwitz, 2011).

CRITICAL EVALUATION

Conflict theories have several weaknesses. First, from 1992 to 2012, racial-ethnic diversity increased about 56 percent at community colleges and four-year public and private colleges (Zweifler, 2013). Attaining a bachelor's degree or higher has also risen for minorities (see *Figure 13.2* on page 251). Such data show that racial-ethnic inequality isn't as fixed as many conflict theorists assert.

Second, a hidden curriculum doesn't necessarily determine job placement because many students aren't passive recipients of educational systems. Below the bachelor's degree level, for example, nearly 25 million U.S. adults have acquired professional certificates and licenses in wide-ranging fields (e.g., nursing, computer and information sciences, business management). These and other "alternative credentials" have labor market value that increases employment and earnings (Ewert and Kominski, 2014). Conflict theorists may denounce credentialism and credential inflation, but employers often view credentials as useful initial screening tools that signal job candidates' discipline and drive.

Third, do conflict theorists overstate the importance of social class achievement gaps? A study tracked more than 5,200 Asian American and white students from kindergarten through high school and found that the former, regardless of social class, performed better than their white counterparts. The researchers attributed the Asian American students' greater success to cultural beliefs that achievement is learned rather than innate, parental pressures to succeed, and, especially among recent immigrants, ethnic community resources such as private tutoring and vital information about navigating the education system (Hsin and Xie 2014; see also Chapter 10).

13-2c Feminist Theories: Is There a Gender Gap in Education?

Since 1982, women have been graduating from college at higher rates than men. Some view this trend as an indicator of greater gender equality; others are alarmed about the supposed "male crisis" in and "feminization" of higher education (Mullen, 2012; see also Chapter 9). Has women's progress come at the cost of men? Or are there still gender gaps, particularly in higher education?

WHO'S GETTING DEGREES

Almost equal numbers of girls and boys are high school graduates. Women earn more associate's, bachelor's, and master's degrees than men, but their percentage of professional and doctoral degrees drops considerably (see *Figure 13.3*).

Across all racial-ethnic groups, students from affluent families are the most likely to earn a bachelor's degree, but men are more likely to do so than women (Mullen, 2012). In 2012, Asian females (50 percent) and Asian males (54 percent) had the highest rates of attaining a bachelor's degree or higher, followed by white males (36 percent) and white females (34 percent). In contrast, black women and Latinas had higher rates than their male counterparts (Snyder and Dillow, 2013; Kena et al., 2014). Such data contradict the description of higher education as feminized, but why do women achieve more degrees than men up to the master's level?

In both high school and college, women spend more time than men studying, earn better grades, hold more leadership positions, and are more involved in student clubs and community volunteer work. Women (81 percent) are also more likely than men (67 percent) to value higher education (Goldin et al., 2006). This may be one reason why women are less likely to drop out of college and enter the labor force. Regardless of the reasons for women's success, many colleges have been giving men preferential treatment in admissions to avoid large gender imbalances in their student bodies (Gewertz, 2009; Kahlenberg, 2010; Dwyer et al., 2013).

MAJORS AND CAREERS

A major gender gap is women's underrepresentation in the high-paying fields of science, technology, engineering, and mathematics (STEM). In elementary, middle, and high school, girls and boys take math and science courses in roughly equal numbers, and about the same numbers leave high school planning to pursue STEM majors in college. After a few years, however, men outnumber women in nearly every STEM field, and in some—mathematics, engineering, and computer science—women earn only 10 to 20 percent of the bachelor's degrees. Their numbers decline further at the graduate level and yet again in the workplace (Hill et al., 2010; Landivar, 2013; Snyder and Dillow, 2013).

Why is there a gradual attrition? In college, females are initially as persistent as men in a STEM major and earn

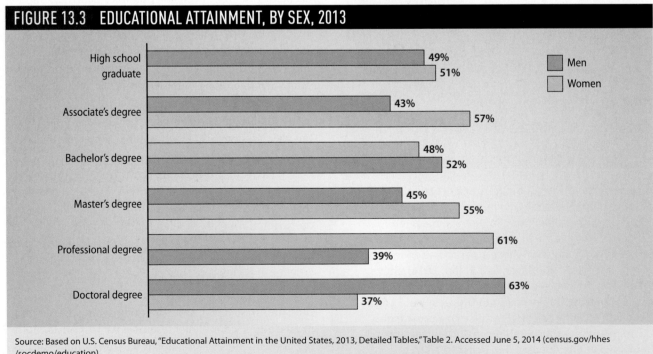

FIGURE 13.3 EDUCATIONAL ATTAINMENT, BY SEX, 2013

	Men	Women
High school graduate	49%	51%
Associate's degree	43%	57%
Bachelor's degree	48%	52%
Master's degree	45%	55%
Professional degree	61%	39%
Doctoral degree	63%	37%

Source: Based on U.S. Census Bureau, "Educational Attainment in the United States, 2013, Detailed Tables," Table 2. Accessed June 5, 2014 (census.gov/hhes /socdemo/education).

higher grades. However, they're less satisfied than men with the core courses and more likely to doubt their ability to succeed in a male-dominated discipline. As a result, women's self-confidence falters and they change majors (Shapiro and Williams, 2012; Jagacinski, 2013). The exit from a STEM major is also associated with having few female faculty role models, and some science professors' beliefs that female students won't benefit from mentoring because they're less competent than men (Moss-Racusin et al., 2012; Williams and Ceci, 2012).

CRITICAL EVALUATION

A common criticism is that feminist scholars address women's education barriers but not their choices. Socialization, gender stereotypes, and teachers' expectations affect our behavior (see Chapters 4 and 9), but it's still not clear why many women choose fields of study that they know are on the lower end of the pay scale (e.g., health, education).

Some fault feminist theorists, who support gender equality, for being more interested in women's than men's gender gaps, especially at the associate's, bachelor's, and master's levels (Baenninger, 2011). There's also the question of why feminist scholars devote little attention to issues such as why boys are more disruptive in elementary school, which leads to suspensions and decreases the chance of attending college by at least 16 percent for each suspension; whether the general public's image of men as underachievers affects men's motivation; and

Over a five-year college career, nearly 20 percent of female students experience rape or other sexual assault on campus. About 6 percent of college men rape or attempt to rape women. Of this group, 63 percent are repeat offenders, committing an average of 6 rapes (Lisak and Miller, 2002; Krebs et al., 2007). In mid-2014, the U.S. Department of Education announced that it would investigate 55 universities for possible violations in handling sexual assault accusations.

why parents are more likely to pay for their daughters' (40 percent) than their sons' (29 percent) college education (Bertrand and Pan, 2011; Wang and Parker, 2011).

13-2d Symbolic Interaction: How Do Social Contexts Affect Education?

None of us is born a student or a teacher. Instead, these roles, like others, are socially constructed (see Chapters 3–5). For symbolic interactionists, education is an active *process* that includes students, teachers, peers, and parents and involves tracking, labeling, and student engagement.

TRACKING

Beginning in kindergarten, practically all schools sort students by aptitude. Such sorting results in **tracking** (also called *streaming* or *ability grouping*), assigning students to specific educational programs and classes on the basis of test scores, previous grades, or perceived ability.

Some educators believe that tracking is beneficial because students learn better in groups with others like themselves, and it allows teachers to develop curricula for students with similar ability. Many interactionists maintain, however, that tracking creates and reinforces inequality.

▶ High-track students take classes that involve critical thinking, problem solving, and creativity that high-status occupations require. Low-track students take classes that are limited to simple skills (e.g., punctuality and conformity) that usually characterize lower status jobs.

▶ High-track students have more homework, better quality instruction, and more enthusiastic teachers. One result is that high-track students are more likely to see themselves as "bright," whereas low-track students see themselves as "dumb" or "slow."

▶ The effects of tracking are usually cumulative and lasting. Teachers tend to have low expectations for low-track students, who therefore fall further behind every year in reading, mathematics, and interaction skills (Oakes, 1985; Hanushek and Woessman, 2005).

Middle school and high school become even more stratified as high-track students are sorted into gifted, honors, and advanced courses. In college, students continue to be tracked and sorted into honors programs and accelerated undergraduate courses.

tracking (also called *streaming* or *ability grouping*) assigning students to specific educational programs and classes on the basis of test scores, previous grades, or perceived ability.

LABELING

Tracking often leads to labeling, a serious problem because "there's a widespread culture of disbelief in the learning capacities of many of our children, especially children of color and the economically disadvantaged" (Howard, 2003: 83). Labeling, in turn, can result in a *self-fulfilling prophecy*. That is, students live up or down to teachers' expectations and evaluations that are influenced by a student's social class, skin color, hygiene, accent, and test scores (see Chapter 5).

Parents affect their children's academic outcomes through conscious or unintentional labeling. Only half of parents with annual incomes of less than $25,000 expect their child to attain a four-year college degree compared with 88 percent of parents with incomes over $75,000. Because higher-income parents expect their children to graduate from college, they talk about what's going on in the classroom, provide out-of-school learning opportunities, foster positive attitudes toward school, and applaud academic achievement. As a result, the children, regardless of innate abilities, internalize personal goals that include earning a college degree (Kagan, 2011; Child Trends Data Bank, 2012).

mevans/iStockphoto.com

STUDENT ENGAGEMENT

Many schools assess performance not only through tests but also in terms of *student engagement*, how involved students are in their own learning. A team of researchers who spent thousands of hours in more than 2,500 elementary school classrooms concluded that the typical U.S. child has only a 1 in 14 chance of being in a school that encourages her or his engagement. Because of standardized tests, teachers in public elementary schools spend most of the day on basic reading and math drills and little time on problem solving, reasoning, science, and social studies. As a result, many classrooms are often "dull and bleak places" where kids don't get much teacher feedback or face-to-face interaction with their peers (Pianta et al., 2007).

U.S. high school students aren't as engaged in their education as they could be: 26 percent admit that they usually don't do their homework, and 10 percent drop out. The students who are most likely to be disengaged are from low-income families, African American or Latino, don't live with both biological parents, attend financially strapped urban public high schools that are typically overcrowded and understaffed, and have fewer resources such as computers and even textbooks (Finn, 2006; Planty et al., 2008).

In college, the typical undergraduate spends more time partying, socializing, and working for pay than studying. On average, even full-time college students study only 14 hours a week compared with 24 hours a week in 1961

(Babcock and Marks, 2011). This is well below the 24 to 30 hours faculty members say students should be spending on class preparation if they're taking three courses.

According to one college instructor, "Education is the only business in which the clients want the least for their money" (Perlmutter, 2001: B1). However, some analysts also blame professors for students' not studying. Many faculty have watered down their required readings, say nothing when students don't prepare for classes or do assignments, and often give easy exams to avoid complaints about low grades. Only 55 percent of first-year college students and 61 percent of seniors say their courses "challenged them to do their best work" (National Survey of Student Engagement, 2013: 9). When faculty dilute their course requirements and tests, some students don't study because they're bored by courses that don't stimulate them (Benton, 2011; Glenn, 2011).

CRITICAL EVALUATION

One weakness is that interactionists gloss over the power of individuals to change the course of events, including overcoming the effects of labeling. For example, a number of low-income parents, despite language barriers among recent immigrants, contact school personnel on a regular basis. Others go to college open houses to speak to college representatives, with their children translating the questions and answers (Tornatzky et al., 2002).

Another limitation is that interactionists, because of their micro-level analysis, neglect the macro-level structural constraints that are built into society. Black and Latino male students enter community colleges with higher aspirations than their white peers, but white males are six times more likely to graduate in three years with a certificate or degree. A major reason for the difference is that minority students tend to enter college with weaker academic skills because many have attended underfunded and understaffed high schools (Center for Community College Student Engagement, 2014; see also Rios, 2012).

13-3 SOME CURRENT ISSUES IN U.S. EDUCATION

In an open-ended question about "the most important problem facing this country today," only 5 percent of Americans named education (Riffkin, 2014). A dysfunctional government and weak economy ranked the highest, but the U.S. education system has problems at all levels, some more controversial than others.

13-3a Elementary Schools

Advanced economies, including ours, rely not on physical labor but on "cognitive labor" that requires formal analytical abilities, written communications, and specific technical knowledge—skills that people acquire and cultivate beginning in preschool (Autor, 2014). Three of the ongoing issues regarding elementary education are low test scores, achievement gaps, and effective teaching.

LOW TEST SCORES AND ACHIEVEMENT GAPS

Children who read proficiently by the end of the third grade are more likely to do well in other subjects, including mathematics, and to graduate from high school. Between 1992 and 2013, the average reading and mathematics scores of fourth- and eighth-grade students increased, but there are considerable achievement gaps by race and ethnicity. Asian and white children have the highest proficiency rates in both reading and math, but only about a third of U.S. students are proficient in both subjects (see *Table 13.2*).

Social class explains some of the variations. In 2012, 75 percent of 3- to 5-year-olds whose parents had either a graduate or professional degree, compared with 60 percent of those whose parents had only a high school diploma, were enrolled in kindergarten, preschool, and nursery school programs that provided educational experiences (Kena et al., 2014). In 2013, 80 percent of fourth graders in low-income families were below proficiency in reading compared with 49 percent of higher-income children (National Center for Education Statistics, 2013).

Social class alone doesn't explain student performance, however. Homework also affects test scores. Many elementary and middle school children aren't getting or doing their homework, but parents and the public often blame teachers for the students' poor academic performance (Loveless, 2014).

EFFECTIVE TEACHING

A multi-year project concluded that the most accurate way to evaluate elementary school teachers is to use a three-pronged approach that includes student test scores, multiple classroom observations, and student evaluations (Cantrell and Kane, 2013), but 40 states use only student test scores to evaluate teachers. Much research shows that test scores are invalid and unreliable measures of teacher effectiveness. Other factors affect test scores, including a teacher's quality of undergraduate education, years of experience, being assigned to low-track or high-track classrooms, whether or not students study, and school funding (Kalogrides et al., 2013; Amrein-Beardsley, 2014; Polikoff and Porter, 2014).

13-3b High Schools

High schools, like elementary schools, must grapple with low achievement scores and effective teaching. In addition, the SAT is becoming an increasingly controversial issue.

INTERNATIONAL TEST SCORES

Among 65 countries, U.S. 15-year-olds rank 30th in math, 23rd in science, and 20th in reading. These rankings have slipped since 2009, and many other countries' best students outperform ours. Of 32 nations assessed on computer-based mathematics literacy, the United States comes in only 13th (Banchero, 2013; Kelly et al., 2013; OECD, 2013).

The United States is one of the wealthiest countries in the world; spends more per student from the ages of 6 to 15 than 30 other industrialized countries; ranks above average in the share of high-SES high school students; and has fewer non-English speaking students than other Western countries, including Canada (Carnoy and Rothstein, 2013; OECD, 2013). Why, then, are U.S. students doing so poorly? A major difference between the United States and other developed nations is teacher preparation and quality.

TABLE 13.2   READING AND MATHEMATICS PROFICIENCY, BY RACE AND ETHNICITY, 2013

	National Average	African American	American Indian	Latino	White	Asian and Pacific Islander
Fourth graders who scored at or above proficient in reading	34%	17%	22%	19%	45%	51%
Eighth graders who scored at or above proficient in math	34%	14%	21%	21%	44%	60%

Note: "Proficient" indicates the ability to handle challenging subject matter.

Source: Based on Annie E. Casey Foundation, 2014, Table 1.

About 7 percent of Americans ages 16 to 24 are high school dropouts, down from 12 percent in 1990. The dropout rate is slightly higher for males than females, and highest among foreign-born Latinos and youth whose families are in the bottom 25 percent of all family incomes (Kena et al., 2014).

TEACHER PREPARATION AND QUALITY

American teachers spend, on average, 20 to 50 percent more hours teaching in class than do their Australian, Canadian, and Finnish counterparts, all of whose students outrank Americans in international tests (Lawrence, 2013). A combination of structural factors helps explain why, despite their individual efforts, many U.S. teachers aren't as effective as they could be.

First, the top-performing countries accept only the top applicants for education programs. Finland, Singapore, and South Korea recruit the top 5 to 30 percent of high school graduates, but accept only 10 percent who have high standardized test scores in science, math, and reading (Auguste et al., 2010; Sawchuck, 2012). Several recent reports described U.S. teacher preparation programs as "an industry of mediocrity" that has low or no academic entry standards, easy coursework, and gives out many easy A's (Greenberg et al., 2013; Putnam et al., 2014). In top-performing countries, 100 percent of teachers are in the top third of their college graduating classes. In the United States, 47 percent of kindergarten through twelfth grade teachers come from the bottom third. "In other words, we hire lots of our lowest performers to teach, and then we scream when our kids don't excel" (Cloud, 2010: 48).

Second, compared with teachers in top-performing countries, a higher number of U.S. teachers don't have a bachelor's degree or higher in the subjects they teach. For example, almost 30 percent of public high school teachers who teach math didn't major in the field, and 54 percent who teach chemistry didn't major in the subject (Hill, 2011).

Third, the top-performing countries offer teachers competitive salaries. In South Korea and Singapore, teachers on average earn more than lawyers and engineers. In these countries, Finland, and many others, teachers are highly regarded and enjoy the same prestige as physicians and other high-status professions and have lifelong careers, resulting in very low teacher attrition. In contrast, 40 to 50 percent of new U.S. teachers leave the profession within their first five years on the job. High turnover rates are especially harmful to students in low-performing schools: Replacing a teacher—especially a good one—is costly, disruptive, and demoralizing for both students and parents (OECD, 2011; Ronfeldt et al., 2011; Kafka et al., 2014).

THE GROWING CONTROVERSY OVER SATs, ACTs, AND AP COURSES

In 2013, about 3.5 million high school students took either the SAT or the ACT, standardized college entrance examinations. The administrators/owners of both exams say that the scores can indicate where students are falling behind in college readiness, measure students' capability to do college-level work, and even predict college success (College Board, 2013; ACT, Inc., 2014).

A growing number of critics, including college officials, contend that the tests are little more than gatekeeping tools that exclude lower socioeconomic students from higher education, don't predict college success, and that students who submit test scores and those who don't perform similarly in college. As a result, 850 four-year colleges don't require applicants to submit SAT or ACT scores (FairTest, 2013; Botstein, 2014; Fraire, 2014; Hiss and Franks, 2014).

Advanced Placement (AP) offers high school students college-level curricula and exams. About 78 percent of degree-granting colleges and universities give course credit to students who get high scores on the exams (Berrett, 2014). In 2013, 30 percent of high school students took at least one AP exam. Of those that took the tests, 27 percent were low income, and only 48 percent of this group scored high enough to get college credit (College Board, 2014).

Like the ACT and SAT, AP is becoming more controversial as critics point to a number of problems. First, AP courses are rarely as demanding as college courses and don't offer the same content. Second, to increase a high school's prestige, superintendents and principals may pressure teachers to offer AP courses and students to take them. Both groups may be unprepared to do so, and failure can make students and teachers feel inferior. Finally, only about 45 percent of the nation's public high schools offer AP courses, and those that do so are typically in high-income and predominantly white neighborhoods (Jackson, 2012; Simon, 2013; Berrett, 2014).

13-3c Colleges and Universities

Public postsecondary institutions often feel embattled by many constituencies and external pressures. Many colleges and universities are grappling with affirmative action issues, grade inflation and cheating, low graduation rates, and rising student loan debt.

AFFIRMATIVE ACTION

In higher education, *affirmative action* refers to admission policies that provide equal access for groups, particularly women and minorities, that have been historically excluded or underrepresented. The enrollment rates of these groups has increased steadily since the 1970s, but using race and ethnicity in college admission decisions has been, and continues to be, a hotly debated topic (see "The Affirmative Action Debate in Higher Education").

In 2006, Michigan voters approved an amendment to the state's constitution to ban using race as a factor in deciding who's admitted to the state's public universities. In 2014, the U.S. Supreme Court (*Schuette v. Coalition to Defend Affirmation Action*) ruled that Michigan's ban was constitutional. The court's decision legitimated similar laws in seven other states (Schmidt, 2014). Affirmative action opponents praised the decision for reinforcing **meritocracy**, a system that rewards people because of their individual accomplishments. We say things like "work hard and you'll get ahead" and "pull yourself up by your bootstraps," but is that what happens? As you've seen, the best predictor of who goes to college is not ability but income.

GRADE INFLATION AND CHEATING

Grade inflation is widespread across all education levels, and some children start cheating on tests as early as the third grade. The consequences of grade inflation and cheating in postsecondary education are more serious, however, because someone who cheated her or his way through college or graduate school may be your dentist, lawyer, doctor, or tax accountant.

Grade Inflation. Has an A replaced the C of the 1970s? In 1971, only 26 percent of first-year college students

> **WOULD YOU WANT A MILITARY OFFICER, A FINANCIAL PLANNER, A DENTIST, OR A DOCTOR WHO CHEATED HER OR HIS WAY THROUGH COLLEGE OR GRADUATE SCHOOL?**

expected to earn at least a B average in college, compared with 52 percent in 1995, and a whopping 70 percent in 2010 (Pryor, 2011), and even though students are studying less than in the past.

Many faculty give high grades because it decreases student complaints, involves less time and thought in grading exams and papers, and reduces the chances of students' challenging a grade. Some faculty believe that they can get favorable course evaluations from students by handing out high grades, and others accept students' view of high grades as a reward for simply showing up in class. Inflating grades decreases student attrition and satisfies administrators, especially when state legislators base funding on graduation rates (Bartlett and Wasley, 2008; Gillespie, 2014).

Cheating. Despite grade inflation, cheating is common in college and graduate school. Recently, New York authorities

> **meritocracy** a system that rewards people because of their individual accomplishments.

The Affirmative Action Debate in Higher Education

Supporters argue that affirmative action programs...

▸ give disadvantaged and underrepresented students a needed boost in overcoming past and current discrimination

▸ have doubled or tripled the number of minority applications, resulting in student populations that are more representative of the surrounding community

▸ have increased minorities' upward social and economic mobility

▸ promote diversity and encourage people to work together effectively in a multicultural society

Opponents argue that affirmative action programs...

▸ are discriminatory because they're based on race rather than academic achievement

▸ benefit primarily middle- and upper-class minorities

▸ are condescending because they imply that minorities can't succeed without preferential treatment

▸ admit students who often can't live up to an institution's academic standards, resulting in high failure rates

What else would you add to either side of the debate?

Sources: Messerli, 2012; Krishnamurthy and Edlin, 2014.

charged 20 students from an affluent Long Island community with paying impersonators up to $3,600 per SAT to increase their chances of getting into a selective school (Winter, 2011). Between 50 and 75 percent of college students admit having cheated at least once on tests and assignments. Numerous studies have uncovered cheating at the Air Force Academy and in many graduate programs, including business administration and dentistry (McCabe et al., 2006; Frosch, 2007; Wasley, 2007; Barrett, 2014).

Cheating is prevalent because most students aren't caught. If caught, the sanctions are usually mild. In 2013, for example, Harvard University found that 125 undergraduates cheated on a government take-home exam. More than half had to withdraw for a few semesters, some received probation, and others weren't punished at all (Levitz, 2013). Despite widespread grade inflation and cheating, millions of students don't graduate from college.

GRADUATION RATES

The United States has one of the lowest college graduation rates in the industrialized world. In 2012, only 59 percent of full-time students at four-year colleges completed a bachelor's degree within 6 years, and 31 percent at 2-year colleges earned a certificate or associate's degree within 3 years (OECD, 2013; Kena et al., 2014).

Once on campus, 36 percent must take one or more remedial courses and, as you saw earlier, many students admit that they don't study very much. Dropout rates are higher for males than females, part-time than full-time students, and students from lower-income families. At four-year colleges, the more selective the institution, the higher the graduation rate: 33 percent at colleges with open admissions policies, 60 percent at colleges that accept 50 percent or more students, and 86 percent at colleges that accept only 25 percent of applicants (Kena et al., 2014).

Family and work commitments can reduce the likelihood of completing college. Fully three-quarters of undergraduates are managing some combination of family responsibilities, jobs, and commuting to class (Princiotta et al., 2014). Students also drop out because they get sick, accept attractive job offers, or decide that college isn't for them, but cost is a major reason for attrition.

STUDENT LOAN DEBT

Student loan debt has almost tripled since 2004, is now more than $1 trillion, and is second only to mortgage debt (Lee, 2013). About 71 percent of students who graduated from four-year colleges in 2012 had an average debt of almost $30,000, up from $9,500 in 1993. Those at private nonprofit colleges and for-profit colleges (e.g., University of Phoenix) owed much more—$32,300 and almost $40,000, respectively (Reed and Cochrane, 2013; Institute for College Access & Success, 2014).

There's no simple answer for the rising student loan debt because much depends on factors like an institution's endowment, in-state and out-of-state tuition prices, and the growing number and costs of full-time administrators (*Chronicle of Higher Education*, 2014). However, a major reason for the rise is that, since 2000, public spending on higher education has dropped 30 percent even though enrollment at public colleges has jumped 34 percent (Kena et al., 2014). As state funding has decreased, tuition has increased, and so has borrowing to cover tuition, fees, and, for many students, the costs for room and board (see *Figure 13.4*).

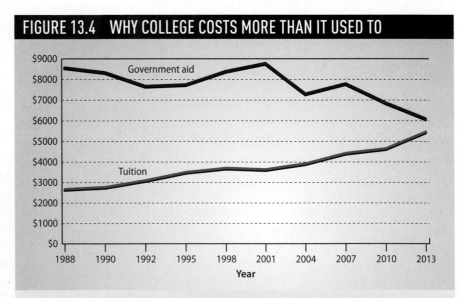

FIGURE 13.4   WHY COLLEGE COSTS MORE THAN IT USED TO

Note: These are public higher-education revenues per student, in 2013 dollars.

Source: Based on State Higher Education Executive Officers, 2014, Figure 3.

13-4 RELIGION AND SOCIETY

Like education, religion has a profound effect on society. Some form of religion exists in all societies and cultures. Throughout history, religion has been a central part of the human experience, affecting the family, economy, and other social institutions. Why is religion important to many people? And does it always benefit society? You'll see that the four sociological perspectives answer these and other questions differently.

13-4a What Is Religion?

Religion is a social institution that involves shared beliefs, values, and practices related to the supernatural. It unites believers into a community, but customs and practices differ across cultures and groups. For example, being Catholic involves confessing (telling one's sins to a priest), a practice not followed by Protestants, Jews, Muslims, and other religious groups.

Émile Durkheim (1961) distinguished between the sacred and the profane. **Sacred** refers to anything that people see as awe-inspiring, supernatural, holy, and not part of the physical world. In contrast, **profane** refers to the ordinary and everyday elements of life that aren't related to religion. Sociologists differentiate religion from religiosity and spirituality.

13-4b Religion, Religiosity, and Spirituality

Religion is a belief system, but religious expression can vary. When sociologists examine **religiosity**, the ways people demonstrate their religious beliefs, they find that religion and religiosity differ. For example, 58 percent of Americans say that religion is "very important" in their lives, but only 37 percent attend worship services once a week or more (Lugo et al., 2012).

Spirituality is a personal quest to feel connected to a reality greater than oneself. About 37 percent of Americans describe themselves as "spiritual but not religious." Within this group, however, 44 percent pray at least once a day, and 55 percent are "absolutely certain" that there's a God (Lugo et al., 2012). Thus, religious people are spiritual, but spiritual people aren't necessarily religious.

13-5 RELIGIOUS ORGANIZATION AND MAJOR WORLD RELIGIONS

People manifest their religious beliefs most commonly through organized groups, including cults, sects, denominations, and churches. The major world religions differ in their membership and beliefs.

13-5a Cults (New Religious Movements)

A **cult** is a religious group that is devoted to beliefs and practices that are outside of those accepted in

---

religion a social institution that involves shared beliefs, values, and practices related to the supernatural.

sacred anything that people see as awe-inspiring, supernatural, holy, and not part of the natural world.

profane the ordinary and everyday elements of life that aren't related to religion.

religiosity the ways people demonstrate their religious beliefs.

cult a religious group that is devoted to beliefs and practices that are outside of those accepted in mainstream society.

Hare Krishna is the popular name for the International Society of Krishna Consciousness (ISKCON), a new religious movement based in Hinduism. Established in the United States in 1965, its practices include an austere and simple life, vegetarianism, abstinence from drugs and alcohol, chanting, and evangelism.

mainstream society. Many sociologists use **new religious movement (NRM)** rather than *cult* because the media have used the latter term in derogatory ways to describe any unfamiliar, new, or seemingly bizarre religious group (Roberts, 2004).

NRMs usually organize around a **charismatic leader** (like Jesus) whom followers see as having exceptional or superhuman powers and qualities (see Chapter 11). Some NRMs have become established religions. The early Christians were a renegade group that broke away from Judaism, and Islam, the world's second largest religion, began as a cult around Muhammad. Most contemporary cults are fragmentary, loosely organized, and temporary, but others (e.g., the Church of Scientology) have developed into lasting and highly bureaucratic international organizations.

new religious movement (NRM) term used instead of *cult* by most sociologists.

charismatic leader someone that followers see as having exceptional or superhuman powers and qualities.

sect a religious group that has broken away from an established religion.

denomination a subgroup within a religion that shares its name and traditions and is generally on good terms with the main group.

church a large established religious group that has strong ties to mainstream society.

13-5b Sects

A **sect** is a religious group that has broken away from an established religion. People who begin sects are usually dissatisfied members who believe that the parent religion has become too secular and has abandoned key original doctrines. Like cults, some sects are small and disappear after a time, whereas others become established and persist. Examples of sects that have persisted include the Amish, the Jewish Hassidim, Jehovah's Witnesses, Quakers, and Seventh-Day Adventists (Bainbridge, 1997). Some sects develop into denominations.

13-5c Denominations

A **denomination** is a subgroup within a religion that shares its name and traditions and is generally on good terms with the main group. Denominations can form slowly or develop rapidly, depending on factors like geography, immigration, and a country's birth rate. Some scholars describe a denomination as somewhere between a sect and a church. Like sects, denominations have a professional ministry. Unlike sects, denominations view other religious groups as valid and don't make claims that only they possess the truth (Hamilton, 2001). Denominations typically accommodate themselves to the larger society instead of trying to dominate or change it. As a result, people may belong to the same denomination as did their grandparents or even great-grandparents. Denominations exist in all religions, including Christianity, Judaism, and Islam. In the United States, the many Protestant denominations include Episcopalians, Baptists, Lutherans, Methodists, and Evangelicals.

13-5d Churches

A **church** is a large established religious group that has strong ties to mainstream society. Because leadership is attached to an office rather than a specific leader, new generations of believers replace previous ones, and members follow tradition or authority rather than a charismatic leader. As in a denomination, people are usually born into a church, but may later decide to leave it.

Churches (e.g., Roman Catholic Church, Anglican Church in North America) are typically bureaucratically organized, have formal worship services and trained clergy, and often maintain some degree of control over political or educational institutions (see Chapter 11). Because churches are an integral part of the social order, they often become dependent on, rather than critical of, the ruling classes (Hamilton, 1995).

13-5e Some Major World Religions

Worldwide, the largest religious group is Christians, followed by Muslims. If the world's population is represented as an imaginary village of 100 people, it has about:

▶ 32 Christians

▶ 23 Muslims

▶ 16 Unaffiliated (people who may or may not believe in God but don't identify with any particular religious group)

▶ 15 Hindus

▶ 7 Buddhists

▶ 6 Folk religionists (includes followers of African traditional religions, Chinese folk traditions, Native American religions, and Australian aboriginal religions)

▶ 1 Other (includes Jews, Baha'is, Sikhs, Jains, Shintoists, and many others) (based on Pew Research Center, 2014)

Note that no religious group comes close to being a global majority, that the third largest group is religiously unaffiliated, and that non-Christians outnumber Christians 2 to 1. Five religious groups in particular have had a worldwide impact on economic, political, and social issues. *Table 13.3* provides a brief overview of these groups.

TABLE 13.3 CHARACTERISTICS OF FIVE MAJOR WORLD RELIGIONS

Religion	Date of Origin	Founder	Number of Followers
Christianity Onfokus/iStockphoto.com	0 C.E.	Jesus Christ	2.2 billion
Jesus, the son of God, sacrificed his life to redeem humankind. Those who follow Christ's teachings will enter the Kingdom of Heaven. Sinners who don't repent will burn in hell for eternity.			
Islam kickstand/iStockphoto.com	600 C.E.	Muhammad	1.6 billion
God is creator of the universe, omnipotent, omniscient, just, forgiving, and merciful. Those who sincerely repent and submit (the literal meaning of Islam) to God will attain salvation; the wicked will burn in hell.			
Hinduism Ferenc Cegledi/iStockphoto.com	Between 4000 and 1500 B.C.E.	No specific founder	887 million
Life in all its forms is an aspect of the divine. The aim of every Hindu is to use pure acts, thoughts, and devotion to escape a cycle of birth and rebirth (samsara) determined by the purity or impurity of past deeds (karma).			
Buddhism Pablo Salgado Barrientos/iStockphoto.com	525 B.C.E.	Siddhartha Gautama	386 million
Life is misery and decay with no ultimate reality. Meditation and good deeds will end the cycle of endless birth and rebirth, and the person will achieve nirvana, a state of liberation and bliss.			
Judaism Howard Sandler/iStockphoto.com	2000 B.C.E.	Abraham	15 million
God is the creator and the absolute ruler of the universe. By obeying the divine law God gave them, Jews bear special witness to God's mercy and justice.			

Note: C.E. (Common Era) is the nondenominational abbreviation for A.D. (Anno Domini, Latin for "In the year of our Lord") and B.C.E. (Before the Common Era) is the nondenominational abbreviation for B.C. (Before Christ).

Sources: Based on a number of sources including the Center for the Study of Global Christianity, 2007; "Religions of the World...," 2007; and "Global Christianity...," 2011.

13-6 RELIGION IN THE UNITED STATES

Among U.S. adults, 90 percent believe in God, 8 percent don't, and 2 percent don't know or don't care (Lugo et al., 2013). For sociologists, religiosity is a better measure of being religious than simply asking people whether they believe in God (or a universal spirit) and which religion they follow. Religiosity includes a number of variables, but the most common are religious belief, affiliation, and participation.

13-6a Religious Belief

Some 58 percent of Americans say that religion is very important in their lives, down from 75 percent in 1952 (Newport, 2013). Not surprisingly, religion isn't important for *agnostics* (those who say that it's impossible to know whether there's a God), *atheists* (those who believe that there's no God), or others who are skeptics. Nonetheless, about 25 percent of Americans, including some atheists and agnostics, embrace the tenets of some Eastern religions or elements of New Age spirituality that include reincarnation, meditation, astrology, and the evil eye (casting curses and evil spells) (Lugo et al., 2012).

13-6b Religious Affiliation

Since 2007, the greatest declines have occurred among evangelical and mainstream white Protestants. In contrast, 20 percent of Americans say that they have no religious preference or affiliation (see *Table 13.4*), up from only 8 percent in 1990 (Kosmin and Keysar, 2009). "Nones," a popular term for the religiously unaffiliated, include Asians, young people, males, political independents, those living in the Pacific and New England regions, college graduates, and people in higher-income brackets (Lugo et al., 2012; Saad, 2013; Newport, 2014).

About half of U.S. adults—especially those who were raised as Catholics or Protestants—have changed their religious affiliation since childhood. Some have done so more than once because of intermarriage or finding a religion they liked better. The most important reasons for becoming unaffiliated include not believing the teachings, seeing many religious people as hypocritical or judgmental, losing respect for religious leaders who focus on power and money, and negative teachings or treatment of lesbians and gays (Pew Forum on Religion & Public Life, 2009; Cooperman et al., 2014; Taylor et al., 2014).

13-6c Religious Participation

Turning to the third measure of religiosity, religious participation, among Americans who say that religion is

TABLE 13.4   RELIGIOUS AFFILIATION IN THE UNITED STATES, 2012

	Percentage	Change Since 2007
CHRISTIAN	73%	−5%
Protestant	48	−5
Catholic	22	−1
Mormon	2	—
OTHER FAITH	7	+2
UNAFFILIATED	20	+4.3
Atheist	2.4	+0.8
Agnostic	3.3	+1.2
Nothing in particular	14	+2.3

Source: Based on Lugo et al., 2012.

important to them, 27 percent seldom or never attend religious services. Thus, many people are more likely to believe in a religion than to practice it by attending services regularly. Mormons (81 percent), Protestants (64 percent), Muslims (64 percent), and Catholics (60 percent) have higher attendance rates at religious services than do Jews (34 percent) or non-Christians (Pew Forum on Religion and Public Life, 2008; Newport, 2012).

13-6d Some Characteristics of Religious Participants

Americans differ in their beliefs and affiliations. Religious participation also varies by sex, age, race and ethnicity, and social class.

SEX

Across all age and faith groups in 145 countries, women tend to be more religious than men in believing in God, praying, attending services, and saying that religion is very important in their lives. In the United States, atheists and agnostics are much more likely to be male (64 percent) than female (36 percent) (Taylor et al., 2009; Lugo et al., 2012).

It may be that women are expected to be more pious and spiritual because, especially as nurturers, they transmit religious values to their children (see Chapter 9). Women may turn to religion because they have a harder life than most men. As you saw in previous chapters, women worldwide are more likely than men to experience assaults and poverty. Also, women, on average, live longer than men, and religiosity increases with age (Newport, 2012).

AGE

Generally, Americans age 65 and older are more likely than younger people to describe themselves as religious, to say that religion is very important in their lives, and to attend services at least weekly. These age-related differences may reflect several factors: Older Americans grew up decades ago when church attendance was higher, they seek spiritual comfort as elderly friends and relatives die, they want to lessen a sense of isolation or loneliness, and they're preparing for death (Taylor et al., 2009).

"Millennials" (those born after 1980) are less religious than their elders: They're less likely to pray, to regularly attend worship services, or to identify themselves with a religious group. For example, 32 percent of adults under 30 have no religious affiliation compared with just 9 percent who are 65 and older. Also, young adults are much more likely to be unaffiliated than previous generations were at a similar stage in their lives (Lugo et al., 2012; Martínez and Lipka, 2014).

RACE AND ETHNICITY

Blacks are more religious than any other U.S. racial or ethnic group. They're the most likely to report a religious affiliation, to belong to a church, to pray every day, and to attend religious services every week. The vast majority (78 percent) are Protestant; within this group, 40 percent are Baptist (Sahgal and Smith, 2009; Newport, 2012).

Latinos' religious landscape has been shifting. About 55 percent are Catholic (down from 67 percent in 2010), 22 percent are Protestant and primarily evangelical rather than mainline, and 18 percent are unaffiliated (up from 10 percent in 2010). Some Catholic and evangelical Protestant churches have been especially successful in attracting recent Latino immigrants because the churches offer services in Spanish, the ceremonies are expressive rather than formal, and the clergy are more responsive to their members' social and economic needs (Dias, 2013; Cooperman et al., 2014; "Select-a-Faith," 2014).

Asian Americans may be the most diverse religious group in America. Nationally, 42 percent are Christian, 26 percent are unaffiliated, 24 percent are Hindu or Buddhist, and 7 percent are Muslim, Sikh, or some other religion. There's also considerable variation across subgroups. For example, 65 percent of Filipinos are Catholic, 52 percent of Chinese Americans are unaffiliated, and 61 percent of Korean Americans are Protestant (Funk et al., 2012).

SOCIAL CLASS

People with less education are generally more religious than those with higher educational levels. Because education and income are highly correlated, as education increases, the importance of religion generally decreases (see *Figure 13.5*). Among the "nones," for instance, 38 percent of atheists and agnostics have an annual family income of at least $75,000 compared with 29 percent of those who have a religious affiliation (Lugo et al., 2012).

13-6e Secularization: Is Religion Declining or Thriving?

Industrialized nations have been experiencing **secularization**, a process in which religion loses its social and cultural influence. Some social scientists, including sociologists, maintain that secularization is increasing; others contend that religiosity is flourishing.

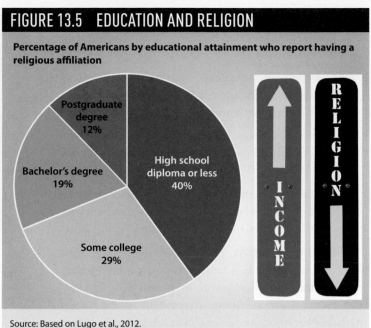

FIGURE 13.5   EDUCATION AND RELIGION

Percentage of Americans by educational attainment who report having a religious affiliation

- Postgraduate degree 12%
- Bachelor's degree 19%
- Some college 29%
- High school diploma or less 40%

INCOME ↑ RELIGION ↓

Source: Based on Lugo et al., 2012.

IS SECULARIZATION INCREASING?

You've seen that the number of "nones" is growing, attendance at religious services has dropped, and that fewer Americans say that religion is "very important" in their lives. There's other evidence that secularization is rising in the United States and globally. For example:

▶ 45 percent of Americans have a "great deal" of confidence in the church and organized religion, down from 68 percent in 1975 (Riffkin, 2014)

▶ 70 percent of Americans attended religious services on Christmas Eve or Day when they were children; only 54 percent do so in adulthood (Cooperman et al., 2013)

▶ 68 percent of people worldwide describe themselves as religious, down from 77 percent in 2005 (WIN-Gallup International, 2013)

There's also evidence of greater "forced secularization" worldwide. About 64 percent of the world's population—up from 58 percent in 2007—lives in countries with high government restrictions on religion. Some of the laws and policies ban particular faiths, prohibit conversions, or limit preaching. Other laws and practices allow religion-related conflict or terrorism, mob violence against a religious minority, and harassment over religious attire (Grim and Cooperman, 2014; "Trouble in Trappes," 2014).

secularization a process in which religion loses its social and cultural influence.

Shelly_Au/iStockphoto.com

IS RELIGION THRIVING?

Many sociologists contend that the extent of secularization has been greatly exaggerated. For example:

▸ 61 percent of Americans say that it's important for members of Congress to have strong religious beliefs ("Growing Number of Americans…," 2010)

▸ 83 percent support Christmas displays on government property (Pew Research Center for the People & the Press, 2010)

▸ 93 to 99 percent of people in the Middle East, Asia/Pacific, Latin America, and Africa say that believing in God is essential to morality (Simmons et al., 2014)

Fundamentalism, the belief in the literal meaning of a sacred text (e.g., Christian Bible, Muslim Qur'an, Jewish Torah), is widespread in practically all countries except Europe. About 75 percent of Americans say that everything or almost everything in the Bible should be taken literally because it's the word of God (Saad, 2014).

The U.S. Constitution established a separation of church and state. Over the years, that separation has become fuzzier. Several southern states have passed laws that allow public schools and government buildings to display traditional Christmas scenes and symbols; many public elementary and high schools allow prayer before class, permit students to leave classes for religious instruction, support religious student clubs that faculty sponsor, and offer after-school evangelical

Pascal Deloche/Terra/Corbis

programs. A growing number of public colleges and universities has approved building privately funded "religious dorms" (Bruinius, 2013; Campo-Flores, 2013; Lawrence, 2013).

One of the most divisive issues has been over which explanations for human origins should be taught in public schools. For scientists, it's "a fact" that human beings (and other creatures) evolved from earlier animals through a process known as "natural selection" over several million years (National Science Board, 2014). In contrast, *creationism*, based on a fundamentalist interpretation of the Bible, argues that God created humans in their present form about 10,000 years ago. About 42 percent of Americans espouse creationism. Because of such beliefs, Republicans in conservative states (e.g., Louisiana, Missouri, Tennessee, Oklahoma) have introduced bills for public schools to teach creationism in science classes ("Faith and Reason," 2014; Newport, 2014).

---

fundamentalism the belief in the literal meaning of a sacred text.

AP Images/Harry Cabluck

BRENDAN SMIALOWSKI/Getty Images

In 2005, the U.S. Supreme Court ruled that a 6-foot-high monument containing the Ten Commandments outside the Texas state capitol was constitutional (left), even though it's on government property. In 2014, the court ruled that prayer in civic life, including legislative bodies, is not only constitutional, but can use explicit Christian or other religious language (right). Do such decisions illustrate greater secularization or religiosity?

"WE ESTABLISH NO RELIGION IN THIS COUNTRY. WE COMMAND NO WORSHIP. WE MANDATE NO BELIEF, NOR WILL WE EVER. CHURCH AND STATE ARE AND MUST REMAIN SEPARATE." RONALD REAGAN, REPUBLICAN, 40TH U.S. PRESIDENT (1911–2004).

PaulMaguire/iStockphoto.com

Another indicator that religion is booming in the United States and elsewhere is the prevalence of **civil religion** (sometimes called *secular religion*), integrating religious beliefs into secular life. The Pledge of Allegiance was written in 1892; the phrase "under God" was added in 1954. Other examples of civil religion include the phrase "In God We Trust" on U.S. currency, prayer in public schools, a legally mandated National Day of Prayer held on the first Thursday of May, and many local jurisdictions' closing public schools, public libraries, and government offices on Good Friday, a Christian holy day.

13-7 SOCIOLOGICAL PERSPECTIVES ON RELIGION

You've seen that religion plays an important role in many people's lives. Why? How does society affect religion? And how does religion affect society? *Table 13.5* summarizes the major sociological perspectives on religion.

13-7a Functionalism: Religion Benefits Society

Durkheim (1961: 13, 43) described religion as a "permanent aspect of humanity" whose "essential task is to maintain, in a positive manner, the normal course of life." Religion can also be dysfunctional, but first let's consider its benefits.

RELIGION IS A SOCIAL GLUE

Religion fulfills a variety of important functions at individual, community, and societal levels. All of the following, according to functionalists, contribute to a society's survival, stability, and solidarity:

▶ *Belonging and identity.* Communal worship and rituals increase social contacts, develop a sense of acceptance and identity, and reinforce people's feeling of belonging to a group. During Christmas, 86 percent of Americans, including "nones," attend holiday activities with family or friends (Cooperman et al., 2013).

▶ *Purpose and emotional comfort.* Religion provides meaning in life and offers hope for the future. Religiosity is highest among the world's poorest countries because religion helps people cope with daily struggles to survive (Crabtree, 2010; WIN-Gallup International, 2013).

▶ *Well-being.* Very religious Americans report being healthier and experiencing less depression and worry than those who aren't religious (Newport et al., 2012; Newport and Himelfarb, 2013).

▶ *Social service.* Religious groups raise millions of dollars for the victims of natural disasters, distribute food and clothing, and find shelter for displaced families. About 88 percent of Americans, including 78 percent of "nones," agree that churches and other religious organizations are important because they help the poor and sick (Lugo et al., 2012).

▶ *Social control.* Because many religious people fear punishment in the afterlife, they try to follow societal rules. Doing so suppresses deviant and antisocial behavior and promotes societal cohesion (Durkheim, 1961; Shariff and Aknin, 2014).

RELIGION PROMOTES SOCIAL CHANGE

Religion can also spearhead social change. Mohandas Gandhi (1869–1948), a spiritual leader in India, worked for his country's independence from Great Britain through nonviolence and peaceful negotiations. In the

> **civil religion** (sometimes called *secular religion*) integrating religious beliefs into secular life.

TABLE 13.5 SOCIOLOGICAL EXPLANATIONS OF RELIGION

Theoretical Perspective	Level of Analysis	Key Points
Functionalist	Macro	Religion benefits society by providing a sense of belonging, identity, meaning, emotional comfort, and social control over deviant behavior.
Conflict	Macro	Religion promotes and legitimates social inequality, condones strife and violence between groups, and justifies oppression of poor people.
Feminist	Macro and micro	Religion subordinates women, excludes them from decision-making positions, and legitimizes patriarchal control of society.
Symbolic Interactionist	Micro	Religion provides meaning and sustenance in everyday life through symbols, rituals, and beliefs, and binds people together in a physical and spiritual community.

United States, religious leaders, especially the Reverend Martin Luther King, Jr., were at the forefront of the civil rights movement during the late 1960s.

In 2011, after decades of debate, the Presbyterian Church, one of the nation's largest Protestant denominations, approved ordaining people with same-sex partners as ministers, elders, and deacons. Numerous Lutheran churches have also welcomed gay pastors. In 2014, and despite many of its congregations' threats of leaving, the Presbyterian Church voted to allow same-sex marriages. Other religious groups (e.g., United Church of Christ, Episcopal Church, Quakers, Reform and Conservative Judaism) have taken similar steps (Goodstein, 2014).

Max Weber (1920) asserted that religion sparks economic development. His study of Calvinism, a Christian sect that arose in Europe during the sixteenth century, led to his coining the term **Protestant ethic**, which he described as a belief that hard work, diligence, self-denial, frugality, and economic success would lead to salvation in the afterlife. According to Weber, the harder the early Calvinists worked and saved, the more likely they were to accumulate money, to become successful, and to drive the growth of capitalism.

Was Weber right about the relationship between the Protestant ethic and the rise of capitalism? The data are mixed. Some studies show that people are diligent not because of religious beliefs but simply because amassing savings provides resources when disaster strikes (as when droughts wipe out crops). On the other hand, a study of 59 industrialized and developing countries found that religious beliefs that encourage hard work and thrift spur economic growth (Cohen, 2002; Barro and McCleary, 2003).

IS RELIGION DYSFUNCTIONAL?

Functionalists emphasize its benefits, but also recognize that religion can be dysfunctional when it harms individuals, communities, and societies. Religious intolerance can spark conflict between groups (including vandalizing churches, mosques, and synagogues) and attacks on religious minorities. The United States has a long history, beginning with the Pilgrims in 1620, of people discriminating against other religious groups even though they themselves sought religious freedom (Davis, 2010).

Recently, a scathing United Nations report accused the Roman Catholic Church and the Vatican of "systematically" protecting predator priests for at least five decades, allowing tens of thousands of children to be abused, humiliating victims' families into

Megachurches are Christian congregations that have a regular weekly attendance of more than 2,000 and tend to be evangelical. They represent only 0.5 percent of all U.S. churches, but their number has almost tripled (to more than 1,600) since 2000 (Bird and Thumma, 2011). From a functionalist perspective, why do you think megachurches are so popular?

Protestant ethic a belief that hard work, diligence, self-denial, frugality, and economic success will lead to salvation in the afterlife.

silence, and still not removing perpetrators from their posts (United Nations Committee on the Rights of the Child, 2014).

CRITICAL EVALUATION

One weakness is that functionalists, by emphasizing religion's benefits, imply that religion is indispensable to leading a good life. People who aren't religious do many good things for society, and some religious people commit heinous acts. Primarily because of priests' sexual abuse of young children, Americans' rating of the clergy's honesty and ethics fell from 67 percent in the mid-1980s to 47 percent in 2013 (Swift, 2013).

A second limitation is functionalists' exaggerating religion's positive effect on physical and emotional health. Jews, who are one of the least religious U.S. groups, report the highest rates of well-being, and "nones," including atheists, report higher rates of well-being than do religious groups (Newport et al., 2012; Leurent et al., 2014).

A third, and major, criticism is that functionalists gloss over numerous dysfunctional aspects of religion that generate wars, terrorism, and genocide. Conflict theory addresses these issues.

13-7b Conflict Theory: Religion Promotes Social Inequality

From a conflict perspective, religion promotes and reinforces social inequality. Throughout history and currently, religion has created discord and divisiveness within groups, between groups, and in the larger society.

"THE OPIUM OF THE PEOPLE"

Much conflict theory reflects the work of Karl Marx (1845/1972), who described religion as "the sigh of the oppressed creature" and "the opium of the people" because it encouraged passivity and acceptance of social and economic inequality. Marx viewed religion as a form of **false consciousness**, an acceptance of a system of beliefs that prevents people from protesting oppression. Contemporary conflict theorists don't view religion as an opiate, but they agree with Marx that religion often justifies violence and promotes social inequality.

RELIGION JUSTIFIES INTOLERANCE AND VIOLENCE

For thousands of years, many governments and religious leaders have condoned or perpetrated widespread violence in the name of religion. Religion promotes conflict when religious groups differentiate between "we" and "they" ("We're right and they're wrong," "Our God is the *real* God'"). Such self-righteousness condones aggression, oppression, and brutality within and across societies. For example:

▶ The "eternal war" between Iraq's Sunnis and Shiites, two major Islamic sects, dates back more than 1,300 years, and there are long-standing conflicts and wars between Muslims and Jews in Israel and Palestine (Crowley, 2014).

▶ Because of widespread violence against Christians, they now make up only 5 percent of the Middle East's population, down from 20 percent in 1914 (Baker, 2014).

▶ In nearly 75 percent of the world's countries, religious groups experience harassment and intimidation, including physical or verbal assaults, desecration of holy sites, and discrimination in employment, education, and housing (Theodorou and Henne, 2014).

▶ In several of Asia's Buddhist-majority nations, monks have incited bigotry and violence—mostly against Muslims (Beech, 2013).

RELIGION PROMOTES SOCIAL INEQUALITY

Conflict theorists see religion as a tool dominant groups use to control society, protect their own interests, and derail social change. U.S. Roman Catholic bishops have a long history of helping the poor and fostering social justice. Most recently, however, they have opposed comprehensive health care because it would fund abortions, have publicly chastised pro-choice Catholic politicians, and have campaigned against same-sex marriage—all of which fuel social inequality and conflict (Doyle, 2011).

Conflict theorists also point out that private evangelical schools, colleges, and universities receive massive state and federal financial aid that diverts much-needed tax revenues from public schools and colleges. Many of the schools' textbooks describe homosexuals and abortion rights supporters as "evil," and are hostile toward other religions, including nonevangelical Protestants, Jews, and Catholics (Berkowitz, 2011; Tabachnick, 2011).

CRITICAL EVALUATION

Functionalists may overemphasize consensus and harmony, but conflict theorists, according to some critics,

---

false consciousness an acceptance of a system of beliefs that prevents people from protesting oppression.

"Maybe You should tell them that You and Allah are the same person."

often ignore religion's role in challenging homophobia and uniting people. Nationally, for instance, the share of American religious congregations allowing an openly gay or lesbian couple to become full-fledged members grew from 37 percent in 2006 to 48 percent in 2012 (Chaves and Anderson, 2014).

A second criticism is that conflict theorists attach too much credence to Marx's concept of false consciousness. Instead of passively submitting to religious oppression, many people revolt, complain, or resist. The EEOC, for instance, has filed lawsuits against a trucking company that requires its Muslim drivers to deliver alcohol, which Islam prohibits. Others have sued employers, sometimes successfully, who denied requests to not work on holy days, and whose secular dress codes don't allow head scarves, turbans, or long, shaggy beards (Trottman, 2013). Whether religious discrimination is real or imagined, many Americans protest, demonstrate, and seek legal remedies.

13-7c Feminist Theories: Religion Subordinates and Excludes Women

Feminist theorists agree with conflict theorists that religion can foster violence and inequality. They go further, however, by criticizing organized religions as sexist, patriarchal, and shutting women out of leadership positions.

SEXISM, PATRIARCHY, AND SUBORDINATION

"Religion is probably the most important force in continuing the oppression of women worldwide," a feminist blogger writes, and "religiosity and sexism go hand in hand" (Marcotte, 2014). From a feminist perspective, most religions are patriarchal: They emphasize men's experiences and a male point of view and see women as subordinate to men. According to the apostle Paul, "Wives should submit to their husbands in everything" (*Ephesians* 5: 24). The idea that Eve was created out of Adam's rib is often used to justify men's domination of women. Many religious teachings and institutions continue to propagate such beliefs (Ferguson, 2014).

In Orthodox Judaism, a man's daily prayers include this line: "Blessed art thou, O Lord, our God, King of the Universe, that I was not born a woman." The Qur'an tells Muslims that men are in charge of women and that women should obey men. Almost all contemporary religions worship a male deity, and none of the major world religions treat women and men equally (Gross, 1996; Jeffreys, 2011).

Feminist scholars offer alternatives to patriarchal interpretations of Scripture. Some contend that Jesus was a feminist. For example, he defended women against men, encouraged women's intellectual pursuits when it wasn't the norm, and there are numerous passages in the Bible about women spreading Jesus' teachings (Gross, 1996; Ferguson, 2014).

Muslim feminists, similarly, note that men have interpreted sacred Islamic texts to ensure male dominance and control. Women were among some of Muhammad's earliest converts, and the Qur'an has numerous passages that establish women's equal rights in inheritance and family roles (Menissi, 1991, 1996). Thus, women's subordination in many Islamic societies isn't due to religious tenets but to men's interpretations of Scripture to maintain their power and privilege (Smith, 1994).

In the Middle East and North Africa, 35 percent of the countries have religious police who, among other things, enforce strict segregation of the sexes and punish women who are perceived as behaving or dressing improperly. Worldwide, harassing women over "indecent" attire increased from 7 percent of countries in 2007 to 25 percent in 2011 (Stencel and Grim, 2013; Theodorou, 2014). Thus, it's women's fault if men have lustful thoughts or assault women.

EXCLUSION OF WOMEN FROM LEADERSHIP POSITIONS

In 2011, women earned 32 percent of theology degrees, a dramatic increase from only 2 percent in 1970. Still, women make up only 21 percent of the nation's clergy, lead only 8 percent of congregations, and earn 20 percent less than their male counterparts (Chaves et al., 2009; Snyder and Dillow, 2013; BLS Reports, 2014).

Roman Catholicism, Orthodox Judaism, and the Mormon Church don't allow women to be ordained because "women should serve and not lead." U.S. Catholic women contribute $6 billion a year during Sunday Masses, but "the presence of women anywhere within the institutional power structure is virtually nil" (Miller, 2010: 39; Gibson, 2012). Mormon women can be missionaries and provide "welfare and compassionate service," but are excluded from the priesthood and the all-male

In 2014, the Mormon Church excommunicated Kate Kelly (left), leader of the Ordain Women group. According to some U.S. Catholic bishops, the ideas of Sister Simone Campbell (right), and other nuns who advocate ordaining women, border on heresy. Both women are human rights lawyers.

central leadership that governs the church (Kantor and Goodstein, 2014).

Some Protestant denominations justify women's exclusion from leadership positions based on biblical passages such as "I permit no woman to teach or have authority over men; she is to keep silent" (I *Timothy* 2: 11–12). Many Protestant groups—including Southern Baptists and evangelical born-again Christians—interpret this and similar passages to mean that women should never, under any circumstances, instruct men, within or outside of religious institutions. Even in liberal Protestant congregations, female clergy tend to be relegated to specialized ministries with responsibilities for music, youth, or Bible studies (Banerjee, 2006; Richie, 2013).

What happens when religious women openly protest their exclusion? In 2013, Pope Benedict XVI accused the Leadership Conference of Women Religious, which represents 80 percent of U.S. nuns, of promoting "radical feminism" for challenging the Catholic Church's policies of not ordaining women, and set up re-education programs "to bring the nuns back into line." His successor, Pope Francis, recently praised American nuns for their service and sacrifices, but said nothing about ordaining women (Carballo, 2014).

CRITICAL EVALUATION

Feminist perspectives on religion have three weaknesses. First, some feminist Muslim scholars have criticized Western feminists for misreading Islamic and other sacred scriptures, and reducing practically all discussions of gender to the *hijab* (a veil or scarf that Muslim women wear) instead of focusing on justice for both women and men in marriage, employment, and other areas (Fakhraie, 2009; Daneshpour, 2013).

Second, religious women aren't as oppressed as many feminist scholars claim. For instance, a Muslim recently won a lawsuit against Abercrombie & Fitch after she was fired for wearing a head scarf instead of conforming to the company's dress code (Ballhaus, 2013). Even in the most conservative religions, women don't blindly submit to religious dogma. Instead, they challenge existing doctrines, stop volunteering at church, don't attend weekly services, and establish their own parishes (Padgett, 2010; Miller, 2012).

Third, some feminist scholars overlook the progress and variation of women's leadership in organized religion. In the U.S. Presbyterian Church, for instance, 36 percent of the active ministers are female, up from only 3 percent in 1977 (Hartford Institute for Religion Research, 2012; Presbyterian Mission Agency, 2013). Between 2002 and 2012, the number of female clergy in the Church of England rose from 7 to 21 percent and, in 2014, its lawmaking body cast a final vote to ordain women bishops (Bennhold, 2014).

13-7d Symbolic Interaction: Religion Is Socially Constructed

For symbolic interactionists, religion isn't innate but socially constructed. As a result, people can learn and interpret the same religion differently across cultures and over time. Symbols, rituals, and beliefs are three of the most common means of learning and internalizing religion.

SYMBOLS

A *symbol* is anything that stands for or represents something else to which people attach meaning (see Chapter 3). Many religious symbols are objects (a cross, a steeple, a Bible), but also include behaviors (kneeling or bowing one's head), words ("Holy Father," "Allah," "the Prophet"), and physical appearance (e.g., wearing head scarves, skull caps, turbans, clerical collars).

Religious symbols, like all symbols, are shorthand communication tools. Some interactionists define a religion as "a system of symbols" because it's a community that's unified by its symbols (Berger and Luckmann, 1966; Geertz, 1966). Religious symbols can also be divisive (e.g., a Ten Commandments monument on government property that alienates non-Christians and atheists).

RITUALS

A *ritual* (sometimes called a *rite*) is a formal and repeated behavior that unites people (see Chapter 3). Religious rituals, like secular ones, strengthen a participant's self-identity (Reiss, 2004). Religious rites of passage—*bat mitzvah* for girls and *bar mitzvah* for boys in the Jewish community, and first communion and confirmation for Catholic children—reinforce the individual's sense of belonging to a particular religious group. A group's rituals symbolize its spiritual beliefs and include a wide range of practices—praying, chanting, fasting, singing, dancing, and offering sacrifices.

All religions have rituals like birth and marriage that mark significant life events (see Chapters 3, 5, and 12). Death rituals are probably the most elaborate and sacred worldwide. They vary across religious groups and societies, but all of them comfort the living and show respect for the dead.

BELIEFS

Rituals and symbols come from *beliefs*, convictions about what people think is true. Religious beliefs can be passive (believing in God but never attending services) or active (participating in rituals and ceremonies). Beliefs bind people together into a spiritual community.

One of the strongest beliefs worldwide is that prayer is important. Islam requires prayer five times a day. In the United States, 55 percent of U.S. adults say that they pray every day, and for a variety of reasons, including feeling close to God, as well as requesting better health, more money, and cures for sick pets. Even 21 percent of the "nones" pray at least once a day (Wicker, 2009; Cooperman et al., 2014). Prayer offers psychological and spiritual benefits such as comfort and a sense of unity among those who pray together, but depending on God can also diminish people's motivation to actively shape their own lives (Schieman, 2010).

CRITICAL EVALUATION

A common criticism is that interactionists' focus on micro-level behavior ignores the ways that religion promotes social inequality at the macro level. Conflict theorists and feminist scholars, especially, maintain that people often use religion to justify violence and women's subordination.

Some critics also wonder if interactionists paint too rosy a picture of religion even on a micro level because people's beliefs can wreak considerable havoc. The Islamic zealots who targeted the World Trade Center on 9/11 didn't see themselves as "crazed terrorists" but as "true believers": They were carrying out "God's will" by imposing their religious beliefs on others or destroying their "religious enemies" (Juergensmeyer, 2003).

STUDY TOOLS 13

READY TO STUDY? IN THE BOOK YOU CAN:

☐ Check your understanding of what you've read with the Test Your Learning Questions provided on the chapter review card at the back of the book.

☐ Rip out the chapter review card for a handy summary of the chapter and key terms.

ONLINE AT CENGAGEBRAIN.COM YOU CAN:

☐ Prepare for tests with quizzes.

☐ Review the key terms with Flash Cards.

☐ Play games to master concepts.

© Anelina/Shutterstock.com

WHY CHOOSE?

Every 4LTR Press solution comes complete with a visually engaging textbook in addition to an interactive eBook. Go to CourseMate for **SOC4** to begin using the eBook. Access at **www.cengagebrain.com**

Complete the Speak Up survey in CourseMate at **www.cengagebrain.com**

 Follow us at **www.facebook.com/4ltrpress**

14 | Health and Medicine

LEARNING OBJECTIVES

 14-1 Describe social epidemiology and explain why there are global health gaps.

14-2 Explain how and why environmental, demographic, and lifestyle factors affect U.S. health and illness. Illustrate your explanation with specific examples.

14-3 Compare U.S. with other countries' health care systems in terms of access, costs, and outcomes.

14-4 Compare and evaluate the theoretical explanations of health and medicine.

After you finish
this chapter go to
PAGE 288 for
STUDY TOOLS.

Most of us would probably agree with American poet Ralph Waldo Emerson, who wrote, in 1860, "Health is the first wealth." **Health** is the state of physical, mental, and social well-being. We can usually determine physical health by using objective measures like weight and blood pressure. Mental and social well-being are more difficult to gauge because they rely on subjective definitions that change over time.

14-1 GLOBAL HEALTH AND ILLNESS

How does health differ around the world? And why? Social epidemiology addresses both questions.

14-1a Social Epidemiology

Health varies among and across societies, but all people experience *disease*, a disorder that impairs a person's normal physical and/or mental condition. **Social epidemiology** examines how societal factors affect

health the state of physical, mental, and social well-being.

social epidemiology examines how societal factors affect the distribution of disease within a population.

What do you think?

The government should be responsible for making sure that all Americans have health care.

1 2 3 4 5 6 7
strongly agree   strongly disagree

the distribution of disease within a population. Why, for example, do people live much longer in some countries than in others?

In answering this and other questions, epidemiologists look at two factors. One is *incidence*, the number

of new cases of a health problem that occur in a given population during a given time period (e.g., in 2012, there were 1.7 million new diabetes cases among Americans aged 20 or older). The other measure is *prevalence*, the total number of cases (extent) of an illness or health problem within a population or at a particular point in time (e.g., in 2012, 12 percent of the U.S. population aged 20 or older had diabetes) (*National Diabetes Statistics...*, 2014). Epidemiological studies show large health differences across countries.

14-1b Global Health Disparities

The World Bank classifies countries into four broad economic groups based on gross national income (GNI) per capita. In 2013, the GNI per capita was the following: $1,045 or less for *low-income countries*, between $1,046 and $4,125 for *lower-middle-income countries*, between $4,126 and $12,745 for *upper-middle income countries*, and $12,746 or more for *high-income countries* (World Bank, 2014).

Figure 14.1 shows the location of these countries and compares them on life expectancy, infant mortality, and per capita health expenditures. These three variables are among the most frequently used to measure health status because they indicate a population's living standards, people's average socioeconomic status, and the quality and financial resources that a country devotes

to **health care**, the prevention, treatment, and management of illness.

The 75 high-income countries make up only 13 percent of the world's population, but have the greatest access to clean water, sanitation, food, and health services. In contrast, chronic malnutrition is the most common killer of children under 5 years old in lower-income countries, and these death rates increased from 37 percent in 1990 to 44 percent in 2012. More than a third of the world's population lacks basic sanitation, almost half are at risk of dying of malaria, and 44 percent of people in low- and lower-middle-income countries can't afford medicine to treat infections and diseases (World Health Organization, 2014).

High income populations live longer, have low infant mortality rates, and spend more on health (see *Figure 14.1*), but experience "diseases of wealth" (e.g., diabetes, heart disease, various cancers). Compared with 16 other high-income countries, the United States has a large and growing "health disadvantage" and ranks last or near-last in nine key health areas including drug-related deaths, chronic lung and heart disease, obesity, and the prevalence of HIV and AIDS (Woolf and Aron, 2013).

health care the prevention, management, and treatment of illness.

FIGURE 14.1 WORLDWIDE HEALTH GAPS

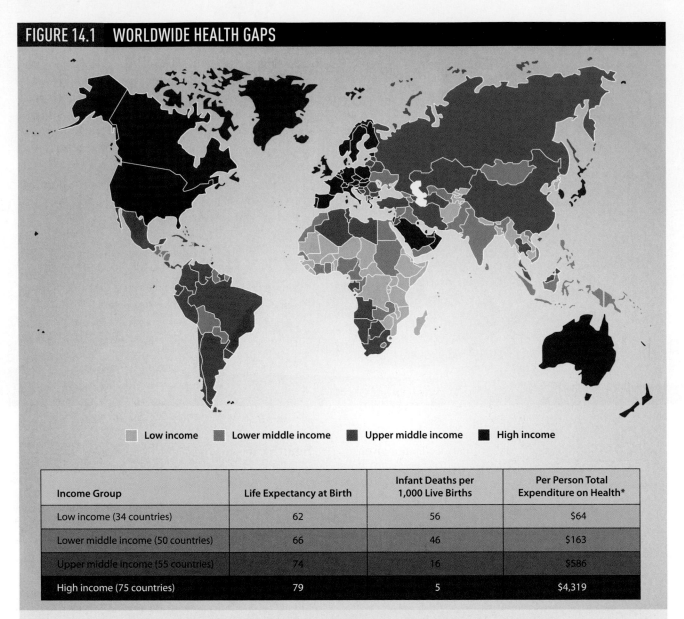

Low income ■ Lower middle income ■ Upper middle income ■ High income

Income Group	Life Expectancy at Birth	Infant Deaths per 1,000 Live Births	Per Person Total Expenditure on Health*
Low income (34 countries)	62	56	$64
Lower middle income (50 countries)	66	46	$163
Upper middle income (55 countries)	74	16	$586
High income (75 countries)	79	5	$4,319

*These figures include private, employer, and government expenditures in U.S. dollars.

Sources: Based on Kaiser Family Foundation, 2014; World Bank, 2014; World Health Organization, 2014, Tables 1 and 7.

14-2 HEALTH AND ILLNESS IN THE UNITED STATES

Why do Americans, compared with their peers in other high-income countries, have a health disadvantage that includes high disability rates? Compared with other industrialized nations, non-medical social determinants contribute to large health disparities in the United States.

disability any physical or mental impairment that limits a person's ability to perform a basic life activity.

14-2a Understanding Disability

A **disability** is any physical or mental impairment, temporary or permanent, that limits a person's ability to perform a basic life activity. These activities include walking; doing errands alone; bathing, dressing, or feeding oneself; and working.

Almost 30 percent of Americans aged 18 and over have a disability. Age is a major predictor of disability, affecting 59 percent of adults age 65 and older compared with 26 percent of Americans ages 18 to 64. Disability also increases with age—from 26 percent for people aged 65 to 74 to 73 percent for those age 85 and older. Besides age, disability rates are higher for women (33 percent)

Christophe Calais/Corbis News/Corbis

than men (25 percent), for people living in poverty (38 percent versus 22 percent who aren't poor), and are highest among American Indians (34 percent) (He and Larsen, 2014; National Center for Health Statistics, 2014; West et al., 2014).

The number of Americans experiencing one or more disabilities rose from 61 million in 1997 to almost 74 million in 2012 (National Center for Health Statistics, 2014). Why the increase? First, as you'll see shortly, behavior previously considered normal has been medicalized. Second, medical advances help many people survive diseases, car accidents, and wars, but they may have lifelong health ailments. Third, as the proportion of the population age 65 and older increases, so does the proportion living with disabilities.

14-2b Social Determinants of Health and Illness

No single variable explains well-being. Genes affect all of us, but environmental, demographic, and lifestyle factors—all of which are interrelated—have a significant impact on health and illness.

ENVIRONMENTAL FACTORS

Many environmental hazards affect our bodies. For example:

▸ There's a strong association between early-life exposure to air pollution and autism, schizophrenia, and learning disabilities (Currie et al., 2013; Roberts et al., 2013; Allen et al., 2014).

▸ Many toxic chemicals in food and everyday products (e.g., dishwashing detergents, shampoos, shaving products, and makeup) have been linked to asthma, some types of cancer, and children's behavioral disorders (Sathyanarayana et al., 2013; Blake, 2014; Grandjean and Landrigan, 2014; Rosenberg, 2014).

▸ Norovirus—a very contagious virus—is the leading cause of disease outbreaks from contaminated food.

About 21 million Americans get sick from norovirus each year, 71,000 are hospitalized, and about 800 die. Outbreaks on cruise ships account for only 1 percent of all norovirus; 70 percent of cases are due to food workers who are infected, don't wash their hands, or go to work when sick (Hall et al., 2014).

Access to health care can prolong life, but an estimated 440,000 Americans die each year because of preventable medical errors (e.g., doctors operating on the wrong patient or body part, patient infections). Medication errors—misdiagnosing an illness, prescribing the wrong drugs, or giving patients drugs that interact dangerously—injure about 1.3 million Americans and kill about 7,000 every year. Moreover, thousands of patients a year leave operating rooms with surgical sponges in their bodies that cause infections, lifetime digestive problems, removal of intestines, and even death (Brown and McGann, 2011; Eisler, 2013; James, 2014).

A growing problem is antibiotic resistance, which kills an estimated 23,000 Americans each year. Bacteria have become more resistant because of the overuse of antibiotics: People take them for illnesses where antibiotics aren't effective, and farmers feed them to animals to promote growth. No new antibiotics have been discovered since 1987, largely because pharmaceutical companies reap much higher profits by developing very expensive and specialized medicines for small populations with chronic conditions, including multiple sclerosis, cancer, HIV, and hepatitis C ("The Drugs Don't Work," 2014: 54; Tozzi, 2014b).

DEMOGRAPHIC FACTORS

Religion, family size, marital status, urban-rural residence, and other variables affect health. Four of the most important demographic factors, however, are age, sex, social class, and race and ethnicity.

Age. Not surprisingly, age is the single best predictor of illness and death. Death rates drop sharply shortly after birth, begin to rise at about age 45, and escalate, particularly after age 78.

From birth through elementary school, children have health problems over which they have no control. Nearly 1 in 100 U.S. babies are born with *fetal alcohol spectrum disorders*, a range of permanent birth defects caused by a mother drinking alcohol during pregnancy. The disorders include intellectual infirmities, speech and language delays, poor social skills, mental retardation, and physical abnormalities like congenital heart defects (National Organization on Fetal Alcohol Syndrome, 2012; Weinhold, 2012).

Many adolescents have health problems because of their own choices. In 2013, 21 percent of twelfth graders took drugs without a doctor's prescription, 28 percent smoked, 47 percent drank alcohol, and 60 percent texted or emailed while driving (Kann et al., 2014). Such risky behavior may be due to peer pressure, the fact that some parts of the teen brain are still developing, and adolescents' becoming more independent as they break away from parental supervision (see Chapter 4).

Health problems emerge and increase during one's late thirties because of genes, lifestyle choices (which we'll address shortly), and because physical decline is normal and inevitable as we age. No matter how well tuned we keep our bodies, the parts start wearing down: Reflexes slow, hearing and eyesight dim, and stamina and muscle strength decrease.

After age 65, chronic rather than acute diseases comprise the majority of health problems, including disability. **Chronic diseases** (e.g., asthma and high blood pressure) are long-term or lifelong illnesses that develop gradually or are present from birth. In contrast, **acute diseases** (e.g., chicken pox) are illnesses that strike suddenly and often disappear rapidly but can cause incapacitation and sometimes death. Chronic diseases increase as people age. *Dementia*, a loss of mental abilities, most commonly occurs during one's 70s. The most debilitating form of dementia is **Alzheimer's disease**, a progressive, degenerative disorder that attacks the brain and impairs memory, thinking, and behavior.

Sex. The gender gap in life expectancy has decreased since 1975, but women, on average, still live longer than men (see Chapter 12). Men of all racial-ethnic groups are two to three times more likely than women to die in motor vehicle crashes, to be victims of homicide, to smoke and drink alcohol, to abuse drugs, and to work in dangerous occupations (e.g., construction, law enforcement) (National Center for Health Statistics, 2014).

Depression is the single most common mental health problem among adolescents, usually begins at about age 14, and affects 29 percent of U.S. high school students. Girls (36 percent) are more likely than boys (22 percent) to report being depressed, but the reasons are similar: Substance abuse, bullying (either as a victim or perpetrator), sexual or physical abuse, parental divorce, and a family history of mental disorders. In adulthood, and across all ages, women are more likely than men to be taking antidepressant medications (Murphey et al., 2013; National Center for Health Statistics, 2014).

Depression may lead to suicide. Among high school students, females are twice as likely as males to attempt suicide. Nationally, however, men are about four times more likely than women to die by suicide, and the highest rates are among white men aged 75 and older (Kann et al., 2014; National Center for Health Statistics, 2014).

Social Class. Social class has a strong effect on health. As early as age 45, the lower the family income, the greater the likelihood that adults will experience two or more chronic diseases. Living in poor neighborhoods increases the likelihood of experiencing stress (which affects diet, smoking, and alcohol/drug usage), having a sedentary lifestyle because of limited recreational facilities, and less access to health care (Beckles and Truman, 2013; Hummer and Hernandez, 2013).

© chiakto/Shutterstock.com

Dangerous occupations, including mining, can decrease life expectancy because of accidents, disease, and a higher risk of death.

---

chronic diseases long-term or lifelong illnesses that develop gradually or are present from birth.

acute diseases illnesses that strike suddenly and often disappear rapidly but can cause incapacitation and sometimes death.

Alzheimer's disease a progressive, degenerative disorder that attacks the brain and impairs memory, thinking, and behavior.

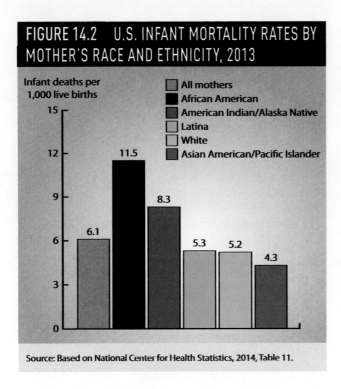

FIGURE 14.2 U.S. INFANT MORTALITY RATES BY MOTHER'S RACE AND ETHNICITY, 2013

Infant deaths per 1,000 live births

- All mothers
- African American
- American Indian/Alaska Native
- Latina
- White
- Asian American/Pacific Islander

All mothers: 6.1
African American: 11.5
American Indian/Alaska Native: 8.3
Latina: 5.3
White: 5.2
Asian American/Pacific Islander: 4.3

Source: Based on National Center for Health Statistics, 2014, Table 11.

People in lower social classes are also more likely to have dangerous jobs. In 2012, more than 4,600 U.S. workers died due to occupation-related injuries, and almost 3.8 million experienced workplace injuries or illnesses caused, among other things, by falls, overexertion, and accidents using equipment ("Workplace Injury and Illness Summary," 2013; U.S. Bureau of Labor Statistics, 2014).

Race and Ethnicity. A good measure of a population's health is its *infant mortality rate*, the number of babies under age 1 who die per 1,000 live births in a given year. Infants born to black women are more than twice as likely to die before age 1 as those born to white, Asian, and Hispanic women (see *Figure 14.2*). No one knows why for sure, but the higher black infant mortality rate may be due to a combination of factors: More babies born before 24 weeks who don't survive, less access to quality health care, and black mothers' greater likelihood of having some conditions (high blood pressure, diabetes, obesity) that are associated with negative birth outcomes (Williams, 2011; Chen et al., 2014; MacDorman et al., 2014).

You'll recall that Latinos, despite their generally low education levels and household incomes, have the highest life expectancy rates (see *Table 12.3* on page 241 and the related text). Some researchers speculate that this "Hispanic health paradox" may be due to several reasons: (1) a greater likelihood of being undercounted than other groups because of their unauthorized status, therefore leading to a false appearance of greater longevity; (2) those who migrate to the United States

tend to be healthier than those who don't migrate; and (3) compared with other racial/ethnic groups, Latinos are the least likely to engage in unhealthy behavior and to die from illicit drugs and prescription drug abuse (Frieden, 2011; Riosmena et al., 2013).

The longer Latinos live in the United States, however, the worse their rates of heart disease, high blood pressure, some types of cancer, and diabetes. Acculturation may result in improved access to health care and preventive services, but also to adopting unhealthy behaviors (e.g., smoking, excessive alcohol consumption), less physical activity, and unhealthier diets (Siegel et al., 2012; Scommegna, 2013c; Tavernise, 2013).

LIFESTYLE FACTORS

Our lifestyle choices can improve or impair our health. The top three preventable lifestyle health hazards, in order of priority, are smoking, obesity, and substance abuse. Sexually transmitted diseases also cause infections and illness.

Smoking. Worldwide and in the United States, tobacco use, primarily cigarette smoking, is the leading cause of preventable disease, disability, and death. Smoking declined from 42 percent in 1965 to 18 percent in 2013 among adults, and from 36 percent in 1997 to 16 percent in 2013 among high school students (Agaku et al., 2014; Kann et al,. 2014). Smoking continues to be higher among males than female across all age groups, and hasn't changed much for the oldest population (see *Figure 14.3*).

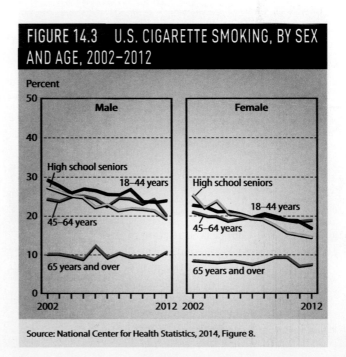

FIGURE 14.3 U.S. CIGARETTE SMOKING, BY SEX AND AGE, 2002–2012

Percent

Male
- High school seniors
- 18–44 years
- 45–64 years
- 65 years and over

Female
- High school seniors
- 18–44 years
- 45–64 years
- 65 years and over

Source: National Center for Health Statistics, 2014, Figure 8.

There are now more former than current smokers, but tobacco use is responsible for about 1 in 5 deaths annually, and, on average, smokers die about 10 years earlier than nonsmokers. Because tobacco harms nearly every human organ, smoking is linked to cancer, heart disease, stroke, Type 2 diabetes, macular degeneration, erectile dysfunction, lung diseases (including emphysema and bronchitis), and birth defects (U.S. Department of Health and Human Services, 2014).

Prevention is difficult because the tobacco industry spends almost $10 billion a year to market its products, half of all movies for children under 13 contain scenes of tobacco use, half of the states continue to allow smoking in public places, and images and messages normalize tobacco use in magazines, on the Internet, and at retail stores frequented by youth. During 2014, states collected more than $25 billion from tobacco taxes and legal settlements, but spent only 2 percent of the money on tobacco control programs (Agaku et al., 2014).

Public health experts are debating whether electronic cigarettes, or e-cigarettes, are safer than tobacco cigarettes. E-cigarettes deliver nicotine in a liquid smoke-free vapor (often called "vaping"), and don't contain the same carcinogens and other toxic chemicals as regular cigarettes. One of the earliest studies found that e-cigarettes were "modestly effective" in helping smokers quit tobacco products, but the results were similar to wearing a nicotine patch (Bullen et al., 2013). Since then, researchers have found that adolescents who use e-cigarettes are more likely to smoke tobacco cigarettes as well, and to become regular smokers (Dutra and Glantz, 2014; Grana et al., 2014).

In 2012, 7 percent of U.S. students in grades 6 to 12 had tried e-cigarettes, double the number from the previous year (U.S. Department of Health and Human Services, 2014). To attract adolescents, the e-cigarette industry uses celebrities, cartoon characters, and "kid-friendly" flavors like chocolate, gummy bears, and mint candy (Zimmerman, 2014).

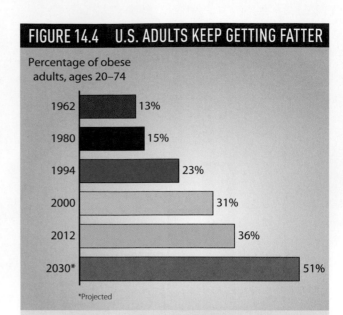

FIGURE 14.4 U.S. ADULTS KEEP GETTING FATTER

Percentage of obese adults, ages 20–74

Year	Percentage
1962	13%
1980	15%
1994	23%
2000	31%
2012	36%
2030*	51%

*Projected

Sources: Based on Finkelstein et al., 2012, and National Center for Health Statistics, 2014, Table 69.

Obesity. Obesity is the second leading and preventable cause of disease, disability, and death. As you saw in Chapter 6, childhood and adult obesity has increased since the early 1970s. Preschoolers who are overweight or obese are 5 times more likely than normal-weight children to be overweight or obese as adults, increasing their risk for heart disease, high blood pressure, strokes, diabetes, osteoporosis, and several types of cancer (Cunningham et al., 2014).

In 1990, not a single state had an obesity rate above 15 percent. Now *all* states have obesity rates above 20 percent and 15 states have obesity rates equal to or greater than 30 percent ("Adult Obesity Facts," 2014). Some researchers predict that, at the current rate, 51 percent of American adults will be obese in 2030 (see *Figure 14.4*), and 9 percent of that group will be severely obese (Finkelstein et al., 2012).

Why is obesity more prevalent than in the past? People may be getting fatter because of poor eating habits, large food portions, and not exercising, but structural factors also affect obesity rates. Healthy food is less expensive than during the 1950s, more widely available than ever before, and easy to prepare. However, a smaller share of income now buys many more calories, including sugar-sweetened beverages and prepared food that are high in unhealthy carbohydrates or fat. Thus, since 1970, the average American's consumption of calories has risen by about 20 percent, fueling an "obesity epidemic" that has affected all groups (Sturm and An, 2014).

Low-fat diets rarely reduce weight because much of our food contains sugar that's added during preparation, processing, or at the table. Products with added sugar include those with packaging statements like "made with whole grain," "excellent source of calcium," "fat-free," "good source of Vitamin D," or "100% juice." However, a small container of Greek yogurt can have three or more teaspoons of added sugar.

Substance Abuse. The nation's third leading lifestyle-related and preventable cause of death is **substance abuse**, an overindulgence in and dependence on a drug or other chemical that harms a person's physical and mental health. Nearly 85,000 Americans die each year because of excessive drinking, and more than 4 million visit emergency rooms for alcohol-related problems. Excessive alcohol use includes *binge drinking*, having 4 or more drinks on a single occasion for women or 5 or more drinks on a single occasion for men, generally within about 2 hours. *Heavy drinking* involves consuming 15 or more drinks per week for men, and 8 or more drinks for women. Between 1993 and 2012, the prevalence of binge drinking increased from 14 to 23 percent, and from 3 to 7 percent of heavy drinking (Substance Abuse and Mental Health Services Administration, 2013; "Alcohol and Public Health," 2014).

Those ages 12 to 20 consume more than 90 percent of alcohol by binge drinking. Every year, 38 percent of young people under age 21 who are involved in fatal motor vehicle crashes are drunk (Crowe et al., 2012; Kann et al., 2014). More than 80 percent of college students drink alcohol. Compared with part-time college students and their peers not enrolled in college, full-time college students are more likely to binge drink, to do so more often, to be heavy drinkers, and to have a higher incidence of drunk driving (National Institute on Alcohol Abuse and Alcoholism, 2013, 2014; Substance Abuse and Mental Health Services Administration, 2013). *Table 14.1* summarizes some of the consequences of abusive college drinking on students, their families, and the larger community.

Excessive alcohol use results in myriad short- and long-term health risks that include unintentional injuries, violence, risky sexual behaviors, miscarriage and stillbirth, physical and mental birth defects, alcohol poisoning, unemployment, psychiatric problems, heart disease, several types of cancer, and liver disease. For teenagers and young adults under age 25, alcohol consumption can irreversibly damage a part of the brain that involves memory, learning, and social interaction (National Institute on Alcohol Abuse and Alcoholism, 2010; Crowe et al., 2012).

Illicit (illegal) *drugs* include marijuana/hashish, cocaine and crack, heroin, hallucinogens (e.g., LSD, PCP), inhalants, and any prescription-type psychotherapeutic

TABLE 14.1 SOME CONSEQUENCES OF ABUSIVE COLLEGE DRINKING

Researchers estimate that, each year, among college students aged 18 to 24…	
1,825	Die from alcohol-related unintentional injuries, including motor vehicle crashes
599,000	Are unintentionally injured while drunk
696,000	Are assaulted by another student who's been drinking
97,000	Are victims of alcohol-related sexual assault or date rape
400,000	Have unprotected sex while intoxicated, and more than 100,000 were too drunk to remember if they consented to having sex
3.4 million	Drive while drunk
25%	Have academic problems because of their drinking, including missing classes, doing poorly on exams or papers, earning lower grades overall, and flunking out of college

Sources: Based on National Institute on Alcohol Abuse and Alcoholism, 2013, 2014.

drug, including stimulants and sedatives, used nonmedically. More than 9 percent of Americans (almost 24 million people) age 12 and older use illegal drugs. Illicit drug use is more common among men than women, most common among high school dropouts and people aged 18 to 25, and least common among Asian Americans (Substance Abuse and Mental Health Services Administration, 2013).

Since 2012, Alaska, Colorado, Oregon, and Washington have legalized marijuana for recreational use, and 23 states (so far) and the District of Columbia allow marijuana for medical use. Federal laws still forbid possessing, using, and selling marijuana, but the Department of Justice isn't challenging state laws that enact strict regulations to control its use and sale. Marijuana is the most commonly used illicit drug (see *Figure 14.5*). Even before legalization in some states, the number of marijuana users rose from 14.5 million in 2007 to 19 million in 2012 (Substance Abuse and Mental Health Services Administration, 2013).

Tighter restrictions on pain prescriptions such as Oxycontin and Vicodin have resulted in a surge of heroin use, nearly doubling among persons age 12 and older between 2007 and 2012 (Substance Abuse and Mental Health Services Administration, 2013). Prescription pain pills containing narcotics sell for up to $80 for an 80 mg

substance abuse an overindulgence in and dependence on a drug or other chemical that harms a person's physical and mental health.

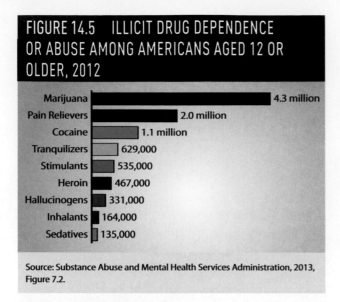

FIGURE 14.5 ILLICIT DRUG DEPENDENCE OR ABUSE AMONG AMERICANS AGED 12 OR OLDER, 2012

Marijuana	4.3 million
Pain Relievers	2.0 million
Cocaine	1.1 million
Tranquilizers	629,000
Stimulants	535,000
Heroin	467,000
Hallucinogens	331,000
Inhalants	164,000
Sedatives	135,000

Source: Substance Abuse and Mental Health Services Administration, 2013, Figure 7.2.

pill compared with only $9 for a dose of heroin. Heroin has become so easy to get that, in many parts of the country, "dealers deliver to the suburbs and run specials to attract young, professional, upper-income customers." According to the substance abuse director at a North Carolina medical center, "Our heroin patients come from the five best neighborhoods" (Leger, 2013).

Illicit drugs have the same health risks as excessive alcohol use but can also have immediate life-threatening consequences, including death. In addition, illicit drug users, compared with non-users, have higher rates of mental illness, suicidal thought and behavior, major depressive episodes, and impaired ability to function at school, home, or work (Hedegaard et al., 2014).

Sexually Transmitted Diseases. There are nearly 19 million new sexually transmitted disease (STD) cases in the United States every year. People infected with STDs are at least two to five times more likely than uninfected individuals to contract HIV, the virus that causes AIDS. More than 1.2 million Americans are living with HIV, including 14 percent who aren't aware of their infection. Men who have sex with men (MSM)—whether straight, gay, or bisexual—represent only 4 percent of the male population but 78 percent of new HIV infections. Another high-risk group is people who inject drugs. HIV-infected people live much longer than in the past, but 14,000 Americans died of AIDS in 2011 (CDC Fact Sheet, 2014; "HIV in the United States...," 2014).

medicine a system of individuals, organizations, and institutions that provide scientific diagnosis, treatment, and prevention of illness, injury, and other health impairments.

We've looked at some of the macro and micro reasons that help explain why some people are healthier than others. Who gets health care? And who pays for it?

14-3 HEALTH CARE: UNITED STATES AND GLOBAL

Health care encompasses a number of components. One of the most important is **medicine**, a system of individuals, organizations, and institutions that provide scientific diagnosis, treatment, and prevention of illness, injury, and other health impairments.

14-3a U.S. Health Care Coverage

In late 2013, 13 percent of Americans (42 million people) had no health insurance, down from 16 percent in 2010. The least likely to be insured were Latinos, those ages 19 to 34, and people with household incomes of less than $50,000 a year (Smith and Medalia, 2014).

In 2010, Congress passed the Patient Protection and Affordable Care Act, also called the Affordable Care Act (ACA) or "Obamacare." The ACA's goal is to give more Americans age 64 and under access to affordable, quality health insurance, and to reduce the growth in U.S. health care spending.

Despite numerous computer problems of Health-Care.gov, the federal website, 5 percent of nonelderly Americans were newly insured by early 2014 through federal and state *exchanges*—marketplaces where people can compare and purchase health insurance plans (Ander and Newport, 2014). Still, more than 10 percent of 18- to 64-year-olds will be uninsured in 2016 if people live in states where Medicaid doesn't cover the "nearly poor," are undocumented immigrants, or have incomes above the federal poverty level but can't afford the programs offered by employers or private insurers (Congressional Budget Office, 2014a; Eibner and Saltzman, 2014).

14-3b Who Pays for Health Care?

About 66 percent of Americans have health coverage through private insurance, primarily employer-based programs that cover most or all costs for workers and their family members, but also through private insurance not provided by employers. Some 25 percent of Americans have government health insurance funded by taxpayers; others have no insurance (see *Table 14.2*).

TABLE 14.2 HOW AMERICANS GET THEIR HEALTH INSURANCE, 2014

Percentage of Americans aged 18 to 64 who have coverage from...	
Private health insurance	66%
Current or former employer	43
Paid for by self or family member	20
Other (e.g., union)	3
Government health insurance	25
Medicaid	9
Medicare	7
Military health care	5
Other (e.g., state health plan)	4
No insurance	16

Notes: The percentages include coverage by more than one type of health insurance, and people who receive Medicare benefits before age 65.

Source: Based on Levy and Witters, 2014.

Employer-based health insurance coverage has deteriorated. Some employers with 50 or more full-time workers are considering stopping health insurance because the ACA's penalties will be less expensive than the insurance costs. Many large companies (Target, Home Depot, Walgreens) continue to maximize their record profits by ending health plans for part-time workers and increasing the premiums of full-time employees. Since 1999, workers at both large and small firms have been paying higher premiums, deductibles, copayments, and other out-of-pocket costs, whereas wages, including those of middle-class families, have stagnated (Kellerman and Auerbach, 2013; Pear, 2013; Eskow, 2014; Jilani, 2014).

Medicare pays most of the medical costs of Americans age 65 and over, regardless of income. Thus, even billionaires are eligible for Medicare. Medicaid, another government program, provides medical care, regardless of age, for people living below the poverty level (see Chapter 12).

14-3c The United States Compared with Other Countries

The United States spends more on health care than any other nation in the world—more, in fact, than the next 10 highest-spending developed countries combined. U.S. health care costs have risen steadily—per person, nationally, and as a percentage of the gross domestic product. We spend more on health care than 34 other high-income countries, but rank last among the top 11 according to criteria that include quality of care, access to care, efficiency, and health outcomes (Davis et al., 2014; OECD, 2014c).

By 2022, health care expenses will consume $1 of every $5 in the economy (see *Table 14.3*). Why, then, are many Americans worse off than their counterparts in other high-income countries? One reason is that from a third to almost half of U.S. health care spending is wasteful or inefficient (Redberg, 2012; Newhouse et al., 2013).

In Canada, Germany, Great Britain, France, Japan, and Sweden, the government picks up most of the health care bill. In these and other countries, patients and health care workers don't have to submit bills to several insurance providers, resulting in considerable administrative savings. The government, nonprofit organizations, or large groups (cities and industries) have considerable power in keeping down drug costs and setting prices for health care providers and services. There are also private hospitals, but most are nonprofit (OECD, 2014b).

In the United States, "hospitals are the most expensive part of the world's most expensive health system." Even nonprofit hospitals are businesses: Up to 26 percent of their revenues are profits, executives can earn almost $6 million a year, and they sell as many services as possible at the highest price (Brill, 2013; "Prescription for Change," 2013).

TABLE 14.3 THE INCREASING COST OF U.S. HEALTH CARE, 1980–2022

	1980	1990	2000	2010	2022
Average cost per person	$1,100	$2,864	$4,878	$8,402	$14,664
National health expenditure	$256 billion	$724 billion	$1.4 trillion	$2.6 trillion	$5 trillion
Percent of gross domestic product (GDP)	9%	13%	14%	18%	20%

Notes: The "average cost per person" includes medical care, supplies, drugs, and health insurance. All numbers are in current dollars; those for 2022 are projected.

Sources: Based on Centers for Medicare & Medicaid Services, 2010, Table 1, and 2014, Table 1.

With its high prescription prices, the United States spends far more per capita on medicines than other developed countries. Unlike other countries where the government sets a national wholesale price for each drug, the United States leaves prices to market competition among pharmaceutical companies. Congressional lawmakers—heavily lobbied by the pharmaceutical industry, the American Hospital Association, and other groups—have forbidden Medicare, the nation's largest medical insurer and the world's largest buyer of prescription drugs, to negotiate drug prices or the cost of health products. Thus, a hospital can charge $24.00 for each niacin (Vitamin B3) pill that drugstores sell for about a nickel apiece, and $1.50 for an acetaminophen tablet (which relieves pain and fever) that costs $1.49 for a box of 100 at Amazon—a 10,000 percent markup (Brill, 2013).

Medical technology has also increased health care costs. Technology can save lives, but U.S. doctors, compared with those in many other countries, are more likely to adopt new and expensive machines that are highly profitable for both physicians and hospitals. The more machines there are, the more doctors use them, increasing a patient's, insurer's, and nation's medical costs. For instance, U.S. doctors do 71 percent more CT scans than do doctors in Germany (Brill, 2013). *Figure 14.6* provides more examples of why U.S. health care costs are so high.

14-4 SOCIOLOGICAL PERSPECTIVES ON HEALTH AND MEDICINE

Sociologists focus on different aspects of health, health care, and medicine. *Table 14.4* summarizes these perspectives.

14-4a Functionalism: Good Health and Medicine Benefit Society

For functionalists, good health and medicine are critical for a society's survival and stability. Thus, countries try to develop a medical care system that, ideally, benefits the entire population regardless of age, sex, race, ethnicity, social class, or other characteristics.

HEALTH CARE AND OTHER INSTITUTIONS

To function effectively, health care systems are integrally related to other institutions. Despite ongoing debates about the ACA and the government's requiring people to buy health insurance, millions of individuals and families are now eligible for government assistance to pay for coverage, and can get free preventive health services (e.g., diabetes tests, colonoscopies, routine vaccinations).

FIGURE 14.6 SHOULD YOU TRAVEL ABROAD FOR MEDICAL CARE?

Would you be better off living outside the United States when it comes to health care costs? Except for Argentina, all of the countries below have higher life expectancy rates than the United States.

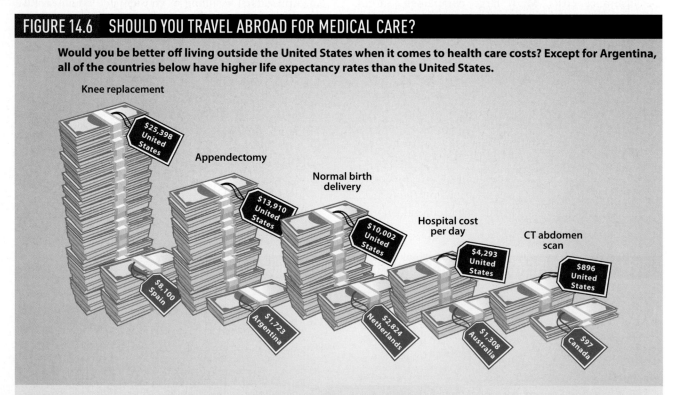

Knee replacement — $25,398 United States — $8,100 Spain
Appendectomy — $13,910 United States — $1,723 Argentina
Normal birth delivery — $10,002 United States — $2,824 Netherlands
Hospital cost per day — $4,293 United States — $1,308 Australia
CT abdomen scan — $896 United States — $97 Canada

Source: Based on International Federation of Health Plans, 2014. See also Howard, 2014, for a comparison of surgery costs in the United States and other countries.

TABLE 14.4 SOCIOLOGICAL PERSPECTIVES ON HEALTH AND MEDICINE

Perspective	Level of Analysis	Key Points
Functionalism	Macro	• Health and medicine are critical in ensuring a society's survival and are closely linked to other institutions. • Illness is dysfunctional because it prevents people from performing expected roles. • Sick people are expected to seek professional help and get well.
Conflict	Macro	• There are gross inequities in the health care system. • The medical establishment is a powerful social control agent. • A drive for profit ignores people's health needs.
Feminist	Macro and micro	• Women are less likely than men to receive high-quality health care. • Gender stratification in medicine and the health care industry reduces women's earnings. • Men control women's health.
Symbolic Interaction	Micro	• Illness and disease are socially constructed. • Labeling people as ill increases their likelihood of being stigmatized. • Medicalization has increased the power of medical associations, parents, and mental health advocates and the profits of pharmaceutical companies.

The government is deeply involved in health care in other ways. It funds much of the scientific research at universities as well as federal agencies like the Centers for Disease Control and Prevention and the National Institutes of Health, both of which deal with prevention, treatment, and health care policy. At both the national and state levels, numerous government agencies are responsible for passing and enforcing regulations regarding new drugs, medical procedures, and access to medical care.

Health care systems are also linked to the economy. The Congressional Budget Office (2014b) estimates that because of health care reform, Americans will work about 2 percent fewer hours between 2017 and 2024, the equivalent of 2.5 million full-time workers. Some will take part-time jobs without benefits or retire early because the ACA guarantees individual health coverage. On the other hand, the ACA allows people to quit jobs they hate, start their own business, or take care of family members.

Even if some people quit their jobs, health care is the nation's largest employer. "Demand for health services will rise as Obamacare expands insurance and Americans grow older, fatter, and sicker" ("The Health Paradox," 2013: 27). Our huge health care spending threatens funding for other public programs, including education and transportation, but functionalists would point out that the trillions spent on health care have generated millions of jobs at hospitals, drug companies, insurance companies, nursing homes, and information-technology firms (see Chapter 11).

THE SICK ROLE

Talcott Parsons (1951), an influential sociologist, introduced the concept of the **sick role**, a social role that excuses people from normal obligations because of illness. In Parsons' model, sick people aren't responsible for their condition and, therefore, have legitimate reasons for not performing their usual social roles ("I missed the exam because I had the flu").

Parsons emphasized that the sick role is temporary, and that people must seek medical help to hasten their recovery. Otherwise, people will view them as hypochondriacs, slouchers, and malingerers who aren't living up to their responsibilities. Thus, from a functionalist perspective, the sick role is legitimate if it's short-lived, but dysfunctional if people feign illness to shirk their duties long-term in the family, the workplace, or other groups.

THE PHYSICIAN AS GATEKEEPER

From a functionalist perspective, physicians play a key gatekeeping role in limiting the sick role. They verify a person's condition as sick and provide an excuse for temporarily not performing necessary roles. They also designate a patient as "recovered" and ready to meet societal role expectations again. Doctors' specialized knowledge gives them considerable authority in defining health and illness that's unmatched by other health care providers, such as nurses and pharmacists.

sick role a social role that excuses people from normal obligations because of illness.

© Iakov Filimonov/Shutterstock.com

How sick is too sick to miss work or class? Are menstrual cramps enough? What about a severe cold? A hangover? And when does being sick become dysfunctional—after a day, a week, a month?

CRITICAL EVALUATION

Functionalist theories are limited for several reasons. First, health care policies can be divisive. For example, 56 percent of Americans say that health care coverage isn't the government's responsibility (Wilke, 2013). Moreover, most of the job growth in the health care sector has been in low-paying occupations (hospital orderlies, nursing home assistants, lab technicians) (Ross et al., 2014). Thus, many people aren't benefiting from the booming health care industry.

Second, people often view those with chronic illnesses like arthritis and diabetes as healthy enough to perform their expected roles regardless of how they sick they feel. Many people can't assume even a short-lived sick role, even if an illness is life threatening, because they don't have the resources to see doctors or may lose a portion of their earnings if they miss workdays. Almost one in five Americans with health insurance has put off treatment for a serious condition because of high out-of-pocket costs (Brown, 2013).

Third, because physicians are powerful gatekeepers, many prosper by extending sick roles. The more times that doctors see patients or visit them at the hospital, and the more tests that doctors order, the higher

medical-industrial complex a network of business enterprises that influences medicine and health care.

their salaries. Conflict theorists, especially, maintain that much of health care industry is dysfunctional because access to health care varies considerably, and only the wealthy don't have to worry about receiving and paying for the best available medical care.

14-4b Conflict Theory: Health Care and Medicine Don't Benefit Everyone

For conflict theorists, medicine and the health care industry benefit some groups much more than others. In contrast to functionalists, conflict theorists argue that the medical system reinforces social inequality, exerts social control to maintain the status quo, and is often driven by a profit motive rather than a concern for people's well-being.

SOCIAL CONTROL OF MEDICINE AND HEALTH CARE

From a conflict perspective, those at the top of the medical hierarchy have considerable control. Physicians, for instance, have almost absolute power in diagnosing an illness, providing treatment, and deciding on medical procedures.

The American Medical Association (AMA) and health care industry can maintain the status quo because of the **medical-industrial complex**, a network of business enterprises that influences medicine and health care. The medical-industrial complex includes many groups—doctors, nurses, hospitals, lawyers, nursing homes and hospices, insurance companies, drug manufacturers, accountants, banks, and real estate and construction businesses.

Who benefits from the medical-industrial complex? All of those involved, especially hospitals and drug companies. Big hospitals have been buying up previously independent cancer clinics and doctors' offices. Doing so gives them the power to raise prices for private insurers and government programs, particularly Medicare. As a result, they can charge higher rates, sometimes three times as much, for cancer drugs and outpatient visits and services (Tozzi, 2014a).

The percentage of Americans taking five or more prescription drugs increased from 4 percent in 1994 to more than 10 percent in 2010, and consumption of narcotic painkillers, or opioids, jumped 300 percent between 1999 and 2010. The prices for dozens of drugs that treat everything from blood pressure to multiple sclerosis have doubled since 2007, and the prices of generic drugs increased ten-fold during 2013 alone (Katz, 2013; Langreth, 2014; National Center for Health Statistics, 2014).

Direct-to-consumer prescription drug marketing ("Ask your doctor about…"), which began in the 1980s, increases a drug's sales. A recent study of 100,000 adults found that people who saw drug advertising were more likely to request their doctors a particular brand-name medication. Doctors may feel pressure to accommodate their patients' requests, which, in turn, increases a drug's price and sales (Timmermans and Oh, 2010; Niederdeppe et al., 2013).

HEALTH, PROFIT, AND WASTE

From a conflict perspective, health care is big business. In 2013, the average compensation of an health insurance CEO was almost $14 million (a 19 percent increase over the previous year) compared with $35,000 for the average hospital worker. Large hospitals are crowded with "senior VPs, VPs of this, that, and the other" whose salaries are two to three times higher than those of surgeons. A Wisconsin surgeon discovered that a brief outpatient appendectomy he had performed for a fee of $1,700 generated over $12,000 in hospital bills. Thus, "the biggest bucks are earned not through the delivery of care, but from overseeing the business of medicine" (Healthcare-NOW! 2014; Rosenthal, 2014: SR4).

Most physicians are concerned about quality care, but Medicare is a source of wealth for many. In 2012, 2 percent of doctors received 23 percent of all Medicare fees, including a Florida ophthalmologist who took in more than $26 million to treat fewer than 900 patients. In the same year, Medicare also paid $7 million to at least eight doctors whose medical licenses had been revoked or suspended (Terhune et al., 2014; Tozzi et al., 2014).

About 37 percent of the Medicaid and Medicare spending is wasteful. A physician's Medicare fee is based partly on a drug's price. Thus, there's little financial incentive to use a drug that costs $50 instead of $2,000 a shot. About 10 to 20 percent of all surgeries are unnecessary, and more than 2,200 doctors nationwide are suspected of writing elderly and disabled patients prescriptions, particularly opioids, that are expensive, addictive, and medically unnecessary. Other wasteful spending includes the following: Expensive services that have the same health outcomes as lower-priced alternatives, preventable hospital injuries and readmissions that cost Medicare almost $5 billion a year, unnecessary tests and treatments to guard against liability in malpractice lawsuits, and bureaucratic procedures and forms that result in needlessly complex and time-consuming billing work for physicians and their staff (Lallemand, 2012; Eisler and Hansen, 2013; Office of Inspector General, 2013; Whoriskey et al., 2014).

CRITICAL EVALUATION

The most common criticism is that conflict theorists often ignore the contributions of medical and health care systems. Without them, people would suffer more, die at a younger age, and have a lower quality of life. Second, not all medical providers are motivated by profit. For example, doctors in family practices earn, on average, only about half as much ($208,000) as surgeons (Bureau of Labor Statistics, 2014–2015).

Third, doctors, medical scientists, and some medical associations (e.g., American Academy of Family Physicians) have been among the most vocal critics of unneeded surgery, tests, and procedures. Examples include cardiac stress tests, antibiotic prescriptions for sinusitis that usually resolves itself within a few weeks, colonoscopies, pap smears, routine prostate cancer screening because the risks can outweigh the advantages, and taking vitamins and other supplements that don't benefit health. Others have urged ending expensive end-of-life care that simply prolongs the dying process (Kellerman, 2012; Guallar et al., 2013; Agnvall, 2014; Makary, 2014).

Finally, most people aren't simply victims of a malicious medical-industrial complex. Many make unhealthy lifestyle choices (unprotected sex, drug abuse), and complain rather than change their behavior when employers impose health insurance penalties for smoking and obesity. Others join forces with drug companies to lobby the FDA to decrease current restrictions of potent narcotic painkillers (Meier and Lipton, 2013).

14-4c Feminist Theories: Health and Medicine Benefit Men More Than Women

Both feminist and conflict theorists emphasize the connection between health and inequality. Feminist theorists go further, however, by addressing issues of the health costs of being a woman, gender stratification in medicine and health care, and men's control over women's choices.

THE HEALTH COSTS OF BEING A WOMAN

Women, compared with men, have greater health care needs and risks, especially during their reproductive years. Almost 33 percent of women who give birth in the United States have a cesarean section (C-section), up from only 5 percent in 1970. The main reason for the increase is hospital administrators' and maternity clinicians' fear of lawsuits if a vaginal birth leads to bad outcomes (Morris, 2014). Because C-sections are major surgical procedures, the risks include infection, blood clots, and injury to other organs, particularly the bladder.

Is Being Fashionable Hazardous to Women's Health?

Squeezing into skinny jeans and tight pants can cause nerve compression, numbness, and digestive problems. Narrow-toed shoes and stilettos wreak havoc: Blisters, bunions, corns, calluses hammertoes, nerve damage, back and neck pain, painful inflammation, stress fractures, and ankle sprains. Still, according to a podiatrist, "Women will wear their high-heeled shoes until their feet are bloody stumps" (Ianzito, 2013: 3; Gleiber, 2014).

Many cosmetics contain chemicals that can trigger skin problems (e.g., itching, rashes, acne). Others—contaminated with bacteria, yeasts, or molds—can lead to a range of problems from simple rashes to serious infections that can cause swelling and breathing difficulties (Dahl, 2011). A recent study that tested

32 commonly sold lipsticks and lip glosses found that they contained lead, cadmium, chromium, aluminum, and five other metals— some at potentially toxic levels that are linked to stomach tumors. The more often and heavily that women use lipstick, the greater the health risks (Liu et al., 2013).

Except for some hair color additives, the FDA doesn't require companies to test products for safety or list toxic ingredients. Japan, Canada, and the European Union have banned more than 500 cosmetic products that contain toxic ingredients that are sold in the United States (Rano and Houlihan, 2012). Skin Deep (ewg.org/skindeep) ranks the safety of a range of cosmetic products.

maron/Age Fotostock

After following nearly 90,000 women aged 40 to 59 for 25 years, the researchers found that women who got mammograms every year for five years were no less likely to die of breast cancer than those who didn't, and recommended cutting back on the screenings (Miller et al., 2014). Mammograms can save women's lives, but also have health risks. If a tumor is benign and there's no history of breast cancer, patients could be subjected to unnecessary procedures—surgery, chemotherapy, and radiation—that could do more harm than good given their complications. Nonetheless, many major organizations, including the American Cancer Society (2013), continue to recommend annual screenings beginning at age 40.

On average, women live longer than men but are more likely than men, particularly at later ages, to suffer from stress, obesity, hypertension, and chronic illnesses. Women are also less likely than men to receive high-intensity treatments like organ transplants and coronary bypasses (Weitz, 2013).

Social class affects every person's health, but women experience higher health costs simply because of their sex. Women earn less than men (see Chapter 11) but use more health care services. Less income, greater usage of health care services, and rising medical costs jeopardize many women's health. In 2010, 48 percent of adult women

(compared with 34 percent in 2001) reported that they didn't fill a prescription because of cost; skipped a recommended test, treatment, or follow-up; or didn't see a specialist when needed. In addition, 44 percent of women, compared with 35 percent of men, had problems paying their medical bills or debts (Robertson and Collins, 2011).

GENDER STRATIFICATION IN HEALTH CARE OCCUPATIONS

Economists expect job growth in health care occupations to surge until at least 2022 because of a graying population and corporations' inability to offshore most medical services (Kolet and Chandra, 2012; see also Chapter 11 on offshoring). Many of these jobs, however, including home health aides and dental assistants (not dental hygienists), are low-paying and dominated by women.

In health care jobs, men consistently earn more than women (see *Table 14.5*). Among physical therapists, for example, males earn almost $75,800 a year compared with $67,600 for females. Thus, in only 10 years, male therapists earn almost $82,000 more than females, even when both have the same level of education, years of experience, and work responsibilities (see also Chapters 8 and 11).

Feminist scholars attribute much of the gender wage gap to male doctors' gatekeeping. Because many registered

TABLE 14.5 WOMEN EARN LESS THAN MEN IN HEALTH CARE OCCUPATIONS, 2013

	Men	Women	Percentage of Women in This Occupation
All health care practitioners	**$68,224**	**$51,378**	**75**
Pharmacists	$108,784	$93,704	54
Physicians and surgeons	$108,524	$77,844	36
Physical therapists	$75,764	$67,600	66
Registered nurses	$64,272	$56,472	89
Emergency medical technicians and paramedics	$43,264	$40,820	41
Support technologists and technicians	$34,528	$31,876	81

Source: Based on Bureau of Labor Statistics, "Median Weekly Earnings of Full-Time Wage and Salary Workers by Detailed Occupation and Sex," Table 39. Accessed July 30, 2014 (www.bls.gov).

nurses, pharmacists, and physical therapists (occupations that have more women than men) now receive doctoral degrees, they want to use the honorific title of "doctor." Doing so would win more respect from patients, help women land top administrative jobs that pay more, provide more autonomy in treating patients, and bring higher fees from health insurers. Physicians oppose the idea for several reasons: They want to maintain their prestige in the health care industry; they treat patients first, whereas nurses and others play only secondary roles; and they have considerably more education and training than others to diagnose and treat illness and disease (Harris, 2011).

MEN'S CONTROL OF WOMEN'S HEALTH

Feminist scholars maintain that in patriarchal societies, including ours, men control many aspects of women's health. Catholic bishops, all of whom are men, condemn contraceptives as sinful, even though 98 percent of U.S. Catholic women have used birth control (Jones and Dreweke, 2011).

Most recently, the Supreme Court ruled, 5-4 (*Burwell v. Hobby Lobby Stores*), that privately held for-profit corporations don't have to pay for their employees' contraceptive coverage, as the ACA requires, if doing so conflicts with the owners' personal religious beliefs. In effect, a religious objection can trump a federal law, and businesses can shift insurance costs to taxpayers or private insurers. According to some Johns Hopkins doctors, "The Supreme Court decision has started us down a dangerous path on which the religious beliefs of a third party enter the examination room and interfere with the doctor-patient relationship" (Singal et al., 2014: 17; see also Bassett and Reilly, 2014; and Rosenfeld, 2014). All of the justices who ruled against women's contraceptive coverage were males.

In many states, conservative lawmakers have passed laws to restrict or block women's ability to get abortions (see Chapter 9). They aren't passing similar laws to limit the availability of Viagra and other erectile dysfunction medications for men because "Viagra is a wonderful drug" (Beadle, 2012). The Hobby Lobby chain doesn't have to pay for IUDs and emergency contraceptive pills, but covers Viagra and other erectile dysfunction drugs.

CRITICAL EVALUATION

Critics have questioned feminist theories for three reasons. First, feminist scholars sometimes gloss over the

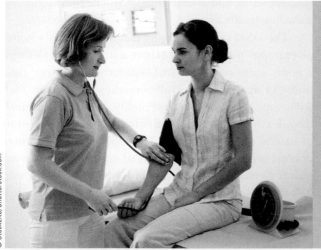

© StockLite/Shutterstock.com

Nurse practitioners (NPs), 89 percent of whom are women, complete a master's or doctoral program that includes diagnosing and treating patients, managing acute and chronic illnesses, and prescribing medications. Because of a projected shortage of doctors, NPs are pressing 34 states to relax laws requiring doctors to oversee their work. Many doctors' groups, led by the AMA, are fighting the NPs, arguing that less supervision would jeopardize patient safety. Several national studies have found no evidence that NPs provide lower-quality care than a physician (Beck, 2013; Pettypiece, 2013). How might you explain the AMA's resistance from a feminist perspective?

fact that social class, rather than gender, has a big effect on people's health and receiving health care services (Kindig and Cheng, 2013; Montez and Zajacova, 2013).

Second, it's not entirely clear that men can be blamed for gatekeeping because 54 percent of physicians and surgeons are women. Also, perhaps the gatekeeping by physicians (men or women) is justified because a doctoral degree in nursing, pharmacy, or physical therapy requires considerably less schooling and training than getting a Doctor of Medicine (M.D.) degree (Harris, 2011).

Third, and like conflict theorists, feminist scholars rarely address lifestyle choices that affect people's health. For example, 21 percent of women aged 18 to 25 use illicit drugs, 36 percent aged 20 and older are obese, and 29 percent aged 45 and older smoke (National Center for Health Statistics, 2014).

14-4d Symbolic Interaction: The Social Construction of Health and Illness

Symbolic interactionists focus on how we define and construct views about health, illness, and medicine and then implement these definitions in everyday life. Social constructions include labeling, stigmatizing behavior, and medicalizing attitudes and behaviors as normal or sick.

© Monkey Business Images/Shutterstock.com

THE SOCIAL CONSTRUCTION OF ILLNESS

Medical models assume that illness is anything that deviates from normal biological functioning, and that an illness has specific features that a doctor can recognize. In contrast, interactionists view illness and medicine as social constructions that can change over time. In 1956, the AMA declared that alcoholism is a treatable disease that's due to a genetic predisposition rather than a person's lacking moral character or self-discipline. And, in 2013, the AMA officially designated obesity as a disease—rather than a lifestyle choice—that requires research funding and medical intervention, including surgery and drugs (Newberry, 2013; Pollack, 2013).

medicalization a process that defines a nonmedical condition or behavior as an illness, disorder, or disease that requires medical treatment.

With or without medical intervention, individuals construct and manage their illnesses differently. Some people's social worlds shrink when they become immersed in the day-to-day aspects of managing a chronic illness like rheumatoid arthritis: They become increasingly cut off from everyday routines if they're unable to work, spend less time with family and friends, or can't move about freely. Others create new self-identities by describing themselves as survivors (of breast cancer, for instance), exchange information about treatment options, and participate in local or national fundraising events (Conrad and Barker, 2010).

Social class also affects social constructions of illness. As educational level rises, people report poorer health, presumably because they're more aware of medical knowledge and evaluate their own physical well-being more critically (Schnittker, 2009).

THE MEDICALIZATION OF HEALTH

In 1952, the American Psychiatric Association (APA) published the first edition of the *Diagnostic and Statistical Manual of Mental Disorders* (DSM), which has become the global "bible of mental illness." The DSM relies on subjective definitions and criteria, but is influential in labeling (or unlabeling) mental illness. The number of disorders increased from 106 in 1952 to nearly 400 in 2013. Some of the new mental illnesses include "binge eating" (frequent overeating), "bereavement" (mourning the loss of a loved one), "caffeine intoxication" (drinking too much coffee), "disruptive mood dysregulation disorder" (children's temper tantrums), and "hoarding" (accumulating too much stuff) (American Psychiatric Association, 2013).

The DSM's ever-changing diagnoses and labels are an example of **medicalization**, a process that defines a nonmedical condition or behavior as an illness, disorder, or disease that requires medical treatment. The APA once classified homosexuality as a psychological disorder, but dropped sexual orientation from its roster of mental illnesses in 1973. The DSM affects patients, doctors, insurers, pharmaceutical companies, and taxpayers because psychiatrists and other physicians use it to bill insurance companies. DSM-based diagnoses also determine whether people get special services at school, qualify for disability benefits, are stigmatized, and even whether they're able to adopt children.

Medicalization is a lucrative business. Pharmaceutical corporations have reaped enormous profits because everyday normal anxieties, discomforts, and stresses (frustration with traffic, boredom with routine housekeeping chores, feeling sad or personally insecure) can be "fixed" by popping a Prozac or other pill (Herzberg, 2009). The more behaviors that the DSM defines as mental illness, the more likely psychiatrists are to increase their number of patients. Mental rights advocates and parents also benefit from medicalization, as when children who are diagnosed with ADHD get insurance coverage or special treatment, including more attention from teachers and health specialists (Conrad and Barker, 2010; Rochman, 2012).

According to critics, the DSM encourages misdiagnosis, overdiagnosis, the medicalization of normal behavior, and the prescription of a large number of unnecessary drugs, especially in the case of ADHD. In 2014, more than 10,000 2-to-3-year-olds (that's right, 2-to-3-year-olds!), particularly those on Medicaid, were being medicated for ADHD. About 11 percent of children aged 4 to 17 have been diagnosed with ADHD. Between 2003 and 2012, ADHD rates in this age group climbed by 42 percent, and more than two-thirds of those with a current diagnosis were receiving prescriptions for stimulants (e.g., Ritalin, Adderall). ADHD diagnoses among high-school students are particularly high— 10 percent for girls and 19 percent for boys. Such numbers alarm many physicians because ADHD medications have many side effects that include delayed growth, sleep problems, decreased appetite, headaches, stomachaches, moodiness, and irritability (Rettew, 2014; Schwarz, 2014; Visser et al., 2014).

ADHD is diagnosed about 25 times more often in the United States than in the United Kingdom. Attitudes vary by country, but many European parents, teachers, and doctors are reluctant to use medication for what they consider routine childhood behavioral problems. Europeans also don't believe that ADHD is a disease because "parents are loath to get their child labeled" (Kelley, 2013: 26). Thus, America's high ADHD rates and drug sales illustrate cultural differences in the social construction of health and illness.

MEDICALIZATION, LABELING, AND STIGMA

From an interactionist perspective, conditions and behaviors that are diagnosed and labeled as illness or disease change over time, differ among groups, and vary across countries. Medicalization and labeling stigmatize some illnesses and diseases more than others. Male impotence, which commonly increases as men age and was stigmatized, is now called "erectile dysfunction," and

treated with drugs, regardless of age, to enhance a man's sexual experience. Ads and commercials ("Cialis is ready when you are") have flooded magazines, newspapers, and online sites.

Obese women have higher rates of gynecological cancers than nonobese women. They avoid routine exams, however, because of the stigma of obesity and the possible negative attitudes of health care professionals toward overweight people (Amy et al., 2006). Western stigmas about fat people seem to be spreading to other nations. Some countries that had traditionally viewed plumper female bodies as attractive or associated with being wealthy now stigmatize being overweight. The stigmatizing often leads to ridicule, criticism, and a negative self-image (Brewis et al., 2011).

CRITICAL EVALUATION

Symbolic interaction theories are limited for several reasons. The most common criticism is that they don't address structural factors such as laws and government policies. One reason for Europe's low ADHD rates is that the governments prohibit direct-to-consumer drug ads (Kelley, 2013). After Quebec expanded insurance

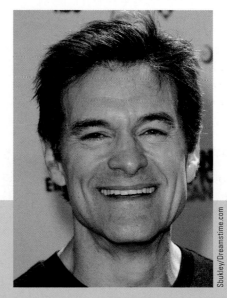

How do authority figures affect people's health perceptions and behavior? Dr. Mehmet Oz is a highly respected and published cardiothoracic surgeon, and host of the popular *Dr. Oz Show*. Products he endorses on his show "are almost guaranteed to fly off the shelves" (Weathers, 2014a, 2014b). Members of a Senate committee recently reprimanded Dr. Oz, however, for falsely portraying some of the diet pills as a "miracle" and a "magic weight-loss cure."

coverage for prescription medications in 1997, there was a significant increase in Ritalin use compared with the rest of Canada, which didn't provide similar insurance coverage (Currie et al., 2014).

Second, people often reject medical definitions of health and illness and treatment. Many deaf people have ignored recommendations to get cochlear implants (devices that can increase hearing). They don't see deafness as a medical disability but as a social reality that helps them form a community and identify with other deaf people (Conrad and Barker, 2010). A "fat pride" movement argues that obesity's health risks are exaggerated and focuses, instead, on changing discrimination against overweight people (see Chapter 16).

Third, interactionists emphasize that health and illness are social constructions, but there are many serious health problems whether or not they're medicalized. People die from unknown diseases, for example. Others suffer from "contested" illnesses, including chronic fatigue syndrome, that many physicians question because the symptoms are difficult to diagnose and treat (Walker, 2012).

STUDY TOOLS 14

READY TO STUDY? IN THE BOOK YOU CAN:

☐ Check your understanding of what you've read with the Test Your Learning Questions provided on the chapter review card at the back of the book.

☐ Rip out the chapter review card for a handy summary of the chapter and key terms.

ONLINE AT CENGAGEBRAIN.COM YOU CAN:

☐ Prepare for tests with quizzes.

☐ Review the key terms with Flash Cards.

☐ Play games to master concepts.

© Anelina/Shutterstock.com

ONE APPROACH.
70 UNIQUE SOLUTIONS.

15 | Population, Urbanization, and the Environment

LEARNING OBJECTIVES

15-1 Explain how and why populations change, and evaluate the population growth theories.

15-2 Explain how and why global and U.S. urbanization are changing, describe the consequences of urbanization, and compare and evaluate the theoretical explanations of urbanization.

15-3 Describe and illustrate the major environmental issues, and discuss whether sustainable development is achievable.

After you finish this chapter go to **PAGE 310** for **STUDY TOOLS.**

In 2014, the world hit a population milestone of 7.2 billion people, and is projected to increase to 9.6 billion by 2050. Our planet has a record number of inhabitants, and many are living longer than ever before. This chapter examines how such changes affect the planet's population, urbanization, and environment. Let's begin with population.

15-1 POPULATION CHANGES

Global population has grown rapidly since 1800, and more than tripled since 1900 (see *Figure 15.1*). By 2100, the world population will climb to almost 11 billion, and nearly all of the growth will occur in eight developing African countries (United Nations Population Division, 2013).

The world adds almost 393,000 people each day (Haub and Kaneda, 2014), but population changes vary across countries. By 2050, India, with almost 1.7 billion people, will surpass China as the world's largest country, Nigeria will replace the United States as the third largest country, and three African nations will be the most populous (see *Table 15.1* on page 292).

What do you think?

Individuals can do little to prevent global warming.

| 1 | 2 | 3 | 4 | 5 | 6 | 7 |

strongly agree   strongly disagree

FIGURE 15.1 WORLD POPULATION GROWTH THROUGH HISTORY

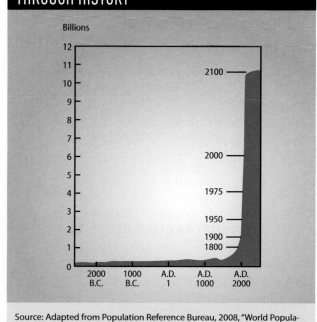

Billions

| 2100 |
| 2000 |
| 1975 |
| 1950 |
| 1900 |
| 1800 |

2000 B.C. 1000 B.C. A.D. 1 A.D. 1000 A.D. 2000

Source: Adapted from Population Reference Bureau, 2008, "World Population Projections to 2100," accessed at www.prb.org.

Information about these and other changes comes from **demography**, the scientific study of human populations. A **population** is a group of people who share a geographic territory. A territory can be as small as a town or as vast as the planet, depending on a researcher's focus. Demographers analyze populations in terms of size, composition, distribution, and change. They also study personally relevant topics, including your probability of getting married or divorced, the kind of job you'll probably have, how many times you'll move, and how long you'll probably live (McFalls, 2007).

15-1a Why Populations Change

Populations are never static. Their growth or decline involves three key factors: fertility (how many people are born), mortality (how many die), and migration (how many move from one area to another).

demography the scientific study of human populations.

population a group of people who share a geographic territory.

TABLE 15.1 WORLD'S TEN LARGEST COUNTRIES, 2014 AND 2050

2014		2050		Estimated Population Change Between 2014 and 2050
Country	Population (In Millions)	Country	Population (In Millions)	
China	1,364	China	1,312	−4%
India	1,296	India	1,657	+28%
United States	318	United States	395	+24%
Indonesia	251	Indonesia	365	+46%
Brazil	203	Brazil	226	+11%
Pakistan	194	Pakistan	348	+79%
Nigeria	177	Nigeria	396	+124%
Bangladesh	158	Bangladesh	202	+28%
Russia	144	Democratic Republic of Congo	194	(Not applicable)
Japan	127	Ethiopia	165	(Not applicable)

Source: Based on Haub and Kaneda 2014, and author's calculations.

FERTILITY: ADDING NEW PEOPLE

The study of population changes begins with **fertility**, the number of babies born during a specified period in a particular society. Demographers use several fertility measures depending on the level of specificity needed. One of the most commonly used measures is the **crude birth rate** (also called the *birth rate*), the number of live births for every 1,000 people in a population in a given year. "Crude" implies that the rate is a general measure of a society's childbearing. It's based on the total population rather than more specific measures (e.g., a woman's age or marital status). In 2013, the crude birth rate was 20 worldwide, 36 for Africa, 13 for the United States, and 11 for Europe (Haub and Kaneda, 2014).

---

fertility the number of babies born during a specified period in a particular society.

crude birth rate (also called the *birth rate*) the number of live births per 1,000 people in a population in a given year.

total fertility rate (TFR) the average number of children born to a woman during her lifetime.

mortality the number of deaths in a population during a specified period.

crude death rate (also called the *death rate*) the number of deaths per 1,000 people in a population in a given year.

Another standard measure, the **total fertility rate (TFR)**, is the average number of children born to a woman during her lifetime. In the early 1900s, U.S. women had an average TFR of 3.5 compared with 1.8 in 2014. Worldwide, TFRs are much higher in the least developed nations (4.3) than in developed nations (1.6), ranging from a low of under 1.4 in some Asian and European countries to 6.0 and higher in 10 African nations (Haub and Kaneda, 2014; Martin et al., 2014).

To maintain a stable population, a woman must have an average of two children, the *replacement rate* for herself and her partner. TFRs above 2.1 indicate that a country's population is growing and getting younger. Rates below 2.1 mean that a country's population is decreasing and growing older. The older a country's population, the more difficult it is to support people who are living into their eighties and longer and to provide resources for the young (United Nations Population Division, 2013; see also Chapter 12 on the old-age dependency ratio).

MORTALITY: SUBTRACTING PEOPLE

The second factor in population change is **mortality**, the number of deaths in a population during a specified period. Demographers typically measure mortality by the **crude death rate** (also called the *death rate*), the number of deaths per 1,000 people in a population in a given year. In 2013, the death rate was 6 for South America,

Africa's Niger and South Sudan have the world's highest TFRs, 7.6 and 7.0 respectively (left). The reasons for the high fertility include teen pregnancies, poverty, women's limited access to education, and cultural norms that exclude females from family planning decisions. In contrast, some European towns are facing "slow death" due to low fertility and immigration rates. Germany is trying to boost its shrinking TFR (1.4 compared with 2.5 in the 1960s) by offering families tax breaks and free day care for all children age 1 and older (right) (Daley and Kulish, 2013; Williamson, 2013; Rossi and Jucca, 2014).

7 for Asia, 8 worldwide and for the United States, 11 for Europe, but 15 or higher for 6 African nations (Haub and Kaneda, 2014).

A death rate isn't necessarily the best measure of a population's health. Developed countries, compared with most developing countries, also have large proportions of people age 65 and older who are no longer employed but require costly medical services (Colby and Ortman, 2014; see also Chapter 14).

A better measure of a population's health is the *infant mortality rate*, the number of deaths of infants younger than 1 year per 1,000 live births. Generally, as the standard of living improves—access to clean water, adequate sanitation, and medical care—the infant mortality rate decreases. In 2013, the infant mortality rate was 5 in developed countries, 64 in the least developed countries, and 90 or higher in 6 African nations (Haub and Kaneda, 2014).

Lower infant mortality greatly raises *life expectancy*, the average number of years that people who were born at about the same time can expect to live. Worldwide, the average life expectancy is 71—69 for males and 73 for females. Again, however, there are considerable variations across countries—from a high of 84 in Hong Kong and 83 in Japan to a low of 44 in several African nations. The United States, with a life expectancy of 79, ranks below at least 25 other industrialized countries and only slightly higher than less developed countries such as Cuba and Uruguay (Haub and Kaneda, 2014). In almost every society, people with a higher socioeconomic status live longer and healthier lives because their occupations are

physically safe, and their resources include clean drinking water, sanitation, and medical services.

MIGRATION: ADDING AND SUBTRACTING PEOPLE

The third demographic factor in population change is **migration**, the movement of people into or out of a specific geographic area. There are two types of migration: international and internal. *International migration*, movement to another country, includes *emigrants* (people who move out of a country) and *immigrants* (people who move into a country). You may be a product of international migration if your great-great-grandparents emigrated from Ireland and immigrated to the United States.

International Migration. The number of international migrants reached an all-time high of 232 million in 2013, and is expected to double by 2050. The largest flows have moved from one developing country to another (as from Indonesia to Saudi Arabia) or from a developing to an industrialized country (as from Mexico to the United States). Also, most emigrants move to a neighboring country (from Mexico to the United States, for example, rather than from Mexico to Canada). The United States has 20 percent of the world's migrants, more than any other country in the world, and is the only industrialized

migration the movement of people into or out of a specific geographic area.

nation where one-quarter of migrants are unauthorized (Martin, 2013).

Some of the most common reasons for international migration include religious and political freedom, better schools, lower crime rates, war, natural disasters, and, especially, economic opportunities (Kohut and Wike, 2013). Thus, the reasons for international migration today are similar to those of European immigrants who came to the United States during the twentieth century.

Internal Migration. International migration has increased, but 92 percent of adults worldwide who change residences do so because of *internal migration*, movement within a country. Over a five-year period, 24 percent of U.S. adults move within the country, similar to rates reported in other advanced economies, including New Zealand, Finland, and Norway (Esipova et al., 2013).

Between 2012 and 2013, 36 million people living in the United States (12 percent of the population) moved to a different residence. Why do people move? The most common reasons (48 percent) are housing-related (e.g., "Wanted new or better home/apartment," eviction), 30 percent are family-related (e.g., change in marital status), and 20 percent are job-related (e.g., new job, job transfer, to look for work) (Ihrke, 2014).

Social class and race/ethnicity also affect internal migration. Between 1980 and 2013, approximately 381,000 Puerto Ricans relocated from the island to the mainland, but 38 percent have done so since 2010. The top reasons for moving stateside were job-related (42 percent) and family-related (38 percent), but many also cited crumbling schools and pervasive crime, particularly drug trafficking (Alvarez, 2014; Cohn et al., 2014). After Hurricanes Katrina and Rita devastated much of the U.S. Gulf coast in 2005, 75 percent of New Orleans' population—primarily low-income blacks—moved to other states; they didn't have home insurance policies that covered the cost of the damages. As a result, blacks now make up only 59 percent of the city's population compared with 67 percent before Katrina (Bremner and Hunter, 2014).

15-1b Population Composition and Structure

Demographers study age and sex to understand a population's composition and structure. Two of the most common measures are *sex ratios* and *population pyramids*.

sex ratio the proportion of men to women in a population.

So far, more than 9 million Syrians (over one-third of the population) have fled the civil war that began in 2011 or have been displaced from their homes inside the country. "Many have been on the run for a year or more…moving from village to village, up to as many as 20 times, before they finally made it across an international border" (Bengali, 2014: 10).

SEX RATIOS

A **sex ratio** is the proportion of men to women in a population. A sex ratio of 100 means that there are equal numbers of men and women; a ratio of 95 means that there are 95 men for every 100 women (fewer males than females). Worldwide, without human intervention, 102 to 107 boys are born for every 100 girls. In the United States, the sex ratio is 105 at birth, but skewed in many countries: 115 in Curaçao, 114 in Armenia, 112 in India and Vietnam, 111 in China and Albania, and as high as 138 in some of these countries' rural regions (Gilles and Feldman-Jacobs, 2012; *CIA World Factbook*, 2014).

These sex ratio imbalances are attributed primarily to *female infanticide* (sometimes called *gendercide*)—the intentional killing of baby girls. In many Asian countries, including China and India, there's a preference for boys because males are expected to carry on the family name, care for elderly parents, inherit property, and play a central role in family rituals. Consequently, and particularly in rural areas, hundreds of thousands of female infants die every year because of neglect, abandonment, and starvation. Others are aborted after ultrasound scanners reveal the child's sex ("Sex-Selective Abortion…," 2013).

China is a good example of some of the unintended negative consequences of lopsided sex ratios. In 1979, China passed a one-child policy to combat poverty and overpopulation. The policy succeeded in limiting population growth because of at least 600 million abortions and sterilizations. The country's TFR fell from 6 in the late 1960s to 1.5 by 2010, but well below the 2.1 replacement rate to maintain a constant population (Eberstadt, 2013; "Family Planning…," 2014).

Because of the unbalanced sex ratio for the last 30 years, by 2020 China will have about 35 million more young men than women. That means a large number of men won't be able to find wives—a problem that has already increased the illegal trafficking of women from poorer neighboring countries, including Cambodia, Myanmar, and Vietnam—and there'll be millions of aging bachelors (Tsai, 2012; Larson, 2014).

To reduce the burden of elderly care, the government eased the policy in 2013 by allowing married couples to have two children if one spouse was an only child. The urban, educated middle class—the group that China's leaders want to see increase its family size—hasn't shown much interest in doing so because of high costs of living and expenses in raising a second child. Thus, China's working-age population is expected to start declining rapidly in 2015, resulting in fewer workers to support a growing elderly population (Beech, 2013; Burkitt, 2013; "Reforming the One-Child Policy...," 2013).

POPULATION PYRAMIDS

A **population pyramid** is a graphic depiction of a population's age and sex distribution at a given point in time. As *Figure 15.2* shows, Mexico is a young country: Much of its population is under age 45 (which also means that many women are in their childbearing years), and there are relatively few people 65 years and older. In contrast, Italy is an old country, and the United States is somewhere in the middle.

The shape of the pyramid (a triangle for Mexico, a rectangle for the United States, and a diamond for Italy) has future implications for young and old countries. Italy has a relatively small number of women ages 15 to 44 (in their reproductive years) and a bulge of people ages 45 to 79. This suggests that there'll be a scarcity of workers to support an aging population and a greater need for social services for older people than for children and adolescents (Kochhar and Oates, 2014; see also Chapter 12). Thus, population pyramids give us a snapshot of a country's demographic profile and indicate some of the problems that countries are likely to face in the future.

15-1c Population Growth: A Ticking Bomb?

Some of the world's largest countries, many of them in the developing world, will grow even more by 2050 (see *Table 15.1* on p. 296). So, has population growth gotten out of hand? There are many views on this question, but

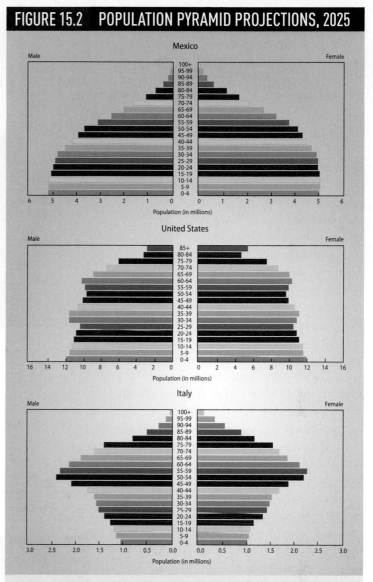

FIGURE 15.2   POPULATION PYRAMID PROJECTIONS, 2025

Source: U.S. Census Bureau, International Database, www.census.gov/ipc/www/idb /pyramids.html. Accessed January 20, 2007.

two of the most influential have been Malthusian theory (which argues that the world can't sustain its unprecedented population surge), and demographic transition theory (which maintains that population growth is slowing).

MALTHUSIAN THEORY

For many demographers, population growth is a ticking bomb. They subscribe to **Malthusian theory**, which maintains that the population is growing faster than the food supply needed to sustain it. This theory is named

population pyramid   a graphic depiction of a population's age and sex distribution at a given point in time.

Malthusian theory   maintains that population is growing faster than the food supply needed to sustain it.

after Thomas Malthus (1766–1834), an English economist, clergyman, and college professor who argued that humans are multiplying faster than the earth's ability to produce sufficient food.

According to Malthus (1798/1965), population grows at a *geometric rate* (2, 4, 8, and so on), whereas the food supply grows at an *arithmetic rate* (1, 2, 3, 4, and so on). That is, two parents can have 4 children and 16 grandchildren within 50 years. The available number of acres of land, farm animals, and other sources of food can increase in that time period, but certainly not quadruple. In effect, then, the food supply will not keep up with population growth. Because there are millions of parents, the results could be catastrophic, with masses of

Hulton Archive/Stringer/Getty images

people living in poverty or dying of starvation.

Malthus posited that two types of checks affect population size. *Positive checks* (famine, disease, war) limit reproduction, raise the death rate, and lower the overall population. *Preventive checks* (contraception, postponing marriage, abortion, premarital and extramarital sex) also limit reproduction by reducing birth rates and, consequently, ensure a higher standard of living for all (Malthus, 1872/1991).

Malthusian theory has had a lasting influence. *Neo-Malthusians* (or New Malthusians) agree that the population is exploding beyond food supplies. The world's population reached its first billion in 1800. In the 200 years that followed, the world added another 5 billion people (see *Figure 15.1* on p. 295). As a result of this growth, according to some influential neo-Malthusians, the earth has become a "dying planet"—a world with insufficient food and a rapidly expanding population that pollutes the environment (Ehrlich, 1971; Ehrlich and Ehrlich, 2008).

An estimated one in eight people in the world suffer from chronic hunger that prevents them from working and having a normal life. Resources and technical knowledge are available to increase food production by 70 percent

by 2050, but poverty and environmental problems plague the countries that need food the most (FAO et al., 2013).

DEMOGRAPHIC TRANSITION THEORY

Some demographers are more optimistic than neo-Malthusians. **Demographic transition theory** maintains that population growth is kept in check and stabilizes as countries experience economic and technological development, which, in turn, affects birth and death rates. According to this theory, population growth changes as societies undergo industrialization, modernization, technological progress, and urbanization. During these processes, a nation goes through four stages (see *Figure 15.3*), from high birth and death rates to low birth and death rates.

▶ **Stage 1: Preindustrial society.** In this initial stage, there's little population growth. The birth rate is high because people rarely use birth control: They want as many children as possible to provide unpaid agricultural labor and support parents in old age, but a high death rate offsets the high birth rate. No country exists in this current stage of the demographic transition.

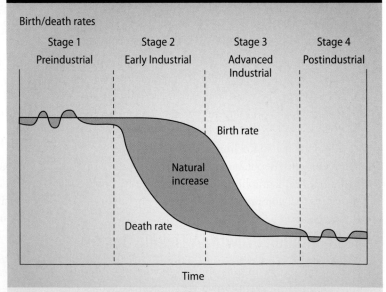

FIGURE 15.3 THE CLASSICAL DEMOGRAPHIC TRANSITION MODEL

Birth/death rates

| Stage 1 | Stage 2 | Stage 3 | Stage 4 |
| Preindustrial | Early Industrial | Advanced Industrial | Postindustrial |

Birth rate

Natural increase

Death rate

Time

Source: U.S. Census Bureau, International Database, www.census.gov/ipc/www/idb/pyramids.html. Accessed January 20, 2007.

demographic transition theory maintains that population growth is kept in check and stabilizes as countries experience economic and technological development.

▶ **Stage 2: Early industrial society.** There's significant population growth because the birth rate is higher than the death rate. The birth rate may even increase over what it was in Stage 1 because mothers and their children enjoy improved health care. Couples may still have large numbers of children because they fear that many of them will die, but the death rate declines because of better sanitation, better nutrition, and medical advances (e.g., immunizations and antibiotics). Most of the world's poorest countries—including sub-Saharan Africa, Afghanistan, and Yemen—are currently in Stage 2.

▶ **Stage 3: Advanced industrial society.** As the infant mortality rate declines, parents have fewer children. Effective birth control reduces family size. The decrease in child care responsibilities, in turn, enables women to work outside the home. Brazil, Mexico, and some Asian countries are currently in Stage 3.

▶ **Stage 4: Postindustrial society.** In this stage, the demographic transition is complete, and the society has low birth and death rates. Women tend to be well educated and to have full-time jobs or careers. If there's little immigration, the population may even decrease because the birth rate is low. This is the case today in Canada, Japan, Singapore, Hong Kong, Australia, New Zealand, the United States, and Europe.

CRITICAL EVALUATION

The dire predictions of Malthus and his successors that global population growth would lead to worldwide famine, disease, and poverty haven't come true. Still, 1.2 billion people (22 percent of the world's population) live in abject poverty, subsisting on less than $1.25 a day (United Nations Development Program, 2014; Malik, 2014; see also Chapter 8).

Despite neo-Malthusians' fears, global fertility is half of what it was in 1972. The population of some industrialized countries is declining because people aren't having enough babies to replace themselves. These countries are experiencing **zero population growth (ZPG)**, a stable population level that occurs when each woman has no more than two children.

Fearing that there won't be enough young workers to pay for social security and the rising cost of health care for aging populations, some low-birth nations with fertility rates below ZPG are paying women to have more children. Russia gives mothers with one child $12,500 for each additional baby; Japan has expanded its day care facilities and offers families a monthly allowance of $145 per child younger than 15; Germany and France have liberal parental leave; South Korea turns off its office lights once a month so staff have more time to devote to "childbirth and upbringing," and offers other incentives, including cash gifts, for staff who produce more than two babies; and China is encouraging married couples to have two children ("S. Korea Orders...," 2010; Hales, 2014; Nechepurenko, 2014).

Some neo-Malthusians maintain, however, that it's irresponsible for *any* country to encourage higher fertility rates. They worry about the consequences of adding 3 billion more inhabitants to the planet in less than 50 years, especially for many developing countries "with desperate economic outlooks" (Sachs, 2005; Shorto, 2008).

One result of population growth is urban growth. Cities attract people because of jobs and cultural activities, but urbanization has also created numerous problems.

15-2 URBANIZATION

If you've flown over the United States, you've probably noticed that people tend to cluster in and around cities. After sunset, some areas glow with lights, whereas others are engulfed in darkness. The average person, in the United States and worldwide, is more likely to live in a city than a rural area, and this trend is rising.

A *city* is a geographic area where a large number of people live relatively permanently and make a living primarily through nonagricultural activities. **Urbanization**, which increases the size of cities, is the movement of people from rural to urban areas. Most of this discussion focuses on U.S. cities, but let's begin with a brief look at global urbanization.

15-2a Urbanization: A Global View

In 2008, for the first time in history, a majority of the world's population lived in urban areas. By 2050, 66 percent of the world's population is projected to be urban (United Nations, 2014).

zero population growth (ZPG) a stable population level that occurs when each woman has no more than two children.

urbanization people's movement from rural to urban areas.

During the Industrial Revolution, many cities, like this one in Hamburg, Germany, constructed canals that provided an inexpensive means of transportation and distributing goods.

North Wind Picture Archives

TABLE 15.2 URBANIZATION AROUND THE WORLD

Percentage of People Living in Urban Areas

	1950	2014	2050 (Projected)
World	**29**	**54**	**66**
Africa	14	40	56
Asia	18	48	64
Latin America and the Caribbean	41	80	86
North America	64	81	87
Europe	51	73	82
Oceania	62	71	74

Source: Based on United Nations Department of Economic and Social Affairs 2012, Table 2, and United Nations 2014, Table 1.

ORIGIN AND GROWTH OF CITIES

Cities are one of the most striking features of modern life, but they've existed for centuries. About 7,000 years ago, for example, people built small cities in the Middle East and Latin America to protect themselves from attackers and to increase trade. By 1800, 56 cities in Western Europe had a population of 40,000 or more (Chandler and Fox, 1974; De Long and Shleifer, 1992).

Before the Industrial Revolution, which began in the late eighteenth century, urban settlements in Europe, India, and China developed largely because people figured out how to use natural resources (e.g., mining coal and transporting water for irrigation and consumption). The Industrial Revolution spurred ever-increasing numbers of people to move to cities in search of jobs, schooling, and improved living conditions. As a result, the urban population surged—from 3 percent of the world's population in 1800 to 14 percent in 1900 (Sjoberg, 1960; Mumford, 1961). As industrialization advanced, urbanization increased.

WORLD URBANIZATION TRENDS

Between 1920 and 2014, the world's urban population increased from 270 million to 3.9 billion, and is expected to rise to 6.3 billion by 2050. Africa and Asia are home to nearly 90 percent of the world's rural population, but both are urbanizing faster than other regions (see *Table 15.2*). Between 2014 and 2050, just three

countries—India, China, and Nigeria—are expected to account for 37 percent of the world's urban population growth (United Nations, 2014).

Many of the world's largest cities are **megacities**, metropolitan areas with at least 10 million inhabitants. In 1950, the world had only two megacities: Tokyo (11.3 million) and New York-Newark (12.3 million). By 2030, there'll be 41 megacities. Only two of them will be in the United States—New York-Newark (19.9 million), and Los Angeles-Long Beach-Santa Ana (13.3 million)—and both will be much smaller than Tokyo (37.2 million), Delhi (36.1 million), Shanghai (30.8 million), and other Asian cities. Thus, the number of megacities is growing, and they're much bigger than in the past (United Nations, 2014).

Should the explosive growth of cities and megacities concern us? Generally, cities provide jobs, offer better health services, and have more educational opportunities, but not everyone benefits from such advantages. The urban poor are often crowded into slums where children are less likely to be enrolled in school, sanitation is inadequate, and the economic gap between haves and have-nots is widening (Laneri, 2011).

15-2b Urbanization in the United States

Like many other countries, the United States is becoming more urban. How, specifically, has the urban landscape changed? And what are some of consequences?

HOW URBAN AMERICA IS CHANGING

During the Industrial Revolution, millions of Americans in agricultural areas migrated to cities to find jobs. As a result, between 1900 and 2000, the urban population

megacities metropolitan areas with at least 10 million inhabitants.

MANAN VATSYAYANA/AFP/Getty Images

surged from 39 to 79 percent (Riche, 2000; U.S. Census Bureau, 2010).

Suburbanization. As cities grew more crowded, dirtier, and noisier, urban growth sparked *suburbanization*, people's moving to communities just outside a city. In 1920, about 15 percent of Americans lived in the suburbs compared with more than half today (Palen, 2014).

During the 1950s, two-thirds of urban dwellers moved to suburbs. The federal government, fearing a return to the economic depression of the 1930s, underwrote the construction of much new housing in the suburbs. The general public obtained low-interest mortgages, veterans were offered the added incentive of being able to purchase a home with a $1 down payment, and massive highway construction programs enabled commuting by car. As a result, suburbs mushroomed (Rothman, 1978).

Suburban life is appealing. Some of the benefits include more privacy and space, one's own yard and garages, better schools, safer streets, and, if there are jobs nearby, shorter commutes. Suburban elites have sprawling "McMansions," considerable acreage, and gated communities. Especially since 1980, however, much of the social and physical separation between cities and suburbs has blurred. One in three poor Americans now live in suburbs and crime rates, although lower than in cities, have increased (Kneebone and Berube, 2013). Suburbs are also attracting fewer young adults, especially those with young children. This means that many communities no longer have the taxes that support schools, services for older people, and public recreational facilities (Frey, 2011; El Nasser and Overberg, 2011). For these and other reasons, as you'll see shortly, many big cities are growing faster than suburbs.

Edge Cities and Exurbs. Originally, most suburbs were bedroom communities for commuters with jobs in the city. Over the last few decades, suburbanization has generated **edge cities**, business centers that are within or close to suburban residential areas and include offices, schools, shopping, entertainment, malls, hotels, and medical facilities.

People have also created **exurbs**, areas of new development beyond the suburbs that are more rural but on the fringe of urbanized areas. About 6 percent of Americans live in exurbs. The average exurbanite is white, a middle-income earner, married with children, a "super commuter" (one who travels two or more hours a day for work), and owns a large house outside of an expensive metropolitan suburb (Berube et al., 2006; Lalasz, 2006).

Metropolitan Statistical Areas. Together, suburbs, edge cities, and exurbs form **metropolitan statistical areas** (MSAs), also called *metro areas*, that consist of a central city of at least 50,000 people and the urbanized areas linked to it. The vast majority of Americans (84 percent) live in 366 MSAs. With a population of nearly 20 million, New York-Newark-Jersey City is the largest MSA; Carson City, Nevada, with a population of 55,000, is the smallest. Between 2000 and 2013, the fastest-growing metro areas were in the South and West (e.g., Dallas-Fort Worth, Atlanta, Los Angeles), but six were in the Midwest (e.g., Bismarck, North Dakota, and Casper, Wyoming), where mining, quarrying, and oil and gas extraction fueled job opportunities (Wilson et al., 2012; U.S. Census Bureau Newsroom, 2014).

Since 2011, and for the first time in nearly a hundred years, metro areas—particularly those with more than 1 million people—grew faster than suburbs (Frey, 2014). Two groups—young professionals and Baby Boomers—have driven the trend in metro living. Because many young professionals are postponing marriage and parenthood (see Chapter 12), they don't

edge cities business centers that are within or close to suburban residential areas.

exurbs areas of new development beyond the suburbs that are more rural but on the fringe of urbanized areas.

metropolitan statistical area (MSA, also called *metro area*) a central city of at least 50,000 people and urban areas linked to it.

Micro-apartments, sometimes smaller than college dorm rooms, are cropping up in many U.S. metro areas to accommodate the growing number of single professionals, students, and older people. The rents are high, however. In San Francisco and other metro areas, renting a 17-by-17-foot apartment ranges from $939 to $1,500 a month (Wong, 2013).

have to worry about the quality of schools. Moreover, many work in cities, want to reduce commute times, and can't afford the down payment for a house in the suburbs. A number of Baby Boomers, particularly those who are retiring, are moving to metro areas for a variety of reasons (e.g., better public transportation, greater access to cultural events, social services for older people, no longer being healthy enough to maintain a suburban house) (Kneebone and Berube, 2013; Westcott, 2014).

SOME CONSEQUENCES OF URBANIZATION

Cities offer many benefits, including a vast array of culturally diverse restaurants, shops, and activities, but urbanization also creates problems. Some of the drawbacks include urban sprawl, increased traffic congestion, a scarcity of affordable housing, and residential segregation.

Urban Sprawl. **Urban sprawl**—the rapid, unplanned, and uncontrolled spread of development into regions adjacent to cities—is widespread. According to some

> **urban sprawl** the rapid, unplanned, and uncontrolled spread of development into regions adjacent to cities.
>
> **gentrification** a process in which upper-middle-class and affluent people buy and renovate houses and stores in downtown urban neighborhoods.

estimates, urbanization in the Southeast will increase by up to 190 percent by 2060. Development on that scale will result in losing 15 percent of agricultural land, 12 percent of grasslands, and 10 percent of forests (Terando et al., 2014).

Urban sprawl has created rapid *job sprawl*, which occurs when companies move jobs from metropolitan areas to suburbs. The more distant the suburb, the less likely low-income people are to hear about employment opportunities through informal networks, to afford houses in these areas, and to have transportation to the jobs (Raphael and Stoll, 2010).

Some large metro areas (e.g., Washington, D.C., New York, Boston, Atlanta, Miami, and Denver) are implementing "walkable urbanism" to curb urban sprawl. That is, the developments are neighborhoods where everyday destinations (e.g., work, school, restaurants) are concentrated and within walking distances. For the most part, however, most urban development is built around driving rather than walking (Leinberger and Lynch, 2014).

Traffic Congestion. Generally, the only way to get around in urban sprawl areas is by automobile. This means that most suburban households face the costs of buying, fueling, insuring, and maintaining multiple cars. The U.S. population has increased by 23 percent over the last 25 years, but total highway miles have increased by only 5 percent, resulting in greater traffic congestion within and outside cities. More than 10 million Americans (8 percent of all people who don't work at home) now travel 60 minutes or longer one-way, a proportion that has increased by 95 percent since 1990. Traffic snarls and long commutes increase air pollution and stress and decrease the time that people have for family and leisure activities (McKenzie, 2013).

Lack of Affordable Housing. In some cases, the poor are pushed out by **gentrification**, a process in which upper-middle-class and affluent people buy and renovate houses and stores in downtown urban neighborhoods. Governments in many older cities encourage gentrification to increase dwindling populations, to revitalize urban areas, and to augment tax revenues. Rent increases, however, have displaced many low-income residents and small businesses.

Residential Segregation. Since 1970, racial residential segregation has decreased. Among metropolitan areas with a population of 500,000 or more, the least segregated are in the South and West, and the most segregated are mainly in the Northeast and Midwest. There may be a "continued easing of the color line" because racial segregation has waned even in long-standing "hypersegregated" cities like Chicago and New York. The nation's

Gentrification improves old city neighborhoods and increases property values, but also displaces low-income residents.

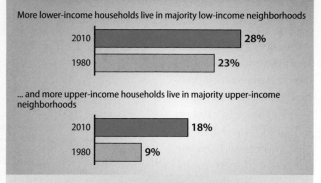

FIGURE 15.4   LOWER-INCOME AND UPPER-INCOME HOUSEHOLDS WHO LIVE MAINLY AMONG THEMSELVES, 1980 AND 2010

More lower-income households live in majority low-income neighborhoods

2010   28%

1980   23%

... and more upper-income households live in majority upper-income neighborhoods

2010   18%

1980   9%

Note: Lower-income households had an annual income of less than $34,000 (in 2010 dollars); upper-income households had an annual income of $104,000 and above.

Source: Based on Fry and Taylor 2012: 1.

20 most multiethnic metropolitan regions are now "global neighborhoods" that have substantial numbers of whites, blacks, Latinos, and Asians (Scommegna, 2011; Iceland et al., 2013: 119).

Whereas racial residential segregation has decreased, social class residential segregation has increased (see *Figure 15.4*). The share of families living in middle-income neighborhoods dropped from 65 to 42 percent between 1970 and 2009. The residential segregation gap between affluent and all other social classes is more pronounced than in the past, and the most residentially segregated are low-income blacks and Latinos (Fry and Taylor, 2012; Bischoff and Reardon, 2013).

Residential segregation along income lines has increased primarily because of the long-term rise in income inequality (see Chapters 8 and 11). The geographic isolation of affluent neighborhoods has led to their residents' lower investments in resources and services, including public schools and parks, that would benefit more people. Some wealthy neighborhoods

have even fought off state laws that mandate constructing affordable housing for low- and middle-income families. According to the high-income homeowners, building less expensive houses would raise crime rates, lower property values, and "There's plenty of affordable housing *in neighboring communities*" (McCabe, 2014: 39, emphasis added).

15-2c Sociological Explanations of Urbanization

How and why do cities change? In answering this and other questions, functionalists underscore urban development, conflict theorists emphasize the impact of capitalism and big business, feminist scholars focus on women's safety and space, and symbolic interactionists examine the quality of city life (*Table 15.3* summarizes these perspectives).

TABLE 15.3   SOCIOLOGICAL EXPLANATIONS OF URBANIZATION

Perspective	Level of Analysis	Key Points
Functionalist	Macro	Cities serve many important social and economic functions, but urbanization can also be dysfunctional.
Conflict	Macro	Driven by greed and profit, large corporations, banks, developers, and other capitalist groups shape cities' growth or decline.
Feminist	Macro and micro	Whether they live in cities or suburbs, women generally experience fewer choices and more constraints than men.
Symbolic Interactionist	Micro	City residents differ in their types of interaction, lifestyles, and perceptions of urban life.

FUNCTIONALISM: HOW AND WHY CITIES CHANGE

Urban ecology studies the relationships between people and urban environments. Over the years, social scientists have revised urban ecology theories (see *Figure 15.5*).

Sociologists Robert Park and Ernest Burgess (1921) proposed *concentric zone theory* to explain the distribution of social groups within urban areas. According to this model, a city grows outward from a central point in a series of rings. The innermost ring, the central business district, is surrounded by a zone of transition, which contains industry and poor-quality housing. The third and fourth rings have housing for the working and middle classes. The outermost ring is occupied by people who live in the suburbs and commute daily to work in the central business district.

Economist Homer Hoyt's (1939) *sector theory* refined concentric zone theory. He proposed that cities develop in sectors (instead of rings) that radiate from the central business district depending on various economic and social activities. Some sectors are predominantly industrial, some contain stores and offices, and others, generally farther away from the central business district, are middle- and upper-class residential areas.

Geographers Chauncey Harris and Edward Ullman (1945) developed another influential model, *multiple-nuclei theory*, which proposed that a city contains more than one center around which activities revolve. For example, a "minicenter" often includes an outlying business district with stores and offices that are accessible to middle- and upper-class residential neighborhoods, whereas airports typically attract hotels and warehouses.

As cities grew after World War II, these models no longer described urban spaces. Thus, Harris (1997) proposed a *peripheral* theory of urban growth, which emphasized the development of suburbs around a city but away from its center. As suburbs and edge cities burgeon, highways that link the city's central business district to outlying areas and beltways that loop around the city provide relatively easy access to airports, the downtown, and surrounding areas.

Functionalists also examine urban dysfunctions, including overcrowding, poverty, deviant behavior, and environmental destruction. From 2000 to 2010, for instance, the number of vacant housing units—many of them in older industrial cities like Baltimore and Detroit—increased by 4.5 million, or 44 percent. Abandoned buildings can increase crime rates and a neighborhood's social disorganization (Williams, 2013).

CONFLICT THEORY: THE IMPACT OF CAPITALISM AND BIG BUSINESS

For functionalists, people's choices shape urban changes. In contrast, **new urban sociology**, a perspective heavily influenced by conflict theory, views urban changes as largely the result of decisions made by powerful capitalists and high-income groups. That is, economic and political factors favorable to the rich, not ordinary citizens, determine urban growth or decline. When a local government wants to rejuvenate parts of the inner city, it typically offers tax breaks, changes zoning laws, and allows real estate, construction, and banking industries to seek profits with little regard for the needs of low-income households or

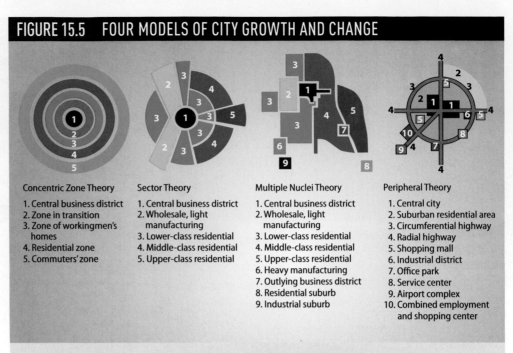

FIGURE 15.5   FOUR MODELS OF CITY GROWTH AND CHANGE

Concentric Zone Theory
1. Central business district
2. Zone in transition
3. Zone of workingmen's homes
4. Residential zone
5. Commuters' zone

Sector Theory
1. Central business district
2. Wholesale, light manufacturing
3. Lower-class residential
4. Middle-class residential
5. Upper-class residential

Multiple Nuclei Theory
1. Central business district
2. Wholesale, light manufacturing
3. Lower-class residential
4. Middle-class residential
5. Upper-class residential
6. Heavy manufacturing
7. Outlying business district
8. Residential suburb
9. Industrial suburb

Peripheral Theory
1. Central city
2. Suburban residential area
3. Circumferential highway
4. Radial highway
5. Shopping mall
6. Industrial district
7. Office park
8. Service center
9. Airport complex
10. Combined employment and shopping center

Source: Based on Park and Burgess 1921, Hoyt 1939, Harris and Ullman 1945, and Harris 1997.

urban ecology studies the relationships between people and urban environments.

new urban sociology view that urban changes are largely the result of decisions made by powerful capitalists and high-income groups.

the homeless (Feagin and Parker, 1990; Macionis and Parrillo, 2007).

For conflict theorists, urban space is a commodity that's bought and sold for profit. It's not the average American, they argue, but bankers, developers, politicians, and influential businesspeople who determine how to use urban space. Because increasing a property's value is a higher priority than community needs, poor and low-income people are crowded into dilapidated neighborhoods (Logan and Molotch, 1987; Gottdiener and Hutchison, 2000).

Some analysts are especially critical of the tech sector for being "bad urbanists." To illustrate, in 2011 Twitter received generous tax incentives to move its new headquarters into one of San Francisco's poorest neighborhoods. City officials expected that the company's presence would revitalize the community. Instead, Twitter employees eat all their meals in the dining area, rarely leave the building to shop because small businesses can't afford the area's skyrocketing rents, and the neighborhood's evictions increased 38 percent by 2013 because longtime residents couldn't pay the soaring rent increases (Arieff, 2013; Butler, 2014; Steinmetz, 2014).

FEMINIST THEORIES: GENDER ROLES, SPACE, AND SAFETY

Feminist theories emphasize gender-related constraints. Whether they live in cities or suburbs, women generally experience more problems than men because living spaces are usually designed by men who have tended to ignore women's changing roles.

Many women fear the city, especially urban public spaces like streets, parks, and public transportation

Slums are one of the most dysfunctional by-products of urbanization. Megacities—particularly those in India, Egypt, Pakistan, Kenya, and Mexico—have the world's largest slums.

(Domosh and Seager, 2001). They see these places as risky for their physical safety, despite the fact that most violence against women occurs at home. A few cities provide public transportation (e.g., minivans that operate seven nights a week) to prevent crimes against women, usually minority women, who must travel to work after 8:00 p.m. and return home before dawn. For the most part, however, such services are rare (Hayden, 2002).

In the suburbs, both women's and men's physical mobility is limited because of the scarcity of public transportation systems. Consequently, many suburban households have two or more cars, but women living in the suburbs usually experience greater problems than men. For example, there may be fewer job opportunities (especially if mothers want to be at home when children return from school), and women often spend much of their time maintaining a single-family home, decreasing their time for educational or leisure activities (Cichocki, 1981; Hayden, 2002).

SYMBOLIC INTERACTION THEORY: HOW PEOPLE EXPERIENCE CITY LIFE

Symbolic interactionists are most interested in the impact of urban life on city residents. In a classic essay, sociologist Louis Wirth (1938: 14) described the city as a place where "our physical contacts are close, but our social contacts are distant."

Wirth defined the city as a large, dense, and socially and culturally diverse area. These characteristics produce *urbanism*, a way of life that differs from that of rural dwellers. Wirth saw urbanites as more tolerant than residents of small towns or rural areas of a variety of lifestyles, religious practices, and attitudes.

He also believed that urbanism has negative consequences, including alienation, friction because of physical congestion, impersonal relationships, and a disintegration of kinship and friendship ties. Some studies have supported Wirth's theory of urbanism (see Guterman, 1969), but others have challenged his views. In a recent national study, people living in large metropolitan areas scored higher than those living in small towns and rural areas on physical and emotional health, access to basic necessities, and being satisfied with life (Witters, 2010).

CRITICAL EVALUATION

Functionalists tend to overlook urbanization's negative political and economic impact, especially when profit and greed guide urban planning. Conflict theory seems to assume that residents are helpless victims as developers and corporations raze low-income houses. In fact, environmental groups have had considerable success in pushing through legislation to maintain and even increase open public spaces and build energy-saving homes in low-income neighborhoods (Moore, 2008).

Feminist sociologists have made important contributions through studies of the everyday lives of low-income women, especially in central cities (see Chapter 11), but urbanization has received much less attention. Urbanites are more diverse than some symbolic interactionists claim. People living in cities aren't necessarily more self-centered or isolated than those in small towns or rural areas. Instead, many have close family bonds, friends, and satisfying relationships with coworkers (Crothers, 1979; Wilson, 1993). Also, symbolic interaction doesn't show how political, educational, religious, and economic factors shape people's experiences of city life (Hutter, 2007).

You've seen that the world's population is growing rapidly and becoming more urbanized. Both population growth and urbanization are taxing the planet's limited resources.

Among developing countries, 25 percent of girls aren't in school compared with 17 percent of boys. One reason for this difference is that girls are more likely than boys to be responsible for collecting the family's water, making it difficult for them to attend school (Global WASH Fast Facts, 2013).

15-3 THE ENVIRONMENT

Consider the following:

▶ Each year, Americans throw away 40 percent of their food—50 percent more than in the 1970s—that ends up in landfills (Gunders, 2012).

▶ Up to 3.5 million Americans get sick every year from polluted beach water that causes a wide range of diseases, including ear-nose-eye infections, hepatitis, skin rashes, and respiratory illnesses (Dorfman and Haren, 2014).

▶ The United States uses 1.2 billion pounds of pesticides a year, but only 0.01 percent reaches the intended target. The other 99.99 percent contaminates the food, air, and water (Hsu et al., 2014).

Such environmental problems threaten our **ecosystem**, a community of living and non-living organisms that share a physical environment. Plants, animals, and humans depend on each other for survival. Because the ecosystem is interconnected worldwide, what happens in one country affects others. Water, air pollution, and climate change are three interrelated factors that are endangering the global ecosystem.

15-3a Water

An expanding world population, extreme weather patterns, and industrial pollution are jeopardizing already limited water supplies. More than 768 million people worldwide don't have clean water, and 1.8 billion drink contaminated water. Contaminated water, inadequate availability of water, and lack of access to sanitation together contribute to 88 percent of deaths from diarrheal diseases. In some developing countries, families spend up to 25 percent of their income to purchase water, and many women and children spend up to 6 hours a day carrying it home (Prüss-Üstün et al., 2008; World Health Organization and UNICEF, 2014; WWAP, 2014).

AVAILABILITY AND CONSUMPTION

Just 3 percent of the earth's water is fresh, and two-thirds of it is locked up in the ground, glaciers, and ice caps. That leaves about 1 percent of the earth's water for the world's more than 7 billion people. The world's demand for water has tripled over the last half century. By 2030, total global water supply will meet just 60 percent of the demand. By 2050, more than 40 percent of the global population is projected to be living in areas with severe water shortages (Wheeler, 2012; World Health Organization and UNICEF, 2014; WWAP, 2014).

Some refer to water as "blue gold" because it's becoming one of the earth's most precious and scarce commodities. Water shortages—rather than oil or diamonds—are behind conflicts and even wars in a number

ecosystem a community of living and non-living organisms that share a physical environment.

How Much Water Does It Take to Make...

In 1950, Americans used 150 billion gallons of water every day compared with 400 billion gallons today. It takes water to make everything. For example, it takes:

2,600 gallons to make a pair of blue jeans

© Jacob Kearns/Shutterstock.com

713 gallons to make a cotton shirt

© Stocksnapper/Shutterstock.com

634 gallons to make an average hamburger

© Christopher Elwell/Shutterstock.com

53 gallons to produce a cup of to-go latte coffee

magicoven/Shutterstock.com

1.5 gallons to produce an average 18 ounces of bottled water

© Evgeny Karandaev/Shutterstock.com

Sources: Based on Connell 2011; Postel 2012.

---

of countries. According to a past Secretary-General of the United Nations, "Too often, where we need water, we find guns..." (World Water Assessment Program, 2009: 20; Reynolds, 2012).

Industrialized nations not only have greater access to clean water than the developing world, they use more and pay less for it. The average person in the United States uses about 163 gallons of water per day compared with 40 gallons in Germany, 30 in Denmark, 23 in China, and less than 3 in Mozambique. On average, Americans pay only $.48 per gallon compared with $1.65 a gallon in Denmark. In the developing world, people typically pay five times as much as Europeans (Maxwell, 2012). Thus, in many countries, clean water is a luxury rather than a basic human right.

THREATS TO WATER SUPPLIES

Precipitation (in the form of rain, snow, sleet, or hail) is the ecosystem's main source of water. Clean water has been depleted for many reasons, including pollution, privatization, and waste.

Pollution. Every year, more than 860 billion gallons of sewage, pesticides, fertilizers, automotive chemicals, and trash spoil the country's freshwater. Industrial sites and farm fertilizers are two of the major pollution sources. As a result, 20 percent of the nation's lakes and 55 percent of its rivers and streams can't provide drinking water or support healthy aquatic life (EPA, 2013, 2014b).

In 2014, a chemical spill in West Virginia left 300,000 people without access to running water and unable to use their faucets to drink, cook, or bathe for five days. The chemical industry maintains that such spills are rare, but they occur every day. In 2013, for instance, 76 U.S. companies had almost 3,900 spills. From 2001 to 2010, there were 992 oil and gas fluid spills in three Colorado counties alone (Chemi, 2014; Walsh, 2014).

Few corporate polluters are fined, and many are exempt from major environmental laws, including the 1974 Safe Water Drinking Act. Consider hydraulic fracturing, or *fracking*, a process that injects water, sand, and chemicals at high pressure to extract gas and oil from rock that lies deep underground. Many U.S. lawmakers—some of them with deep ties to industry that finances political campaigns—endorse fracking to increase energy supplies (Blake, 2014).

Fracking has risks because gases can escape into drinking water. Scientists have linked some of the 750 chemicals injected into the ground with higher risks of cancer, infertility, and other health problems (Jackson et al., 2013; Kassotis et al., 2014). Others suspect that fracking has increased the number and magnitude of earth tremors from Colorado to the Atlantic coast. Cities and counties nationwide have passed 430 measures to control or ban fracking, but companies typically prevail despite opposition (Keranen et al., 2014; Shauk and Olson, 2014).

Privatization. Water is a big business because of *privatization*, transferring some or all of the assets or operations of public systems into private hands. Perrier, Evian, Coca-Cola, PepsiCo—and particularly the French giants Vivendi and Suez—have been buying the rights to extract water in the United States and other countries at will from aquifers (underground layers of rock that hold water), then bottling and selling it around the world.

About 70 percent of Americans say they drink bottled water (Mui, 2009). Bottling water is lucrative for corporations, but there are many environmental drawbacks. It depletes local water supplies, whether the water comes from municipal sources (40 percent) or local springs (60 percent). Moreover, about 86 percent of the empty plastic bottles in the United States clog landfills instead of being recycled (Velasquez-Manoff, 2009; Food & Water Watch, 2013).

Waste. Most water problems are due to human mismanagement and waste, not nature. Of all available water worldwide, agriculture consumes about 70 percent, industry uses 20 percent, and 10 percent is residential (WWAP, 2014). In agriculture, many irrigation systems are inefficient, farmers and agribusiness (large agriculture companies) often grow water-hungry crops like cotton and sugarcane in arid areas, and pesticide and chemical fertilizer runoff from fields pollutes streams, rivers, and lakes. Some Americans are trying to conserve water, but they're a minority. For example, nearly 75 percent of residential water use in California—which has experienced numerous water emergencies—goes to outdoor purposes, mostly landscaping (Goodale, 2009; Pickert, 2014).

In the United States, a significant water pipe bursts, on average, every 2 minutes somewhere in the country, losing 7 billion gallons of water a day. Most of the nation's pipes, especially in cities along the eastern seaboard, are nearly 200 years old, and some are made of wood. A water pipe that leaks or bursts wastes huge amounts of drinking water, damages streets and homes, and seeps dangerous pollutants into drinking water. Despite such hazards, replacing a city's substandard pipes has decreased. From 1890 to 1920, for instance, Chicago replaced aging water and sewage pipes at a rate of about 75 miles a year. That number has declined steadily over the years, dropping to 30 miles a year in 2003 (Duhigg, 2010; Pearlstine, 2013).

gmuttu/iStockphoto.com

15-3b Air Pollution, Global Warming, and Climate Change

Global warming is a serious environmental problem. Let's begin with air pollution, the major cause of climate change and global warming.

AIR POLLUTION: SOME SOURCES AND CAUSES

There are many reasons for air pollution, but four are among the most common. First, a major source of air pollution is the burning of *fossil fuels*, substances obtained from the earth, including coal, petroleum, and natural gas. The exhaust gases of cars, trucks, and buses contain poisons—sulfur dioxide, nitrogen oxide, carbon dioxide (CO_2), and carbon monoxide. Power plants that produce electricity by burning coal or oil also spew pollutants.

Second, manufacturing plants that produce consumer goods pour pollutants into the air. Formaldehyde-based vapors that can lead to cancer and respiratory problems are emitted by many household and personal care products: Pressed wood (often used in furniture), plastic grocery bags, waxed paper, latex paints, detergents, nail polish, cosmetics, and shampoos ("Formaldehyde," 2012; see also Chapter 14).

Third, winds blow contaminants in the air across borders and oceans. Air pollution from Asia, particularly China, has affected the air quality in the Sequoia and Kings Canyon national parks in California. Air pollution originating in Europe has been tracked to Asia, the Arctic, and even rural areas in the western United States where there's little industry or automobile traffic (Lamb, 2009; Cooper et al., 2010).

A fourth reason for air pollution is that government policies and enforcement vary from one administration to another. Between 2002 and 2006, Justice Department lawsuits against polluters declined by 70 percent, and criminal and civil fines for polluting decreased by more than half compared with the period from 1996 to 2000. The Bush administration blocked the efforts of 18 states to cut emissions from cars and trucks because it believed that tougher regulations would hurt the U.S. economy (Environmental Integrity Project, 2007; "No Action on Greenhouse Gases," 2008).

President Obama promised to decrease air pollution, but environmental groups have accused his administration of being too soft on the fossil-fuel industry by not limiting their CO_2 emissions. The rules also exclude Indian reservations—which have some of the most

China is the planet's biggest air polluter. Sixteen of the world's 20 most polluted cities are in China. The smog is so thick on some days that visibility drops to a few yards, and cities close schools, airports, and highways (Denyer, 2013; Cendrowski, 2014).

polluting coal-fired power plants in the country—from meeting state goals to reduce CO_2 emissions by 2030 (Eilperin and Bernstein, 2014; Magill, 2014).

GLOBAL WARMING AND THE GREENHOUSE EFFECT

Global warming is an increase in the average temperature of earth's atmosphere. The warming has resulted from numerous factors. Some are natural, such as changes in solar radiation, the earth's orbit, and the frequency or intensity of volcanic activity. At least 95 percent of global warming is due to human activities, however, and not natural climate swings (Intergovernmental Panel on Climate Change, 2014).

Global warming begins with the **greenhouse effect**, the heating of earth's atmosphere because of the presence of certain atmospheric gases. When the sun's heat enters the atmosphere, some of it is absorbed by earth's surface, and some of it is reflected back to space. Greenhouse gases in the atmosphere trap some of this heat. Heat is necessary to support life, but when greenhouse gases increase, earth becomes warmer than it would be otherwise, endangering public health and the welfare of current and future generations (U.S. Environmental Protection Agency, 2009).

Air pollutants, primarily CO_2, ignite the greenhouse effect. Between 2000 and 2010, greenhouse-gas emissions grew at 2.2 percent a year—almost twice as fast as in the previous 30 years. During this decade, 90 percent of the world's CO_2 emissions came from burning fossil fuels, primarily from coal-fired power plants. The volume of carbon dioxide hit a record in 2013, the largest year-to-year increase since 1984, when reliable global record keeping began (Intergovernmental Panel on Climate Change, 2014; World Meteorological Organization, 2014b).

China is now emitting almost twice as much CO_2 as the next-biggest polluter, the United States. At current rates, it will produce 500 billion tons of carbon dioxide between 1990 and 2050—as much as the whole world produced between the start of the Industrial Revolution and 1970. China is the world's top coal producer, consumer, and importer, accounting for about half of global coal usage. It burns nearly as much coal every year (more than 4 billion tons) as the rest of the world combined, an important factor in its carbon dioxide emissions ("China's Environment...," 2014; U.S. Energy Information Administration, 2014).

In the United States, about 80 percent of all greenhouse gases come from industrial coal-fired power plants like Southern Company and its subsidiaries in Georgia, Alabama, Indiana, Ohio, and West Virginia owned by America Electric Power. At least 10 other states also have high-polluting power plants (Food and Agriculture Organization of the United Nations, 2011; Cappiello, 2012).

SOME EFFECTS OF CLIMATE CHANGE

Climate change is a change in overall temperatures and weather conditions over time. Global warming probably began thousands of years ago when humans began changing the planet by cutting down and burning forests to grow food (Fischman, 2009). Such *deforestation*, clearing massive amounts of trees, affects global climate changes because forests recycle carbon dioxide into oxygen.

Regardless of when it began, climate change is "unequivocal," has "moved firmly into the present," and is having a profound impact on every ecosystem worldwide. Since such record keeping began in 1850, the years 2001 to 2012 have been the hottest on record. Scientists predict that extreme weather, especially heat waves, will produce heavier rainfall in some regions; more floods, landslides, and uncontained wildfires; stronger hurricanes and tornadoes; and more intense droughts around the world (Howard, 2014; Intergovernmental Panel on Climate Change, 2014; NOAA National Climactic Data Center, 2014).

global warming the increase in the average temperature of earth's atmosphere.

greenhouse effect the heating of earth's atmosphere because of the presence of certain atmospheric gases.

climate change a change in overall temperatures and weather conditions over time.

Changes have already occurred because of global warming and climate change. For example:

▶ Higher CO_2 levels are turning oceans more acidic, resulting in some shellfish (e.g., clams and crabs) dying or not developing (Spotts, 2009).

▶ Global weather- and climate-related disasters rose from 743 between 1971–1980 to 3,496 between 2001–2010 (World Meteorological Organization, 2014a).

▶ Excessive heat in the United States claims more lives each year than floods, lightning, tornadoes, and hurricanes combined (National Oceanic and Atmospheric Administration, 2014).

▶ The world's largest ice sheets in Antarctica and western Greenland are melting at a rate of 120 cubic miles a year, the highest speed in the last 20 years. The melting glaciers could contribute to sea levels rising 2 to 3 feet by 2100 depending on the region (Helm et al., 2014; Intergovernmental Panel on Climate Change, 2014).

Such data show that our planet is ailing. Nonetheless, people can at least slow some of earth's devastation through better environmental policies and practices, particularly by endorsing sustainable development.

15-3c Is Sustainable Development Possible?

Sustainable development refers to economic activities that don't threaten the environment. There are reasons to be both pessimistic and optimistic about achieving sustainable development.

> **sustainable development** economic activities that don't threaten the environment.

REASONS TO BE PESSIMISTIC

Which country is the greenest? Not the United States. For example:

▶ A study of environmental performance ranked 178 countries on factors like water quality, air pollution, and protecting the ecosystem. Switzerland, Luxembourg, Australia, Singapore, and the Czech Republic were the top five. The United States ranked 33rd, and below a number of developing countries, including Slovakia and Serbia (Hsu et al., 2014).

▶ In a global study of 132 nations, the United States ranked only 69th in ecosystem sustainability (Porter and Stern, 2014).

▶ Among the world's 16 largest economies, Germany ranks first in energy-efficiency policies and programs (e.g., fuel economy standards for vehicles, energy efficiency standards for appliances). China is 4th, India is 11th, and the United States is only 13th (Young et al., 2014).

Why does the United States rank so low in environmental performance? There are many reasons, but three are especially important. First, many Americans' worries about global warming have waned and they see economic growth as more important than protecting the environment (see *Table 15.4*). People in many other nations, including developing countries, are consistently more concerned than Americans about climate change, pollution by companies, and other environment issues (see "Climate Change and Financial Instability...," 2013; Grunwald, 2014; Ipsos MORI, 2014). About 85 percent of Americans say that they recycle, but only 34 percent really do so (EPA, 2012).

Second, environmental problems are a low priority. Despite the World Bank's emphasis on the environment, of the 60 projects it financed during 2012–2013,

TABLE 15.4 IS ECONOMIC GROWTH OR THE ENVIRONMENT MORE IMPORTANT?		
Percentage of Americans Who...		
	Worry about Global Warming	**Believe That Economic Growth Is More Important Than Protecting the Environment**
1999	68%	28%
2007	64%	37%
2014	56%	41%

Note: The numbers may not sum to 100 percent because of no responses and rounding.

Sources: Based on Newport 2014 and Swift 2014.

75 percent provided no assessment of climate risks, and only 12 percent considered greenhouse gas emissions (Polycarp et al., 2014). Among almost 2,000 U.S. companies, 67 percent of businesspeople say that social and environmental matters are very important but only 10 percent report that their companies are developing sustainability policies (Kiron et al., 2014). Chevron Corporation pulled back from key efforts to produce clean energy because the profits from non-renewable oil and gas sales were much higher. Some environmentalists are especially dismayed by *greenwashers*, companies and other organizations that pollute the planet while presenting an environmentally responsible public image. In 2000, for instance, BP (British Petroleum) rebranded itself "Beyond Petroleum" and changed its logo to a

Fluffy and ultra-plush toilet paper is made by chopping down and grinding up the pulp of trees that are decades or even a century old. U.S. corporations have introduced "earth-friendly" toilet paper, but "customers are unwavering in their desire for the softest paper possible" (Fahrenthold, 2009: A1).

ARE AMERICANS ECO-FRIENDLY ONLY WHEN IT'S CONVENIENT?

green-and-yellow sunburst, but the majority of its business produces fossil fuels (Elgin, 2014).

Third, lawmakers often accommodate business. Oil and utilities industries have been successful in lobbying Congress to drop laws that would've required them to develop renewable wind and solar energy sources and to pay $13 billion more in taxes for using only fossil fuels. Corporations and many small businesses profit from government subsidies for agriculture and exports, but taxpayers bear the costs of air and water pollution (Philpott, 2013; Saperstein, 2014; see also Chapter 11 on the close ties between the government and corporations).

Americans say that they want cleaner energy, but most oppose wind farms because the 40-foot windmills would "disturb my view of the landscape" (Dunlap, 2007). Thus, are Americans eco-friendly only when it's convenient?

REASONS TO BE OPTIMISTIC

There's been considerable progress since 1970, when the United States celebrated its first Earth Day. Most cars no longer burn leaded gasoline, ozone-destroying chlorofluorocarbons (CFCs) have been generally phased out, and total emissions of the six major air pollutants declined by 54 percent during the same period as the U.S. population increased by 47 percent. The U.S. recycling rate increased from less than 10 percent in 1980 to over 34 percent in 2012. As a result, landfill trash decreased from 89 to 54 percent during this interval (Sperry, 2008; EPA, 2014a).

Many companies have found that being green is good for their profits, image, and environment, and are now more eco-friendly. For example:

▶ Each pair of Levi's new Waste<Less jeans is composed of eight recycled plastic bottles (Berfield, 2012).

▶ Ford, Honda, and Toyota use recycled materials—jeans, sweaters, plastic bottles, and sugar cane by-products—for dashboards, tires, seat covers, and cushions (Fleming, 2014).

▶ SABMiller, the world's second-largest brewer, is working with India's wheat farmers to use water more efficiently to ensure that the company has water to brew beer ("A New Green Wave," 2014).

▸ Coca-Cola uses water-conservation technologies, and Nike is using more synthetic material that's less dependent on weather conditions (Davenport, 2014).

Economic self-interest rather than a concern for the environment sometimes drives such eco-friendly practices. Some corporations are eager to reduce their reliance on water-intensive products such as cotton, for instance. Regardless of motives, enacting environmentally friendly policies promotes sustainability.

Current efficiency standards for appliances, lighting, and other equipment that were implemented during the early 1980s will save the United States the equivalent of two years of energy use ($1.1 trillion by 2035) and slash greenhouse gas emissions. Also, local governments in southern California and some Texas cities are recycling wastewater for drinking (Lowenberger et al., 2012; Preston, 2014).

In other cases, progress has been mixed. For instance, Walmart says that it now reuses or recycles 80 percent of the waste in domestic stores, and hopes to achieve zero waste. Some environmentalists contend, however, that Walmart's manufacturing low-quality products increases landfills and accelerates consumer consumption (Clifford, 2012).

In 2014, and despite protests by industry and some state governments, a federal appeals court upheld the EPA's first limits on air toxics—including emissions of mercury, arsenic, and acid gases—produced by coal plants. The same year, the Supreme Court upheld an EPA regulation requiring Midwestern states with coal-fired power plants to reduce harmful emissions that drift to the East Coast, but also ruled in favor of an electronics manufacturing corporation accused of contaminating a North Carolina community (Savage and Banerjee, 2014; Peeples, 2014).

STUDY TOOLS 15

READY TO STUDY? IN THE BOOK YOU CAN:

☐ Check your understanding of what you've read with the Test Your Learning Questions provided on the chapter review card at the back of the book.

☐ Rip out the chapter review card for a handy summary of the chapter and key terms.

ONLINE AT CENGAGEBRAIN.COM YOU CAN:

☐ Prepare for tests with quizzes.

☐ Review the key terms with Flash Cards.

☐ Play games to master concepts.

© Anelina/Shutterstock.com

4LTR Press solutions are designed for today's learners through the continuous feedback of students like you. Tell us what you think about **SOC4** and help us improve the learning experience for future students.

YOUR FEEDBACK MATTERS.

Complete the Speak Up survey in CourseMate at
www.cengagebrain.com

 Follow us at
www.facebook.com/4ltrpress

16 | Social Change: Collective Behavior, Social Movements, and Technology

LEARNING OBJECTIVES

16-1 Compare, illustrate, and evaluate the major types of collective behavior, their theories, and functions.

16-2 Compare, illustrate, and evaluate the major types of social movements, their theories, and functions.

16-3 Describe and illustrate recent technological advances, their benefits, costs, and ethical controversies.

After you finish this chapter go to **PAGE 329** for **STUDY TOOLS.**

For the most part, as you've seen throughout this textbook, sociologists study behavior and social processes that are relatively institutionalized, routine, stable, and predictable. This chapter examines **social change**, the transformations of societies and social institutions over time. Some collective behavior is short-lived with few long-term changes; others, particularly social movements, can have lasting effects. Technology has also played a critical role in sparking social change. Let's begin with collective behavior.

16-1 COLLECTIVE BEHAVIOR

Do you have several boxes crammed with collectibles like Barbie dolls or baseball cards? A tattoo? Have you ever signed a petition? Joined a club? Texted? Taken a selfie? If so, you've engaged in collective behavior.

social change transformations of societies and social institutions over time.

What do you think?

I sometimes feel overwhelmed in keeping up with texting, e-mail, Twitter, and Facebook.

1	2	3	4	5	6	7

strongly agree   strongly disagree

16-1a What Is Collective Behavior?

Collective behavior is the spontaneous and unstructured behavior of a large number of people. Collective behavior encompasses a wide range of actions, including riots, fads, fashion, panic, rumors, and responses to disasters.

Collective behavior has two important characteristics. First, it's an act rather than a state of mind. You may *feel* panic when a tornado threatens your town, but you don't engage in collective behavior until you actually *leave* your home and head for a safer location.

Second, collective behavior varies in its degree of spontaneity and structure. Panic, the least structured form of collective behavior, is typically short-lived. Fads (e.g., diets, "ugly" Christmas sweaters) are more structured. They may last several years, are planned, and may be expensive. Other forms of collective behavior, including pro- and anti-abortion groups, become highly institutionalized social movements that include a staff, budget, and lobbying.

16-1b Why Collective Behavior Occurs

Over the years, sociologists have proposed many explanations of collective behavior. Four of the most influential

are contagion, convergence, emergent norm, and structural strain theory. Each has strengths and weaknesses.

CONTAGION THEORY

Contagion theory proposes that individuals act emotionally and irrationally due to a crowd's almost hypnotic influence. This perspective is rooted in the work of French scholar Gustave Le Bon (1841–1931), who asserted that people in crowds, particularly mobs and riots, undergo a radical transformation. The anonymity of the crowd, a feeling of power, and an infectious "mob mind" embolden people to abandon personal responsibility and to engage in antisocial, and often violent, behavior (Le Bon, 1896/1968).

American sociologists later refined Le Bon's theory. They emphasized "milling," a process of people moving about, talking to one another, and becoming increasingly excited. Milling produces a "circular reaction" in which the participants become more intense and more unified in attacking a target (Park and Burgess, 1921; Blumer,

> **collective behavior** the spontaneous and unstructured behavior of a large number of people.

1969). Because people feel anonymous in crowds, they abandon self-control and adopt behavior that they'd reject in other settings (Brown and Goldin, 1973).

CONVERGENCE THEORY

Contagion theory asserts that crowds are highly suggestible and out of control. In contrast, *convergence theory* proposes that crowds consist of like-minded people who deliberately assemble in a place to pursue a common goal. Thus, collective behavior occurs when people who share similar values, beliefs, attitudes, emotions, and goals come together in, or converge on, a certain location. Instead of the crowd affecting individuals, as contagion theory claims, convergence theory argues that individuals affect a crowd. Examples of convergence theory include marches, rallies, and acts of civil disobedience that oppose a government's domestic or foreign policies.

EMERGENT NORM THEORY

Emergent norm theory emphasizes that social norms shape crowd behavior. When people perceive a situation as unusual, they create rules that guide their behavior in rational ways. Those who are the most committed may become the leaders and pacesetters; those who disagree or have a different agenda, like looting, are typically ignored. In effect, then, people aren't spontaneously "infected" with the emotions of others, as contagion theory postulates. Instead, crowd members develop norms as a situation unfolds. Unionized workers may orchestrate a strike against an airline or hospital. If, however, the negotiators can't reach an agreement, the strikers may become more intense and develop new norms that include sit-ins and boycotts (Brown and Goldin, 1973; Turner and Killian, 1987).

In 2014, from Los Angeles to New York City, hundreds of fast-food workers and their supporters were arrested as they escalated a fight for better pay with strikes, rallies, and sit-ins. Which collective behavior theory would you use to explain the protests? Why?

STRUCTURAL STRAIN THEORY

Structural strain theory (also called *value-added theory*) proposes that collective behavior occurs only if six conditions are present (Smelser, 1962). These conditions are "value-added" in the sense that each condition leads to the next one, ending in an episode of collective behavior:

1. *Structural conduciveness* refers to the social conditions that foster collective behavior. When channels for expressing a grievance either aren't available or fail, like-minded people may resort to protests.

2. *Structural strain* occurs when societal problems (e.g., discrimination, war) make people angry, frustrated, or interfere with their everyday lives.

3. In the *growth and spread of a generalized belief* stage, people begin to see a situation as a widespread problem instead of their own fault, attribute the problem to a person or group, and believe that something should be done.

4. *Precipitating factors* are incidents or events that trigger action.

5. During the *mobilization for action* stage, leaders emerge.

6. In the *social control* stage, opposing groups prevent, interrupt, or repress those advocating social change. Government officials, the police, community and business leaders, courts, the mass media, and other social control agents—all of whom benefit from the status quo—may ridicule or quash the emerging collective behavior.

CRITICAL EVALUATION

Over time, scholars found little evidence for contagion theory's claim that crowds are typically irrational or out of control. Crowds can sway emotions, but can also be effective in meeting their objectives, and even mobs can be focused and rational (Couch, 1968; McPhail, 1991).

Convergence theories are useful in describing crowds that are deliberate and planned, but have several limitations. They don't explain why some people who share the same perspectives join a protest, rally, or demonstration while others don't, and don't account for the shifts in crowd behavior, no matter how well planned (Berk, 1974; McPhail, 1991).

Emergent norm theory helps explain why many crowds are orderly and focused rather than disorderly and mob-like. In the case of protestors, those who join a crowd quickly learn "the rules" ("Ignore hecklers" or "Don't chat with spectators"). Nonetheless, emergent norm theory doesn't explain which norms emerge, why, and how they differ in crowds. In the case of large crowds—those numbering in the thousands, for instance—how are emergent norms disseminated so quickly and accepted by the participants? It's also not clear why some participants

A day after the CDC confirmed the first U.S. Ebola case in Texas, and emphasized that the risk of infection is low, there were hundreds of social media rumors that the disease had spread to Iowa and other states. Why did the rumors proliferate? Almost 67 percent of Americans believed that Ebola spreads easily (it doesn't), 40 percent expected a major outbreak in the United States (highly unlikely), and 26 percent were personally terrified that they or someone in their immediate family would get the Ebola virus (also highly unlikely) (Harvard School of Public Health, 2014).

Facts *about* **Ebola** in the U.S.

You CAN'T get Ebola through AIR

You CAN'T get Ebola through WATER

You CAN'T get Ebola through FOOD grown or legally purchased in the U.S.

Source: Centers for Disease Control

conform to emergent norms while others ignore them and act irrationally (as looting stores and vandalizing property) (Brown and Goldin, 1973; Berk, 1974).

Structural strain theory helps to predict when and where collective behavior might occur, and offers insights on why, at every stage, collective behavior may either fade or escalate. However, the model doesn't explain all forms of collective behavior. In the case of fads and rumors, as you'll see shortly, all six stages don't necessarily occur. A second limitation is that the sequence of stages isn't necessarily the same as Smelser outlined. Moreover, the theory doesn't say how much structural strain must exist to spark collective behavior (Rule, 1988; Locher, 2002).

16-1c Types of Collective Behavior

There are many forms of collective behavior; some are more fleeting or harmful than others (Turner and Killian, 1987). Let's begin with rumors, one of the most common types of collective behavior.

RUMORS

There were widespread rumors that on January 1, 2000, a glitch (Y2K) in operating systems would cause computers around the world to crash, leading to global power outages, banks losing all of their customers' statements, and even airplanes falling from the skies. None of this occurred.

A **rumor** is unfounded information that people spread quickly. Through modern communication technology, a rumor can spread to millions of people within seconds, and incite riots, panic, or widespread anxiety. Because of the Y2K rumor, thousands of people built underground shelters, and millions of others stocked up on bottled water, canned food, batteries, and medical supplies.

Most rumors (that rock stars Elvis Presley and John Lennon are alive, for example) are harmless. Others can be damaging. After a woman claimed that she found part of a human finger in her cup of Wendy's beef chili, the

restaurant's business dropped by half nationally, and rumors warning people to stop eating fast food altogether spread over the Internet (Richtel and Barrionuevo, 2005). The woman admitted, ultimately, that she had planted the finger to try to get a lucrative settlement. Nonetheless, some customers are still leery about eating at fast-food restaurants.

Rumors are typically false, so why do so many people believe them? First, rumors often deal with an important subject about which—especially during uncertain economic times or natural disasters—people are anxious, insecure, or stressed. This makes people especially suggestible. In Hurricane Katrina's aftermath, the media reported numerous rumors of carjacking, murders, thefts, and rapes, the overwhelming majority of which subsequently proved to be false.

Second, there's often little factual information to counter a rumor, or people distrust the sources of information. During Y2K, the people who stockpiled groceries and so forth didn't believe computer scientists or federal officials who assured the general public that there wouldn't be a calamity. Third, rumors offer entertainment, diversion, and drama in our otherwise mundane daily lives (Campion-Vincent, 2005; Heath, 2005). Gossip and urban legends are two of the most common types of rumor.

Gossip. Gossip is the act of spreading rumors, often negative, about other people's personal lives. Someone once said that "no one gossips about other people's virtues." Because of its tendency to be derogatory, gossip makes us feel superior ("Did you know that Margie just had breast

---

rumor unfounded information that people spread quickly.

gossip rumors, often negative, about other people's personal lives.

implants? Isn't she pathetic?"). Gossip is also interesting, entertaining, and exposes hypocrisy (Epstein, 2011).

Sometimes, individuals gossip to control other people's behavior and to reinforce a community's moral standards. Comments about someone's drug abuse or marital infidelity strengthen norms of what's deviant or unacceptable (see Chapter 4). In other cases, people gossip because they resent or envy someone's physical appearance or accomplishments, or because the gossip reinforces what people think they already know (Sunstein, 2009; Watson, 2012).

Urban Legends. Another form of rumor is **urban legends** (also called *contemporary legends* and *modern legends*), stories—funny, horrifying, or just odd—that supposedly happened somewhere. Some of the most common and enduring urban legends, but with updated variations, deal with food contamination, like the finger at Wendy's. Others have targeted politicians. We still hear, for example, that President Obama wasn't born in the United States even though he produced his birth certificate.

Health-related urban legends also persist. In 1998, medical researchers showed conclusively that there's no relationship between autism and vaccinations for measles, mumps, and rubella. Despite the scientific evidence, some parents still don't vaccinate their children because of "the ongoing antivaccine noise by celebrities like Jenny McCarthy who erroneously links vaccines to autism" (Sifferlin, 2014: 7). As a result, there have been mumps and measles outbreaks in the United States and Europe (Whalen and McKay, 2013; see also Chapters 2 and 14).

---

urban legends (also called *contemporary legends* and *modern legends*) rumors about stories that supposedly happened somewhere.

panic a collective flight, often irrational, from a real or perceived danger.

mass hysteria an intense, fearful, and anxious reaction to a real or imagined threat by large numbers of people.

Why do urban legends persist much longer than gossip? First, they reflect contemporary anxieties and fears—about contaminated food, unscrupulous companies, and corrupt and unresponsive governments. Second, urban legends are cautionary tales that warn us to watch out in a dangerous world. There have been tales that sunscreens cause blindness, and that women have died sniffing perfume samples sent to them in the mail. Third, we tend to believe urban legends because we hear them from people we trust, especially family members, coworkers, and friends. Fourth, rejecting slanted information and outright lies is hard work because doing so requires rethinking already-held beliefs (Lewandowsky et al., 2012). Finally, urban legends—like the one about alligators living in New York City's sewer system—are fun to tell and "too beguiling to fade away" (Brunvand, 2001; Ellis, 2005).

PANIC AND MASS HYSTERIA

In 2003, an indoor fireworks display to kick off a heavy metal concert in West Warwick, Rhode Island, set off a fire that killed 100 people and injured 200 others. As thick black smoke poured through the audience, hundreds of patrons stampeded for the front door (even though there were three other exits), trampling and crushing those who had fallen beneath them.

Most of the deaths in West Warwick weren't due to the fire but to **panic**—a collective flight, often irrational, from a real or perceived danger. The danger seems so overwhelming that people desperately jam an escape route, jump from high buildings, leap from a sinking ship, or sell off their stock. Fear drives panic: "Each person's concern is with his [or her] own safety and personal security, whether the danger is physical, psychological, social, or financial" (Lang and Lang, 1961: 83).

Panic is similar to **mass hysteria**—an intense, fearful, and anxious reaction to a real or imagined threat by large numbers of people. In mid-2009, the World Health

sx70/iStockphoto.com; Canon_Bob/iStockphoto.com

STRDEL/Getty Images

In 2013, 110 people were killed and more than 100 injured during a pilgrimage to a temple in India. As thousands crossed a bridge, one of the railings snapped because of overcrowding. Some people shouted that the bridge was collapsing, which set off panic and a stampede as people rushed to safety (Naveen and Singh, 2013).

Organization issued an alert that H1N1 (the "swine flu") would become a worldwide epidemic and result in numerous deaths. Millions of Americans stood in line for hours, sometimes overnight, because the vaccine was initially in short supply, but the swine flu never materialized on the predicted scale. Unlike panic, which usually subsides quickly, mass hysteria may last longer because warnings—especially about health and food—reinforce our general fears and anxieties about life's dangers.

FASHIONS

Fashion is a popular way of dressing during a particular time or among a particular group of people. Over the years, black women's hairstyles have changed—from Afros in the late 1960s, to straightened hair during the 1980s, to braids, cornrows, dreadlocks, hair extensions, and coloring more recently. All reflect gender politics, racial solidarity, generational differences, identity, and changing images of beauty (Banks, 2000; Desmond-Harris, 2009).

Fashion involves periodic changes in the popularity of clothes, furniture, music, language usage, books, automobiles, sports, recreational activities, the names parents give their children, and even the dogs that people own. Teenagers who want to be fashionable buy clothes with prominent labels, but what's fashionable changes from year to year. Thus, U.S. adults regularly wear only about 20 percent of their wardrobe (Smith, 2013).

Why do fashions, particularly clothes, change fairly quickly? One reason is that designers, manufacturers, and retailers must continuously create demand for new products to maintain a profit. Second, many people keep up with fashion because they don't want to seem different, out-of-date, or dowdy. Third, shopping for new clothes and other products decreases the boredom of everyday routine. Also, clothes and other merchandise are status symbols that signal being an insider: "Others . . . will admire me for . . . making stylish choices" (Best, 2006: 85–86;

for classic analyses of fashion and collective behavior, see Veblen, 1899/1953; Packard, 1959; and Bourdieu, 1984).

FADS

A **fad** is a fashion that spreads rapidly and enthusiastically but lasts only a short time (Turner and Killian, 1987; Lofland, 1993). Fads include *products* (e.g., bean bag chairs, Crocs), *activities* (e.g., twerking, diet books), and *popular personalities and television characters* (e.g., the Lone Ranger during the 1950s, the Kardashians more recently).

Why do fads pop up? A major reason is profit. Because children and adolescents are especially likely to adopt fads, manufacturers create numerous products and activities that they hope will catch on. The products include toys, sportswear, and new cereals. The hottest fads are usually the must-have Christmas toys that children plead for every year. A few months later, the toy may be thrown away or stuffed in the back of a closet.

In the early 1980s, Cabbage Patch Kids dolls were extremely popular. Barely a decade later, "the dolls had fallen victim to the toy industry's perennial challenge: winning a new crop of fans every few years when kids outgrow their toys." To pump up sagging sales and profit, in 2014 the toymaker reinvigorated the dolls by introducing Cabbage Patch Kids wearing new outfits and Twinkle Toes, a popular line of sneakers that sparkle and light up (Townsend, 2014: 27).

Some people dismiss fads as "ridiculous" or "silly," but they serve several functions. In a mass society, where people often feel anonymous, a fad can develop strong in-group feelings and a sense of belonging, especially among people who have similar interests and attitudes. Fads can also be fun, promise to resolve a nagging problem (e.g., being overweight), and help us keep up with technological changes (Marx and McAdam, 1994; Best, 2006).

Most fads are soon forgotten, but some become established. Pez candy dispensers, which originated in 1952

fashion a popular way of dressing during a particular time or among a particular group of people.

fad a fashion that spreads rapidly and enthusiastically but lasts only a short time.

BETTY CROCKER MAKEOVER

1936 **1955** **1965** **1968**

1972 **1980** **1986** **1996**

AP Images/General Mills

In 1921, General Mills created Betty Crocker, a fictitious woman, to answer thousands of questions about baking that came in from consumers every year. As fashions and hairstyles changed, so did Betty Crocker's image. The original image of a stern, gray-haired older woman has changed over the years so that, by 1996, she had a darker complexion and wore casual attire. Can you think of other brands that have changed their image over the years to keep up with changing trends?

PUBLICS, PUBLIC OPINION, AND PROPAGANDA

A **public** is a group of people, not necessarily in direct contact with each other, who are interested in a particular issue. A public is different than the general public, which consists of everyone in a society.

There are as many publics as there are issues—abortion, gun control, pollution, health care, and same-sex marriage, to name just a few. Even within one organization or institution, there may be several publics that are concerned about entirely different issues. At a college, students may be most concerned about the cost of tuition, faculty may spend much time discussing instructional technology, and maintenance employees may be most interested in wages.

The interaction within a public is often indirect rather than face-to-face—as through the mass media, social media, blogs, or professional journals. Because publics aren't organized groups with memberships, they're often

and cost 49 cents, are still inexpensive (under $2.00), are sold in more than 60 countries, and have been continuously updated to include popular television characters like the Simpsons (Paul, 2002). Other fads, like streaking (running around nude in public places), reemerge from time to time ("Streaking," 2005).

DISASTERS

A **disaster** is an unexpected event that causes widespread damage, destruction, distress, and loss. Some disasters are due to *social causes* like war, genocide, terrorist attacks, and civil strife. Some are due to *technological causes*, including oil spills, nuclear accidents, burst dams, building collapses. Others—like floods, landslides, earthquakes, hurricanes, and tsunamis—have natural causes (Marx and McAdam, 1994).

Disasters often inspire organized behavior rather than chaos. Instead of panicking, most people are rational, cooperative, and altruistic. They often care for family members instead of fleeing, and thousands of volunteers offer financial, medical, and other help.

disaster an unexpected event that causes widespread damage, destruction, distress, and loss.

public a group of people, not necessarily in direct contact with each other, who are interested in a particular issue.

Rachel Weill/FoodPix/Jupiter Images

transitory. Publics expand or contract as people lose or develop interest in an issue. A public may surge during a highly publicized and controversial incident, like removing a patient's life support, but then evaporate quickly. In other cases, publics organize and become enduring social movements (a topic we'll examine shortly).

Some publics express themselves through **public opinion**, widespread attitudes on a particular issue. Public opinion has three components: (1) it's a verbalization rather than an action; (2) about a matter that concerns many people; and (3) involves a controversial issue (Turner and Killian, 1987). Like publics, public opinions wax and wane over time. People's interest in crime and education drops, for example, when they're more concerned about pressing issues such as jobs and income.

Public opinion can be swayed by **propaganda**, spreading information (or misinformation) to influence people's attitudes or behavior. Propaganda isn't a type of collective behavior, but it affects collective behavior in several important ways. First, it can create attitudes that will arouse collective outbursts (e.g., strikes or riots). Second, propaganda can try to or succeed in preventing collective outbursts, as when corporations convince employees that job losses are due to a weak economy rather than to offshoring (see Chapter 11). Third, propaganda tries to gain followers for a cause, whatever it might be (Smelser, 1962). Propaganda is institutionalized in advertising, political campaign literature, and government policies. It's conveyed in many ways: The mass media, social media, political speeches, religious groups, rumor, and symbols (e.g., flags, bumper stickers).

CROWDS

Much collective behavior is scattered geographically, but crowds are concentrated in a limited physical space. A **crowd** is a temporary gathering of people who share a common interest or participate in a particular event. Regardless of size—whether it's a few dozen people or millions—crowds come together for a specific reason, including a religious leader's death, a concert, or a riot.

Crowds differ in their motives, interests, and emotional level:

▶ A *casual crowd* is a loose collection of people who have little in common except for being in the same place at the same time and participating in a common activity or event. There's little if any interaction, the gathering is temporary, and there's little emotion. Examples include people watching a street performer, spectators at the scene of a fire, and shoppers at a busy mall.

▶ A *conventional crowd* is a group of people that assembles for a specific purpose and follows established norms. Unlike casual crowds, conventional crowds are structured, their members may interact, and they conform to rules that are appropriate for the situation. Examples include people attending religious services, funerals, graduation ceremonies, and parades.

▶ An *expressive crowd* is a group of people who show strong emotions toward some object or event. The feelings—which can range from joy to grief—pour out freely as the crowd reacts to a stimulus. Examples include attendees at religious revivals, revelers during Mardi Gras, and enthusiastic fans at a football game.

▶ An *acting crowd* is a group of people who have intense emotions and a single-minded purpose. The event may be planned, but acting crowds can also be spontaneous. Examples include people fleeing a burning building, soccer fans storming a field, and college students having a water balloon fight.

▶ A *protest crowd* is a group of people who assemble to achieve a specific goal. Protest crowds demonstrate their support of or opposition to an idea or event. Most demonstrations—like antiwar protests, civil rights

One of the best-known examples of propaganda in the United States is this Uncle Sam poster, designed in 1917, and used to recruit soldiers for World Wars I and II. Opponents of the Vietnam War used it as an anti-war poster during the 1960s and 1970s.

public opinion widespread attitudes on a particular issue.

propaganda spreading information (or misinformation) to influence people's attitudes or behavior.

crowd a temporary gathering of people who share a common interest or participate in a particular event.

"Public mourning" includes strangers coming together to show their respect to the dead. Less than a day after Apple's co-founder Steve Jobs' death in 2011, fans across the globe created memorials in Japan and other countries. Is this kind of collective behavior an example of a conventional, expressive, or acting crowd? Why?

marches, boycotts, and labor strikes—are usually peaceful. Peaceful protesters can become aggressive, however, resulting in destruction and violence (Blumer, 1946; McPhail and Wohlstein, 1983; Pell, 2011).

One type of crowd can easily change into another. A conventional crowd at a nightclub can turn into an acting crowd if a fire causes people to panic and flee for safety. Any of the five types of crowds—from casual to protest—can become a mob or a riot.

MOBS

A **mob** is a highly emotional and disorderly crowd that uses force, the threat of force, or violence against a specific target. The target can be a person, a group, or a property. Mobs often arise when people are demanding radical societal changes, like removing a corrupt government official. In other cases, especially when authority breaks down, people who take advantage of a situation may engage in mob behavior. After a 2010 earthquake in Port-au-Prince, Haiti, most of the city's 3 million survivors focused on clearing the streets of debris and pulling bodies out of the rubble. However, dozens of armed men—some wielding machetes, others with sharpened pieces of wood—"dodged from storefront to storefront, battering down doors and hauling away whatever they

mob a highly emotional and disorderly crowd that uses force, the threat of force, or violence against a specific target.

riot a violent crowd that directs its hostility at a wide and shifting range of targets.

social movement a large and organized group of people who want to promote or resist a particular social change.

could carry…" (Romero and Lacey, 2010: A1). After attacking, a mob tends to dissolve quickly.

RIOTS

Compared with mobs, riots usually last longer. A **riot** is a violent crowd that directs its hostility at a wide and shifting range of targets. Unlike mobs, which usually have a specific target, rioters unpredictably attack whomever or whatever gets in their way. Most riots arise out of long-standing anger, frustration, or dissatisfaction that may have smoldered for years or even decades. Some of these long-term tensions are due to discrimination, poverty, unemployment, economic deprivation, or other unaddressed grievances.

There are numerous protests in the United States every year, but race riots have been the most violent and destructive. More than 150 U.S. cities experienced race riots after the assassination of Martin Luther King Jr. in 1968. In 1992, riots broke out in 11 cities after four white police officers were acquitted of beating Rodney King, a black motorist, in Los Angeles. The violence resulted in deaths and considerable property damage because of fires and looting. In 2014, riots erupted in Ferguson, a small suburb in Missouri, after a white police officer shot and killed an 18-year-old unarmed black male. Looting accompanied the riots.

Riots are usually expressions of deep-seated hostility, but this isn't always the case. In the 2010 Championship Series Sports, riots erupted after the Los Angeles Lakers beat the Boston Celtics. Such "celebration riots" are due to extreme enthusiasm and excitement rather than anger or frustration, and can lead to "an orgy of gleeful destruction" (Locher, 2002: 95).

Much collective behavior, like a mob or riot, is spontaneous and short-lived. Collective behavior that's structured and enduring brings long-term social changes. Social movements are important vehicles for creating or suppressing societal changes.

16-2 SOCIAL MOVEMENTS

There are hundreds of social movements in the United States alone. Why are they so prevalent? And why do they matter?

16-2a What Is a Social Movement?

A **social movement** is a large and organized group of people who want to promote or resist a particular social change. "Social movements are as American as apple pie. The abolition of slavery, women's right to vote, unions, open admissions to public colleges and student aid, and Head

Start are all changes in our society that were won through social movements" (Ewen, 1998: 81–82).

Unlike other forms of collective behavior, social movements are goal-oriented, deliberate, structured, and can have a lasting impact on a society. And in contrast to many other forms of collective behavior (like crowds, mobs, and riots), the people who make up a social movement are dispersed over time and space, and usually have little face-to-face interaction (Turner and Killian, 1987; Lofland, 1996).

Some U.S. social movements, like white supremacists, are relatively small. Others are large and have subgroups that appeal to different segments of the population. The U.S. environmental movement has at least 50 subgroups, including Earth First!, Greenpeace, the National Audubon Society, the Union of Concerned Scientists, and the Wilderness Society.

16-2b Types of Social Movements

Sociologists generally classify social movements according to their goals (changing some aspect of society or resisting change) and the amount of change they seek (limited or widespread). Some social movements are perceived as more threatening than others because they challenge the existing social order (see *Table 16.1*).

Alternative social movements focus on changing some people's attitudes or behavior in a specific way. They typically emphasize spirituality, self-improvement, or physical well-being. These movements are the least threatening to the status quo because they seek limited change and only for some people. For instance, millions of non-Asian Americans, influenced by Asian religions, have embraced yoga, meditation, and healing practices like acupuncture (Cadge and Bender, 2004).

Redemptive social movements (also called *religious* or *expressive movements*) propose a dramatic change, but only for some people. They're usually based on spiritual or supernatural beliefs, promising to renew people from within and guarantee some form of salvation or rebirth. Examples include any religious movements that actively seek converts (e.g., Jehovah's Witnesses, some Christian evangelical groups) (see Chapter 13).

Reformative social movements want to change everyone, but only in a specific way. These movements, the most common type in U.S. society, don't want to replace the existing economic, political, or social class arrangements, but to change society in some specific way. Examples include groups that champion the rights of the disabled, gays, crime victims, fat people, and animals.

Resistance social movements (also called *reactionary movements*) try to preserve the status quo by blocking change or undoing change that has already occurred. Resistance movements are often called *countermovements* because they usually form immediately after an earlier movement has succeeded in creating change. For example, antiabortion groups that arose in the United States shortly after the Supreme Court decision in *Roe v. Wade* (1973), which legalized abortion, seek to reverse that decision.

Revolutionary social movements want to completely destroy the existing social order and replace it with a new one. These movements range from utopian groups that withdraw from society and try to create their own to terrorists who use violence and intimidation. Examples of the latter include Al Qaeda in the Middle East and Boko Haram in Africa.

16-2c Why Social Movements Emerge

A social movement is "an answer either to a threat or a hope" (Touraine, 2002: 89), but not everyone who feels threatened or hopeful joins a social movement. Why not? Let's look at four explanations, beginning with the oldest.

MASS SOCIETY THEORY

Early on, sociologists believed that the people who formed social movements felt powerless, insignificant, and isolated in modern mass societies, which are impersonal, industrialized, and highly bureaucratized. Thus, according to *mass society theory*, social movements offer a sense of belonging to people who feel alienated and disconnected from others (Kornhauser, 1959).

CRITICAL EVALUATION

Mass society theory may explain why some people form extreme political movements like Fascism and Nazism, but subsequent research has shown that movement organizers are typically

TABLE 16.1	FIVE TYPES OF SOCIAL MOVEMENTS	
Movement	**Goal**	**Examples**
Alternative	Change some people in a specific way	Alcoholics Anonymous, transcendental meditation
Redemptive	Change some people, but completely	Jehovah's Witnesses, born-again Christians
Reformative	Change everyone, but in specific ways	Gay rights advocates, Mothers Against Drunk Driving (MADD)
Resistance	Preserve status quo by blocking or undoing change	Antiabortion groups, white supremacists
Revolutionary	Change everyone completely	Right-wing militia groups, Communism, Islamic State in Iraq and Syria (ISIS)

not isolated but well-integrated into their families and communities. Also, historically, many political activists in the United States, including those behind the civil rights and women's rights movements during the late 1960s, weren't powerless but came from relatively privileged backgrounds (McAdam and Paulsen, 1994).

RELATIVE DEPRIVATION THEORY

Relative deprivation theory is broader than mass society theory. **Relative deprivation** is a gap between what people have and what they think they should have compared with others in a society. What people think they should have includes money, social status, power, or privilege.

Relative deprivation theorists note two other elements. First, people often feel that they *deserve* better than they have ("I've worked hard all my life"). Second, they believe that they *cannot attain their goals through conventional channels* ("I've written people in Congress, and they just ignore me"). Thus, shared beliefs combined with unfulfilled expectations can trigger change-oriented social movements (Davies, 1979; Morrison, 1971).

CRITICAL EVALUATION

Relative deprivation theory helps explain why some social movements emerge. Critics point out, however, that there's a certain degree of relative deprivation in all societies, but people don't always react by forming social movements. Relative deprivation theory also doesn't explain why some people join movements even though they don't see themselves as deprived, and don't expect to gain anything personally if the movement succeeds (Gurney and Tierney, 1982; Johnson and Klandermans, 1995).

RESOURCE MOBILIZATION THEORY

It takes more than feeling alienated (mass society theory) or disadvantaged (relative deprivation theory) to sustain a social movement. Instead, according to *resource mobilization theory*, a social movement will succeed if it can put together (or mobilize) an organization and leadership dedicated to advancing its cause (Oberschall, 1995; Gamson 1990). Other important resources include money, devoted volunteers, paid staff, access to the media, effective communication systems, special technical or legal knowledge and skills, equipment, physical space, alliances with like-minded groups, and lobbyists who finance campaign elections (see Chapter 11).

Consider U.S. gun-rights and gun-control movements. In 2014, Georgia passed a law that allows people with firearm licenses to carry guns into bars, churches,

schools, some airport areas, and government buildings not protected by security guards during business hours. The gun control movement has seen some victories, including 16 states strengthening gun-control laws in 2013. Still, the National Rifle Association (NRA), a powerful leader of the gun-rights movement, defeated a federal bill mandating tougher background checks, something 85 percent of Americans favored. The NRA has brushed off gun-control efforts because it has "millions of members…who will give us small amounts of money [that] add up to an annual budget of more than $300 million" ("Gun Laws: A Shot and a Beer," 2014).

CRITICAL EVALUATION

A major contribution of resource mobilization theory is its emphasis on structural factors (like organization and leadership) in explaining why some social movements thrive whereas others shrivel. A major criticism, however, is that resource mobilization theory largely ignores the role of relative deprivation in a social movement's formation. If there aren't large numbers of dissatisfied people to initiate a movement, even plentiful resources won't be able to sustain it (Klandermans, 1984; Buechler, 2000).

NEW SOCIAL MOVEMENTS THEORY

New social movements theory, which became prominent during the 1970s, emphasizes the linkages among culture, politics, and ideology. Unlike the earlier perspectives, new social movements theory proposes that many recent movements (like those that work for peace and environmental protection) promote the rights and welfare of *all* people rather than specific groups in particular countries (Laraña et al., 1994; Melucci, 1995). Thus, new social movements theory is especially interested in "the struggle to liberate the voices of the dispossessed" (Schehr, 1997: 6).

For these theorists, recent social movements differ from older ones in several ways. They attract a disproportionate number of people who are well-educated and

IF THERE AREN'T LARGE NUMBERS OF DISSATISFIED PEOPLE TO INITIATE A MOVEMENT, EVEN PLENTIFUL RESOURCES WON'T BE ABLE TO SUSTAIN IT.

relative deprivation a gap between what people have and what they think they should have compared with others in a society.

relatively affluent, who represent a wide array of professions (e.g., educators, scientists, actors, businesspeople, political leaders), and who share a broad goal—improving the quality of life for all people around the world. Moreover, recent social movements pursue goals or advance values that may not personally benefit its members, such as eradicating measles or tuberculosis in developing countries (Obach, 2004).

CRITICAL EVALUATION

Unlike earlier perspectives, new social movements theory contributes to our understanding of collective behavior that crosses international boundaries. According to some critics, however, neither this perspective nor the groups that it examines are novel. Some social movements (like feminism and environmentalism) have been around for a long time and still focus on the same basic issues, like women's second-class citizenship and population growth. In addition, some scholars point out that educated middle-class or wealthy activists were as common in the old social movements as in more recent ones (Rose, 1997; Buechler, 2000; Sutton, 2000).

Critics also contend that new social movements theory often overstates people's altruistic motivations. Many people who join environmental groups do so for reasons referred to as NIMBY (not in my backyard). That is, they're concerned about their own community's environment, but show little interest in ecological threats to people elsewhere.

These four theories, despite their limitations, help us understand social movements because "no single theory is sufficient to explain the complexities of any social movement" (Blanchard, 1994: 8). The next question is why some social movements flourish and others collapse.

16-2d The Stages of Social Movements

Most social movements are short-lived. Some never really get off the ground; others meet their goals and disband. Social movements generally go through four stages: Emergence, organization, institutionalization, and decline (see *Figure 16.1*) (Tilly, 1978).

EMERGENCE

During *emergence*, the first stage of a social movement, a number of people are distressed about some condition and want to change it. One or more individuals, serving

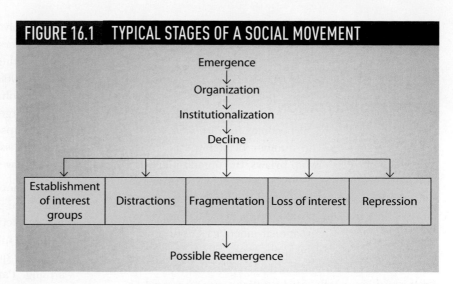

FIGURE 16.1   TYPICAL STAGES OF A SOCIAL MOVEMENT

Emergence → Organization → Institutionalization → Decline → [Establishment of interest groups | Distractions | Fragmentation | Loss of interest | Repression] → Possible Reemergence

as agitators or prophets, emerge as leaders. They verbalize the feelings of the discontented, crystallize the issues, and push for action. If leaders don't get much support, the movement may die.

If, on the other hand, the discontent resonates among growing numbers of people, public awareness increases and the movement attracts like-minded people. In India, women's groups have had some success in combatting pervasive sexual violence because growing numbers of both women and men are protesting rape and sexual assaults. Some Indian cities have passed new laws that include sanctions for police who refuse to register rape allegations, but many offenders still aren't charged or prosecuted (Lahiri and Rana, 2013).

ORGANIZATION

Once people's consciousness has been raised, the second stage is *organization*. The most active members form alliances, seek media coverage, develop strategies and tactics, recruit members, and acquire the necessary resources. The organization establishes a division of labor in which leaders make policy decisions and followers perform necessary tasks like preparing mass mailings, developing websites, and responding to phone calls and emails. At this stage, the movement may develop chapters at local, regional, national, and international levels.

The fast-food workers of the strikes that began in 2012 have demanded a pay raise to $15.00 per hour and the right to unionize. According to some analysts, the protesters may be spearheading a broader "living wage movement" for several reasons: They have been organized by Fast Food Forward (a powerful labor group in New York City), have been financed by the Service Employees International Union, and have received considerable (and supportive) media coverage (Weissmann, 2014). It remains to be seen, however, whether, as predicted, the strikes will attract millions of unhappy retail and other low-income earners, and whether a living wage movement will become institutionalized.

INSTITUTIONALIZATION

As a movement grows, it becomes *institutionalized* and more bureaucratic: The number of staff positions increases, members draw up bylaws, the organization may hire outsiders (writers, attorneys, and lobbyists) to handle some of the necessary tasks, and the leaders may spend more of their time on speaking tours, in media interviews, and at national or international meetings. As the social movement grows and becomes more bureaucratic and self-sufficient, the original leaders may move on to better paying and more influential positions in government or the private sector.

DECLINE

Almost all social movements end sooner or later. This *decline*, the last stage, may take several forms:

▶ If a social movement is successful, it can become an *interest group* and a part of society's fabric. A small antismoking movement that began in the mid-1970s now has the enthusiastic support of numerous prestigious organizations, including the American Cancer Society, the American Heart Association, the American Medical Association, the World Health Organization, and governing bodies within and outside the United States (Wolfson, 2001).

Food movements have burgeoned worldwide, especially in urban centers. Many U.S. cities have institutionalized food carts and trucks by passing or revising regulations that support food vending. Doing so encourages neighborhood entrepreneurs to use available city spaces, attracts "foodies," and offers residents and tourists a variety of ethnic, organic, and gourmet "culinary experiences" (Hanser and Hyde, 2014: 47).

▶ Those involved in a social movement may become *distracted* because the group loses sight of its original goals, their enthusiasm diminishes, or both. In the mid-1960s, when Ralph Nader first criticized the automobile industry for car safety defects, he gained a large following. As Nader and his consumer rights groups expanded their focus to include environmental issues and corporate crimes, many of the initial followers lost interest.

▶ A social movement may experience *fragmentation* because the participants disagree about goals, strategies, or tactics. The environmental movement encompasses numerous groups that focus on different issues—air quality, marine life, land use, and global warming. Participants may also drift away from a movement because of time constraints, health problems, or similar reasons.

▶ Social movements may also decline because of *repression*. Many autocratic governments quash dissent. A government can crush an emerging social movement by arresting protestors and imprisoning or even executing leaders (see Chapter 11).

Social movements can wane for a combination of reasons. In 2011, the Occupy Wall Street (OWS) movement protested socioeconomic inequality, corporate greed, and corporate power over the U.S. government. The group's slogan, "We are the 99%," referred to the income inequality between the top 1 percent and the rest of the population. Despite similar demonstrations in 70 major U.S. cities and over 600 communities, and 900 cities worldwide, the movement faded away. Why? OWS raised awareness of the injustice of this nation's inequality, but ultimately failed because it lacked leadership, clear objectives, solutions, organization, and a long-term commitment (Madrick, 2013; Sandbu, 2013).

A social movement that declines can sometimes experience a resurgence. For example, there have been several waves of the women's rights movement in the United States since the mid-nineteenth century. The first wave ensured women's right to vote in 1920; the second wave expanded employment and educational rights during the 1960s and 1970s; and the third wave, during the 1990s, focused on economic and other inequalities experienced by women of different social classes, sexual orientations, and nationalities. Despite considerable backlash over the past hundred years, American feminism has appealed to diverse groups, agitated in a multitude of spheres ranging from athletics to religion, and has used a variety of strategies in pressing for social change (Cobble et al., 2014).

16-2e Why Social Movements Matter

On an *individual level*, many of us enjoy a variety of rights as workers, consumers, voters, and even victims—rights that we owe to highly dedicated people who were determined to change inequitable laws and practices.

On an *institutional level*, social movements can change general practices. Shopping for healthy food is much easier today than it was before the 1990s, when veggie burgers, tofu, and nutritional labeling were practically nonexistent. Now, mainstream grocery stores have large organic food sections, a variety of fruits and vegetables, and breads and cereals made with whole grains, nuts, and less salt, sugar, and chemical additives. The "farm-to-table" movement, with its emphasis on locally grown produce and meat, has benefitted people's health and regional economies. In effect, then, vegetarian and consumer groups have changed the way many farms operate and loosened corporate control of food products.

On a *societal level*, social movements have had a major effect in the United States and globally. Most of the world's great religions began as protest movements (see Chapter 13). Also, democratic forms of government in the United States and other countries grew out of the activities of revolutionary groups that sought greater political and economic freedom (Giugni et al., 1999; della Porta and Diani, 1999).

You've seen that social movements can generate or resist change. However, "what all of these social movements have in common is the desire by ordinary citizens to have a say in the operation of their society" (Locher, 2002: 246). Technology is another important source of societal change.

16-3 TECHNOLOGY AND SOCIAL CHANGE

This brief "history" of how to cure an earache has circulated on the Internet:

▶ 2000 B.C.—Here, eat this root.

▶ 1000 A.D.—That root is heathen, say this prayer.

▶ 1850 A.D.—That prayer is superstition, drink this potion.

▶ 1940 A.D.—That potion is snake oil, swallow this pill.

▶ 1985 A.D.—That pill is ineffective, take this antibiotic.

▶ 2000 A.D.—That antibiotic is artificial. Here, eat this root.

As this anecdote suggests, despite technological progress over the centuries, some of the old remedies are enjoying renewed popularity (peoplespharmacy.com, for instance, offers home treatments for everything from arthritis to whooping cough). Still, technological advances have brought enormous social changes, including benefits, costs, and ethical concerns.

16-3a Some Recent Technological Advances

Technology, the application of scientific knowledge for practical purposes, is a vital aspect of human life: "For good or ill, [technologies] are woven inextricably into the

technology the application of scientific knowledge for practical purposes.

fabric of our lives, from birth to death, at home, in school, in paid work" (MacKenzie and Wajcman, 1999: 3).

In the next 50 years, technology is likely to change our lives more dramatically than ever before. Several companies are working on "smart" pills to improve mental ability and restore brains that have been impaired by disease or injury. Our houses may be built with sensors that automatically test for anthrax, environmental contaminants, allergens, and radioactivity, and with devices that can defend against the release of chemical and biological agents (Rubin, 2004; Murphy, 2005). Whether such predictions will become reality is anyone's guess, but General Motors and Nissan plan to sell driverless cars by 2020. In the meantime, technological advances are changing our lives. Let's look briefly at a few of the most influential.

COMPUTER TECHNOLOGY

In 1887, English mathematician Charles Babbage designed the first programmable computer. Since then, computers have gone through seven generations of evolution. *Artificial intelligence (AI)*, a branch of computer science, is developing theories and computer systems to perform tasks that normally require human intelligence (e.g., speech recognition, decision making, and understanding). AI advances have spurred the development of *robots*, machines that are programmed to perform humanlike functions. Industrial robots, known as "collaborative robots" or "cobots," already work alongside people in doing tasks like stacking heavy packages. By 2025, according to some experts, AI and robots will affect nearly every aspect of daily life (Smith et al., 2014; Swan, 2014).

Robots have performed repetitive, dirty, or dangerous human tasks, including auto assembly, toxic waste cleanup, mining, minefield sweeping, underwater and space exploration, testing military equipment, and locating bombs. In time, robots may be able to repair damaged brain cells, print body parts, discuss stock market investment strategies, give you advice about a personal problem, read a book to you in any desired language or in a voice of either sex, cook and serve your meals, and be emotionally savvy companions who cheer you up when you're unhappy (Schwartz, 2014b).

BIOTECHNOLOGY

Biotechnology is a broad term that applies to all practical uses of living organisms. It covers anything from the use of microorganisms (e.g., yeast) to ferment beer to *genetic engineering*, sophisticated techniques that can change the makeup of cells and move genes across species to produce new organisms.

In agriculture, biotechnology produces genetically modified (GM) crops like corn, soybeans, and cotton.

Between 1996 and 2012, the total earth's land cultivated with GM crops rose from 4.2 to 395 million acres. U.S. farmers continue to be the chief users of GM crops, producing 43 percent of the world's total (Atici, 2014). Biotech crops increase production and make agriculture less costly because new varieties of plants can tolerate herbicides and resist uncontrollable diseases that destroy millions of tons of food each year. Critics contend, however, that GM crops may not be safe, may harm soil quality, and may disrupt ecosystems (Nestle, 2012; Caplan, 2013).

In medicine, biotechnology includes genetic testing/screening that can determine a child's biological parents, a person's ancestry, inherited disorders, and a person's chance of developing or passing on a genetic disease. You'll see shortly, however, that genetic screening also raises worrisome ethical questions.

NANOTECHNOLOGY

Another promising recent technology is *nanotechnology*, the ability to build objects one atom or molecule at a time. The key characteristic of these objects is tiny size. Nanotechnology is based on structures measured in nanometers, a unit of measurement equal to 1 billionth of a meter, or 1/80,000th the width of a human hair.

In medicine, researchers foresee the day when thousands of nanotubes will be packed into a hairlike capsule the size of a splinter and painlessly implanted under the skin. The nanotubes would continuously monitor blood sugar, cholesterol, and hormone levels, and destroy tumors without also frying adjacent healthy cells, one of the negative effects of current chemotherapy and radiation treatments for cancer (Weiss, 2004, 2005). Scientists are also working on nanosensors that would live in the bloodstream and send messages to smartphones about signs of infection, an impending heart attack, or other issues that are early warnings for disease or death. Swallowing little pills that contain nanosensors—tiny robots with legs, propellers, and cameras—may become a routine part of doctors' office visits in the future (Alkhatib et al., 2014; Cha, 2014).

These and other technological innovations promise longer and healthier lives in the future, but what about the present? What are some of the current benefits and costs of technological changes?

16-3b Some Benefits and Costs of Recent Technologies

The ever-accelerating pace of change means that most U.S. children born after 2000 are using technologies that didn't exist just a decade ago. Indeed, much of this "iGeneration" views even those in their 20s as outdated in

FIGURE 16.2   SOME TECHNOLOGY IS HARD TO PUT DOWN

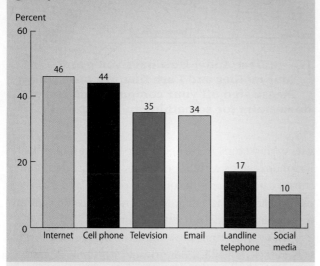

Percentage of U.S. adults who said, in 2014, that these technologies would be "very hard or impossible to give up"

Source: Based on Fox and Rainie, 2014: 6.

their tech skills (Rosen, 2010). Technology may be creating greater generation gaps than in the past, but a majority of Americans (59 percent) believes that technological advancements will improve people's lives (Smith, 2014). Large numbers also say that it'd be almost impossible to give up some of their hi-tech gadgets (see *Figure 16.2*).

Is the average American too upbeat about new technology? Recently, a computer scientist noted that every technology has a "dark side" because practically every benefit also has a cost (Lohr, 2011: A1). Here are a few contemporary examples.

COLLABORATIVE CONSUMPTION

▶ *Benefits*: Technology has spurred "collaborative consumption" (also called a "sharing economy" and "access economy") that enables people to earn and save money by sharing goods and services. Airbnb matches budget travelers and people with a spare room or other lodging for rent, RelayRides allows people to rent your car, and taxi-like services (e.g., Lyft, Uber, Sidecar) bring drivers and passengers together.

▶ *Costs*: The sharing economy has created conflict: Airbnb drivers have sometimes ignored zoning laws, don't always pay city or state taxes, and neighbors have complained about noise and a steady stream of strangers. Taxicab companies have protested that ridesharing has decreased their business because the drivers have avoided all regulations, inspections, fees, and insurance requirements ("All Eyes on the Sharing Economy," 2013; Goodale, 2013).

3-D PRINTING

▶ *Benefits*: 3-D printing—a process for making a physical object from a three-dimensional digital model—is revolutionizing manufacturing. Because the process can reduce manufacturing costs by 25 percent or more, the CEO of an aluminum corporation describes 3-D printing as "the beginning of…a second Industrial Revolution" (Geier, 2014: 76). Among other things, 3-D printing produces medical implants (as knee replacements), synthetic heart valves, prosthetics that are easy to use, models of buildings for architects, and even cars that are safe and fuel efficient. Desktop 3-D printers produce a wide range of objects like toys, clocks, ceramics, and wrenches.

▶ *Costs*: As with any new technology, 3-D printers have negative impacts: They consume 50 to 100 times more electrical energy than traditional molds, increase usage of plastics that aren't recycled, and emit toxic particles when used in the home that can pose health risks. Moreover, people can print weapons, including plastic guns, that metal detectors or x-ray scanners can't spot (Bodley, 2014; Fung, 2014; Gilpin, 2014).

HEALTH AND MEDICINE

▶ *Benefits*: People who use health-related mobile gadgets and apps can monitor their blood pressure, calories, and physical fitness. The number of robotic procedures increases by 30 percent each year. Robotic surgery reduces physician fatigue because she/he doesn't have to stand over the patient for hours, and a robot's "hands" can reach into tight spots and move in ways that human hands can't. In addition, minimally invasive robotic surgeries usually result in less blood loss and faster recoveries because there's a smaller incision to heal.

AnthonyRosenberg/iStockphoto.com

▶ *Costs*: Health-related apps encourage "IPochondria" among people who are anxious about their health which, in turn, increases doctors' workloads and insurance costs. Robotic surgery is expensive, the outcomes are generally similar to traditional operations, there are no national training standards for robotic surgery, one company dominates the market, and adverse events increased 34 percent between 2011 and 2012 alone (Langreth, 2013; Wright et al., 2013; Howard, 2013/2014; "M-health...," 2014).

MONITORING SYSTEMS

▸ *Benefits*: New technology allows older people to live at home rather than in an institutional setting. Sensors can track people's medications and falls. Adult children can monitor an aging parent or relative remotely via cellular connections, seeing if they go outside or get home safely after short drives.

▸ *Costs*: Medicaid pays for some aging-at-home technology, but Medicare doesn't. Even if families can afford the technology, some older people feel intimidated by even one-button devices and don't use them (Abrahms, 2014; Tsukayama, 2014).

Some policy analysts fear that technology will produce more jobs for robots than for humans, even in occupations like teaching, sales, nursing, and stock trading. For example, driverless cars could displace millions of truck drivers, bus drivers, and others who drive for a living. Others contend that computers will never replace jobs that require people skills like teamwork, creativity, managing diverse employees, brainstorming, leadership, and decision-making (Von Drehle, 2013; Colvin, 2014; "Rise of the Robots," 2014).

There's considerably more consensus that almost every technological advance reduces privacy and increases data breaches. Many "data mining" companies routinely collect information about people as they click from site to site. Much of this web tracking is done anonymously, but a new crop of "snooper" sites is making it easier than ever before for anyone with Internet access to assemble and sell personal information, including your name, Social Security number, address, what you buy, whom you love, and which sites you've visited on almost any topic, product, or service.

The data come from a variety of sources, including public records on campaign contributions, property sales, and court cases; networking sites where people provide information about themselves, their jobs, relatives, and friends; cell phone companies; and hospitals that share their medical records with state health agencies, which in turn sell the information to private data-mining companies (Leber, 2013; Robertson, 2013; Suich, 2014).

Google compiles enough data to build comprehensive portfolios of most users—who they are, where they go, and what they do. To create a new stream of profit, Facebook sells marketers detailed information about its

"Dear Andy: How have you been? Your mother and I are fine. We miss you. Please sign off your computer and come downstairs for something to eat. Love, Dad."

© Randy Glasbergen

users. Marketers, in turn, can flood millions of websites and mobile apps with targeted ads. At many department stores, cameras have become so sophisticated that companies can analyze what shoppers are looking at, and even their mood (Efrati, 2013; Rosen, 2013; Goel, 2014).

Many Americans are vulnerable because they're not very knowledgeable about privacy laws. Only 22 percent know that if a website has a privacy policy, the site can share information about you with other companies without your permission (Turow et al., 2009). Just 21 percent of U.S. adults have "a lot of trust" in the businesses or companies they regularly patronize to keep their personal information secure. However, millions provide information—including gender, age, and income—for as little as a "$1 off" online coupon or a $5.00 to $10.00 prepaid gift card, and agree to be tracked over GPS, Wi-Fi, and cellular networks. The companies then sell the data to storeowners, online retailers, and app developers (Clifford and Hardy, 2013; Anderson and Rainie, 2014; Fleming and Kampf, 2014).

16-3c Some Ethical Issues

DNA testing has given millions of people information about their genetic predispositions for diseases like cystic fibrosis, cancer, and Huntington's disease (an incurable neurological disorder). Having such information has helped doctors and patients make better health care decisions, but the expanding use of genetic testing is having an unforeseen consequence: More people who don't show any symptoms are being told they have genes for potentially fatal diseases. For instance, a pediatric lung specialist tells patients and their families: "You have the genetic potential to develop cystic fibrosis, which could be next month, next year, when you are 60 years old, or never" (Marcus, 2013: D1).

Those affected become "patients-in-waiting" who undergo continuous screening and worry about their condition. Asymptomatic patients with an inherited risk for a genetic heart muscle disease may have a defibrillator implanted, which involves regular maintenance, possible equipment failures, infections, and other complications. Some doctors believe that such preventive measures for patients-in-waiting are unethical because they're unnecessary, risky, and people live in limbo not knowing whether or when a disease will develop (Marcus, 2013).

Stem cell research is also controversial. A *stem cell* is a building block of the human body that can replicate indefinitely and thus serve as a continuous source of new cells. These self-regenerating cells are found in embryos and umbilical cords, but also in parts of adult bodies, including the brain, blood, heart, skin, bone marrow, intestines, and other organs. Embryonic stem cells are more valuable in research because they can produce any cell type of the body, whereas adult stem cells are generally limited to the cell types of their tissue of origin like a liver or kidney (National Institutes of Health, 2006).

Opponents of stem cell research argue that all embryos deserve protection. Proponents maintain that hundreds of thousands of embryos that fertility clinics dispose of every year are a wasted resource: They could be used to treat heart disease, leukemia and other cancers, diabetes, Parkinson's disease, and numerous other health problems. In 2013, scientists for the first time created stem cells by cloning embryos. The experiments' success rekindled debate among bioethicists who have renounced "reproductive cloning" as unethical (Healy, 2013).

There's also concern about parents' and scientists' right to create "designer babies" by choosing genes for a child's physical attributes, as eye and hair color, height, complexion, and even athleticism. Some people believe it's unethical to bioengineer children because better-off parents could use the technology to give their children a competitive edge. On the other hand, 26 percent of U.S. adults believe that prospective parents should be able to alter their children's DNA to produce smarter, healthier, or more athletic offspring (Naik, 2013; Smith, 2014).

Another ethical dilemma is that biotechnological advances, like other technological developments, are most readily available to higher-income people. Screening expectant mothers for fetal abnormalities like Down syndrome is becoming more standard, but not all insurers cover the $1,200 to $2,700 costs of more sophisticated screening tests for high-risk women (Lewis, 2014). Insurance also rarely covers pediatric prosthetics, such as up to $40,000 for a hand that'll have to be replaced about every two years as a child grows (Cohn, 2014).

In 2014, a 36-year-old woman in Sweden who was born without a uterus received a donated womb from a friend in her 60s, and gave birth to a son. Worldwide, the media heralded "the world's first baby born from a transplanted womb." A Swedish charity covered the costs. Because U.S. insurers wouldn't pay about $158,000 for a similar procedure, only affluent women could afford the treatment, operation, follow-up, and drugs (Smith, 2014). Thus, technological advances are fraught with both promise and pitfalls.

STUDY TOOLS 16

READY TO STUDY? IN THE BOOK YOU CAN:

☐ Check your understanding of what you've read with the Test Your Learning Questions provided on the chapter review card at the back of the book.

☐ Rip out the chapter review card for a handy summary of the chapter and key terms.

ONLINE AT CENGAGEBRAIN.COM YOU CAN:

☐ Prepare for tests with quizzes.

☐ Review the key terms with Flash Cards.

☐ Play games to master concepts.

© Anelina/Shutterstock.com

USE THE TOOLS.

- Rip out the Review Cards in the back of your book to study.

Or Visit CourseMate to:

- Read, search, highlight, and take notes in the Interactive eBook
- Review Flashcards (Print or Online) to master key terms
- Test yourself with Auto-Graded Quizzes
- Bring concepts to life with Games, Videos, and Animations!

Go to CourseMate for **SOC4** to begin using these tools.
Access at **www.cengagebrain.com**

Complete the Speak Up survey in CourseMate at **www.cengagebrain.com**

Follow us at **www.facebook.com/4ltrpress**

REFERENCES

The references that are new to this edition are printed in red.

AAA Foundation for Traffic Safety. 2013. "Fact Sheet: Cognitive Distraction." Accessed October 10, 2013 (www.aaafoundation.org).

AARP Magazine. 2014. "You're Old, I'm Not." February 3, 40–42.

AAUW. 2013. *The Simple Truth About the Gender Pay Gap, Fall 2013 Edition*. Washington, DC: American Association of University Women.

Abrahms, Sally. 2014. "Is This the End of the Nursing Home?" *AARP Bulletin*, March, 20, 22.

Abrams, Simon. 2011. "Let's Stay Together." A.V. Club, January 19. Accessed December 6, 2011 (www.avclub.com).

Abudabbeh, Nuha. 1996. "Arab Families." Pp. 333–346 in *Ethnicity and Family Therapy*, 2nd edition, edited by Monica McGoldrick, Joe Giordano, and John K. Pearce. New York: Guilford Press.

Academy of Medical Royal Colleges. 2011. "Induced Abortion and Mental Health: A Systematic Review of the Mental Health Outcomes of Induced Abortion, Including Their Prevalence and Associated Factors." National Collaborating Centre for Mental Health, December. Accessed February 9, 2013 (www.aomrc.org.uk).

Accenture. 2013. "Retail Medical Clinics: From Foe to Friend?" Accessed December 19, 2013 (www.accenture.com).

Acierno, Ron, Melba Hernandez, Amanda B. Amstadter, Heidi S. Resnick, Kenneth Steve, Wendy Muzzy, and Dean G. Kilpatrick. 2010. "Prevalence and Correlates of Emotional, Physical, Sexual, and Financial Abuse and Potential Neglect in the United States: The National Elder Mistreatment Study." *American Journal of Public Health* 100 (February): 292–297.

Acosta, Yesenia D., Luke J. Larsen, and Elizabeth M. Grieco. 2014. "Noncitizens Under Age 35: 2010–2012." U.S. Census Bureau, February. Accessed March 22, 2014 (www.census.gov).

ACT, Inc. 2014. "Improving College and Career Readiness and Success for Everyone." Accessed June 21, 2014 (www.act.org).

Adams, Bert N., and R. A. Sydie. 2001. *Sociological Theory*. Thousand Oaks, CA: Pine Forge Press.

Adamson, David M. 2010. "The Influence of Personal, Family, and School Factors on Early Adolescent Substance Use." Rand Health. Accessed August 20, 2011 (www.rand.org).

"Adding Up the Government's Total Bailout Tab." 2011. *New York Times*, July 24. Accessed September 29, 2011 (www.nytimes.com).

Addy, Sophia, William Engelhardt, and Curtis Skinner. 2013. "Basic Facts About Low-Income Children: Children Under 18 Years, 2011." National Center for Children and Poverty. Accessed February 6, 2014 (www.nccp.org).

Adler, Jerry. 2005. "In Search of the Spiritual." *Newsweek*, August 20–September 5, 44–64.

Administration on Aging. 2013. "Protect Seniors in the Year of Elder Abuse Prevention." Accessed November 30, 2014 (www.aoa.gov).

"Adult Obesity Facts." 2014. Centers for Disease Control and Prevention, March 28. Accessed July 26, 2014 (www.cdc.gov).

"After 39 Wives, Mizo Man Wants to Marry Again." 2013. Matters India, April 27. Accessed May 16, 2014 (www.mattersindia.com).

Agaku, Israel T., et al. 2014. "Tobacco Product Use Among Adults—United States, 2012–2013." *MMWR* 63 (June 27): 542–547.

Agence France-Presse. 2013. "Bill Gates' 'Hand in Pocket' Draws Criticism in South Korea." April 23. Accessed November 10, 2013 (www.rawstory.com).

Agnvall, Elizabeth. 2014. "10 Tests to Avoid." *AARP Bulletin*, March, 12, 14, 16.

Aguila, Raul. 2014. "What We Hate About Holiday Travel." *Consumer Reports*, November, 7.

Aizer, Anna, and Joseph J. Doyle, Jr. 2013. "Juvenile Incarceration, Human Capital and Future Crime: Evidence from Randomly-Assigned Judges." National Bureau of Economic Research, June. Accessed October 19, 2014 (www.nber.org).

Akechi, Hironori, Atsushi Senju, Helen Uibo, Yukiko Kikuchi, Toshikazu Hasegawa, and Jari K. Hietanen. 2013. "Attention to Eye Contact in the West and East: Autonomic Responses and Evaluative Ratings." PLoS One. 8 (3):e59312. Accessed December 5, 2013 (www.plosone.org).

Akers, Ronald L. 1997. *Criminological Theories: Introduction and Evaluation*, 2nd edition. Los Angeles: Roxbury.

"Alcohol and Public Health." 2014. Centers for Disease Control and Prevention, March 14. Accessed July 30, 2014 (www.cdc.gov).

Alexander, Karl L., Doris Entwisle, and Linda Olson. 2014. *The Long Shadow: Family Background, Disadvantaged Urban Youth, and the Transition to Adulthood*. New York: Russell Sage Foundation.

Al-Jassem, Diana. 2011. "Women in Polygamous Marriages Suffering Psychological Torture." ArabNews, March 8. Accessed September 12, 2012 (www.arabnews.com).

Alkhatib, Hasan, et al. 2014. "IEEE CS 2022 Report." IEEE Computer Society, February. Accessed October 8, 2014 (www.computer.org).

"All Eyes on the Sharing Economy." 2013. *Economist Technology Quarterly*, March 9, 13–15.

Allen, Jodie T., and Richard Auxier. 2010. "Ask the Expert." Pew Research Center, December 29. Accessed December 30, 2010 (www.pewresearch.org).

Allen, Joshua, et al. 2014. "Early Postnatal Exposure to Ultrafine Particulate Matter Air Pollution: Persistent Ventriculomegaly, Neurochemical Disruption, and Glial Activation Preferentially in Male Mice." *Environmental Health Perspectives*, June 5, advance publication. Accessed July 30, 2014 (www.ehponline.org).

Allot, Daniel. 2013. "Immigrant Values, American Style." *Christian Science Monitor Weekly*, August 12, 35.

Allport, Gordon W. 1954. *The Nature of Prejudice*. Reading, MA: Addison-Wesley.

Al-Mahmood, Syed Z. 2013. "Bangladesh to Raise Workers' Pay." *Wall Street Journal*, May 13, B4.

Alman, Ashley. 2014. "One Percenter Convicted of Raping Infant Child Dodges Jail Because

He 'Will Not Fare Well.'" *Huffington Post*, March 30. Accessed March 31, 2014 (www.huffingtonpost.com).

Alsever, Jennifer. 2014. "Immigrants: America's Job Creators." *Fortune*, June 16, 56.

Altonji, Joseph G., Sarah Cattan, and Iain Ware. "Identifying Sibling Influence on Teenage Substance Use." National Bureau of Economic Research, October. Accessed July 24, 2011 (www.nber.org).

Alvaredo, Facundo, Anthony B. Atkinson, Thomas Piketty, and Emmanuel Saez. 2013. "The Top 1 Percent in International and Historical Perspective." *Journal of Economic Perspectives* 27 (Summer): 3–20.

Alvarez, Lizette. 2014. "Economy and Crime Spur New Puerto Rican Exodus." *New York Times*, February 9, A1.

___. 2009. "Women at Arms: G.I. Jane Breaks the Combat Barrier." *New York Times*, August 16, A1.

American Association of Suicidology. 2009. "Elderly Suicide Fact Sheet." June 23. Accessed December 17, 2009 (www.suicidology.org).

American Cancer Society. 2013. "Breast Cancer Facts & Figures: 2013–2014." Accessed December 5, 2014 (www.cancer.org).

"American Indian and Alaska Native Heritage Month: November 2014." 2014. U.S. Census Bureau News, November 12. Accessed November 22, 2014 (www.census.gov).

American Psychiatric Association. 2013. *Diagnostic and Statistical Manual of Mental Disorders*, 5th edition. Arlington, VA: American Psychiatric Association.

American Psychological Association. 2011. "About Transgender People, Gender Identity, and Gender Expression." Accessed March 2, 2014 (www.apa.org).

___. 2014. "Are Teens Adopting Adults' Stress Habits?" February 11. Accessed April 12, 2014 (www.stressinamerica.org).

American Society for Aesthetic Plastic Surgery, The. 2014. "Cosmetic Surgery National Data Bank Statistics: 2013." Accessed November 18, 2014 (www.surgery.org).

American Sociological Association. 1999. *Code of Ethics and Policies and Procedures of the ASA Committee on Professional Ethics*. Accessed January 10, 2010 (www.asanet.org).

Amnesty International. 2014. "Death Sentences and Executions 2013." Accessed November 7, 2014 (www.amnesty.org).

Amrein-Beardsley, Audrey. 2014. "Recommended Readings on VAMs." VAMboozled. Accessed June 21, 2014 (www.VAMboozled.com).

Amusa, Malena. 2010. "'Precious' Pushes Past Controversy to Oscar Night." Women's eNews, March 5. Accessed March 8, 2010 (www.womensenew.org).

Amy, N. K., A. Aalborg, P. Lyons, and L. Keranen. 2006. "Barriers to Routine Gynecological Cancer Screening for White and African-American Obese Women." *International Journal of Obesity* 30 (January): 147–155.

Ander, Steve, and Frank Newport. 2014. "After Exchanges Close, 5% of Americans Are Newly Insured." Gallup, June 23. Accessed July 26, 2014 (www.gallup.com).

Andersen, Margaret L., and Patricia Hill Collins. 2010. "Why Race, Class, and Gender Still Matter." Pp. 1–16 in *Race, Class, and Gender: An Anthology,"* 7th edition, edited by Margaret L. Andersen and Patricia Hill Collins. Belmont, CA: Wadsworth.

Anderson, Craig A., et al. 2003. "The Influence of Media Violence on Youth." *Psychological Science in the Public Interest* 4 (December): 81–110.

Anderson, James F., and Laronistine Dyson. 2002. *Criminological Theories: Understanding Crime in America.* Lanham, MD: University Press of America.

Anderson, Jenna, and Lee Rainie. 2014. "The Internet of Things Will Thrive by 2025." Pew Research Center, May 14. Accessed October 8, 2014 (www.pewresearch.org).

Anderson, Sarah, Chuck Collins, Scott Klinger, and Sam Pizzigati. 2011. "Executive Excess 2011: The Massive CEO Rewards for Tax Dodging." Institute for Policy Studies, August 31. Accessed September 22, 2011 (www .ips-dc.org).

Andrzejewski, Adam. 2014. "The Federal Transfer Report." Open the Books, March. Accessed April 12, 2014 (www.openthebooks.com).

Annie E. Casey Foundation. 2014. "Race for Results: Building a Path to Opportunity for All Children." Accessed June 21, 2014 (www. aecf.org).

Arab American Institute Foundation. 2012. "Quick Facts About Arab Americans." Accessed March 22, 2014 (www.aaiusa.org).

Arendt, Hannah. 2004. *The Origins of Totalitarianism.* New York: Schocken.

Arieff, Allison. 2013. "What Tech Hasn't Learned from Urban Planning." *New York Times,* December 13. Accessed September 19, 2014 (www.nytimes.com).

Aris, Phillippe. 1962. *Centuries of Childhood.* New York: Vintage.

Armstrong, Elizabeth A., Laura Hamilton, and Paula England. 2010. "Is Hooking Up Bad for Young Women?" *Contexts* 9 (Summer): 22–27.

Armstrong, Elizabeth A., Paula England, and Alison C. K. Fogarty. 2012. "Accounting for Women's Orgasm and Sexual Enjoyment in College Hookups and Relationships." *American Sociological Review* 77 (June): 435–462.

Arnold, Lindsay W., Deborah C. Girasek, Brian C. Tefft, and Jurek G. Grabowski. 2013. "Temporal Trends in Indicators of Traffic Safety Culture among Drivers in the United States." AAA Foundation for Traffic Safety, August. Accessed October 12, 2013 (www .aaafoundation.org).

ASA Research Department. 2013. "Recruitment and Retention of Sociology Majors." *ASA Footnotes* 41 (January): 1, 4.

Asch, Solomon. 1952. *Social Psychology.* Englewood Cliffs, NJ: Prentice-Hall.

Ashburn, Elyse. 2007. "A Race to Rescue Native Tongues." *Chronicle of Higher Education,* September 28, B15.

Asi, Maryam, and Daniel Beaulieu. 2013. "Arab Households in the United States: 2006–2010." U.S. Census Bureau, May. Accessed March 22, 2014 (www.census.gov).

"Asian/Pacific American Heritage Month: May 2014." 2014. *U.S. Census Bureau News,* April 23. Accessed May 22, 2014 (www.census. gov).

"Asians Fastest-Growing Race or Ethnic Group in 2012, Census Bureau Reports." 2013. Newsroom, U.S. Census Bureau, June 13. Accessed March 22, 2014 (www.census.gov).

Association of Certified Fraud Examiners. 2012. "2012 Report to the Nations on Occupational Fraud and Abuse." Accessed January 12, 2014 (www.acfe.com).

Atchley, Robert C., and Amanda S. Barusch. 2004. *Social Forces and Aging: An Introduction to Social Gerontology,* 10th edition. Belmont, CA: Wadsworth.

Atici, Cemal. 2014. "Climate Change: Genetically Modified Organisms (GMO) as a Mitigating Measure?" Food and Agriculture Organization of the United Nations, May 6. Accessed October 8, 2014 (www.fao.org).

Attinasi, John J. 1994. "Racism, Language Variety, and Urban U.S. Minorities: Issues in Bilingualism and Bidialectalism." Pp. 319–347 in *Race,* edited by Steven Gregory and Roger Sanjek. New Brunswick, NJ: Rutgers University Press.

Aughinbaugh, Alison, Omar Robles, and Hugette Sun. 2013. "Marriage and Divorce: Patterns by Gender, Race, and Educational Attainment." *Monthly Labor Review* (October): 1–19.

Auguste, Byron, Paul Kihn, and Matt Miller. 2010. "Closing the Talent Gap: Attracting and Retaining Top-Third Graduates to Careers in Teaching." McKinsey & Company, September. Accessed March 20, 2012 (www .mckinsey.com).

Aunola, Kaisa, and Jari-Erik Nurmi. 2005. "The Role of Parenting Styles in Children's Problem Behavior." *Child Development* 76 (November/December): 1144–1159.

Auster, Carol J., and Claire S. Mansbach. 2012. "The Gender Marketing of Toys: An Analysis of Color and Type of Toy on the Disney Store Website." *Sex Roles* 67 (October): 375–388.

Austin, Algernon. 2013. "Native Americans and Jobs." Economic Policy Institute, December 17. Accessed March 22, 2014 (www.epi.org).

"Australian Passports Now Offer 'M' for Male, 'F' for Female or 'X.'" 2011. *Baltimore Sun,* September 18, 24.

Autor, David H. 2014. "Skills, Education, and the Rise of Earnings Inequality Among the 'Other 99 Percent.'" *Science* 344 (May): 843–851.

Auxier, Richard. 2010. "Congress in a Wordle." Pew Research Center, March 22. Accessed March 26, 2010 (www.pewresearch.org).

Avellar, Sarah, and Pamela Smock. 2005. "The Economic Consequences of the Dissolution of Cohabiting Unions." *Journal of Marriage and Family* 67 (May): 315–327.

Axtell, Roger E., Tami Briggs, Margaret Corcoran, and Mary Beth Lamb. 1997. *Do's and Taboos Around the World for Women in Business.* New York: John Wiley & Sons.

Babbie, Earl. 2013. *Social Research Counts.* Belmont, CA: Cengage.

Babcock, Philip, and Mindy Marks. 2011. "The Falling Time Cost of College: Evidence from Half a Century of Time Use Data." *The Review of Economics and Statistics* 83 (May): 468–478.

Baenninger, MaryAnn. 2011. "For Women on Campuses, Access Doesn't Equal Success." *Chronicle of Higher Education,* October 7, A26.

Bahadur, Nina. 2013. "'Men Taking Up Too Much Space on the Train Raises Some Interesting Questions About Being Male in Public." *Huffington Post,* September 13. Accessed December 4, 2013 (www.huffingtonpost .com).

Bainbridge, William S. 1997. *The Sociology of Religious Movements.* New York: Routledge.

Baker, Aryn. 2014. "Unholy Choices." *Time,* March 21, 36–41.

Bales, Robert F. 1950. *Interaction Process Analysis.* Reading, MA: Addison-Wesley.

Ballhaus, Rebecca. 2013. "Muslim Worker Wins Bias Case." *Wall Street Journal,* September 10, A3.

Baltimore Sun. 2012. "Elderly Man Dies in Fire at Single-Family Home." December 9, 5.

Banchero, Stephanie. 2013. "Students Slip in Global Tests." *Wall Street Journal,* December 3, A8.

Bandura, Albert, and Richard H. Walters. 1963. *Social Learning and Personality Development.* New York: Holt, Rinehart & Winston.

Bandy, Tawana. 2012. "What Works for Male Children and Adolescents: Lessons from Experimental Evaluations of Programs and Interventions." Child Trends Research Brief, August. Accessed September 22, 2012 (www .childtrends.org).

Banerjee, Abhijit, Esther Duflo, Maitreesh Ghatak, and Jeanne Lafortune. 2009. "Marry for What? Caste and Mate Selection in Modern India." National Bureau of Economic Research, May. Accessed March 2, 2010 (www.nber.org).

Banerjee, Neela. 2006. "Clergywomen Find Hard Path to Bigger Pulpit." *New York Times,* August 26, A1, A12.

Banfield, Edward C. 1974. *The Unheavenly City Revisited.* Boston: Little, Brown.

Barash, David P. 2012. "The Evolutionary Mystery of Homosexuality." *Chronicle Review,* November 21, B4–B5.

Banks, Duren, and Tracey Kyckelhahn. 2011. "Characteristics of Suspected Human Trafficking Incidents, 2008–2010." Office of Justice Programs, April. Accessed October 25, 2011 (www.ojp.gov).

Banks, Ingrid. 2000. *Hair Matters: Beauty, Power, and Black Women's Consciousness.* New York: New York University Press.

Barker, James B. 1993. "Tightening the Iron Cage: Concertive Control in Self-Managing Teams." *Administrative Science Quarterly* 38 (September): 408–437.

Barker, Megan M. 2011. "Manufacturing Employment Hard Hit During the 2007–09 Recession." *Monthly Labor Review* 134 (April): 28–33.

Barnes, Cynthia. 2006. "China's 'Kingdom of Women.'" *Slate,* November 17. Accessed June 12, 2007 (www.slate.com).

Barnes, Robert. 2014. "Supreme Court Allows Texas to Use Voter ID Law." *Washington Post,* October 18. Accessed November 23, 2014 (www.washingtonpost.com).

___. 2010. "Supreme Court Rules on Employer Monitoring of Cellphone, Computer Conversations." *Washington Post,* June 18, A1.

___. 2011. "Limits on Video Games Rejected." *Washington Post,* June 28, A1.

Barnes, Taylor. 2012. "Watch Your Tongue: Prejudiced Comments Illegal in Brazil." *Christian Science Monitor,* December 4. Accessed November 10, 2013 (www .csmonitor.com).

Barnett, Melissa A., W. R. Mills-Koonce, Hanna Gustafsson, and Martha Cox. 2012. "Mother-Grandmother Conflict, Negative Parenting, And Young Children's Social Development in Multigenerational Families." *Family Relations* 61 (December): 864–877.

Barreto, Michelle, Micheal K. Ryan, and Manuela T. Schmitt, eds. 2009. *The Glass Ceiling in the 21st Century: Understanding Barriers to Gender Equality.* Washington, DC: American Psychological Association.

Barrett, Devlin. 2013. "Overhaul Aims to Cut Prison Population." *Wall Street Journal*, August 12, A3.

Barrett, Paul M. 2014. "Bad Sports." *Bloomberg Businessweek*, March 3–9, 51–55.

Barro, Robert J., and Rachel M. McCleary. 2003. "Religion and Economic Growth across Countries." *American Sociological Review* 68 (October): 760–781.

Barry, Ellen, and Mansi Choksi. 2013. "Gang Rape in India, Routine and Invisible." *New York Times*, October 27, A1.

Barry, Ellen. 2013. "Policing Village Moral Codes as Women Stream to India's Cities." *New York Times*, October 20, A6.

Bart, Pauline B., and Eileen Geil Moran, eds. 1993. *Violence Against Women: The Bloody Footprints*. Thousand Oaks, CA: Sage.

Bartlett, Tom. 2011. "Caffeine Is Definitely Good/Bad for You." *Chronicle of Higher Education*, January 20. Accessed January 25, 2011 (www.chronicle.com/blogs).

Bartlett, Thomas, and Paula Wasley. 2008. "Just Say 'A': Grade Inflation Undergoes Reality Check." *Chronicle of Higher Education*, September 5. Accessed June 21, 2014 (www.chronicle.com).

Barton, Allen H. 1980. "A Diagnosis of Bureaucratic Maladies." Pp. 27–36 in *Making Bureaucracies Work*, edited by Carol H. Weiss and Allen H. Barton. Beverly Hills, CA: Sage.

Baumrind, Diana. 1968. "Authoritarian versus Authoritative Parental Control." *Adolescence* 3 (11): 255–272.

———. 1989. "Rearing Competent Children." Pp. 349–378 in *Child Development Today and Tomorrow*, edited by William Damon. San Francisco: Jossey-Bass.

Bass, Frank, and Dakin Campbell. 2013. "Poor Neighborhoods See Branches Disappear." *Bloomberg Businessweek*, May 13–19, 48–49.

Bassett, Laura, and Ryan J. Reilly. 2014. "Hobby Lobby Case, Dealing Blow to Birth Control Coverage." *Huffington Post*, June 30. Accessed July 26, 2014 (www.huffingtonpost.com).

Basu, Sutapa. 2014. "Sex, Money, and Brutality." *Contexts* 13 (Winter): 16–18.

Bates, Nancy, and Theresa J. DeMaio. 2013. "Measuring Same-Sex Relationships." *Contexts* 12 (Winter): 66–69.

Baum, Sandy, Jennifer Ma, and Kathleen Payea. 2013. "Education Pays 2013: The Benefits of Higher Education for Individuals and Society." College Board. Accessed June 21, 2014 (www.collegeboard.org).

Baum, Sandy, and Kathleen Payea. 2011. "Trends in Student Aid 2011." College Board Advocacy & Policy Center. Accessed April 12, 2012 (www.collegeboard.org).

Bax, Pauline. 2013. "In Ghana, Death Has Become Big Business." *Bloomberg Businessweek*, August 29–September 1, 23–26.

Bazelon, Emily. 2009. "2 Kids + 0 Husbands = Family." *New York Times Magazine*, February 1, 30.

Beadle, Amanda Peterson. 2012. "Minnesota Senator Thinks Abortion Pill Is Wrong, While Viagra Is a 'Wonderful Drug.'" Alternet, May 3. Accessed May 24 (www.alternet.org).

Beck, Allen J., Ramona R. Rantala, and Jessica Rexroat. 2014. "Sexual Victimization Reported by Adult Correctional Authorities, 2009–11." Bureau of Justice Statistics, January. Accessed March 2, 2014 (www.ojp.usdoj.gov).

Beck, Melinda. 2013. "Nurse Practitioners Seek Right to Treat Patients on Their Own." *Wall Street Journal*, August 15, A3.

Becker, Howard, and Ruth H. Useem. 1942. "Sociological Analysis of the Dyad." *American Sociological Review* 7 (February): 3–26.

Becker, Howard S. 1963. *Outsiders: Studies in the Sociology of Deviance*. New York: Free Press.

Beckles, Gloria L., and Benedict I. Truman. 2013. "Education and Income—United States, 2009 and 2011." *MMWR* 62 (November 22): 9–19.

Beech, Hannah. 2013. "The Face of Buddhist Terror." *Time*, July1, 42–50.

———. 2013. "Why China Needs More Children." *Time*, December 2, 36–49.

Begala, Paul. 2011. "Who You Calling Lazy?" *Newsweek*, December 15, 8.

Begley, Sharon. 2010. "Sins of the Grandfathers." *Newsweek*, November 8, 48–50.

Beil, Laura. 2012. "How Much Would You Pay for Three More Months of Life?" *Newsweek*, September 3, 40–44.

Belkin, Douglas, and Caroline Porter. 2012. "Web Profiles Haunt Students." *Wall Street Journal*, October 4, A3.

Belknap, Joanne. 2007. *The Invisible Woman: Gender, Crime, and Justice*, 3rd edition. Belmont, CA: Wadsworth Press.

Bendavid, Naftali. 2013. "Countries Expand Recognition for Alternative 'Intersex' Gender." *Wall Street Journal*, October 31, A9.

Bengali, Shashank. 2014. "Toll of Refugees Fleeing Syrian Strife Tops 3 Million, U.N. Says." *Baltimore Sun*, August 30, 10.

Bennett, Drake. 2010. "This Will Be on the Midterm. You Feel Me?" *Slate*, March 24. Accessed March 26, 2010 (www.slate.com).

Bennett, Jessica. 2009. "Tales of a Modern Diva." *Newsweek*, April 6, 42–43.

Bennett, Jessica, and Jesse Ellison. 2010. "'I Don't': The Case Against Marriage." *Newsweek*, June 21, 42–45.

Bennett-Smith, Meredith. 2013. "Mark and Pam Crawford, Parents of Intersex Child, Sue South Carolina for Sex Assignment Surgery." *Huffington Post*, May 15. Accessed March 2, 2014 (www.huffingtonpost.com).

———. 2013. "More Than 30 Percent of Americans Think Gays Can Become Straight." *Huffington Post*, August 22. Accessed March 2, 2014 (huffingtonpost.com).

Bennhold, Katrin. 2014. "Major Gain for Women in Church of England." *New York Times*, November 18, A4.

Benokraitis, Nijole V. 1997. *Subtle Sexism: Current Practices and Prospects for Change*. Thousand Oaks, CA: Sage.

———. 2014. *Marriages & Families: Changes, Choices, and Constraints*, 8th edition. Upper Saddle River, NJ: Prentice Hall.

Benson, Michael L. 2002. *Crime and the Life Course*. Los Angeles: Roxbury.

Benton, Thomas H. 2011. "A Perfect Storm in Undergraduate Education." *Chronicle of Higher Education*, February 25, A43, A45. Part 2 in April 8, A45–A46.

Berfield, Susan. 2012. "Levi's Has a New Color for Blue Jeans: Green." *Bloomberg Businessweek*, October 22–28, 26–28.

———. 2013. "Fast-Food Workers of the World, Unite!" *Bloomberg Businessweek*, December 9–15, 20–22.

Bergen, Raquel K., Jeffrey L. Edleson, and Claire M. Renzetti, eds. 2005. *Violence Against Women: Classic Papers*. Boston: Allyn & Bacon.

Berger, Peter L., and Thomas Luckmann. 1966. *The Social Construction of Reality: A Treatise in the Sociology of Knowledge*. New York: Doubleday.

Berk, Richard A. 1974. *Collective Behavior*. Dubuque, IA: Wm. C. Brown Company.

Berkos, Kristen M., Terre H. Allen, Patricia Kearney, and Timothy G. Plax. 2001. "When Norms Are Violated: Imagined Interactions as Processing and Coping Mechanisms." *Communication Monographs* 68 (September): 289–300.

Berkowitz, Bill. 2011. "Why Is Jerry Falwell's Evangelical University Getting Filthy Rich Off Your Tax Money?" AlterNet, June 29. Accessed June 29, 2011 (www.alternet.org).

Berlin, Overton B., and Paul Kay. 1969. *Basic Color Terms*. Berkeley: University of California Press.

Bernstein, David E. 2011. "Overt vs. Covert." *New York Times*, May 22. Accessed May 23, 2011 (www.newyorktimes.com).

Bermudez, Esmeralda. 2008. "In L.A., Speaking 'Mexican' to Fit In." *Los Angeles Times*, November 3. Accessed March 22, 2014 (www.latimes.com).

Berrett, Dan. 2014. "Some Elite Colleges Reject AP Credits in Favor of New Core Classes." *Chronicle of Higher Education*, February 21, A13.

Bersani, Bianca E. 2014. "A Game of Catch-Up? The Offending Experience of Second-Generation Immigrants." *Crime & Delinquency* 60 (February): 60–84.

Bertrand, Marianne, and Jessica Pan. 2011. "The Trouble with Boys: Social Influences and the Gender Gap in Disruptive Behavior." National Bureau of Economic Research, October. Accessed April 14, 2012 (www.nber.org).

Berube, Alan, Audrey Sinter, Jill H. Wilson, and William H. Frey. 2006. "Finding Exurbia: America's Fast-Growing Communities at the Metropolitan Fringe." Brookings Institution. Accessed February 2, 2008 (www.brookings.edu).

Best, Joel. 2006. *Flavor of the Month: Why Smart People Fall for Fads*. Berkeley: University of California Press.

Betcher, R. William, and William S. Pollack. 1993. *In a Time of Fallen Heroes: The Re-Creation of Masculinity*. New York: Atheneum.

Bettinger, Eric P., Bridget T. Long, Philip Oreopoulos, and Lisa Sanbonmatsu. 2012. "The Role of Application Assistance and Information in College Decisions: Results from the H&R Block FAFSA Experiment." *Quarterly Journal of Economics* 127 (August): 1205–1242.

Bialik, Carl. 2011. "Irreconcilable Claim: Facebook Causes 1 in 5 Divorces." *Wall Street Journal*, March 12. Accessed November 8, 2011 (www.online.wsj.com).

"Biggest Transnational Companies." 2012. *The Economist*, July 10. Accessed April 12, 2014 (www.economist.com).

Billingsley, Andrew. 1992. *Climbing Jacob's Ladder: The Enduring Legacy of African-American Families*. New York: Simon & Schuster.

Bilton, Nick. 2013. "Minecraft, a Child's Obsession, Finds Use as an Educational Tool." *New York Times*, September 16, B8.

Bird, Chloe E., et al. 2010. "Neighborhood Socioeconomic Status and Biological 'Wear & Tear' in a Nationally Representative Sample of U.S. Adults." *Journal of Epidemiology and Community Health* 64 (October): 860–865.

Bird, Warren, and Scott Thumma. 2011. "A New Decade of Megachurches: 2011 Profile of Large Attendance Churches in the United States." Leadership and Network. Accessed June 21, 2014 (www.leadnet.org).

Bischoff, Kendra, and Sean f. Reardon. 2013. "Residential Segregation by Income,

1970–2009." Russell Sage Foundation, October 16. Accessed September 19, 2014 (www.russellsage.org).

Bitterman, Amy, Rebecca Goldring, and Lucinda Gray. 2013. "Characteristics of Public and Private Elementary and Secondary School Principals in the United States: Results from the 2011–12 Schools and Staffing Survey." National Center for Education Statistics, August. Accessed November 17, 2014 (www.nces.ed.gov).

"Black (African-American) History Month: February 2014." 2014. U.S. Census Bureau News, January 16. Accessed March 22, 2014 (census.gov).

Black, M. C., et al. 2011. "National Intimate Partner and Sexual Violence Survey: 2010 Summary Report." Centers for Disease Control and Prevention. Accessed November 2013 (cdc.gov).

Black, Sandra E., Jane A. Lincove, Jenna Cullinane, and Rachel Veron. 2014. "Can You Leave High School Behind?" National Bureau of Economic Research, January. Accessed June 21, 2014 (www.nber.org).

Black, Thomas. 2010. "More Car Jobs Shift to Mexico." *Bloomberg Businessweek*, June 28–July 4, 10–11.

Black-Owned Firms: 2002. 2006. U.S. Census Bureau. Accessed May 1, 2007 (census.gov).

Blake, Mariah. 2014. "Are Any Plastics Safe?" *Mother Jones*, March/April, 19–25, 60–61.

———. 2014. "The Chevron Communiques: A Trove of Secret Cables Shows How Hillary Clinton's State Department Sold Fracking to the World." *Mother Jones*, September/October, 51–54, 72.

Blanchard, Dallas A. 1994. *The Anti-Abortion Movement and the Rise of the Religious Right: From Polite to Fiery Protest*. New York: Twayne Publishers.

Blau, Peter M. 1986. *Exchange and Power in Social Life*, revised edition. New Brunswick, NJ: Transaction.

Blau, Peter M., and Marshall W. Meyer. 1987. *Bureaucracy in Modern Society*, 3rd edition. New York: Random House.

Bloomberg News. 2014. "Madoff Scam Recovery Hits $10 Billion, Almost 60% of Lost Money, at a Cost of $1 Billion." November 22. Accessed December 9, 2014 (www.investmentnews.com).

BLS News Release. 2013. "Volunteering in the United States—2012." Bureau of Labor Statistics, February 22. Accessed December 19, 2013 (www.bls.gov/cps).

———. 2014. "The Employment Situation—March 2014." Bureau of Labor Statistics, April 4. Accessed April 6, 2014 (www.bls.gov/cps).

———. 2014. "Usual Weekly Earnings of Wage and Salary Workers: Third Quarter 2014." Bureau of Labor Statistics, October 24. Accessed December 2, 2014 (www.bls.gov/cps).

BLS Reports. 2013. "A Profile of the Working Poor, 2011." Bureau of Labor Statistics, April. Accessed February 6, 2014 (www.bls.gov).

———. 2014. "A Profile of the Working Poor, 2012." Bureau of Labor Statistics, March. Accessed November 6, 2014 (www.bls.gov).

———. 2014. "Women in the Labor Force: A Databook." Bureau of Labor Statistics, May. Accessed June 21, 2014 (www.bls.gov).

Blumer, Herbert. 1946. "Collective Behavior." Pp. 65–121 in *New Outline of the Principles of Sociology*, edited by Alfred M. Lee. New York: Barnes & Noble.

———. 1969. *Symbolic Interactionism: Perspective and Method*. Englewood Cliffs, NJ: Prentice Hall.

Board of Trustees, The. 2013. *The 2013 Annual Report of the Board of Trustees of the Federal Old-Age and Survivors Insurance and Federal Disability Insurance Trust Funds*. Accessed July 28, 2013 (www.socialsecurity.gov).

Bobroff-Hajal, Anne. 2006. "Why Cousin Marriage Matters in Iraq." *Christian Science Monitor*, December 26, 9.

Bodley, Michael. 2014. "Custom-Made Knees Using 3-D Printing." *Baltimore Sun*, July 20, 10.

Bogle, Kathleen A. 2008. *Hooking Up: Sex, Dating, and Relationships on Campus*. New York: New York University Press.

Bollag, Burton. 2007. "Credential Creep." *Chronicle of Higher Education*, June 22, A10–A12.

Bohannon, Paul, ed. 1971. *Divorce and After*. New York: Doubleday.

Bonacich, Edna. 1972. "A Theory of Ethnic Antagonism: The Split Labor Market." *American Sociological Review* 37 (October): 547–559.

Bonner, Robert. 2007. "Research—Male Elementary Teachers: Myths and Realities." Men Teach. Accessed June 12, 2008 (www.menteach.org).

Booth, Robert. 2014. "Why the Facebook Experiment Toying with Your Emotions Is So Terrifying." Alternet, June 29 (accessed August 8, 2014).

Borgerhoff Mulder, Monique. 2009. "Serial Monogamy as Polygyny or Polyandry?" *Human Nature* 20 (Summer): 130–150.

Borowiec, Steven. 2014. "Adoptees Try to Lift Stigma." *Christian Science Monitor Weekly*, April 7, 16–17.

Bosk, Charles. 1979. *Forgive and Remember: Managing Medical Failure*. Chicago: University of Chicago Press.

Boslaugh, Sarah. 2007. *Secondary Data Sources for Public Health: A Practical Guide*. New York: Cambridge University Press.

"Boss Sells His Company, Shares Proceeds." 1999. *Baltimore Sun*, September 12, 24A.

Botstein, Leon. 2014. "The SAT Is Part Hoax, Part Fraud." *Time*, March 24, 17.

Bourdieu, Pierre. 1984. *Distinction: A Social Critique of the Judgement of Taste*. Translated by Richard Nice. Cambridge, MA: Harvard University Press.

———. 1986. "The Forms of Capital." Pp. 241–258 in *Handbook for Theory and Research for the Sociology of Education*, edited by J. G. Richardson. Westport, CT: Greenwood.

Bowles, Samuel, and Herbert Gintis. 1977. *Schooling in Capitalist America: Educational Reform and the Contradictions of Economic Life*. New York: Basic Books.

Boyle, Matthew. 2013. "Yes, Real Men Drink Beer and Use Skin Moisturizer." *Bloomberg Businessweek*, October 7–13, 32–33.

Bradshaw, York W., and Michael Wallace. 1996. *Global Inequalities*. Thousand Oaks, CA: Pine Forge Press.

Braga, Anthony A. 2003. "Systematic Review of the Effects of Hot Spots Policing on Crime." Unpublished paper. Accessed May 14, 2005 (www.campbellcollaboration.org).

Braitman, Keli A., Neil K. Chaudhary, and Anne T. McCart. 2011. "Effect of Passenger Presence on Older Drivers' Risk of Fatal Crash Involvement." Insurance Institute for Highway Safety, March. Accessed July 4, 2011 (www.iihs.org).

Bremner, Jason, and Lori M. Hunter. 2014. "Migration and the Environment." *Population Bulletin* 69 (June): 1–11.

Brenner, Joanna, and Aaron Smith. 2013. "72% of Online Adults Are Social Networking Site Users." Pew Research Center, August 5. Accessed December 4, 2013 (www.pewinternet.org).

Brescoll, Victoria L., and Eric L. Uhlmann. 2008. "Can an Angry Woman Get Ahead? Status Conferral, Gender, and Expression of Emotion in the Workplace." *Psychological Science* 19 (March): 268–275.

Brewis, Alexandra A., Amber Wutich, Ashlan Falletta-Cowden, and Isa Rodriguez-Soto. 2011. "Body Norms and Fat Stigma in Global Perspective." *Current Anthropology* 52 (April): 269–276.

Brill, Steven. 2013. "Bitter Pill: How Outrageous Pricing and Egregious Profits Are Destroying Our Health Care." *Time*, March 4, 14–55.

Broadwater, Luke. 2012. "Wells Fargo Settles Bias Suit for $175M." *Baltimore Sun*, July 13, 1, 11.

Brody, Gene H., Shannon Dorsey, Rex Forehand, and Lisa Armistead. 2002. "Unique and Protective Contributions of Parenting and Classroom Processes to the Adjustment of African American Children Living in Single-Parent Families." *Child Development* 73 (January-February): 274–286.

"Broke in the 'Burbs." 2013. *The Economist*, July 20, 30.

Brooks, Arthur C. 2007. "I Love My Work." *The American* (online): September–October. Accessed September 30, 2007 (www.theamericanmag.com).

Brown, Alyssa. 2013. "Costs Still Keep 30% of Americans from Getting Treatment." Gallup, December 9. Accessed July 26, 2014 (www.gallup.com).

Brown, Archie. 2010. "Signposts: Why Did Communism End When It Did?" *History Today*, March. Accessed April 12, 2014 (www.historytoday.com).

Brown, DeNeen L. 2009. "The High Cost of Poverty: Why the Poor Pay More." *Washington Post*, May 18, C1.

Brown, Matthew Hay. 2013. "Breaking the Silence." *Baltimore Sun*, December 15, 1, 24–25.

Brown, Michael, and Amy Goldin. 1973. *Collective Behavior: A Review and Reinterpretation of the Literature*. Pacific Palisades, CA: Goodyear.

Brown, Pamela A., and Elizabeth McGann. 2011. "Medication Error Prevention: A Shared Responsibility." Medscape Medical News, June 14. Accessed April 24, 2012 (www.medscape.com).

Brown, Susan I. 2005. "How Cohabitation Is Reshaping American Families." *Contexts* 4 (Summer): 33–37.

Bruinius, Harry. 2013. "Whose Holidays Are They?" *Christian Science Monitor*, December 16, 21–23.

———. 2014. "A Tarnished American Dream?" *Christian Science Monitor Weekly*, January 13, 21–23.

Brunvand, Jan H. 2001. *The Truth Never Stands in the Way of a Good Story*. Urbana and Chicago: University of Illinois Press.

Buchanan, Patrick J. 2011. "Will Multiculturalism End Europe?" *The American Conservative*, February 14. Accessed October 25, 2014 (www.theamericanconservative.com).

Buchheit, Paul. 2014a. "5 Ways Rich People's 'Entitlements' Cheat You and Me." AlterNet, February 9. Accessed February 14, 2014 (www.alternet.org).

———. 2014b. "4 Money Grabs Show a Few Rich People Are the Ones Getting Wealthier in America." AlterNet, January 19. Accessed February 6, 2014 (www.alternet.org).

Buechler, Steven M. 2000. *Social Movements in Advanced Capitalism: The Political Economy* and Cultural Construction of Social Activism. New York: Oxford University Press.

Buffet, Warren. 2011. "Stop Coddling the Super-Rich." *New York Times,* August 14, B21.

Bullen, Christopher, et al. 2013. "Electronic Cigarettes for Smoking Cessation: A Randomized Controlled Trial." *The Lancet* 382 (November 16): 1629–1637.

Bullock, Karen. 2005. "Grandfathers and the Impact of Raising Grandchildren." *Journal of Sociology and Social Welfare* 32 (March): 43–59.

Burd, Stephen. 2013. "Undermining Pell: How Colleges Compete for Wealthy Students and Leave the Low-Income Behind." New America Foundation, May. Accessed June 21, 2014 (www.newamerica.org).

Bureau of Labor Statistics. 2014–2015. *Occupational Outlook Handbook, 2014–15 Edition.* "Surgeons and Physicians." Accessed July 26, 2014 (www.bls.gov).

___. 2014. "Charting the Labor Market: Data from the Current Population Survey (CPS)." April 4. Accessed April 12, 2014 (www.bls.gov).

Burkitt, Laurie. 2013. "China Eases Limits on Births as Labor Shortage Looms." *Wall Street Journal,* November 16–17, A1, A8.

Bushman, Brad J., Patrick E. Jamieson, Llana Weitz, and Daniel Romer. 2013. "Gun Violence Trends in Movies." *Pediatrics* 32 (5): 1014–18.

Butler, Kiera. 2014. "Google's Magic Bus." *Mother Jones,* May–June, 70–71.

Butterfield, Fox. 2002. "Father Steals Best: Crime in an American Family." *New York Times* (August 15): 1A.

Cabrera, Natasha J., Jay Fagan, and Danielle Farrie. 2008. "Explaining the Long Reach of Fathers' Prenatal Involvement on Later Paternal Engagement." *Journal of Marriage and Family* 70 (December): 1094–1107.

Cadge, Wendy, and Courtney Bender. 2004. "Yoga and Rebirth in America: Asian Religions Are Here to Stay." *Contexts* 3 (Winter): 45–51.

Calderon, Valerie J., and Preety Sidhu. 2014. "Business Leaders Say Knowledge Trumps College Pedigree." Gallup, February 25. Accessed June 21, 2014 (www.gallup.com).

Calderon, Valerie J., and Susan Sorenson. 2014. "Americans Say College Degree Leads to a Better Life." Gallup, April 7. Accessed June 21, 2014 (www.gallup.com).

Cameron, Deborah. 2007. *The Myth of Mars and Venus.* New York: Oxford University Press.

Campo-Flores, Arian. 2013. "Religious Dorms Sprout Up." *Wall Street Journal,* September 3, A3.

Campion-Vincent, Véronique. 2005. "From Evil Others to Evil Elites: A Dominant Pattern in Conspiracy Theories Today." Pp. 103–122 in *Rumor Mills: The Social Impact of Rumor and Legend,* edited by Gary Alan Fine, Veronique Campion-Vincent, and Chip Heath. New Brunswick, NJ: Transaction Publishers.

Cantrell, Steven, and Thomas J. Kane. 2013. "Ensuring Fair and Reliable Measures of Effective Teaching." MET Project, Policy and Practice Brief. Accessed June 21, 2014 (www.metproject.org).

Cantril, Hadley, Hazel Gaudet, and Herta Herzog. 1952. *The Invasion from Mars.* Princeton, NJ: Princeton University Press.

Capaccio, Tony, and Jonathan D. Salant. 2014. "Military: The Pentagon Loves to Pay Top Dollar." *Bloomberg Businessweek, July* 21–27, 26–27.

Cappiello, Dina. 2012. "EPA: Power Plant Are Main Global Warming Culprits." *USA Today,*

January 11. Accessed January 12, 2012 (www.usatoday.com).

"Capital Punishment: The Slow Death of the Death Penalty." 2014. *The Economist,* April 26, 27–29.

Caplan, Arthur L. 2013. "Genetically Modified Food: Good, Bad, Ugly." *Chronicle Review,* September 13, B4–B5.

Carballo, José Rodríguez. 2014. "Final Report on the Apostolic Visitation of Women Religious in the United States of America." Holy See Press Office, December 16. Accessed December 28, 2014 (press.vatican.va)

Card, David, and Laura Giuliano. 2011. "Peer Effects and Multiple Equilibria in the Risky Behavior of Friends." National Bureau of Economic Research, May. Accessed July 24, 2011 (www.nber.org).

Carey, Anne R., and Paul Trap. 2014. "How Much Are 'Mom Jobs' Worth?" *USA Today,* May 5–9, 1A.

Carey, Benedict. 2010. "Revising Book on Disorders of the Mind." *New York Times,* February 10, 1.

Carey, Kevin. 2013. "Too Much 'Merit Aid' Requires No Merit." *Chronicle of Higher Education,* February 22, A20–A21.

Carl, Traci. 2002. "Amid Latte, Mocha Craze, Coffee Growers Go Hungry in Paradise." Accessed September 19, 2002 (www.yahoo.com).

Carnagey, Nicholas L., and Craig A. Anderson. 2005. "The Effects of Reward and Punishment in Violent Video Games on Aggressive Affect, Cognition, and Behavior." *Psychological Science* 16 (November): 882–889.

Carnevale, Anthony P., and Ban Cheah. 2013. "Hard Times: College Majors, Unemployment, and Earnings." Georgetown Center on Education and the Workforce, May. Accessed June 24, 2014 (www.cew.georgetown.edu).

Carney, Ginny. 1997. "Native American Loanwords in American English." *Wicazo SA Review,* 12 (Spring): 189–203

Carnoy, Martin, and Richard Rothstein. 2013. "What Do International Tests Really Show About U.S. Student Performance?" Economic Policy Institute, January 28. Accessed June 21, 2014 (www.epi.org).

Carr, Deborah. 2012. "The Social Stratification of Older Adults' Preparations for End-of-Life Health Care." *Journal of Health and Social Behavior* 53 (3): 297–312.

Carrell, Scott E., Mark Hoekstra, and James E. West. 2010. "Is Poor Fitness Contagious? Evidence from Randomly Assigned Friends." National Bureau of Economic Research, November. Accessed July 24, 2011 (www.nber.org).

Carroll, John. 1956. *Language, Thought & Reality: Selected Writings of Benjamin Lee Whorf.* Cambridge, MA: MIT Press.

Carson, E. Ann. 2014. "Prisoners in 2013." Bureau of Justice Statistics, September. Accessed November 6, 2014 (www.bjs.gov).

Case, Nancy Humphrey. 2010. "When Employees Rule." *Christian Science Monitor.* March 29, 22.

Caston, Richard J. 1998. *Life in a Business-Oriented Society: A Sociological Perspective.* Boston: Allyn & Bacon.

Catalano, Shannan. 2013. "Intimate Partner Violence: Attributes of Victimization, 1993–2011." Bureau of Justice Statistics, November. Accessed November 25, 2014 (www.bjs.gov).

Catalyst. 2014. "Women CEOs of the Fortune 1000." January 15. Accessed February 25, 2014 (www.catalyst.org).

"Catching Up." 2013. *The Economist,* January 12, 67–68.

CDC Fact Sheet. 2014. "Reported STDs in the United States: 2012 National Data for Chlamydia, Gonorrhea, and Syphilis." Centers for Disease Control and Prevention, January. Accessed July 26, 2014 (www.cdc.gov).

CDC/NCHS National Vital Statistics System. 2013. "National Marriage and Divorce Rate Trends." February 19. Accessed May 16, 2014 (www.cdc.gov).

Cech, Erin, Brian Rubineau, Susan Silbey, and Caroll Seron. 2011. "Professional Role Confidence and Gendered Persistence in Engineering." *American Sociological Review* 70 (October): 641–660

Cellini, Stephanie R., Signe-Mary McKernan, and Caroline Ratcliffe. 2008. "The Dynamics of Poverty in the United States: A Review of Data, Methods, and Findings." *Journal of Policy Analysis and Management* 27 (Summer): 577–605.

Cendrowski, Scott. 2014. "Business Created China's Pollution Problem." *Fortune,* April 28, 90–93.

"Census Bureau Reports Minority Business Ownership Increasing at More Than Twice the National Rate." 2010. U.S. Census Bureau Newsroom, July 13. Accessed November 20, 2011 (www.census.gov).

Center for American Women and Politics. 2014. "Women Elective Office 2014." Accessed March 2, 2014 (www.cawp.rutgers.edu).

Center for Community College Student Engagement. 2014. *Aspirations to Achievement: Men of Color and Community Colleges.* Austin, TX: The University of Texas at Austin, Program in Higher Education Leadership.

Center for Responsive Politics. 2014a. "The Money Behind the Elections." Accessed April 12, 2014 (www.opensecrets.org).

___.2014b. "Top PACs." Accessed April 12, 2014 (www.opensecrets.org).

___. 2014c. "2014 Campaign Contribution Limits." Accessed April 12, 2014 (www.opensecrets.org).

Center for the Study of Global Christianity. 2007. "Global Table 5: Status of Global Mission, Presence and Activities, AD 1800–2005." Gordon-Conwell Theological Seminary. Accessed August 24, 2007 (www.gcts.edu).

Centers for Medicare & Medicaid Services. 2010. "NHE Tables for Selected Calendar 1960–2010." April 11. Accessed April 24, 2010 (www.cms.gov).

___. 2014. "National Health Expenditure Projections 2012–2022." May 6. Accessed July 26, 2014 (www.cms.gov).

Cha, Ariana Eunjung. 2014. "'Smart Pills' with Chips, Cameras and Robotic Parts Raise Legal, Ethical Questions." *Washington Post,* May 24. Accessed October 8, 2014 (www.washingtonpost.com).

Chambliss, William J. 1969. *Crime and the Legal Process.* New York: McGraw-Hill.

Chambliss, William J., and Robert B. Seidman. 1982. *Law, Order, and Power,* 2nd edition. Reading, MA: Addison-Wesley.

Chandler, Tertius, and Gerald Fox. 1974. *3000 Years of Urban Growth.* New York: Academic Press.

Chandra, Anjani, William D. Mosher, and Casey Copen. 2011. "Sexual Behavior, Sexual Attraction, and Sexual Identity in the United States: Data from the 2006–2008 National Survey of Family Growth." *National Health Statistics Reports,* No. 36, March 3. Accessed March 2, 2014 (www.cdc.gov/nchs).

Chapman, Tony, and Jenny Hockey, eds. 1999. *Ideal Homes? Social Change and Domestic Life.* New York: Routledge.

Chasin, Barbara H. 2004. *Inequality & Violence in the United States: Casualties of Capitalism,* 2nd edition. Amherst, NY: Humanity Books.

Chaves, Mark, and Shawna L. Anderson. 2014. "Changing American Congregations: Findings from the Third Wave of the National Congregations Study." *Journal for the Scientific Study of Religion* 53 (December): 676–689.

Chaves, Mark, Shawna Anderson, and Jason Byassee. 2009. "American Congregations at the Beginning of the 21st Century." National Congregation Study. Accessed June 21, 2014 (www.soc.duke.edu/natcong).

Chelala, César. 2002. "World Violence Against Women a Great Unspoken Pandemic." *Philadelphia Inquirer,* November 4. Accessed November 7, 2002 (www.commondreams.org).

Chemi, Eric. 2014. "One Year: 3,885 Spills." *Bloomberg Businessweek,* February 3–9, 55.

Chen, Alice, Emily Oster, and Heidi Williams. 2014. "Why is Infant Mortality Higher in the US Than in Europe?" National Bureau of Economic Research, September. Accessed December 5, 2014 (www.nber.org).

Chen, Frances S., Julia A. Minson, Maren Schöne1, and Markus Heinrichs. 2013. "In the Eye of the Beholder: Eye Contact Increases Resistance to Persuasion." *Psychological Science* 24 (November): 2254–2261.

Chen, Juan, and David T. Takeuchi. 2011. "Intermarriage, Ethnic Identity, and Perceived Social Standing among Asian Women in the United States." *Journal of Marriage and Family* 73 (August): 876–888.

Chen, Michelle. 2012. "Health Care System Leaves Patients Frustrated—Nurses Work for a Solution." AlterNet, June 4. Accessed June 5, 2012 (www.alternet.org).

Chesler, Phyllis. 2006. "The Failure of Feminism." *Chronicle of Higher Education,* February 26, B12.

Chesney-Lind, Meda, and Nikki Jones, eds. 2010. *Fighting for Girls: New Perspectives on Gender and Violence.* Albany: State University of New York Press.

Chesney-Lind, Meda, and Lisa Pasko. 2004. *The Female Offender: Girls, Women, and Crime,* 2nd edition. Thousand Oaks, CA: Sage.

Chetty, Raj, Nathaniel Hendren, Patrick Kline, Emmanuel Saez, and Nicholas Turner. 2014. "Is the United States Still a Land of Opportunity? Recent Trends in Intergenerational Mobility." National Bureau of Economic Research, January. Accessed February 6, 2014 (www.nber.org).

"Child Brides in West Africa: Girls Fight Back." 2014. *The Economist,* August 23, 42.

"Child Maltreatment." 2014. Centers for Disease Control and Prevention. Accessed October 26, 2014 (www.cdc.gov).

Child Trends. 2010. "Parental Expectations for Children's Academic Attainment." Accessed July 27, 2011 (www.childtrendsdatabank.org).

Child Trends Data Bank. 2012. "Parental Expectations for Their Children's Academic Attainment: Indicators on Children and Youth." Accessed June 23, 2014 (www.childtrends.org).

___. 2013. "Youth Voting." June. Accessed April 20, 2014 (www.childtrendsdatabank.org).

Children's Bureau. 2013. "Trends in Foster Care and Adoption: FY2002–FY2012." U.S. Department of Health and Human Services. Accessed November 18, 2013 (www.acf.hhs.gov).

Children's Defense Fund. 2014. "The State of America's Children 2014." Accessed February 14, 2014 (www.childrensdefense.org).

"China's Environment: A Small Breath of Fresh Air." 2014. *The Economist,* February 8, 14, 16.

Chittal, Nisha. 2013. "Please Stop Live Tweeting People's Private Conversations." *I.M.H.O.,* November 30. Accessed December 4, 2013 (www.medium.com).

Choice, Pamela, and Leanne K. Lamke. 1997. "A Conceptual Approach to Understanding Abused Women's Stay/Leave Decisions." *Journal of Family Issues* 18 (May): 290–314.

Christakis, Erika. 2012. "The Overwhelming Maleness of Mass Homicide." *Time,* July 24. Accessed November 20, 2013 (www.time.com).

Christina, Greta. 2011. "Wealthy, Handsome, Strong, Packing Endless Hard-Ons: The Impossible Ideals Men Are Expected to Meet." Independent Media Institute, June 20. Accessed June 21, 2011 (www.alternet.org).

Chronicle of Higher Education. 2014. "Data Point." April 18, A23.

Churchill, Ward. 1997. *A Little Matter of Genocide: Holocaust and Denial in the Americas, 1492 to the Present.* San Francisco: City Lights Books.

CIA World Factbook. 2014. "Sex Ratio." March 6. Accessed September 19, 2014 (www.cia.gov).

Cichocki, Mary K. 1981. "Women's Travel Patterns in a Suburban Development." Pp. 151–163 in *New Space for Women,* edited by Gerda R. Wekerle, Rebecca Peterson, and David Morley. Boulder, CO: Westview Press.

Cillizza, Chris. 2014. "Our C-Minus Government (and Why We Deserve It)." *Washington Post,* October 15. Accessed November 23, 2014 (www.brookings.edu).

Citizens for Tax Justice. 2013. "Undocumented Immigrants Pay Taxes, and Will Pay More Under Immigration." July 11. Accessed March 22, 2014 (www.ctj.org).

Clayson, Dennis E., and Debra A. Haley. 2011. "Are Students Telling Us the Truth? A Critical Look at the Student Evaluation of Teaching." *Marketing Education Review* 21 (Summer): 101–112.

Clement, Scott. 2011. "Workplace Harassment Drawing Wide Concern." *Washington Post,* November 16. Accessed February 25, 2012 (www.washingtonpost.com).

Clements, David D. 2012. *Corporations Are Not People: Why They Have More Rights Than You Do and What You Can Do About It.* San Francisco: Berrett-Koehler.

Clifford, Stephanie. 2012. "Unexpected Ally Helps Wal-Mart Cut Waste." *New York Times,* April 13, B1.

Clifford, Stephanie, and Andrew Martin. 2011. "In Time of Scrimping, Fun Stuff Is Still Selling." *New York Times,* September 24, A1.

Clifford, Stephanie, and Quentin Hardy. 2013. "Attention, Shoppers: Store Is Tracking Your Cell." New York Times, July 15, A1.

Clifton, Donna, and Ashley Frost. 2011. "The World's Women and Girls 2011 Data Sheet." Population Reference Bureau. Accessed October 25, 2011 (www.prb.org).

"Climate Change and Financial Instability Seen Top Global Threats." 2013. Pew Research Center, June 24. Accessed September 19, 2014 (www.pewglobal.org).

Cloke, Kenneth, and Joan Goldsmith. 2002. *End of Management and the Rise of Organizational Democracy.* San Francisco: Jossey-Bass.

Cloud, D. S. 2013. "Pentagon Adds Benefits For Same-Sex Couples." *Baltimore Sun,* February 12, 6.

Cloud, John. 2010. "How to Recruit Better Teachers." *Time,* September 20, 47–52.

Clowers, Nicole. 2013. "Improving Personnel Management Is Critical for Agency's Effectiveness." GAO Highlights, GAO-13-621, July 18. Accessed December 19, 2013 (www.gao.gov).

"Clueless." 2014. *The Economist,* June 28, 26.

Cobble, Dorothy Sue, Linda Gordon, and Astrid Henry. 2014. "What 'Lean In' Leaves Out." *Chronicle Review,* September 16, B4–B5.

Cohen, Jere. 2002. *Protestantism and Capitalism: The Mechanisms of Influence.* New York: Aldine de Gruyter.

Cohen, Jon. 2013. "Gay Marriage Support Hits New High in Post-ABC Poll. *Washington Post,* March 18. Accessed May 21, 2013 (www.washingtonpost.com).

Cohen, Stefanie. 2012. "Why Women Writers Still Take Men's Names." *Wall Street Journal,* December 7, D9.

Cohn, D'Vera. 2011. "India Census Offers Three Gender Options." Pew Research Center, February 7. Accessed October 25, 2011 (www.pewresearch.org).

___. 2012. "Divorce and the Great Recession." Pew Research Center, May 2. Accessed May 22, 2014 (www.pewsocialtrends.org).

___. 2013. "Love and Marriage." Pew Research Social & Demographic Trends, February 13. Accessed May 16, 2014 (www.pewsocialtrends.org).

Cohn, D'Vera, Eileen Patten, and Mark Hugo. 2014. "Puerto Rican Population Declines on Island, Grows on U.S. Mainland." Pew Research Center, August 11. Accessed September 19, 2014 (www.pewresearch.org).

Cohn, D'Vera, Gretchen Livingston, and Wendy Wang. 2014. "After Decades of Decline, a Rise in Stay-at-Home Mothers." Pew Research Center, April 8. Accessed April 12, 2014 (www.pewresearch.org).

Cohn, D'Vera, Jeffrey S. Passel, Wendy Wang, and Gretchen Livingston. 2011. "Barely Half of U.S. Adults Are Married—A Record Low." Pew Research Center, December 14. Accessed May 18, 2014 (www.pewsocialtrends.org).

Cohn, D'Vera, Paul Taylor, Mark H. Lopez, Catherine A. Gallagher, Kim Parker, and Kevin T. Maass. 2013. "Gun Homicide Rate Down 49% Since 1993 Peak; Public Unaware." Pew Research Center, May 7. Accessed October 12, 2013 (www.pewresearch.org).

Cohn, Meredith. 2014. "Kids Are Outfitted with New Hands Made on 3-D Printers." *Baltimore Sun,* September 29, 1, 11.

Colapinto, John. 1997. "The True Story of John/Joan." *Rolling Stone* (December 11): 54–73, 92–97.

___. 2001. *As Nature Made Him: The Boy Who Was Raised as a Girl.* New York: Harper Perennial.

___. 2004. "What Were the Real Reasons Behind David Reimer's Suicide?" *Slate,* June 3. Accessed April 24, 2008 (www.slate.com).

Colby, Sandra L., and Jennifer M. Ortman. 2014. "The Baby Boom Cohort in the United States: 2012 to 2060." U.S. Census Bureau, May. Accessed September 19, 2014 (www.census.gov).

College Board. 2013. "2013 College-Bound Seniors: Total Group Profile Report." Accessed June 21, 2014 (www.collegeboard.org).

___. 2014. "The 10th Annual AP Report to the Nation." February 11. Accessed June 21, 2014 (apreport.collegeboard.com).

Collins, Jason, and Franz Lidz. 2013. "The Gay Athlete." *Sports Illustrated*, May 6, 34–41.

Conley, Dalton. 2004. *The Pecking Order: Which Siblings Succeed and Why*. New York: Pantheon.

Conrad, Peter, and Kristin K. Barker. 2010. "The Social Construction of Illness: Key Insights and Policy Implications." *Journal of Health and Social Behavior* 51 (Suppl.): S67–S79.

Colvin, Geoff. 2012. "The Art of the Self-Managing Team." *Fortune*, December 3, 22–23.

Colvin, George. 2014. "In the Future, Will There Be Any Work Left for People to Do?" *Fortune*, June 16, 193–202.

Common Cause. 2014. "McCutcheon v. FEC." Accessed April 12, 2014 (www.commoncause .org).

Congressional Budget Office. 2014a. "Payments of Penalties for Being Uninsured Under the Affordable Care Act: 2014 Update." June. Accessed July 26, 2014 (www.cbo.gov).

———. 2014b. "The Budget and Economic Outlook: 2014 to 2024." February. Accessed July 26, 2014 (www.cbo.gov).

Considine, Austin. 2011. "For Asian-American Stars, Many Web Fans." *New York Times*, July 31, ST6.

Consumer Reports. 2014. "When You Need Medical Care, Stat." August, 13.

ConsumerReports.org. 2009. "6 Top Reasons for Not Having Sex." February. Accessed March 4, 2014 (www.consumerreports.org).

Cook, Thomas D., and Donald T. Campbell. 1979. *Quasi-Experimentation: Design and Analysis Issues for Field Settings*. Chicago: Rand McNally.

Cooley, Charles Horton. 1909/1983. *Social Organization: A Study of the Larger Mind*. New Brunswick, NJ: Transaction Books.

Cooper, O. R., et al. 2010. "Increasing Springtime Ozone Mixing Ratios in the Free Troposphere over Western North America." *Nature* 463 (January 21): 344–348.

Cooperman, Alan, and Jessica H. Martínez. 2013. "U.S. Catholics See Sex Abuse as the Church's Most Important Problem, Charity as Its Most Important Contribution." Pew Research Center, May 6. Accessed June 21, 2014 (www.pewresearch.org).

Cooperman, Alan, Gregory Smith, and Besheer Mohamed. 2013. "Celebrating Christmas and the Holidays, Then and Now." Pew Research Center, December 18. Accessed June 21, 2014 (www.pewresearch.org).

Cooperman, Alan, Mark H. Lopez, Cary Funk, and Jessica H. Martínez. 2014. "The Shifting Religious Identity of Latinos in the United States: Nearly One-in-Four Latinos Are Former Catholics." Pew Research Center, May 7. Accessed June 21, 2014 (www .pewresearch.org).

Copen, Casey E., Anjani Chandra, and Gladys Martinez. 2012. "Prevalence and Timing of Oral Sex with Opposite-Sex Partners Among Females and Males Aged 15–24 Years: United States, 2007–2010." *National Health Statistics Reports*, No. 56, August 16. Accessed March 2, 2014 (www.cdc.gov/nchs).

Copen, Casey E., Kimberly Daniels, and William D. Mosher. 2013. "First Premarital Cohabitation in the United States: 2006–2010 National Survey of Family Growth." *National Health Statistics Reports*, April 4. Accessed May 16, 2014 (www.cdc.gov/nchs).

Copen, Casey E., Kimberly Daniels, Jonathan Vespa, and William D. Mosher. 2012. "First Marriages in the United States: Data from the 2006–2010 National Survey of Family Growth." *National Health Statistics Reports*,

March 22. Accessed May 16, 2014 (www.cdc .gov/nchs).

Corak, Miles. 2013. "Income Inequality, Equality of Opportunity, and Intergenerational Mobility." *Journal of Economic Perspectives* 27 (Summer): 79–102.

Corbett, Christianne, and Catherine Hill. 2012. "Graduating to a Pay Gap: The Earnings of Women and Men One Year After College Graduation." AAUW, October. Accessed March 2, 2014 (www.aauw.org).

Corporation for National & Community Service. 2012. "Volunteering and Civic Life in America 2012." December. Accessed December 19, 2013 (www.nationalservice .gov).

Coser, Lewis. 1956. *The Functions of Social Conflict*. New York: Free Press.

Couch, Carl J. 1968. "Collective Behavior: An Examination of Some Stereotypes." *Social Problems* 15 (Winter): 310–322.

Council for Global Equity. 2014. "The Facts on LGBT Rights in Russia." Accessed March 2, 2014 (www.globalequity.org).

Covenant Eyes, Inc. 2013. "Pornography Statistics." Accessed March 2, 2014 (www.covenanteyes. com).

Cox, Shanna, Karen Pazol, Lee Warner, Lisa Romero, Alison Spitz, Lorrie Gavin, and Wanda Barfield. 2014. "Vital Signs: Births to Teens Aged 15–17 Years—United States, 1991–2012." *MMWR* 63 (April 11): 312–317.

Crabtree, Steve. 2010. "Religiosity Highest in World's Poorest Nations." Gallup, June 3. Accessed August 31, 2012 (gallup.com).

Cressey, Donald R. 1953. *Other People's Money: A Study in the Social Psychology of Embezzlement*. Glencoe, IL: Free Press.

Crockett, Ariel. 2011. "Is the Black Television Series Dead?" Hello Beautiful, June 3. Accessed December 6, 2011 (hellobeautiful. com).

Crothers, Charles. 1979. "On the Myth of Rural Tranquility: Comment on Webb and Collette." *American Journal of Sociology* 84 (May): 1441–1445.

Crowe, Ann H., Tracy G. Mullins, Kimberley A. Cobb, and Nathan C. Lowe. 2012. "Effects and Consequences of Underage Drinking." *Juvenile Justice Bulletin*, Office of Juvenile Justice and Delinquency Prevention, September. Accessed July 30 (www.ojjdp.gov).

Crowley, Michael. 2010. "What $120 Million Buys." *Time*, October 11, 35–37.

———. 2014. "Iraq's Eternal War." *Time*, June 30, 28–34.

Cruikshank, Margaret. 2009. *Learning to Be Old: Gender, Culture, and Aging*. Lanham, MD: Rowman & Littlefield

Cudd, Ann E., and Nancy Holstrom. 2010. *Capitalism, For and Against: A Feminist Debate*. New York: Cambridge University Press.

Cuevas, Carlos A., David Finkelhor, Anne Shattuck, Heather Turner, and Sherry Hamby. 2013. "Children's Exposure to Violence and the Intersection Between Delinquency and Victimization." Office of Juvenile Justice and Delinquency Prevention, October. Accessed January 12, 2014 (www.ojjdp.gov).

Culbert, Samuel A. 2011. "Why Your Boss Is Wrong About You." *New York Times*, March 3, A25.

Culbert, Samuel A., and Lawrence Rout. 2010. *Get Rid of the Performance Review!: How Companies Can Stop Intimidating, Start Managing—and Focus on What Really Matters*. New York: Business Plus.

Cullen, Lisa T., and Coco Masters. 2008. "We Just Clicked." *Time*, January 28, 86–89.

Cumberworth, Erin. 2010. "Homeboy Industries." *Pathways*, Summer, 22–23.

Cunningham, Solveig A., Michael R. Kramer, and K. M. Venkat Narayan. 2014. "Incidence of Childhood Obesity in the United States." *New England Journal of Medicine* 370 (5): 403–411.

Currie, Janet, Joshua S. Graff Zivin, Jamie Mullins, and Matthew J. Neidell. 2013. "What Do We Know About Short and Long Term Effects of Early Life Exposure to Pollution?" National Bureau of Economic Research, October. Accessed July 26, 2014 (www.nber.org).

Currie, E. 1985. *Confronting Crime: An American Challenge*. New York: Pantheon.

Curtiss, Susan. 1977. *Genie*. New York: Academic Press.

Currie, Janet, Mark Stable, and Lauren E. Jones. 2014. "Do Stimulant Medications Improve Educational and Behavioral Outcomes for Children with ADHD?" National Bureau of Economic Research, June. Accessed July 26, 2014 (www.nber.org).

Curtin, Sally C., Stephanie J. Ventura, and Gladys M. Martinez. 2014. "Recent Declines in Nonmarital Childbearing in the United States." National Center for Health Statistics, August. Accessed November 30, 2014 (www .nchs.gov).

Cynkar, Peter, and Elizabeth Mendes. 2011. "More Than One in Six American Workers Also Act as Caregivers." Gallup, July 26. Accessed March 10, 2012 (www.gallup.com).

D'Onofrio, Brian M., and Benjamin B. Lahey. 2010. "Biosocial Influences on the Family: A Decade Review." *Journal of Marriage and Family* 72 (June): 762–782.

Dahl, Robert A. 1961. *Who Governs? Democracy and Power in an American City*. New Haven. CT: Yale University Press.

Dahlin, E. 2011. "There's No 'I' in Innovation." *Contexts* 10 (Fall): 22–27.

Daley, Suzanne, and Nicholas Kulish. 2013. "Germany Fights Population Drop." *New York Times*, August 14, A1.

Daly, Mary C., and Leila Bengali. 2014. "Is It Still Worth Going to College?" Federal Reserve Bank of San Francisco, May 5. Accessed June 21, 2014 (www.frbsf.org).

Daneshpour, Manijeh. 2013. "A Mini-Narrative About My Praxis as a Muslim Feminist." *Family Focus* 57.4 (Winter): F1–F3.

Daniel, Mary-Alice. 2014. "What White People Need to Learn." AlterNet, February 7. Accessed March 22, 2014 (www.alternet.org).

Dank, Meredith, et al. 2014. "Estimating the Size and Structure of the Underground Commercial Sex Economy in Eight Major US Cities." Urban Institute, March. Accessed November 20, 2014 (www.urban.org).

Dao, James. 2011. "In California, Indian Tribes with Casino Money Cast Off Members." *New York Times*, December 11, A20.

———. 2013. "In Debate over Military Sexual Assault, Men Are Overlooked Victims." *New York Times*, June 24, A12.

Davenport, Coral. 2014. "Industry Awakens to Threat of Climate Change." *New York Times*, January 24, A1.

Davey, Monica. 2014. "Immigrants Seen a Way to Refill Detroit Ranks." *New York Times*, January 24, A12.

Davidson, Paul. 2011. "Season of Part-Time Jobs Kicks Off with Holidays." *USA Today*, November 25–27, A1–A2.

Davies, James C. 1979. "The J-Curve of Rising and Declining Satisfaction as a Cause of Revolution and Rebellion." Pp. 413–436 in *Violence in America: Historical and*

Comparative Perspectives, edited by Hugh D. Graham and Ted R. Gurr. Beverly Hills, CA: Sage.

Davis, Karen, Kristof Stremikis, David Squires, and Cathy Schoen. 2014. "Mirror, Mirror on the Wall: How the Performance of the U.S. Health Care System Compares Internationally." The Commonwealth Fund, June. Accessed July 26, 2014 (www.commonwealthfund.org).

Davis, Kenneth C. 2010. "America's True History of Religious Tolerance." Smithsonian, October. Accessed September 28, 2010 (www.smithsonian.com).

Davis, Kingsley, and Wilbert E. Moore. 1945. "Some Principles of Stratification." American Sociological Review 10 (April): 242–249.

De La Haye, Kayla, Garry Robins, Philip Mohr, and Carlene Wilson. 2013. "Adolescents' Intake of Junk Food." Journal of Research on Adolescence 23 (September): 524–536.

de las Casas, Bartolome. 1992. The Devastation of the Indies: A Brief Account. Baltimore, MD: Johns Hopkins University Press.

De Long, J. Bradford, and Andrei Shleifer. 1992. "Princes and Merchants: European City Growth Before the Industrial Revolution." National Bureau of Economic Research, December. Accessed January 10, 2008 (www.nper.org).

de Pommereau, Isabelle. 2013. "A Lift for Europe's Women?" Christian Science Monitor Weekly, August 5,16.

"Deadly Intolerance." 2014. The Economist, March 1, 42.

Death Penalty Information Center. 2009. "The Death Penalty in 2009: Year End Report." Accessed February 25, 2010 (www.deathpenaltyinfo.org).

———. 2014. "Facts about the Death Penalty." January 10. Accessed January 15, 2014 (www.deathpenaltyinfo.org).

Deaton, Angus, and Arthur A. Stone. 2013. "Grandpa and the Snapper: The Wellbeing of the Elderly Who Live with Children." National Bureau of Economic Research, June. Accessed July 10, 2013 (www.nber.org).

Deegan, Mary Jo. 1986. Jane Addams and the Men of the Chicago School, 1892–1918. New Brunswick, NJ: Transaction Books.

Deggans, Eric. 2012. "For Asians and Latinos, Stereotypes Persist in Sitcoms." National Public Radio, February 24. Accessed March 22, 2014 (www.npr.org).

DeGraw, David. 2014. "How the Ultra-Rich .01% Have Sucked Up Even More of America's Wealth Than You Think." AlterNet, August 22. Accessed November 10, 2014 (www.alternet.org).

DeKeseredy, Walter S. 2011. Violence Against Women: Myths, Facts, Controversies. Toronto: University of Toronto Press.

della Porta, Donatella, and Mario Diani. 1999. Social Movements: An Introduction. Malden, MA: Blackwell.

DeNavas-Walt, Carmen, and Bernadette D. Proctor. 2014. Income and Poverty in the United States: 2013. U.S. Census Bureau, Current Population Reports, P60–249, September. Accessed November 9, 2014 (www.census.gov).

DeParle, Jason. 2012. "Harder for Americans to Rise from Lower Rungs." New York Times, January 5, A1.

Dennis, Brady. 2013. "PG-13 Movies Match R Rating, Study Says." Washington Post, November 11. Accessed November 18, 2013 (www.washingtonpost.com).

Denny, C. H., J. Tsai, R. L. Floyd, and P. P. Green. 2009. "Alcohol Use Among Pregnant and Nonpregnant Women of Childbearing Age—United States, 1991–2005." MMWR 58 (May 22): 529–532.

Denyer, Simon. 2013. "Season of Smog Drifting over China." Baltimore Sun, October 23, 13.

Department of Defense. 2013. "Annual Report on Sexual Assault in the Military: Fiscal Year 2012." Department of Defense Sexual Assault Prevention and Response, May. Accessed January 12, 2014 (www.sapr.mil).

Deprez, Esmé E., and William Selway. 2014. "Guns Allowed All Over—Except Near Politicians." Bloomberg Businessweek, May 19–25, 27–28.

DeSilver, Drew. 2014. "5 Facts About Economic Inequality." Pew Research Center, January 7. Accessed February 14, 2014 (www.pewresearch.org).

———. Drew. 2014. "Q/A: What the New York Times' Polling Decision Means." Pew Research Center, July 28. Accessed October 26, 2014 (www.pewresearch.org).

Desmond-Harris, Jenee. 2009. "Why Michelle's Hair Matters." Time, September 7, 55–57.

Dewan, Shaila, and Robert Gebeloff. 2012. "Among the Wealthiest One Percent, Many Variations." New York Times, January 15, A1.

Deveny, Kathleen. 2008. "They're No Baby Einsteins." Newsweek, January 14, 61.

Dholia, Esha. 2014. "Too Many Tropes: Top 3 Stereotypes of South Asians in the Media." Videshi Magazine, March 13. Accessed March 22, 2014 (www.videshimagazine.com).

Diamond, Milton, and H. Keith Sigmundson. 1997. "Sex Reassignment at Birth: Long-Term Review and Clinical Implications." Archives of Pediatrics & Adolescent Medicine 15 (March): 298–304.

Dias, Elizabeth. 2013. "—!Evangelicos!" Time, March 15, 20–28.

Dieter, Richard C. 2013. "The 2% Death Penalty: How Minority of Counties Produce Most Death Cases at Enormous Costs to All." Death Penalty Information Center, October. Accessed January 15, 2014 (www.deathpenaltyinfo.org).

Dilworth-Anderson, Peggye, Linda M. Burton, and William L. Turner. 1993. "The Importance of Values in the Study of Culturally Diverse Families." Family Relations 42 (July): 238–242.

Dimock, Michael, Carroll Doherty, and Jocelyn Kiley. 2014. "Beyond Red vs. Blue: The Political Typology." Pew Research Center, June 26. Accessed August 8, 2014 (www.pewresearch.org).

"DNA Reveals White Supremacist's Black Heritage." 2013. Baltimore Sun, November 14, 8.

DiSalvo, David. 2010. "Are Social Networks Messing with Your Head?" Scientific American Mind 20 (January/February): 48–55.

Dockterman, Eliana. 2013. "The Digital Parent Trap. Should Your Kids Avoid Tech—or Embrace It?" Time, August 19, 54.

Doeringer, Peter B., and Michael J. Piore. 1971. Internal Labor Markets and Manpower Analysis. Lexington, MA: Heath-Lexington Books.

Doherty, Carroll, Juliana M. Horowitz, and Michael Dimock. 2014. "Most See Inequality Growing, but Partisans Differ Over Solutions." Pew Research Center, January 23. Accessed April 12, 2014 (www.pewresearch.org).

Domhoff, G. William. 2013. "Interlocking Directorates in the Corporate Community." Who Rules America, October. Accessed April 12, 2014 (whorulesamerica.net).

Donadio, Rachel. 2013. "When Italians Chat, Hands and Fingers Do the Talking." New York Times, July 1, A6.

"Don't Think, Just Teach." 2014. The Economist, February 15, 40.

Dorfman, Mark, and Angela Haren. 2014. "Testing the Waters." Natural Resources Defense Council, June. Accessed September 19, 2014 (nrdc.org).

Dokoupil, Tony. 2011. "Mad as Hell." Newsweek, June 6, 6–7.

Dolnick, Sam. 2011. "Dance, Laugh, Drink, Save the Date: It's a Ghanaian Funeral." New York Times, April 12, A1.

Domosh, Mona, and Joni Seager. 2001. Feminist Geographers Make Sense of the World. New York: Guilford Press.

Douglas, Susan J., and Meredith W. Michaels. 2004. The Mommy Myth: The Idealization of Motherhood and How It Has Undermined Women. New York: Free Press.

Dovidio, John F. 2009. "Racial Bias, Unspoken but Heard." Science 326 (December 18): 1641–1642.

Doyle, Francis X. 2011. "Bishops' Missed Opportunity." Baltimore Sun, November 14, 15.

Downs, Edward, and Stacy L. Smith. 2010. "Keeping Abreast of Hypersexuality: A Video Game Character Content Analysis." Sex Roles 62 (June): 721–733.

Drake, Bruce. 2013. "Americans See Growing Gap Between Rich and Poor." Pew Research Center, December 5. Accessed February 14, 2014 (www.pewresearch.org).

Draper, Robert. 2013. "Inside the Power of the N.R.A." New York Times, December 15, MM48.

Drexler, Peggy. 2012. "The New Face of Infidelity." Wall Street Journal, October 21, C3.

———. 2013. "The Delicate Protocol of Hugging." Wall Street Journal, September 14–15, C3.

"The Drugs Don't Work." 2014. The Economist, May 3, 54.

Drutman, Lee, Amy Cesal, and Ben Chartoff. 2013. "Where the 1% of the 1% Money Goes." Sunlight Foundation, June 24. Accessed April 20, 2014 (www.sunlightfoundation.com).

Drutman, Lee, and Alexander Furnas. 2014. "K Street Pays Top Dollar for Revolving Door Talent." Sunlight Foundation, January 21. Accessed April 20, 2014 (sunlightfoundation.com).

Du Bois, W. E. B. 1986. The Souls of Black Folk. New York: Library of America.

Dugan, Andrew. 2013. "In U.S., Majority Approves of Unions, but Say They'll Weaken." Gallup, August 30. Accessed April 12, 2014 (www.gallup.com).

———. 2013. "In U.S., Support for Complete Smoking Ban Increases to 22%." Gallup, July 29. Accessed January 12, 2014 (www.gallup.com).

———. 2013. "More Say Crime Is Serious Problem in U.S. Than Locally." Gallup, November 1. Accessed January 12, 2014 (www.gallup.com).

———. 2014. "More Americans Worse Off Financially Than a Year Ago." Gallup, January 15. Accessed April 12, 2014 (www.gallup.com).

Dugan, Andrew, and Brad Hoffman. 2014. "Record Low Say Own Representative Deserves Re-Election." Gallup, January 24. Accessed April 20, 2014 (www.gallup.com).

Dugan, Andrew, and Frank Newport. 2013. "In U.S., Fewer Believe 'Plenty of Opportunity' to Get Ahead." Gallup, November 30. Accessed February 6, 2014 (www.gallup.com).

———. "Americans Most Likely to Say They Belong to the Middle Class." Gallup, November 20. Accessed December 1, 2014 (www.gallup.com).

Duggan, Maeve. 2013. "Cell Phone Activities 2013." Pew Research Center, September 16.

Accessed December 4, 2013 (www.pewinternet.org).

Duggan, Maeve, and Aaron Smith. 2013. "Social Media Update." Pew Research Center, December 30. Accessed March 10, 2014 (www.pewresearch.org).

Duggan, Maeve, and Joanna Brenner. 2013. "The Demographics of Social Media Users—2012." Pew Research Center, February 14. Accessed December 4, 2013 (www.pewinternet.org).

Duhigg, Charles. 2010. "Saving U.S. Water and Sewer Systems Would Be Costly." *New York Times*, March 14, A1.

Dunifon, Rachel, and Lori Kowaleski-Jones. 2007. "The Influence of Grandparents in Single-Mother Families." *Journal of Marriage and Family* 69 (May): 465–481.

Dunlap, Riley E. 2007. "The State of Environmentalism in the U.S." Gallup, April 19. Accessed March 27, 2007 (www.gallup.com).

Durden, Tyler. 2011. "Here Are the 29 Public Companies with More Cash Than the US Treasury." Zero Hedge, July 15. Accessed January 9, 2012 (www.zerohedge.com).

Durkheim, Émile. 1893/1964. *The Division of Labor in Society*. New York: Free Press.

___. 1897/1951. *Suicide: A Study in Sociology*, translated by John A. Spaulding and George Simpson, edited by George Simpson. New York: Free Press.

___. 1898/1956. *Education and Sociology*, translated by Sherwood D. Fox. Glencoe, IL: Free Press.

___. 1961. *The Elementary Forms of the Religious Life*. New York: Collier Books.

Durose, Matthew R., and Patrick A. Langan. 2004. "Felony Sentences in State Courts, 2002." U.S. Department of Justice, Bureau of Justice Statistics.

Duster, Troy. 2005. "Race and Reification in Science." *Science* 307, February 18, 1050–1051.

Dutra, Lauren M., and Stanton A. Glantz. 2014. "Electronic Cigarettes and Conventional Cigarette Use Among US Adolescents: A Cross-Sectional Study." *JAMA Pediatrics* 168 (7): 610–617.

Dwyer, Rachel E., Randy Hodson, and Laura McCloud. 2013. "Gender, Debt, and Dropping Out of College." *Gender & Society* 27 (February): 30–55.

Dye, Thomas R., and Harmon Ziegler. 2003. *The Irony of Democracy: An Uncommon Introduction to American Politics*, 12th edition. Belmont, CA: Wadsworth.

Eagan, Kevin, Jennifer B. Lozano, Sylvia Hurtado, and Matthew H. Case. 2014. *The American Freshman: National Norms Fall 2013*. Los Angeles: Higher Education Research Institute, UCLA.

Eaton, Danice K., et al. 2012. "Youth Risk Behavior Surveillance—United States, 2011." *MMWR* 61 (June 8): 1–162.

Eberstadt, Nicholas. 2009. "Poor Statistics." *Forbes*, March 2, 26.

___. 2013. "China's Coming One-Child Crisis." *Wall Street Journal*, November 27, A15.

Eckenrode, John, Elliott G. Smith, Margaret E. McCarthy, and Michael Dineen. 2014. "Income Inequality and Child Maltreatment in the United States." *Pediatrics* 33 (March): 454–461.

Eckholm, Erik. 2014. " 'Aid in Dying' Movement Takes Hold in Some States." *New York Times*, February 8, A1.

Eckstein, Rick, Rebecca Schoenike, and Kevin Delaney. 1995. "The Voice of Sociology: Obstacles to Teaching and Learning the Sociological Imagination." *Teaching Sociology* 23 (October): 353–363.

Edwards, Ashley N. 2014. "Dynamics of Economic Well-Being: Poverty, 2009–2011." U.S. Census Bureau, January. Accessed February 6, 2014 (www.census.gov).

Efrati, Amir. 2013. "Google's Data-Trove Dance." *Wall Street Journal*, July 31, B1, B4.

Eggebeen, David J. 2005. "Cohabitation and Exchanges of Support." *Social Forces* 83 (May): 1097–1110.

Egley, Arlen, and James C. Howell. 2013. "Highlights of the 2011 National Youth Gang Survey." Office of Juvenile Justice and Delinquency Prevention, September. Accessed January 12, 2014 (www.ojjdp.gov).

Ehrlich, Paul. 1971. *The Population Bomb*, 2nd edition. San Francisco: Freeman.

Ehrlich, Paul R., and Anne H. Ehrlich. 2008. *The Dominant Animal: Human Evolution and the Environment*. Washington, DC: Island Press.

Eibner, Christine, and Evan Saltzman. 2014. "Who Does the Affordable Care Act Leave Behind?" RAND, March 7. Accessed July 26, 2014 (www.rand.org).

Eilperin, Juliet, and Lenny Bernstein. 2014. "Environmental Groups Say Obama Needs to Address Climate Change More Aggressively." *Washington Post*, January 16. Accessed September 19, 2014 (www.washingtonpost.com).

Eisenberg, Abne M., and Ralph R. Smith, Jr. 1971. *Nonverbal Communication*. Indianapolis: Bobbs-Merrill.

Eisenberg, Nancy, et al., 2005. "Relations Among Positive Parenting, Children's Effortful Control, and Externalizing Problems: A Three-Wave Longitudinal Study." *Child Development* 76 (September/October): 1055–1071.

Eisler, Peter. 2013. "What Surgeons Leave Behind Costs Some Patients Dearly." *USA Today*, March 8. Accessed July 26, 2014 (www.usatoday.com).

Eisler, Peter, and Barbara Hansen. 2013. "Doctors Perform Thousands of Unnecessary Surgeries." *USA Today*, June 20. Accessed July 26, 2014 (www.usatoday.com).

Ekman, Paul, and Wallace V. Friesen. 1984. *Unmasking the Face: A Guide to Recognizing Emotions from Facial Clues*. Palo Alto, CA: Consulting Psychologists Press.

Elder, Miriam. 2013. "Russia Passes Law Banning Gay 'Propaganda.' " *The Guardian*, June 11. Accessed March 8, 2014 (www.theguardian.com).

Elgin, Ben. 2014. "An Oil Giant Dims the Lights on Green Power." *Bloomberg Businessweek*, June 2–8, 21–22.

Eliot, Lise. 2012. *Pink Brain, Blue Brain: How Small Differences Grow into Troublesome Gaps—And What We Can Do About It*. Oxford, England: Oneworld Publications.

Ellingwood, Ken. 2009. "Not Being on Time a High Art in Mexico." *Los Angeles Times*, September 12. Accessed September 22, 2009 (www.latimes.com).

Ellis, Bill. 2005. "Legend/AntiLegend: Humor as an Integral Part of the Contemporary Legend Process." Pp. 123–140 in *Rumor Mills: The Social Impact of Rumor and Legend*, edited by Gary Alan Fine, Veronique Campion-Vincent, and Chip Heath. New Brunswick, NJ: Transaction Publishers.

Ellis, Renee. 2013. "Changes in Coresidence of Grandparents and Grandchildren." *Family Focus* (Summer): F13, F15, F17.

Ellis, Renee R., and Tavia Simmons. 2014. "Coresident Grandparents and Their Grandchildren: 2012." U.S. Census Bureau, October. Accessed October 25, 2014 (www.census.gov).

Ellison, Jesse. 2011. "The 2011 Global Women's Progress Report." *Newsweek*, September 26, 27–33.

El Nasser, Haya, and Paul Overberg. 2011. "In Many Neighborhoods, Kids Are Only a Memory." *USA Today*, June 3, 1A, 6A.

"The End of the Beginning." 2014. *The Economist*, March 8, 22–24.

"Ending the Shame." 2013. *The Economist*, September 14, 16,18.

England, Paula. 2005. "Emerging Theories of Care Work." *Annual Review of Sociology* 31(August): 381–399.

England, Paula, and Reuben J. Thomas. 2009. "The Decline of the Date and the Rise of the College Hook Up." Pp. 141–152 in *Family in Transition*, 15th edition, edited by Arlene S. Skolnick and Jerome H. Skolnick. Boston: Pearson Higher Education.

Entmacher, Joan, Lauren Frohlich, Katherine G. Robbins, Emily Martin, and Liz Watson. 2014. "Underpaid & Overloaded: Women in Low-Wage Jobs." National Women's Law Center. Accessed November 23, 2014 (www.nwlc.org).

Environmental Integrity Project. 2007. "Paying Less to Pollute: Environmental Enforcement Under the Bush Administration." Accessed February 12, 2008 (www.environmentalintegrity.org).

Environmental Working Group. 2014. "Children's Cereals: Sugar by the Pound." May. Accessed July 26, 2014 (www.ewg.org).

Epstein, Joseph. 2011. *Gossip: The Untrivial Pursuit*. New York: Houghton Mifflin Harcourt.

Epstein, Leonard H., et al. 2008. "A Randomized Trial of the Effects of Reducing Television Viewing and Computer Use on Body Mass Index in Young Children." *Archives of Pediatrics & Adolescent Medicine* 162 (3): 239–245.

Epstein, Robert. 2008. "Same-Sex Marriage Is Too Limiting." *Los Angeles Times*, December 4. Accessed December 6, 2008 (www.latimes.com)

Equal Employment Opportunity Commission. 2011. "Sexual Harassment Charges EEOC & FEPAs Combined: FY 1997–FY 2011." Accessed March 2, 2014 (www.eeoc.gov).

___. 2014. "Sexual Harassment Charges: FY 2010–FY 2013." Accessed March 2, 2014 (www.eeoc.gov).

Erikson, Kai T. 1966. *Wayward Puritans: A Study in the Sociology of Deviance*. New York: John Wiley & Sons.

Ernst & Young. 2013. "Navigating Today's Complex Business Risks: Europe, Middle East, India and Africa Fraud Survey 2013." Accessed January 12, 2014 (www.ey.com).

Esipova, Neil, Anita Pugliese, and Julie Ray. 2013. "381 Million Adults Worldwide Migrate Within Countries." Gallup, May 15. Accessed September 19, 2014 (www.gallup.com).

Eskow, Richard. 2014. "7 Facts That Show the American Dream Is Dead." AlterNet, October 22. Accessed December 5, 2014 (www.alternet.org).

Esping-Andersen, Gosta. 1990. *The Three Worlds of Welfare Capitalism*. Cambridge, United Kingdom: Polity Press.

Esterl, Mike. 2013. "The Natural Evolution of Food Labels." *Wall Street Journal*, November 6, B1–B2.

Etzioni, Amitai. 1975. *A Comparative Analysis of Complex Organizations*. New York: Free Press.

Evans, Becky. 2013. "Bill Gates 'Disrespects' South Korea's Female President by Shaking Hands While Keeping Other in His Pocket."

MailOnline, April 23. Accessed November 10, 2013 (dailymail.co.uk).

Evans, Harold, Gail Buckland, and David Lefer. 2006. *They Made America: From the Steam Engine to the Search Engine: Two Centuries of Innovators.* New York: Little, Brown.

Ewen, Lynda Ann. 1998. *Social Stratification and Power in America: A View from Below.* Six Hills, NY: General Hall.

Ewert, Stephanie, and Robert Kominski. 2014. "Measuring Alternative Educational Credentials: 2012." U.S. Census Bureau, January. Accessed June 21, 2014 (www.census.gov).

"Extramarital Affairs." 2014. Pew Research Global Attitudes Project. Accessed April 18, 2014 (www.pewglobal.org).

Fahrenthold, David A. 2009. "Environmentalists Seek to Wipe Out Plush Toilet Paper." *Washington Post,* September 24, A1.

FairTest. 2013. "SAT Score Averages Stagnate as Number of Test-Takers Declines: Persistent Race, Gender Gaps Show Failure of Test-Driven Schooling." National Center for Fair & Open Testing, September 26.

"Faith and Reason." 2014. *The Economist,* February 22, 28.

Fakhraie, Fatemeh. 2009. "Feminists Don't Understand Muslim Women." Double X, May 20. Accessed April 26, 2010 (www.doublex.com).

Falah, Ghazi-Wald, and Caroline Nagel, eds. 2005. *Geographies of Muslim Women: Gender, Religion, and Space.* New York: Guilford Press.

"Family Planning: One-Child Proclivity." 2014. *The Economist,* July 19, 40.

Fang, Ferric C., R. Grant Steen, and Arturo Casadevall. 2012. "Misconduct Accounts for the Majority of Retracted Scientific Publications." *Proceedings of the National Academy of Sciences* 109 (42): 17028–17033.

FAO, IFAD, and WFP. 2013. *The State of Food Insecurity in the World 2013: The Multiple Dimensions of Food Security.* Rome: FAO.

Farhi, Paul. 2012. "Adam Lanza, and Others Who Committed Mass Shootings, Were White Males." *Washington Post,* December 20. Accessed November 20, 2013 (www.washingtonpost.com).

Faris, Robert, and Diane Felmlee. 2011. "Status Struggles: Network Centrality and Gender Segregation in Same- and Cross-Gender Aggression." *American Sociological Review* 76 (February): 48–73.

Farley, Melissa. 2001. "Prostitution: The Business of Sexual Exploitation." Pp. 879–891 in *Encyclopedia of Women and Gender,* volume 2, edited by Judith Worrell. New York: Academic Press.

Farley, Sally D., Amie M. Ashcraft, Mark F. Stasson, and Rebecca L. Nusbaum. 2010. "Nonverbal Reactions to Conversational Interruption: A Test of Complementarity Theory and the Status/Gender Parallel." *Journal of Nonverbal Behavior* 34 (December): 193–206.

Farnam, T. W. 2012. "White House Visitor Logs Provide Window into Lobbying Industry." *Washington Post,* May 20. Accessed May 21, 2012 (www.washingtonpost.com).

Feagin, Joe R., and Clairece Booher Feagin. 2008. *Racial and Ethnic Relations,* 8th edition. Upper Saddle River, NJ: Prentice Hall.

Feagin, Joe R., and Melvin P. Sikes. 1994. *Living with Racism: The Black Middle-Class Experience.* Boston: Beacon Press.

Feagin, Joe R., and Robert Parker. 1990. *Building American Cities: The Urban Real Estate Game,* 2nd edition. Englewood Cliffs, NJ: Prentice Hall.

Federal Bureau of Investigation. 2003. "Uniform Crime Reporting Program Releases Hate Crime Statistics for 2002." Accessed February 14, 2014 (www.fbi.gov).

———. 2012. "Financial Crimes Report to the Public." Accessed February 14, 2014 (www.fbi.org).

———. 2013. *Crime in the United States, 2012.* Accessed February 14, 2014 (www.fbi.gov).

Federal Election Commission. 2014. "PAC Count—1974 to Present." January. Accessed April 12, 2014 (www.fec.gov).

Federal Interagency Forum on Aging-Related Statistics. 2012. "Older Americans 2012: Key Indicators of Well-Being." National Center for Health Statistics, June. Accessed May 16, 2014 (www.agingstats.gov).

Federal Interagency Forum on Child and Family Statistics. 2013. *America's Children: Key National Indicators of Well-Being, 2013.* Washington, DC: U.S. Government Printing Office.

Feldman-Jacobs, Charlotte, and Donna Clifton. 2014. "Female Genital Mutilation/Cutting: Data and Trends, Update 2014." Population Reference Bureau. Accessed March 2, 2014 (www.prb.org).

Fellowes, Matt, and Jake Spiegel. 2013. "Debt Savers in Defined Contribution Plans." HelloWallet, October. Accessed April 12, 2014 (www.hellowallet.com).

Ferguson, Christopher J., and Cheryl K. Olson. 2013. "Video Game Violence Use Among 'Vulnerable' Populations: The Impact of Violent Games on Delinquency and Bullying Among Children with Clinically Elevated Depression or Attention Deficit Symptoms." *Journal of Youth and Adolescence* 42 (August), published online.

Ferguson, Emily. 2014. "11 Reasons Christianity Needs Feminism." *Huffington Post,* May 21. Accessed June 21, 2014 (www.huffingtonpost.com).

Fernald, Anne, Virginia A. Marchman, and Adriana Weisleder. 2013. "SES Differences in Language Processing Skill and Vocabulary are Evident at 18 Months." *Developmental Science* 16 (March): 234–248.

Ferree, Myra M. 2005. "It's Time to Mainstream Research on Gender." *Chronicle Review,* August 12, B10.

Fields, Jason, and Lynne. M. Casper. 2001. "America's Families and Living Arrangements: 2000." U.S. Census Bureau, Current Population Reports. Accessed February 12, 2003 (www.census.gov).

File, Thom. 2013. "Computer and Internet Use in the United States." U.S. Census Bureau, May. Accessed December 4, 2013 (www.census.gov).

———. 2013. "The Diversifying Electorate—Voting Rates by Race and Hispanic Origin in 2012 (and Other Recent Elections)." U.S. Census Bureau, May. Accessed April 20, 2014 (www.census.gov).

File, Thom, and Sarah Crissey. 2010. "Voting and Registration in the Election of November 2008: Population Characteristics." U.S. Census Bureau, Current Population Reports. Accessed May 15, 2010 (www.census.gov).

Fingerman, Karen L., Yen-Pi Cheng, Eric D. Wesselmann, Steven Zarit, Frank Furstenberg, and Kira S. Birditt. 2012. "Helicopter Parents and Landing Pad Kids: Intense Parental Support of Grown Children." *Journal of Marriage and Family* 74 (August): 880–896.

Finkelhor, David, Heather Turner, Sherry Hamby, and Richard Ormrod. 2011. "Polyvictimization: Children's Exposure to Multiple Types of Violence, Crime, and Abuse." Office of Justice Programs, October. Accessed March 7, 2012 (www.ojp.usdoj.gov).

Finkelstein, Eric A., et al. 2012. "Obesity and Severe Obesity Forecasts Through 2030." *American Journal of Preventive Medicine* 42 (6). Accessed May 14, 2012 (www.ajponline.org).

Finn, Jeremy D. 2006. *The Adult Lives of At-Risk Students: The Roles of Attainment and Engagement in High School.* Washington, DC: U.S. Department of Education.

Fischman, Josh. 2009. "Global Warming Before Smokestacks." *Chronicle Review,* November 6, B11–B12.

Fisman, Ray, and Tim Sullivan. 2013. *The Org: The Underlying Logic of the Office.* New York: Twelve.

Fitzgerald, Drew. 2013. "Tax-Break Requests Wear Thin in Illinois." *Wall Street Journal,* November 11, B1–B2.

Fitzpatrick, Kathleen. 2013. "#shameonyou." *The Chronicle Review,* April 26, B4–B5.

Flavin, Jeanne. 2001. "Feminism for the Mainstream Criminologist: An Invitation." *Journal of Criminal Justice* 29 (July/August): 271–285.

Fleegler, Eric W., et al. 2013. "Firearm Legislation and Firearm-Related Fatalities in the United State." *JAMA Internal Medicine,* March 6. Accessed January 15, 2014 (www.archinte.jamanetwork.com).

Flegal, Katherine M., Margaret D. Carroll, Cynthia L. Ogden, and Lester R. Curtin. 2010. "Prevalence and Trends in Obesity Among U.S. Adults, 1999–2008." *JAMA* 303 (January 20): 235–241.

Fleishman, Jeffrey, and Amro Hassan. 2009. "Gadget to Help Women Feign Virginity Angers Many in Egypt." *Los Angeles Times,* October 7. Accessed October 28, 2009 (www.latimes.com).

Fleming, Charles. 2014. "Ford Greenlights Recycling." *Baltimore Sun,* July 5, 8.

Fleming, John H., and Elizabeth Kampf. 2014. "Few Consumers Trust Companies to Keep Online Info Safe." Gallup, June 6. Accessed October 8, 2014 (www.gallup.com).

Fletcher, Douglass Scott, and Ian M. Taplin. 2002. *Understanding Organizational Evolution: Its Impact on Management and Performance.* Westport, CT: Quorum Books.

Fletcher, Michael A., and Steven Mufson. 2014. "Why Did GM Take So Long to Respond to Deadly Defect? Corporate Culture May Hold Answer." *Washington Post,* March 30. Accessed April 12, 2014 (www.washingtonpost.com).

Flower, Shawn M. 2010. "Gender-Responsive Strategies for Women Offenders." National Institute of Corrections, November. Accessed August 20, 2011 (www.nicic.gov).

Fogg, Piper. 2005. "Don't Stand So Close to Me." *Chronicle of Higher Education,* April 29, A10–A12.

Fomby, Paula, and Angela Estacion. 2011. "Cohabitation and Children's Externalizing Behavior in Low-Income Latino Families." *Journal of Marriage and Family* 73 (February): 46–66.

Food and Agriculture Organization of the United Nations. 2011. "The State of the World's Land and Water Resources for Food and Agriculture, Managing Systems at Risk: Summary Report." Accessed June 20, 2012 (www.fao.org).

Food and Water Watch. 2013. "Take Back the Tap: Bottled Water Wastes Resources and Money."

July. Accessed September 19, 2014 (www.foodandwaterwatch.org).

"Food Companies Propose Cutting Back on Junk Food Marketing Aimed at Children." 2011. *Washington Post,* July 14. Accessed July 20, (www.washingtonpost.com).

Ford, Peter. 2011. "Why Japan Will Rebound." *Christian Science Monitor,* March 28, 16–29.

___. 2014. "Did China Really Soften?" *Christian Science Monitor Weekly,* January 13, 11–12.

"Formaldehyde." 2012. Environmental Protection Agency. Accessed September 20, 2014 (www.epa.gov).

Foroohar, Rana. 2011. "Whatever Happened to Upward Mobility?" *Time,* November 14, 27–34.

Foroohar, Rana, and Bill Saporito. 2013. "Made in the USA." *Time,* April 22, 22–29.

Fowler, Tom. 2013. "Former Enron CEO's Sentence Cut to 14 Years." *Wall Street Journal,* June 22–23, A3.

Fox, Justin. 2013. "Which Is Worse: Airline Monopolies or Airline Competition?" Harvard Business Review Blog Network, August 15. Accessed April 12, 2014 (www.blogs.hbr.org).

Fox, Susannah, and Lee Rainie. 2014. "The Web at 25 in the U.S." Pew Research Center, February 27. Accessed October 8, 2014 (www.pewresearch.org).

Fox, Susannah, Maeve Duggan, and Kristen Purcell. 2013. "Family Caregivers Are Wired for Health." Pew Research Center, June 20. Accessed July 10, 2013 (www.pewinternet.org).

Fraire, John. 2014. "Why Your College Should Dump the SAT." *Chronicle of Higher Education,* May 2, A44.

Frank, T. A. 2006. "A Brief History of Wal-Mart." CorpWatch. Accessed February 19, 2007 (www.corpwatch.org).

Freedom House. 2014. "Freedom in the World." Accessed April 12, 2014 (www.freedomhouse.org).

Fremstad, Shawn. 2012. "The Poverty Rate Is Higher Than the Federal Government Says It Is." Center for Economic and Policy Research, September 21. Accessed February 6, 2014 (www.cepr.net).

Freese, Jeremy. 2008. "Genetics and the Social Science Explanation of Individual Outcomes." *American Journal of Sociology* 114 (Suppl.): S1–S35.

Frey, Carl B., and Michael A. Osborne. 2013. "The Future of Employment: How Susceptible Are Jobs to Computerization?" University of Oxford, September 17. Accessed April 12, 2014 (www.oxfordmartin.ox.ac.uk).

Frey, William H. 2011. "The Uneven Aging and 'Younging' of America: State and Metropolitan Trends in the 2010 Census." Brookings Metropolitan Policy Program. Accessed June 22, 2012 (www.brookings.edu).

Frieden, Thomas R. 2011. "Foreword." Centers for Disease Control and Prevention. *MMWR* 60 (Suppl): 1–2.

Friedman, Howard S. 2012. "American Voter Turnout Lower Than Other Wealthy Countries." *Huffington Post,* July 10. Accessed April 20, 2014 (www.huffingtonpost.com).

Friedrichs, David O. 2009. *Trusted Criminals: White Collar Crime in Contemporary Society,* 4th edition. Belmont, CA: Wadsworth Publishing.

Friedman, Jaclyn. 2011. *What You Really Really Want: The Smart Girl's Shame-Free Guide to Sex and Safety.* Berkeley, CA: Seal Press.

Frohwirth, Lori, Ann M. Moore, and Renata Maniaci. 2013. "Perceptions of Susceptibility to Pregnancy Among Women Obtaining Abortions." *Social Science & Medicine* 99 (December): 18–26.

Frosch, Dan. 2007. "18 Air Force Cadets Exit Over Cheating." *New York Times,* May 2, 18.

Frueh, Sara. 2014. "Career Choices of Female Engineers." The National Academies Press. Accessed November 17, 2014 (www.nap.edu).

Fry, Richard. 2012. "No Reversal in Decline of Marriage." Pew Research Center, November 20. Accessed February 9, 2013 (www.pewsocialtrends.org).

___. 2013. "Young Adults After the Recession: Fewer Homes, Fewer Cars, Less Debt." Pew Research Center, February 21. Accessed November 20, 2013 (www.pewresearch.org).

Fry, Richard, and D'Vera Cohn. 2011. "Living Together: The Economics of Cohabitation." Pew Research Center, June 27. Accessed March 1, 2012 (www.pewsocialtrends.org).

Fry, Richard, and Jeffrey S. Passel. 2014. "In Post-Recession Era, Young Adults Drive Continuing Rise in Multi-Generational Living." Pew Research Center, July 17. Accessed July 25, 2014 (www.pewresearch.org).

Fry, Richard, and Paul Taylor. 2012. "The Rise of Residential Segregation by Income." Pew Research Center, Social & Demographic Trends, August 1. Accessed September 19, 2014 (www.pewsocialtrends.org).

Fryer, Roland G., Jr., Devah Pager, and Jorg L. Spenkuch. 2011. "Racial Disparities in Job Finding and Offered Wages." National Bureau of Economic Research, September. Accessed December 7, 2011 (www.nber.org).

Fuentes-Nieva, Ricardo, and Nick Galasso. 2014. *Working for the Few.* Oxfam International, January. Accessed February 10, 2014 (www.oxfam.org).

Fung, Brian. 2014. "So, This Exists: A Working Car Has Been 3D-Printed Out of Carbon Fiber Plastic." *Washington Post,* September 19. Accessed October 8, 2014 (www.washingtonpost.com).

Funk, Cary, Luis Lugo, and Alan Cooperman. 2012. "Asian Americans: A Mosaic of Faiths." Pew Research Center, Pew Forum on Religion & Public Life, July 19. Accessed June 21, 2014 (www.pewforum.org).

Gable, Mary. 2013. "Ongoing Joblessness: A National Catastrophe for African American and Latino Workers." Economic Policy Institute, May 29. Accessed April 12, 2014 (www.epi.org).

Gallup Editors. 2013. "Most Americans Practice Charitable Giving, Volunteerism." Gallup, December. Accessed December 19, 2013 (www.gallup.com).

Gallup Historical Trends. 2014. "Gay and Lesbian Rights." Accessed March 2, 2014 (www.gallup.com).

Gallup. 2013. "State of the American Workplace: Employee Engagement Insight for U.S. Business Leaders." Accessed April 12, 2014 (www.gallup.com).

___. 2014. "U.S. Employment (Weekly)." Accessed February 7, 2014 (www.gallup.com).

Galperin, Andrew, et al. 2013. "Sexual Regret: Evidence for Evolved Sex Differences." *Archives of Sexual Behavior* 42 (October): 1145–1161.

Galston, William A. 2013. "The Eroding American Middle Class." *Wall Street Journal,* November 13, A15.

Gamson, William. 1990. *The Strategy of Social Protest,* 2nd edition. Belmont, CA: Wadsworth.

Gambino, Christine P., Edward N. Trevelyan, and John Thomas Fitzwater. 2014. "The Foreign-Born Population from Africa: 2008–2012." U.S. Census Bureau, October. Accessed November 25, 2014 (www.census.gov).

Gans, Herbert J. 1971. "The Uses of Poverty: The Poor Pay All." *Social Policy,* July/August, 78–81.

___. 2005. "Wishes for the Discipline's Future." *Chronicle Review,* August 12, B9.

___. 1999. *Popular Culture & High Culture: An Analysis and Evaluation of Taste.* Revised and updated edition. New York: Basic Books.

Garcia-Diaz, Daniel. 2013. "USDA Needs to Do More to Prevent Improper Payment to Deceased Individuals." GAO Highlights, June 28. Accessed December 19, 2013 (www.gao.gov).

Gardner, Marilyn. 2008. "Happiness Is a Warm 'Thank You.'" *Christian Science Monitor,* January 28, 13, 16.

Garfinkel, Harold. 1956. "Conditions of Successful Degradation Ceremonies." *American Journal of Sociology* 61 (March): 420–424.

___. 1967. *Studies in Ethnomethodology.* Englewood Cliffs, NJ: Prentice-Hall.

Gastañaduy, Paul A., et al. 2014. "Measles—United States, January 1–May 23, 2014." *MMWR* 63 (22): 496–499.

Gates, Gary J. 2011. "How Many People Are Lesbian, Gay, Bisexual, and Transgender?" Williams Institute, April. Accessed March 2, 2014 (www.williamsinstitute.law.ucla.edu).

___. 2013. "LGBT Parenting in the United States." Williams Institute, February. Accessed May 16, 2014 (www.williamsinstitute.law.ucla.edu).

Gates, Gary J., and Frank Newport. 2012. "Special Report: 3.4% of U.S. Adults Identify as LGBT." Gallup, October 18. Accessed March 2, 2014 (www.gallup.com).

Gavin, Lorrie, et al. 2013. "Vital Signs: Repeat Births among Teens—United States, 2007–2010." *MMWR* 62 (April 5): 249–255.

"The Gay Divide." 2014. *The Economist,* October 11, 13.

"Gay Marriage." 2014. Accessed November 17, 2014 (www.gaymarriage.procon.org).

Geary, David C., Mary K. Hoard, Lara Nugent, and Drew H. Bailey. 2013. "Adolescents' Functional Numeracy Is Predicted by Their School Entry Number System Knowledge." *PLOS ONE* 8 (January): e54651.

Geertz, Clifford. 1966. "Religion as a Cultural System." Pp. 1–46 in *Anthropological Approaches to the Study of Religion,* edited by Michael Banton. London: Tavistock.

Geier, Ben. 2014. "Using 3-D Printing to Make Jet Engines." *Fortune,* December 1, 76.

Geller, Adam. 2014. "This Is How Americans Are Grappling with Income Inequality." *Huffington Post,* February 2. Accessed February 4, 2014 (www.huffingtonpost.com).

Gellman, Lindsay. 2013. "Mom, Stop Calling, I'll Text You All Day." *Wall Street Journal,* July 31, D1–D2.

"Gender Equality Universally Embraced, but Inequalities Acknowledged." 2010. Pew Research Center, Global Attitudes Project, July 1. Accessed November 1, 2011 (www.pewglobal.org).

Geoghegan, Tom. 2013. "Why Do So Many Americans Live in Mobile Homes?" *BBC News Magazine,* September 23. Accessed January 12, 2014 (www.bbc.co.uk).

Gerth, H. H., and C. Wright Mills, eds. 1946. *Max Weber: Essays in Sociology.* New York: Oxford University Press.

Gewertz, Catherine. 2009. "Do Men Deserve a Break in College Admissions?" *Education*

Week online, December 9. Accessed December 13, 2010 (www.blogs.edweek.org).

Gibson, David. 2012. "Vatican Orders Crackdown on American Nuns." *USA Today,* April 18. Accessed April 20, 2012 (www.usatoday.com).

Gibson, Megan. 2012. "Celebrities Offering Scientific 'Facts'? Just Say No." *Time,* January 3. Accessed January 24, 2012 (www.newsfeed.time.com).

Gilbert, Dennis. 2011. *The American Class Structure in an Age of Growing Inequality,* 8th edition. Thousand Oaks, CA: Pine Forge Press.

Gilles, Kate, and Charlotte Feldman-Jacobs. 2012. "When Technology and Tradition Collide: From Gender Bias to Sex Selection." Population Reference Bureau, September. Accessed September 19, 2014 (www.prb.org).

Gillespie, Michael Allen. 2014. "Grade Degradation." *Chronicle Review,* January 24, B2.

Gilpin, Lindsey. 2014. "The Dark Side of 3D Printing: 10 Things to Watch." TechRepublic, March 5. Accessed October 8, 2014 (www.techrepublic.com).

Gilson, Dave. 2014. "Don't Tread on Me." *Mother Jones,* January/February, 28–33.

Giugni, Marco, Doug McAdam, and Charles Tilley, eds. 1999. *How Social Movements Matter.* Minneapolis: University of Minnesota Press.

Giuliano, Laura, David I. Levine, and Jonathan Leonard. 2009. "Manager Race and the Race of New Hires." *Journal of Labor Economics* 27 (October): 589–631.

Givhan, Robin. 2010. "Fashion: Michelle Obama Acknowledges Indian Fashion Industry During Trip." *Washington Post,* November 11, C2.

GLAAD. 2013. "Where We Are on TV." Accessed March 2, 2014 (www.glaad.org).

Glaser, Gabrielle. 2013. "Why She Drinks." *Wall Street Journal,* June 22–23, C1–C2.

Glassner, Barry. 2010. "Still Fearful After All These Years." *Chronicle Review,* January 22, B11–B12.

Glauber, Bill. 2011. "Do Monarchies Still Matter?" *Christian Science Monitor,* February 21, 27–31.

Gleiber, Michael A. 2014. "The High Price Our Bodies Pay for High Fashion." *Huffington Post,* October 26. Accessed December 5, 2014 (www.huffingtonpost.com).

Glenn, David. 2011. "For Business Majors, Easy Does It." *Chronicle of Higher Education,* April 22, A1, A3–A5.

Glenn, Norval D., Jeremy Uecker, and Robert W. B. Love, Jr. 2010. "Later First Marriage and Marital Success." *Social Science Research* 39 (September): 787–800.

"Global Christianity: A Report on the Size and Distribution of the World's Christian Population." 2011. Pew Research Center, December 1. Accessed April 8, 2012 (www.pewresearch.org).

Global WASH Fast Facts. 2013. "Information on Water, Sanitation, and Hygiene." Centers for Disease Control and Prevention, November 8. Accessed September 19, 2014 (www.cdc.gov).

Glynn, Sarah Jane, and Audrey Powers. 2012. "The Top 10 facts About the Wage Gap." Center for American Progress, April 16. Accessed November 17, 2014 (www.americanprogress.org).

Goble, Priscilla, Carol L. Martin, Laura D. Hanish, and Richard A. Fabes. 2012." Children's Gender-Typed Activity Choices Across Preschool Social Contexts." *Sex Roles* 67 (October): 435–451.

Godofsky, Jessica, Cliff Zukin, and Carl Van Horn. 2011. "Unfulfilled Expectations: Recent College Graduates Struggle in a Troubled Economy." *Work Trends,* May. Accessed June 5, 2011 (www.heldrich.rutgers.edu).

Goel, Vindu. 2014. "With New Ad Platform, Facebook Opens Gates to Its Vault of User Data." *New York Times,* September 29, B7.

Goffman, Erving. 1959. *The Presentation of Self in Everyday Life.* New York: Doubleday Anchor Books.

___. 1961. *Asylums: Essays on the Social Situation of Mental Patients and Other Inmates.* Garden City, NY: Anchor Books.

___. 1963. *Stigma: Notes on the Management of Spoiled Identity.* Englewood Cliffs, NJ: Prentice-Hall.

___. 1967. *Interaction Ritual: Essays on Face-to-Face Behavior.* New York: Anchor Books.

___. 1969. *Strategic Interaction.* Philadelphia: University of Pennsylvania Press.

Golden, Daniel. 2006. *The Price of Admission: How America's Ruling Class Buys Its Way into Elite Colleges—and Who Gets Left Outside the Gates.* New York: Crown.

Goldin, Claudia, Lawrence F. Katz, and Ilyana Kuziemko. 2006. "The Homecoming of American College Women: The Reversal of the College Gender Gap." National Bureau of Economic Research. Accessed July 25, 2006 (www.nber.org).

Goldring, Rebecca, Lucinda Gray, and Amy Bitterman. 2013. "Characteristics of Public and Private Elementary and Secondary School Teachers in the United States: Results From the 2011–12 Schools and Staffing Survey." National Center for Education Statistics, August. Accessed November 17, 2014 (www.nces.ed.gov).

Goodale, Gloria. 2009. "A Wake-Up Call on Water Use." *Christian Science Monitor,* June 10. Accessed June 12, 2009 (www.csmonitor.com).

___. 2013. "'Sharing' Gets Some Pushback." *Christian Science Monitor Weekly,* October 7, 21–23.

Goode, Erica. 2013. "U.S. Prison Populations Decline, Reflecting New Approach to Crime." *New York Times,* July 26, A11.

Goodnough, Abby. 2010. "Making It Clear That a Clear Parking Space Isn't." *New York Times,* December 28, A10.

Goodstein, Laurie. 2014. "Presbyterians Vote to Allow Same-Sex Marriages." *New York Times,* June 20, A11.

Gopnik, Alison, Andrew N. Meltzoff, and Patricia K. Kuhl. 2001. *The Scientist in the Crib: What Early Learning Tells Us About the Mind.* New York: Perennial.

Gorman, Bill. 2011. "'CSI: Crime Scene Investigation' Is the Most-Watched Drama Series in the World!" TV by the Numbers, June 13. Accessed August 22, 2011 (www.tvbythenumbers.zap2it.com).

Gottdiener, Mark, and Ray Hutchison. 2000. *The New Urban Sociology,* 2nd edition. New York: McGraw-Hill.

Gottlieb, Mark. 2011. "Attention to Duty." *Christian Science Monitor,* June 6, 44.

Gottman, John M. 1994. *What Predicts Divorce? The Relationships Between Marital Processes and Marital Outcome.* Hillsdale, NJ: Lawrence Erlbaum Associates.

Goudreau, Jenna. 2012. "A New Obstacle for Professional Women: The Glass Escalator." *Forbes,* May 21. Accessed December 18, 2013 (www.forbes.com).

Gould, Elise. 2012. "Two in Five Female-Headed Families with Children Live in Poverty." Economic Policy Institute, November 14. Accessed February 6, 2014 (www.epi.org).

Gould, Elise, and Heidi Shierholz. 2012. "Average Worker in 'Right-to-Work' State Earns $1,500 Less Each Year." Economic Policy Institute, December 13. Accessed April 12, 2014 (www.epi.org).

Gouldner, Alvin W. 1962. "Anti-Minotaur: The Myth of a Value-Free Sociology." *Social Problems* 9 (Winter): 199–212.

"Government Resolves to Start Making Sense Under New Law Forbidding Federal Gibberish." 2011. *Washington Post,* May 20. Accessed May 21, 2011 (www.washingtonpost.com).

Governors Highway Safety Association. 2014. "Distracted Driving Laws." Accessed January 12, 2014 (www.ghsa.org).

Goyette, Braden. 2013. "Oprah Talks to Swiss Newspaper About Alleged Racist Encounter." *Huffington Post,* August 14. Accessed March 22, 2014 (www.huffingtonpost.com).

Graif, Corina, and Robert J. Sampson. 2009. "Spatial Heterogeneity in the Effects of Immigration and Diversity on Neighborhood Homicide Rates." *Homicide Studies* 13 (August): 242–260.

Grall, Timothy. 2013. "Custodial Mothers and Fathers and Their Child Support: 2011." U.S. Census Bureau, October. Accessed May 16, 2014 (www.census.gov).

Grana, Rachel A., Lucy Popova, and Pamela M. Ling. 2014. "A Longitudinal Analysis of Electronic Cigarette Use and Smoking Cessation." *JAMA Internal Medicine* 174 (5): 812–813.

Grandjean, Philippe, and Philip J. Landrigan. 2014. "Neurobehavioral Effects of Developmental Toxicity." *The Lancet Neurology* 13 (March): 330–338.

Grauerholz, Liz, and Sharon Bouma-Holtrop. 2003. "Exploring Critical Sociological Thinking." *Teaching Sociology* 31 (October): 485–496.

Graves, Joseph L., Jr. 2001. *The Emperor's New Clothes: Biological Theories of Race at the Millennium.* New Brunswick, NJ: Rutgers University Press.

Gray, Paul S., John B. Williamson, David R. Karp, and John R. Dalphin. 2007. *The Research Imagination: An Introduction to Qualitative and Quantitative Methods.* New York: Cambridge University Press.

Grazian, David. 2008. *On the Make: The Hustle of Urban Nightlife.* Chicago: University of Chicago Press.

Greeley, Brendan. 2013. "How Inequality Became a Household Word." *Bloomberg Businessweek,* December 16–22, 16–17.

Green, Jeff. 2011. "The Silencing of Sexual Harassment." *Bloomberg Businessweek,* November 21–28.

Greenberg, Julie, Arthur McKee, and Kate Walsh. 2013. "Teacher Prep Review: A Review of the Nation's Teacher Preparation Programs 2013." National Council on Teacher Quality. Accessed June 21, 2014 (www.nctq.org).

Grieco, Elizabeth M., et al. 2012. "The Foreign-Born Population in the United States: 2010." U.S. Census Bureau, May. Accessed March 22, 2014 (www.census.gov).

Grier, Peter. 2012. "Briefing: Who Are the '47 Percent'?" *Christian Science Monitor Weekly,* October 1, 12.

Grim, Brian J., and Alan Cooperman. 2014. "Religious Hostilities Reach Six-Year High." Pew Research Center, January 14. Accessed June 21, 2014 (www.pewresearch.org).

Grind, Kirsten. 2013. "Mother, Can You Spare a Room?" *Wall Street Journal,* May 4–5, B1, B10.

Gross, Rita M. 1996. *Feminism and Religion: An Introduction.* Boston: Beacon Press.

Grossman, David. 2011. "Fliers Come to Fisticuffs Over Reclined Seat." *USA Today,* June 6. Accessed June 7, 2011 (www.usatoday.com).

Groves, Robert, and Frank Vitrano. 2011. "The U.S. Decennial Census and the American Community Survey: Looking Back and Looking Ahead." Population Reference Bureau, April. Accessed November 9, 2011 (www.prb.org).

"Growing Number of Americans Say Obama Is a Muslim." 2010. Pew Research Center, August 19. Accessed April 8, 2012 (www.pewresearch.org).

Grunwald, Michael. 2014. "The (Slow) Greening of America: A New Poll Reveals That U.S. Is Reluctant to Recognize and Address Climate Change." *Time,* June 23, 22.

Guallar, Eliseo, Saverio Stranges, Cynthia Mulrow, Lawrence J. Appel, and Edgar R. Miller III. 2013. "Enough Is Enough: Stop Wasting Money on Vitamin and Mineral Supplements." *Annals of Internal Medicine* 159 (12): 850–851.

Guarino, Mark. 2013. "Auto Jobs Are Back, Too, but at Lower Wages." *Christian Science Monitor Weekly,* May 13, 22–23.

Guerra, Nancy G., Carly B. Dierkhising, and Pedro R. Payne. 2013. "How Should We Identify and Intervene With Youth at Risk of Joining Gangs? A Developmental Approach for Children Ages 0–12." Pp. 63–73 in *Changing Course: Preventing Gang Membership,* edited by Thomas R. Simon, Nancy M. Witter, and Reshma R. Mahendra. National Institute of Justice and Centers for Disease Control and Prevention. Accessed December 18, 2013 (www.nij.gov).

"Gun Laws: A Shot and a Beer." 2014. *The Economist,* May 3, 28.

Gunders, Dana. 2012. "Wasted: How America Is Losing Up to 40 Percent of Its Food from Farm to Fork to Landfill." Natural Resources Defense Council, August. Accessed November 10, 2013 (www.nrdc.org).

Gurkoff, Joe, and Anna Ranieri. 2012. *How Can I Help You? What You Can (and Can't) Do to Counsel a Friend, Colleague or Family Member with a Problem.* North Charleston, SC: CreateSpace Independent Publishing Platform.

Gurney, Joan M., and Kathleen T. Tierney. 1982. "Relative Deprivation and Social Movements: A Critical Look at Twenty Years of Theory and Research." *Sociological Quarterly* 23 (Winter): 33–47.

Guterman, Stanley S. 1969. "In Defense of Wirth's 'Urbanism as a Way of Life.'" *American Journal of Sociology* 74 (March): 492–499.

Guttmacher Institute. 2009. "A Real-Time Look at the Impact of the Recession on Women's Family Planning and Pregnancy Decisions." September. Accessed April 12, 2010 (www.guttmacher.org).

___. 2013. "U.S. Women Who Have Abortions." Infographics. Accessed February 11, 2012 (www.guttmacher.org).

___. 2014. "Facts on Induced Abortion in the United States." February. Accessed March 2, 2014 (guttmacher.org).

Haas, Ann P., Phillip L. Rodgers, and Jody L. Herman. 2014. "Suicide Attempts Among Transgender and Gender Non-Conforming Adults." Williams Institute, January. Accessed March 2, 2014 (www.williamsinstitute.law.ucla.edu).

Haas, Steven A., and David R. Schaefer. 2014. "With a Little Help from My Friends? Asymmetrical Social Influence on Adolescent Smoking Initiation and Cessation." *Journal of Health and Social Behavior* 55 (June): 126–143.

Hacker, Jacob S., and Paul Pierson. 2010. *Winner-Take-All Politics: How Washington Made the Rich Richer and Turned Its Back on the Middle Class.* New York: Simon and Schuster.

Hacker, Jacob S. 2002. *The Divided Welfare State: The Battle over Public and Private Social Benefits in the United States.* New York: Cambridge University Press.

Hagan, Frank E. 2008. *Introduction to Criminology: Theories, Methods, and Criminal Behavior,* 6th edition. Thousand Oaks, CA: Sage.

Hagerty, James R. 2013. "Help Wanted. A Lot of It." *Wall Street Journal,* June 11, R7.

Hales, Emily. 2014. "Incentives for Having Kids: Solution for Falling Birth Rates?" *Deseret News,* June 26. Accessed September 21 (www.newsok.com).

"Half of Young Children in the U.S. Are Read to at Least Once a Day, Census Bureau Reports." 2011. U.S. Census Bureau, August 11. Accessed November 18, 2013 (www.census.gov/newsroom).

Hall, Aron J., Mary E. Wikswo, Kimberly Pringle, L. Hannah Gould, and Umesh D. Parashar. 2014. "Vital Signs: Foodborne Norovirus Outbreaks—United States, 2009–2012." *MMWR* 63 (22): 491–495.

Hall, Edward T. 1959. *The Silent Language.* New York: Doubleday.

___. 1966. *The Hidden Dimension.* Garden City, NY: Doubleday.

Halperin, David M. 2012. "How to Be Gay." *Chronicle Review,* September 7, B13–B17.

Halsey, Ashley, III. 2010. "Study: Older People Are Driving More, Having Fewer Accidents." *Washington Post,* June 22, A10.

___. 2011. "Seat in Lap Leads to Scuffle in the Air." *Washington Post,* June 1, A1.

Hamburger, Tom, and Matea Gold. 2014. "Google, Once Disdainful of Lobbying, Now a Master of Washington Influence." *Washington Post,* April 12. Accessed April 20, 2014 (www.washingtonpost.com).

Hamby, Sherry, and David Finkelhor. 2001. "Choosing and Using Child Victimization Questionnaires." *OJJDP Juvenile Justice Bulletin.* Washington, DC: U.S. Department of Justice.

Hamby, Sherry, David Finkelhor, Heather Turner, and Richard Ormrod. 2011. "Children's Exposure to Intimate Partner Violence and Other Family Violence." Office of Justice Programs, October. Accessed May 20, 2014 (www.ojp.usdoj.gov).

Hamedy, Saba. 2014. "Montana Federal Judge Sent Hundreds of Biased Emails, Panel Finds." *Los Angeles Times,* January 18. Accessed November 25, 2014 (www.latimes.com).

Hamilton, Malcolm B. 1995. *The Sociology of Religion: Theoretical and Comparative Perspectives.* New York: Routledge.

___. 2001. *The Sociology of Religion,* 2nd edition. New York: Routledge.

Hamm, Steve. 2007. "The Trouble with India." *Business Week,* March 19, 49–58.

Hammadi, Saad. 2013. "Making Asia's Factories Safer." *Christian Science Monitor Weekly,* November 11, 16.

Hampton, Keith N., Lauren Sessions, Eun Ja Her, and Lee Rainie. 2009. "Social Isolation and New Technology." Pew Internet & American Life Project, November. Accessed February 5, 2010 (www.pewinternet.org).

Hampton, Keith N., Lauren Sessions Goulet, Lee Rainie, and Kristin Purcell. 2011. "Social Networking Sites and Our Lives." Pew Internet & American Life Project, July 16. Accessed July 20, 2011 (www.pewinternet.org).

Hamre, Bridget K., and Robert C. Pianta. 2001. "Early Teacher–Child Relationships and the Trajectory of Children's School Outcomes Through Eighth Grade." *Child Development* 72 (March/April): 625–638.

Hanes, Stephanie. 2010. "Texting Bans Ineffective?" *Christian Science Monitor,* October 11, 20.

___. 2011. "Pretty in Pink?" *Christian Science Monitor,* September 26, 26–31.

___. 2013. "Toddlers on Touch Screens." *Christian Science Monitor Weekly,* October 21, 26–31.

Haney, Craig, Curtis Banks, and Philip Zimbardo. 1973. "Interpersonal Dynamics in a Simulated Prison." *International Journal of Criminology and Psychology* 1: 69–97.

Hansen, Lawrence A. 2010. "Noxious Groupthink." *Chronicle of Higher Education,* November 12, B6–B8.

Hanser, Amy, and Zachary Hyde. 2014. "Foodies Remaking Cities." *Contexts* 13 (Summer): 44–49.

Hanushek, Eric A., and Ludger Woessman. 2005. "Does Educational Tracking Affect Performance and Inequality? Differences-in-Differences Evidence Across Countries." National Bureau of Economic Research, February. Accessed July 25, 2007 (www.nber.org).

Haq, Husna. 2009. "Ethnic Malls Are Buzzing." *Christian Science Monitor,* August 30, 30–31.

Hargrove, Thomas. 2011. "Disappearing Jobs Illustrate U.S. Manufacturing Revolution." Knox News, November 13. Accessed January 9, 2012 (www.knoxnews.com).

Harlow, Harry F., and Margaret K. Harlow. 1962. "Social Deprivation in Monkeys." *Scientific American,* 206 (November): 137–146.

Harnish, Verne. 2014. "5 Ways to Tone Your Operations." *Fortune,* October 27, 46.

Harrell, Erika, and Lynn Langton. 2013. "Victims of Identity Theft, 2012." U.S. Department of Justice, Office of Justice Programs, December. Accessed January 12, 2014 (www.ojp.usdoj.gov).

Harris, Chauncey D. 1997. "'The Nature of Cities' and Urban Geography in the Last Century." *Urban Geography* 18 (1): 15–35.

Harris, Chauncey D., and Edward L. Ullman. 1945. "The Nature of Cities." *Annals* 242: 7–17.

Harris, Elizabeth A., Nicole Perlroth, Nathaniel Popper, and Hilary Stout. 2014. "A Sneaky Path into Target Customers' Wallets." *New York Times,* January 18, A1.

Harris, Gardiner. 2011. "When the Nurse Wants to Be Called 'Doctor.'" *New York Times,* June 6, A23.

___. 2013. "India's New Focus on Rape Shows Only the Surface of Women's Perils." *New York Times,* January 12. Accessed March 4, 2014 (www.nytimes.com).

Harris, Jennifer L., et al. 2013. "Fast Food FACTS 2013: Measuring Progress in Nutrition and Marketing to Children and Teens." Yale Rudd Center for Food Policy & Obesity. Accessed November 10, 2013 (www.fastfoodmarketing.org).

Harter, Jim, and Sangeeta Agrawal. 2013. "'Engaged' Workers Would Keep Jobs upon Winning Lottery." Gallup, December 20. Accessed April 12, 2014 (www.gallup.com).

___. 2014. "Many Baby Boomers Reluctant to Retire." Gallup, January 20. Accessed May 16, 2014 (www.gallup.com).

Hartford Institute for Religion Research. 2012. "Fast Facts About American Religion." Accessed July 1, 2014 (www.hirr.hartsem.edu).

Harvard School of Public Health. 2014. "Poll Finds Many in U.S. Lack Knowledge About Ebola and Its Transmission." August 21. Accessed October 8, 2014 (www.hsph.harvard.edu).

Haskins, Ron. 2008. "Immigration: Wages, Education, and Mobility." Pp. 81–90 in *Getting Ahead or Losing Ground: Economic Mobility in America,* edited by Julia B. Isaacs, Isabel V. Sawhill, and Ron Haskins. Washington DC: Brookings Institution.

Haub, Carl. 2011. "The Caste Census: A Feudal Classification of Society in India?" Population Reference Bureau, September 7. Accessed September 12, 2011 (www.prb.org).

Haub, Carl. 2013. "Rising Trend of Births Outside Marriage." Population Reference Bureau, April. Accessed February 14, 2014 (www.prb.org).

Haub, Carl, and Toshiko Kaneda. 2014. "2014 World Population Data Sheet." Population Reference Bureau. Accessed September 19, 2014 (www.prb.org).

Hausmann, Ricardo, Laura D. Tyson, Yasmina Bekhouche, and Saadia Zahidi. 2014. "The Global Gender Gap Report 2014." World Economic Forum. Accessed December 2, 2014 (www.weforum.org).

Hawai'i Free Press. 2012. "Census Bureau: Hawaii 7th Highest Poverty Rate in US." November 15. Accessed February 14, 2014 (www.hawaiifreepress.com).

Hayden, Dolores. 2002. *Redesigning the American Dream: Gender, Housing, and Family Life.* New York: W.W. Norton

He, Wan, and Luke J. Larsen. 2014. "Older Americans with a Disability: 2008–2012." U.S. Census Bureau, December. Accessed December 5, 2014 (www.census.gov).

"The Health Paradox." 2013. *The Economist,* May 11, 27–28.

"Health-Care Fraud: The $272 Billion Swindle." 2014. *The Economist,* May 31, 26–27.

Healthcare-NOW! 2014. "Health Insurance CEO Pay Skyrockets in 2013." May 5. Accessed July 26, 2014 (www.healthcare-now.org).

Health Resources and Services Administration. 2010. "Assuring Access to Essential Health Care." Accessed July 15, 2011 (www.plainlanguage.gov).

Healy, Melissa. 2013. "1st Stem Cells Made by Cloning Embryos." *Baltimore Sun,* May 16, 8.

Heath, Chip. 2005. "Introduction." Pp. 81–85 in *Rumor Mills: The Social Impact of Rumor and Legend,* edited by Gary Alan Fine, Veronique Campion-Vincent, and Chip Heath. New Brunswick, NJ: Transaction Publishers.

Heath, Jennifer. 2008. *The Veil: Women Writers on Its History, Lore, and Politics.* Berkeley: University of California Press.

Hechter, Michael, and Karl-Dieter Opp. 2001. "Introduction." Pp. xi–xx in *Social Norms,* edited by Michael Hechter and Karl-Dieter Opp. New York: Russell Sage Foundation.

Hedegaard, Holly, and Li-Hui Chen. 2014. "Rates of Drug Deaths Involving Heroin, by Selected Age and Racial/Ethnic Groups—United States, 2002 and 2011." *MMWR* 63 (July 11): 595.

Hedges, Chris. 2011. "Our Public Schools Are Churning Out Drones for the Corporate State." AlterNet, April 11. Accessed April 14, 2011 (www.alternet.org).

Heffer, Simon. 2010. "America Is the Acceptable Face of Cultural Imperialism." *Telegraph Media Group,* July 24. Accessed November 10, 2013 (telegraph.co.uk).

Hegewisch, Ariane, and Stephanie K. Hudiburg. 2014. "The Gender Wage Gap by Occupation and by Race and Ethnicity." Institute for Women's Policy Research, April. Accessed November 2, 2014 (www.iwpr.org).

Heider, Eleanor R., and Donald C. Olivier. 1972. "The Structure of the Color Space in Naming and Memory for Two Languages." *Cognitive Psychology,* 3: 337–354.

Heiman, Julia R., J. Scott Long, Shawna N. Smith, William A. Fisher, Michael S. Sand, and Raymond C. Rosen. 2011. "Sexual Satisfaction and Relationship Happiness in Midlife and Older Couples in Five Countries." *Archives of Sexual Behavior* 40 (4): 741–753.

Heimer, Karen, Stacy Wittrock, and Halime Ünal. 2006. "The Crimes of Poverty: Economic Marginalization and the Gender Gap in Crime." Pp. 115–126 in *Gender and Crime: Patterns of Victimization and Offending,* edited by Karen Heimer and Candace Kruttschnitt. New York: New York University Press.

Heisler, Candace. 2012. "Elder Abuse." Office for Victims of Crime. Accessed May 16, 2014 (www.ovc.ncjrs.gov).

Helliwell, John, Richard Layard, and Jeffrey Sachs (eds). 2013. *World Happiness Report 2013.* United Nations. Accessed February 6, 2014 (www.unsdsn.org).

Helm, V., A. Humbert, and H. Miller. 2014. "Elevation and Elevation Change of Greenland and Antarctica Derived from CryoSat-2." *The Cryosphere* (August): 1539–1559.

Hennessey, Kathleen. 2013. "First Lady Signals a Return to Food Marketing Debate." *Baltimore Sun,* September 19, 8.

Henry, Meghan, Alvaro Cortes, and Sean Morris. 2013. "The 2013 Annual Homeless Assessment Report (AHAR) to Congress." Office of Community Planning and Development, November. Accessed February 6, 2014 (www.onecpd.info).

Herbenick, Debby, Michael Reece, Vanessa Schick, Stephanie A. Sanders, Brian Dodge, and J. Dennis Fortenberry. 2010. "Sexual Behaviors, Relationships, and Perceived Health Status among Adult Women in the United States: Results from a National Probability Sample." *Journal of Sexual Medicine* 7, Supplement 5 (October): 277–290.

Hernandez, Donald J., Jeffrey S. Napierala. 2013. "Diverse Children: Race, Ethnicity, and Immigration in America's New Non-Majority Generation." Foundation for Child Development, July. Accessed March 22, 2014 (www.fcd-us.org).

Herzberg, David. 2009. *Happy Pills in America: From Miltown to Prozac.* Baltimore, MD: Johns Hopkins University Press.

Hetherington, E. Mavis, Ross D. Parke, and Virginia Otis Locke. 2006. *Child Psychology: A Contemporary Viewpoint,* 6th edition. Boston: McGraw-Hill.

Hibbard, Roberta, Jane Barlow, and Harriet MacMillan. 2012. "Psychological Maltreatment." American Academy of Pediatrics. Accessed May 4, 2013 (www.pediatrics.aappublications.org).

Hilbert, Richard A. 1992. *The Classical Roots of Ethnomethodology: Durkheim, Weber, and Garfinkel.* Chapel Hill: University of North Carolina Press.

Hill, Catherine, Christianne Corbett, and Andresse St. Rose. 2010. "Why So Few? Women in Science, Technology, Engineering, and Mathematics." American Association of University Women. Accessed October 25, 2011 (www.aauw.org).

Hill, Jason G. 2011. "Education and Certification Qualifications of Departmentalized Public High School-Level Teachers of Core Subjects: Evidence from the 2007–08 Schools and Staffing Survey." National Center for Education Statistics, May. Accessed June 21, 2014 (www.nces.org).

Hillaker, B. D., H. E. Brophy-Herb, F. A. Villarruel, and B. E. Hass. 2008. "The Contributions of Parenting to Social Competencies and Positive Values in Middle School Youth: Positive Family Communication, Maintaining Standards, and Supportive Family Relationships." *Family Relations* 57 (December): 591–601.

Hillin, Taryn. 2014. "The Divorce Mistakes You Don't Even Know You're Making." *Huffington Post,* March 18. Accessed March 18, 2014 (www.huffingtonpost.com)

Hing, Julianne. 2011. "5 Ways Alabama's New Anti-Immigrant Law Is Even Worse Than Arizona's SB 1070." AlterNet, June 24. Accessed July 2, 2011 (www.alternet.org).

Hinkle, Stephen, and John Schopler. 1986. "Bias in the Evaluation of In-Group and Out-Group Performance." Pp. 196–212 in *Psychology of Everyday Intergroup Relations,* 2nd edition, edited by Stephen Worchel and William G. Austin. Chicago: Nelson-Hall.

Hirsch, Jerry. 2014. "Recall Fever Spreading Fast." *Baltimore Sun,* May 17, 10.

Hirschi, Travis. 1969. *Causes of Delinquency.* Berkeley: University of California Press.

"Hispanic Heritage Month 2014: Sept. 15–Oct. 15." 2014. U.S. Census Bureau News, September 8. Accessed September 22, 2014 (www.census.gov).

Hiss, William C., and Valerie W. Franks. 2014. "Defining Promise: Optional Standardized Testing Policies in American College and University Admissions." National Association for College Admission Counseling, February 5. Accessed June 21, 2014 (www.nacacnet.org).

Hlavka, Heather R. 2014. "Normalizing Sexual Violence: Young Women Account for Harassment and Abuse." *Gender & Society* 20 (June): 1–22.

Hodge-Williams, Erin. 2014. "Black Males Need More Opportunities and Mentors." *Baltimore Sun,* February 18, 11.

Hoecker-Drysdale, Susan. 1992. *Harriet Martineau: First Woman Sociologist.* Providence, RI: Berg.

Hoefer, Michael, Nancy Rytina, and Bryan Baker. 2012. "Estimates of the Unauthorized Immigrant Population Residing in the United States: January 2011." U.S. Department of Homeland Security, Office of Immigration Statistics, March. Accessed March 22, 2014 (www.dhs.gov).

Hoffman, Piper. 2014. "7 Countries That Still Kill 'Witches.'" Care2 Causes, October 21. Accessed October 24, 2014 (www.care2.com).

Hokayem, Charles, and Misty L. Heggeness. 2014. "Living in Near Poverty in the United States: 1966–2012." U.S. Census Bureau, May. Accessed May 16, 2014 (www.census.gov).

Holcomb, Jesse, Jeffrey Gottfried, and Amy Mitchell. 2013. "News Use across Social Media Platforms." Pew Research Center, November 14. Accessed December 4, 2013 (www.pewinternet.org).

Holmes, Elizabeth. 2013. "Grooming Tips from Giant Men." *Wall Street Journal,* November 27, D1.

Homans, George. 1974. *Social Behavior: Its Elementary Forms,* revised edition. New York: Harcourt Brace Jovanovich.

Hondagneu-Sotelo, Pierrette. 2001. *Domestica: Immigrant Workers Cleaning and Caring in the Shadows of Affluence.* Berkeley: University of California Press.

Hooke, Alexander E. 2014. "Speak Not, Lest Ye Be Judged." *Baltimore Sun,* May 27, 13.

Horowitz, Juliana M., Erin Carriere-Kretschmer, and Jacob Poushter. 2010. "Gender Equality Universally Embraced, but Inequalities Acknowledged." Pew Research Center, July 1. Accessed March 2, 2014 (www .pewresearch.org).

Hotz, V. Joseph, and Juan Pantano. 2013. "Strategic Parenting, Birth Order and School Performance." National Bureau of Economic Research, October. Accessed October 26, 2014 (www.nber.org).

Howard, Beth. 2013/2014. "Meet Dr. Robot." *AARP Magazine,* December/January,18, 23.

———. 2014. "Should You Have Surgery Abroad?" *AARP Magazine,* October/November, 26–30.

Howard, Jeff. 2003. "Still at Risk: The Causes and Costs of Failure to Educate Poor and Minority Children for the Twenty-First Century." Pp. 81–97 in *A Nation Reformed? American Education 20 Years After A Nation at Risk,* edited by David T. Gordon and Patricia A. Graham. Cambridge, MA: Harvard Education Press.

Howard, Peter. 2014. "Flammable Planet: Wildfires and the Social Cost of Carbon." Institute for Policy Integrity and National Resources Defense Council, Accessed September 19, 2014 (www.costofcarbon.org).

Hoyt, Homer. 1939. *The Structure and Growth of Residential Neighborhoods in American Cities.* Washington, DC: Federal Housing Administration.

Hsin, Amy, and Yu Xie. 2014. "Explaining Asian Americans' Academic Advantage Over Whites." *Proceedings of the National Academy of Sciences,* Early Edition, May 5. Accessed June 21, 2014 (www.pnas.org).

Hsu, Angel, et al. 2014. "The 2014 Environmental Performance Index: Full Report and Analysis." Yale Center for Environmental Law & Policy. Accessed September 19, 2014 (www.epi.yale.edu).

Hubbard, Ruth. 1990. *The Politics of Women's Biology.* New Brunswick, NJ: Rutgers University Press.

Huff, Rodney, Christian Desilets, and John Kane. 2010. "The 2010 National Public Survey on White Collar Crime." National White Collar Crime Center. Accessed January 12, 2014 (www.nw3c.org).

Huffington Post. 2011. "New American Bible Changes Some Words, Including 'Holocaust.'" March 3. Accessed July 16, 2011 (www.huffingtonpost.com).

Huffman, Matt L. 2013. "Organizations, Managers, and Wage Inequality." *Sex Roles* 68 (February): 216–222.

Hughes, Everett C. 1945. "Dilemmas and Contradictions of Status." *American Journal of Sociology* 50: 353–359.

Huizinga, David, Shari Miller, and the Conduct Problems Prevention Research Group. 2013. "Developmental Sequences of Girls' Delinquent Behavior." Office of Juvenile Justice and Delinquency Prevention, December. Accessed January 12, 2014 (www .ojp.usdoj.gov).

Human Rights Now. 2011. "The World's Worst Places to Be a Woman." Women's Rights Group, June 17. Accessed October 25, 2011 (www.amnestyusa.org).

Hummer, Robert A., and Elaine M. Hernandez. 2013. "The Effect of Educational Attainment on Adult Mortality in the United States." *Population Bulletin* 68 (June): 1–16.

Hunter, Lori M. 2014. "Unmarried Baby Boomers Face Disadvantages as They Grow Older." Population Reference Bureau, February. Accessed May 16, 2014 (www.prb.org).

Hurwitz, Michael. 2011. "The Impact of Legacy Status on Undergraduate Admissions at Elite Colleges and Universities." *Economics of Education Review* 30 (June): 480–492.

Hutter, Mark. 2007. *Experiencing Cities.* Boston: Allyn & Bacon.

Ioannidis, John. 2005. "Contradicted and Initially Stronger Effects in Highly Cited Clinical Research." *JAMA* 294 (July 13): 218–228.

Ianzito, Christina. 2013. "Killer Heels." *Baltimore Sun,* August 1, 3.

Iceland, John, Gregory Sharp, and Jeffrey M. Timberlake. 2013. "Sun Belt Rising: Regional Population Change and the Decline in Black Residential Segregation, 1970–2009." *Demography* 50 (February): 97–123.

Ihrke, David. 2014. "Reason for Moving: 2012 to 2013." U.S. Census Bureau, June. Accessed September 19, 2014 (www.census.gov).

Independent Television Service. 2003. "Race the Power of an Illusion: What Is This Thing Called Race?" Accessed March 22, 2014 (www.itvs.org).

Independent Television Service. 2014. "Economic Empowerment." Public Broadcasting Service. Accessed February 10, 2014 (www.pbs.org /independentlens).

Institute for College Access & Success. 2014. "Quick Facts About Student Debt." March. Accessed June 21, 2014 (www.ticas.org).

Institute on Taxation and Economic Policy. 2013. "Undocumented Immigrants' State and Local Tax Contributions." July. Accessed March 22, 2014 (www.itep.org).

Insurance Institute for Highway Safety. 2010. "Q&A: Older Drivers." Accessed October 28, 2011 (www.iihs.org).

———. 2013. "Status Report." 45 (September 28). Accessed October 18, 2013 (www.iihs.org).

Intergovernmental Panel on Climate Change. 2014. "Climate Change 2014: Impacts, Adaptation and Vulnerability: Summary to Policymakers." Accessed September 19, 2014 (www.ipcc.ch).

International Federation of Health Plans. 2014. "2013 Comparative Price Report: Variation in Medical and Hospice Prices by Country." March 14. Accessed July 26, 2014 (www.ifhp. com).

International Institute for Democracy and Electoral Assistance. 2014. "VAP Turnout." Accessed April 20, 2014 (www.idea.int).

Internet Crime Complaint Center. 2013. "2012 Internet Crime Report." Accessed January 12, 2014 (www.ic3.gov).

Inter-Parliamentary Union. 2013. "Women in National Parliaments." Accessed March 2, 2014 (www.ipu.org).

Ipsos MORI. 2014. "Global Trends 2014: Environment." Accessed September 19, 2014 (www.ipsosglobaltrends.com).

Isaacs, Julia, Katherine Toran, Heather Hahn, Karina Fortuny, and C. Eugene Steurle. 2012. "Kids' Share 2012: Report on Federal Expenditures on Children Through 2011." Urban Institute, July. Accessed May 16, 2014 (www.urban.org).

Isaacs, Julia B. 2008. "International Comparisons of Economic Mobility." Pp. 37–46 in *Getting Ahead or Losing Ground: Economic Mobility in America,* edited by Julia B. Isaacs, Isabel V. Sawhill, and Ron Haskins. Washington, DC: Brookings Institution.

Jackson, C. Kirabo. 2012. "Do College-Prep Programs Improve Long-Term Outcomes?" National Bureau of Economic Research, February. Accessed April 14, 2012 (www.nber.org).

Jackson, D. D. 1998. "'This Hole in Our Heart': Urban Indian Identity and the Power of Silence." *American Indian Culture and Research Journal* 22 (4): 227–254.

Jackson, Henry C. 2010. "Ag Secretary Pushes School Nutrition Plan." *Washington Post,* February 8. Accessed February 15, 2010 (www.washingtonpost.com).

Jackson, Robert B., et al. 2013. "Increased Stray Gas Abundance in a Subset of Drinking Water Wells Near Marcellus Shale Gas Extraction." *Proceedings of the National Academy of Sciences* 110 (July 9): 11250–11255.

Jackson, Shelly L., and Thomas L. Hafemeister. 2011. "Financial Abuse of Elderly People vs. Other Forms of Elder Abuse: Assessing Their Dynamics, Risk Factors, and Society's Response." U.S. Department of Justice, February. Accessed March 5, 2012 (www .ncjrs.gov).

Jackson, Shelly L., and Thomas L. Hafemeister. 2013. "Understanding Elder Abuse." National Institute of Justice, June. Accessed May 16, 2014 (www.nij.gov).

Jacobe, Dennis. 2013. "U.S. Small Businesses Struggle to Find Qualified Employees." Gallup, February 15. Accessed April 12, 2014 (www.gallup.com).

Jacobs, Jerry A. 2005. "Multiple Methods in *ASR.*" *Footnotes* 33 (December): 1, 4.

Jacobs, Ken, Dave Graham-Squire, and Stephanie Luce. 2011. "Living Wage Policies and Big-Box Retail: How a Higher Wage Standard Would Impact Wal-Mart Workers and Shoppers." Center for Labor Research and Education, April. Accessed May 5, 2011 (www.laborcenter.berkeley.edu).

Jacobsen, Linda A., Mary Kent, Marlene Lee, and Mark Mather. 2011. "America's Aging Population." *Population Bulletin* 66 (February): 1–16.

Jacoby, Sanford M. 1997. *Modern Manors: Welfare Capitalism Since the New Deal.* NJ: Princeton University Press.

Jacoby, Susan. 2005. "Sex in America." *AARP* (July/ August): 57–62, 114.

Jagacinski, Carolyn M. 2013. "Women Engineering Students: Competence Perceptions and Achievement Goals in the Freshman Engineering Course." *Sex Roles* 69 (December): 644–657.

Jagger, Alison M., and Paula S. Rothenberg, eds. 1984. *Feminist Frameworks,* 2nd edition. New York: McGraw-Hill.

James, John T. 2013. "A New, Evidence-Based Estimate of Patient Harms Associated with Hospital Care." *Journal of Patient Safety* 9 (September): 122–128.

Jandt, Fred E. 2001. *Intercultural Communication: An Introduction.* Thousand Oaks, CA: Sage.

Janis, Irving L. 1972. *Victims of Groupthink: A Psychological Study of Foreign-Policy Decisions and Fiascoes.* Boston: Houghton Mifflin.

———.1982. *Groupthink: A Psychological Study of Policy Decisions and Fiascoes.* Boston: Houghton Mifflin Company.

Janofsky, Michael. 2003. "Young Brides Stir New Outcry on Utah Polygamy." *New York Times,* February 28, 1.

Jargon, Julie. 2013. "Not-So-Fast Food." *Wall Street Journal,* November 15, B2.

Jantti, Markus, et al. 2006. "American Exceptionalism in a New Light: A

Comparison of Intergenerational Earnings Mobility in the Nordic Countries, the United Kingdom and the United States." Institute for the Study of Labor, January. Accessed February 4, 2012 (www.ffp.iza.org).

Jeffreys, Sheila. 2011. *Man's Dominion: The Rise of Religion and the Eclipse of Women's Rights.* New York: Routledge.

Jernigan, David. 2010. "Alcohol Marketing and Youth: Why It's a Problem and What You Can Do." Center for Alcohol Marketing and Youth, December 14. Accessed April 12, 2012 (www.camy.org).

Jewkes, Rachel, Emma Fulu, Tim Roselli, and Claudia Garcia-Moreno. 2013. "Prevalence of and Factors Associated with Non-Partner Rape Perpetration: Findings from the UN Multi-Country Cross-Sectional Study on Men and Violence in Asia and the Pacific." *Lancet Global Health* 1 (October): e208–218.

Jilani, Zaid. 2011. "Profits at Largest 500 Corporations Grew by 81 Percent in 2010." Think Progress, May 5. Accessed May 7, 2011 (www.alternet.org).

___.2014. "As Walmart's Sales Increase, It Decides to Cut Healthcare for 30,000 Employees." AlterNet, October 8. Accessed December 5, 2014 (www.alternet.org).

Johnson, Dave. 2014. "8 Phony GOP Solutions for Poverty That Will Only Bring More Economic Pain." AlterNet, January 16. Accessed April 12, 2014 (www.alternet.org).

___. 2014. "8 Ways Being Poor Is Wildly Expensive in America." AlterNet, January 29. Accessed February 14, 2014 (www.alternet.org).

Johnson, Ileana. 2010. "Why Has Communism Failed?" *Orthodoxy Today*, October 30. Accessed April 12, 2014 (www.orthodoxytoday.org).

Johnson, Kevin, and H. Darr Beiser. 2013. "Aging Prisoners' Costs Put Systems Nationwide in a Bind." *USA Today*, July 11. Accessed January 15, 2014 (www.usatoday.com).

Johnson, Kirk. 2011. "Between Young and Old, a Political Collision." *New York Times*, June 4, A10.

Jones, Ann. 2013. *How the Wounded Return from America's Wars—The Untold Story.* Chicago, IL: Haymarket Books.

Jones, Rachel K., and Joerg Dreweke. 2011. "Countering Conventional Wisdom: New Evidence on Religion and Contraceptive Use." Guttmacher Institute, April. Accessed April 24, 2012 (www.guttmacher.org).

Johnson, Rachel, James Nunns, Jeffrey Rohaly, Eric Toder, and Roberton Williams. 2011. "Why Some Tax Units Pay No Income Tax." Tax Policy Center, July. Accessed September 22, 2011 (www.taxpolicycenter.org).

Johnston, Pamela. 2005. "Dressing the Part." *Chronicle of Higher Education*, August 10. Accessed August 11, 2005 (www.chronicle.com).

Jones, Jeffrey M. 2012. "Gender Gap in 2012 Vote Is Largest in Gallup's History." Gallup, November 9. Accessed April 20, 2014 (www.gallup.com).

___. 2013. "In U.S., Voter Registration Lags Among Hispanics and Asians." Gallup, November 9. Accessed April 20, 2014 (www.gallup.com).

___. 2013. "U.S. Blacks, Hispanics Have No Preferences on Group Labels." Gallup, July 26. Accessed March 22, 2014 (www.gallup.com).

___. 2014. "Americans Continue to Say a Third Political Party Is Needed." Gallup, September 24. Accessed November 23, 2014 (www.gallup.com).

___. 2014. "In U.S., 14% of Those Aged 24 to 34 Are Living with Parents." Gallup, February 13. Accessed May 22, 2014 (www.gallup.com).

___. 2014. "In U.S., Mobile Tech Aids Interpersonal Communication Most." Gallup, April 28. Accessed April 28, 2014 (www.gallup.com).

___. 2014. "Reports of Alcohol-Related Family Trouble Remain Up in U.S." Gallup, July 29. Accessed November 4, 2014 (www.gallup.com).

___. 2014. "View of Death Penalty as Morally OK Unchanged in U.S." Gallup, May 15. Accessed November 4, 2014 (www.gallup.com).

Jones, Jo, and William D. Mosher. 2013. "Fathers' Involvement with Their Children: United States, 2006–2010." *National Health Statistics Reports*, December 20. Accessed May 16, 2014 (www.cdc.gov/nchs).

Jones, Nicholas A., and Jungmiwha Bullock. 2012. "The Two or More Races Population: 2012." U.S. Census Bureau, September. Accessed March 22, 2014 (www.census.gov).

Jones, R. K., L. Frohwirth, and A. M. Moore. 2013. "More Than Poverty: Disruptive Events Among Women Having Abortions in the USA." *Journal of Family Planning and Reproductive Health Care* 39 (January): 36–43.

Jones, Rachel K., and Jenna Jerman. 2014. "Abortion Incidence and Service Availability in the United States, 2011." *Perspectives on Sexual and Reproductive Health* 46 (1): 3–14.

Jones, Rachel K., Lawrence B. Finer, and Susheela Singh. 2010. "Characteristics of U.S. Abortion Patients, 2008." Guttmacher Institute, May. Accessed March 2, 2014 (www.guttmacher.org).

Jones, Sophia. 2014. "Egyptian Mother on Genital Mutilation: 'I Wake Up at Night Screaming, Just Remembering.'" *Huffington Post*, January 23. Accessed March 2, 2014 (www.huffingtonpost.com).

Jones, Sophia. 2014. "Turkish Activists Say Their Country Is Sliding Backward on Women's Rights." *Huffington Post*, April 18. Accessed April 20, 2014 (www.huffingtonpost.com).

Jonsson, Jan O., David B. Grusky, Reinhard Pollak, and Matthew Di Carlo. 2009. "Recent Trends in Social Mobility in the United States: A New Approach to Modeling Trend in Big Class, Gradational, and Microclass Reproduction." Stanford Center for the Study of Poverty and Inequality, September. Accessed September 22, 2011 (www.inequality.com).

Jonsson, Patrik. 2007. "In US Justice, How Much Bias?" *Christian Science Monitor*, September 21, 1, 10.

Jordan, Miriam, and Mark Peters. 2013. "Tight Market for Farmlands." *Wall Street Journal*, February 20, A3.

Jordan, Reed. 2013. "Poverty's Toll on Mental Health." Metrotrends, November 25. Accessed February 6, 2014 (www.blogs.metrotrends.org).

___. 2014. "Millions of Black Students Attend Public Schools That Are Highly Segregated by Race and by Income." Metrotrends, October 20. Accessed November 7, 2014 (blog.metrotrends.org).

Jordan, William. 2013. "Americans Divided Over Blackface Halloween Make Up." YouGov, November 1. Accessed March 18, 2014 (www.today.yougov.com).

Joshi, Suchi P., Jochen Peter, and Patti M. Valkenburg. 2011. "Scripts of Sexual Desire and Danger in US and Dutch Teen Girl Magazines: A Cross-National Content Analysis." *Sex Roles* 64 (April): 463–474.

Julian, Tiffany. 2012. "Work-Life Earnings by Field of Degree and Occupation for People with a Bachelor's Degree: 2011." U.S. Census Bureau, October. Accessed June 21, 2014 (www.census.gov).

Juergensmeyer, Mark. 2003. "Thinking Globally About Religion." Pp. 3–13 in *Global Religions: An Introduction,* edited by Mark Juergensmeyer. New York: Oxford University Press.

Kafka, Stephanie, Jordan Jeromcheck, and Susan Sorenson. 2014. "States in West and Midwest Lead Nation in Teacher Respect." Gallup, April 10. Accessed June 21, 2014 (www.gallup.com).

Kagan, Jerome. 2011. "Want Better Students? Teach Their Parents." *Christian Science Monitor,* February 21, 34.

Kahan, Dan M., Hank Jenkins-Smith, and Donald Braman. 2011. "Cultural Cognition of Scientific Consensus." *Journal of Risk Research* 14 (2): 147–174.

Kahlenberg, Richard D. 2010. "10 Myths About Legacy Preferences in College Admissions." *Chronicle of Higher Education,* October 1, A23, A25.

Kaiser Family Foundation. 2014. "Global Health Facts." Accessed July 26, 2014 (www.kff.org).

Kalev, Alexandra, Frank Dobbin, and Erin Kelly. 2006. "Best Practices or Best Guesses? Assessing the Efficacy of Corporate Affirmative Action and Diversity Policies." *American Sociological Review* 71 (August): 589–617.

Kalish, Rachel, and Michael Kimmel. 2011. "Hooking Up." *Australian Feminist Studies* 26 (March): 137–151.

Kalogrides, Demetra, Susanna Loeb, and Tara Beteille. 2013. "Systematic Sorting: Teacher Characteristics and Class Assignments." *Sociology of Education* 86 (April): 103–123.

Kann, Laura, et al. 2014. "Youth Risk Behavior Surveillance—United States, 2013." *MMWR* 63 (June 13): 1–168.

Kantor, Jodi, and Laurie Goodstein. 2014. "From Mormon Women, a Flood of Requests and Questions on Their Role in the Church." *New York Times*, March 7, A2.

Kantrowitz, Mark. 2011. "The Distribution of Grants and Scholarships by Race." September 2. Accessed March 20, 2012 (www.finaid.org).

Kaplan, David A. 2010. "The Best Company to Work For." *Fortune,* February 8, 57–72.

Kapp, Diana. 2013. "Can New Building Toys for Girls Improve Math and Science Skills?" *Wall Street Journal,* April 17, D1, D3.

Kassotis, Christopher D., Donald E. Tillitt, J. Wade Davis, Annette M. Hormann, and Susan C. Nagel. 2014. "Estrogen and Androgen Receptor Activities of Hydraulic Fracturing Chemicals and Surface and Ground Water in a Drilling-Dense Region." *Endocrinology* 155 (March): 897–907.

Katz, Alan. 2013. "Generic, but Not Cheap." *Bloomberg Businessweek*, December 16–22, 23–24.

Katz, Ian, and Alex Tanzi. 2013. "A Girl Can Be Anything She Wants, as Long as She Wants to Be a Waitress." *Bloomberg Businessweek*, September 30–October 6, 19–20.

Katz, Jackson. 2006. *The Macho Paradox: Why Some Men Hurt Women and How All Men Can Help.* Naperville, IL: Sourcebooks.

Katzenbach, Jon. 2003. *Why Pride Matters More Than Money: The Power of the World's Greatest Motivational Force.* New York: Crown Business.

Kavanaugh, Megan L., Jenna Jerman, Kathleen Ethier, and Susan Moskosky. 2013.

"Meeting the Contraceptive Needs of Teens and Young Adults: Youth-Friendly and Long-Acting Reversible Contraceptive Services in U.S. Family Planning Facilities." *Journal of Adolescent Health* 52 (March): 284–292.

Kay, Aaron C., Martin V. Day, Mark P. Zanna, and A. David Nussbaum. 2013. "The Insidious (and Ironic) Effects of Positive Stereotypes." *Journal of Experimental Social Psychology* 49 (March): 287–291.

Kearney, Melissa S., and Phillip B. Levine. 2014a. "Media Influences on Social Outcomes: The Impact of MTV's *16 and Pregnant* on Teen Childbearing." Brookings Institution, January. Accessed May 16, 2014 (www.brookings.edu).

___. 2014b. "Teen Births Are Falling: What's Going On?" Brookings Institution, March. Accessed May 16, 2014 (www.brookings.edu).

Kearney, Melissa S., Benjamin H. Harris, Elisa Jácome, and Lucie Parker. 2014. "Ten Economic Facts About Crime and Incarceration in the United States." The Hamilton Project, May. Accessed November 4, 2014 (www.hamiltonproject.org).

Keels, Micere. 2014. "Choosing Single Motherhood." *Contexts* 13 (Spring): 70–72.

Keeter, Scott. 2009. "New Tricks for Old—and New Dogs: Challenges and Opportunities Facing Communications Research." Pew Research Center Publications, March 3. Accessed December 31, 2009 (www.pewresearch.org).

___. 2010. "Ask the Expert." Pew Research Center, December 29. Accessed December 30, 2010 (www.pewresearch.org).

Kelleher, James B. 2013. "With New Year Comes a Raft of New State Laws." *Baltimore Sun*, January 1, 15.

Kellermann, Arthur L. 2012. "Should We Ration End-of-Life Care?" The RAND Blog, October 30. Accessed July 26, 2014 (www.rand.org).

Kellermann, Arthur R., and David I. Auerbach. 2013. "Health Care Cost Growth Is Hurting Middle-Class Families." RAND Health Affairs Blog, January 8. Accessed July 26, 2014 (www.rand.org).

Kelley, Trista. 2013. "The Big Bucks in Keeping Kids Focused." *Bloomberg Businessweek*, October 14–20, 25–27.

Kelly, Dana, Holly Xie, Christine Winquist Nord, Frank Jenkins, Jessica Ying Chan, and David Kastberg. 2013. "Performance of U.S. 15-Year-Old Students in Mathematics, Science, and Reading Literacy in an International Context." National Center for Education Statistics, December. Accessed June 21, 2014 (www.nces.org).

Kemp, Alice Abel. 1994. *Women's Work: Degraded and Devalued*. Englewood Cliffs, NJ: Prentice Hall.

Kena, Grace, Susan Aud, Frank Johnson, Xiaolei Wang, Jijun Zhang, Amy Rathbun, Sidney Wilkinson-Flicker, and Paul Kristapovich. 2014. "The Condition of Education 2014." National Center for Education Statistics, May. Accessed June 21, 2014 (www.nces.ed.gov).

Kendall, Diana. 2002. *The Power of Good Deeds: Privileged Women and the Social Reproduction of the Upper Class*. Lanham, MD: Rowman & Littlefield.

Kennedy, Sean, and Luke Gelber. 2012. "2012 Congressional Pig Book Summary." Citizens Against Government Waste. Accessed April 20, 2014 (www.cagw.org).

Keranen, K. M., M. Weingarten, G. A. Abers, and B. A. Bekins, S. Ge. 2014. "Sharp Increase in Central Oklahoma Seismicity Since 2008 Induced by Massive Wastewater Injection." *Science* 345 (6195): 448–451.

Kettner, Peter M., Robert M. Moroney, and Lawrence L. Martin. 1999. *Designing and Managing Programs: An Effectiveness-Based Approach*, 2nd ed. Thousand Oaks, CA: Sage.

Khadaroo, Stacy Teicher. 2013. "Ending U.S. Child Sex Trafficking." *Christian Science Monitor Weekly*, November 25, 12.

Khan, Shamus R. 2012. *Privilege: The Making of an Adolescent Elite at St. Paul's School*. Princeton, NJ: Princeton University Press.

Khaw, Lyndal B. L., and Jennifer L. Hardesty. 2009. "Leaving an Abusive Partner: Exploring Boundary Ambiguity Using the Stages of Change Model." *Journal of Family Theory & Review* 1 (March): 38–53.

Kidd, Dustin. 2007. "Harry Potter and the Functions of Popular Culture." *The Journal of Popular Culture* 40 (February): 69–89.

Kilmann, R. H. 2011. *Quantum Organizations: A New Paradigm for Achieving Organizational Success and Personal Meaning*. Newport Coast, CA: Kilmann Diagnostics.

Kindig, David L., and Erika R. Cheng. 2013. "Even as Mortality Fell in Most US Counties, Female Mortality Nonetheless Rose in 42.8 Percent of Counties from 1992 to 2006." *Health Affairs* 32 (March): 451–458.

Kindy, Kimberly. 2013. "Tribes' Fight with Government Shakes Contractors." *Baltimore Sun*, December 29, 8.

Kington, Tom. 2013. "Pope Reaffirms Crackdown Against Group of U.S. Nuns." *Baltimore Sun*, March 16, 6.

Kinsey Institute. 2011. "Continuum of Human Sexuality." Accessed November 24, 2012 (www.indiana.edu/~kinsey).

Kinsey, Alfred C., Wardell B. Pomeroy, and Clyde E. Martin. 1948. *Sexual Behavior in the Human Male*. Philadelphia: Saunders.

Kirişci, Kemal. 2014. "Turkey, the Twitter Ban, and Upcoming Local Elections." Brookings Institution, March 26. Accessed April 12, 2014 (www.brookings.edu).

Kirkland, Allegra. 2014. "There Goes the Neighborhood: 6 Billionaires Who Moved Somewhere and Ruined It." AlterNet, September 26. Accessed November 10, 2014 (www.alternet.org).

Kirkpatrick, David D. 2010. *The Facebook Effect: The Inside Story of the Company That Is Connecting the World*. New York: Simon & Schuster.

Kirkova, Deni. 2013. "Trash-the Dress-Photography." *Mail Online*, August. Accessed January 12, 2014 (www.dailymail.co.uk).

Kiron, David, Nina Kruschwitz, Holger Rubel, Martin Reeves, and Sonja-Katrin Fuisz-Kehrbach. 2013. "Sustainability's Next Frontier: Walking the Talk on the Sustainability Issues That Matter Most." *MIT Sloan Management Review* 55 (December): 1–28.

Kivisto, Peter, and Dan Pittman. 2001. "Goffman's Dramaturgical Sociology: Personal Sales and Service in a Commodified World." Pp. 311–334 in *Illuminating Social Life: Classical and Contemporary Theory Revisited*, 2nd edition. Thousand Oaks, CA: Pine Forge Press.

Klandermans, Bert. 1984. "Mobilization and Participation: Social Psychological Explanations of Resource Mobilization Theory." *American Sociological Review* 49 (October): 583–600.

Klein, Allison, and Josh White. 2011. "Technology, Police Tactics Are Reining in Car Thefts." *Washington Post*, July 24, A1.

Klein, Karen E., and Nick Leiber. 2013. "$9 an Hour Doesn't Sound So Bad." *Bloomberg Businessweek*, February 25– March 3, 46–48.

Klein, Ezra. 2011. "Do We Still Need Unions? Yes: Why They're Worth Fighting For." *Newsweek*, March 7, 18.

Klos, Diana Mitsu. 2013. "The Status of Women in the U.S. Media 3013." Women's Media Center. Accessed March 2, 2014 (www.womensmediacenter.com).

Knafo, Saki. 2013. "America Has More Prisoners Than High School Teachers." *Huffington Post*, November 5. Accessed January 15, 2014 (www.huffingtonpost.com).

Kneebone, Elizabeth, and Alan Berube. 2013. *Confronting Suburban Poverty in America*. Washington, DC: Brookings Institution.

___. 2014. "Does the Suburbanization of Poverty Mean the War on Poverty Failed?" Brookings, January 8. Accessed February 6, 2014 (www.brookings.edu).

Koba, Mark. 2013. "Workers Unite! (So You Can Become Capitalists)." CNBC, December 13. Accessed December 19, 2013 (www.nbcnews.com).

Kochhar, Rakesh. 2007. "1995–2005: Foreign-Born Latinos Make Progress on Wages." Pew Hispanic Center, August 21. Accessed October 10, 2007 (www.pewhispanic.org).

Kochhar, Rakesh, Richard Fry, and Paul Taylor. 2011. "Twenty-to-One: Wealth Gaps Rise to Record Highs Between Whites, Blacks and Hispanics." Pew Research Center, July 26. Accessed September 22, 2011 (www.pewsocialtrends.org).

Kochhar, Rakesh, and Rich Morin. 2014. "Despite Recovery, Fewer Americans Identify as Middle Class." Pew Research Center, January 27. Accessed February 6, 2014 (www.pewresearch.org).

Kochhar, Rakesh, and Russ Oates. 2014. "Attitudes About Aging: A Global Perspective." Pew Research Center, January 30. Accessed September 19, 2014 (www.pewresearch.org).

Kocieniewski, David. 2011. "G.E.'s Strategies Let It Avoid Taxes Altogether." *New York Times*, March 24, A1.

Kohut, Andrew, and Richard Wike. 2013. "Mexicans and Salvadorans Have Positive Picture of Life in U.S." Pew Research Center, October 24. Accessed September 19, 2014 (www.pewresearch.org).

Kohut, Andrew, Carroll Doherty, Michael Dimock, Alan Cooperman, Scott Keeter, and Gregory Smith. 2012. "Two-Thirds of Democrats Now Support Gay Marriage." Pew Research Center, July 31. Accessed March 2, 2014 (www.pewresearch.org).

Kohut, Andrew, Carroll Doherty, Michael Dimock, and Scott Keeter. 2011. "Beyond Red vs. Blue Political Typology." Pew Research Center for the People & the Press, May 4. Accessed July 9, 2011 (www.people-press.org).

Kolet, Ilan, and Shobhana Chandra. 2012. "Health Care's Jobs Boom." *Bloomberg Businessweek*, February 6–12, 20–21.

Kopczuk, Wojciech, Emmanuel Saez, and Jae Song. 2009. "Earnings Inequality and Mobility in the United States: Evidence from Social Security Data Since 1937." February 3. Accessed September 25, 2011 (www.elsa.berkeley.edu).

Korenman, Sanders, and Dahlia Remler. 2013. "Rethinking Elderly Poverty: Time for a Health Inclusive Poverty Measure?" National Bureau of Economic Research, March. Accessed February 6, 2014 (www.nber.org).

Kornhauser, William. 1959. *The Politics of Mass Society*. New York: Free Press.

Koseff, Alexei, and John Fritze. 2013. "IRS Recognizes Same-Sex Marriages." *Baltimore Sun*, August 30, 8.

Kosmin, Barry A., and Ariela Keysar. 2009. *American Religious Identification Survey [ARIS 2008]*. Summary Report, March. Accessed April 25, 2010 (www .americanreligionsurvey-aris.org).

Kosova, Weston, and Pat Wingert. 2009. "Crazy Talk." *Newsweek*, June 8, 54–62.

Kozol, Jonathan. 2005. *The Shame of the Nation: The Restoration of Apartheid Schooling in America*. New York: Crown.

Krache, D. 2008. "How to Ground a 'Helicopter Parent.'" CNN, August 19. Accessed July 10, 2009 (www.cnn.com).

Kramer, Adam D. I., Jamie E. Guillory, and Jeffrey T. Hancock. 2014. "Experimental Evidence of Massive-Scale Emotional Contagion through Social Networks." *Proceedings of the National Academy of Sciences* 111 (24): 8788–8790.

Kraska, Peter B. 2004. *Theorizing Criminal Justice: Eight Essential Orientations*. Long Grove, IL: Waveland Press.

Krasnova, Hanna, Helena Wenninger, Thomas Widjaja, and Peter Buxmann. 2013. "Envy on Facebook: A Hidden Threat to Users' Life Satisfaction?" Unpublished paper presented at the 11th International Conference on Wirtschaftinformatik, February 27–March 1, Leipzig, Germany. Accessed December 5, 2013 (www.aisel.aisnet.org).

Kreager, Derek A., and Jeremy Staff. 2009. "The Sexual Double Standard and Adolescent Peer Acceptance." *Social Psychology Quarterly* 72 (June): 143.

Krebs, Christopher P., Christine H. Lindquist, Tara D. Warner, Bonnie S. Fisher, and Sandra L. Martin. 2007. "The Campus Sexual Assault (CSA) Study." U.S. Department of Justice, December. Accessed June 21, 2014 (www .doj.gov).

Kreeger, Karen Y. 2002a. "Sex-Based Differences Continue to Mount." *The Scientist* 16 (February 18). Accessed November 20, 2013 (www.the-scientist.com).

———. 2002b. "X and Y Chromosomes Concern More Than Reproduction." *The Scientist* 16 (February 4). Accessed November 20, 2013 (www.the-scientist.com).

Kreider, Rose M., and Renee Ellis. 2011. "Number, Timing, and Duration of Marriages and Divorces: 2009." U.S. Census Bureau, Current Population Reports, May. Accessed March 10, 2012 (www.census.gov).

Krishnamurthy, Prasad, and Aaron Edlin. 2014. "Affirmative Action and Stereotypes in Higher Education Admissions." National Bureau of Economic Research, October. Accessed December 5, 2014 (www.nber.org).

Kristof, Nicholas. 2014. "A Nation of Takers?" *New York Times*, March 26, A31.

Krogstad, Jens Manuel. 2014. "Asian American Voter Turnout Lags Behind Other Groups; Some Non-Voters Say They're 'Too Busy'." Pew Research Center, April 9. Accessed April 20, 2014 (www.pewresearch.org).

Krogstad, Jens Manuel, and Mark Hugo Lopez. 2014. "Hispanic Nativity Shift: U.S. Births Drive Population Growth as Immigration Stalls." Pew Research Center, April 29. Accessed November 25, 2014 (www .pewresearch.org).

Kroll, Luisa, and Kerry A. Dolan. 2013. "Forbes Billionaires." *Forbes*, March 25, 85–144, 164–174, 190–200.

Kromer, Braedyn, and David Howard. 2013. "Labor Force Participation and Work Status

of People 65 Years and Older." U.S. Census Bureau, January. Accessed May 16, 2014 (www.census.gov).

Kross, Ethan, et al. 2013. "Facebook Use Predicts Declines in Subjective Well-Being in Young Adults." *PLoS ONE* 8 (8): e69841. Accessed December 5, 2013 (www.plosone.org).

Krueger, Alan B., Judd Cramer, and David Cho. 2014. "Are the Long-Term Unemployed on the Margins of the Labor Market?" Paper presented at the Brookings Panel on Economic Activity, March 20–21. Accessed April 12, 2014 (www.brookings.edu).

Kuhn, Manford H. and Thomas S. McPartland. 1954. "An Empirical Investigation of Self-Attitudes." *American Sociological Review* 19(1): 68–76.

Kulczycki, Andrei, and Arun P. Lobo. 2002. "Patterns, Determinants, and Implications of Intermarriage Among Arab Americans." *Journal of Marriage and the Family* 64 (February): 202–210.

Küntzle, Julia, and Paul Blondé. 2013. "The Unhappy Fate of Ghanaian Witches." VICE United States, October 22. Accessed February 14, 2014 (www.vice.com).

Kurlantzick, Joshua. 2014. "The Rise of Elected Autocrats." *Bloomberg Businessweek*, January 27–February 2, 8–9.

Kusenbach, Margarethe. 2009. "Salvaging Decency: Mobile Home Residents' Strategies of Managing the Stigma of 'Trailer' Living." *Qualitative Sociology* 32 (December): 399–428.

Kutner, Lawrence, and Cheryl Olson. 2008. *Grand Theft Childhood*. New York: Simon & Schuster.

Kwok, Alvin C., et al. 2011. "The Intensity and Variation of Surgical Care at the End of Life: A Retrospective Cohort Study." *The Lancet*, October 6. Accessed April 15, 2012 (www .lancet.com).

Kyckelhahn, Tracey. 2013. "State Corrections Expenditures, FY 1982–2010." Bureau of Justice Statistics, December 11. Accessed January 15, 2014 (www.bjs.gov).

Kymlicka, Will. 2012. *Multiculturalism: Success, Failure, and the Future*. Washington, DC: Migration Policy Institute.

"Labor Day 2014: Sept. 1. 2014." U.S. Census Bureau News, July 8. Accessed September 12, 2014 (www.census.gov).

La France, Betty H., David D. Henningsen, Aubrey Oates, and Christina M. Shaw. 2009. "Social-Sexual Interactions? Meta-Analyses of Sex Differences in Perceptions of Flirtatiousness, Seductiveness, and Promiscuousness." *Communication Monographs* 76 (September): 263–285.

La Puma, John. 2014. "Don't Ask Your Doctor About 'Low T.'" *New York Times*, February 4, A21.

Lahart, Justin. 2013. "Worry Over Inequality Occupies Wall Street." *Wall Street Journal*, November 11, C6.

Lahiri, Tripti, and Preetika Rama. 2013. "In India, New Anger at Police Response to Rapes." *Wall Street Journal*, May 1, A13.

Lakoff, Robin. T. 1990. *Talking Power: The Politics of Language*. New York: Basic Books.

Lalasz, Robert. 2006. "Americans Flocking to Outer Suburbs in Record Numbers." Population Reference Bureau. Accessed January 24, 2008 (www.prg.org).

Lallemand, Nicole C. 2012. "Reducing Waste in Health Care." Health Policy Brief, December 13. Accessed July 26, 2014 (www .healthaffairs.org).

Lamb, Gregory M. 2009. "National Parks Face New Threats." *Christian Science Monitor*, September 27, 36–37.

Lamb, Sharon, and Lyn Mikel Brown. 2007. *Packaging Girlhood: Rescuing Our Daughters from Marketers' Schemes*. New York: St. Martin's Press.

Landivar, Liana C. 2013. "Disparities in STEM Employment by Sex, Race, and Hispanic Origin." U.S. Census Bureau, September. Accessed June 21, 2014 (www.census.gov).

Landsberger, Henry A. 1958. *Hawthorne Revisited*. Ithaca, NY: Cornell University Press.

Laneri, Raquel. 2011. "Slumdog Millions." *Forbes*, May 9, 100–101.

Lang, James M. 2013. "How Orwell and Twitter Revitalized My Course." *Chronicle of Higher Education*, November 1, A35–A37.

Lang, Kurt, and Gladys Engel Lang. 1961. *Collective Dynamics*. New York: Thomas Y. Crowell.

Langreth, Robert. 2013. "Do Robot Surgeons Do No Harm?" *Bloomberg Businessweek*, March 14, 17–18.

———. 2014. "Big Pharma's Favorite Prescription: Higher Prices." *Bloomberg Businessweek*, May 12–18, 22–24.

Lannutti, Pamela J., Melanie Laliker, and Jerold L. Hale. 2001. "Violations of Expectations and Social-Sexual Communication in Student/Professor Interactions." *Communication Education* 50 (January): 69–82.

Laraña, Enrique, Hank Johnston, and Joseph R. Gusfield, eds. 1994. *New Social Movements: From Ideology to Identity*. Philadelphia: Temple University Press.

Larence, Eileen. 2011. "Private Sector and Law Enforcement Collaborate to Deter and Investigate Theft." U.S. Government Accountability Office, GAO-11-675, June. Accessed January 12, 2014 (www.gao.gov).

LaRossa, Ralph, and Donald C. Reitzes. 1993. "Symbolic Interactionism and Family Studies." Pp. 135–163 in *Sourcebook of Family Theories and Methods: A Contextual Approach*, edited by Pauline G. Boss, William J. Doherty, Ralph LaRossa, Walter R. Schumm, and Suzanne K. Steinmetz. New York: Plenum Press.

Larson, Christina. 2014. "China Needs Girls. More Are on the Way." *Bloomberg Businessweek*, August 4–10, 13–14.

Lauerman, John. 2013. "Colleges Slow to Investigate Assaults." *Baltimore Sun*, June 20, 10.

Laumann, Edward O., Sara A. Leitsch, and Linda J. Waite. 2008. "Elder Mistreatment in the United States: Prevalence Estimates from a Nationally Representative Study." *Journal of Gerontology* 63B (4): S248–S254.

Lawless, Jennifer L., and Richard L. Fox. 2012. "Men Rule: The Continued Under-Representation of Women in U.S. Politics." Women & Politics Institute, January. Accessed April 21, 2014 (www.american.edu).

Lawless, Jennifer L., and Richard L. Fox. 2013. "Girls Just Wanna Not Run: The Gender Gap in Young Americans' Political Ambition." Women & Politics Institute, March. Accessed April 21, 2014 (www.american.edu).

Lawn, Richard. 2013. *Experiencing Jazz*, 2nd edition. New York: Routledge.

Lawrence, Lee. 2013. "Was God Expelled?" *Christian Science Monitor*, June 17, 26–32.

Lazear, Edward P., Kathryn L. Shaw, and Christopher T. Stanton. 2013. "The Value of Bosses." National Bureau of Economic Research, June. Accessed December 10, 2013 (www.nber.org).

Leaper, Campbell, and Melanie M. Ayres. 2007. "A Meta-Analytic Review of Gender Variations in Adults' Language Use: Talkativeness,

Affiliative Speech, and Assertive Speech." *Personality and Social Psychology Review* 11 (November): 328–363.

Le Bon, Gustave. 1896/1968. *The Crowd: A Study of the Popular Mind,* 2nd edition. Dunwoody, GA: Norman S. Berg.

Leaf, Clifton. 2013. "Do Clinical Trials Work?" *New York Times,* July 14, SR1.

Leber, Jessica. 2013. "How Wireless Carriers Are Monetizing Your Movements." *MIT Technology Review,* April 12. Accessed October 8, 2014 (www.technologyreview.com).

Ledger, Kate. 2009. "Sociology and the Gene." *Contexts* 8 (Summer): 16–20.

Lee, Donghoon. 2013. "Household Debt and Credit: Student Debt." Federal Reserve Bank of New York, February 28. Accessed June 21, 2014 (ny.frb.org).

Lee, Marlene. 2009. "Aging, Family Structure, and Health." Population Reference Bureau, October. Accessed April 10, 2010 (www.prb.org).

Legatum Prosperity Index. 2013. "The Prosperity Index 2013." Accessed February 6, 2014 (www.prosperity.com).

Leger, Donna L. 2013. "OxyContin a Gateway to Heroin for Upper-Income Addicts." *USA Today,* April 25. Accessed July 28, 2014 (www.usatoday.com).

Leinberger, Christopher B., and Patrick Lynch. 2014. "Foot Traffic Ahead: Ranking Walkable Urbanism in America's Largest Metros." Smart Growth America. Accessed September 19, 2014 (www.smartgrowthAmerica.org).

Lemert, Edwin M. 1951. *Social Pathology: A Systematic Approach to the Theory of Sociopathic Behavior.* New York: McGraw-Hill.

———. 1967. *Human Deviance, Social Problems and Social Control.* Englewood Cliffs, NJ: Prentice Hall.

Lencioni, Patrick. 2002. *The Five Dysfunctions of a Team: A Leadership Fable.* San Francisco: Jossey-Bass.

Lenhart, Amanda, and Maeve Duggan. 2014. "Couples, the Internet, and Social Media." Pew Research Center, February 11. Accessed February 12, 2014 (www.pewresearch.org).

Lengermann, Patricia Madoo, and Jill Niebrugge-Brantley. 1992. "Contemporary Feminist Theory." Pp. 308–357 in *Contemporary Sociological Theory,* 3rd edition, edited by George Ritzer. New York: McGraw-Hill.

Leonardsen, Dag. 2004. *Japan as a Low-Crime Nation.* New York: Palgrave Macmillan.

Leung, Maxwell. 2013. "Jeremy Lin's Model Minority Problem." *Contexts* 12 (Summer): 53–56.

Leurent, B., et al. 2013. "Spiritual and Religious Beliefs as Risk Factors for the Onset of Major Depression: An International Cohort Study." *Psychological Medicine* 43 (10): 2109–2120.

LeVay, Simon. 2011. *Gay, Straight, and the Reason Why: The Science of Sexual Orientation.* New York: Oxford University Press.

Levin, Diane E., and Jean Kilbourne. 2009. *So Sexy So Soon: The New Sexualized Childhood and What Parents Can Do to Protect Their Kids.* New York: Ballantine Books.

The Levin Institute. 2013. "Pop Culture." State University of New York. Accessed November 10, 2013 (www.globalization101.org).

Levine, Lawrence W. 1998. *Highbrow/Lowbrow: The Emergence of Cultural Hierarchy in America.* Cambridge, MA: Harvard University Press.

Levine, Linda. 2012. "An Analysis of the Distribution of Wealth Across Households, 1989–2010." Congressional Research Service, July 17. Accessed February 6, 2014 (www.crs.gov).

Levitz, Jennifer. 2013. "Harvard Punishes Dozens of Students for Cheating." *Wall Street Journal,* February 3, A2.

Levy, Jenna, and Dan Witters. 2014. "More in U.S. Have Self-Funded Health Coverage, Medicaid." Gallup, May 9. Accessed July 26, 2014 (www.gallup.com).

Lewandowsky, Stephan, Ullrich K. H. Ecker, Colleen M. Seifert, Norbert Schwarz, and John Cook. 2012. "Misinformation and Its Correction: Continued Influence and Successful Debiasing." *Psychological Science in the Public Interest* 13 (December): 106–131.

Lewin, Kurt, Ronald R. Lippit, and Ralph K. White. 1939. "Patterns of Aggressive Behavior in Experimentally Created 'Social Climates.'" *Journal of Social Psychology* 10 (2): 271–301.

Lewin, Tamar. 2013. "New Milestone Emerges: Baby's First iPhone App." *New York Times,* October 26, A17.

Lewis, David Levering. 1993. *W. E. B. Du Bois: Biography of a Race, 1868–1919.* New York: Henry Holt.

Lewis, James A., and Stewart Baker. 2013. "The Economic Impact of Cybercrime and Cyber Espionage." McAfee Center for Strategic and International Studies, July. Accessed January 12, 2014 (www.csis.org).

Lewis, Oscar. 1966. "The Culture of Poverty." *Scientific American* 115 (October): 19–25.

Lewis, Ricki. 2014. "Noninvasive Prenatal DNA Test Okay for Low-Risk Pregnancies." Medscape, February 27. Accessed October 8, 2014 (www.medscape.com).

Lichter, Daniel T., Zhenchao Qian, and Leanna M. Mellott. 2006. "Marriage or Dissolution? Union Transitions Among Poor Cohabiting Women." *Demography* 43 (May): 223–240.

"Life Size Barbie Shows Young Girls the Dangers of Unrealistic Body Expectations." 2011. Diets in Review, May 3. Accessed October 25, 2011 (www.dietsinreview.com).

Lilly, J. Robert, Francis T. Cullen, and Richard A. Ball. 1995. *Criminological Theory: Context and Consequences,* 2nd edition. Thousand Oaks, CA: Sage.

Lindau, Stacy T., L. P. Schumm, Edward O. Laumann, Wendy Levinson, Colm A. O'Muircheartaigh, and Linda J. Waite. 2007. "A Study of Sexuality and Health Among Older Adults in the United States." *New England Journal of Medicine* 357 (August 23): 762–774.

Lindner, Andrew M. 2012. "An Old Tool with New Promise." *Contexts* 11 (Winter): 70–72.

Lindo, Jason M., Jessamyn Schaller, and Benjamin Hansen. 2013. "Economic Conditions and Child Abuse." National Bureau of Economic Research, April. Accessed May 16, 2014 (www.nber.org).

Lindsey, Linda L. 2005. *Gender Roles: A Sociological Perspective,* 4th edition. Upper Saddle River, NJ: Prentice Hall.

Lino, Mark. 2014. "Expenditures on Children by Families, 2013." U.S. Department of Agriculture, Center for Nutrition Policy and Promotion. Accessed November 18, 2014 (www.cnpp.usda.gov).

Linton, Ralph. 1936. *The Study of Man.* New York: Appleton-Century-Crofts.

———. 1964. *The Study of Man: An Introduction.* New York: Appleton-Century-Crofts.

Lipka, Michael. 2014. "5 Facts About Abortion." Pew Research Center, January 22. Accessed March 2, 2014 (www.pewresearch.org).

Lipka, Michael, and Sandra Stencel. 2013. "Racial and Ethnic Groups View 'Radical Life Extension' Differently." Pew Research Center, August 6. Accessed May 16, 2014 (www.pewresearch.org).

Lisak, David, and Paul M. Miller. 2002. "Repeat Rape and Multiple Offending Among Undetected Rapists." *Violence and Victims* 17 (1): 73–84.

Liu, Sa, S. Katherine Hammond, and Ann Rojas-Cheatham. 2013. "Concentrations and Potential Health Risks of Metals in Lip Products." *Environmental Health Perspectives* 121 (June): 705–710.

Livingston, Gretchen. 2014. "Four-in-Ten Couples Are Saying 'I Do,' Again." Pew Research Center, November 14. Accessed November 20, 2014 (pewresearch.org).

Livingston, Gretchen, and Anna Brown. 2014. "Birth Rate for Unmarried Women Declining for First Time in Decades." Pew Research Center, August 11. Accessed August 14, 2014 (www.pewresearch.org).

Livingston, Gretchen, and D'Vera Cohn. 2013. "Long-Term Trend Accelerates Since Recession: Record Share of New Mothers Are College Educated." Pew Research Center, May 10. Brookings Institution, March. Accessed May 16, 2014 (pewresearch.org).

Livingston, Gretchen, and Kim Parker. 2011. "A Tale of Two Fathers." Pew Social & Demographic Trends, June 15. Accessed May 20, 2014 (pewsocialtrends.org).

Llana, Sara M., and Sibylla Brodzinsky. 2012. "Latin America's Silent Scourge." *Christian Science Monitor,* November 26, 18–20.

"Lobbying Database." 2014. Center for Responsive Politics. Accessed April 12, 2014 (www.opensecrets.org).

Locher, David A. 2002. *Collective Behavior.* Upper Saddle River, NJ: Prentice Hall.

Lodge, Amy C., and Debra Umberson. 2012. "All Shook Up: Sexuality of Mid- To Later Life Married Couples." *Journal of Marriage and Family* 74 (June): 428–443.

Lofland, John. 1993. "Collective Behavior: The Elementary Forms." Pp. 70–75 in *Collective Behavior and Social Movements,* edited by Russell L. Curtis, Jr. and Benigno E. Aguirre. Boston: Allyn & Bacon.

———. 1996. *Social Movement Organizations: Guide to Research on Insurgent Realities.* New York: Aldine De Gruyter.

Lofquist, Daphne, Terry Lugaila, Martin O'Connell, and Sarah Feliz. 2012. "Households and Families: 2010." Census Briefs, April. Accessed May 15, 2012 (www.census.gov).

Logan, John, and Harvey Molotch. 1987. *Urban Fortunes: The Political Economy of Place.* Berkeley: University of California Press.

Lohr, Steve. 2011. "Computers That See You and Keep Watch Over You." *New York Times,* January 1, A1.

Lopez, Mark H., and Ana Gonzalez-Barrera. 2013. "What Is the Future of Spanish in the United States?" Pew Research Center, September 5, 2013. Accessed November 10, 2013 (www.pewresearch.org).

Lorber, Judith. 2005. *Gender Inequality: Feminist Theories and Politics,* 3rd edition. Los Angeles: Roxbury.

Lorber, Judith, and Lisa Jean Moore. 2007. *Gendered Bodies: Feminist Perspectives.* Los Angeles: Roxbury.

Loveless, Tom. 2014. "The 2014 Brown Center Report on American Education: How Well Are American Students Learning?" Brown Center on Education Policy, March. Accessed June 21, 2014 (www.brookings.edu).

Lowenberger, Amanda, Joanna Mauer, Andrew deLaski, Marianne DiMascio, Jennifer Amann, and Steven Nadel. 2012. "The Efficiency Boom: Cashing in on the Savings from Appliance Standards." American Council for an Energy-Efficient Economy, and Appliance Standards Awareness Project, March. Accessed June 22, 2012 (www.aceee.org).

Lowenkamp, Christopher T., and Edward J. Latessa. 2005. "Developing Successful Reentry Programs: Lessons Learned from the 'What Works' Research." Corrections Today 67 (April): 72–77.

Lowery, Wesley. 2014. "Senate Republicans Unanimously Reject Equal Pay Bill." Washington Post, April 9. Accessed April 21, 2014 (www.washingtonpost.com).

Lowrey, Annie. 2013. "Top 10% Took Home Half of U.S. Income in 2012." New York Times, September 11, B4.

Lublin, Joann S. 2010. "CEO Pay in 2010 Jumped 11%." Wall Street Journal, May 6. Accessed September 27, 2011 (www.wsj.com).

___. 2013. "Tyco's Former CEO Set for 2014 Parole." Wall Street Journal, December 4, B2.

Luckey, John R. 2008. "CRS Report for Congress: The United States Flag: Federal Law Relating to Display and Associated Questions." Congressional Research Service, April 14. Accessed July 20, 2011 (www.senate.gov).

Lugaila, Terry. A. 1998. "Marital Status and Living Arrangements: March 1998 (Update)." U.S. Census Bureau, Current Population Reports. Accessed August 8, 2000 (www.census.gov).

Lugo, Luis, Alan Cooperman, and Cary Funk. 2013. "Abortion Viewed in Moral Terms: Fewer See Stem Cell Research and IVF as Moral Issues." Pew Research Center, August 15. Accessed June 21, 2014 (www.pewresearch.org).

___. 2013. "Living to 120 and Beyond: Americans' Views on Aging, Medical Advances, and Radical Life Extension." Pew Research Center, August 6. Accessed May 16, 2014 (www.pewresearch.org).

Lugo, Luis, Alan Cooperman, and Gregory A. Smith. 2013. "A Portrait of Jewish Americans." Pew Research Center, October 1. Accessed June 21, 2014 (www.pewresearch.org).

Lugo, Luis, Alan Cooperman, Cary Funk, and Gregory A. Smith. 2012. 'Nones' on the Rise: One-in-Five Adults Have No Religious Affiliation." Pew Forum on Religion & Public Life, October 9. Accessed June 21, 2014 (www.pewforum.org).

Luling, Todd Van. 2013. "Hey Politicians, It's the 21st Century. Time to Stop Being Sexist Idiots." Huffington Post, October 8. Accessed March 2, 2014 (www.huffingtonpost.com).

Lundberg, Shelly, and Robert A. Pollak. 2007. "The American Family and Family Economics." Journal of Economic Perspectives 21 (Spring): 3–26.

Lutz, William. 1989. Doublespeak. New York: Harper & Row.

Lynch, Eleanor W., and Marci J. Hanson, eds. 1999. Developing Cross-Cultural Competence: A Guide for Working with Children and Their Families, 2nd edition. Baltimore, MD: Paul H. Brookes.

Lynn, David B. 1969. Parental and Sex Role Identification: A Theoretical Formulation. Berkeley, CA: McCutchen.

Lynn, M., and M. Todoroff. 1995. "Women's Work and Family Lives." Pp. 244–271 in Feminist Issues: Race, Class, and Sexuality, edited by Nancy Mandell. Scarborough, Ontario: Prentice Hall Canada.

"M-Health: Health and Appiness." 2014. The Economist, February 1, 56–57.

Macartney, Suzanne, Alemayehu Bishaw, and Kayla Fontenot. 2013. "Poverty Rates for Selected Detailed Race and Hispanic Groups by State and Place: 2007–2011." U.S. Census Bureau, February. Accessed February 6, 2014 (www.census.gov).

Maccoby, Eleanor E., and John A. Martin. 1983. "Socialization in the Context of the Family: Parent-Child Interaction." Pp. 1–101 in Socialization, Personality, and Social Development: Vol. 4. Handbook of Child Psychology, edited by E. Mavis Hetherington. New York: Wiley.

MacDorman, Marian F., T. J. Mathews, Ashna D. Mohangoo, and Jennifer Zeitlin. 2014. "International Comparisons of Infant Mortality and Related Factors: United States and Europe, 2010." National Vital Statistics Reports 63 (5): 1–7.

Mach, Annie l., and Ada S. Cornell. 2013. "Federal Employees Health Benefits Program (FEHBP): Available Health Insurance Options." Congressional Research Service, November, 13. Accessed November 17, 2014 (www.crs.gov).

Macionis, John J., and Vincent N. Parrillo. 2007. Cities and Urban Life, 4th edition. Upper Saddle River, NJ: Prentice Hall.

MacKenzie, Donald, and Judy Wajcman, eds. 1999. The Social Shaping of Technology, 2nd edition. Philadelphia: Open University Press.

Madden, Mary, Amanda Lenhart, Sandra Cortesi, Urs Gasser, Maeve Duggan, and Aaron Smith. 2013. "Teens, Social Media, and Privacy." Pew Research Center, May 21. Accessed December 4, 2013 (www.pewinternet.org).

Madigan, Nick. 2009. "Seniors Increasingly Targeted by Con Artists; Police 'Struggling to Keep Up.'" Baltimore Sun, January 14, 3.

Madrick, Jeff. 2013. "The Anti-Economist: The Fall and Rise of Occupy Wall Street." Harper's Magazine, March, 9–11.

Magill, Bobby. 2014. "Clean Power Plan Exempts Major CO_2 Emitters." Climate Central, June 17. Accessed September 19, 2014 (www.climatecontrol.org).

Magnier, Mark. 2009. "In Northern India, Village Elders Order 'Honor Killings.'" Los Angeles Times, September 26. Accessed September 28, 2009 (www.latimes.com).

Maher, Kris. 2013. "Unions Target Home Workers." Wall Street Journal, June 20, A3.

Mahr, Krista. 2013. "India's Shame." Time, January 14, 12.

Maines, David R. 2001. The Faultline of Consciousness: A View of Interactionism in Sociology. New York: Aldine De Gruyter.

Major, Brenda, Mark Appelbaum, Linda Beckman, Mary Ann Dutton, Nancy Felipe Russo, and Carolyn West. 2008. "Report of the APA Task Force on Mental Health and Abortion." American Psychological Association, August 13. Accessed September 5, 2008 (www.apa.org).

Major, William. 2011. "Thoreau's Cellphone Experiment." Chronicle of Higher Education, January 21, A33, A35.

Makary, Marty. 2014. "The Cost of Chasing Cancer." Time, March 10, 24.

Makinen, Julie. 2013. "Law on Elder Visits in Effect." Baltimore Sun, August 6, 6.

___. 2014. "Report Shows 'Brutal' N. Korea." Baltimore Sun, February 18, 5.

Malik, Khalid. 2014. "Human Development Report 2014." United Nations Development Program. Accessed September 20, 2014 (hdr.undp.org).

Maltby, Emily. 2013. "Tattoo Timeline." Time, October 28, 55.

Malthus, Thomas Robert. 1798/1965. An Essay on Population. New York: Augustus Kelley.

___. 1872/1991. An Essay on the Principle of Population, 7th edition. London: Reeves & Turner.

Mangan, Katherine. 2013. "Comanche Nation College Tries to Rescue a Lost Tribal Language." Chronicle of Higher Education, June 14, A18–A19.

___. 2013. "Life After Steel." Chronicle of Higher Education, April 26, A33–A35.

Manning, Jennifer E. 2013. "Membership of the 113th Congress: A Profile." Congressional Research Service, August 2013. Accessed June 21, 2014 (www.crs.gov).

Manning, Wendy D., and Jessica A. Cohen. 2012. "Premarital Cohabitation and Marital Dissolution: An Examination of Recent Marriages." Journal of Marriage and Family 74 (April): 377–387.

Manning, Wendy D., and Pamela J. Smock. 2005. "Measuring and Modeling Cohabitation: New Perspectives from Qualitative Data." Journal of Marriage and Family 67 (November): 989–1002.

"Manufacturing Day: Oct. 4, 2013." 2013. U.S. Census Bureau News, October 1. Accessed April 12, 2014 (www.census.gov).

Manyika, James, Michael Chui, Jacques Bughin, Richard Dobbs, Peter Bisson, and Alex Marrs. 2013. "Disruptive Technologies: Advances That Will Transform Life, Business, and the Global Economy." McKinsey Global Institute, May. Accessed April 12, 2014 (www.mckinsey.com).

Marcotte, Amanda. 2014. "Is Religion Inherently Oppressive?" AlterNet, April 9. Accessed June 21, 2014 (www.alternet.org).

Marcus, Amy Dockser. 2013. "Genetic Testing Leaves More Patients Living in Limbo." Wall Street Journal, November 19, D1–D2.

Marquand, Robert. 2011. "Europe Rejects Multiculturalism." Christian Science Monitor, February 21, 14.

Martin, Joyce A., Brady E. Hamilton, and Michelle J. K. Osterman. 2014. "Births in the United States, 2013." NCHS Data Brief, December. Accessed December 10, 2014 (www.cdc.gov/nchs).

Martin, Joyce A., Brady E. Hamilton, Michelle J. K. Osterman, Sally C. Curtin, and T. J. Mathews. 2013. "Births: Final Data for 2012." National Vital Statistics Reports 62 (December 30): 1–87.

Martin, Molly A. 2008. "The Intergenerational Correlation in Weight: How Genetic Resemblance Reveals the Social Role of Families." American Journal of Sociology 114 (Suppl.): S67–S105.

Martin, Philip. 2013. "The Global Challenge of Managing Migration." Population Bulletin 68 (November): 1–16.

Martin, Steven, Nan Astone, and Elizabeth Peters. 2014. "Fewer Marriages, More Divergence: Marriage Projections for Millennials to Age 40." Urban Institute, April 29. Accessed July 25, 2014 (www.urban.org).

Martinez Gladys, Casey E. Copen, and Joyce C. Abma. 2011. "Teenagers in the United States: Sexual Activity, Contraceptive Use, and Childbearing, 2006–2010 National Survey of Family Growth." Vital and Health Statistics, 2011, No. 31. October. Accessed December 14, 2012 (www.cdc.gov/nchs).

Martínez, Jessica, and Michael Lipka. 2014. "Hispanic Millennials Are Less Religious Than Older U.S. Hispanics." Pew Research

Center, May 8. Accessed June 21, 2014 (www.pewresearch.org).

Martinez, Ruben O., ed. 2011. *Latinos in the Midwest*. East Lansing, MI: Michigan State University Press.

Maruschak, Laura M., and Thomas P. Bonczar. 2013. "Probation and Parole in the United States, 2012." Bureau of Justice Statistics, December. Accessed January 15, 2014 (www.bjs.gov).

Marx, Gary T., and Douglas McAdam. 1994. *Collective Behavior and Social Movements: Process and Structure*. Upper Saddle River, NJ: Prentice Hall.

Marx, Karl. 1844/1964. *Economic and Philosophic Manuscripts of 1844*. New York: International Publishers.

___. 1845/1972. "The German Ideology." Pp. 110–164 in *The Marx-Engels Reader*, edited by Robert C. Tucker. New York: W. W. Norton.

___. 1867/1967. *Capital*, edited by Friedrich Engels. New York: International Publishers.

___. 1934. *The Class Struggles in France*. New York: International Publishers.

___. 1964. *Karl Marx: Selected Writings in Sociology and Social Philosophy*, translated by T. B. Bottomore. New York: McGraw-Hill.

Masci, David, Elizabeth Sciupac, and Michael Lipka. 2014. "Gay Marriage Around the World." Pew Research Religion & Public Life Project, February 5. Accessed March 2, 2014 (www.pewforum.org).

Masis, Julie. 2010. "Leashed for Safekeeping." *Christian Science Monitor*, July 26, 7.

Massey, Douglas S. 2007. *Categorically Unequal: The American Stratification System*. New York: Russell Sage Foundation.

Masters, Jonathan. 2011. "Militant Extremists in the United States." Council on Foreign Relations, February 7. Accessed July 19, 2011 (www.cfr.org).

Matjasko, Jennifer L., Phyllis H. Niolon, and Linda A. Valle. 2013. "The Role of Economic Factors and Economic Support in Preventing and Escaping from Intimate Partner Violence." *Journal of Policy Analysis and Management* 32 (October): 122–141.

Mather, Mark, Keven Pollard, and Linda A. Jacobsen. 2011. *First Results from the 2010 Census*. Washington, DC: Population Reference Bureau.

Matthews, Hannah, and Stephanie Schmit. 2014. "Child Care Assistance Spending and Participation in 2012." CLASP, February. Accessed May 16, 2014 (www.clasp.org).

Maxwell, Steve. 2012. "Water Is *Still* Cheap: Demonstrating the True Value of Water." *American Water Works Association* 104 (May): 31–36.

Mayo, Elton. 1945. *The Problems of an Industrial Civilization*. Cambridge, MA: Harvard University Press.

Mayo, Michael. 2012. "Home for the Holidays: Convicted Killers Goodman, LeVin." *Sun Sentinel*, December 19. Accessed January 12, 2014 (www.articles.sun-sentinel.com).

McAdam, Doug, and Ronnelle Paulsen. 1994. "Specifying the Relationship Between Social Ties and Activism." *American Journal of Sociology* 99 (November): 640–667.

McCabe, Brian J. 2014. "When Property Values Rule." *Contexts* 13 (Winter): 38–43.

McCabe, Donald L., Kenneth D. Butterfield, and Linda K. Trevino. 2006. "Academic Dishonesty in Graduate Business Programs: Prevalence, Causes, and Proposed Action." *Academy of Management Learning Education* 5 (September): 294–305.

McCabe, Janice, Karin L. Brewster, and Kathryn H. Tillman. 2011. "Patterns and Correlates of Same-Sex Sexual Activity Among U.S. Teenagers and Young Adults." *Perspectives on Sexual and Reproductive Health* 43 (September): 15–21.

McCarthy Justin. 2014. "Same-Sex Marriage Support Reaches New High at 55%." Gallup, May 21. Accessed November 17, 2014 (www.gallup.com).

___. 2014. "No Improvement for Congress' Job Approval Rating." Gallup, April 10. Accessed April 20, 2014 (www.gallup.com).

___. 2014. "Seven in 10 Americans Back Euthanasia." Gallup, June 18. Accessed June 21, 2014 (www.gallup.com).

McCoy, J. Kelly, Gene H. Brody, and Zolinda Stoneman. 2002. "Temperament and the Quality of Best Friendships: Effect of Same-Sex Sibling Relationships." *Family Relations* 51 (July): 248–255.

McCoy, Kevin. 2013. "Ex-Enron CEO Skilling's Resentenced to 14 Years." *USA Today*, June 21. Accessed February 14, 2014 (www.usatoday.com).

McDermott, Rose, James H. Fowler, and Nicholas A. Christakis. 2013. "Breaking Up Is Hard to Do, Unless Everyone Else Is Doing It Too: Social Network Effects on Divorce in a Longitudinal Sample." *Social Forces* 92 (2): 491–519.

McDonald, Kim A. 1999. "Studies of Women's Health Produce a Wealth of Knowledge on the Biology of Gender Differences." *Chronicle of Higher Education*, June 25, A19, A22.

McDonough, Katie. 2014. "The GOP's Other War on Women: 5 Gender Battlegrounds Beyond Abortion and Contraception." AlterNet, March 10. Accessed April 21, 2014 (www.alternet.org).

McEwan, Melissa. 2012. "Unbelievable: Man Beats Wife, Judge Orders Him to Take Her Out to Red Lobster and the Bowling Alley." AlterNet, February 9. Accessed February 10, 2012 (www.alternet.org).

McFadden, Robert D., and Angela Macropoulos. 2008. "Wal-Mart Employee Trampled to Death." *New York Times*, November 29, A16.

McFalls, Joseph A., Jr. 2007. "Population: A Lively Introduction, 5th edition." *Population Bulletin* 62 (March): 1–31.

McGregor, Jena, and Steve Hamm. 2008. "Managing the Workforce." *BusinessWeek*, January 28, 34–43.

McHale, Susan M. 2001. "Free-Time Activities in Middle Childhood: Links with Adjustment in Early Adolescence." *Child Development* 76 (November/December): 1764–1778.

McHale, Susan M., Kimberly A. Updegraff, and Shawn D. Whiteman. 2012. "Sibling Relationships and Influences in Childhood and Adolescence." *Journal of Marriage and Family* 74 (October): 913–930.

McIntosh, Peggy. 1995. "White Privilege and Male Privilege: A Personal Account of Coming to See Correspondences Through Work in Women's Studies." Pp. 76–87 in *Race, Class, and Gender: An Anthology*, 2nd edition, edited by Margaret L. Andersen and Patricia Hill Collins. Belmont, CA: Wadsworth.

McKenzie, Brian. 2013. "Out-of-State and Long Commutes: 2011." U.S. Census Bureau, February. Accessed September 19, 2014 (www.census.gov).

McKernan, Signe-Mary, Caroline Ratcliffe, Eugene Steuerle, and Sisi Zhang. 2013. "Less Than Equal: Racial Disparities in Wealth Accumulation." Urban Institute, April. Accessed February 6, 2014 (www.urban.org).

McKernan, Signe-Mary, Caroline Ratcliffe, Margaret Simms, and Sisi Zhang. 2012. "Do Financial Support and Inheritance Contribute to the Racial Wealth Gap?" Urban Institute, September. Accessed March 22, 2014 (www.urban.org).

McKinnon, Mark. 2011. "Do We Still Need Unions? No: Let's End a Privileged Class." *Newsweek*, March 7, 19.

McLane, Adam. 2013. "Why You Should Delete Snapchat." August 22. Accessed December 5, 2013 (www.adammclane.com).

McNeil, David G. 2010. "U.S. Apologizes for Syphilis Program in Guatemala." *New York Times*, October 1, A1.

McNeill, David. 2012. "Japanese Fraud Case Highlights Weaknesses in Scientific Publishing." *Chronicle of Higher Education*, October 12, A16–A17.

McPhail, Clark. 1991. *The Myth of the Madding Crowd*. New York: deGruyter.

McPhail, Clark, and Ronald T. Wohlstein. 1983. "Individual and Collective Behavior within Gatherings, Demonstrations, and Riots." *Annual Review of Sociology* 9 (August): 579–600.

McRae, Susan. 1999. "Cohabitation or Marriage?" Pp. 172–190 in *The Sociology of the Family*, edited by Graham Allan. Malden, MA: Blackwell Publishers.

McWhirter, Cameron, and Gary Fields. 2012. "Crime Migrates to Suburbs." *Wall Street Journal*, December 31, A3.

Mead, George Herbert. 1934. *Mind, Self, and Society*. Chicago: University of Chicago Press.

___. 1964. *On Social Psychology*. Chicago: University of Chicago Press.

Medrano, Lourdes. 2013. "On Illegal Immigration, More Cities Are Rolling Out a Welcome Mat." *Christian Science Monitor*, November 28. Accessed March 22, 2014 (www.csmonitor.com).

Mehl, Matthias R., Simine Vazire, Nairan Ramirez-Esparza, Richard B. Slatcher, and James W. Pennebaker. 2007. "Are Women Really More Talkative Than Men?" *Science* 317 (July): 82.

Meier, Barry, and Eric Lipton. 2013. "F.D.A. Shift on Painkillers Was Years in the Making." *New York Times*, October 28, A1.

Melucci, Alberto. 1995. "The New Social Movements Revisited: Reflections on a Sociological Misunderstanding." Pp. 107–119 in *Social Movements and Social Classes: The Future of Collective Action*, edited by Louis Maheu. Thousand Oaks, CA: Sage.

Melton, Glennon. 2013. "5 Reasons Social Media Is Dangerous for Me." *Huffington Post*, October 1. Accessed December 4, 2013 (www.huffingtonpost.com).

Mencher, Steve. 2013. "Visit Your Parents or Be Sued: Is That Un-American?" AARP, July 8. Accessed November 2, 2013 (www.aarp.org).

Mendelberg, Tali, and Christopher F. Karpowitz. 2012. "More Women, but Not Nearly Enough." *New York Times*, November 9, A31.

Mendelsohn, Oliver, and Maria Vicziany. 1998. *The Untouchables, Subordination, Poverty and the State in Modern India*. New York: Cambridge University Press.

Mendes, Elizabeth. 2013. "Americans Favor Giving Illegal Immigrants a Chance to Stay." Gallup, April 12. Accessed March 22, 2014 (www.gallup.com).

Mendes, Elizabeth, and Joy Wilke. 2013. "Americans' Confidence in Congress Falls to Lowest on Record." Gallup, June 13. Accessed December 19, 2013 (www.gallup.com).

Meng, Liu. 2011. "Chinese College Drops Plan to Discourage Kissing and Other 'Uncivilized

Behavior' on Campus." *Global Times*, April 22. Accessed October 25, 2011 (www.china .globaltimes.cn).

Menissi, Fatima. 1991. *The Veil and the Male Elite: A Feminist Interpretation of Women's Rights in Islam*. Reading, MA: Addison-Wesley.

———. 1996. *Women's Rebellion and Islamic Memory*. Atlantic Highlands, NJ: Zed Books.

Merryman, Ashley. 2013. "Losing Is Good for You." *New York Times*, September 25, A29.

Merton, Robert K. 1938. "Social Structure and Anomie." *American Sociological Review* 3 (December): 672–682.

———. 1948/1996. "The Self-Fulfilling Prophecy." Pp. 183–201 in *Robert K. Merton: On Social Structure and Science*, edited by Piotr Sztompka. Chicago: University of Chicago Press.

———. 1949. "Discrimination and the American Creed." Pp. 99–126 in *Discrimination and National Welfare*, edited by Robert M. MacIver. New York: Harper.

———. 1968. *Social Theory and Social Structure*. New York: Free Press.

Merton, Robert K., and Alice K. Rossi. 1950. "Contributions to the Theory of Reference Group Behavior." Pp. 40–105 in *Continuities in Social Research*, edited by Robert K. Merton and Paul L. Lazarsfeld. New York: Free Press.

Messerli, Joe. 2012. "Should Affirmative Action Policies, Which Give Preferential Treatment Based on Minority Status, Be Eliminated?" BalancedPolitics.org, January 7. Accessed June 21, 2014 (www.balancedpolitics.org).

Meston, Cindy M., and David M. Buss. 2007. "Why Humans Have Sex." *Archives of Sexual Behavior* 36 (August): 477–507.

Meteyer, Karen B., and Maureen Perry-Jenkins. 2009. "Dyadic Parenting and Children's Externalizing Symptoms." *Family Relations* 58 (July): 289–302.

Mettler, Suzanne. 2014. "Equalizers No More: Politics Thwart Colleges' Role in Upward Mobility." *Chronicle Review*, March 7, B7–B11.

Michels, Robert. 1911/1949. *Political Parties*. Glencoe, IL: Free Press.

Mider, Zachary R., and Jeff Green. 2012. "Heads or Tails, Some CEOs Win the Pay Game." *Bloomberg Businessweek*, October 8–14, 23–24.

Mikkelson, Barbara, and David Mikkelson. 2005. "Super Bull Sunday." Accessed August 14, 2009 (www.snopes.com).

Milan, Lynn. 2012. "Characteristics of Doctoral Scientists and Engineers in the United States: 2008." National Science Foundation, December. Accessed May 15, 2013 (www .nsf.gov).

Milgram, Stanley. 1963. "Behavioral Study of Obedience." *Journal of Abnormal and Social Psychology* 67 (4): 371–378.

———. 1965. "Some Conditions of Obedience and Disobedience to Authority." *Human Relations* 18 (February): 57–76.

Miller, Anthony B., Claus Wall, Cornelia J. Barnes, Ping Sun, Teresa To, and Steven A. Narod. 2014. "Twenty-Five Year Follow-Up for Breast Cancer Incidence and Mortality of the Canadian National Breast Screening Study: Randomised Screening Trial." *BMJ* 348 (February): 1–5.

Miller, Claire C. 2013. "Angry Over U.S. Surveillance, Tech Giants Bolster Defenses." *New York Times*, November 1, A1.

Miller, Claire C., Somini Sengupta. 2013. "Selling Secrets of Phone Users to Advertisers." *New York Times*, October 6, A1.

Miller, D. W. 2001. "DARE Reinvents Itself—With Help From Its Social-Scientist Critics." *Chronicle of Higher Education*, October 16, A12–A14.

Miller, Lisa. 2010. "A Woman's Place Is in the Church." *Newsweek*, April 12, 34–41.

———. 2012. "Feminism's Final Frontier? Religion." *Washington Post*, March 8. Accessed March 9, 2012 (www.washingtonpost.com).

Miller, Sean J. 2012. "When Prison Doors Swing Open." *Christian Science Monitor Weekly*, May 21, 25–31.

"Millionaires' Club: For First Time, Most Lawmakers Are Worth $1 Million-Plus." 2014. Center for Responsive Politics, January 9. Accessed March 30, 2014 (www.opensecrets.gov).

Millman, Joel. 2012a. "Many Apples, Few Pickers." *Wall Street Journal*, October 10, A3.

———. 2012b. "Tribes Clash Over Gambling." *Wall Street Journal*, November 10, A3.

Mills, C. Wright. 1956. *The Power Elite*. New York: Oxford University Press.

———. 1959. *The Sociological Imagination*. New York: Oxford University Press.

Mills, Theodore M. 1958. "Some Hypotheses on Small Groups from Simmel." *American Journal of Sociology* 63 (May): 642–650.

Mincy, Ronald, ed. 2006. *Black Males Left Behind*. Washington, DC: Urban Institute Press.

Mischel, Walter. 1966. "A Social Learning View of Sex Differences." Pp. 57–81 in *The Development of Sex Differences*, edited by Eleanor E. Maccoby. Stanford, CA: Stanford University Press.

Mishel, Lawrence, and Alyssa Davis. 2014. "CEO Pay Continues to Rise as Typical Workers Are Paid Less." Economic Policy Institute, June 12. Accessed November 23, 2014 (www .epi.org).

Mishel, Lawrence, Josh Bivens, Elise Gould, and Heidi Shierholz. 2012. *The State of Working America*, 12th edition. Ithaca, NY: Cornell University Press.

Mokhiber, Russell. 2007. "Twenty Things You Should Know About Corporate Crime." *Corporate Crime Reporter* 21 (June 12). Accessed February 25, 2010 (www .corporatecrimereporter.com).

Molin, Anna. 2012. "In Sweden, Playtime Goes Gender-Neutral for Holidays." *Wall Street Journal*, November 29, D1–D2.

Money, John, and Anke A. Ehrhardt. 1972. *Man & Woman, Boy & Girl: The Differentiation and Dimorphism of Gender Identity from Conception to Maturity*. Baltimore, MD: Johns Hopkins University Press.

Montanaro, Domenico, Rachel Wellford, and Simone Pathe. 2014. "Money Is Pretty Good Predictor of Who Will Win Elections." Public Broadcasting Service, November 11. Accessed November 23, 2014 (www.pbs.org).

Montez, Jennifer Karas, and Anna Zajacova. 2013. "Explaining the Widening Education Gap in Mortality Among U.S. White Women." *Journal of Health and Social Behavior* 54 (June): 165–181.

Monto, Martin and Anna Carey. 2014. "A New Standard of Sexual Behavior? Are Claims Associated with the 'Hookup Culture' Supported by General Social Survey Data?" *Journal of Sex Research* 51(6): 605–615.

Moore, Kathleen. 2008. "Low-Income Homes Green—and Affordable." *Daily Gazette*, July 1. Accessed September 21, 2008 (www .dailygazette.com).

Morales, Lymari. 2011. "U.S. Adults Estimate That 25% of Americans Are Gay or Lesbian." Gallup, May 27. Accessed November 24, 2012 (www.gallup.com).

Morello, Carol. 2010. "An Unexpected Result for Some Census Takers: The Wrath of Irate Americans." *Washington Post*, June 20, A4.

———. 2014. "Census to Change the Way It Counts Gay Married Couples." *Washington Post*, May 26. Accessed October 23, 2014 (www .washingtonpost.com).

Morello, Carol, and Scott Clement. 2014. "'Happy Days' No More: Middle-Class Families Squeezed as Expenses Soar, Wages Stall." *Washington Post*, April 26. Accessed March 27, 2014 (www.washingtonpost.com).

Morgenson, Gretchen. 2004. "No Wonder C.E.O.'s Love Those Mergers." *New York Times*, July 18, C1.

Morin, Richard. 2009. "What Divides America?" Pew Research Center, September 24. Accessed March 2, 2010 (www.pewresearch.org).

———. 2012. "Rising Share of Americans See Conflict Between Rich and Poor." Pew Research Center, January 11. Accessed January 15, 2012 (www.pewsocialtrends .org).

Morin, Rich, and Seth Motel. 2012. "A Third of Americans Now Say They Are in the Lower Class." Pew Social & Demographic Trends, September 10. Accessed February 6, 2014 (www.pewsocialtrends.org).

Morris, David. 2014. "How the Private Sector Is Destroying Our Personal Space." AlterNet, January 8. Accessed January 19, 2014 (www .alternet.org).

Morris, Desmond. 1994. *Bodytalk: The Meaning of Human Gestures*. New York: Crown Trade Paperbacks.

Morris, Theresa. 2014. "C-Section Epidemic." *Contexts* 13 (Winter): 70–72.

Morrison, Denton E. 1971. "Some Notes Toward Theory on Relative Deprivation, Social Movements, and Social Change." *American Behavioral Scientist* 14 (May–June): 675–690.

Moss-Racusin, Corrine A. John F. Dovidio, Victoria L. Brescoli, Mark, J. Graham, and Jo Handelsman. 2012. "Science Faculty's Subtle Gender Biases Favor Male Students." *Proceedings of the National Academy of Sciences* 109 (October 9): 16474–16479.

Motel, Seth, and Eileen Patten. 2013. "Statistical Portrait of the Foreign-Born Population in the United States, 2011." Pew Research Hispanic Trends Project, January 29. Accessed March 22, 2014 (www.pewhispanic .org).

Motivans, Mark. 2004. "Intellectual Property Theft, 2002." Washington, DC: U.S. Department of Justice, Bureau of Justice Statistics.

———. 2013. "Federal Justice Statistics 2010 – Statistical Tables." Bureau of Justice Statistics, December. Accessed January 12, 2014 (www.bjs.gov).

Moyer, Imogene L. 2001. *Criminological Theories: Traditional and Nontraditional Voices and Themes*. Thousand Oaks, CA: Sage.

———. 2003. "Jane Addams: Pioneer in Criminology." *Women & Criminal Justice* 14 (3/4): 1–14.

Mui, Ylan Q. 2009. "Bottled Water Boom Appears Tapped Out." *Washington Post*, August 13, A10.

Mukherjee, Sy. 2013. "Congresswoman Opposes Pay Equity Laws because Women 'Don't Want the Decisions Made in Washington.'" *Nation of Change*, June 3. Accessed April 21, 2014 (www.nationofchange.org).

Mullen, Ann. 2012. "The Not-So-Pink Ivory Tower." *Contexts* 4 (Fall): 34–38.

Mumford, Lewis. 1961. *The City in History: Its Origins, Transformations, and Its Prospects*. New York: Harcourt, Brace.

Murdock, George P. 1940. "The Cross-Cultural Survey." *American Sociological Review*, 5: 361–370.

Murphey, David, Megan Barry, and Brigitte Vaughn. 2013. "Mental Health Disorders." Child Trends, January. Accessed July 26, 2014 (www.childtrends.org).

Murphy, Cait. 2005. "Fast-Forward to the Future." *Fortune*, September 19, 271.

Murphy, Caryle. 2009. "Behind the Veil." *Christian Science Monitor,* December 13, 12–17.

Murphy, John. 2004. "S. Africa's New Goal: Economic Equality." *Baltimore Sun,* April 27, 1A, 4A.

Murphy, Kevin. 2013. "Hallmark Responds to 'Gay' Ornament Controversy with Regret." *Huffington Post*, November 1. Accessed November 1, 2013 (www.huffingtonpost.com).

Murray, Sara. 2013. "Businesses Push for More Low-Skill Visas." *Wall Street Journal*, August 9, A3.

Mutzabaugh, Ben. 2011. "Reports: Thai Airline Recruits 'Third-Sex' Attendants." *USA Today*, January 29. Accessed October 25, 2011 (www.usatoday.com).

Myers, John P. 2007. *Dominant–Minority Relations in America: Convergence in the New World,* 2nd edition. Boston: Allyn & Bacon.

Naik, Gautam. 2013. "New Advance Toward 'Designer Babies.'" *Wall Street Journal*, October 4, A3.

Naili, Hajer. 2011. "Study Details Sex-Traffic in Post-Saddam Iraq." Women's e-News, November 9. Accessed February 25, 2012 (www.womensenews.org).

Nash, Elizabeth, Rachel Benson Gold, Andrea Rowan, Gwendolyn Rathbun, and Yana Vierboom. 2014. "Laws Affecting Reproductive Health and Rights: 2013 State Policy Review." Guttmacher Institute. Accessed March 2, 2014 (www.guttmacher.org).

National Academy of Sciences. 2014. *New Directions in Child Abuse and Neglect Research.* Washington, DC: The National Academies Press.

National Center for Education Statistics. 2013. "The Nation's Report Card: A First Look: 2013 Mathematics and Reading." U.S. Department of Education, November. Accessed June 21, 2014 (www.nationsreprtcard.gov).

National Center for Health Statistics. 2014. "Health, United States, 2013." May. Accessed July 26, 2014 (www.cdc.gov/nchs).

National Center on Elder Abuse. 2012. "Statistics/Data." Accessed May 16, 2014 (www.ncea.aoa.gov).

National Conference of State Legislatures. 2014. "State Laws Regarding Marriages Between First Cousins." Accessed May 16, 2014 (www.ncsl.org).

National Diabetes Statistics Report: Estimates of Diabetes and Its Burden in the United States. 2014. Centers for Disease Control and Prevention. Atlanta, GA: U.S. Department of Health and Human Services.

"National Grandparents Day 2014: Sept. 7." 2014. U.S. Census Bureau, July 8. Accessed October 20, 2014 (www.census.gov).

National Institute of Mental Health. 2011. "The Teen Brain: Still Under Construction." Accessed November 20, 2013 (www.nimh.nih.gov).

National Institute on Alcohol Abuse and Alcoholism. 2010. "Beyond Hangovers: Understanding Alcohol's Impact on Your Health." Accessed July 30, 2014 (www.niaaa.nih.gov).

___. 2013. "College Drinking." July. Accessed July 26, 2014 (www.niaaa.nih.gov).

___. 2014. "Alcohol Facts and Statistics." May. Accessed July 30, 2014 (www.niaaa.nih.gov).

National Institute on Drug Abuse. 2012. "Principles of Drug Addiction Treatment: A Research-Based Guide." Accessed January 15, 2014 (www.drugabuse.gov).

National Institutes of Health. 2006. "Stem Cell Information." Accessed August 6, 2008 (www.nih.gov).

National Oceanic and Atmospheric Administration. 2014. "Heat Wave: A Major Summer Killer." NOAA Watch. Accessed September 19, 2014 (www.noaawatch.gov).

National Organization on Fetal Alcohol Syndrome. 2012. "FASD: What Everyone Should Know." Accessed July 25, 2014 (www.nofas.org).

National Public Radio. 2010. "Black Male Privilege?" Interview transcript, March 4. Accessed March 10, 2010 (www.npr.org).

National Retail Foundation. 2013. "Retailers Estimate Holiday Return Fraud Will Cost Them $3.4 Billion, According to NRG Survey." December 6. Accessed January 15, 2014 (www.nrf.com).

National Science Board. 2012. *Science and Engineering Indicators 2012.* Arlington, VA: National Science Foundation.

___. 2014. *Science and Engineering Indicators 2014.* Arlington VA: National Science Foundation.

National Survey of Student Engagement. 2013. "A Fresh Look at Student Engagement–Annual Results 2013." Bloomington, IN: Indiana University Center for Postsecondary Research. Accessed June 21, 2014 (www.nsse.iub.edu).

National White Collar Crime Center. 2013. "Cyberstalking." May. Accessed December 4, 2013 (www.nw3c.org).

National Women's Law Center. 2014. "Women's Unemployment Rises Despite Job Gains, Growth Concentrated in Low-Wage Sectors, NWLC Analysis Shows." April 4. Accessed April 12, 2014 (www.nwlc.org).

Naumann, Rebecca B., and Ann M. Dellinger. 2013. "Mobile Device Use While Driving—United States and Seven European Countries, 2011." *MMWR* 62 (March 15): 177–182.

Naveen, P., and Amarjeet Singh. 2013. "110 Pilgrims Killed in Stampede on Bridge Leading to Ratangarh Temple in MP." *The Times of India*, October 14. Accessed October 2, 2014 (timesofindia.com).

Nechepurenko, Ivan. 2014. "Russia Reverses Birth Decline—But for How Long?" *The Moscow Times*, June 22. Accessed September 21 (www.themoscowtimes.com).

Neelakantan, Shailaja. 2006. "In India, Conservatives Want Women Under Wraps." *Chronicle of Higher Education,* May 26, A47–A48.

___. 2011. "In India, Caste Discrimination Still Plagues University Campuses." *Chronicle of Higher Education*, December 16, A12–A15.

Neider, Linda L., and Chester A. Schriesheim, eds. 2005. *Understanding Teams.* Greenwich, CT: Information Age.

Niewyk, Donald L., and Francis R. Nicosia. 2000. *The Columbia Guide to the Holocaust.* New York: Columbia University Press.

Nelson, Colleen M. 2013. "Poll: Most Women See Bias in the Workplace." *Wall Street Journal*, April 12, A4.

Nestle, Marion. 2012. "Genetically Modified Food Myths and Truths—A Critical Review of the Science." AlterNet, June 21. Accessed June 26, 2012 (www.alternet.org).

Neuhaus, Jessamyn. 2010. "Marge Simpson, Blue-Haired Housewife: Defining Domesticity on *The Simpsons.*" *The Journal of Popular Culture* 43 (August): 761–781.

"The New American Father." 2013. Pew Research Center, June 14. Accessed May 16, 2014 (www.pewresearch.org).

"A New Green Wave." 2014. *The Economist*, August 30, 61.

Newberry, Sydne. 2013. "What's in a Name? Calling Obesity a Disease Could Help Improve Chronic Disease Outcomes." The RAND Blog, October 16. Accessed July 26, 2014 (www.rand.org).

Newhouse, Joseph P., Alan M. Garber, Robin P. Graham, Margaret A. McCoy, Michelle Mancher, and Ashna Kibria, eds. 2013. *Variation in Health Care Spending: Target Decision Making, Not Geography.* Washington, DC: The National Academies Press.

Newkirk, Margaret, and Gigi Douban. 2012. "Legal Immigrants Wanted for Dirty Jobs." *Bloomberg Businessweek*, October 8–14, 34, 36.

Newman, John. 2014. "Coaches, Not Presidents, Top Public-College Pay List." *Chronicle of Higher Education*, May 16. Accessed June 14, 2014 (www.chronicle.com).

Newport, Frank. 2011. "Americans Prefer Boys to Girls, Just as They Did in 1941." Gallup, June 23. Accessed October 25, 2011 (www.gallup.com).

Newport, Frank. 2012. "Americans Like Having a Rich Class, as They Did 22 Years Ago." Gallup. May 11. Accessed February 14, 2014 (www.gallup.com).

___. 2012. "Seven in 10 Are Very or Moderately Religious." Gallup, December 4. Accessed June 21, 2014 (www.gallup.com).

___. 2013. "In U.S., 87% Approve of Black-White Marriage, vs. 4% in 1958." Gallup, July 25. Accessed March 22, 2014 (www.gallup.com).

___. 2013. "In U.S., Four in 10 Report Attending Church in Last Week." Gallup, December 24. Accessed June 21, 2014 (www.gallup.com).

___. 2014. "Americans Show Low Levels of Concern on Global Warming." Gallup, April 4. Accessed September 19, 2014 (www.gallup.com).

___. 2014. "In U.S., 42% Believe Creationist View of Human Origins." Gallup, June 4. Accessed June 21, 2014 (www.gallup.com).

___. 2014. "Mississippi Most Religious State, Vermont Least Religious." Gallup, February 3. Accessed June 21, 2014 (www.gallup.com).

Newport, Frank, and Brandon Busteed. 2013. "Americans Still See College Education as Very Important." Gallup, December 17. Accessed June 21, 2014 (www.gallup.com).

Newport, Frank, Dan Witters, and Sangeeta Agrawal. 2012. "In U.S., Very Religious Americans Have Higher Wellbeing Across All Faiths." Gallup, December 1. Accessed April 8, 2012 (www.gallup.com).

Newport, Frank, and Igor Himelfarb. 2013. "In U.S., Strong Link Between Church Attendance, Smoking." Gallup, August 5. Accessed June 21, 2014 (www.gallup.com).

Newport, Frank, and Joy Wilke. 2013. "Desire for Children Still Norm in U.S." Gallup, September 25. Accessed May 16, 2014 (www.gallup.com).

___. 2014. "Americans Rate Economy as Top Priority for Government." Gallup, January 16. Accessed April 12, 2014 (www.gallup.com).

Newport, Frank, Jeffrey M. Jones, and Lydia Saad. 2014. "State of the Union: The Public Weighs In On 10 Key Issues." Gallup, January 31. Accessed March 22, 2014 (www.gallup.com).

Nichols, James. 2014. "Cecil Chao, Hong Kong Billionaire, Doubles Reward for Any Man

Who Can Make Lesbian Daughter Straight." *Huffington Post*, January 24. Accessed March 2, 2014 (www.huffingtonpost.com).

Niederdeppe, Jeff, Sahara Byrne, Rosemary J. Avery, and Jonathan Cantor. 2013. "Direct-To-Consumer Television Advertising Exposure, Diagnosis with High Cholesterol, and Statin Use." *Journal of General Internal Medicine* 28 (July): 886–893.

Nielsen Company, The. 2008. "College Spring Break Study." February 27. Accessed April 10, 2008 (www.alcoholstats.com).

Niquette, Mark, and Richard Rubin. 2014. "States Target Corporate Cash Stashed Overseas." *Bloomberg Businessweek*, April 21–27, 27–28.

NOAA National Climatic Data Center. 2014. "State of the Climate: Global Analysis for June 2014." Accessed September 19, 2014 (www.noaa.gov).

"No Action on Greenhouse Gases." 2008. *Baltimore Sun*, July 12, 2A.

Norris, Tina, Paula L. Vines, and Elizabeth M. Hoeffel. 2012. "The American Indian and Alaska Native Population: 2010." U.S. Census Bureau. 2010 Census Briefs, January. Accessed March 20, 2012 (www.census.gov).

Norton, Michael I., and Samuel R. Sommers. 2011. "Whites See Racism as a Zero-Sum Game That They Are Now Losing." *Perspectives on Psychological Science* 6 (3): 215–218.

Nossiter, Adam. 2011. "Hinting at an End to a Curb on Polygamy, Interim Libyan Leader Stirs Anger." *New York Times*, October 30, A6.

"Number of Jobs Held, Labor Market Activity, and Earnings Growth Among the Youngest Baby Boomers: Results from a Longitudinal Survey." 2012. Bureau of Labor Statistics, News Release, July 25. Accessed November 18, 2013 (www.bls.gov).

Nyseth, Hollie, Sarah Shannon, Kia Heise, and Suzy Maves McElrath. 2011. "Embedded Sociologists." *Contexts* 10 (Spring): 44–50.

Oakes, Jeannie. 1985. *Keeping Track: How Schools Structure Inequality*. New Haven, CT: Yale University Press.

Obach, Brian K. 2004. *Labor and the Environmental Movement: The Quest for Common Ground*. Cambridge, MA: MIT Press.

Oberschall, Anthony. 1995. *Social Movements: Ideologies, Interests, and Identities*. New Brunswick, NJ: Transaction.

O'Brien, Jodi, and Peter Kollock. 2001. *The Production of Reality: Essays and Readings on Social Interaction*, 3rd edition. Thousand Oaks, CA: Pine Forge Press.

O'Connor, Anahad. 2014. "New Concern About Testosterone and Heart Risks." *New York Times*, January 30, A13.

O'Donnell, Victoria. 2007. *Television Criticism*. Thousand Oaks, CA: Sage.

OECD. 2011. "Building a High-Quality Teaching Profession." Accessed June 21, 2014 (www.oecd.org).

———. 2013. "Education at a Glance 2013." Accessed June 21, 2014 (www.oecd.org).

———. 2013. "Results from PISA 2012: United States." Accessed June 22, 2014 (www.oecd.org).

———. 2014a. "Income Inequality Update." June. Accessed November 10, 2014 (www.oecd.org).

———. 2014b. *OECD Factbook 2014: Economic, Environmental and Social Statistics*. OECD Publishing. Accessed July 26, 2014 (www.oecd.org).

———. 2014c. "Society at a Glance 2014: OECD Social Indicators." Accessed July 26, 2014 (www.oecd.org).

Office of the Deputy Chief of Staff for Intelligence. 2006. "Arab Cultural Awareness: 58 Factsheets." U.S. Army Training and Doctrine Command, Ft. Leavenworth, Kansas, January. Accessed February 15, 2006 (www.fas.org).

Office of Inspector General. 2013. "Prescribers with Questionable Patterns in Medicare Part D." June. Accessed July 26, 2014 (www.oig.hhs.gov).

Office of Justice Programs. 2011. "Adult Crime Solutions." Accessed August 25, 2011 (www.crimesolutions.gov).

Office of Juvenile Justice and Delinquency Prevention. 2010. "Best Practices to Address Community Gang Problems: OJJDP's Comprehensive Gang Model," 2nd edition, October. Accessed August 21, 2011 (www.ncjrs.gov).

Office of the United Nations High Commissioner for Human Rights. 2010. "The Right to Water. Fact Sheet No. 35." United Nations, August. Accessed November 10, 2014 (www.un.org).

Ogburn, William F. 1922. *Social Change with Respect to Culture and Original Nature*. New York: Dell.

Offutt, Susan. 2013. "Insights Gained from Efforts to Quantify the Effects of Counterfeit and Pirated Goods in the U.S. Economy." U.S. Government Accountability Office, GAO-13-762T, July 9. Accessed January 12, 2014 (www.gao.gov).

Ogunro, Nola. 2012. "10 Great Paying Jobs You Can Get Without a 4 Year College Degree." Alternet, September 21. Accessed June 23, 2014 (www.alternet.org).

Ogunwole, Stella U., Malcolm P. Drewery, Jr., and Merarys Rios-Vargas. 2012. "The Population with a Bachelor's Degree or Higher by Race and Hispanic Origin: 2006–2010." U.S. Census Bureau, May. Accessed March 22, 2014 (www.census.gov).

O'Keefe, Ed. 2014. "Women Are Wielding Notable Influence in Congress." *Washington Post*, January 16. Accessed April 21, 2014 (www.washingtonpost.com).

Ohlemacher, Stephen. 2011. "Social Security Makes $6.5B in Overpayments." *USA Today*, June 14. Accessed June 15, 2011 (www.usatoday.com).

Olinto, Pedro, Kathleen Beegle, Carlos Sobrado, and Hiroki Uematsu. 2013. "The State of the Poor: Where Are the Poor, Where Is Extreme Poverty Harder to Find, and What Is the Current Profile of the World's Poor?" World Bank, October. Accessed February 6, 2014 (www.worldbank.org).

Olson, Jonathan R. 2010. "Choosing Effective Youth-Focused Prevention Strategies: A Practical Guide for Applied Family Professionals." *Family Relations* 59 (April): 207–220.

Olson, Theodore B. 2010. "The Conservative Case for Gay Marriage." *Newsweek*, January 18, 48–53.

Oppel, Richard A., Jr. 2011. "Steady Decline in Major Crime Baffles Experts." *New York Times*, May 4, A17.

Orenstein, Peggy. 2011. "Should the World of Toys Be Gender-Free?" *New York Times*, December 30, A23.

Oreopoulos, Philip, and Uros Petronijevic. 2013. "Making College Worth It: A Review of Research on the Returns to Higher Education." Accessed June 21, 2014 (www.ticas.org).

Oropesa, R. Salvatore, and Nancy S. Landale. 2004. "The Future of Marriage and Hispanics." *Journal of Marriage and Family* 66 (November): 901–920.

Ortman, Jennifer M., Victoria A. Velkoff, and Howard Hogan. 2014. "An Aging Nation: The Older Population in the United States." U.S. Census Bureau, May. Accessed May 16, 2014 (www.census.gov).

Ovide, Shira, and Rachel Feintzeig. 2013. "Microsoft Abandons Dreaded 'Stack.'" *Wall Street Journal*, November 13, B1, B5.

Owen, Daniela J., Amy M. S. Slep, and Richard E. Heyman. 2012. "The Effect of Praise, Positive Nonverbal Response, Reprimand, and Negative Nonverbal Response on Child Compliance: A Systematic Review." *Clinical Child and Family Psychology Review* 15 (December): 364–385.

Oxfam. 2013. "The Cost of Inequality: How Wealth and Income Extremes Hurt Us All." January 18. Accessed February 6, 2014 (www.oxfam.org).

———. 2014. "Even It Up: Time to End Extreme Inequality." October. Accessed November 10, 2014 (www.oxfam.org).

Packard, Vance. 1959. *The Status Seekers*. New York: David McKay.

Padgett, Tim. 2010. "Robes for Women." *Time*, September 27, 53–55.

Palen, J. John. 2014. *The Urban World*, 10th edition. New York: Oxford University Press.

Paletta, Damian, and Caroline Porter. 2013. "Use of Food Stamps Swells Even as Economy Improves." *Wall Street Journal*, March 28, 1, A12.

Palmeri, Christopher, and Peter S. Green. 2010. "Former Executives Bomb at the Ballot Box." *Bloomberg Businessweek*, November 8–14, 35–36.

Park, Robert, and Ernest Burgess. 1921. *Introduction to the Science of Sociology*. Chicago: University of Chicago Press.

Parker, Kim. 2012. "The Boomerang Generation." Pew Research Center, March 15. Accessed May 22, 2014 (www.pewsocialtrends.org).

———. 2012. "The Boomerang Generation: Feeling OK About Living with Mom and Dad." Pew Social & Demographic Trends, March 15. Accessed November 20, 2013 (pewsocialtrends.org).

Parker, Kim, and Eileen Patten. 2013. "The Sandwich Generation: Rising Financial Burdens for Middle-Aged Americans." Pew Research Center, January 30. Accessed May 22, 2014 (www.pewsocialtrends.org).

Parker, Kim, Richard Fry, D'Vera Cohn, and Wendy Wang. 2011. "Is College Worth It? College Presidents, Public Assess Value, Quality and Mission of Higher Education." Pew Research Center, May 16. Accessed April 15, 2012 (www.pewsocialtrends.org).

Parker, Kim, and Wendy Wang. 2013. "Modern Parenthood." Pew Research Center, March 14. Accessed March 10, 2014 (www.pewresearch.org).

Parker-Pope, Tara. 2011. "Web of Popularity, Achieved by Bullying." *New York Times*, February 14, A1.

Parramore, Lynn Stuart. 2014. "11 Jobs Where an Honest Day's Work Earns You Poverty." AlterNet, January 21. Accessed February 6, 2014 (www.alternet.org).

Parsons, Talcott. 1951. *The Social System*. Glencoe, IL: Free Press.

———. 1954. *Essays in Sociological Theory*, revised edition. New York: Free Press.

———. 1959. "The School Class as a Social System: Some of Its Functions in American Society." *Harvard Educational Review* 29 (Fall): 297–313.

___. 1960. *Structure and Process in Modern Societies.* New York: Free Press.

Parsons, Talcott, and Robert F. Bales, eds. 1955. *Family, Socialization, and Interaction Process.* New York: Free Press.

Partnership for Public Service. 2011. "Scores by Effective Leadership." Accessed August 13, 2011 (www.bestplacestowork.org).

___. 2013. "Federal Leadership on the Decline." April. Accessed December 19, 2013 (www.ourpublicservice.org).

Passel, Jeffrey S., D'Vera Cohn, and Ana Gonzalez-Barrera. 2013. "Population Decline of Unauthorized Immigrants Stalls, May Have Reversed." Pew Research Center, September 23. Accessed March 22, 2014 (www.pewresearch.org).

Passel, Jeffrey S., Gretchen Livingston, and D'Vera Cohn. 2012. "Explaining Why Minority Births Now Outnumber White Births." Pew Research Center, May 17. Accessed May 18, 2012 (www.pewsocialtrends.org).

Passel, Jeffrey S., and D'Vera Cohn. 2008. "U.S. Population Projections: 2005–2050." Pew Research Center, February 11. Accessed November 15, 2011 (www.pewresearch.org).

Passel, Jeffrey S., D'Vera Cohn, and Mark Hugo Lopez. 2011. "Hispanics Account for More Than Half of Nation's Growth in Past Decade." Pew Research Center, March 24. Accessed December 7, 2011 (www.pewhispanic.org).

Patchin, Justin W. 2013. "Cyberbullying Research: 2013 Update." Cyberbullying Research Center, November 20. Accessed December 4, 2013 (www.cyberbullying.us).

Paul, Annie Murphy. 2010. *Origins: How the Nine Months Before Birth Shape the Rest of Our Lives.* New York: Free Press.

Paul, Noel. C. 2002. "The Birth of a Would-Be-Fad." *Christian Science Monitor,* September 23, 11, 14–16.

Paul, Richard, and Linda Elder. 2007. *The Miniature Guide to Critical Thinking: Concepts and Tools.* Dillon Beach, CA: Foundation for Critical Thinking.

Paulson, Amanda. 2014. "Changes of Address." *Christian Science Monitor Weekly,* May 26, 12–13.

Payne, K. K., and J. Copp. 2013. "Young Adults in the Parental Home and the Great Recession." National Center for Family & Marriage Research. Accessed November 20, 2013 (www.hcfmr.bgsu.edu).

Pazol, Karen, Andrea A. Creanga, Kim D. Burley, Brenda Hayes, and Denise J. Jamieson. 2013. "Abortion Surveillance—United States, 2010." *MMWR* 62 (November 29): 1–48.

Pear, Robert. 2013. "On Health Exchanges, Premiums May Be Low, but Other Costs Can Be High." *New York Times,* December 5, A18.

Pearlstine, Norman. 2013. "Fix This: Water." *Bloomberg Businessweek,* March 25–31, 46–51.

Pedersen, Paul. 1995. *The Five Stages of Culture Shock: Critical Incidents around the World.* Westport, CT: Greenwood Press.

Peeples, Lynne. 2014. "Supreme Court Decision Is a Boon for Polluters, Critics Say." *Huffington Post,* June 10. Accessed September 19, 2014 (www.huffingtonpost.com).

Pell, Nicholas. 2011. "Beyond Occupy Wall Street: 11 American Uprisings You've Never Heard of That Changed the World." AlterNet, October 21. Accessed October 22, 2011 (www.alternet.org).

Perlmutter, David D. 2001. "Students Are Blithely Ignorant; Professors Are Bitter." *Chronicle of Higher Education,* July 27, B20.

Perry, Gina. 2013. *Behind the Shock Machine: The Untold Story of the Notorious Milgram Psychology Experiments.* New York: New Press.

Persell, Caroline Hodges, and Peter W. Cookson, Jr. 1985. "Chartering and Bartering: Elite Education and Social Reproduction." *Social Problems* 33 (December): 114–129.

Peter, Tom A. 2012. "Mistreatment of Afghan Women Caused by Far More Than Taliban." *Christian Science Monitor,* January 31. Accessed March 4, 2014 (www.csmonitor.com).

Peterman, Amber, Tia Palermo, and Caryn Bredenkamp. 2011. "Estimates and Determinants of Sexual Violence Against Women in the Democratic Republic of Congo." *American Journal of Public Health* 101 (June): 1060–1067.

Petersen, Andrea. 2012. "Smarter Ways to Discipline Kids." *Wall Street Journal,* December 26, D1, D3.

___. 2013. "Gap Between Those Who Use Internet and Those Who Don't Is Widening." *Washington Post,* November 13. Accessed December 4, 2013 (www.washingtonpost.com).

Peterson, James L., Josefina J. Card, Marvin B. Eisen, and Bonnie Sherman-Williams. 1994. "Evaluating Teenage Pregnancy Prevention and Other Social Programs: Ten Stages of Program Assessment." *Family Planning Perspectives* 26 (May): 116–120, 131.

Peterson, Scott. 2008. "In Iran, Barbie Seen as Cultural Invader." *Christian Science Monitor,* September 15, 4.

Petrosino, Anthony, Carolyn Turpin-Petrosino, and John Buehler. 2003. "Scared Straight and Other Juvenile Awareness Programs for Preventing Juvenile Delinquency: A Systematic Review of the Randomized Experimental Evidence." *Annals of the American Academy of Political and Social Science* 589 (September): 41–62.

Pettypiece, Shannon. 2013. "Anything You Can Do, I Can Do Better." *Bloomberg Businessweek,* March 11–17, 27–29.

Pew Center on the States. 2010. "Prison Count 2010." April. Accessed August 24, 2011 (www.pewcenteronthestates.org).

___. 2011. "State of Recidivism: The Revolving Door of America's Prisons." April. Accessed August 24, 2011 (www.pewcenteronthestates.org).

Pew Forum on Religion & Public Life. 2008. "U.S. Religious Landscape Survey: Religious Beliefs and Practices: Diverse and Politically Relevant." June. Accessed June 26, 2008 (www.religions.pewforum.org).

___. 2009. "Faith in Flux: Changes in Religious Affiliation in the U.S." April. Accessed April 25, 2010 (www.pewforum.org).

Pew Hispanic Center. 2012. "When Labels Don't Fit: Hispanics and Their Views of Identity." Pew Research Center, April 4. Accessed March 22, 2014 (www.pewhispanic.org).

___. 2013. "The Path Not Taken." Pew Research Center, February 4. Accessed March 22, 2014 (pewhispanic.org).

___. 2014. "Global Religious Diversity: Half of the Most Religiously Diverse Countries Are in Asia-Pacific Region." April 4. Accessed June 21, 2014 (pewresearch.org/religion).

___. 2014. "Most See Inequality Growing, But Partisans Differ over Solutions." January. Accessed February 14, 2014 (www.pewresearch.org).

Pew Research Center for the People & the Press. 2010. "83%—Support Christmas Displays in Public." Accessed April 26, 2010 (www.pewresearch.org).

___. 2014. "Thirteen Years of the Public's Top Priorities." January 27. Accessed March 22, 2014 (www.people-press.org).

Pewewardy, Cornel. 1998. "Fluff and Feathers: Treatment of American Indians in the Literature and the Classroom." *Equity & Excellence in Education* 31 (April): 69–76.

Pfeffer, Fabian T., Sheldon Danziger, and Robert F. Schoeni. 2014. "Wealth Levels, Wealth Inequality, and the Great Recession." Russell Sage Foundation, June. Accessed November 10, 2014 (www.russellsage.org).

Pflaumer, Alicia. 2011. "Texting Bride Video Goes Viral on the Web." *Christian Science Monitor,* November 1. Accessed November 3, 2011 (www.csmonitor.com).

Pflanz, Mike. 2014. "Briefing: Africa's Stance on Gays." *Christian Science Monitor Weekly,* March 17, 13.

Phelps, Glenn, and Steve Crabtree. 2013a. "More Than One in Five Worldwide Living in Extreme Poverty." Gallup, December 23. Accessed February 6, 2014 (www.gallup.com).

___. 2013b. "Worldwide, Median Household Income About $10,000." Gallup, December 16. Accessed February 6, 2014 (www.gallup.com).

Phelps, Glenn, and Steve Crabtree. 2014. "Worldwide, Richest 3% Hold One-Fifth of Collective Income." Gallup, January 3. Accessed February 6, 2014 (www.gallup.com).

Phelps, Timothy M. 2014. "Holder: Treat Gay Marriages Equally in Court, Elsewhere." *Baltimore Sun,* February 9, 12.

___. 2014. "Justice Dept. Policy to Let Tribes Grow, Sell Cannabis." *Baltimore Sun,* December 11, 6.

Phillips, Matthew. 2012. "Manufacturing: A Rebound, Not a Renaissance." *Bloomberg Businessweek,* December 17–23, 9–10.

___. 2014. "Welders, America Needs You." *Bloomberg Businessweek,* March 24–April 6, 19–21.

Philpott, Tom. 2013. "Are We Becoming China's Factory Farm?" *Mother Jones,* November/December, 72.

Pianta, Robert C., Jay Belsky, Renate Houts, Fred Morrison, and The National Institute of Child Health and Human Development (NICHD) Early Child Care Research Network. 2007. "Teaching: Opportunities to Learn in America's Elementary Classrooms." *Science* 315 (March 30): 1795–1796.

Pickert, Kate. 2014. "Turning Off the Tap." *Time,* August 4, 18–19.

Pierce, Lamar, Daniel Snow, and Andrew McAfee. 2013. "Cleaning House: The Impact of Information Technology Monitoring on Employee Theft and Productivity." Social Science Research Network, August 24. Accessed October 24, 2013 (www.ssrn.com).

Pilkington, Ed. 2013. "Research Exposes Racial Discrimination in America's Death Penalty Capital." *The Guardian,* March 13. Accessed January 17, 2014 (www.theguardian.com).

Pinsky, Paul G. 2014. "Maryland Cedes Millions in Corporate Taxes." *Baltimore Sun,* February 12, 19.

Pipkin, Whitney. 2013. "Are Asian Americans Taking Over the Internet?" *Asian Fortune,* November 29. Accessed December 4, 2013 (www.asianfortunenews.com).

Pitz, Will. 2005. "Closing the Gap: Solutions to Race-Based Health Disparities." Applied Research Center & Northwest Federation of

Community Organizations, June. Accessed April 20, 2007 (www.arc.org).

Planty, M., W. Hussar, T. Snyder, S. Provasnik, G. Kena, R. Dinkes, A. KewalRamani, and J. Kemp. 2008. *The Condition of Education 2008*. Washington, DC: National Center for Education Statistics, U.S. Department of Education.

Plateris, Alexander A. 1973. *100 Years of Marriage and Divorce Statistics: 1867–1967*. Rockville, MD: National Center for Health Statistics.

Polgreen, Lydia. 2010. "One Bride for Two Brothers: A Custom Fades in India." *New York Times*, July 16, A4.

"Policing Philadelphia: Boots on the Street." 2013. *The Economist*, August 24, 33.

Polikoff, Morgan S., and Andrew C. Porter. 2014. "Instructional Alignment as a Measure of Teaching Quality." *Educational Evaluation and Policy Analysis* 36 (December): 399–416.

Pollack, Andrew. 2013. "A.M.A. Recognizes Obesity as a Disease." *New York Times*, June 19, B1.

Pollard, Kelvin, and Paola Scommegna. 2013. "The Health and Life Expectancy of Older Blacks and Hispanics in the United States." *Today's Research on Aging*, No. 28, June, 1–8.

Pollick, Michael. 2014. "Why Didn't Communism Work in Eastern Europe?" Conjecture Corporation, April 1. Accessed April 12, 2014 (www.wisegeek.org).

Polsby, Nelson W. 1959. "Three Problems in the Analysis of Community Power." *American Sociological Review* 24 (December): 796–803.

Polyak, Llana. 2014. "Sudden Wealth Can Leave You Broke." CNBC, October 1. Accessed November 10, 2014 (www.cnbc.com).

Polycarp, Clifford, Milap Patel, and Joonkyung Seong. 2014. "Designed for the Future? Assessing Principles of Sustainable Development and Governance in the World Bank's Project Plans." World Resources Institute. Accessed September 19, 2014 (www.wri.org).

Ponemon Institute. 2013. "2013 Cost of Cyber Crime Study: United States." October. Accessed January 12, 2014 (www.ponemon.org).

Pong, Suet-ling, Lingxin Hao, and Erica Gardner. 2005. "The Roles of Parenting Styles and Social Capital in the School Performance of Immigrant Asian and Hispanic Adolescents." *Social Science Quarterly* 86 (December): 928–950.

Popper, Nathaniel, and Somini Sengupta. 2013. "U.S. Says Ring Stole 160 Million Credit Card Numbers." *New York Times*, July 26, B7.

Porter, Eduardo. 2013. "America's Sinking Middle Class." *New York Times*, September 19, A1.

———. 2014. "A Global Boom, but Only for Some." *New York Times*, March 19, B1.

Porter, Michael E., and Scott Stern. 2014. "Social Progress Index 2014: Executive Summary." Social Progress Imperative. Accessed September 19, 2014 (www.socialprogressimperative.org).

Poushter, Jacob. 2014. "Russia's Moral Barometer: Homosexuality Unacceptable, but Drinking, Less So." Pew Research Center, February 6. Accessed March 2, 2014 (www.pewresearch.org).

———. 2014. "What's Morally Acceptable? It Depends on Where in the World You Live." Pew Research Center, April 15. Accessed May 16, 2014 (www.pewresearch.org).

"The Power of a Party." 2013. *The Economist*, August 13, 29–30.

Powers, Charles H. 2004. *Making Sense of Social Theory: A Practical Introduction*. Lanham, MD: Rowman & Littlefield.

Presbyterian Mission Agency. 2013. "The Top Most Frequently Asked Questions About the PC (USA)." Accessed July 2, 2014 (www.presbyterianmission.org).

"Prescription for Change." 2013. *The Economist*, June 29, 61–62.

Preston, Darrell. 2014. "Cooling Off with a Nice, Tall Glass of Toilet Water." *Bloomberg Businessweek*, April 28–May 4, 37.

Price, Barbara Raffel, and Natalie J. Sokoloff, eds. 2004. *The Criminal Justice System and Women: Offenders, Prisoners, Victims, & Workers*, 3rd edition. New York: McGraw-Hill.

Princiotta, Daniel, Laura Lippman, Renee Ryberg, Hannah Schmitz, David Murphey, and Mae Cooper. 2014. "Social Indicators Predicting Postsecondary Success." Child Trends, April 1. Accessed June 23, 2014 (www.childtrends.org).

Prins, Nomi. 2011. "Guess How Much More Wall Street Spends on Bonuses Than on Penalties for Torpedoing the Economy?" AlterNet, June 27. Accessed June 28, 2012 (www.alternet.org).

Prior, Markus. 2009. "The Immensely Inflated News Audience: Assessing Bias in Self-Reported News Exposure." *Public Opinion Quarterly* 73 (Spring): 130–143.

Project on Government Oversight. 2010. "Letter to NIH on Ghostwriting Academics." November 29. Accessed July 3, 2011 (www.pogo.org).

Protess, Ben. 2013. "Big Banks Get Break in Rules to Limit Risks." *New York Times*, May 16, A1.

Prüss-Üstün, Annette, Robert Bos, Fiona Gore, and Jamie Bartram. 2008. "Safer Water, Better Health: Costs, Benefits and Sustainability of Interventions to Protect and Promote Health." World Health Organization. Accessed September 19, 2014 (www.who.int).

Pryor, John H. 2011. "The Changing First-Year Student: Challenges for 2011." Higher Education Research Institute at UCLA, January 27. Accessed April 10, 2012 (www.heri.ucla.edu).

Puddington, Arch. 2014. "The Democratic Leadership Gap." Freedom House. Accessed April 12, 2014 (www.freedomhouse.org).

Purcell, Kristen. 2011. "Search and Email Still Top the List of Most Popular Online Activities." Pew Research Center, August 9. Accessed August 10, 2011 (www.pewinternet.org).

Putnam, Hannah, Julie Greenberg, and Kate Walsh. 2014. "Easy A's and What's Behind Them." National Council on Teacher Quality, November. Accessed December 5, 2014 (www.nctq.org).

Qian, Zhenchao. 2012. "During the Great Recession, More Young Adults Lived with Parents." Census Brief prepared for Project US. 2010, August. Accessed November 20, 2013 (www.s4.brown.edu/us2010).

Quinney, Richard. 1980. *Class, State, and Crime*. Boston: Little, Brown.

Rackin, Heather, and Christina M. Gibson-Davis. 2012. "The Role of Pre- and Postconception Relationships for First-Time Parents." *Journal of Marriage and Family* 74 (June): 389–398.

Rainie, Lee, Aaron Smith, and Maeve Duggan. 2013. "Coming and Going on Facebook." Pew Research Center, February 5. Accessed December 4, 2013 (www.pewinternet.org).

Rainie, Lee, Kristen Purcell, and Aaron Smith. 2011. "The Social Side of the Internet." Pew Internet & American Life Project, January 1. Accessed August 10, 2011 (www.pewinternet.org).

Rampell, Catherine. 2011. "Companies Spend on Equipment, Not Workers." *New York Times*, June 10, A1.

———. 2014. "The Safety Net Catches the Middle Class More Than the Poor." *Washington Post*, April 7. Accessed March 8, 2014 (www.washingtonpost.com).

Rank, Mark R. 2013. "Poverty in America Is Mainstream." *New York Times*, November 3, SR12.

Rano, Jason, and Jane Houlihan. 2012. "Myths on Cosmetic Safety." Skin Deep Cosmetics Database. Accessed May 10, 2012 (www.ewg.org).

Rape Crisis Center. 2014. "Get the Facts." Accessed February 14, 2014 (www.rccmsc.org).

Raphael, Steven, and Michael A. Stoll. 2010. "Job Sprawl and the Suburbanization of Poverty." Brookings Institute, Metropolitan Policy Program, March. Accessed May 1, 2010 (www.brookings.edu).

Rattray, Sharon. 2010. "2010 Toy Sales." June 23. Accessed July 20, 2011 (www.suite101.com).

Ray, Chris. 2013. "Why Conglomerates Have Gone Out of Fashion." Hot Shot Trader, November 13. Accessed April 12, 2014 (www.hotshottrader.com).

Ray, Rebecca, Milla Sanes, and John Schmitt. 2013. "No-Vacation Nation Revisited." Center for Economic and Policy Research, May. Accessed April 12, 2014 (cepr.net).

Rayasam, Renuka. "Immigrants: The Unsung Heroes of the U.S. Economy." *U.S. News & World Report*, February 26, 58.

Reaney, Patricia. 2012. "Average Cost of U.S. Weddings Hits $27,021." Reuters, May 23. Accessed June 2, 2012 (www.reuters.com).

Reardon, Sean F. 2011. "The Widening Academic Achievement Gap Between the Rich and the Poor: New Evidence and Possible Explanations." Pp. 91–116 in *Whither Opportunity? Rising Inequality and the Uncertain Life Chances of Low-Income Children*, edited by R. Murnane and G. Duncan. New York: Russell Sage Foundation Press.

Redberg, Rita. 2012. "Less Is More." Institute of Medicine, September. Accessed July 26, 2014 (www.iom.edu).

Redd, Zakia, Tahilin Sanchez Karver, David Murphey, Kristin Anderson Moore, and Dylan Knewstub. 2011. "Two Generations in Poverty: Status and Trends among Parents and Children in the United States, 2000–2010." Child Trends Research Brief, November. Accessed January 12, 2012 (www.childtrends.org).

Reddy, Sumathi. 2013. "'I Don't Smoke, Doc,' and Other Patient Lies." *Wall Street Journal*, February 19, D3.

Reece, Michael, Debby Herbenick, Vanessa Schick, Stephanie A. Sanders, Brian Dodge, and J. Dennis Fortenberry. 2010. "Sexual Behaviors, Relationships, and Perceived Health among Adult Men in the United States: Results from a National Probability Sample." *Journal of Sexual Medicine* 7, Supplement 5 (October): 291–304.

Reed, Matthew, and Debbie Cochrane. 2013. "Student Debt and the Class of 2012." Institute for College Access & Success, December. Accessed June 21, 2014 (www.ticas.org).

Reeves, Jay. 2013. "Alabama Woman's Body Removed from Front Yard Grave." *U.S. News & World Report*, November 15. Accessed August 10, 2014 (www.us.news.com).

Reeves, Richard V., and Kerry Searle Grannis. 2013. "Five Strong Starts for Social Mobility." Brookings Institution, January 9. Accessed February 14, 2014 (www.brookings.edu).

"Reforming the One-Child Policy." 2013. *The Economist*, March 16, 45–46.

Reger, Jo. 2012. *Everywere and Nowhere: Contemporary Feminism in the United States*. New York: Oxford University Press.

Regnerus, Mark, and Jeremy Uecker. 2011. *Premarital Sex in America: How Young Americans Meet, Mate, and Think About Marrying*. New York: Oxford University Press.

Reich, Robert. 2014. "The Four Biggest Myths About Income Inequality." *Christian Science Monitor*, May 5. Accessed June 21, 2014 (www.csmonitor.com).

___. 2014. "Why Widening Inequality Is Hobbling Equal Opportunity." AlterNet, February 6. Accessed February 10, 2014 (www.alternet .org).

Reichert, Tom, Courtney Carpenter Childers, and Leonard N. Reid. 2012. "How Sex in Advertising Varies by Product Category: An Analysis of Three Decades of Visual Sexual Imagery in Magazine Advertising." *Journal of Current Issues & Research in Advertising* 33 (January): 1–19.

Reilly, Katie. 2013. "Sesame Street Reaches Out to 2.7 Million American Children with an Incarcerated Parent." Pew Research Center, June 21. Accessed November 18, 2013 (www .pewresearch.org).

Reiman, Jeffrey, and Paul Leighton. 2010. *The Rich Get Richer and the Poor Get Prison: Ideology, Class, and Criminal Justice*, 9th edition. Upper Saddle River, NJ: Prentice Hall.

Rein, Lisa. 2011. "Federal Workers Tell Us What Should Be Cut from the Budget." *Washington Post*, April 14, B4.

Reinhold, Steffen 2010. "Reassessing the Link Between Premarital Cohabitation and Marital Instability." *Demography* 47 (August): 719–733.

"Religions of the World: Number of Adherents, Names of Houses of Worship. . . ." 2007. Religious Tolerance. Accessed August 24, 2007 (www.religioustolerance.org).

Restaurant Opportunities Centers United, et al. 2014. "The Glass Floor: Sexual Harassment in the Restaurant Industry." New York, NY: Restaurant Opportunities Centers United.

Rettew, David. 2014. "Do You Have a Disorder or Just a Trait?" *Psychology Today*, January/February, 42–43.

Reynolds, Kelly A. 2012. "The Price of Drinking Water." *Water Conditioning & Purification* 54 (July): 50–52.

Rheault, Magali, and Kyley McGeeney. 2011. "Education Is a Key Predictor of Emotional Health After 65." Gallup, August 19. Accessed April 3, 2012 (www.gallup.com).

Ricciardelli, Rosemary, Kimberley A. Clow, and Philip White. 2010. "Investigating Hegemonic Masculinity: Portrayals of Masculinity in Men's Lifestyle Magazines." *Sex Roles* 63 (March): 64–78.

Rich, Motoko. 2013. "Subtract Teachers, Add Pupils: Math of Today's Jammed Schools." *New York Times*, December 22, A1.

Richburg, Keith B. 2009. "States Seek Less Costly Substitutes for Prison." *Washington Post*, July 13, A1.

Riche, Martha Farnsworth. 2000. "America's Diversity and Growth: Signposts for the 21st Century." *Population Bulletin* 55 (June): 1–41.

Richie, Christina. 2013. "The Scandal of the (Female) Evangelical Mind." *Chronicle of Higher Education*, June 14, A37–A38.

Richtel, Matt, and Alexei Barrionuevo. 2005. "Wendy's Restaurants." *New York Times*, April 22, A9.

Rideout, Victoria. 2013. "Zero to Eight: Children's Media Use in America 2013." Common Sense Media, Fall. Accessed November 18, 2013 (www.commonsensemedia.org).

Rideout, Victoria, Ulla G. Foehr, and Donald F. Roberts. 2010. *Generation M2: Media in the Lives of 8- to 18-Year-Olds*. Kaiser Family Foundation, January. Accessed January 30, 2010 (www.kff.org).

Riesman, David. 1953. *The Lonely Crowd*. New York: Doubleday.

Riffkin, Rebecca. 2014. "Americans Still Prefer a Male Boss to a Female Boss." Gallup, October 14. Accessed November 17, 2014 (www.gallup.com).

___. 2014. "Jobs, Government, and Economy Remain Top U.S. Problems." Gallup, May 19. Accessed June 21, 2014 (www.gallup.com).

___. 2014. "New Record Highs in Moral Acceptability." Gallup, May 30. Accessed August 10, 2014 (www.gallup.com).

___. 2014. "Public Faith in Congress Falls Again, Hits Historic Low." Gallup, June 19. Accessed June 21, 2014 (www.gallup.com).

Rios, Victor. 2012. "Reframing the Achievement Gap." *Contexts* 4 (Fall): 8–10.

Riosmena, Fernando, Rebecca Wong, and Alberto Palloni. 2013. "Migration Selection, Protection, and Acculturation in Health: A Binational Perspective on Older Adults." *Demography* 50 (June): 1039–1064.

"Rise of the Robots." 2014. *The Economist*, March 29, 13.

Ritzer, George. 1992. *Contemporary Sociological Theory*, 3rd edition. New York: McGraw-Hill.

___. 1996. *The McDonaldization of Society: An Investigation into the Changing Character of Contemporary Social Life*. Thousand Oaks, CA: Pine Forge Press.

___. 2008. *The McDonaldization of Society*, 5th edition. Los Angeles: Pine Forge Press.

Robers, Simone, Jana Kemp, and Jennifer Truman. 2013. "Indicators of School Crime and Safety: 2012." National Center for Education Statistics, June. Accessed December 4, 2013 (www.nces.ed.gov).

Robert Wood Johnson Foundation. 2010. "California Nurse Ratio Law Saves Lives, Improves Nurse Morale, Study Finds." May 16. Accessed June 6, 2012 (www.rwjf.org).

Roberts, Andrea L., et al. 2013. "Perinatal Air Pollutant Exposures and Autism Spectrum Disorder in the Children of Nurses' Health Study II Participants." *Environmental Health Perspectives* 121 (August): 978–984.

Roberts, James A., Luc Honore Petnji Yaya, and Chris Manolis. 2014. "The Invisible Addiction: Cell-Phone Activities and Addiction Among Male and Female College Students." *Journal of Behavioral Addictions* 3 (December): 254–265.

Roberts, Keith A. 2004. *Religion in Sociological Perspective*, 4th edition. Belmont, CA: Wadsworth.

Robertson, Jordan. 2013. "Your Not-So-Secret Medical History." *Bloomberg Businessweek*, August 12–25, 41–42.

Robertson, Ruth, and Sara R. Collins. 2011. "Realizing Health Reform's Potential." Commonwealth Fund, May. Accessed April 24, 2012 (www.commonwealthfund.org).

Robinson, Laurie O., and Jeff Slowikowski. 2011. "Scary—and Ineffective." *Baltimore Sun*, February 1, 11.

Rochman, Bonnie. 2012. "The End of an Epidemic?" *Time*, February 6, 16.

___. 2013. "Hover No More: Helicopter Parents May Breed Depression and Incompetence in Children." *Time*, February 22. Accessed

November 20, 2013 (www.healthland.time .com).

Rodrigue, Edward P., and Richard V. Reeves. 2014. "Horatio Alger Goes to Washington: Representation and Social Mobility." Brookings, October 2. Accessed November 23, 2014 (www.brookings.edu).

Roethlisberger, F. J., and William J. Dickson. 1939/1942. *Management and the Worker: An Account of a Research Program Conducted at the Western Electric Company, Hawthorne Works, Chicago*. Cambridge, MA: Harvard University Press.

Rohrlich, Justin. 2010. "Why White-Collar Criminals Don't Fear Getting Caught." Minyanville Media, Inc., November 12. Accessed January 12, 2014 (www.minyanville.com).

Roman, Caterina Gouvis, et al. 2012. "Social Networks, Delinquency, and Gang Membership: Using a Neighborhood Framework to Examine the Influence of Network Composition and Structure in a Latino Community." The Urban Institute, February. Accessed January 12, 2014 (www .urban.org).

Romano, Andrew, and Allison Samuels. 2012. "Is Obama Making It Worse?" *Newsweek*, April 6, 40–45.

Romer, Dan. 2011. "After 11 Years of Setting the Record Straight, Stories About Holiday Suicides Still Outnumber Those Debunking the Myth." Annenberg Public Policy Center, December 13. Accessed December 14, 2011 (www.annenbergpublicpolicycenter.org).

Romero, Simon, and Marc Lacey. 2010. "Looting Flares Where Authority Breaks Down." *New York Times*, January 17, A1.

Ronfeldt, Matthew, Hamilton Lankford, Susanna Loeb, and James Wyckoff. 2011. "How Teacher Turnover Harms Student Achievement." National Bureau of Economic Research, June. Accessed April 14, 2012 (www.nber.org).

Roscigno, Vincent J. 2010. "Ageism in the American Workplace." *Contexts* 9 (Winter): 16–21.

Rose, Fred. 1997. "Toward a Class Cultural Theory of Social Movements: Reinterpreting New Social Movements." *Sociological Forum* 12 (September): 461–494.

Rose, Peter I. 1997. *They and We: Racial and Ethnic Relations in the United States*, 5th edition. New York: McGraw-Hill.

Rose, Stephen J., and Scott Winship. 2009. "Ups and Downs: Does the American Economy Still Promote Upward Mobility?" Economic Mobility Project. Accessed September 20, 2011 (www.economicmobility.org).

Rosen, David. 2013. "Customer Beware: You Are Being Tracked." AlterNet, January 12. Accessed October 8, 2014 (www.alternet.org).

Rosen, Larry D. 2010. *Rewired: Understanding the iGeneration and the Way They Learn*. New York: Palgrave Macmillan.

Rosenberg, Martha. 2014. "6 Nasty Drugs Your Meat Is On." AlterNet, February 7. Accessed July 26, 2014 (www.alternet.org).

Rosenberg, Paul. 2014. "10 Things You Might Not Know About Poverty." AlterNet, January 8. Accessed February 6, 2014 (www.alternet.org).

Rosenbloom, Stephanie. 2013. "Bad Manners Are in the Air." *New York Times*, November 3, TR1.

Rosenfeld, Steven. 2014. "The Supreme Court's Radical Right Wing Majority: Waging War on Women and Boosting Corporate Power." AlterNet, June 30. Accessed July 26, 2014 (www.alternet.org).

Rosenthal, Elizabeth. 2014. "Medicine's Top Earners Are Not the M.D.s." *New York Times*, May 18, SR4.

Rosenthal, Robert, and Lenore Jacobsen. 1968. *Pygmalion in the Classroom: Teacher Expectations and Pupils' Intellectual Development.* New York: Holt, Rinehart, and Winston.

Rosin, Hanna. 2012. "Boys on the Side." *The Atlantic,* September. Accessed January 7, 2013 (www.theatlantic.com).

Ross, Jeffrey Ian. 2013. "New Laws Aren't Enough." *Baltimore Sun,* September 22, 29.

Ross, Martha, Nicole P. Svajlenka, and Jane Williams. 2014. "Part of the Solution: Pre-Baccalaureate Healthcare Workers in a Time of Health System Change." Brookings, July. Accessed December 5, 2014 (www.brookings.edu).

Ross, Terris, Grace Kena, Amy Rathbun, Angelina KewalRamani, Jijun Zhang, Paul Kristapovich, and Eileen Manning. 2012. "Higher Education: Gaps in Access and Persistence Study." National Center for Education Statistics, August. Accessed June 21, 2014 (www.nces.ed.gov).

Rossano, Matt J. 2012. "The Essential Role of Ritual in the Transmission and Reinforcement of Social Norms." *Psychological Bulletin* 138 (May): 529–549.

Rossi, Max, and Lisa Jucca. 2014. "Villages Face 'Slow Death' as Italy's Population Ages." *Baltimore Sun,* March 2, 14.

Rothlin, Phillippe, and Peter R. Werder. 2008. *Boreout! Overcoming Workplace Demotivation.* Philadelphia: Kogan Page.

Rothman, Sheila M. 1978. *Women's Proper Place: A History of Changing Ideals and Practices, 1870 to the Present.* New York: Basic Books.

Rowe-Finkbeiner, Kristin. 2004. *The F-Word: Feminism in Jeopardy, Women, Politics, and the Future.* Emeryville, CA: Seal Press.

___. 2012. "It's Not a 'Mommy War', It's a War on Moms." Moms Rising, April 14. Accessed April 12, 2014 (www.momsrising.org).

Rowland, Lauren. 2011. "Top 20 Toys for Christmas 2011." Kidspot. Accessed October 22, 2012 (www.kidspot.com).

Rubin, Alissa J. 2013. "Afghan Policewomen Say Sexual Harassment Is Rife." *New York Times,* September 17, A4.

Rubin, Kenneth, William Bukowski, and Jeffrey G. Parker. 1998. "Peer Interactions, Relationships, and Groups." Pp. 619–700 in *Handbook of Child Psychology: Vol. 3. Social, Emotional, and Personality Development,* edited by William Damon and Nancy Eisenberg. New York: Wiley.

Rubin, Rita. 2004. "'Smart Pills' Make Headway." *USA Today,* July 7, 1D.

Ruetschlin, Catherine, and Dedrick Asante-Muhammad. 2013. "The Challenge of Credit Card Debt for the African American Middle Class." Demos & NAACP, December 4. Accessed March 22, 2014 (www.demos.org).

Rugh, Jacob S., and Douglas S. Massey. 2010. "Racial Segregation and the American Foreclosure Crisis." *American Sociological Review* 75 (October): 629–651.

Ruiz, Rebecca. 2013. "'Trash the Dress' Trend Gone Too Far? Bride Burns Wedding Gown—While Wearing It." *Today,* July 19. Accessed January 12, 2014 (www.today.com).

Rule, James B. 1988. *Theories of Civil Violence.* Berkeley: University of California Press.

Ruth, Jennifer. 2014. "Non-Critical Thinking in China." *Chronicle Review,* February 28, B20.

Rutter, Virginia, and Pepper Schwartz. 2012. *The Gender of Sexuality: Exploring Sexual Possibilities,* 2nd edition. Lanham, MD: Rowman & Littlefield.

Ryan, Camille. 2013. "Language Use in the United States: 2011." U.S. Census Bureau, August. Accessed March 22, 2014 (www.census.gov).

"S. Korea Orders Lights Out to Boost Birthrate." 2010. *Asia One,* January 20. Accessed September 25 (www.asiaone.com.sg).

Saad, Lydia. 2001. "Majority Considers Sex Before Marriage Morally Okay." *Gallup Poll Monthly,* No. 428 (May): 46–48.

___. 2006. "Families of Drug and Alcohol Abusers Pay an Emotional Toll." Gallup News Service, August 25. Accessed August 27, 2006 (www.gallup.com).

—. 2010. "Nearly 4 in 10 Americans Still Fear Walking Alone at Night." Gallup, November 5. Accessed August 22, 2011 (www.gallup.com).

___. 2013. "Americans' Abortion Views Steady Amid Gosnell Trial." Gallup, May 10. Accessed March 2, 2014 (www.gallup.com).

___. 2013. "Besides Pay, Women as Satisfied as Men with Job Aspects." Gallup, August 21. Accessed April 12, 2014 (www.gallup.com).

___. 2013. "Half in U.S. Support Publicly Financed Federal Campaigns." Gallup, June 24. (Accessed April 16, 2014 (www.gallup.com).

___. 2013. "In U.S., Rise in Religious 'Nones' Slows in 2012." Gallup, January 10. Accessed June 21, 2014 (www.gallup.com).

___. 2013. "Majority of U.S. Workers Say Job Doesn't Require a Degree." Gallup, September 9. Accessed June 21, 2014 (www.gallup.com).

___. 2013. "U.S. Support for Euthanasia Hinges on How It's Described." Gallup, May 29. Accessed May 16, 2014 (www.gallup.com).

___. 2014. "More Women Than Men in U.S. Workforce Are Irked by Pay." Gallup, October 16. Accessed November 23, 2014 (www.gallup.com).

___. 2014. "The '40-Hour' Workweek is Actually Longer—by Seven Hours." Gallup, August 29. Accessed November 23, 2014 (www.gallup.com).

___. 2014. "Three in Four in U.S. Still See the Bible as Word of God." Gallup, June 4. Accessed June 21, 2014 (www.gallup.com).

Sachs, Jeffrey S. 2005. "Confusion over Population: Growth or Dearth?" *Pop!ulation Press* 11 (Winter/Spring): 17.

Sacks, Vanessa, David Murphey, and Kristin Moore. 2014. "Adverse Childhood Experiences: National and State-Level Prevalence." Child Trends Research Brief, July. Accessed November 25, 2014 (www.childtrends.org).

Saez, Emmanuel. 2013. "Striking It Richer: The Evolution of Top Incomes in the United States (Updated with 2012 Preliminary Estimates)." UC Berkeley, September 3. Accessed February 6, 2014 (www.elsa.berkeley.edu).

Saez, Emmanuel, and Gabriel Zucman. 2014. "Wealth Inequality in the United States Since 1913: Evidence from Capitalized Income Tax Data." National Bureau of Economic Research, October. Accessed November 10, 2014 (www.nber.org).

Sagarin, Edward. 1975. *Deviants and Deviance.* New York: Praeger.

Sahgal, Neha. 2014. "Coke, 'America the Beautiful,' and the Language of Diversity." Pew Research Center, February 3. Accessed November 25, 2014 (www.pewresearch.org).

Sahgal, Neha, and Greg Smith. 2009. "A Religious Portrait of African-Americans." Pew Research Center, Pew Forum on Religion & Public Life Project, January 30. Accessed June 21, 2014 (www.pewforum.org).

Sahgal, Neha, and Tim Townsend. 2014. "Four-in-Ten Pakistanis Say Honor Killing of Women Can Be at Least Sometimes Justified." Pew Research Center, May 30. Accessed November 17, 2014 (www.pewresearch.org).

Salamone, Frank A. 2005. "Jazz and Its Impact on European Classical Music." *The Journal of Popular Culture* 38 (May): 732–743.

Salant, Jonathan D. 2014. "$350M NASA Tower Shows How Congress Protects Pork." *Baltimore Sun,* January 12, 12.

Salvy, S-J., K. de la Haye, J. C. Bowker, and R. C. J. Hermans. 2012. "Influence of Peers and Friends on Children's and Adolescents' Eating and Activity Behaviors." *Physiology & Behavior* 106 (June): 369–378.

Sanburn, Josh. 2010. "Brief History: Secret Medical Testing." *Time,* October 18, 30.

Sánchez, Erika L. 2013. "The Challenge of Defining Muslim Feminism." *Huffington Post,* February 10. Accessed October 12, 2013 (www.huffingtonpost.com).

Sandberg, Sheryl. 2013. "Why I Want Women to Lean In." *Time,* March 18, 44–45.

Sandbu, Martin. 2013. "Talkin' 'Bout a Revolution." *Financial Times,* April 19. Accessed October 8, 2014 (www.ft.com).

Sandholtz, Nathan, Lynn Langton, and Michael Planty. 2013. "Hate Crime Victimization, 2003–2011." Bureau of Justice Statistics, March. Accessed January 12, 2014 (www.bjs.gov).

Sandstrom, Kent L., Daniel D. Martin, and Gary Alan Fine. 2006. *Symbols, Selves, and Social Reality: A Symbolic Interactionist Approach to Social Psychology and Sociology,* 2nd edition. Los Angeles: Roxbury.

Sang-Hun, Choe. 2009. "Group Resists Korean Stigma for Unwed Mothers." *New York Times,* October 8, 6.

Saperstein, Guy. 2014. "Cows, Rice Fields and Big Agriculture Consumes Well Over 90% of California's Water." AlterNet, May 17. Accessed September 21, 2014 (www.alternet.org).

Sapir, Edward. 1929. "The Status of Linguistics as a Science." *Language* 5 (4): 207–214.

Sassler, Sharon, and Amanda J. Miller. 2011. "Class Differences in Cohabitation Processes." *Family Relations* 60 (April): 163–177.

Sassler, Sharon, Fenaba R. Addo, and Daniel T. Lichter. 2012. "The Tempo of Sexual Activity and Later Relationship Quality." *Journal of Marriage and Family* 74 (August): 708–725.

Sathyanarayana, Sheela, et al. 2013. "Unexpected Results in a Randomized Dietary Trial to Reduce Phthalate and Bisphenol A Exposures." *Journal of Exposure Science and Environmental Epidemiology* 23 (February): 378–384.

Saul, Michael Howard. 2013. "Judge Halts New York Ban on Large Sodas." *Wall Street Journal,* March 12, A3.

Saulny, Susan. 2011. "Counting by Race Can Throw Off Some Numbers." *New York Times,* February 10, A1.

Savage, David G., and Neela Banerjee. 2014. "Justices Uphold Midwest Coal Plant Curbs." *Baltimore Sun,* April 30, 14.

Sawchuck, Stephen. 2012. "Teacher Quality, Status Entwined Among Top-Performing Nations." *Education Week* 31 (16): 12–16.

Sawhill, Isabel V., and Joanna Venator. 2014. "Families Adrift: Is Unwed Childbearing the New Norm?" Brookings, October 13. Accessed October 14, 2014 (www.brookings.edu).

Scarlett, W. George, Sophie Naudeau, Dorothy Salonius-Pasternak, and Iris Ponte. 2005. *Children's Play.* Thousand Oaks, CA: Sage.

Schachter, Jason P. 2004. "Geographical Mobility: 2002 to 2003." U.S. Census Bureau, Current

Population Reports. Accessed May 15, 2007 (www.census.gov).

Schaeffer, Robert K. 2003. *Understanding Globalization: The Social Consequences of Political, Economic, and Environmental Change*, 2nd edition. Lanham, MD: Rowman & Littlefield.

Schehr, Robert C. 1997. *Dynamic Utopia: Establishing Intentional Communities as a New Social Movement*. Westport, CT: Bergin & Garvey.

Schieman, Scott. 2010. "Socioeconomic Status and Beliefs About God's Influence in Everyday Life." *Sociology of Religion* 71 (Spring): 25–51.

Schiesel, Seth. 2011. "Supreme Court Has Ruled; Now Games Have a Duty." *New York Times*, July 29, C1.

Schiffrin, Holly H., Miriam Liss, Haley Miles-McLean, Katherine A. Geary, Mindy J. Erchull, and Taryn Tashner. 2014. "Helping or Hovering? The Effects of Helicopter Parenting on College Students' Well Being." *Journal of Child and Family Studies* 23 (April): 548–557.

Schlesinger, Izchak M. 1991. "The Wax and Wane of Whorfian Views." Pp. 7–44 in *Influence of Language on Culture & Thought*, edited by Robert Cooper and Bernard Spolsky. New York: Mounton de Gruyter.

Schlesinger, Robert. 2011. "Two Takes: Collective Bargaining Rights for Public Sector Unions?" *U.S. News Weekly*, February 25, 15–16.

Schmall, Emily. 2007. "The Cult of Chick-fil-A." *Forbes*, July 23, 80, 83.

Schmidt, Peter. 2008. "2 Studies Raise Questions about Research Based on Student Surveys." *Chronicle of Higher Education*, November 6. Accessed November 9, 2008 (www.chronicle.com).

———. 2014. "Supreme Court Exposes Affirmative Action at Colleges to Continued Political Assault." *Chronicle of Higher Education*, May 2, A3–A4.

Schnall, Marianne. 2012. "Letting Girls Be Girls—A Global Campaign." Women's Media Center, January 25. Accessed January 2012 (www.womensmediacenter.com).

Schneider, Mark. 2013. "Higher Education Pays: But a Lot More for Some Graduates Than for Others." College Measures. Accessed June 21, 2014 (www.collegemeasures.org).

Schneider, Monica C., and Angela L. Bos. 2014. "Measuring Stereotypes of Female Politicians." *Political Psychology* 25 (April): 245–266.

Schnittker, Jason. 2008. "Happiness and Success: Genes, Families, and the Psychological Effects of Socioeconomic Position and Social Support." *American Journal of Sociology* 114 (Suppl.): S233–S259.

———. 2009. "Mirage of Health in the Era of Biomedicalization: Evaluating Change in the Threshold of Illness, 1972–1996." *Social Forces* 87 (June): 2155–2182.

Schramm, J. B., Chad Aldeman, Andrew Rotherham, Rachael Brown, and Jordan Cross. 2013. "Smart Shoppers: The End of the 'College for All' Debate." College Summit. Accessed June 21, 2014 (www.collegesummit.org).

Schuck, Peter. 2013. "The ObamaCare Website Failure Was Inevitable." *Wall Street Journal*, November 5, A15.

Schultz, Ellen E. 2011. *Retirement Heist: How Companies Plunder and Profit from the Nest Eggs of American Workers*. New York: Penguin.

Schur, Edwin M. 1968. *Law and Society: A Sociological View*. New York: Random House.

Schurman-Kauflin, Deborah. 2000. *The New Predator: Women Who Kill*. New York: Algora.

Schutz, Alfred. 1967. *The Phenomenology of the Social World*. Evanston, IL: Northwestern University Press.

Schwalbe, Michael. 2001. *The Sociologically Examined Life: Pieces of the Conversation*, 2nd edition. Mountain View, CA: Mayfield.

Schwartz, Larry. 2014b. "10 Mind-Blowing Medical Gadgets That Are Likely in Your Future." AlterNet, August 31. Accessed October 8, 2014 (www.alternet.org).

———. 2014a. "8 Mistakes We're Making About Ebola That We Also Made when AIDS Appeared." AlterNet, October 17. Accessed October 18, 2014 (www.alternet.org).

Schwartz, Nelson D. 2011. "Bank Closings Tilt Toward Poor Areas." *New York Times*, February 23, B1.

Schwarz, Alan. 2014. "Thousands of Toddlers Are Medicated for A.D.H.D., Report Finds, Raising Worries." *New York Times*, May 17, A11.

Schweizer, Peter. 2013. *Extortion: How Politicians Extract Your Money, Buy Votes, and Line Their Own Pockets*. New York: Houghton Mifflin Harcourt.

Schwyzer, Hugo. 2011. "How Our Sick Culture Makes Girls Think They Have to Be Gorgeous to Be Loved." AlterNet, April 12. Accessed October 29, 2012 (www.alternet.org).

Scommegna, Paola. 2011. "Least Segregated U.S. Metros Concentrated in Fast-Growing South and West." Population Reference Bureau, September 11. Accessed June 23, 2012 (www.prb.org).

———. 2011. "U.S. Parents Who Have Children with More Than One Partner." Population Reference Bureau, June. Accessed March 6, 2012 (www.prb.org).

———. 2013a. "Aging U.S. Baby Boomers Face More Disability." Population Reference Bureau, March. Accessed May 16, 2014 (www.prb.org).

———. 2013b. "Elderly Immigrants in the United States." *Today's Research on Aging* 29 (October): 1–9.

———. 2013c. "Exploring the Paradox of U.S. Hispanics' Longer Life Expectancy." Population Reference Bureau, July. Accessed July 26, 2014 (www.prb.gov).

Scott, Robert E. 2012. "The China Toll." EPI Briefing Paper 345, Economic Policy Institute, August 23. Accessed April 12, 2014 (www.epi.org).

"Second-Generation Americans." 2013. Pew Research Center, February 7. Accessed March 22, 2014 (www.pewresearch.org).

"Select-a-Faith." 2014. *The Economist*, May 17, 30.

Selingo, Jeffrey J. 2013. "The Diploma's Vanishing Value." *Wall Street Journal*, April 27–28, C3.

Seltzer, Sarah. 2012. "Skinny Minnie? Our Culture's Bizarre Obsession with Stick-Thin Women." AlterNet, October 16. Accessed May 15, 2013 (www.alternet.org).

———. 2012. "The 5 Most Sexist Moments of the Campaign." AlterNet, November 9. Accessed April 21, 2014 (www.alternet.org).

Semba, Richard D., et al. 2014. "Resveratrol Levels and All-Cause Mortality in Older Community-Dwelling Adults." *JAMA Internal Medicine* 174 (7): 1077–1084.

Semuels, Alana. 2010. "Can a Prison Save a Town?" *Los Angeles Times*, May 3. Accessed May 4, 2010 (www.latimes.com).

Sengupta, Somini. 2013. "The Information-Gathering Paradox." *New York Times*, October 27, SR4.

———. 2013. "What You Didn't Post, Facebook May Still Know." *New York Times*, March 26, B1.

Sentencing Project, The. 2013. "Report of the Sentencing Project to the United Nations Human Rights Committee." August. Accessed January 15, 2014 (www.sentencing-project.org).

Senter, Mary S., Nicole Van Vooren, and Roberta Spalter-Roth. 2014. "Sociology, Criminology Concentrations, and Criminal Justice: Differences in Reasons for Majoring, Skills, Activities, and Early Outcomes?" American Sociological Association, April. Accessed October 25, 2014 (www.asanet.org).

"75%—A Nation of Flag Wavers." 2011. Pew Research Center. Accessed July 15, 2011 (www.pewresearch.org).

Seward, Zachary M. 2012. "58 Countries with Better Voter Turnout Than the United States." Quartz, November 6. Accessed April 20, 2014 (www.qz.com).

"Sex-Selective Abortion: Gendercide in the Caucasus." 2013. *The Economist*, September 21, 54–55.

Shaefer, H. Luke, and Kathryn Edin. 2014. "The Rise of Extreme Poverty in the United States." *Pathways* (Summer): 28–32.

Shamoo, Adil E., and Bonnie Bricker. 2011. "The Threat of Bad Science." *Baltimore Sun*, January 11, 13.

Shanahan, Michael J., Shawn Bauldry, and Jason Freeman. 2010. "Beyond Mendel's Ghost." *Contexts* 3 (Fall): 34–39.

Shapira, Ian. 2014. "In Arizona, a Navajo High School Emerges as a Defender of the Washington Redskins." *Washington Post*, October 26. Accessed October 27, 2014 (www.washingtonpost.com).

Shapiro, Jenessa R., and Amy M. Williams. 2012. "The Role of Stereotype Threats in Undermining Girls' and Women's Performance and Interest in STEM Fields." *Sex Roles* 66 (February): 175–183.

Shariff, Azim F., and Lara B. Aknin. 2014. "The Emotional Toll of Hell: Cross-National and Experimental Evidence for the Negative Well-Being Effects of Hell Beliefs." *PLoS ONE* 9 (January): e85251.

Sharifzadeh, Virginia-Shirin. 1997. "Families with Middle Eastern Roots." Pp. 441–482 in *Developing Cross-Cultural Competence: A Guide for Working with Children and Families*, edited by Eleanor W. Lynch and Marci J. Hanson. Baltimore, MD: Paul H. Brookes.

Sharpe, Lindsey. 2013. "U.S. Obesity Rate Climbing in 2013." Gallup, November 1. Accessed December 19, 2013 (www.gallup.com).

Shattuck, Rachel M., and Rose M. Kreider. 2013. "Social and Economic Characteristics of Currently Unmarried Women with a Recent Birth: 2011." U.S. Census Bureau, May. Accessed May 16, 2014 (www.census.gov).

Shauk, Zain, and Bradley Olson. 2014. "Where Frackers Are Friendly Neighbors." *Bloomberg Businessweek*, September 8–14, 37–38.

Shea, Christopher. 2014. "The Liar's 'Tell.'" *The Chronicle Review*, October 17, B6–B9.

Shellenbarger, Sue. 2013. "Finding the Just-Right Level of Self-Esteem for a Child." *Wall Street Journal*, February 27, D1, D3.

Shepherd, Julianne Escobedo. 2011. "Alabama Latinos Flee Immigration Law, Leaving Insufficient Workforce for Tuscaloosa Tornado Cleanup." AlterNet, June 30. Accessed July 2, 2011 (www.alternet.org).

———. 2011. "Hard-Partying Rich Boy Kills Two in Hit-and-Run, Buys Himself Out of a Prison

Sentence." *AlterNet,* June 6. Accessed June 10, 2011 (www.alternet.org).

Sheppard, Kate. 2011. "The Hackers and the Hockey Stick." *Mother Jones,* May/June, 33–45.

___. 2013. "Scientific Misconceptions: The Junk Science Behind Anti-Abortion Advocates' Wildest Claims." *Mother Jones,* January/February, 14.

Sherwood, Jessica Holden. 2010. *Wealth, Whiteness, and the Matrix of Privilege: The View from the Country Club.* Lanham, MD: Lexington Books.

Shibutani, Tamotsu. 1986. *Social Process: An Introduction to Sociology.* Berkeley: University of California Press.

Shilo, Guy, and Riki Savaya. 2011. "Effects of Family and Friend Support on LGB Youths' Mental Health and Sexual Orientation Milestones." *Family Relations* 60 (July): 318–330.

Shorrocks, Anthony, James B. Davies, Rodrigo Lluberas, Markus Stierli, and Antonios Koutsoukis. 2014. *Global Wealth Report 2014.* Credit Suisse Research Institute, October. Accessed November 14, 2014 (www.credit-suisse.com).

Shorto, Russell. 2008. "No Babies?" *New York Times,* June 29, 34.

Shriver, Maria. 2014. *The Shriver Report: A Woman's Nation Pushes Back from the Brink.* New York: Rosetta Books.

Shteir, Rachel. 2013. "Feminism Fizzles." *The Chronicle Review,* February 1, B6–B9.

Siegel, Andrea F. 2009. "Dog Killer Gets 3 Years." *Baltimore Sun,* August 21, 10.

Siegel, Rebecca, Deepa Naishadham, and Ahmedin Jemal. 2012. "Cancer Statistics for Hispanics/Latinos, 2012." *CA: A Cancer Journal for Clinicians* 62 (September/October): 283–298.

Siegman, Aron W., and Stanley Feldstein, eds. 1987. *Nonverbal Behavior and Communication.* Hillsdale, NJ: Lawrence Erlbaum Associates.

Sifferlin, Alexandra. 2014. "Rise of the Mumps: What's Behind the New Cases." *Time,* April 7, 24.

Silverman, Rachel E. 2003. "Provisions Boost Rights of Couples Living Together." *Wall Street Journal,* March 5, D1.

___. 2013. "Tracking Sensors Invade the Workplace." *Wall Street Journal,* March 7, B1, B2.

Simmel, Georg. 1902. "The Number of Members as Determining the Sociological Form of the Group." *American Journal of Sociology* 8 (July): 1–46.

Simmons, Katie, James Bell, Richard Wike, and Alan Cooperman. 2014. "Worldwide, Many See Belief in God as Essential to Morality: Richer Nations Are Exception." Pew Research Center, May 27. Accessed June 21, 2014 (www.pewresearch.org).

Simon, Paula. 2013. "AP Is Not for Everyone." *Baltimore Sun,* August 24, 17.

Singal, Pooja, Adi Rattner, and Meghana Desale. 2014. "Court Ruling Limits a Doctor's Options." *Baltimore Sun,* July 17, 17.

Singer, Natasha. 2013. "Group Criticizes Learning Apps for Babies." *New York Times,* August 8, B8.

___. 2013. "They Loved Your G.P.A. Then They Saw Your Tweets." *New York Times,* November 10, BU3.

Sjoberg, Gideon. 1960. *The Preindustrial City: Past and Present.* Glencoe, IL: Free Press.

Skoning, Gerald D. 2013. "How Congress Puts Itself Above the Law." *Wall Street Journal,* April 16, A15.

Slater, Dan. 2013. "Darwin Was Wrong About Dating." *New York Times,* January 13, SR1.

"A Smarter War on Poverty." 2014. *Bloomberg Businessweek,* January 13–19, 8.

Smedley, Audrey. 2007. *Race in North America: Origin and Evolution of a Worldview,* 3rd edition. Boulder, CO: Westview Press.

Smelser, Neil J. 1962. *Theory of Collective Behavior.* New York: Free Press.

___. 1988. "Social Structure." Pp. 103–129 in *Handbook of Sociology,* edited by Neil J. Smelser. Newbury Park, CA: Sage.

Smith, Aaron. 2010. "Neighbors Online." Pew Internet & American Life Project, June 9. Accessed June 12, 2010 (www.pewinternet.org).

___. 2012. "The Best (and Worst) of Mobile Connectivity." Pew Research Center, November 30. Accessed December 4, 2013 (www.pewinternet.org).

___. 2014. "5 Facts About Online Dating." Pew Research Center, February 13. Accessed March 2, 2014 (www.pewresearch.org).

___. 2014. "U.S. Views of Technology and the Future: Science in the Next 50 Years." Pew Research Center, April 17. Accessed October 8, 2014 (www.pewresearch.org).

Smith, Aaron, and Maeve Duggan. 2013. "Online Dating & Relationships." Pew Research Center, October 21. Accessed December 4, 2013 (www.pewinternet.org).

Smith, Aaron, Janna Anderson, and Lee Rainie. 2014. "AI, Robotics, and the Future of Jobs." Pew Research Center, August 6. Accessed October 8, 2014 (www.pewresearch.org).

Smith, Adam. 1776/1937. *An Inquiry into the Nature and Causes of the Wealth of Nations.* New York: Modern Library.

Smith, Dorothy E. 1987. *The Everyday World as Problematic: A Feminist Sociology.* Toronto: University of Toronto Press.

Smith, Elliot Blair, and Phil Kuntz. 2013. "Some CEOs Are More Equal Than Others." *Bloomberg Businessweek,* May 6–12, 70–71.

Smith, Hedrick. 2014. "'The Divide: American Injustice in the Age of the Wealth Gap' by Matt Taibbi." *Washington Post,* April 11. Accessed April 13, 2014 (www.washingtonpost.com).

Smith, Jane I. 1994. "Women in Islam." Pp. 303–325 in *Today's Woman in World Religions,* edited by Arvind Sharma. Albany: State University of New York Press.

Smith, Jessica C., and Carla Medalia. 2014. "Health Insurance Coverage in the United States: 2013." U.S. Census Bureau, September. Accessed December 5, 2014 (www.census.gov).

Smith, N., and A. Leiserowitz. 2013. "American Evangelicals and Global Warming." *Global Environmental Change* 23 (October): 1009–1017.

Smith, Ray A. 2013. "A Closet Filled with Regrets." *Wall Street Journal,* April 18, D1, D4.

Smith, Rebecca. 2014. "The First Baby Has Been Born Following a Womb Transplant." *The Telegraph,* October 3. Accessed October 8, 2014 (telegraph.co.uk).

Smith, Stacy L., Katherine M. Pieper, Amy Granados, and Marc Choueiti. 2010. "Assessing Gender-Related Portrayals in Top-Grossing G-Rated Films." *Sex Roles* 62 (June): 774–786.

Smith, Stacy L., and Marc Choueiti. 2011. "Gender Inequality in Cinematic Content? A Look at Females on Screen & Behind-the-Camera in Top-Grossing 2008 Films." Annenberg School for Communication & Journalism, University of Southern California." April 22.

Accessed October 25, 2011 (www.annenberg.usc.edu).

Smith, Tom W., Peter V. Marsden, and Michael Hout. 2011. *General Social Survey, 1972–2010, Cumulative File Codebook.* Ann Arbor, MI: Inter-University Consortium for Political and Social Research. Accessed February 13, 2014 (www.icpsr.umich.edu).

Snider, Laureen. 1993. "Regulating Corporate Behavior." Pp. 177–210 in *Understanding Corporate Criminality,* edited by Michael B. Blankenship. London: Garland Press.

Snipp, C. Matthew. 1996. "A Demographic Comeback for American Indians." *Population Today* 24 (November): 4–5.

Snyder, Thomas D., and Sally A Dillow. 2013. *Digest of Education Statistics 2012.* U.S. Department of Education, National Center for Education Statistics, December. Accessed June 21, 2014 (www.nces.ed.gov).

Soguel, Dominique. 2009. "Wage Gap Study Arrives in Time for Equal Pay Day." Women's eNews, April 28. Accessed April 28, 2009 (www.womensenews.org).

Sommers, Dixie, and James C. Franklin. 2012. "Overview of Projections to 2020." *Monthly Labor Review* 135 (January): 3–20.

Sonfield, Adam, and Kathryn Kost. 2013. "Public Costs from Unintended Pregnancies and the Role of Public Insurance Programs in Paying for Pregnancy Infant Care: Estimates for 2008." Guttmacher Institute, October. Accessed May 16, 2014 (www.guttmacher.org).

Sorenson, Susan. 2013. "Lower Your Health Costs While Boosting Performance." *Gallup Business Journal,* September 19. Accessed December 19, 2013 (www.businessjournal.gallup.com).

Sorenson, Susan, and Keri Garman. 2013. "More Educated, Less Engaged." *Gallup Business Journal,* August 13. Accessed April 12, 2014 (www.gallup.com).

Southern Poverty Law Center. 2013. "Active U.S. Hate Groups." Accessed July 20, 2014 (www.splcenter.org).

Spalter-Roth, Roberta, Nicole Van Vooren, and Mary S. Senter. 2013. "Social Capital for Sociology Majors: Applied Activities and Peer Networks." American Sociological Association. Accessed October 12, 2013 (www.asanet.org).

Sparshott, Jeffrey. 2013. "TARP Firms' Pay Unchecked." *Wall Street Journal,* January 29, C3.

Speigel, Lee. 2013. "48 Percent of Americans Believe UFOs Could Be ET Visitations." *Huffington Post,* September 11. Accessed September 12, 2013 (www.huffingtonpost.com).

Sperry, Shelley. 2008. "Ozone Defense." *National Geographic* 214 (October): no page given.

Spencer, Herbert. 1862/1901. *First Principles.* New York: P. F. Collier & Son.

Spetz, Joanne, Stephen T. Parente, Robert J. Town, and Dawn Bazarko. 2013. "Scope-of-Practice Laws for Nurse Practitioners Limit Cost Savings That Can Be Achieved in Retail Clinics." *Health Affairs* 32 (November): 1977–1984.

Spitznagel, Eric. 2013. "Men on the Run." *Bloomberg Businessweek,* March 25–31, 82.

Spotts, Peter N. 2009. "New Climate Change Signal: Oceans Turning Acidic." *Christian Science Monitor,* December 9. Accessed December 12, 2010 (www.csmonitor.com).

Spivak, Howard R., E. Lynn Jenkins, Kristi VanAudenhove, Debbie Lee, Mim Kelly, and John Iskander. 2014. "CDC Grand Rounds:

A Public Health Approach to Prevention of Intimate Partner Violence." *MMWR* 63 (January 17): 38–41.

Spradley, J. P., and M. Phillips. 1972. "Culture and Stress: A Quantitative Analysis." *American Anthropologist* 74 (3): 518–529.

Sprigg, Peter. 2011. "Marriage's Public Purpose." *Baltimore Sun*, February 2, 27.

Srivastava, Mehul. 2013. "In Bangladesh, Outside Inspectors Are Still MIA." *Bloomberg Businessweek*, November 4–10, 28–30.

Srivastava, Mehul, and Kartikay Mehrotra. 2013. "India's Second-Class Citizens." *Bloomburg Businessweek*, January 14–20, 10–12.

Starr, Christine R., and Gail M. Ferguson. 2012. "Sexy Dolls, Sexy Grade-Schoolers? Media and Maternal Influences on Young Girls' Self-Sexualization." *Sex Roles* 67 (October): 463–476.

State Higher Education Executive Officers (SHEEO). 2014. "State Higher Education Finance FY 2013." Accessed June 21, 2014 (www.sheeo.org).

Steffensmeier, Darrell J., Jennifer Schwartz, and Michael Roche. 2013. "Gender and Twenty-First-Century Corporate Crime: Female Involvement and the Gender Gap in Enron-Era Corporate Frauds." *American Sociological Review* 78 (June): 448–476.

Steinberg, Julia R., and Lawrence B. Finer. 2011. "Examining the Association of Abortion History and Current Mental Health: A Reanalysis of the National Comorbidity Survey Using a Common-Risk-Factors Model." *Social Science & Medicine* 72 (1): 72–82.

Steinhauer, Jennifer. 2013. "As Fund-Raisers in Congress, Women Break the Cash Ceiling." *New York Times*, November 30, A11.

Steinhauser, Paul. 2014. "CNN Poll: Pathway to Citizenship Trumps Border Security." Cable News Network, February 6. Accessed March 22, 2014 (www.cnn.com).

Steinmetz, Katy. 2014. "Disrupted: The Tech Wealth Transforming San Francisco Has Unemployment Down, Evictions Up and Tensions Flaring." *Time*, February 10, 32–35.

Stencel, Sandra, and Brian J. Grim. 2013. "Headscarf Incident in Sudan Highlights a Global Trend." Pew Research Center, September 18. Accessed June 21, 2014 (www.pewresearch.org).

Sternbergh, Adam. 2008. "Why White People Like 'Stuff White People Like'." *New Republic*, March 17. Accessed November 6, 2008 (www.tnr.com).

Steuerle, Eugene, Signe-Mary McKernan, Caroline Ratcliffe, and Sisi Zhang. 2013. "Lost Generations? Wealth Building Among Young Americans." Urban Institute, March. Accessed February 6, 2014 (www.urban.org).

"Steve Jobs: Adopted Child Who Never Met His Biological Father." 2011. *Telegraph*, October 6. Accessed March 22, 2014 (telegraph.co.uk).

Stevenson, Betsey, and Justin Wolfers. 2013. "Subjective Well-Being and Income: Is There Any Evidence of Satiation?" *American Economic Review* 103 (May): 598–604.

Stillars, Alan L. 1991. "Behavioral Observation." Pp. 197–218 in *Studying Interpersonal Interaction*, edited by B. M. Montgomery and S. Duck. New York: Guilford Press.

Stilwell, Victoria. 2013. "Body Parts for Sale by Desperate U.S. Workers." *Bloomberg Businessweek*, October 21–27, 22, 24.

Stokes, Bruce. 2013a. "U.S. Stands Out as a Rich Country Where a Growing Minority Say They Can't Afford Food." Pew Research Center, May 24. Accessed February 14, 2014 (www.pewresearch.org).

———. 2013b. "The U.S.'s High Income Gap Is Met with Relatively Low Public Concern." Pew Research Center. Accessed February 14, 2014 (www.pewresearch.org).

Stolley, Giordano, and Somchai Taphaneeyapan. 2002. "Stomaching Bugs in Thailand." *Baltimore Sun*, June 20, 2A.

Strasburger, Victor C., et al. 2006. "Children, Adolescents, and Advertising." *Pediatrics* 118 (December 6): 2563–2569.

Strasburger, Victor C., and Marjorie J. Hogan. 2013. "Policy Statement: Children, Adolescents, and the Media." *Pediatrics* 132 (5): 958–961.

Straus, Murray A. 2011. "Gender Symmetry and Mutuality in Perpetration of Clinical-Level Partner Violence: Empirical Evidence and Implications for Prevention and Treatment." *Aggression and Violent Behavior* 16 (July–August): 279–288.

———. 2014. "Addressing Violence by Female Partners Is Vital to Prevent or Stop Violence Against Women: Evidence From the Multisite Batterer Intervention Evaluation." *Violence Against Women* 20 (7): 889–899.

"Streaking." 2005. Wikipedia Encyclopedia. Accessed September 8, 2005 (www.wikipedia.org).

Strean, William B. 2009. "Remembering Instructors: Play, Pain, and Pedagogy." *Qualitative Research in Sport and Exercise* 1 (November): 210–220.

Streib, Jessi. 2011. "Class Reproduction by Four Year Olds." *Qualitative Sociology* 34 (June): 337–352.

Stroebe, Margaret, Maarten van Son, Wolfgang Stroebe, Rolf Kleber, Henk Schut, and Jan van den Bout. 2000. "On the Classification and Diagnosis of Pathological Grief." *Clinical Psychology Review* 20 (January): 57–75.

Strom, Stephanie. 2014. "Coca-Cola to Remove an Ingredient Questioned by Consumers." *New York Times*, May 6, B6.

"Study Shows Positive Results from Early Head Start Program." 2002. U.S. Department of Health and Human Services, June 3. Accessed July 10, 2002 (www.hhs.gov).

Strumpf, Dan. 2013. "With Fewer to Lock Up, Prisons Shut Doors." *Wall Street Journal*, February 11, A3.

Sturm, Roland, and Ruopeng An. 2014. "Obesity and Economic Environments." *CA: A Cancer Journal for Clinicians* 65 (5): 337–350.

Substance Abuse and Mental Health Services Administration. 2013. *Results from the 2012 National Survey on Drug Use and Health: Summary of National Findings.* Accessed July 26, 2014 (www.store.samhsa.gov).

Sudarkasa, Niara 2007. "African American Female-Headed Households: Some Neglected Dimensions." In *Black Families*, 4th edition., edited by Harriette Pipes McAdoo, 172–183. Thousand Oaks, CA: Sage.

Suddath, Claire. 2013. "So Hard to Say Goodbye." *Bloomberg Businessweek*, November 4–10, 79–81.

———Suddath, Claire. 2013. "You Get a D+ in Teamwork." *Bloomberg Businessweek*, November 11–17, 91.

Suich, Alexander. 2014. "Little Brother." *The Economist*, September 13, 3–16.

Sullivan, Andrew. 2011. "Why Gay Marriage Is Good for Straight America." *Newsweek*, July 25, 12–14.

Sulzberger, A. G. 2011. "Hispanics Reviving Faded Towns on the Plains." *New York Times*, November 14, A1.

Summers, Nick. 2010. "Do Fines Ever Make Corporations Change?" *Newsweek*, November 13, 56.

Sumner, William G. 1906. *Folkways.* New York: Ginn.

Sunstein, Cass R. 2009. *On Rumors: How Falsehoods Spread, Why We Believe Them, What Can Be Done.* New York: Farrar, Straus and Giroux.

Sutherland, Edwin H. 1949. *White Collar Crime.* New York: Holt, Rinehart, and Winston.

Sutherland, Edwin H., and D. R. Cressey. 1970. *Criminology*, 8th edition. Philadelphia: Lippincott.

Sutton, Philip W. 2000. *Explaining Environmentalism: In Search of a New Social Movement.* Burlington, VT: Ashgate.

Swan, Noelle. 2014. "Rise of the People-Friendly Machines." *Christian Science Monitor Weekly*, October 6, 21–23.

Swift, Art. 2013. "Americans' Satisfaction with Life Similar to Levels in 1998." Galllup, November 1. Accessed February 10, 2014 (www.gallup.com).

———. 2013. "For First Time, Americans Favor Legalizing Marijuana." Gallup, October 22. Accessed January 12, 2014 (www.gallup.com).

———. 2013. "Honesty and Ethics Rating of Clergy Slides to New Low." Gallup, December 16. Accessed April 20, 2014 (www.gallup.com).

———. 2014. "Americans Again Pick Environment Over Economic Growth." Gallup, March 20. Accessed September 19, 2014 (www.gallup.com).

———. 2014. "Less Than Half of Americans Support Stricter Gun Laws." Gallup, October 31. Accessed November 4, 2014 (www.gallup.com).

Szarota, Piotr. 2010. "The Mystery of the European Smile: A Comparison Based on Individual Photographs Provided by Internet Users." *Journal of Nonverbal Behavior* 34 (December): 249–256.

Tabachnick, Rachel. 2011. "The 'Christian' Dogma Pushed by Religious Schools That Are Supported by Your Tax Dollars." AlterNet, May 23. Accessed July 12, 2011 (www.alternet.org).

Taibbi, Matt. 2014. *The Divide: American Injustice in the Age of the Wealth Gap.* New York: Spiegel & Grau.

Tajfel, Henri. 1982. "Social Psychology of Intergroup Relations." *Annual Review of Psychology* 33 (February): 1–39.

Tannen, Deborah. 1990. *You Just Don't Understand: Women and Men in Conversation.* New York: Ballantine.

Tannenbaum, Frank. 1938. *Crime and the Community.* New York: Columbia University Press.

Taormina-Weiss, Wendy. 2013. "Psychological and Social Aspects of Disability." Disabled World, June 22. Accessed November 18, 2013 (www.disabled-world.com).

"Tattoos in the Workplace: Ink Blots." 2014. *The Economist*, August 2, 22.

Tavernise, Sabrina. 2013. "The Health Toll of Immigration." *New York Times*, May 19, A1.

Tavris, Carol. 2013. "How Psychiatry Went Crazy." *Wall Street Journal*, May 18, C5.

———. 2013. "The Experiments That Still Shock." *Wall Street Journal*, September 6, C1, C7.

Taylor, Carl S., and Pamela R. Smith. 2013. "The Attraction of Gangs: How Can We Reduce It?" Pp. 19–29 in *Changing Course: Preventing Gang Membership*, edited by Thomas R. Simon, Nancy M. Witter, and Reshma R. Mahendra. National Institute of Justice and Centers for Disease Control and Prevention. Accessed December 18, 2013 (www.nij.gov).

Taylor, Frederick W. 1911/1967. *The Principles of Scientific Management.* New York: W. W. Norton & Company.

Taylor, Jay. 1993. *The Rise and Fall of Totalitarianism in the Twentieth Century.* New York: Paragon House.

Taylor, Paul, et al. 2012. "The Lost Decade of the Middle Class." Pew Research Center, August 22. Accessed February 14, 2014 (www.pewsocialtrends.org).

Taylor, Paul, et al. 2013. "A Survey of LGBT Americans: Attitudes, Experiences and Values in Changing Times." Pew Research Center, June 13. Accessed March 2, 2014 (www.pewresearch.org).

Taylor, Paul, et al. 2013. "The Rise of Asian Americans." Pew Research Center, April 4. Accessed May 16, 2014 (www.pewsocialtrends.org).

Taylor, Paul, and Mark H. Lopez. 2013. "Skepticism About the Census Voter Turnout Finding." Pew Research Center, May 15. Accessed August 8, 2014 (www.pewresearch.org).

Taylor, Paul, Carroll Doherty, Rich Morin, and Kim Parker. 2014. "Millennials in Adulthood: Detached from Institutions, Networked with Friends." Pew Research Center, March 7. Accessed June 21, 2014 (www.pewresearch.org).

Taylor, Paul, Rich Morin, Kim Parker, and D'Vera Cohn. 2009. "Growing Old in America: Expectations vs. Reality." Pew Research Center, June 29. Accessed April 30, 2010 (www.pewsocialtrends.org).

Taylor, Susan C. 2003. *Brown Skin: Dr. Susan Taylor's Prescription for Flawless Skin, Hair, and Nails.* New York: HarperCollins.

Terando, Adam J., Jennifer Costanza, Curtis Belyea, Robert R. Dunn, Alexa McKerrow, and Jaime A. Collazo. 2014. "The Southern Megalopolis: Using the Past to Predict the Future of Urban Sprawl in the Southeast U.S." *PLoS ONE* 9 (July): e102261.

Teranishi, Robert. 2011. "Asian Americans and Pacific Islanders: Facts, Not Fiction—Setting the Record Straight." CARE and College Board. Accessed December 7, 2011 (www.nyu.edu/projects/care).

Terhune, Chad, Noam N. Levy, and Doug Smith. 2014. "Medicare Unveils 2012 Payouts: 2% of U.S. Providers Got 23% of Fees." *Baltimore Sun*, April 9, 1, 16.

Tétreault, Mary Ann. 2001. "A State of Two Minds: State Cultures, Women, and Politics in Kuwait." *International Journal of Middle East Studies* 33 (May): 203–220.

Tharenou, Phyllis. 2013. "The Work of Feminists Is Not Yet Done: The Gender Pay Gap—A Stubborn Anachronism." *Sex Roles* 68 (February): 198–206.

Theodorou, Angelina, and Peter Henne. 2014. "How Religious Harassment Varies by Region Across the Globe." Pew Research Center, May 2. Accessed June 21, 2014 (www.pewresearch.org).

___. 2014. "Religious Police Found in Nearly One-in-Ten Countries Worldwide." Pew Research Center, March 19. Accessed June 21, 2014 (www.pewresearch.org).

Thio, Alex D., Jim D. Taylor, and Martin D. Schwartz. 2012. *Deviant Behavior*, 11th edition. Upper Saddle River, NJ: Pearson.

Thibaut, John W., and Harold H. Kelley. 1959. *The Social Psychology of Groups.* New York: Wiley.

Thomas, Anita Jones, Jason Daniel Hacker, and Denada Hoxha. 2011. "Gendered Racial Identity of Black Young Women." *Sex Roles* 64 (April): 530–542.

Thomas, W. I., and Dorothy Swaine Thomas. 1928. *The Child in America.* New York: Alfred A. Knopf.

Thompson, Arienne. 2010. "16, Pregnant...and Famous: Teen Moms Are Newest Stars." *USA Today,* November 23. Accessed November 24, 2011 (www.usatoday.com).

Thompson, Mark. 2014. "A Troubled Marine's Final Fight." *Time*, February 10, 35–42.

Thompson, Robert S., et al. 2006. "Intimate Partner Violence: Prevalence, Types, and Chronicity in Adult Women." *American Journal of Preventive Medicine* 30 (June): 447–457.

Thorpe-Moscon, Jennifer, and Alixandra Pollock. 2014. "Feeling Different: Being the 'Other' in US Workplaces." Catalyst. Accessed March 22, 2014 (www.catalyst.org).

Tilly, Charles. 1978. *From Mobilization to Revolution.* Reading, MA: Addison-Wesley.

Timberg, Craig, and Ellen Nakashima. 2013. "Amid NSA Spying Revelations, Tech Leaders Call for New Restraints on Agency." *Washington Post,* November 1. Accessed November 10, 2013 (www.washingtonpost.com).

Timberlake, Cotten. 2013. "Don't Even Think About Returning This Dress." *Bloomberg Businessweek*, September 30–October 6, 29–31.

Timmermans, Stefan, and Hyeyoung Oh. 2010. "The Continued Social Transformation of the Medical Profession." *Journal of Health and Social Behavior* 51 (March): S94–S109.

Tjaden, Patricia, and Nancy Thoennes. 2006. *Extent, Nature, and Consequences of Rape Victimization: Findings from the National Violence Against Women Survey.* Washington, DC: U.S. Department of Justice, Office of Justice Statistics.

Tobar, Hector. 2009. "Language as a Bridge and an Identity." *Los Angeles Times,* September 22. Accessed September 24, 2009 (www.latimes.com).

Tormey, Simon. 1995. *Making Sense of Tyranny: Interpretations of Totalitarianism.* New York: Manchester University Press.

Tornatzky, Louis G., Richard Cutler, and Jongho Lee. 2002. "College Knowledge: What Latino Parents Need to Know and Why They Don't Know It." The Tomas Rivera Policy Institute. Accessed January 2, 2005 (www.trpi.org).

Toth, Emily. 2011. "No Girls Aloud." *Chronicle of Higher Education*, April 1, A26.

Touraine, Alain. 2002. "The Importance of Social Movements." *Social Movement Studies* 1 (April): 89–96.

Townsend, Matt. 2014. "The Cabbage Patch Kids Get a Makeover." *Bloomberg Businessweek*, June 9–22, 27–28.

Tozzi, John. 2014a. "When Hospitals Buy Clinics, Prices Go Up." *Bloomberg Businessweek*, July 28–August 3, 27–28.

___. 2014b. "Who Should Pay the Bill for Wonder Drugs?" *Bloomberg Businessweek*, July 10–16, 27–28.

Tozzi, John, David Armstrong, and Caroline Chen. 2014. "Why Medicare Keeps Paying Sketchy Docs." *Bloomberg Businessweek*, May 5–11, 32–34.

"A Trillion in the Trough." 2014. *The Economist*, February 8, 27–28.

Trottman, Melanie. 2013. "Religious Discrimination Claims on the Rise." *Wall Street Journal*, October 28, B1, B4.

Trottman, Melanie, and Kris Maher. 2013. "Organized Labor Loses Members." *Wall Street Journal*, January 24, A6.

"Trouble in Trappes." 2013. *The Economist*, July 27, 44.

Truman, Jennifer L., and Erica L. Smith. 2012. "Prevalence of Violent Crime among Households with Children, 1993–2010." Bureau of Justice Statistics, September. Accessed May 16, 2014 (www.bjs.gov).

Truman, Jennifer L., and Lynn Langton. 2014. "Criminal Victimization, 2013." Bureau of Justice Statistics, September. Accessed November 6, 2014 (www.bjs.gov).

Truman, Jennifer L., and Rachel E. Morgan. 2014. "Nonfatal Domestic Violence, 2003–2012." Bureau of Justice Statistics, April. Accessed May 16, 2014 (www.bjs.gov).

Truman, Jennifer, Lynn Langton, and Michael Planty. 2013. "Criminal Victimization, 2012." Bureau of Justice Statistics, October. Accessed January 12, 2014 (www.bjs.gov).

Trumbull, Mark. 2014. "The Wage War Now Widens." *Christian Science Monitor Weekly*, January 13, 11.

TrustLaw. 2011. "The World's Five Most Dangerous Countries for Women." Women's Rights Homepage, June 17. Accessed November 20, 2013 (www.trust.org).

Tsai, Tyjen. 2012. "China Has Too Many Bachelors." Population Reference Bureau, January. Accessed September 19, 2014 (www.prb.org).

Tsukayama, Hayley. 2014. "High-Tech Upgrades May Let Aging Boomers Live Independently in Their Own Homes Longer." *Washington Post*, January 20. Accessed October 8, 2014 (www.washingtonpost.com).

Tumin, Melvin M. 1953. "Some Principles of Stratification: A Critical Analysis." *American Sociological Review* 18 (August): 387–393.

Turk, Austin T. 1969. *Criminality and the Legal Order.* Chicago: Rand-McNally.

___. 1976. "Law as a Weapon in Social Conflict." *Social Problems* 23 (February): 276–291.

Turner, Jennifer. 2013. "A Living Death: Life Without Parole for Nonviolent Offenses." American Civil Liberties Union. Accessed January 15, 2014 (www.aclu.org).

Turner, Margery A., Rob Santos, Diane K. Levy, Doug Wissoker, Claudia Aranda, and Rob Pitingolo. 2013. "Housing Discrimination Against Racial and Ethnic Minorities 2012." The Urban Institute, June. Accessed March 22, 2014 (www.huduser.org).

Turner, Ralph H., and Lewis M. Killian. 1987. *Collective Behavior*, 3rd edition. Englewood Cliffs, NJ: Prentice Hall.

Turow, Joseph, Jennifer King, Chris Jay Hoofnagle, Amy Bleakley, and Michael Hennessy. 2009. "Americans Reject Tailored Advertising." Social Science Research Network, September. Accessed May 11, 2010 (www.ssrn.com).

Twenge, Jean M. 2007. *Generation Me: Why Today's Young Americans Are More Confident, Assertive, Entitled—and More Miserable Than Ever Before.* New York: Simon & Schuster.

U.S. Bureau of Labor Statistics American Time Use Survey. 2013. "American Time Use Survey—2012 Results." USDL-13-1178, June 20. Accessed June 30, 2013 (www.bls.gov).

U.S. Bureau of Labor Statistics. 2014. "Labor Force Participation Rate." Accessed February 13, 2014 (www.bls.gov).

___. 2014. "Number of Fatal Work Injuries, 1992–2012." Accessed July 26, 2014 (www.bls.gov).

U.S. Census Bureau. 2012. *Statistical Abstract of the United States: 2012*, 131st edition. Washington, DC: Government Printing Office.

U.S. Census Bureau News. 2008. "An Older and More Diverse Nation by Midcentury." August 14. Accessed August 18, 2008 (www.census.gov).

U.S. Census Bureau Newsroom. 2014. "Energy Boom Fuels Rapid Population Growth in

Parts of Great Plains; Gulf Coast Also Has High Growth Areas, Says Census Bureau." U.S. Census Bureau, March 27. Accessed September 19, 2014 (www.census.gov).

U.S. Census Bureau, Current Population Survey, Annual Social and Economic Supplements. 2013. Tables F-2 and H-3 on income. Accessed February 13, 2014 (www.census .gov).

U.S. Census Bureau, Current Population Survey. 2013. "Annual Social and Economic Supplements." November. Accessed October 20, 2014 (www.census.gov).

U.S. Census Bureau Population Division. 2012. "Table 3. Annual Estimates of the Resident Population by Sex, Race, and Hispanic Origin for the United States: April 1, 2010 to July 1, 2011 (NC–EST2011–03)." Accessed May 15, 2012 (www.census.gov).

"U.S. Census Bureau Projections Show a Slower Growing, Older, More Diverse Nation a Half Century from Now." 2012. United States Census Bureau, December 12. Accessed May 16, 2014 (www.census.gov).

U.S. Census Bureau, Voting and Registration. 2013. "Voting and Registration in the Election of November 2012—Detailed Tables." Census Bureau, May 8. Accessed April 18, 2014 (www.census.gov).

U.S. Department of Commerce. 1993. "We the American Foreign Born." U.S. Census Bureau. Washington, DC: U.S. Government Printing Office.

U.S. Department of Health and Human Services. 2013. Child Maltreatment 2012. Administration for Children and Families, Children's Bureau. Accessed May 16, 2014 (www.acf.hhs.gov).

___. 2014. "The Health Consequences of Smoking—50 Years of Progress. A Report of the Surgeon General." Office of the Surgeon General. Accessed July 30, 2014 (www.cdc .gov).

U.S. Department of Justice. 1996. "Policing Drug Hot Spots." National Institute of Justice, January. Accessed May 3, 2005 (www.ncjrs.org).

___. 2011. "U.S. Attorney Announces Drug Endangered Children Task Force." Press Notice, May 31. Accessed July 25, 2011 (www.usdoj.gov).

U.S. Department of State. 2006. Trafficking in Persons Report. June. Accessed November 1, 2006 (www.state.gov).

___. 2013. "Trafficking in Persons Report: June 2013." Accessed March 2, 2014 (www.state .gov).

U.S. Energy Information Administration. 2014. "China." February 4. Accessed September 19, 2014 (www.eia.gov).

U.S. Environmental Protection Agency. 2009. "EPA's Endangerment Finding." December 7. Accessed May 1, 2010 (www.epa.gov).

___. 2012. "Municipal Solid Waste." Environmental Protection Agency, April 3. Accessed June 22, 2012 (www.epa.gov).

___. 2013. "The National Rivers and Streams Assessment 2008–2009: A Collaborative Survey." March. Accessed September 19, 2014 (www.epa.gov).

___. 2014a. "Municipal Solid Waste Generation, Recycling, and Disposal in the United States: Facts and Figures for 2012." February. Accessed September 19, 2014 (www.epa.gov).

___. 2014b. "Nutrient Pollution." August 26. Accessed September 19, 2014 (www.epa.gov).

U.S. Office of Personnel Management. 2014. "2014 Federal Employee Viewpoint Survey Results." Accessed November 2, 2014 (www .opm.gov/fevs).

U.S. Senate. 2004. Report on the U.S. Intelligence Community's Prewar Intelligence Assessments on Iraq. Accessed September 13, 2006 (www.gpoaccess.gov).

U.S. Senate Special Committee on Aging, American Association of Retired Persons, Federal Council on the Aging, and U.S. Administration on Aging. 1991. Aging America: Trends and Projections, 1991. Washington, DC: U.S. Department of Health and Human Services.

U.S. Travel Association. 2014. "Working for Free: U.S. Workforce Forfeits $52.4 Billion in Time Off Benefits Annually." October 21. Accessed November 23, 2014 (www.ustravel.org).

UN Women. 2014. "Facts and Figures: Ending Violence Against Women." Accessed March 2, 2014 (www.unwomen.org).

___. 2014. "Promoting Equality, Including Social Equity, and Gender Equality and Women's Empowerment." February 5. Accessed February 10, 2014 (www.unwomen.org).

UNICEF. 2013. "Female Genital Mutilation/ Cutting: A Statistical Overview and Exploration of the Dynamics of Change." Unite for Children. Accessed March 2, 2014 (www.childinfo.org).

"Union Members—2013." 2014. Bureau of Labor Statistics, January 24. Accessed January 26, 2014 (www.bls.gov).

United Human Rights Council. 2004. "History of Genocide." Accessed April 29, 2004 (www .unitedhumanrights.org).

United Nations Committee on the Rights of the Child. 2014. "Concluding Observations on the Second Periodic Report of the Holy See." February 25. Accessed June 21, 2014 (www .ohchr.org).

United Nations Department of Economic and Social Affairs. 2012. "World Urbanization Prospects: The 2011 Revision, Highlights." March. Accessed June 23, 2012 (www.esa.un .org).

United Nations Development Program. 2013. "Humanity Divided: Confronting Inequality in Developing Countries." November. Accessed February 6, 2014 (www.undp.org).

___. 2014. "Human Development Report 2014." Accessed September 19, 2014 (www.undp.org).

United Nations Population Division. 2013. "World Population Prospects: The 2012 Revision." Department of Economic and Social Affairs. Accessed September 19, 2014 (www.un.org).

United Nations. 2014. World Urbanization Prospects: The 2014 Revision. Department of Economics and Social Affairs, Population Division. Accessed September 19, 2014 (www.esa.un.org).

United States Conference of Mayors. 2013. "Hunger and Homelessness Survey: A Status on Hunger and Homelessness in America's Cities." December. Accessed February 6, 2014 (www.usmayors.org).

Urbina, Ian. 2006. "In Online Mourning, Don't Speak Ill of the Dead." New York Times, November 5, 1.

"Vacation Habits Around the World." 2013. Expedia 2013 Vacation Deprivation Study, November 18. Accessed April 12, 2014 (www.viewfinder .expedia.com).

Vallotton, Claire, and Catherine Ayoub. 2011. "Use Your Words: The Role of Language in the Development of Toddlers' Self-Regulation." Early Childhood Research Quarterly 26 (2): 169–181.

van Agtmael, Peter. 2013. "Laugh After Death." Time, November 18, 44–49.

Van Vooren, Nicole, and Roberta Spalter-Roth. 2010. "Tracking Master's Students Through Programs and into Careers." Footnotes (September/October): 10–11.

Vartabedian, Ralph, and Ken Bensinger. 2010. "Toyota Faces $16.4 Million Penalty." Baltimore Sun, April 6, 6.

Veblen, Thorstein. 1899/1953. The Theory of the Leisure Class. New York: New American Library.

Velasco, Schuyler. 2013. "The Hidden Cost of Fast Food." Christian Science Monitor Weekly, November 11, 43.

___. 2014. "Food Additives: More Fuss Than Harm?" Christian Science Monitor, May 26, 43.

Velasquez-Manoff, Moises. 2009. "Pressure Builds over Bottled Water." Christian Science Monitor, October 18, 36–37.

Venkatesh, Sudhir. 2008. Gang Leader for a Day: A Rogue Sociologist Takes to the Streets. New York: Penguin Press.

Vespa, Jonathan, Jamie M. Lewis, and Rose M. Kerner. 2013. "America's Families and Living Arrangements: 2012." U.S. Census Bureau, August. Accessed May 16, 2014 (www.census .gov).

Visser, Susanna N., et al. 2014. "Trends in the Parent-Report of Health Care Provider-Diagnosed and Medicated Attention-Deficit/ Hyperactivity Disorder: United States, 2003–2011." Journal of the American Academy of Child & Adolescent Psychiatry 53 (January): 34–46.

Vitello, Paul. 2006. "The Trouble when Jane Becomes Jack." New York Times (August 20): H1, H6.

Vold, George B. 1958. Theoretical Criminology. New York: Oxford University Press.

Volz, Matt. 2012. "$3.4 Billion Indian Land Royalty Settlement Upheld." Salon, May 22. Accessed March 22, 2014 (www.salon.com).

Von Drehle, David. 2013. "The Robot Economy." Time, September 9, 42–47.

___. 2014. "Breaking the Marriage Bans." Time, March 3, 12.

Voosen, Paul. 2014a. "As People Shun Pollsters, Researchers Put Online Surveys to the Test." Chronicle of Higher Education, September 5, A8, A10.

___. 2014b. "In Backlash Over Facebook Research, a Trove of Data Is at Risk." Chronicle of Higher Education, July 18, A5.

"Wages Maximus." 2014. Christian Science Monitor Weekly, March 10, 33.

"The Waiting Wounded." 2013. The Economist, March 23, 33–34.

Wald, Matthew L. 2012. "Woman Becomes First Openly Gay General." New York Times, August 13, A8.

___. 2014. "U.S. Agency Knew About G.M. Flaw but Did Not Act." New York Times, March 31, A1.

Waldron, Travis. 2011. "Pastor at Kentucky Church That Banned Interracial Couples Calls for Vote to Reverse Decision." AlterNet, December 3. Accessed December 7, 2011 (www.alternet.org).

Walker, Jacob. 2012. "Researching Contested Illnesses: The Case of Chronic Fatigue Syndrome (CFS)." July 4. Accessed July 26, 2014 (www.in-training.org).

Walker, Lenore E. 2000. The Battered Woman Syndrome, 2nd edition. New York: Springer.

Wallace, Bruce. 2006. "Japanese Schools to Teach Patriotism." Los Angeles Times, December 16. Accessed December 18, 2006 (www .latimes.com).

Wallis, Cara. 2011. "Performing Gender: A Content Analysis of Gender Display in Music Videos." Sex Roles 64 (February): 160–172.

Walls, Mark. 2011. "High Court Strikes Down Calif. Law on Violent Video Games." Education

Week, June 27. Accessed June 28, 2011 (www.edweek.org).

Walmsley, Roy. 2013. "World Prison Population List (Tenth Edition)." International Center for Prison Studies. Accessed January 15, 2014 (www.prisonstudies.org).

Walsh, Bryan. 2014. "After the Spill." *Time*, January 27, 14.

Walsh, Janet. 2011. "Failing Its Families: Lack of Paid Leave and Work-Family Supports in the US." Human Rights Watch. Accessed April 7, 2013 (www.hrw.org).

Wan, William. 2011. "Chinese Dog Eaters and Dog Lovers Spar Over Animal Rights." *Washington Post*, May 28, A14.

Wang, Wendy. 2012. "The Rise of Intermarriage." Pew Research Center, Social & Demographic Trends, February 16. Accessed March 22, 2014 (www.pewsocialtrends.org).

Wang, Wendy. 2013. "Parents' Time with Kids More Rewarding Than Paid Work—and More Exhausting." Pew Research Center, October 8. Accessed March 2, 2014 (www.pewresearch.org).

Wang, Wendy, and Kim Parker. 2011. "Women See Value and Benefits of College; Men Lag on Both Fronts, Survey Finds." Pew Research Center, August 17. Accessed April 15, 2012 (www.pewsocialtrends.org).

___. 2014. "Record Share of Americans Have Never Married: As Values, Economic and Gender Patterns Change." Pew Research Center, September 24. Accessed October 26, 2014 (www.pewresearch.org).

Wang, Wendy, Kim Parker, and Paul Taylor. 2013. "Breadwinner Moms." Pew Research Center, May 29. Accessed May 18, 2014 (www.pewresearch.org).

Warren, Kenneth R. 2012. "NIH Statement on International FASD Awareness Day." National Institutes of Health, September 5. Accessed November 20, 2013 (www.nih.gov).

Wartik, Nancy. 2005. "The Perils of Playing House." *Psychology Today* (July/August): 42–52.

Wasley, Paula. 2007. "46 Students Are Disciplined for Cheating at Indiana University's Dental School." *Chronicle of Higher Education*, May 9. Accessed May 10, 2007 (www.chronicle.com).

___. 2008. "The Syllabus Becomes a Repository of Legalese." *Chronicle of Higher Education* 54, March 14, A1, A8–A11.

Watson, David C. 2012. "Gender Differences in Gossip and Friendship." *Sex Roles* 67 (November): 494–502.

Wealth-X and UBS. 2014. "Billionaire Census 2014." Accessed November 10, 2014 (www.billionairecensus.com).

Weathers, Cliff. 2014a. "4 of the Biggest Quacks Plaguing America with False Claims About Science." AlterNet, July 17. Accessed July 26, 2014 (www.alternet.org).

___. 2014b. "Research Behind Dr. Oz-Endorsed Diet Supplement Was Bogus, Authors Admit." AlterNet, October 22. Accessed December 5, 2014 (www.alternet.org).

Webb, Alex. 2013. "Germany's Coalition Looks to Gender Quotas." *Bloomberg Businessweek*, December 16–22, 17–18.

Weber, Lynn, Tina Hancock, and Elizabeth Higginbotham. 1997. "Women, Power, and Mental Health." Pp. 380–396 in *Women's Health: Complexities and Differences*, edited by Sheryl B. Ruzek, Virginia L. Olesen, and Adele E. Clarke. Columbus: Ohio State University Press.

Weber, Max. 1920/1958. *The Protestant Ethic and the Spirit of Capitalism*, translated by Talcott Parsons. New York: Charles Scribner's Sons.

___. 1925/1947. *The Theory of Social and Economic Organization*. New York: Free Press.

___. 1925/1978. *Economy and Society*, edited by Guenther Roth and Claus Wittich. Berkeley: University of California Press.

___. 1946. *From Max Weber: Essays in Sociology*, translated and edited by H. H. Gerth and C. Wright Mills. Berkeley: University of California Press.

Weinhold, Bob. 2012. "A Steep Learning Curve: Decoding Epigenetic Influences on Behavior and Mental Health." *Environmental Health Perspectives* 120 (October): A396–A401.

Weintraub, Arlene. 2010. "Break That Hovering Habit Early." *U.S. News & World Report*, September, 42–43.

Weiss, Carol H. 1998. *Evaluation: Methods for Studying Programs and Policies*, 2nd edition. Upper Saddle River, NJ: Prentice Hall.

Weiss, Rick. 2004. "Nanomedicine's Promise Is Anything but Tiny." *Washington Post*, January 31, A8.

___. 2005. "The Power to Divide." *National Geographic* 208 (July): 3–27.

Weissman, Jordan. 2014. "The Fast-Food Strikes Have Been a Stunning Success for Organized Labor." *Slate*, September 7. Accessed October 8, 2014 (www.slate.com).

Weitz, Rose. 2013. *The Sociology of Health, Illness, and Health Care: A Critical Approach*, 6th edition. Boston: Wadsworth.

Welch, David. 2011. "For the UAW, a Bargaining Dilemma." *Bloomberg Businessweek*, September 19–25.

Wenneras, Christine, and Agnes Wold. 1997. "Nepotism and Sexism in Peer Review." *Nature* 387 (May 22): 341–343.

West, Candace, and Don H. Zimmerman. 2009. "Accounting for Doing Gender." *Gender & Society* 23 (February): 112–122.

West, Loraine A., Samantha Cole, Daniel Goodkind, and Wan He. 2014. "65+ in the United States: 2010." Current Population Reports, June. Accessed July 25, 2014 (www.census.gov).

Westcott, Lucy. 2014. "More Americans Moving to Cities, Reversing the Suburban Exodus." The Wire, March 27. Accessed September 19, 2014 (www.thewire.com).

Whelan, David. 2007. "The Sentencing Game." *Forbes*, February 12, 40.

Whalen, Jeanne, and Betsy McKay. 2013. "Fifteen Years After Autism Panic, a Plague of Measles Erupts." *Wall Street Journal*, July 20–21, A11–A12.

Wheeler, William. 2012. "Troubled Waters." *Christian Science Monitor Weekly*, December 3, 26–31.

White, Alexander, and Carolyn Witkus. 2013. "How Legal Marijuana May Affect Troubled Families." *Christian Science Monitor Weekly*, March 11, 36.

White, James W. 2005. *Advancing Family Theories*. Thousand Oaks, CA: Sage.

White, Nicole, and Janet L. Lauritsen. 2012. "Violent Crime Against Youth: 1994–2010." Bureau of Justice Statistics, December. Accessed May 4, 2013 (www.bjs.gov).

Whiteaker, Chloe, Laurie Meisler, Yvette Romero, and Karen Weise. 2014. "Voter ID States That Make Voting Super Simple—or Really Hard." *Bloomberg Businessweek*, October 27–November 2, 36.

Whitehead, Jaye C. 2011. "The Wrong Reasons for Same-Sex Marriage." *New York Times*, May 16, A21.

Whitelaw, Kevin. 2000. "But What to Call It?" *U.S. News & World Report*, October 16, 42.

Whiteman, Shawn, D., Susan M. McHale, and Anna Soli. 2011. "Theoretical Perspectives on Sibling Relationships." *Journal of Family Theory & Review* 3 (June): 124–139.

Whitesides, John. 2014. "Anti-Union Group Pledges to Remain on the Offensive." *Baltimore Sun*, February 18, 2.

Whitson, Sarah Leah. 2013. "Restricting Speech Restricts Arab Freedom." *Contexts* 12 (Spring): 20–21.

"Who Is 'Black' in America?" 2011. U.S. Census Bureau. Accessed March 22, 2014 (www.census.gov).

Whorf, Benjamin Lee. 1956. *Language, Thought, and Reality*. Cambridge, MA: MIT Press.

Whoriskey, Peter, Dan Keating, and Lena H. Sun. 2014. "Data Uncover Nation's Top Medicare Billers." *Washington Post*, April 9. Accessed July 26, 2014 (www.washingtonpost.com).

"Who Votes, Who Doesn't, and Why: Regular Voters, Intermittent Voters, and Those Who Don't." 2006. Pew Research Center for the People & the Press, October 18. Accessed May 3, 2007 (www.peoplepress.org).

Wicker, Christine. 2009. "How Spiritual Are We?" *Parade*, October 4, 4–5.

Wightman, Patrick, Robert Schoeni, and Keith Robinson. 2012. "Familial Financial Assistance to Young Adults." Paper prepared for the annual meetings of the Population Association of America, May 3. Accessed February 13, 2014 (www.npc.umich.edu).

Wiik, Kenneth A., Renske Keizer, and Trude Lappegård. 2012. "Relationship Quality in Marital and Cohabiting Unions Across Europe." *Journal of Marriage and Family* 74 (June): 389–398.

___. 2012. "Relationship Quality in Marital and Cohabiting Unions Across Europe." *Journal of Marriage and Family* 74 (June): 389–398.

Wilcox, W. Bradford, and Elizabeth Marquardt. 2011. "The State of Our Unions, Marriage in America 2011: When Baby Makes Three." Institute for American Values. Accessed March 2, 2014 (www.stateofourunions.org).

Wildsmith, Elizabeth, Nicole R. Steward-Streng, and Jennifer Manlove. 2011. "Childbearing Outside Marriage: Estimates and Trends in the United States." Child Trends, November. Accessed March 4, 2012 (www.childtrends.org).

Wilke, Joy. 2013. "Majority in U.S. Say Healthcare Not Gov't Responsibility." Gallup, September 18. Accessed July 26, 2014 (www.gallup.com).

Wilke, Joy, and Frank Newport. 2014. "Americans' Satisfaction with Economy Sours Most Since 2001." Gallup, January 16. Accessed April 12, 2014 (www.gallup.com).

Wilkinson, Charles. 2006. *Blood Struggle: The Rise of Modern Indian Nations*. New York: W. W. Norton & Company.

Wilkinson, Lindsey, and Jennifer Pearson. 2013. "High School Religious Context and Reports of Same-Sex Attraction and Sexual Identity in Young Adulthood." *Social Psychology Quarterly* 76 (June): 180–202.

Williams, Claudia, and Barbara Gault. 2014. "Paid Sick Days Access in the United States: Differences by Race/Ethnicity, Occupation, Earnings, and Work Schedule." Institute for Women's Policy Research, March. Accessed April 12, 2014 (www.iwpr.org).

Williams, Frank P., III, and Marilyn D. McShane. 2004. *Criminological Theory*, 4th edition. Upper Saddle River, NJ: Prentice Hall.

Williams, Lauren. 2014. "21 Things You Can't Do While Black." AlterNet, February 12. Accessed March 22, 2014 (www.alternet.org).

Wilson, Meagan M. 2014. "Hate Crime Victimization, 2004–2012—Statistical Tables." Bureau of Justice Statistics, February. Accessed March 2, 2014 (www.bjs.gov).

Williams, Robin M., Jr. 1970. *American Society: A Sociological Interpretation*, 3rd edition. New York: Knopf.

Williams, Timothy. 2011. "Tackling Infant Mortality Rates Among Blacks." *New York Times*, October 14, A10.

Williams, Timothy. 2013. "Blighted Cities Prefer Razing to Rebuilding." *New York Times*, November 13, A15.

Williams, Wendy H., and Stephen J. Ceci. 2012. "When Scientists Choose Motherhood." *American Scientist* 100 (March/April). Accessed February 20, 2012 (www.americanscientist.org).

Williamson, Nancy. 2013. "Motherhood in Childhood: Facing the Challenge of Adolescent Pregnancy." United Nations Population Fund. Accessed September 19, 2014 (www.unfpa.org).

Wilson, Steven G., David A. Plane, Paul J. Mackun, Thomas R. Fischetti, and Justyna Goworowska. 2012. "Patterns of Metropolitan and Micropolitan Population Change: 2000 to 2010." U.S. Census Bureau, September. Accessed September 19, 2014 (www.census.gov).

Wilson, Thomas C. 1993. "Urbanism and Kinship Bonds: A Test of Four Generalizations." *Social Forces* 71 (March): 703–712.

Winerip, Michael. 2013. "Pushed Out of a Job Early." *New York Times*, December 6. Accessed May 16, 2014 (www.nytimes.com).

Wing, Nick, and Carly Schwartz. 2014. "Here's the Painful Truth About What It Means to Be 'Working Poor' in America." *Huffington Post*, May 19. Accessed May 19, 2014 (www.huffingtonpost.com).

Wingfield, Adia Harvey. 2010. "Are Some Emotions Marked 'Whites Only'? Racialized Feeling Rules in Professional Workplaces." *Social Problems* 57 (May): 258–268.

WIN-Gallup International. 2013. "Global Index of Religiosity and Atheism, 2012." Accessed June 21, 2014 (www.gallup-international.com).

Winship, Scott. 2011. "Mobility Impaired." *National Review*, November 7. Accessed February 13, 2014 (www.nationalreview.com).

Winter, Michael. 2011. "13 More Charged in SAT Cheating Scandal in N.Y." *USA Today*, November 23. Accessed November 24, 2011 (www.usatoday.com).

Wirth, Louis. 1938. "Urbanism as a Way of Life." *American Journal of Sociology* 44 (July): 1–24.

Witters, Dan. 2010. "Large Metro Areas Top Small Towns, Rural Areas in Wellbeing." Gallup, May 17. Accessed May 20, 2010 (www.gallup.com).

"Wives Who Earn More Than Their Husbands, 1978–2012." 2014. U.S. Bureau of Labor Statistics, Labor Force Statistics from the Current Population Survey, March 24. Accessed April 9, 2014 (www.bls.gov/cps).

Wolff, Edward N. 2012. "The Asset Price Meltdown and the Wealth of the Middle Class." National Bureau of Economic Research, November. Accessed February 6, 2014 (www.nber.org).

Wolfson, Mark. 2001. *The Fight Against Big Tobacco: The Movement, the State, and the Public's Health*. New York: Aldine de Gruyter.

Wondergem, Taylor R., and Mihaela Friedlmeier. 2012. "Gender and Ethnic Differences in Smiling: A Yearbook Photographs Analysis from Kindergarten Through 12th Grade." *Sex Roles* 67 (October): 403–411.

Wong, Curtis M. 2013. "Hallmark's 'Holiday Sweater' Keepsake Ornament Omits 'Gay' from 'Deck the Halls' Lyrics." *Huffington Post*, October 28. Accessed November 10, 2013 (www.huffingtonpost.com).

Wong, Venessa. 2013. "Living Small in the Big City." *Bloomberg Businessweek*, March 18–24, 46–47.

Wood, Graeme. 2012. "Who Are the Millionaires?" *National Review*, December 19. Accessed February 6, 2014 (www.nationalreview.com).

Wood, Julia T. 2011. *Gendered Lives: Communication, Gender, & Culture*, 9th edition. Belmont, CA: Wadsworth.

Woolf, Steven H., and Laudon Aron. 2013. *U.S. Health in International Perspective: Shorter Lives, Poorer Health*. National Research Council and Institute of Medicine of the National Academies. Accessed July 26, 2014 (www.national-academies.org).

World Bank. 2012. *Gender Equality and Development*. Washington, DC. Accessed September 22, 2011 (www.worldbank.org).

___. 2014. "Country and Lending Groups." Accessed July 26, 2014 (www.worldbank.com).

World Bank and Collins. 2013. *Atlas of Global Development: A Visual Guide to the World's Greatest Challenges*, 4th edition. Washington, DC, and Glasgow: World Bank and Collins.

World Economic Forum. 2013. "Outlook on the Global Agenda." Accessed February 6, 2014 (www.weforum.org).

World Health Organization. 2005. *WHO Multi-Country Study on Women's Health and Domestic Violence Against Women: Summary Report of Initial Results on Prevalence, Health Outcomes and Women's Responses*. Geneva: World Health Organization.

___. 2013. "Global and Regional Estimates of Violence Against Women: Prevalence and Health Effects of Intimate Partner Violence and Non-Partner Sexual Violence." Accessed March 2, 2014 (www.who.int).

___. 2014. "World Health Statistics 2014." Accessed July 26, 2014 (www.who.int).

World Health Organization and UNICEF. 2014. "Progress on Drinking Water and Sanitation." May. Accessed September 19, 2014 (www.unicef.org).

World Meteorological Organization. 2014a. "Atlas of Mortality and Economic Losses from Weather, Climate and Water Extremes (1970–2012)." Accessed September 19, 2014 (wmo.int).

___. 2014b. "The State of Greenhouse Gases in the Atmosphere Based on Global Observations Through 2013." WMO Greenhouse Gas Bulletin, September 9. Accessed September 19, 2014 (www.wmo.int).

World Water Assessment Program. 2009. *The United Nations World Water Development Report 3: Water in a Changing World*. Paris: UNESCO and London: Earthscan.

"Workplace Injury and Illness Summary." 2013. U.S. Bureau of Labor Statistics, November 7. Accessed July 26, 2014 (www.bls.gov).

Worley, Heidi. 2014. "Top 10 Countries Closing Gender Gap." Population Reference Bureau, February. Accessed March 2, 2014 (www.prb.org).

Wozniak, Jesse, and Bob Groves. 2011. "A National Ceremony." *Contexts* 10 (Winter): 12–18.

Wright, Jason D., et al. 2013. "Robotically Assisted vs. Laparoscopic Hysterectomy Among Women with Benign Gynecologic Disease." *JAMA* 309 (February): 689–698.

Wright, Wynne, and Elizabeth Ransom. 2005. "Stratification on the Menu: Using Restaurant Menus to Examine Social Class." *Teaching Sociology* 33 (July): 310–316.

WWAP. 2014. *The United Nations World Water Development Report 2014: Water and Energy*. United Nations World Water Assessment Program. Accessed September 19, 2014 (www.unesco.org).

Wynnyckj, Mychailo. 2014. "Crimea—Putin's Sudetenland." Project Maidan, March 3. Accessed April 12, 2014 (www.projectmaidan.com).

Wysong, Earl, Robert Perrucci, and David Wright. 2014. *The New Class Society: Goodbye American Dream?* 4th edition. Lanham, MD: Rowman & Littlefield.

Xie, Yu, James Raymo, Kimberly Goyette, and Arland Thornton. 2003. "Economic Potential and Entry into Marriage and Cohabitation." *Demography* 40 (May): 351–367.

Yabiku, Scott T., and Constance T. Gager. 2009. "Sexual Frequency and the Stability of Marital and Cohabiting Unions." *Journal of Marriage and Family* 71 (November): 983–1000.

Yaccino, Steven, and Lizette Alvarez. 2014. "New G.O.P. Bid to Limit Voting in Swing States." *New York Times*, March 30, A1.

Yen, Hope. 2013. "U.S. Census Surveys Will No Longer Use the Term 'Negro.'" *Huffington Post*, February 25. Accessed March 22, 2014 (www.huffingtonpost.com).

Yglesias, Matthew. 2013. "The End of Cheap Airfare." Slate, February 19. Accessed April 12, 2014 (www.slate.com).

Yokota, Fumise, and Kimberly M. Thompson. 2000. "Violence in G-rated Animated Films." *JAMA* 283 (May 24/31): 2716–2720.

You, Danzhen, Phillip Jingxian Wu, and Tessa Wardlaw. 2013. "Levels & Trends in Child Mortality: Report 2013." UN Inter-Agency Group for Child Mortality Estimation. Accessed February 6, 2014 (www.childmortality.org).

Young, Jeffrey R. 2011. "Programmed for Love." *Chronicle of Higher Education*, January 21, B6–B11.

Young, Rachel, et al. 2014. "The 2014 International Energy Efficiency Scorecard." American Council for an Energy-Efficient Economy. July. Accessed September 19, 2014 (www.aceee.org).

Yu, Roger. 2007. "Indian-Americans Book Years of Success." *USA Today*, April 18, 1B–2B.

Zahedi, Ashraf. 2008. "Concealing and Revealing Female Hair: Veiling Dynamics in Contemporary Iran." Pp. 250–265 in *The Veil: Women Writers on Its History, Lore, and Politics*, edited by Jennifer Heath. Berkeley: University of California Press.

Zahn, Margaret A., et al. 2010. "Causes and Correlates of Girls' Delinquency." Office of Juvenile Justice and Delinquency Prevention, April. Accessed January 12, 2014 (www.ojjdp.gov).

Zezima, Katie. 2010. "For Many, 'Washroom' Seems to Be Just a Name." *New York Times*, September 13, A14.

Zhang, S. X. 2012. "Trafficking of Migrant Laborers in San Diego County: Looking for a Hidden Population." San Diego State University, San Diego, CA.

Zhang, Yuanyuan, Travis L. Dixon, and Kate Conrad. 2010. "Female Body Image as a Function of Themes in Rap Music Videos: A Content Analysis." *Sex Roles* 62 (June): 787–797.

Zickuhr, Kathryn. 2013. "Who's Not Online and Why." Pew Research Center, September 25. Accessed December 4, 2013 (www .pewinternet.org).

Ziliak, James P. 2013. "Why Are So Many Americans on Food Stamps? The Role of Economy, Policy, and Demographics." University of Kentucky Center for Poverty Research, September. Accessed November 10, 2014 (www.ukcpr.org).

Zimbardo, Philip. G. 1975. "Transforming Experimental Research into Advocacy for Social Change." Pp. 33–66 in *Applying Social Psychology: Implications for Research,* *Practice, and Training,* edited by Morton Deutsch and Harvey A. Hornstein. Hillsdale, NJ: Erlbaum.

Zimbardo, Philip G., Christina Maslach, and Craig Haney. 2000. "Reflections on the Stanford Prison Experiment: Genesis, Transformations, Consequences." Pp. 193–237 in *Obedience to Authority: Current Perspectives on the Milgram Paradigm,* edited by Thomas Blass. Mahwah, NJ: Lawrence Erlbaum Associates.

Zimmerman, Gregory M., and Greg Pogarsky. 2011. "The Consequences of Parental Underestimation and Overestimation of Youth Exposure to Violence." *Journal of Marriage and Family* 73 (February): 194–208.

Zimmerman, Jonathan. 2014. "Are E-Cigarettes Ensnaring a New Generation of Smokers?" *Christian Science Monitor Weekly,* March 31, 35.

Zweifler, Seth. 2013. "Elite Institutions: Far More Diverse Than They Were 20 Years Ago." *Chronicle of Higher Education,* November 1, B16.

Zweigenhaft, Richard L., and G. William Domhoff. 2006. *Diversity in the Power Elite: How It Happened, Why It Matters.* Lanham, MD: Rowman & Littlefield.

NAME INDEX

Mehl, Matthias R., 91
Mehrotra, Kartikay, 58
Meier, Barry, 283
Melton, Glennon, 98
Melucci, Alberto, 322
Mencher, Steve, 49
Mendelberg, Tali, 221
Mendelson, Danial A., 140
Mendes, Elizabeth, 182, 226
Meng, Liu, 177
Menissi, Fatima, 266
Merkel, Angela, 221
Merryman, Ashley, 76
Merton, Robert K., 14, 69, 84, 89, 103, 109, 126–127, 186–187, 208
Mestral, George de, 54–55
Meteyer, Karen, 70
Mettler, Suzanne, 249
Meyer, Marshall W., 109
Michels, Robert, 109
Mider, Zachary, 155
Mikkelson, Barbara, 21
Milan, Lynn, 165
Milgram, Stanley, 105, 115
Miller, Anthony B., 56
Miller, Claire C., 229
Miller, D. W., 35, 251
Miller, Lisa, 267
Miller, Robert K., 137
Millman, Joel, 182
Mills, C. Wright, 5, 12, 103, 220
Mincy, Ronald, 189
Mishel, Lawrence, 147, 205, 209
Mokhiber, Russell, 129
Molin, Anna, 72
Molotch, Harvey, 303
Money, John, 63
Montanaro, Domenico, 215
Montez, Jennifer Karas, 286
Monto, Martin, 169
Moore, Elizabeth S., 75
Moore, Kathleen, 304
Moore, Lisa Jean, 160
Moore, Wilbert, 153–154
Morales, Lymari, 160
Moran, Eileen Geil, 176
Morello, Carol, 29, 148
Morgan, Rachel E., 233
Morgenson, Gretchen, 201
Morin, Richard, 145, 147–148, 194
Morris, David, 283
Morris, Desmond, 92
Morris, Theresa, 200
Morrison, Denton F., 322
Mosher, William D., 232
Moss-Racusin, Corrine A., 251
Motel, Seth, 145, 148, 193
Motivans, Mark, 123
Moyer, Imogene, 13, 130
Mukherjee, Sy, 221
Mullen, Ann, 250
Mumford, Lewis, 298
Murdock, George P., 39
Murphey, David, 274
Murphy, John, 183
Murphy, Sherry L., 326
Murray, Sara, 182

N

Nader, Ralph, 324
Nagel, Caroline, 95
Naik, Gautam, 329
Naili, Jaher, 176
Nakashima, Ellen, 56
Napierala, Jeffrey S., 183
Nash, Elizabeth, 170
Naumann, Rebecca, 2
Naveen, P., 317
Nechepurenko, Ivan, 297
Neelakantan, Shailaja, 140, 175
Nelson, Colleen, 220
Nestle, Marion, 326
Neuhaus, Jessamyn, 58
Newberry, Sydne, 286
Newhouse, Joseph P., 279
Newkirk, Margaret, 193
Newport, Frank, 148, 157, 160, 196, 198, 202, 203, 230, 246, 259–263, 265, 278
Nichols, Austin, 161
Nicosia, Francis, 184
Niebrugge-Brantley, Jill, 16
Niederdeppe, Jeff, 283
Niewyk, Donald, 184
Niquette, Mark, 201
Norris, TIna, 191
Nossiter, Adam, 226
Nurmi, Jari-Erik, 70
Nyseth, Holly, 6

O

Oakes, Jennie, 251
Oates, Russ, 295
Obach, Brian K., 323
Obama, Barack, 184–185, 196, 205, 217, 278, 306–307, 316
Obama, Michelle, 50, 117, 214
Oberschall, Anthony, 322
O'Brien, Jodi, 17
O'Donnell, Victoria, 53
Offutt, Susan, 124
Ogburn, William F., 40
Ogunro, Nola, 247
Oh, Hyeyoung, 283
Ohlemacher, Stephen, 114
O'Keefe, Ed, 221
Olivier, Donald C., 42
Olson, Cheryl, 74
Olson, Jonathan, 305
Opp, Karl-Dieter, 47
Oppel, Richard A. Jr., 122
Orenstein, Peggy, 72
Oreopoulos, Philip, 246
Oropesa, B. Salvatore, 228
Ortman, Jennifer, 237–238, 293
Osborne, Michael A., 204
Overberg, Paul, 299
Ovide, Shira, 112
Owen, Daniela, 76

P

Packard, Vance, 317
Padgett, Tim, 267
Palen, J. John, 299
Paletta, Damian, 148
Palmeri, Christopher, 220
Pan, L., 251
Pantano, Joan, 66
Park, Robert, 302, 313
Parker, Amy W., 45
Parker, Kim, 78, 164, 225, 227–228, 230, 246, 251
Parker, Robert, 303
Parker-Pope, Tara, 22
Parramore, Lynn Stuart, 148
Parrillo, Vincent N., 303
Parsons, Talcott, 14, 173, 208, 223, 245, 281
Pasko, Lisa, 64
Passel, Jeffrey S., 6, 179, 181, 238
Patten, Eileen, 193
Paul, Annie Murphy, 6
Paul, Noel C., 318
Paul, Richard, 64
Paulsen, Ronelle, 322
Payea, Kathleen, 249
Payne, K. K., 78
Pazol, Karen, 169
Pearlstein, Norman, 306
Pearson, Jennifer, 177, 279
Pedersen, Paul, 51
Peeples, Lyn, 310
Pell, Nicholas, 320
Perlmutter, David D., 252
Perry, Gina, 105
Perry-Jenkins, 70
Peter, Tom A., 48, 172
Peterman, Amber, 172
Petersen, Andrea, 76
Peterson, James, 35
Peterson, Scott, 98
Petronijevec, Uros, 246
Petrosino, Anthony, 135
Pettypiece, Shannon, 285
Peweardy, Cornel, 191
Pflammer, Alicia, 58
Pflanz, Mike, 173
Phelps, Glenn, 151, 152
Phelps, Timothy, 192
Phillips, M., 51, 210
Philpott, Tom, 309
Pianta, Robert C., 73, 252
Pickert, Kate, 306
Pierce, Lamar, 30
Pierson, Paul, 143, 155
Pilkington, Ed, 137
Pinsky, Paul C., 155
Piore, Michael J., 194
Pittman, Dan, 115
Pitts, Will, 180
Pogarsky, Greg, 235
Polgreen, Lydia, 226
Polikoff, Morgan S., 253
Pollack, William S., 173, 183, 286
Pollak, Robert A., 240
Pollard, Kelvin, 237
Pollick, Michael, 200
Polsby, Nelson W., 218
Pong, Suet-ling, 70
Popper, Karl, 124
Porter, Eduardo, 99

Porter, Michael, 148, 149, 203, 253
Poushter, Jacob, 173, 227
Powers, Audry, 165
Powers, Charles H., 69
Preston, Darrell, 310
Price, Barbara Raffel, 131
Princiotta, Daniel, 256
Prins, Nomi, 129
Prior, Markus, 29, 255
Proctor, Bernadette D., 141, 145, 147, 188, 189
Protess, Ben, 216
Prüss-Üstin, Annette, 304
Pryor, Richard, 52
Puddington, Arch, 212
Purcell, Kristin, 97
Putin, Vladimir, 212
Putnam, Robert D., 254

Q

Qian, Zenchao, 78
Quinney, Richard, 128

R

Rainie, Lee, 96, 106–107, 327, 328
Rampell, Catherine, 113, 148
Rank, Mark R., 155
Rano, Jason, 284
Raphael, Steven, 300
Ray, Rebecca, 202, 206
Reagan, Ronald, 214, 263
Reardon, Sean F., 248, 301
Redberg, Rita, 279
Redd, Zakia, 232
Reddy, Samathi, 29, 93
Reece, Michael, 168
Reed, Matthew, 256
Reeves, Jay, 40
Reeves, Richard V., 148
Reger, Jo, 16
Regnerus, Mark, 167
Reich, Robert, 155
Reichert, Tom, 176
Reilly, Kate, 76
Reimer, David, 63
Reinhold, Steffen, 229
Reitzes, Donald C., 18
Remler, Dahlia, 147
Rettew, David, 287
Reynolds, Kelly A., 305
Rheault, Magali, 246
Ricciardelli, Rosemary, 75
Richburg, Keith B., 136
Riche, Martha Farnsworth, 299
Richie, Christina, 267
Rideout, Victoria, 73–75, 97
Riesman, David, 218
Riffkin, Rebecca, 40, 252, 261
Rios, Victor, 252
Riosmena, Fernando, 275
Ritzer, George, 5, 69, 111
Roberts, Andrea J., 97
Roberts, Keith A., 258, 273
Robertson, Ruth, 328
Robinson, Laurie O., 135

Eye contact, nonverbal communication and, 93

F

Facebook, 22, 97–98
  ethical research and, 36–37
  gender identity on, 161–162
Face-to-face interviews, 28–29
Facial expression, as nonverbal communication, 93
Facts
  social facts, 9
  sociology and, 4
Fads, 317–318
False consciousness, 265–266
Families
  alcohol and drug use impact on, 125
  authority and power in, 225–226
  child abuse in, 234–235
  cohabitation and, 228–229
  conflict theory and, 174–175, 240–241
  courtship and mate selection, 226
  defined, 222
  divorce trends and, 227
  elder abuse and neglect, 235–236
  extended, 224–225
  female-headed households, 232–233
  feminist theories and, 241
  functionalist perspective on, 239–240
  functions of, 223–224
  gender and, 163–164
  global comparisons of, 223–225
  marriage trends and, 227
  nonmarital childbearing, 230–231
  nonverbal communication in, 93
  nuclear, 224–225
  online interaction and, 97
  personal space and, 95
  poverty and, 147
  as primary group, 101
  relational status in, 84
  residence patterns, 225
  same-sex marriage and, 170–171
  single-parent, 120, 147, 150, 225, 231–232
  as social institution, 116
  as socialization agent, 69–70
  social mobility and, 150–151, 223
  symbolic interactionist perspective on, 242
  touch in, 94
  violence and conflict in, 233–236
Farm-to-table movement, 325

Fashion
  as collective behavior, 317
  health and, 284
Fast Food Forward, 323–324
Fast-food industry
  advertising by, 52
  economy and jobs in, 210
  global presence of, 53
  labor unrest in, 314, 323–324
Female genital mutilation/cutting (FGM/C), 172–173
Female-headed households, increase in, 232–233
Female infanticide, 294
Feminist theories
  basic principles, 15–17, 19
  culture and, 58
  defined, 15–16
  deviance and, 130–132
  economy and, 210–211
  education and, 250–251
  gender and sexuality and, 173, 175–176
  health and health care and, 283–286
  politics and power and, 220–221
  race and ethnicity and, 194–195
  religion and, 266–267
  social groups and, 114–115
  social interaction and, 91
  social stratification and, 156
  urbanization and, 303
Feminization of poverty, 147–148, 156
Fertility, 292
Fetal alcohol spectrum disorders, 274
Fetal origins theory, 64
Fictive kin, 232
Field research, 29–30
Filial piety, 49
Financial aid for education, 248–249
Finland
  crime rates in, 129
  internal migration in, 294
  social mobility in, 149
  teacher training in, 254
Firearm laws, 135
Flags, as symbols, 18, 41
Flexible norms, 46
Folk religions, 259
Folkways, 46–47
Food, cost of, 146
Food poisoning, 273
Foot patrol policing, 135
Forbes magazine, 145
"Forced secularization," 261
Foreign-born population
  origins of, 181
  in United States, 181–183
Formal behavior, 85–86
Formal deviance, 120
Formal organizations, 107–112
  bureaucracies, 108–112
  characteristics of, 107–108

coercive organizations, 107–108
  normal organizations, 107
  utilitarian organizations, 107
Formal social controls, 134
Fortune magazine, 115
Fossil fuels, 306, 309
Foster care, 77
Fragmentation of social movements, 324
France
  health care in, 279
  population policies in, 297
Fraud
  in bureaucracies, 114
  in health care, 282–283
  in research, 36–37
  as white-collar crime, 128–129
  by women, 130
Freedom
  as cultural value, 44
  global measurement of, 211–212
Friends
  online interaction with, 98
  as primary group, 101
Front stage, social interaction performance, 90
Functionalism, theory of
  basic principles, 13–14, 19
  on culture, 56–57
  defined, 13–14
  on deviance, 125–128
  on education, 245–247
  families and aging and, 240
  gender and sexuality and, 173–174
  health and medicine and, 280–282
  politics and power and, 218–220
  on poverty, 147–148
  racial-ethnic factors of inequality and, 193–194
  religion and, 263–265
  on social groups and organizations, 112–113
  on social stratification, 153–154
  urbanization and, 301–302
  work and the economy and, 208–209
Fundamentalism, 262
Fundamentalist Church of Jesus Christ of Latter-Day Saints, 226

G

Gambling
  American Indian casinos and, 191–192
  legalization of, 121
Game stage
  self-development, 67–68
Gangs
  as groups, 103
  research on, 29–30
Gay bashing, 164

Gays, 160–161
  acceptance of, 163–164
  language referring to, 42
Gender. *See also* Feminist theories; Glass ceiling; Homosexuality; Sex; Sexuality
  abuse and neglect and, 235–236
  capitalism and inequality in, 174–175
  child abuse and, 234–235
  conflict theory and, 174–175
  content analysis research and, 30–32
  crime and, 122–123, 127–128, 130–132
  culture and, 58, 131–132, 161–162, 171–173
  defined, 159
  development and health differences, 63
  deviance and, 130–132
  discrimination, 164–167
  dropout rate and, 254
  education and, 164–165, 250–251
  environment and, 304
  expression, 162
  facial expressions and, 93
  family and, 163–165
  female-headed households, 232–233
  feminist theory and, 16–17, 175–176
  functionalism and, 173–174
  global stratification and, 151–152
  health and illness and, 274
  health care and, 283–284
  higher education and, 164–165
  identity, 63, 161–162
  incarceration rates and, 136–137
  inequality, 164–167, 171–172
  intimate partner violence and, 233–234
  language and, 42
  life expectancy and, 236–237
  media images of, 52–53
  nonmarital childbearing and, 230–231
  norms of, 46
  online interaction, 96
  pay gap, 115, 152, 165–166, 210–211
  personal space and, 94–95
  play and, 71–72
  politics and, 165–166, 220–221
  poverty and, 147–148, 156
  power and, 220–221
  racism and, 194–195
  religion and, 260, 266–267
  roles, 86–87, 162, 173–174, 241
  same-sex marriage and, 170–171
  sex and, 158–159
  sex ratios and, 294–295

Human immunodeficiency virus (HIV), 278
Humanitarianism, as cultural value, 44
Hunger, global poverty and, 152
Hurricane Katrina, 315
  internal migration and, 294
  selective perception of, 195
Hurricane Rita, 294
Hydraulic fracturing (fracking), 305
Hypersexuality, of women and girls, 168
Hypotheses
  formulation of, 27
  in scientific method, 22–23

I

Ideal culture, 48–49
Ideal types, social groups, 102
Identity
  gender identity, 161–162
  language and, 43
  religion and, 263
  socialization and, 61
Illicit drugs, 277–278
Illness
  contested illnesses, 287–288
  social construction of, 286–288
Immigration. *See also* Migration
  contact hypothesis and, 195–196
  gendered racism and, 194
  interracial dating and marriage and, 197
  labeling and, 195–196
  labor exploitation and, 128
  language and, 43
  life expectancy and, 237
  multiculturalism and, 51
  negative assimilation and, 231
  selective perception of, 195–196
  social mobility and, 149–150
  subcultures and, 50
  unauthorized immigrants, 182–183
  in United States, 181–183
Impersonality of bureaucracies, 108
Impotence drugs, 176
Impression management
  development and, 68
Incest taboo, 223–224, 226
Income
  of Native Americans, 191–192
  capitalism and, 154–155
  cohabitation trends and, 229
  deviance and, 133–134
  distribution of, 141, 151
  dropout rate and, 254
  education and, 150, 246–247, 252

elder abuse and neglect and, 236
female-headed households and, 232–233
gender pay gap, 156, 165–166, 206–208
global stratification and, 151
health disparities and, 271–272, 275
Internet use and, 96–97
life expectancy and, 237
marriage and divorce and, 227–228
mean household income, 141
of Middle Eastern Americans, 192–193
multigenerational households and, 238
nonmarital childbearing and, 230–231
normalization of deviance and, 121
politics and power and, 220
poverty and, 145–146
by race and ethnicity, 188–189
religion and, 261
residential segregation and, 301
segregation by, 248
social mobility and, 148–151
social stratification and, 140–141
test scores and, 254
voting rates and, 217
Inconsistent status, 85–86, 142
Independent variables, 23
Independent voters, 215
India
  caste system in, 139–140
  culture and rituals in, 57–58
  female infanticide in, 294
  gang rape in, 58
  McDonald's in, 53
  offshoring to, 203–204
  panic and mass hysteria in, 317
  polyandry in, 226
  population projections in, 290–292
  social movements in, 3223
  tent cities in, 152
  urbanization in, 298–299
  violence against women in, 172
Indirect learning, 65
Individual discrimination, 186
Individualism, as cultural value, 45
Inducement, social control and, 134
Inductive reasoning, in scientific method, 23–24
Industrialization, demographic transition theory and, 297
Industrial Revolution, 10
  urbanization and, 298
Inequality
  conflict theory and, 15, 154–155, 194

feminist theory and, 175–176
functionalism, 193–194
gender, 164–167, 171–172, 174–175
global stratification and, 151–153
racial-ethnic factors in, 193–196
religion and, 265–266
sexism and, 162–163
sexual inequality, 172–173
social mobility and, 149–150
universality of, 152–153
wealth and income inequality, 141
Infancy
  electronic use in, 37
  socialization in, 75–76
Infant mortality rate, 151–152, 273–275
  population health and, 293
Inflated language, 89
Informal behavior, 85–86
Informal deviance, 120
Informal group networks, 110
Informal social controls, 134
Informed consent, in social research, 36–37
Informed decisions, sociological research and, 6
In-group, 102–103
Inherited wealth, 143, 201
Initiation, 65
Innovation
  adaptation and, 127
  culture and, 54
Institutional discrimination, 186
  racial-ethnic inequality and, 193–194
Institutionalization
  emotional effects, 62
  involuntary resocialization and, 80
  of social movements, 324
Institutionalized means, social strain theory and, 126–128
Institutional Review Boards (IRBs), 36–37
Instrumental needs, 102
Instrumental norms, 46
Instrumental roles in sexuality, 173–174
Instrumental tasks, 104
Insurance, bureaucracies in, 109
Intellectual property theft, 123
Interest groups, 324
Intergenerational mobility, 148–149
Interlocking directories, 202
Internalization, socialization and, 61
Internal migration, 294
Internal Revenue Service (IRS), 43
International Brotherhood of Electrical Workers, 216

International migration, 293–294
International Red Cross and Red Crescent Movement, 41
International Society of Krishna Consciousness, 258
Internet
  income and use of, 96
  organ black market on, 58
  relationships on, 97–98
  socialization and, 73–74
  social networks and, 106–107
Internet Crime Complaint Center, 124
Interracial marriage, 224
Intersexuality
  defined, 160
  sex reassignment and, 63
Intervention
  in crime and deviance, 135
  social, 35
Interviews, research using, 28–29
Intimate partner violence (IPV), 233–234
Intolerance, religion and, 265
Intragenerational mobility, 148–149
Invention, culture and, 54
Involuntary resocialization, 80
Iran, executions in, 137
Iranian hostage crisis, 106
Iraq, executions in, 137
Iraq War, 106
Irish Catholics, 187
Iron law of oligarchy, 109–110
Islam
  characteristics of, 259, 266
  symbols in, 41
  violence in, 265
  women in, 58, 266–267
Isolation, emotional effects of, 62
Italy, birth rate in, 295

J

Japan
  crime rates in, 129
  cultural change in, 45, 50
  eye contact in, 93
  health care in, 279
  monarchy in, 213
  population policies in, 297
  rape in, 58
  verbal gestures in, 93
  work conditions in, 206
Jazz, cultural influence of, 52–54
Job satisfaction, 205–206
John F. Kennedy Center for the Performing Arts, 52
Journal of the American Medical Association (JAMA), 36
Judaism, 259–266
  women and, 266–267

Norway
   internal migration in, 294
   monarchy in, 213
   social mobility in, 149
   wealth disparity in, 151
Not in my backyard
   (NIMBY), 323
Nouveau riche, 143
Nuclear family, 224–225
Nurse practitioners, 285
Nurture, 64

O

Obesity
   genetics and, 65
   health and illness and,
     276–277
   life expectancy and, 237
   popular culture and, 52
   research integration
     concerning, 116–117
   stigmatization of, 121,
     287–288
Objective approach to social
   class, 142
Observation, 65
Observational studies, 29–30
Occupations and work. See also
   Employment; Labor
   bureaucracies in, 108–109
   bureaucratic segregation of,
     108
   changes in, 204–208
   conflict theory and, 209–210
   as cultural value, 44
   deindustrialization and
     globalization and, 203
   educational levels and, 142
   education and, 246–247
   feminist theories and,
     210–211
   functionalist perspective
     on, 112–113
   functionalist view of,
     208–209
   gender and, 152, 156,
     164–165
   in health care, 281–282
   health care occupations,
     284–285
   health coverage and, 279
   informal group networks,
     110
   job satisfaction and stress,
     205–206
   minorities and, 206–207
   modern work teams, 110
   off-shoring and labor unions,
     203–204
   part-time work, 205
   performance reviews, 112
   prestige and, 141–142
   reference groups in, 103
   relational status in, 84
   sex segregation in, 165
   social class and, 143–145
   social interaction at, 91
   social mobility and, 150–151

sociological research and,
   208–211
STEM (science, technology,
   engineering, and math)
   educational fields, 165
subcultures and, 50
symbolic interactionism
   and, 211
underemployed workers,
   205
unemployment and
   discourages workers, 205
in United States, 202–208
women and, 206–207
work roles, 78
Occupy Wall Street (OWS),
   11, 324
Offshoring, 203–204
Old-age dependency ratio, 238
Oligarchy, iron law of,
   109–110
Oligopoly, 199–200
Online dating sites, 98
Online interaction, 96–99
   activities, 97
   benefits and drawbacks,
     97–98
   connections, compulsions,
     and complications, 98
   demographics of, 96
   family ties and, 97
   harassment and public
     shaming, 98–99
   privacy and, 99
   relationships and, 97–98
   social networking, 97
Open-ended questionnaires, 28
Open stratification system,
   139–140
Ordain Women group, 267
Organizations. See also Social
   groups
   conflict theory perspective on,
     113–114
   functionalist perspective on,
     112–113
Organized crime, 124–125
Organ markets, 58
Otherness
   generalized other, 68
   significant others, 67–68
Out-group, 102–103
Overgeneralization
   research and avoidance
     of, 21

P

Pain prescriptions, substance
   abuse and, 277–278
Pakistan, sexual violence in, 172
Panic, 316–317
Parenting
   education and, 252
   functionalist perspective and,
     240
   helicopter parents, 77
   socialization and, 76–77
   styles of, 70

test scores and achievement
   gaps and, 253
Participant observation, 29–30
Part-time work, 205
   social class and, 144–145
Patient Protection and
   Affordable Care Act, 278
Patriarchy
   economy and, 211
   in family system, 225–226
   gender discrimination and,
     156, 241
   health care and, 285
   power and politics and,
     220–221
   religion and, 266–267
   violence against women and,
     131
Patrilocal residence pattern, 225
Pearl Harbor attack, 106
Peer groups
   differential association theory
     and, 132–133
   socialization and, 71–72
Perception
   interpretation of, 66
   of self, 66
Performance
   learning and, 65
   reviews, 112
   in social interaction, 90
Peripheral theory of
   urbanization, 302
Permissive parenting, 70
Personal space
   nonverbal communication
     and, 94–95
   power and, 95
Peru, social mobility in, 149
Pez candy, 317–318
Physical contact, as sexual
   harassment, 166
Physical disability, status and, 85
Physicians
   assisted suicide and, 239
Physicians
   assisted suicide and, 239
   as gatekeepers, 281–282
   gender and 284–825
Plain Writing Act, 43
Play
   gender and, 71–72
   self-development and stages
     of, 67–68
   socialization and, 71
Pledge of Allegiance, 263
Pluralism, 184
   multiculturalism, 51
   in politics, 218–219
Police, control of crime and
   deviance and, 135
Political action committees
   (PACs), 215–216
Political parties, 214–215
Politics
   conflict theory and, 220
   economy and, 198
   family background and, 150
   feminist theories and,
     220–221

functionalist perspective on,
   218–220
gender and, 156, 165–166,
   172
global political systems,
   211–213
lobbyists and, 216
online interaction and, 98
power and, 213–218
public opinion concerning,
   218
situational and structural
   factors in, 218
social institutions in,
   116–117
special-interest groups in,
   215–216
subcultures and, 50
totalitarianism and dictator-
   ships, 212
in United States, 214–218
voting patterns and, 216–217
Pollution, water supply and, 305
Polyandry, 226
Polygamy, 226
Polygyny, 226
Popular culture, 52–54
   globalization of, 53–54
   obesity and, 52
   socialization and, 73–74
Population
   aging of, 237–238
   changes in, 290–292
   composition and structure,
     294–295
   defined, 291
   fertility and, 292
   growth of, 295–297
   migration and, 293–294
   mortality and, 292–293
   sampling, 24
   sex ratios and, 294–295
Population pyramid, 295
Pornography, 119, 176
Positive checks on population,
   296
Positive sanctions, 48, 134
Postindustrial society, demographic transition
   theory and, 297
Poverty, 145–148
   achievement gap and, 248
   age and, 146–147
   causes of, 147–148
   child abuse and, 240–241
   costs of, 146
   crime and, 123, 128
   demographics, 146–147
   disability and, 272–273
   elderly and, 79, 240–241
   female-headed households
     and, 232–233
   functionalist approach to,
     147–148
   gender and family structure,
     147, 156
   global stratification and,
     152–153
   government assistance and,
     155

CHAPTER 1 SUMMARY

1-1 What Is Sociology?

Sociology is the systematic study of human behavior in society, behavior that takes place among individuals, small groups (such as families), large organizations (such as Apple), and entire societies (such as between the United States and other countries). Sociology goes well beyond common sense and conventional wisdom because it examines claims and beliefs critically, considers many points of view, and enables us to move beyond established ways of thinking.

1-2 What Is a Sociological Imagination?

The *sociological imagination* emphasizes the connection between personal troubles (biography) and structural (public and historical) issues in understanding the social world. *Microsociology* examines the relationships between individuals, whereas *macrosociology* focuses on large-scale patterns and processes that characterize society as a whole. Macro-level systems shape society, often limiting our personal options on the micro level.

1-3 Why Study Sociology?

Regardless of your major, this course will help you (1) make more informed decisions, (2) understand diversity, (3) increase your input in shaping social policies and practices, (4) think critically, and (5) expand your career opportunities.

1-4 Some Origins of Sociological Theory

Sociologists use *theories* to explain why a phenomenon occurs among people, institutions, and societies. Theories produce knowledge, but can also offer solutions to everyday social problems. Some of the most influential theorists have included Auguste Comte, Harriet Martineau, Émile Durkheim, Karl Marx, Max Weber, Jane Addams, and W. E. B. Du Bois. All of these early thinkers shaped contemporary sociological theories.

1-5 Contemporary Sociological Theories

Sociologists typically use more than one theory to explain human behavior. The fullest understanding of society comes from using all four of these perspectives:

• *Functionalism maintains that society is a complex system of interdependent parts that work together to ensure a society's survival.* Critics contend that functionalism is so focused on order and stability that it often ignores social change, social inequality, and social conflict.

• *Conflict theory sees disagreement and the resulting changes in society as natural, inevitable, and even desirable.* Critics argue that conflict theory ignores the importance of harmony and cooperation.

• *Feminist theories*, which build on conflict theory, *emphasize women's social, economic, and political inequality.* Critics claim that these theories are too narrowly focused.

• *Symbolic interaction focuses on individuals' everyday behavior through the communication of knowledge, ideas, beliefs, and attitudes.* Critics maintain that this theoretical perspective overlooks the impact of macro-level factors on our everyday behavior.

Example: Critical Thinking versus Common Sense and Conventional Wisdom When thinking critically, it's important to differentiate between common sense and conventional wisdom myths and data-based facts. Here are a few examples:

Myth: Older people make up the largest group of poor people.

Fact: Children younger than 6, not older people, make up the largest group of poor people.

Myth: Divorce rates are higher today than ever before.

Fact: Divorce rates are lower today than they were between 1980 and 1990.

Now, based on the material you read in this chapter, construct your own "myth" and "fact."

KEY TERMS

sociology the systematic study of human behavior in society.

sociological imagination the ability to see the relationship between individual experiences and larger social influences.

microsociology examines the patterns of individuals' social interaction in specific settings.

macrosociology examines the large-scale patterns and processes that characterize society as a whole.

theory a set of statements that explains why a phenomenon occurs.

empirical information that is based on observations, experiments, or other data collection rather than on ideology, religion, intuition, or conventional wisdom.

social facts aspects of social life, external to the individual, that can be measured.

social solidarity social cohesiveness and harmony.

division of labor an interdependence of different tasks and occupations, characteristic of industrialized societies, that produces social unity and facilitates change.

capitalism an economic system in which the ownership of the means of production is in private hands.

alienation the feeling of separation from one's group or society.

value free separating one's personal values, opinions, ideology, and beliefs from scientific research.

functionalism (structural functionalism) maintains that society is a complex system of interdependent parts that work together to ensure a society's survival.

CHAPTER REVIEW 1

dysfunctions social patterns that have a negative impact on a group or society.

manifest functions purposes and activities that are intended and recognized; they're present and clearly evident.

latent functions purposes and activities that are unintended and unrecognized; they're present but not immediately obvious.

conflict theory examines how and why groups disagree, struggle over power, and compete for scarce resources.

feminist theories examine women's social, economic, and political inequality.

symbolic interaction theory (*interactionism*) examines individuals' everyday behavior through the communication of knowledge, ideas, beliefs, and attitudes.

social interaction a process in which people take each other into account in their own behavior.

TEST YOUR LEARNING

1. _____ looks at the relationship between individual characteristics; _____ examines the relationships between institutional characteristics.
   a. Microsociology; macrosociology
   b. Macrosociology; microsociology
   c. Metasociology; macrosociology
   d. Metasociology; microsociology

2. Which social class, as identified by Karl Marx, includes the ruling elite who own the means of production?
   a. Capitalists
   b. Communists
   c. Power elite
   d. Proletariat

3. James sees Julie laughing in the hallway with Sam and assumes that Julie likes Sam. James is using Weber's
   a. explanatory understanding.
   b. surveillance understanding.
   c. common understanding.
   d. direct observational understanding.

4. Jeremy views society as a system of interrelated parts, but Thomas sees society as composed of groups competing for scarce resources. Jeremy would be considered a _____ theorist, and Thomas would be seen as a _____ theorist.
   a. symbolic interactionist; functionalist
   b. conflict; functionalist
   c. functionalist; symbolic interactionist
   d. functionalist; conflict

5. Many people buy designer clothes that they can't afford. The clothes are an example of a status symbol that reflects a
   a. latent function.
   b. manifest function.
   c. dysfunction.
   d. social system.

6. *True or False* In the definition of sociology, "systematic" means behavior that is built into the larger social structure of society.

7. *True or False* Émile Durkheim saw sociology as the scientific study of two aspects of society: social statics and social dynamics.

8. *True or False* Jane Addams was an early sociologist who published extensively on topics such as social disorganization, immigration, and urban neighborhoods.

9. *True or False* Much contemporary functionalism grew out of the work of Auguste Comte and Émile Durkheim.

10. *True or False* Feminist theorists see society as cooperative and harmonious.

11. What does C. Wright Mills mean when he says that there's a connection between personal troubles and structural issues? Use an example, but not unemployment, to explain and illustrate the sociological imagination concept.

12. Which theory do you think better explains how the United States operates—functionalist, conflict, feminist, or symbolic interactionist? Why?

1. a 2. a 3. d 4. d 5. a 6. False 7. False 8. False 9. True 10. False

TABLE 1.2 LEADING CONTEMPORARY PERSPECTIVES IN SOCIOLOGY

THEORETICAL PERSPECTIVE	FUNCTIONALISM	CONFLICT	FEMINIST	SYMBOLIC INTERACTION
Level of Analysis	Macro	Macro	Macro and Micro	Micro
Key Points	• Society is composed of interrelated, mutually dependent parts. • Structures and functions maintain a society's or group's stability, cohesion, and continuity. • Dysfunctional activities that threaten a society's or group's survival are controlled or eliminated.	• Life is a continuous struggle between the haves and the have-nots. • People compete for limited resources that are controlled by a small number of powerful groups. • Society is based on inequality in terms of ethnicity, race, social class, and gender.	• Women experience widespread inequality in society because, as a group, they have little power. • Gender ethnicity, race, age, sexual orientation, and social class—rather than a person's intelligence and ability—explain many of our social interactions and lack of access to resources. • Social change is possible only if we change our institutional structures and our day-to-day interactions.	• People act on the basis of the meaning they attribute to others. Meaning grows out of the social interaction that we have with others. • People continuously reinterpret and reevaluate their knowledge and information in their everyday encounters.
Key Questions	• What holds society together? How does it work? • What is the structure of society? • What functions does society perform? • How do structures and functions contribute to social stability?	• How are resources distributed in a society? • Who benefits when resources are limited? Who loses? • How do those in power protect their privileges? • When does conflict lead to social change?	• Do men and women experience social situations in the same way? • How does our everyday behavior reflect our gender social class, age, race, ethnicity, sexual orientation, and other factors? • How do macro structures (such as the economy and the political system) shape our opportunities? • How can we change current structures through social activism?	• How does social interaction influence our behavior? • How do social interactions change across situations and between people? • Why does our behavior change because of our beliefs, attitudes, values, and roles? • How is "right" and "wrong" behavior defined, interpreted, reinforced, or discouraged?
Example	• A college education increases one's job opportunities and income.	• Most low-income families can't afford to pay for a college education.	• Gender affects decisions about a major and which college to attend.	• College students succeed or fail based on their degree of academic engagement.

CHAPTER 2 SUMMARY

2-1 How Do We Know What We Know?

Much of our knowledge is based on tradition and authority. In contrast, sociologists rely on *research methods* to get information about a particular topic.

2-2 Why Is Sociological Research Important in Our Everyday Lives?

In contrast to tradition and authority, sociological research challenges overgeneralizations, exposes myths, helps explain why people behave as they do, influences social policies, and sharpens critical thinking skills about issues that affect our everyday lives.

2-3 The Scientific Method

The *scientific method* incorporates careful data collection, exact measurement, accurate recording and analysis of findings, thoughtful interpretation of results, and, when appropriate, a generalization of the findings to a larger group. A research question or a *hypothesis* examines the association between an *independent variable* and a *dependent variable*. Sociologists use both *qualitative* and *quantitative* approaches, and are always concerned about the *reliability* and *validity* of their measures.

2-4 The Research Process: The Basics

Research involves choosing a topic, summarizing the pertinent research, formulating a hypothesis or asking a research question, describing the data collection methods, collecting the data, presenting the findings, and analyzing and explaining the results.

2-5 Some Major Data Collection Methods

Five data collection methods are especially common in sociology (see Table 2.3 on the next page). In designing their studies, sociologists weigh the advantages and limitations of each data collection method. *Evaluation research*, which uses one or more data collection methods, examines whether a social intervention has produced the intended result. Because sociologists don't conduct research in a cultural vacuum, many groups use the findings to change current policies and practices.

Example: The Uneasy Relationship between Research and Practice

Many supporters of the DARE (Drug Abuse Resistance Education) program were unhappy when more than 30 research studies showed that DARE had negligible long-term effects in reducing teen drug use. About 75 percent of U.S. school districts revised the DARE curriculum, but continued to use it because they believed that the program built a positive relationship between police, students, parents, and educators (see text).

2-6 Ethics and Social Research

Sociological research demands a strict code of ethics to avoid mistreating participants. For example, participants must give informed consent and must not be harmed, humiliated, abused, or coerced; researchers must honor their guarantees of privacy, confidentiality, and/or anonymity. Still, sociologists often encounter pressure from policy makers and others to limit their research to topics that won't stir controversy on sensitive issues.

KEY TERMS

research methods organized and systematic procedures to gain knowledge about a particular topic.

scientific method a body of objective and systematic techniques for investigating phenomena, acquiring knowledge, and testing hypotheses and theories.

concept an abstract idea, mental image, or general notion that represents some aspect of the world.

variable a characteristic that can change in value or magnitude under different conditions.

independent variable a characteristic that has an effect on the dependent variable.

dependent variable the outcome that may be affected by the independent variable.

control variable a characteristic that is constant and unchanged during the research process.

hypothesis a statement of the expected relationship between two or more variables.

deductive reasoning an inquiry process that begins with a theory, prediction, or general principle that is then tested through data collection.

inductive reasoning an inquiry process that begins with a specific observation, followed by data collection, a conclusion about patterns or regularities, and the formulation of hypotheses that can lead to theory construction.

reliability the consistency with which the same measure produces similar results time after time.

validity the degree to which a measure is accurate and really measures what it claims to measure.

population any well-defined group of people (or things) about whom researchers want to know something.

sample a group of people (or things) that is representative of the population researchers wish to study.

probability sample each person (or thing) has an equal chance of being selected because the selection is random.

nonprobability sample there is little or no attempt to get a representative cross section of the population.

qualitative research examines nonnumerical material and interprets it.

CHAPTER REVIEW 2

quantitative research focuses on a numerical analysis of people's responses or specific characteristics.

causation a relationship in which one variable is the direct consequence of another.

correlation the relationship between two or more variables.

survey a method for collecting that data that includes questionnaires, face-to-face or telephone interviews, or a combination.

questionnaire a series of written questions that ask for information.

interview a researcher directly asks respondents a series of questions.

field research data collected by observing people in their natural surroundings.

content analysis a data collection method that systematically examines some form of communication.

experiment a controlled artificial situation that allows researchers to manipulate variables and measure the effects.

experimental group the participants who are exposed to the independent variable.

control group the participants who aren't exposed to the independent variable.

secondary analysis examination of data that have been collected by someone else.

evaluation research the process of determining whether a social intervention has produced the intended result.

TEST YOUR LEARNING

1. Sociological research is important in our daily lives for a number of reasons. Which of the following is NOT one of those reasons?
   a. It exposes myths.
   b. It challenges the findings of psychology researchers.
   c. It affects social policy.
   d. It sharpens our critical thinking skills.

2. What is a hypothesis?
   a. A theory based on one's opinions.
   b. A measure of how accurate a study is.
   c. A statement of a relationship between two or more variables.
   d. An analysis of multiple studies that researchers use to make generalizations.

3. Alexandra notices that she performs best on tests that are given in the afternoon. She then begins to collect data of her test grades and asks close friends and classmates to do the same. Alexandra is using
   a. intuition.
   b. inferences.
   c. deductive reasoning.
   d. inductive reasoning.

4. Which of the following examines and interprets nonnumerical data?
   a. Qualitative research
   b. Surveys
   c. Secondary analysis
   d. Quantitative research

5. There are three golden rules in sociological research. Which of the following is NOT one of them?
   a. Do no harm.
   b. Be fully transparent with the subject.
   c. Get the subject's informed consent.
   d. Protect the subject's confidentiality.

6. **True or False** Social researchers continuously question the quality of existing studies.

7. **True or False** If researchers use a nonprobability sample, they can generalize the results to a larger population.

8. **True or False** Validity is the consistency with which the same measure produces similar results time after time.

9. **True or False** Observing second graders interact with one another in their classroom is an example of field research.

10. **True or False** Correlation is pretty much the same as causation.

11. State a hypothesis about your GPA, the dependent variable. What three independent variables might you use? Explain why.

12. Suppose you want to examine the effects of student employment on class attendance. Which data collection method(s) would you use? Why?

1. b 2. c 3. d 4. a 5. b 6. True 7. False 8. False 9. True 10. False

TABLE 2.3 SOME DATA COLLECTION METHODS IN SOCIOLOGICAL RESEARCH

Method	Example	Advantages	Disadvantages
Surveys	Sending questionnaires and/or interviewing students on why they succeeded in college or dropped out	Questionnaires are fairly inexpensive and simple to administer; interviews have high response rates; findings are often generalizable	Mailed questionnaires may have low response rates; respondents tend to be self-selected; interviews are usually expensive
Secondary analysis	Using data from the National Center for Education Statistics (or similar organizations) to examine why students drop out of college	Usually accessible, convenient, and inexpensive; often longitudinal and historical	Information may be incomplete; some documents may be inaccessible; some data can't be collected over time
Field research	Observing classroom participation and other activities of first-year college students with high and low grade-point averages (GPAs)	Flexible; offers deeper understanding of social behavior; usually inexpensive	Difficult to quantify and to maintain observer/subject boundaries; the observer may be biased or judgmental; findings are not generalizable
Content analysis	Comparing the transcripts of college graduates and dropouts on variables such as gender, race/ethnicity, and social class	Usually inexpensive; can recode errors easily; unobtrusive; permits comparisons over time	Can be labor intensive; coding is often subjective (and may be distorted); may reflect social class bias
Experiments	Providing tutors to some students with low GPAs to find out if such resources increase college graduation rates	Usually inexpensive; plentiful supply of subjects; can be replicated	Subjects aren't representative of a larger population; the laboratory setting is artificial; findings can't be generalized

CHAPTER 3 SUMMARY

3-1 Culture and Society

Culture is learned, transmitted from one generation to another, adaptive, and always changing. A *society* shares a culture and sees itself as a social unit. People construct a *material culture* (such as buildings) and *nonmaterial culture* (such as rules for behavior) that influence each other (such as forbidding smoking in public buildings).

3-2 The Building Blocks of Culture

The following are some of the fundamental building blocks of culture:

• *Symbols* take many forms, distinguish one culture from another, can change over time, and can unify or divide a society.

• *Language* makes us human; it can change over time, and affects perceptions about gender, race, social class, and ethnicity.

• *Values* provide general guidelines for behavior; they sometimes conflict, are usually emotion laden, vary across cultures, and change over time.

• *Norms*—whether folkways, mores, or laws—regulate our behavior; they vary across cultures, and range from mild to severe sanctions.

• *Rituals* transmit and reinforce norms that unite people and strengthen their relationships.

Example: Sanctions for Violating the Dead

Sanctions are more severe for violating laws than folkways. Legacy.com, which carries a death notice or obituary for almost all of the roughly 2.4 million Americans who die each year, dedicates at least 30 percent of its budget to weeding out comments (a relatively mild punishment) that "diss the dead" (Urbina, 2006). In contrast, in many states, vandalizing a tombstone, a property crime, can result in a fine of up to $1,000, a jail sentence of up to a year, or both.

3-3 Some Cultural Similarities

Customs and specific behaviors vary across countries, but *cultural universals*, such as

food taboos, are common to all societies. *Ideal culture* often differs from *real culture*. *Ethnocentrism* has its benefits but can also lead to conflict and discrimination. *Cultural relativism*, the opposite of ethnocentrism, encourages cross-cultural understanding and respect.

3-4 Some Cultural Variations

Subcultures and *countercultures* account for some of the complexity within a society. Subcultures differ from the larger society in some ways, whereas countercultures oppose or reject some of the dominant culture's basic beliefs, values, and norms. In *multiculturalism*, several cultures coexist without trying to dominate one another. People who encounter an unfamiliar way of life or environment may experience *culture shock* because everyone is culture bound to some degree.

3-5 High Culture and Popular Culture

High culture refers to the cultural expression of a society's highest social classes, whereas *popular culture*, which is widely shared among a population, Includes music, social media, sports, hobbies, fashions, the food we eat, the people with whom we spend time, the gossip we share, and the jokes we pass along. *Cultural capital* affects social class boundaries, but our everyday life can contain elements of both popular and high culture. Popular culture is typically spread through *mass media*, including television and the Internet, and has enormous power in shaping our attitudes and behavior.

3-6 Cultural Change and Technology

Some societies are relatively stable because of *cultural integration*, but all societies change over time because of diffusion, innovation, invention, discovery, and external pressures. *Cultural lags* can create confusion, ambiguity about what's right and wrong, conflict, and a feeling of helplessness. They also expose contradictory values and behavior.

3-7 Sociological Perspectives on Culture

See Table 3.3 on the next page.

KEY TERMS

culture the learned and shared behaviors, beliefs, attitudes, values, and material objects that characterize a particular group or society.

society a group of people who share a culture and defined territory.

material culture the physical objects that people make, use, and share.

nonmaterial culture the ideas that people create to interpret and understand the world.

symbol anything that stands for something else and has a particular meaning for people who share a culture.

language a system of shared symbols that enables people to communicate with one another.

values the standards by which people define what is good or bad, moral or immoral, proper or improper, desirable or undesirable, beautiful or ugly.

norms specific rules of right and wrong behavior.

folkways norms that involve everyday customs, practices, and interaction.

mores norms that people consider very important because they maintain moral and ethical behavior.

taboos strong prohibitions of any act that is forbidden because it's considered to be extremely offensive.

laws formally defined norms about what is permissible or illegal.

sanctions rewards for good or appropriate behavior and/or penalties for bad or inappropriate behavior.

rituals formal and repeated behaviors that unite people.

cultural universals customs and practices that are common to all societies.

ideal culture the beliefs, values, and norms that people say they hold or follow.

real culture people's actual everyday behavior.

ethnocentrism the belief that one's own culture, society, or group is inherently superior to others.

cultural relativism the belief that no culture is better than another and should be judged by its own standards.

CHAPTER REVIEW 3

subculture a group within society that has distinctive norms, values, beliefs, lifestyle, or language.

counterculture a group within society that openly opposes and/or rejects some of the dominant culture's norms, values, or laws.

multiculturalism (sometimes called *cultural pluralism*) the coexistence of several cultures in the same geographic area, without one culture dominating another.

culture shock confusion, disorientation, or anxiety that accompanies exposure to an unfamiliar way of life.

high culture the cultural expression of a society's highest social classes.

popular culture the beliefs, practices, activities, and products that are widespread among a population.

cultural capital resources such as knowledge, verbal and social skills, education, and other assets that give a group advantages.

mass media forms of communication designed to reach large numbers of people.

cultural imperialism the cultural values and products of one society influence or dominate those of another.

cultural integration the consistency of various aspects of society that promotes order and stability.

cultural lag the gap that occurs when material culture changes faster than nonmaterial culture.

TEST YOUR LEARNING

1. Society and culture are mutually
   a. dependent.
   b. exclusive.
   c. destructive.
   d. diversified.

2. _____ stand for something else and have a particular meaning for people who share a culture.
   a. Mores
   b. Symbols
   c. Values
   d. Norms

3. At a party, George was eating nachos and salsa. He took a bite of his chip and then dipped it back into the salsa bowl. His friend Brett gave him a disgusted look. Which of the following did George break?
   a. Mores
   b. A taboo
   c. A folkway
   d. A law

4. Which of the following is an example of a subculture?
   a. A group of adolescent Goths
   b. Roman Catholics
   c. A 40-and-over women's bridge club
   d. All of the above
   e. None of the above

5. Anaz is an 8-year-old Iranian girl who loves the female fashions she sees in U.S. films and television shows. Lately she's been questioning why her mother wears a burka. Anaz's questioning of her family's traditions could best be explained by
   a. cultural relativism.
   b. multiculturalism.
   c. cultural imperialism.
   d. ethnocentrism.

6. **True or False** Culture is adapting and always changing.

7. **True or False** Language helps communicate ideas, but it's not considered a symbol.

8. **True or False** In most societies, the real culture matches the ideal culture.

9. **True or False** The legal controversy over file sharing and downloading music off the Web is an example of cultural lag.

10. **True or False** Symbolic interactionists rely on a macro approach in examining culture.

11. List five examples of material culture and five of nonmaterial culture that affect you. Explain (a) how they intersect and (a) how they shape your attitudes and behavior.

12. Provide an example of each type of norm (folkways, mores, and laws) that you or someone you know has violated. Describe the negative sanctions for each violation.

1. a 2. b 3. c 4. d 5. c 6. True 7. False 8. False 9. True 10. False

---

TABLE 3.3 SOCIOLOGICAL PERSPECTIVES OF CULTURE

THEORETICAL PERSPECTIVE	FUNCTIONALIST	CONFLICT	FEMINIST	SYMBOLIC INTERACTIONIST
Level of Analysis	**Macro**	**Macro**	**Macro and Micro**	**Micro**
Key Points	• Similar beliefs bind people together and create stability. • Sharing core values unifies a society and promotes cultural solidarity.	• Culture benefits some groups at the expense of others. • As powerful economic monopolies increase worldwide, the rich get richer and the rest of us get poorer.	• Women and men often experience culture differently. • Cultural values and norms can increase inequality because of sex, race/ethnicity, and social class.	• Cultural symbols forge identities (that change over time). • Culture (such as norms and values) helps people merge into a society despite their differences .
Examples	• Speaking the same language (English in the United States) binds people together because they can communicate with one another, express their feelings, and influence one another's attitudes and behaviors.	• Much of the English language reinforces negative images about gender ("slut"), race ("honky"), ethnicity ("jap"), and age ("old geezer") that create inequality and foster ethnocentrism.	• Using male language (e.g., "congressman," "fireman," and "chairman") conveys the idea that men are superior to and dominant over women, even when women have the same jobs.	• People can change the language they create as they interact with others. Many Americans now use "police officer" instead of "policeman," and "single person" instead of "bachelor" or "old maid."

CHAPTER 4 SUMMARY

4-1 Socialization: Its Purpose and Importance

Socialization, a lifelong process, fulfills four key purposes: It establishes our social identity, teaches us roles, controls our behavior (through *internalization*), and transmits culture to the next generation. The research on children who are isolated or institutionalized shows that socialization is critical to our social and physical development.

4-2 Nature and Nurture

Biologists tend to focus on the role of heredity (or genetics) in human development. In contrast, most social scientists, including sociologists, underscore the role of learning, socialization, and culture. This difference of opinion is often called the *nature–nurture debate*. Sociobiologists argue that genetics (nature) can explain much of our behavior, whereas most sociologists maintain that socialization and culture (nurture) shape even biological inputs.

4-3 Sociological Explanations of Socialization

Sociologists have offered many explanations of socialization, but two of the most influential, both at the micro level, have been social learning and symbolic interaction theories (see Table 4.2 on the next page). The theories help us understand social development processes, but also have limitations.

4-4 Primary Socialization Agents

Parents are the first and most important *socialization agents*, but siblings, grandparents, and other family members also play important roles. Other important socialization agents include play and peer groups, teachers and schools, popular culture, and the media. Electronic media and advertising are especially powerful socialization sources.

Example: Are Parents Raising Self-Centered Children?

Many parents tell their offspring that they're beautiful, can achieve anything they want, and are very intelligent. In fact, only about 5 percent of American kids can be considered "gifted" (endowed with significantly higher than average intellectual or other abilities), even though many are enrolled in gifted classes. Some educators maintain that telling average—or even above average—children that they're superior does them a disservice: It gives them false expectations on how the world will treat them, encourages being self-centered, and increases anger and unhappiness when they don't succeed in college or the workplace (Deveny, 2008).

4-5 Socialization Throughout Life

As we progress through the life course, we learn culturally approved norms, values, and roles. Infants are born with an enormous capacity for learning that parents and other caregivers can enrich and shape. In adolescence, these and other socialization agents teach children how to form relationships on their own, to get along with others, and to develop their social identity through play and peer groups. In adulthood, people must learn new roles that include singlehood, marriage, parenthood, divorce, work, and experiencing the death of a loved one. Socialization continues in later life when many people learn still new roles such as grandparents, retirees, older workers, and being widowed.

4-6 Resocialization

Resocialization can be voluntary or involuntary. Voluntary resocialization includes entering a religious order, joining a religious cult, seeking treatment in a drug abuse rehabilitation facility, or serving in the military. The changes can be short-term and pleasant, but also long, difficult, and intense. Most involuntary resocialization takes place in *total institutions*, where people are isolated from the rest of society, stripped of their former identities, and required to conform to new rules and behavior.

KEY TERMS

socialization the lifelong process through which people learn culture and become functioning members of society.

internalization the process of learning cultural behaviors and expectations so deeply that we accept them without question.

social learning theories maintain that people learn new attitudes, beliefs, and behaviors through social interaction, especially during childhood.

role models people we admire and whose behavior we imitate.

self an awareness of one's social identity.

looking-glass self a self-image based on how we think others see us.

role taking learning to take the perspective of others.

significant others the people who are important in one's life, such as parents (or other primary caregivers), siblings, and grandparents.

anticipatory socialization the process of learning how to perform a role one does not yet occupy.

generalized other people who don't have close ties to a child but who influence her or his internalization of society's norms and values.

impression management the process of providing information and cues to others to present oneself in a favorable light while downplaying or concealing one's less appealing characteristics.

reference groups groups of people who shape an individual's self-image, behavior, values, and attitudes in different contexts.

socialization agents the individuals, groups, or institutions that teach us how to participate effectively in society.

CHAPTER REVIEW 4

multigenerational households homes in which three or more generations live together.

peer group people who are similar in age, social status, and interests.

resocialization the process of unlearning old ways of doing things and adopting new attitudes, values, norms, and behavior.

total institutions settings where people are isolated from the rest of society, stripped of their former identities, and required to conform to new rules and behavior.

TEST YOUR LEARNING

1. Mom taught Emma that girls can do anything that boys can do, including being good in math and science. This is an example of the ___ process.
   a. multicultural
   b. feminist
   c. socialization
   d. hereditary

2. "Human development is fairly fixed." This statement is an example of which side of the nature/nurture debate?
   a. Nature side of the debate
   b. Nurture side of the debate
   c. Neither the nature nor nurture side of the debate
   d. Both the nature and nurture side of the debate

3. According to social learning theory, the greatest impact of social interaction occurs during
   a. childhood.
   b. adolescence.
   c. early adulthood.
   d. middle adulthood.

4. What are the three stages of Mead's role-taking theory?
   a. Pre-work stage, work stage, post-work stage
   b. Anal stage, phallic stage, post-phallic stage
   c. Mirror stage, active stage, passive stage
   d. Preparatory stage, play stage, game stage

5. According to Mead, children learn how to perform a role they don't yet occupy. Mead referred to this process as
   a. a developing self.
   b. anticipatory socialization.
   c. role taking.
   d. the generalized other.

6. **True or False** Institutionalization is the process of learning cultural behaviors and expectations so deeply that we assume they are correct and accept them without question.

7. **True or False** Charles Horton Cooley proposed that the looking-glass self develops in five phases.

8. **True or False** According to Erving Goffman, social life mirrors a theatre.

9. **True or False** Healthy child development is most likely in authoritative homes.

10. **True or False** Resocialization is the process of presenting ourselves in a favorable light while downplaying or concealing our less appealing characteristics.

11. Think about your early socialization. Were social learning theories, symbolic interaction theories, or both important in your socialization? Explain and illustrate your answer using specific examples.

12. Have you ever engaged in impression management? Describe when, where, how, and why.

1. c 2. a 3. a 4. d 5. b 6. False 7. False 8. True 9. True 10. False

TABLE 4.2 | KEY ELEMENTS OF SOCIALIZATION THEORIES

Social Learning Theories	Symbolic Interaction Theories
• Social interaction is important in learning appropriate and inappropriate behavior.	• The self emerges through social interaction with significant others.
• Socialization relies on direct and indirect reinforcement.	• Socialization includes role taking and controlling the impression we give to others.
Example: Children learn how to behave when they are scolded or praised for specific behaviors.	*Example:* Children who are praised are more likely to develop a strong self-image than those who are always criticized.

CHAPTER 5 SUMMARY

5-1 Social Structure

Social interaction, central to all social activity, affects people's behavior. Our interaction is part of the *social structure*, which guides our actions and gives us a feeling that life is orderly and predictable. Every society has a social structure that encompasses statuses and roles.

5-2 Status

A *status* is a social position that a person occupies in a society. Every person has many statuses that form her or his *status set*, which include both ascribed and achieved statuses. An *ascribed status* is a social position that a person is born into and can't control, change, or choose (such as age, race, and being male or female). An *achieved status* is a social position that a person attains through personal effort or assumes voluntarily (such as college student or wife). A *master status* is usually immediately apparent, makes the biggest impression, affects others' perceptions, and, consequently, often shapes a person's entire life.

Because we hold many statuses, some clash. People experience *status inconsistency* when they occupy social positions that are ranked differently (such as being a low-paid college professor).

5-3 Role

A *role* defines how we're expected to behave in a particular status, but people vary considerably in fulfilling the responsibilities associated with their roles. These differences reflect *role performance*, the actual behavior of a person who occupies a status. A *role set* encompasses different roles attached to a single status (such as a parent who is a teacher, chauffeur, and PTA member). Playing many roles often leads to *role conflict* because it's difficult to meet the requirements of two or more statuses, and to *role strain*, the stress that arises because of incompatible demands among roles within a single status.

To deal with role conflict and role strain, some people deny that there's a problem. More effective ways to minimize role conflict and role strain include compromising, negotiating, setting priorities, compartmentalizing our roles, refusing to take on more roles, and exiting one or more current roles.

Example: Exiting a Marriage

Divorce is a good example of role exit, but it often involves a long process of five stages that may last several decades (Bohannon 1971):

• The "emotional divorce" begins when one or both partners feel disillusioned or unhappy.

• The "legal divorce" is the formal dissolution of the marriage during which the partner who doesn't want the divorce may try to stall the end of the marriage.

• During the "economic divorce" stage, the partners may argue about who should pay past debts, property taxes, and unforeseen expenses (such as moving costs).

• The "coparental divorce" stage involves parents' agreeing on issues such as child support and visitation rights.

• During the "community divorce" stage, partners inform friends, family, and others that they are no longer married. Finally, the couple goes through a "psychic divorce" in which the partners separate from each other emotionally. In many cases, one or both spouses never complete this stage because they can't let go of their pain, anger, and resentment—even if they remarry.

5-4 Explaining Social Interaction

See Table 5.2 on next page.

KEY TERMS

social interaction the process by which we act toward and react to people around us.

social structure an organized pattern of behavior that governs people's relationships.

status a social position that a person occupies in a society.

status set a collection of social statuses that a person occupies at a given time.

ascribed status a social position that a person is born into.

achieved status a social position that a person attains through personal effort or assumes voluntarily.

master status a status that overrides other statuses and forms an important part of a person's social identity.

status inconsistency the conflict that arises from occupying social positions that are ranked differently.

role the behavior expected of a person who has a particular status.

role performance the actual behavior of a person who occupies a status.

role set the different roles attached to a single status.

role conflict the frustrations and uncertainties a person experiences when confronted with the requirements of two or more statuses.

role strain the stress that arises from incompatible demands among roles within a single status.

self-fulfilling prophecy a situation in which if we define something as real and act on it, it can, in fact, become real.

ethnomethodology the study of how people construct and learn to share definitions of reality that make everyday interactions possible.

CHAPTER REVIEW 5

dramaturgical analysis examines social interaction as if occurring on a stage where people play different roles and act out scenes for the audiences with whom they interact.

social exchange theory posits that social interaction is based on each person's trying to maximize rewards (or benefits) and minimize punishments (or costs).

nonverbal communication messages that are sent without using words.

social media websites that enable users to create, share, and/or exchange information and ideas.

social networking site a website that connects people who share similar personal or professional interests.

5-5 Nonverbal Communication

Our *nonverbal communication* includes gestures, facial expressions, eye contact, and silence. Touching and how we use space are also important forms of nonverbal communication because they send powerful messages about our feelings and power. Nonverbal communication varies from society to society, which can lead to cross-cultural misinterpretation and misunderstanding.

5-6 Online Interaction

Many people interact in *cyberspace*, an online world of computer networks that includes *social media* and *social networking sites*. Internet usage varies by sex, age, race, ethnicity, and social class. Online interaction can be impersonal, socially isolating, and jeopardizes our privacy, but can also save time, foster closer ties among family members and friends, and facilitate working from home.

TEST YOUR LEARNING

1. Jenna is the youngest of three girls in her family. Being the youngest sister is an example of a(n)
   a. achieved status.
   b. ascribed status.
   c. problematic status.
   d. interchangeable status.

2. Jake has a full-time course load and works 30 hours a week. Given the demands of both work and school, Jake is likely to experience
   a. role set.
   b. role strain.
   c. role conflict.
   d. role exchange.

3. According to ___, social interaction is based on trying to maximize one's rewards while minimizing costs.
   a. ethnomethodology theory
   b. feminist theory
   c. conflict theory
   d. social exchange theory

4. Gestures, eye contact, and silence are all examples of
   a. emotional language.
   b. punishment.
   c. healthy relationship interaction.
   d. nonverbal communication.

5. Which one of the following statements is false?
   a. Almost equal numbers of women and men use the Internet.
   b. The higher a family's income, the greater the likelihood that its members are Internet users.
   c. A major disadvantage of cyberspace communication is that it's more impersonal than face-to-face interaction.
   d. White Americans are the most wired group in the United States.

6. ***True or False*** In sociology, status signifies prestige.

7. ***True or False*** A master status is based on one's achieved status.

8. ***True or False*** Cyberstalking is more common among adolescents than adults.

9. ***True or False*** If we define something as real and act on it, it can, in fact, become real. This is known as the self-fulfilling prophecy.

10. ***True or False*** Being an African American college student is an example of having two ascribed statuses.

11. List your social statuses. Which ones are ascribed, and which ones are achieved? Do you have one or more master statuses, and how do they affect you? After describing the difference between role conflict and role strain, choose an example from each and explain how you cope with or have resolved a role conflict or role strain.

12. Describe three benefits and three costs of online interaction that you've encountered. Include any of the topics you read about in the chapter, but illustrate your answer with specific examples.

1. b 2. c 3. d 4. d 5. d 6. False 7. False 8. False 9. True 10. False

TABLE 5.2	SOCIOLOGICAL EXPLANATIONS OF SOCIAL INTERACTION
Perspective	**Key Points**
Symbolic Interactionist	• People create and define their reality through social interaction. • Our definitions of reality, which vary according to context, can lead to self-fulfilling prophecies.
Social Exchange	• Social interaction is based on a balancing of benefits and costs. • Relationships involve trading a variety of resources, such as money, youth, and good looks.
Feminist	• Females and males act similarly in many interactions but may differ in communication styles and speech patterns. • Men are more likely to use speech that's assertive (to achieve dominance and goals), whereas women are more likely to use language that connects with others.

CHAPTER 6 SUMMARY

6-1 Social Groups

A *social group* (such as friends or work groups) gives us a common identity and a sense of belonging. Social groups include *primary groups* (such as family members) that shape our social and moral development and *secondary groups* (such as the students in your sociology class) that pursue a specific goal or activity.

Members of an *in-group* share a sense of identity and "we-ness," whereas *out-groups* are viewed and treated negatively because they're seen as having values, beliefs, and other characteristics that differ from those of the in-group. We also have *reference groups* that influence who we are, what we do, and who we'd like to be in the future. Groups often form a *social network* that may be tightly knit and interact every day or may include large numbers of people whom we don't know personally and with whom we interact only rarely or indirectly.

Example: Secondary Groups Can Replace Primary Groups

In 1864, an alcoholic who had ruined a promising career on Wall Street because of his constant drunkenness cofounded Alcoholics Anonymous (AA), a program that would enable people to stop drinking by undergoing a spiritual awakening and seeking help from a buddy to stay sober. Initially, AA was a secondary group that tried to beat alcoholism by encouraging its members to attend regular meetings where alcoholics talked about their accomplishments in staying sober. Over the years, however, AA has become a primary group for many members because it offers a relatively small group of people who engage in face-to-face interaction over an extended period, especially when their family and friends have rejected them.

6-2 Formal Organizations

We depend on a variety of *formal organizations* to provide goods and services in a stable and predictable way. Utilitarian, normative, and coercive organizations differ in their general characteristics and membership, but a single formal organization can fall into all three categories. Bureaucracies (such as your college), are supposed to accomplish goals and tasks in the most efficient and rational way possible, but many experience shortcomings such as weak reward systems, rigid rules, goal displacement, alienation, communication problems, and dehumanization.

6-3 Sociological Perspectives on Social Groups and Organizations

See Table 6.3 on the next page.

6-4 Social Institutions

A *social institution* meets a society's basic survival needs. Social institutions are abstractions, but they have an organized purpose, weave together norms and values, and, consequently, guide behavior. Because social institutions are linked to one another, they can tell us a lot about how a society functions and how we're connected to one another.

KEY TERMS

social group two or more people who share some attribute and interact with one another.

primary group a small group of people who engage in intimate face-to-face interaction over an extended period.

secondary group a large, usually formal, impersonal, and temporary collection of people who pursue a specific goal or activity.

ideal types general traits that describe a social phenomenon rather than every case.

in-group people who share a sense of identity and "we-ness" that typically excludes and devalues outsiders.

out-group people who are viewed and treated negatively because they're seen as having values, beliefs, and other characteristics different from those of an in-group.

reference group people who shape our behavior, values, and attitudes.

dyad a group with two members.

triad a group with three members.

authoritarian leader gives orders, assigns tasks, and makes all major decisions.

democratic leader encourages group discussion and includes everyone in the decision-making process.

laissez-faire leader offers little or no guidance to group members and allows them to make their own decisions.

groupthink a situation in which in-group members make faulty decisions because of group pressures, rather than critically testing, analyzing, and evaluating ideas and evidence.

social network a web of social ties that links individuals or groups to one another.

CHAPTER REVIEW 6

formal organization a complex and structured secondary group designed to achieve specific goals in an efficient manner.

bureaucracy a formal organization designed to accomplish goals and tasks in an efficient and rational way.

goal displacement a preoccupation with rules and regulations rather than achieving the organization's objectives.

alienation a feeling of isolation, meaninglessness, and powerlessness.

iron law of oligarchy the tendency of a bureaucracy to become increasingly dominated by a small group of people.

glass ceiling workplace attitudes or organizational biases that prevent women from advancing to leadership positions.

glass escalator men who enter female-dominated occupations and receive higher wages and faster promotions than women.

social institution an organized and established social system that meets one or more of a society's basic needs.

TEST YOUR LEARNING

1. A(n) _____ is a small group of people who engage in frequent and intimate face-to-face interaction.
   a. out-group
   b. reference group
   c. secondary group
   d. primary group

2. _____ conducted experiments that used "teachers" who administered electric shocks to "learners."
   a. Solomon Asch
   b. Stanley Milgram
   c. Philip Zimbardo
   d. Irving Janis

3. Max Weber outlined the characteristics of an efficient and productive bureaucracy. Which of the following was NOT one of those characteristics?
   a. High degree of specialization
   b. Explicit rules and regulations
   c. Qualifications-based employment
   d. Decentralized authority

4. Regarding social groups and organizations, symbolic interaction maintains that
   a. some people benefit more than others.
   b. cooperation works.
   c. men benefit more than women.
   d. people define and shape their situations.

5. _____ contend that organizations are based on vast differences in power and control.
   a. Functionalists
   b. Conflict theorists
   c. Symbolic interactionists
   d. Exchange theorists

6. **True or False** A high school football team is a good example of a primary group.

7. **True or False** Groupthink occurs most often when in-group members discuss a diversity of ideas.

8. **True or False** People join normative organizations because of shared interests and to pursue goals that they consider personally worthwhile or rewarding.

9. **True or False** The iron law of oligarchy states that bureaucracies have a tendency to become increasingly dominated by a small group of people.

10. **True or False** The glass ceiling refers to organizational barriers, but not attitudes, in the workplace.

11. Describe yourself using the following concepts: primary group, secondary group, in-group, out-group, reference group, and social network. Illustrate your self-description with specific examples.

12. In the United States, we live in a "McDonaldized" society. What does this concept mean? What are some of the benefits and costs of a "McDonaldized" society? Illustrate your answer with specific examples from your everyday life.

1. d 2. b 3. d 4. d 5. b 6. False 7. False 8. True 9. True 10. False

TABLE 6.3 VOLUNTEERING IN THE UNITED STATES, 2012

Percentage of adults who volunteer	27
Total number of volunteers age 16 and older	65 million
Median annual hours per volunteer	50
Total dollar value of volunteer time	$171 billion

Sources: Based on Corporation for National & Community Service, 2012, and BLS News Release, 2013.

CHAPTER 7 SUMMARY

7-1 What Is Deviance?

Deviance, the violation of social norms, includes *crime*. Perceptions of deviance and crime vary across and within societies, can change over time. People with authority or power decide what's right or wrong, and a *stigma* accompanies deviant behavior.

Example: Deviance and College Drinking

According to many college presidents, alcohol abuse is the most serious problem on campus. Alcohol abuse results in alcohol poisoning and blackouts and leads to sexual assault, violent behavior, injuries, and academic problems. Because drinking laws are rarely enforced, some college presidents have proposed that the drinking age be lowered from 21 to 18. Others argue that doing so would increase traffic fatalities and drinking problems. Young people can get a driver's license at 16 and vote and enlist in the military at 18. Should they be the ones, then, to decide whether drinking laws should be changed?

7-2 Types of Deviance and Crime

Sociologists study *noncriminal deviance*—such as suicide, alcoholism, lying, mental illness, and adult pornography—and *criminal deviance*, behavior that violates laws. Two of the most important sources of U.S. crime statistics are the FBI's *Uniform Crime Report*, and *victimization surveys*. The media usually focuses on street crimes, but Americans are much more likely to be victimized by *hate crimes*, *white-collar crimes*, *corporate crimes*, *cybercrime*, and *organized crime*. *Victimless crimes* violate laws, but the parties involved don't consider themselves victims.

7-3 to 7-6 Sociological Explanations of Deviance and Crime

See Table 7.2 on the next page.

7-7 Controlling Deviance and Crime

The purpose of *social control* is to eliminate, or at least reduce, deviance and crime. Formal social control is administered by those in authority or power. Informal social control is internalized from childhood. Most people conform because of positive and negative *sanctions*. The *criminal justice system* relies on three major approaches in controlling crime: prevention and intervention, punishment, and rehabilitation. A *crime control model* supports a tough approach toward criminals in sentencing, imprisonment, and capital punishment. In contrast, many people believe that *rehabilitation* can change offenders into productive and law-abiding citizens.

KEY TERMS

deviance a violation of social norms.

crime a violation of society's formal laws.

stigma a negative label that devalues a person and changes her or his self-concept and social identity.

victimization survey interviews people about being crime victims.

hate crime a criminal act motivated by the offender's bias regarding race, religion, ethnicity, sexual orientation, gender, or disability.

white-collar crime illegal activities committed by high-status people in the course of their occupations.

corporate crimes (also known as *organizational crimes*) illegal acts committed by executives to benefit themselves and their companies.

cybercrime (also called *computer crime*) illegal activities that are conducted online.

organized crime activities of individuals and groups that supply illegal goods and services for profit.

victimless crimes acts that violate laws but those involved don't consider themselves victims.

anomie the condition in which people are unsure of how to behave because of absent, conflicting, or confusing social norms.

strain theory posits that people may engage in deviant behavior when they experience a conflict between goals and the means available to obtain the goals.

patriarchy a hierarchical system in which cultural, political, and economic structures are controlled by men.

rape culture an environment in which sexual violence is prevalent, pervasive, and perpetuated by the media and popular culture.

CHAPTER REVIEW 7

differential association theory asserts that people learn deviance through interaction, especially with significant others.

labeling theory holds that society's reaction to behavior is a major factor in defining oneself or others as deviant.

primary deviance the initial act of breaking a rule.

secondary deviance rule-breaking behavior that people adopt in response to others' reactions.

medicalization of deviance diagnosing and treating a violation of social norms as a medical disorder.

social control the techniques and strategies that regulate people's behavior in society.

control theory proposes that deviant behavior decreases when people have strong social bonds with others.

sanctions rewards or punishments for obeying or violating a norm.

criminal justice system government agencies that are charged with enforcing laws, judging offenders, and changing criminal behavior.

crime control model proposes that crime rates increase when offenders don't fear apprehension or punishment.

rehabilitation a social control approach which maintains that appropriate treatment can change offenders into productive, law-abiding citizens.

TEST YOUR LEARNING

1. Which of the following statements is false?
   a. Deviance can be a trait or a behavior.
   b. Informal deviance violates laws.
   c. Deviance is usually accompanied by social stigmas.
   d. Deviance varies across situations.

2. Who of the following is most likely to be the victim of a crime?
   a. A black man
   b. A black woman
   c. A white man
   d. A white woman

3. When 3-year-old Kyle colored the kitchen table blue with his new crayons, his mother frowned and scolded him. Kyle's mother used _____ to control his behavior?
   a. punishment
   b. negative sanctions
   c. behavior modification
   d. positive sanctions

4. Merton proposed four deviant modes by which people adapted to social strain. Which of the following was NOT one of Merton's modes?
   a. Innovation
   b. Retreatism
   c. Ritualism
   d. Recidivism

5. What is a fundamental question that a conflict theorist would ask regarding crime?

   a. "Why do some people commit crimes whereas others do not?"
   b. "Why are some acts defined as criminal whereas others are not?"
   c. "Why do men commit more violent crimes than women?"
   d. "How does one's social context impact deviant behavior?"

6. **True or False** Female crime rates have decreased.

7. **True or False** Differential association theory claims that people learn deviant behaviors through interaction with others.

8. **True or False** Labeling theory contends that society's reaction to a behavior is a major factor in defining oneself or others as deviant.

9. **True or False** Capital punishment decreases crime.

10. **True or False** The crime control model maintains that rehabilitation is the best way to decrease the frequency of crime.

11. What's the difference, if any, between deviance and crime? Include the positive and negative functions, if any, of deviance and crime in your answer.

12. Which sociological perspective do you think provides the best explanation of deviance and crime? Explain why, illustrating your answer with specific examples.

1. b 2. a 3. b 4. d 5. b 6. False 7. True 8. True 9. False 10. False

TABLE 7.2 | SOCIOLOGICAL EXPLANATIONS OF DEVIANCE

Theoretical Perspective	Level of Analysis	Key Points
Functionalist	Macro	• Deviance is both functional and dysfunctional. • Anomie increases the likelihood of deviance. • People are deviant when they experience blocked opportunities to achieve the culturally approved goal of economic success.
Conflict	Macro	• There's a strong association between capitalism, social inequality, power, and deviance. • The most powerful groups define what's deviant. • Laws rarely punish the illegal activities of the powerful.
Feminist	Macro and micro	• There's a large gender gap in deviant behavior. • Women's deviance reflects their general oppression due to social, economic, and political inequality. • Many women are offenders or victims because of patriarchal beliefs and practices and living in a rape culture.
Symbolic Interactionist	Micro	• Deviance is socially constructed. • People learn deviant behavior from significant others such as parents and friends. • If people are labeled or stigmatized as deviant, they're likely to develop negative self-concepts and engage in criminal behavior.

CHAPTER 8 SUMMARY

8-1 Social Stratification Systems and Bases

Social stratification is a society's ranking of people who have different access to valued resources. A *closed stratification system* differs from an *open stratification system* because the latter allows movement from one social class to another. Stratification includes *wealth, prestige,* and *power*. People are more likely to experience status inconsistency if they rank differently on these three dimensions, such as a football player who has great wealth but little power.

8-2 Social Class in America

A good indicator of social class is *socioeconomic status (SES)*, an overall rank based on a person's income, education, and occupation. Using SES and other variables (such as values, power, and *conspicuous consumption*), most sociologists agree that there are at least four social classes in the United States: upper, middle, working, and lower. These groups can be divided further into upper-upper, lower-upper, upper-middle, lower-middle, and the working class. The lower class includes the *working poor* and the *underclass*. A major outcome of social stratification is *life chances*.

Example: Restaurant Menus and Stratification

Two sociologists—in Iowa and Virginia—asked their students in introductory sociology classes to do a content analysis (see Chapter 1) of 10 menus that represented a sampling of restaurants by social class. The students found that the restaurants that catered to upper-class clientele had higher than average entrée prices, described the entrées in foreign languages, used fancy sauces, recommended expensive wines, and had few illustrations. Middle-class menus emphasized "value for the dollar," presented photos of entrées

with "bountiful plates overflowing with appetizing food," and popular items such as quesadillas. Menus at lower-class restaurants featured low prices ($3 to $10 entrées), the items were numbered, none of the entrées had "pretentious names," and the typesetting was simple (Wright and Ransom, 2005). In effect, then, even menus denote social class and social status.

8-3 Poverty

Absolute poverty is a more serious social problem than *relative poverty* because millions of Americans live below the *poverty line*. Poverty varies by age, gender, family structure, race, and ethnicity. One explanation for poverty maintains that people are poor because of individual failings; another contends that societal structures create and sustain poverty.

8-4 Social Mobility

Social mobility can be *horizontal, vertical, intragenerational,* or *intergenerational*. There has been a sharp decline in the share of middle-income adults since 1971 because many have slid into a lower class. Structural, demographic, and individual factors affect a person's social mobility. Much social mobility depends on structural, demographic, and family background factors, all of which are interrelated.

8-5 Global Stratification

Global inequality is widespread, and all societies are stratified, but some countries are much wealthier than others, inequality is increasing across the globe. Historically, currently, and across all nations, women and children experience the greatest poverty. Sociologists use *modernization theory, dependency theory,* and *world-system theory* to explain why inequality is universal.

8-6 Sociological Explanations: Why There Are Haves and Have-Nots

See Table 8.2 on the next page.

KEY TERMS

social stratification a society's ranking of people based on their access to valued resources such as wealth, power, and prestige.

slavery system people own others as property and have almost total control over their lives.

caste system people's positions are ascribed at birth and largely fixed.

class system people's positions are based on both birth and achievement.

social class people who have a similar standing or rank in a society based on wealth, education, power, prestige, and other valued resources.

wealth economic assets that a person or family owns.

income the money a person receives, usually through wages or salaries, but can also include other earnings.

prestige respect, recognition, or regard attached to social positions.

power the ability to influence or control the behavior of others despite opposition.

socioeconomic status (SES) an overall ranking of a person's position in society based on income, education, and occupation.

working poor people who work at least 27 weeks a year but whose wages fall below the official poverty level.

underclass people who are persistently poor and seldom employed, residentially segregated, and relatively isolated from the rest of the population.

life chances the extent to which people have positive experiences and can secure the good things in life because they have economic resources.

absolute poverty not having enough money to afford the basic necessities of life.

relative poverty not having enough money to maintain an average standard of living.

CHAPTER REVIEW 8

poverty line the minimal income level that the federal government considers necessary for basic subsistence (also called the *poverty threshold*).

feminization of poverty the disproportionate number of the poor who are women.

social mobility movement up or down the social class hierarchy.

intragenerational mobility movement up or down the class hierarchy over one's lifetime.

intergenerational mobility movement up or down the class hierarchy relative to the position of one's parents.

global stratification worldwide inequality patterns that result from differences in wealth, power, and prestige.

infant mortality rate the number of babies under age 1 who die per 1,000 live births in a given year.

Davis–Moore thesis the functionalist view that social stratification benefits a society.

meritocracy a belief that social stratification is based on people's accomplishments.

bourgeoisie those who own and control capital and the means of production.

proletariat workers who sell their labor for wages.

corporate welfare subsidies, tax breaks, and assistance that the government has created for businesses.

TEST YOUR LEARNING

1. Slavery and castes are ____ systems.
   a. capitalist
   b. open stratification
   c. closed stratification
   d. socialist

2. A person's socioeconomic status (SES) is based on
   a. income, education, and occupation.
   b. wealth, prestige, and power.
   c. wealth, prestige, and lifestyle.
   d. education, income, and age.

3. The economic gap between the wealthy and the poor
   a. fluctuates from year to year.
   b. is about the same as during the 1990s.
   c. is decreasing.
   d. is increasing.

4. Which of the following is NOT a major structural factor in social mobility?
   a. Changes in the economy
   b. Immigration patterns
   c. Consumer confidence
   d. Number of available positions in given occupations

5. ____ claim that social stratification ultimately benefits society.
   a. Functionalists
   b. Conflict theorists
   c. Feminist theorists
   d. Symbolic interactionists

6. **True or False** One of the criticisms of conflict theory is that it ignores structural factors in explaining stratification.

7. **True or False** Conspicuous consumption displays one's social status and enhances one's prestige.

8. **True or False** Intergenerational mobility refers to moving up or down the class hierarchy over one's lifetime.

9. **True or False** The Davis–Moore thesis is a symbolic interactionist perspective.

10. **True or False** According to world-system theory, the global economic system helps richer countries stay rich while poorer countries remain poor.

11. Describe your current social class and explain why you place yourself in this class. Ten years from now, do you expect to experience upward or downward mobility? Explain why.

12. Should public assistance recipients be required to work, regardless of their education level? If no, why not? If yes, should the government subsidize preschool and afterschool care for their children while the parent(s) is/are at work? In your answer, incorporate the sociological perspectives on social stratification.

1. c 2. a 3. d 4. c 5. a 6. False 7. True 8. False 9. False 10. True

TABLE 8.2 SOCIOLOGICAL EXPLANATIONS OF SOCIAL STRATIFICATION

Perspective	Level of Analysis	Key Points
Functionalist	Macro	• Fills social positions that are necessary for a society's survival • Motivates people to succeed and ensures that the most qualified people will fill the most important positions
Conflict	Macro	• Encourages workers' exploitation and promotes the interests of the rich and powerful • Ignores a wealth of talent among the poor
Feminist	Macro and micro	• Constructs numerous barriers in patriarchal societies that limit women's achieving wealth, status, and prestige • Requires most women, not men, to juggle domestic and employment responsibilities that impede upward mobility
Symbolic Interactionist	Micro	• Shapes stratification through socialization, everyday interaction, and group membership • Reflects social class identification through symbols, especially products that signify social status

CHAPTER 9 SUMMARY

9-1 Sex, Gender, and Culture

Sex refers to biological characteristics, whereas *gender* refers to learned attitudes and behaviors. *Gender roles* differ depending on whether people perceive themselves as masculine or feminine and because a society expects women and men to think and behave differently. Many Americans still have *gender stereotypes* about how people will look, act, think, and feel based on their sex.

Our sexual identity incorporates a *sexual orientation* that can be *homosexual*, *heterosexual*, *bisexual*, or *asexual*. Many biological theories maintain that sexual orientation has a strong genetic basis, but social constructionists argue that culture, not biology, plays a large role in forming people's *gender identity*.

9-2 Contemporary Gender Stratification and Inequality

Sexism is widespread because of *gender stratification*, which can lead to inequality in the family (particularly child care and housework), education, the workplace (as in the case of *occupational sex segregation* and a *gender pay gap*), and politics. *Sexual harassment* and workplace bullying are also common in the workplace.

9-3 Sexuality

Sexual attitudes and behavior can vary from situation to situation and change over time, including why we have sex. Contrary to some stereotypes, adolescents aren't sexually promiscuous and older people aren't asexual. All of us have internalized *sexual scripts*, and *sexual double standards* persist.

Example: Gender Roles and Hooking Up—Are Women the Losers?

Hooking up refers to physical encounters, no strings attached, and can mean anything from kissing and genital fondling to oral sex and sexual intercourse. The prevalence of hooking up has increased only slightly since the late 1980s, but is now more common than dating at many high schools and colleges. Hooking up has its advantages. It's much cheaper than dating. It's also assumed that hooking up requires no commitment of time or emotion. In addition, hookups remove the stigma from those who can't get dates but can experience sexual pleasure, and they make people feel sexy and desirable (Bogle, 2008; Monto and Carey, 2014). Hooking up also has disadvantages, especially for women, because women who hook up generally get a bad reputation as being "easy" (England and Thomas, 2009). In effect, then, and despite the advantages of hooking up, the sexual double standard persists.

9-4 Some Current Social Issues about Sexuality

Two of the most controversial and politically contested issues continue to be *abortion* and *same-sex marriage*. Almost equal percentages of Americans support or condemn abortion. Those who favor same-sex marriage argue that people should have the same legal rights regardless of sexual orientation; those who oppose same-sex marriage contend that such unions are immoral and contrary to religious beliefs.

9-5 Gender and Sexuality across Cultures

Worldwide, women have fewer rights and opportunities than men. In all countries and regions, the greatest gender gaps are in economic opportunity and participation and political leadership. People in some countries are more accepting of homosexuality than in the past, but *heterosexism* prevails.

9-6 Sociological Explanations of Gender and Sexuality

See Table 9.4 on the next page.

KEY TERMS

sex the biological characteristics with which we are born.

gender learned attitudes and behaviors that characterize women and men.

intersexuals people whose medical classification at birth isn't clearly either male or female.

sexual identity our awareness of ourselves as male or female and the ways that we express our sexual values, attitudes, feelings, and beliefs.

sexual orientation a preference for sexual partners of the same sex, of the opposite sex, of both sexes, or neither sex.

homosexuals those who are sexually attracted to people of the same sex.

heterosexuals those who are sexually attracted to people of the opposite sex.

bisexuals those who are sexually attracted to both sexes.

asexuals those who lack any interest in or desire for sex.

gender identity a perception of oneself as either masculine or feminine.

transgender an umbrella term for people whose gender identity and behavior differ from the sex to which they were assigned at birth.

gender expression the way a person communicates gender identity to others through behavior, clothing, hairstyles, voice, or body characteristics.

gender roles the characteristics, attitudes, feelings, and behaviors that society expects of females and males.

gender stereotypes expectations about how people will look, act, think, and feel based on their sex.

sexism an attitude or behavior that discriminates against one sex, usually females, based on the assumed superiority of the other sex.

CHAPTER REVIEW 9

heterosexism belief that heterosexuality is the only legitimate sexual orientation.

homophobia a fear and hatred of lesbians and gays.

gender stratification people's unequal access to wealth, power, status, prestige, and other valued resources because of their sex.

occupational sex segregation (sometimes called *occupational gender segregation*) the process of channeling women and men into different types of jobs.

gender pay gap the overall income difference between women and men in the workplace (also called the *wage gap, pay gap,* and *gender wage gap*).

sexual harassment any unwanted sexual advance, request for sexual favors, or other conduct of a sexual nature that makes a person uncomfortable and interferes with her or his work.

sexual script specifies the formal and informal norms for acceptable or unacceptable sexual behavior.

sexual double standard a code that permits greater sexual freedom for men than women.

abortion the expulsion of an embryo or fetus from the uterus.

same-sex marriage (also called *gay marriage*) a legally recognized marriage between two people of the same biological sex and/or gender identity.

TEST YOUR LEARNING

1. ____ refers to learned attitudes and behaviors, whereas ____ refers to biological characteristics with which we are born.
   a. Sex; gender
   b. Gender; sex
   c. Sex; gender roles
   d. Sexual identity; gender stereotypes

2. Generally, the gender pay gap increases as the level of educational attainment
   a. increases.
   b. decreases.
   c. neither of the above because it depends on how hard a person works
   d. neither a nor b because such data are not available

3. Sean, age 16, lies to his friends about the frequency of his sexual activity and new sexual exploits every week. Sean is
   a. being a normal male.
   b. following a sexual script.
   c. falling into a pathologic life course.
   d. challenging gender norms.

4. ____ is the belief that heterosexuality is superior to and more natural than homosexuality or bisexuality.
   a. Heterobiology
   b. Homosexism
   c. Heterosexism
   d. Bisexism

5. ____ posit that gender inequality and sexuality are socially constructed.
   a. Functionalists
   b. Conflict theorists
   c. Feminist theorists
   d. Symbolic interactionists

6. **True or False** Sexism can only be enacted by men against women.

7. **True or False** Sexual harassment is the fastest growing type of employment discrimination.

8. **True or False** U.S. abortion rates have been increasing.

9. **True or False** A majority of Americans support same-sex marriage.

10. **True or False** Conflict theorists believe that capitalism increases gender inequality.

11. How do sex and gender differ? How do they affect each other? "Our sexual behavior is spontaneous." Explain why you agree or disagree with this statement.

12. Describe how gender stratification affects women and men in family life, education, the workplace, and politics. Provide examples to illustrate your answer.

1. b 2. a 3. b 4. c 5. d 6. False 7. True 8. False 9. False 10. True

TABLE 9.4 | SOCIOLOGICAL EXPLANATIONS OF GENDER AND SEXUALITY

Theoretical Perspective	Level of Analysis	Key Points
Functionalist	Macro	• Gender roles are complementary, equally important for a society's survival, and affect human capital. • Agreed-on sexual norms contribute to a society's order and stability.
Conflict	Macro	• Gender roles give men power to control women's lives. • Most societies regulate women's, but not men's, sexual behavior.
Feminist	Macro and micro	• Women's inequality reflects their historical and current domination by men, especially in the workplace. • Many men use violence—including sexual harassment, rape, and global sex trafficking—to control women's sexuality.
Symbolic Interactionist	Micro	• Gender is a social construction that emerges and is reinforced through everyday interactions. • The social construction of sexuality varies across cultures because of societal norms and values.

CHAPTER 10 SUMMARY

10-1 U.S. Racial and Ethnic Diversity

Perhaps the most multicultural country in the world, the United States includes about 150 distinct ethnic or racial groups among more than 312 million inhabitants. By 2025, only 58 percent of the U.S. population will be white—down from 86 percent in 1950.

10-2 The Social Significance of Race and Ethnicity

Race refers to physical characteristics, whereas an *ethnic group* identifies with a common national origin or cultural heritage. A *racial-ethnic group* has both distinctive physical and cultural characteristics.

10-3 Our Changing Immigration Mosaic

In 1900, almost 85 percent of immigrants came from Europe; now immigrants come primarily from Asia and Latin America. Many Americans are ambivalent about immigrants, especially those who are in the country illegally. Many scholars argue, however, that, in the long run, both legal and undocumented immigrants bring more benefits than costs.

10-4 Dominant and Minority Groups

A *dominant group* has more economic and political power than a *minority*. The latter may be larger in number than a dominant group but is often subject to differential and unequal treatment because of its physical, cultural, or other characteristics. Patterns of dominant-minority group relations include *genocide*, *segregation*, *acculturation*, *assimilation*, and *pluralism*.

10-5 Some Sources of Racial-Ethnic Friction

Racism justifies and preserves the social, economic, and political interests of dominant groups. *Prejudice* is an attitude; *discrimination* is an act that occurs at both the individual and institutional level. All of us can be prejudiced, but minorities are typically targets of *stereotypes* and *ethnocentrism* that often lead to *scapegoating*.

Example: Stuff White People Like

The popular blog *Stuff White People Like* has generated clones (e.g., *Stuff Educated Black People Like* and *Stuff Asian People Like*). Why are these sites so popular? Many fans say that the descriptions are funny because they're true. According to some critics, however, by poking fun at privileged upper-middle-class white people, the sites fuel stereotypes instead of having painfully frank discussions about U.S. race and racism (Sternbergh, 2008). Do you agree or disagree?

10-6 Major Racial and Ethnic Groups in the United States

European Americans, who settled the first colonies, are declining in population, whereas Latinos now comprise 17 percent of the population. Other large racial and ethnic populations are African Americans (13 percent), Asian Americans (6 percent), and American Indians (almost 2 percent). Middle Eastern Americans, who comprise less than 0.05 percent of the population, come from more than 30 countries. All of these groups have experienced prejudice and discrimination, but they have enhanced U.S. society and culture.

10-7 Sociological Explanations of Racial-Ethnic Inequality

See Table 10.4 on the next page.

10-8 Interracial and Interethnic Relationships

About 97 percent of Americans report being only one race, but the numbers of biracial children are increasing because of interracial dating and marriage. The rise of racial-ethnic intermarriage reflects many micro and macro factors such as greater interethnic and interracial contact, changing attitudes, and acculturation.

KEY TERMS

racial group people who share visible physical characteristics that members of a society consider socially important.

ethnic group people who identify with a common national origin or cultural heritage.

racial-ethnic group people who have distinctive physical and cultural characteristics.

dominant group any physically or culturally distinctive group that has the most economic and political power, the greatest privileges, and the highest social status.

minority a group of people who may be treated differently and unequally because of their physical, cultural, or other characteristics.

genocide the systematic effort to kill all members of a particular ethnic, religious, political, racial, or national group.

segregation the physical and social separation of dominant and minority groups.

acculturation the process of adopting the language, values, beliefs, and other characteristics of the host culture.

assimilation conforming to the dominant group's culture, adopting its language and values, and intermarrying with that group.

pluralism minority groups maintain many aspects of their original culture while living peacefully with the host culture.

racism a set of beliefs that one's own racial group is inherently superior to other groups.

prejudice an attitude that prejudges people, usually in a negative way.

stereotype an oversimplified or exaggerated generalization about a group of people.

CHAPTER REVIEW 10

scapegoats individuals or groups whom people blame for their own problems or shortcomings.

discrimination behavior that treats people unequally or unfairly because of their group membership.

individual discrimination harmful action on a one-to-one basis by a dominant group member against someone in a minority group.

institutional discrimination unequal treatment and fewer opportunities that minority groups members experience because of the everyday operations of a society's laws, rules, policies, practices, and customs.

gendered racism the overlapping and cumulative effects of inequality due to racism *and* sexism.

contact hypothesis posits that the more people get to know members of a minority group personally, the less likely they are to be prejudiced against that group.

miscegenation marriage or sexual relations between a man and a woman of different races.

TEST YOUR LEARNING

1. What's the difference between race and ethnicity?
   a. Race is a negative term that refers to difference; ethnicity is the politically correct term that refers to difference.
   b. Race refers to people's physical characteristics; ethnicity refers to people's cultural heritage.
   c. Race refers to people around the world; ethnicity refers only to a group of people within a specific nation.
   d. There is no difference between race and ethnicity; the terms are interchangeable.

2. Today most immigrants to the United States come from which of the following countries?
   a. China and Mexico
   b. Mexico and Canada
   c. England and Canada
   d. Japan and Russia

3. _____ is the physical and social separation of dominant and minority groups.
   a. Racism
   b. Internal colonialism
   c. Segregation
   d. Institutional discrimination

4. Tyrone and Jeanette were up for the same job. Tyrone told his friends that he didn't get the job not because Jeanette was more qualified but because "women are always hired first." This is an example of
   a. scapegoating.
   b. pluralism.
   c. stereotyping.
   d. prejudice.

5. Laws against miscegenation were overturned nationally in
   a. 1807.
   b. 1867.
   c. 1907.
   d. 1967.

6. **True or False** On the continuum of dominant and minority group relations, genocide reflects the least tolerance, whereas pluralism illustrates the greatest tolerance.

7. **True or False** Ethnocentrism is the belief that one's own racial group is naturally inferior to others.

8. **True or False** Asian Americans are the fastest growing U.S. minority group.

9. **True or False** American Indians are on the verge of vanishing from the American landscape.

10. **True or False** The lower the educational level, the greater the likelihood that a person will have an interracial marriage.

11. What's the difference between individual discrimination and institutional discrimination? Is one more powerful than the other in its effect on minority groups? If so, which one and why? If not, why not?

12. Choose any two minorities discussed in this chapter. Compare them in terms of factors such as population size, diversity, constraints, and economic success.

1. b 2. a 3. c 4. a 5. d 6. True 7. False 8. False 9. False 10. False

TABLE 10.4 SOCIOLOGICAL EXPLANATIONS OF RACIAL-ETHNIC INEQUALITY

Theoretical Perspective	Level of Analysis	Key Points
Functionalist	Macro	Immigration provides needed workers; acculturation and assimilation increase social solidarity; racial-ethnic inequality can be dysfunctional, but benefits dominant groups.
Conflict	Macro	There's ongoing strife between dominant and minority groups; powerful groups maintain their advantages primarily through economic exploitation; race is a more important factor than social class in perpetuating racial-ethnic inequality.
Feminist	Macro and micro	Minority women suffer from the combined effects of racism and sexism; gendered racism occurs within and across racial-ethnic groups.
Symbolic Interactionist	Micro	Because race and ethnicity are socially constructed, social interaction can increase or reduce racial and ethnic hostility; antagonistic attitudes toward minorities, which are learned, can be lessened through cooperative interracial and interethnic contacts.

CHAPTER 11 SUMMARY

11-1 Global Economic Systems

The *economy* determines how a society produces, distributes, and consumes goods and services. *Capitalism* frequently spawns *monopolies* and *oligopolies* that dominate the market and discourage competition. *Socialism* and *communism* promise cooperation and collective ownership of property, but all countries have mixed economies that allow private profits.

11-2 Corporations and the Economy

A *corporation*, usually created for profit, often forms a *conglomerate*. Both corporations and conglomerates are governed by *interlocking directorates* that have become more powerful than ever because of the growth of *transnational corporations* and *transnational conglomerates*.

11-3 Work in U.S. Society Today

Work produces goods or services. Many Americans have been casualties of *deindustrialization* and *globalization*. Others have lost their jobs to *offshoring*. Widespread *downsizing* has resulted in unemployment, *underemployment*, discouraged workers, and low-paid jobs.

11-4 Sociological Explanations of Work and the Economy

See Table 11.3 on the next page.

11-5 Global Political Systems

In a *democracy*, ideally, citizens have a high degree of control over the state. *Totalitarianism* controls people's lives; *authoritarianism* generally permits some degree of individual freedom. A *monarchy* is the oldest type of authoritarian regime.

11-6 Politics, Power, and Authority

Politics includes *power* and *authority*. *Legitimate power* comes from having a role, position, or title that people accept as legal and appropriate; *coercive power* relies on force or threat of force to impose one's will on others. Authority can be based on *tradition, charisma, rational-legal power*, or a combination of these sources.

11-7 Politics and Power in U.S. Society

A *political party* tries to influence and control government. Whereas political parties include diverse individuals, *special-interest groups* are usually made up of people who are similar in social class and political objectives. The most powerful special-interest groups are *political action committees* and *lobbyists*. The voting rate is much higher among some groups than others. Situational and structural factors, such as voter registration, requiring proof of citizenship, and convenience can also encourage or discourage voting.

11-8 Sociological Perspectives on Politics and Power

See Table 11.7 on the next page.

KEY TERMS

economy determines how a society produces, distributes, and consumes goods and services.

politics individuals and groups acquire and exercise power and authority and make decisions.

capitalism an economic system based on private ownership of property and competition in producing and selling goods and services.

monopoly domination of a particular market or industry by one person or company.

oligopoly domination of a market by a few large producers or suppliers.

socialism an economic system based on the public ownership of the production of goods and services.

communism a political and economic system in which property is communally owned and all people are considered equal.

welfare capitalism (also called *state capitalism*) an economic system that combines private ownership of property, market competition, and a government's regulation of many programs and services.

corporation an organization that has legal rights, privileges, and liabilities apart from those of its members.

conglomerate a corporation that owns a collection of companies in different industries.

interlocking directorate the same people serve on the boards of directors of several companies or corporations.

transnational corporation (also called a *multinational corporation* or an *international corporation*) a large company that's based in one country but operates across international boundaries.

transnational conglomerate (also called a *multinational conglomerate*) a corporation that owns a collection of different companies in various industries in a number of countries.

work a physical or mental activity that produces goods or services.

deindustrialization a process of social and economic change resulting from the reduction of industrial activity, especially manufacturing.

globalization the growth and spread of investment, trade, production, communication, and new technology around the world.

offshoring sending work or jobs to another country to cut a company's costs at home.

discouraged workers people who stop searching for employment because they believe that job hunting is futile.

underemployed people who have part-time jobs but want full-time work or whose jobs are below their experience, skill, and education levels.

motherhood penalty (also called *motherhood wage penalty* or *mommy penalty*) a pay gap between women who are and aren't mothers.

government a formal organization that has the authority to make and enforce laws.

democracy a political system in which, ideally, citizens have control over the state and its actions.

totalitarianism the government controls almost every aspect of people's lives.

authoritarianism the state controls the lives of citizens but permits some degree of individual freedom.

monarchy power is allocated solely on the basis of heredity and passes from generation to generation.

power the ability of a person or group to influence others, even if they resist.

authority the legitimate use of power.

traditional authority power based on customs that justify the ruler's position.

charismatic authority power based on exceptional individual abilities and characteristics that inspire devotion, trust, and obedience.

CHAPTER REVIEW 11

rational-legal authority power based on the belief that laws and appointed or elected political leaders are legitimate.

political party an organization that tries to influence and control government by recruiting, nominating, and electing its members to public office.

special-interest group (also called an *interest group*) a group of people that seeks or receives benefits or special treatment.

political action committee (PAC) a special-interest group that raises money to elect one or more candidates to public office.

lobbyist someone hired by a special-interest group to influence legislation on the group's behalf.

pluralism a political system in which power is distributed among a variety of competing groups in a society.

power elite a small group of influential people who make the nation's major political decisions.

TEST YOUR LEARNING

1. _____ is a market system dominated by a few large producers or suppliers.
   a. Capitalism
   b. Monopoly
   c. Oligopoly
   d. Socialism

2. Felipe serves on the board of directors of several U.S. companies. He's part of
   a. an interlocking directorate.
   b. a transnational corporation.
   c. a conglomerate.
   d. a monopoly.

3. Which of the following is NOT a major reason for the surge in women's employment since the 1970s?
   a. Men's falling wages
   b. Women's increased education
   c. Rising cost of home ownership
   d. Decrease in sexism

4. According to Max Weber, which one of the following types of authority is most characteristic of the majority of U.S. presidents?
   a. Charismatic
   b. Absolute
   c. Traditional
   d. Rational-legal

5. _____ claim that the U.S. political structure is pluralistic.
   a. Functionalists
   b. Conflict theorists
   c. Feminist theorists
   d. Symbolic interactionists

6. *True or False* Compared with low-wage U.S. jobs, high-wage jobs are relatively safe from offshoring.

7. *True or False* Deindustrialization has spread because of globalization.

8. *True or False* Asian Americans have higher voting rates than Latinos.

9. *True or False* Some of the best-paid lobbyists are retired corporate executives (CEOs).

10. *True or False* In the United States, most Republicans believe that the federal government should provide social programs for its citizens.

11. Have you personally benefited or been harmed by globalization? Be specific in explaining your answer.

12. Suppose someone asked you to explain U.S. political power. Are you more likely to use a functionalist, conflict, or feminist perspective? A combination? Or none? Explain your answer.

1. c 2. a 3. a 4. d 5. a 6. False 7. True 8. False 9. False 10. False

---

TABLE 11.3 SOCIOLOGICAL EXPLANATIONS OF WORK AND THE ECONOMY

Theoretical Perspective	Level of Analysis	Key Points
Functionalist	Macro	Capitalism benefits society; work provides an income, structures people's lives, and gives them a sense of accomplishment.
Conflict	Macro	Capitalism enables the rich to exploit other groups; most jobs are low-paying, monotonous, and alienating; productivity isn't always rewarded.
Feminist	Macro and micro	Gender roles structure women's and men's work experiences differently and inequitably.
Symbolic Interactionist	Micro	How people define and experience work in their everyday lives affects their workplace behavior and relationships with coworkers and employers.

TABLE 11.7 SOCIOLOGICAL EXPLANATIONS OF POLITICAL POWER

	Functionalism: A Pluralist Model	Conflict Theory: A Power Elite Model	Feminist Theories: A Patriarchal Model
Who has political power?	The people	Rich upper-class people—especially those at top levels in business, government, and the military	White men in Western countries; most men in traditional societies
How is power distributed?	Very broadly	Very narrowly	Very narrowly
What is the source of political power?	Citizens' participation	Wealthy people in government, business corporations, the military, and the media	Being white, male, and very rich
Does one group dominate politics?	No	Yes	Yes
Do political leaders represent the average person?	Yes, the leaders speak for a majority of the people.	No, the leaders are most concerned with keeping or increasing their personal wealth and power.	No, the leaders are rarely women who have decision-making power.

CHAPTER 12 SUMMARY

12-1 What Is a Family?

Among other activities, the members of a *family* care for one another and any children. Worldwide, however, families vary in characteristics such as structure (a *nuclear* or an *extended* family), living arrangements (*patrilocal*, *matrilocal*, or *neolocal*), who has authority (*matriarchal*, *patriarchal*, or *egalitarian*), and how many marriage mates a person can have (*monogamy* or *polygamy*).

12-2 How U.S. Families Are Changing

The United States has one of the highest marriage and divorce rates in the world. *Divorce* is easier to obtain than in the past because all states now have *no-fault divorce* laws, but divorce rates vary by race, ethnicity, and social class. The number of single people has risen greatly, primarily because many people are postponing marriage. There has also been a striking increase in *cohabitation* and nonmarital births. Female-headed single-parent families are increasingly common.

12-3 Family Conflict and Violence

We're more likely to experience *intimate partner violence (IPV)* than assault by a stranger. IPV causes are complex and often the product of multiple individual, demographic, and societal factors. *Child maltreatment* is widespread, and parents

are the most common offenders. Similarly, in *elder abuse*, most of the offenders are adult children, spouses, or other family members. Both micro- and macro-level reasons help explain IPV, child maltreatment, and elder abuse and neglect.

12-4 Our Aging Society

How people define "old" varies across societies. There are also significant differences between the *young-old*, the *old-old*, and the *oldest-old* in their health, ability to live independently, and to work. Historically and currently, women live longer than men, but life expectancy varies by race, ethnicity, and social class. Our *old-age dependency ratio* has increased, placing a larger burden on the working-age population. Some important current aging issues include the rise of *multigenerational households*, physician-assisted suicide, and competition for scarce resources.

Example: "He Gets Prettier; I Get Older"

When comparing her own public image with that of her actor husband, the late Paul Newman, actress Joanne Woodward once remarked, "He gets prettier; I get older." Was she right? If aging gracefully is acceptable, why do many companies tout antiaging products? And why are the products targeted primarily at women?

12-5 Sociological Explanations of Family and Aging

See Table 12.4 on the next page.

KEY TERMS

family an intimate group consisting of two or more people who (1) have a committed relationship, (2) care for one another and any children, and (3) share close emotional ties and functions.

incest taboo cultural norms and laws that forbid sexual intercourse between close blood relatives.

marriage a socially approved mating relationship that people expect to be stable and enduring.

endogamy (often used interchangeably with *homogamy*) cultural practice of marrying within one's group.

exogamy (often used interchangeably with *heterogamy*) cultural practice of marrying outside one's group.

nuclear family a family form composed of married parents and their biological or adopted children.

extended family a family form composed of parents, children, and other kin.

patrilocal residence pattern newly married couples live with the husband's family.

matrilocal residence pattern newly married couples live with the wife's family.

neolocal residence pattern a newly married couple sets up its own residence.

boomerang generation young adults who move back into their parents' home or never leave it in the first place.

matriarchal family system the oldest females control cultural, political, and economic resources and, consequently, have power over males.

patriarchal family system the oldest males control cultural, political, and economic resources and, consequently, have power over females.

egalitarian family both partners share power and authority fairly equally.

marriage market prospective spouses compare the assets and liabilities of eligible partners and choose the best available mate.

arranged marriage parents or relatives choose the children's spouses.

monogamy one person is married exclusively to another person.

serial monogamy individuals marry several people, but one at a time.

polygamy a man or woman has two or more spouses.

CHAPTER REVIEW 12

cohabitation two unrelated people aren't married but live together and are in a sexual relationship.

dating cohabitation a couple that spends a great deal of time together decides to move in together.

premarital cohabitation a couple lives together before getting married.

fictive kin nonrelatives who are accepted as part of a family.

intimate partner violence (IPV) abuse that occurs between people in a close relationship.

child maltreatment (also called *child abuse*) a broad range of behaviors that place a child at serious risk, including physical and sexual abuse, neglect, and emotional mistreatment

elder abuse (sometimes called *elder mistreatment*) any knowing, intentional, or negligent act by a caregiver or other person that causes harm to people age 65 or older.

life expectancy the average expected number of years of life remaining at a given age.

baby boomers people born between 1946 and 1964.

old-age dependency ratio the number of people age 65 and older who aren't in the labor force relative to the number of working-age adults ages 18 to 64.

activity theory proposes that many older people remain engaged in numerous roles and activities, including work.

exchange theory posits that people seek through their social interactions to maximize their rewards and minimize their costs.

ageism discrimination against older people.

continuity theory posits that older adults can substitute satisfying new roles for those they've lost.

TEST YOUR LEARNING

1. This chapter defines a *family* in terms of three criteria. Which of the following is NOT one of the criteria?
   a. Relationships are legally sanctioned by the state
   b. Live together in a committed relationship
   c. Care for one another and any children
   d. Share close emotional ties and bonds

2. The family serves many important functions in a society. Which of the following functions does the family NOT serve?
   a. Sexual regulation
   b. Social placement
   c. Peer approval
   d. Economic security

3. Worldwide, the most common residence pattern is
   a. patriarchal.
   b. matriarchal.
   c. neolocal.
   d. egalitarian.

4. Which of the following is NOT a common factor in intimate partner violence (IPV)?
   a. Income level
   b. Employment status
   c. Drug abuse
   d. Number of children

5. _____ use both a macro and a micro approach in examining families and aging.
   a. Functionalists
   b. Conflict theorists
   c. Feminist theorists
   d. Symbolic interactionists

6. **True or False** Jack marries several people, but one at a time. Jack is a serial monogamist.

7. **True or False** U.S. divorce rates have decreased since 1980.

8. **True or False** College-educated women are less likely to cohabit than women with a high school diploma.

9. **True or False** Most child maltreatment offenders are relatives and unrelated caregivers such as foster parents and boyfriends.

10. **True or False** Baby boomers are the fastest growing segment of the U.S. population.

11. "U.S. families are falling apart." Explain whether you agree or disagree with this statement, using functionalist, conflict, feminist, and symbolic interactionist perspectives.

12. What do you think your life will be like at age 85? How might your options be different because of your sex, social class, race, and ethnicity? Support your answer with specific examples.

TABLE 12.4 SOCIOLOGICAL PERSPECTIVES ON FAMILIES AND AGING

Theoretical Perspective	Level of Analysis	Key Points
Functionalist	Macro	Families are important in maintaining societal stability and meeting family members' needs.
		Older people who are active and engaged are more satisfied with life.
Conflict	Macro	Families promote social inequality because of social class differences.
		Many corporations view older workers as disposable.
Feminist	Macro and micro	Families both mirror and perpetuate patriarchy and gender inequality.
		Women have an unequal burden in caring for children as well as older family members and relatives.
Symbolic Interactionist	Micro	Families construct their everyday lives through interaction and subjective interpretations of family roles.
		Many older family members adapt to aging and often maintain previous activities.

CHAPTER 13 SUMMARY

13-1 Education and Society

U.S. *education* and *schooling* have changed in four important ways: Universal education has expanded, community colleges have flourished, public higher education has burgeoned, and student diversity has greatly increased.

13-2 Sociological Perspectives on Education

See Table 13.1 on the next page.

13-3 Some Current Issues in U.S. Education

Compared with their counterparts in a number of other countries, many U.S. students are performing poorly in elementary and high schools—especially in mathematics and the sciences. U.S. teachers' salaries are low, many are poorly prepared, out of field, and have less control over curricula than ever before. A growing number of critics contend that standardized tests (such as SATs and ACTs) are little more than gatekeeping tools that exclude lower socioeconomic students from access to higher education, and that the tests scores don't predict college success. Despite grade inflation, high school and college dropout rates are high, and cheating is widespread.

13-4 Religion and Society

Some form of religion exists in all societies and cultures. *Religion* unites believers into a community. Every known society distinguishes between *sacred* and *profane (secular)* activities. Religion, *religiosity*, and spirituality differ; for example, people who describe themselves as religious may not attend services.

Example: Sacred vs. Secular

The corporate mission of Chick-fil-A (a franchise that prepares sandwiches), as stated on a plaque at company headquarters, is "to glorify God." Chick-fil-A is the only national fast-food chain that closes on Sunday so employees can go to church, and prospective employees are asked about their religious activities. Some franchise operators are delighted with the religious emphasis; others believe that a business should stay out of its workers' personal lives (Schmall, 2007). What do *you* think?

13-5 Religious Organization and Major World Religions

People express their religious beliefs most commonly through organized groups, including *cults* (also called *new religious movements [NRMs]*), *sects, denominations,* and *churches.* Some NRMs, which usually organize around a *charismatic leader,* have been short-lived, whereas others have become established religions with highly organized bureaucracies. Worldwide, the largest religious group is Christians, followed by Muslims, but no religious group comes close to being a global majority. The third largest group is people who don't have a religious affiliation.

13-6 Religion in the United States

Religion in the United States is complex and diverse. About half of U.S. adults have changed their religion since childhood, many opting for no religion at all. Mainline Protestant groups have declined whereas evangelicals have surged. Religious participation varies by gender, age, race, ethnicity, and social class. Some sociologists maintain that *secularization* is increasing rapidly in the United States; others contend that this claim is greatly exaggerated, especially as witnessed by the growth of *fundamentalism* and the prevalence of *civil religion.*

13-7 Sociological Perspectives on Religion

See Table 13.5 on the next page.

KEY TERMS

education a social institution that transmits attitudes, knowledge, beliefs, values, norms, and skills.

schooling formal training and instruction provided in a classroom setting.

achievement gap the difference in academic performance that shows up in grades, standardized test scores, and college completion rates.

hidden curriculum school practices that transmit nonacademic knowledge, values, attitudes, norms, and beliefs.

credentialism an emphasis on certificates or degrees to show that people have certain skills, educational attainment levels, or job qualifications.

tracking (also called *streaming* or *ability grouping*) assigning students to specific educational programs and classes on the basis of test scores, previous grades, or perceived ability.

meritocracy a system that rewards people because of their individual accomplishments.

religion a social institution that involves shared beliefs, values, and practices related to the supernatural.

sacred anything that people see as awe-inspiring, supernatural, holy, and not part of the natural world.

profane the ordinary and everyday elements of life that aren't related to religion.

religiosity the ways people demonstrate their religious beliefs.

cult a religious group that is devoted to beliefs and practices that are outside of those accepted in mainstream society.

new religious movement (NRM) term used instead of *cult* by most sociologists.

CHAPTER REVIEW 13

charismatic leader someone that followers see as having exceptional or superhuman powers and qualities.

sect a religious group that has broken away from an established religion.

denomination a subgroup within a religion that shares its name and traditions and is generally on good terms with the main group.

church a large established religious group that has strong ties to mainstream society.

secularization a process in which religion loses its social and cultural influence.

fundamentalism the belief in the literal meaning of a sacred text.

civil religion (sometimes called *secular religion*) integrating religious beliefs into secular life.

Protestant ethic a belief that hard work, diligence, self-denial, frugality, and economic success will lead to salvation in the afterlife.

false consciousness an acceptance of a system of beliefs that prevents people from protesting oppression.

TEST YOUR LEARNING

1. _____ maintain that the education system creates and perpetuates social inequality.
   a. Functionalists
   b. Conflict theorists
   c. Feminist theorists
   d. Symbolic interactionists

2. An emphasis on certificates or degrees is called
   a. meritocracy.
   b. tracking.
   c. the hidden curriculum.
   d. credentialism.

3. Today, the largest educational achievement gap is between
   a. males and females.
   b. blacks and whites.
   c. social classes.
   d. none of the above.
   e. all of the above.

4. A _____ is a religious group that is devoted to beliefs and practices that are outside of those accepted in mainstream society.
   a. cult
   b. sect
   c. denomination
   d. church

5. According to _____, religion isn't innate but taught.
   a. functionalists
   b. conflict theorists
   c. feminist theorists
   d. symbolic interactionists

6. **True or False** Cultural innovation is one of the latent functions of education.

7. **True or False** U.S. high school dropout rates have been increasing.

8. **True or False** Both religion and religiosity are social institutions.

9. **True or False** The Protestant work ethic is a belief that hard work, diligence, self-denial, frugality, and economic success will lead to salvation in the afterlife.

10. **True or False** One of the criticisms of functionalist theories on religion is that they often ignore the role that religion plays in creating social cohesion and harmony.

11. How does education reinforce existing inequalities in the United States based on gender, social class, race, and ethnicity? Illustrate your answer with specific examples.

12. Explain how religious and secular rituals differ and overlap. Be specific. Also, why do people who say that they have no religion practice civil religion?

1. b 2. d 3. c 4. a 5. d 6. False 7. False 8. False 9. True 10. False

TABLE 13.1 SOCIOLOGICAL EXPLANATIONS OF EDUCATION

Theoretical Perspective	Level of Analysis	Key Points
Functionalist	Macro	Contributes to society's stability, solidarity, and cohesion and provides opportunities for upward mobility
Conflict	Macro	Reproduces and reinforces inequality and maintains a rigid social class structure
Feminist	Macro and micro	Produces inequality based on gender
Symbolic Interactionist	Micro	Teaches roles and values through everyday face-to-face interaction and behavior

TABLE 13.5 SOCIOLOGICAL EXPLANATIONS OF RELIGION

Theoretical Perspective	Level of Analysis	Key Points
Functionalist	Macro	Religion benefits society by providing a sense of belonging, identity, meaning, emotional comfort, and social control over deviant behavior.
Conflict	Macro	Religion promotes and legitimates social inequality, condones strife and violence between groups, and justifies oppression of poor people.
Feminist	Macro and micro	Religion subordinates women, excludes them from decision-making positions, and legitimizes patriarchal control of society.
Symbolic Interactionist	Micro	Religion provides meaning and sustenance in everyday life through symbols, rituals, and beliefs, and binds people together in a physical and spiritual community.

CHAPTER 14 SUMMARY

14-1 Global Health and Illness

Health varies among individuals and societies, but all people experience *disease*. In addressing the question of why some people are healthier than others, health and medical practitioners, researchers, and sociologists examine *social epidemiology*, which includes both the incidence and prevalence of illness or health problems within a population during a specific time period. Worldwide, high income populations live longer, have low infant mortality rates, and spend more on health, but experience "diseases of wealth" such as diabetes, heart disease, and various cancers.

14-2 Health and Illness in the United States

Many people live with a *disability*; others die at an earlier age than expected for a variety of macro- and micro-level reasons. The three most important reasons for illness and early death are environmental, demographic, and lifestyle choices. After age 65, people are more likely to experience *chronic diseases* rather than *acute diseases*. Women tend to live longer than men but have more chronic diseases, such as arthritis, asthma, cancer, and mental illness, and experience higher depression rates than men. Among U.S. racial-ethnic groups, black babies have the *highest infant mortality rate*, and Latinos are the least likely to die from illicit drugs and prescription drug abuse. Our lifestyle choices also improve or impair health. The top three preventable health hazards, in order of priority, are smoking, obesity, and substance abuse. Sexually transmitted diseases are another important source of preventable health hazards.

14-3 Health Care: United States and Global

Medicine is a vital part of *health care* in diagnosing, treating, and preventing illness, injury, and other health impairments. The United States is one the richest nations in the world, but only the very wealthy don't have to worry about receiving and paying for the best medical care available. About 14 percent of Americans have no health insurance. Since 2000, employer-based health insurance coverage has deteriorated. Consequently, workers at both large and small firms have been making higher contributions to premiums and paying higher deductibles or copayments. *Medicare* pays many of the medical costs of Americans age 65 and over, regardless of income. *Medicaid*, another government program, provides medical care for those living below the poverty level.

The United States spends more on health care than any other nation in the world, but ranks only 34th worldwide in life expectancy. Compared with other high-income countries, the United States covers a smaller percentage of the total population, and has poorer health outcomes. Unlike people in other high-income countries, Americans have fewer and shorter doctor visits, fewer days of inpatient hospital care, and pay considerably more for prescription drugs and routine medical services (such as doctor's office visits and normal birth deliveries).

In 2010, Congress passed the Patient Protection and Affordable Care Act, also called the Affordable Care Act (ACA) or "Obamacare." The ACA's goal is to give more Americans age 64 and under access to affordable, quality health insurance, and to reduce the growth in U.S. health care spending. Some key features of the ACA prohibit health care insurers from denying coverage to children with a preexisting condition and imposing lifetime spending limits. ACA also requires insurers to provide free preventive health services, such as tests for diabetes, mammograms, colonoscopies, and routine vaccinations; and permits people to choose a primary

KEY TERMS

health the state of physical, mental, and social well-being.

social epidemiology examines how societal factors affect the distribution of disease within a population.

health care the prevention, management, and treatment of illness.

disability any physical or mental impairment that limits a person's ability to perform a basic life activity.

chronic diseases long-term or lifelong illnesses that develop gradually or are present from birth.

acute diseases illnesses that strike suddenly and often disappear rapidly but can cause incapacitation and sometimes death.

Alzheimer's disease a progressive, degenerative disorder that attacks the brain and impairs memory, thinking, and behavior.

substance abuse an overindulgence in and dependence on a drug or other chemical that harms a person's physical and mental health.

medicine a system of individuals, organizations, and institutions that provide scientific diagnosis, treatment, and prevention of illness, injury, and other health impairments.

sick role a social role that excuses people from normal obligations because of illness.

medical-industrial complex a network of business enterprises that influences medicine and health care.

medicalization a process that defines a nonmedical condition or behavior as an illness, disorder, or disease that requires medical treatment.

CHAPTER REVIEW 14

care doctor outside of a health plan's provider network. In 2012, however, the U.S. Supreme Court allowed states to opt out of the ACA's "Medicaid expansion" provision that covers the "nearly poor" who live just above the poverty level and don't qualify for Medicaid. Thus, "nearly poor" people in about half of the states must pay for their own health insurance.

14-4 Sociological Perspectives on Health and Medicine

See Table 14.5 below.

TEST YOUR LEARNING

1. The sick role is most closely associated with which theoretical perspective?
   a. Functionalist
   b. Conflict
   c. Feminist
   d. Symbolic interactionist

2. ____ is the top preventable lifestyle health hazard.
   a. Obesity
   b. Drug abuse
   c. Sexually transmitted diseases
   d. Smoking

3. Some have hailed the ____ as an important milestone in U.S. health care reform.
   a. AMA
   b. DSM
   c. ACA
   d. HMO

4. Which one of the following statements about U.S. health and medicine is *false*?
   a. Illness and death rates vary considerably, especially by social class.
   b. According to functionalists, the medical profession and health care industry maintain the status quo because of the medical-industrial complex.
   c. Women use more health care services than men do.
   d. According to symbolic interactionists, physicians, mental rights advocates, and parents all benefit from medicalization.

5. One of the most common criticisms of ____ theories is that they often ignore the contributions of medical and health care systems.
   a. functionalist
   b. conflict
   c. feminist
   d. symbolic interaction

6. *True or False* Acute diseases increase as people age.

7. *True or False* Chronically ill people are responsible for almost 50 percent of all U.S. health care spending.

8. *True or False* Compared with other groups, Latinos are the most likely to die from illegal drugs and prescription drug abuse.

9. *True or False* All states have obesity rates equal to or greater than 30 percent.

10. *True or False* Medicalization is a network of business enterprises that influences medicine and health care.

11. Is government-paid health care a basic human right? Or should it be limited to taxpayers? Explain your answer. Be as specific as possible, and incorporate the material in this chapter such as health care in other countries and the ACA.

12. Identify any two limitations of functionalist, conflict, feminist, and symbolic interaction theories in explaining health and medicine. (Extra points if you add a third limitation for each theory that's not discussed in this chapter.)

1. a 2. d 3. c 4. b 5. b 6. False 7. False 8. True 9. False 10. False

TABLE 14.5 WOMEN EARN LESS THAN MEN IN HEALTH CARE OCCUPATIONS, 2013

	Men	Women	Percentage of Women in This Occupation
All health care practitioners	**$68,224**	**$51,378**	**75**
Pharmacists	$108,784	$93,704	54
Physicians and surgeons	$108,524	$77,844	36
Physical therapists	$75,764	$67,600	66
Registered nurses	$64,272	$56,472	89
Emergency medical technicians and paramedics	$43,264	$40,820	41
Support technologists and technicians	$34,528	$31,876	81

Source: Based on Bureau of Labor Statistics, "Median Weekly Earnings of Full-Time Wage and Salary Workers by Detailed Occupation and Sex," Table 39. Accessed July 30, 2014 (www.bls.gov).

CHAPTER 15 SUMMARY

15-1 Population Changes

Demography examines the interplay among *fertility, mortality,* and *migration.* The *crude birth rate, total fertility rate, crude death rate,* and infant mortality rate measure a population's life expectancy and health. Push and pull factors affect international and internal migration. Demographers also use *sex ratios* and *population pyramids* to understand a population's composition and structure.

Demographers who believe that population growth is a ticking bomb subscribe to *Malthusian theory,* which argues that the world's food supply won't keep up with population growth. *Demographic transition theory,* in contrast, maintains that population growth is kept in check and stabilizes as countries experience greater economic and technological development.

15-2 Urbanization

Globally and in the United States, cities and *urbanization* mushroomed during the twentieth century and are expected to increase. As more people move from rural to urban areas, many of the world's largest cities are becoming *megacities.* In the United States, urban growth has led to suburbanization, *edge cities, exurbs, gentrification,* and *urban sprawl.*

Sociologists offer several perspectives on how and why cities change, and the consequences (see Table 15.3 on the next page).

15-3 The Environment

Population growth and urbanization are changing the planet's *ecosystem* and, many argue, endangering plants, animals, and humans. Water and air pollution and global warming are good examples of threats to the ecosystem in the United States and globally. Clean water has been depleted for many reasons, including pollution, privatization, waste, and mismanagement.

Four of the most common sources and causes of air pollution are burning fossil fuels, manufacturing plants that spew pollutants into the air, winds that carry contaminants across borders and oceans, and lax governmental policies. Air pollution, which can lead to the *greenhouse effect,* is a major cause of *climate change* and *global warming.*

The rise of environmental problems has sparked discussions about *sustainable development.* Those who are pessimistic about achieving sustainable development show that, worldwide, the United States has one of the worst records on environmental performance, largely because of the close ties between government officials and corporations. Others are optimistic about achieving sustainable development and point to examples such as decreases in the emission of major air pollutants and some large U.S. corporations' switching to practices that decrease pollution and energy consumption.

KEY TERMS

demography the scientific study of human populations.

population a group of people who share a geographic territory.

fertility the number of babies born during a specified period in a particular society.

crude birth rate (also called the *birth rate*) the number of live births per 1,000 people in a population in a given year.

total fertility rate (TFR) the average number of children born to a woman during her lifetime.

mortality the number of deaths in a population during a specified period.

crude death rate (also called the *death rate*) the number of deaths per 1,000 people in a population in a given year.

migration the movement of people into or out of a specific geographic area.

sex ratio the proportion of men to women in a population.

population pyramid a graphic depiction of a population's age and sex distribution at a given point in time.

Malthusian theory maintains that population is growing faster than the food supply needed to sustain it.

demographic transition theory maintains that population growth is kept in check and stabilizes as countries experience economic and technological development.

zero population growth (ZPG) a stable population level that occurs when each woman has no more than two children.

urbanization people's movement from rural to urban areas.

megacities metropolitan areas with at least 10 million inhabitants.

CHAPTER REVIEW 15

edge cities business centers that are within or close to suburban residential areas.

exurbs areas of new development beyond the suburbs that are more rural but on the fringe of urbanized areas.

metropolitan statistical area (MSA, also called *metro area*) a central city of at least 50,000 people and urban areas linked to it.

urban sprawl the rapid, unplanned, and uncontrolled spread of development into regions adjacent to cities.

gentrification a process in which upper-middle-class and affluent people buy and renovate houses and stores in downtown urban neighborhoods.

urban ecology studies the relationships between people and urban environments.

new urban sociology view that urban changes are largely the result of decisions made by powerful capitalists and high-income groups.

ecosystem a community of living and non-living organisms that share a physical environment.

global warming the increase in the average temperature of earth's atmosphere.

greenhouse effect the heating of earth's atmosphere because of the presence of certain atmospheric gases.

climate change a change in overall temperatures and weather conditions over time.

sustainable development economic activities that don't threaten the environment.

TEST YOUR LEARNING

1. ____ is the study of human populations.
   a. Sociology
   b. Demography
   c. Ecology
   d. Zero population growth (ZPG)

2. Emma is comparing the number of deaths per 1,000 people in Spain and Russia for 2011. She is measuring
   a. fertility.
   b. mortality.
   c. the crude birth rate.
   d. the crude death rate.

3. Which one of the following is NOT one of the phases of demographic transition theory?
   a. Preindustrial society
   b. Early industrial society
   c. Advanced industrial society
   d. Declining industrial society

4. Maleek and Shonda live in a newly developed area that's rural but on the fringe of an urbanized region. They live in
   a. an exurb.
   b. a suburb.
   c. a gentrified area.
   d. an edge city.

5. Why do many feminist scholars maintain that women suffer more problems than men when living in urban areas?
   a. Women don't have the natural aggressive instinct that men have.

   b. Husbands tend to maintain tighter controls over their wives in urban areas.
   c. Urban areas have typically been designed by men and for men.
   d. All of the above explain why women suffer more problems in urban areas.

6. **True or False** Malthusian theory claims that population growth is kept in check and stabilizes as developing countries experience economic and technological development.

7. **True or False** Population pyramids are graphic depictions of a population's makeup in terms of the age and sex of its members at a given point in time.

8. **True or False** Concentric zone theory emphasizes the development of suburbs around a city but away from its center.

9. **True or False** A sex ratio of 115 means that there are 115 women for every 100 men in a population.

10. **True or False** Global warming begins with the greenhouse effect.

11. Does Malthusian theory or demographic transition theory come closer to your perspective on population growth? Why?

12. How do functionalist, conflict, feminist, and symbolic interaction theories differ in their explanations of urbanization? Be specific in illustrating your answer.

1. b 2. d 3. d 4. a 5. c 6. False 7. True 8. False 9. False 10. True

---

TABLE 15.3 | SOCIOLOGICAL EXPLANATIONS OF URBANIZATION

Perspective	Level of Analysis	Key Points
Functionalist	Macro	Cities serve many important social and economic functions, but urbanization can also be dysfunctional.
Conflict	Macro	Driven by greed and profit, large corporations, banks, developers, and other capitalist groups shape cities' growth or decline.
Feminist	Macro and micro	Whether they live in cities or suburbs, women generally experience fewer choices and more constraints than men.
Symbolic Interactionist	Micro	City residents differ in their types of interaction, lifestyles, and perceptions of urban life.

CHAPTER 16 SUMMARY

16-1 Collective Behavior

Social scientists offer several explanations of *collective behavior*. Contagion theory proposes that individuals act emotionally and irrationally due to a crowd's almost hypnotic influence. According to convergence theory, crowds consist of like-minded people who deliberately assemble in a place to pursue a common goal. Emergent norm theory emphasizes the importance of social norms in shaping crowd behavior. Structural strain theory proposes that collective behavior occurs only if six conditions are present.

There are many types of collective behavior, some more short-lived or harmful than others. *Rumors*, *gossip*, and *urban legends* are typically untrue, but many people believe and pass them on for a number of reasons, such as anxiety or to reinforce a community's moral standards. In contrast, *panic* and *mass hysteria* can have dire consequences, including death. *Fashions* and *fads* are harmless because they usually last only a short time and change over time. In contrast, a *disaster* is an unexpected event due to social, technological, or natural causes that result in widespread damage and destruction.

Publics, *public opinion*, and *propaganda* also affect large numbers of people. *Crowds* vary in their motives, interests, and emotional level. A casual crowd, for example, has little, if any, interaction, the gathering is temporary, and there is little emotion. On the other hand, protest crowds, especially in *mobs* and *riots*, can wreak considerable havoc on property and result in death.

Example: Crowds Can Be Deadly

On Thanksgiving Day, 2008, crowds started gathering at 9:00 p.m. outside the Wal-Mart store in Valley Stream, New York,

for a bargain-hunting ritual known as Black Friday, the day after Thanksgiving. By 4:55 a.m., the crowd had grown to more than 2,000 people and could no longer be held back. Suddenly, according to witnesses, the glass doors shattered and "the shrieking mob surged through in a blind rush for holiday bargains." A 34-year-old male temporary worker, who had been hired for the holiday season, was trampled to death, and four other people, including a 28-year-old woman who was eight months pregnant, were treated for injuries (McFadden and Macropoulos, 2008). Review the types of crowds in this chapter. Which type of crowd do you think this Wal-Mart incident most nearly represents?

16-2 Social Movements

Unlike collective behavior, *social movements* are typically organized and have long-lasting effects. Some of the most common social movements are alternative, redemptive, reformative, resistance, and revolutionary (see Table 16.1 on the next page). Sociologists have offered several explanations for the emergence of social movements, including mass society theory, relative deprivation theory, resource mobilization theory, and new social movements theory. Each theory has strengths and weaknesses in explaining social movements.

Social movements generally go through four stages: emergence, organization, institutionalization, and decline. Decline occurs when a social movement is successful and becomes a part of society's fabric; when the members become distracted because the group loses sight of its original goals and/or their enthusiasm wanes; when the membership fragments because the participants disagree about goals, strategies, or tactics; or when a government quashes dissent. Social movements are important because

KEY TERMS

social change transformations of societies and social institutions over time.

collective behavior the spontaneous and unstructured behavior of a large number of people.

rumor unfounded information that people spread quickly.

gossip rumors, often negative, about other people's personal lives.

urban legends (also called *contemporary legends* and *modern legends*) rumors about stories that supposedly happened somewhere.

panic a collective flight, often irrational, from a real or perceived danger.

mass hysteria an intense, fearful, and anxious reaction to a real or imagined threat by large numbers of people.

fashion a popular way of dressing during a particular time or among a particular group of people.

fad a fashion that spreads rapidly and enthusiastically but lasts only a short time.

disaster an unexpected event that causes widespread damage, destruction, distress, and loss.

public a group of people, not necessarily in direct contact with each other, who are interested in a particular issue.

public opinion widespread attitudes on a particular issue.

propaganda spreading information (or misinformation) to influence people's attitudes or behavior.

crowd a temporary gathering of people who share a common interest or participate in a particular event.

mob a highly emotional and disorderly crowd that uses force, the threat of force, or violence against a specific target.

CHAPTER REVIEW 16

riot a violent crowd that directs its hostility at a wide and shifting range of targets.

social movement a large and organized group of people who want to promote or resist a particular social change.

relative deprivation a gap between what people have and what they think they should have compared with others in a society.

technology the application of scientific knowledge for practical purposes.

they can create or resist change at the individual, institutional, and societal levels.

16-3 Technology and Social Change

Technology also generates changes. Some of the most important technological advances have included computer technology, biotechnology, and nanotechnology—all of which have changed our lives. Technology has both benefits and costs, however. For example, the Internet and other forms of telecommunication technology can bring people together, but can also intrude on our privacy. In addition, technological advances raise numerous ethical questions, such as their greater availability to higher-income people.

TEST YOUR LEARNING

1. According to ____, crowds are highly suggestible and out of control.
   a. contagion theory
   b. convergence theory
   c. emergent norm theory
   d. structural strain theory

2. A major difference between panic and mass hysteria is that
   a. panics are typically not as severe as mass hysteria.
   b. panics don't usually last as long as mass hysteria.
   c. panics stem from imagined events, whereas mass hysteria arises from real events.
   d. panics tend to occur less today, whereas mass hysteria occurs more frequently.

3. Cabbage Patch Kids dolls are an example of a
   a. fashion.
   b. fad.
   c. craze.
   d. disaster.

4. ____ maintains that social movements emerge because there's a gap between what people have and what they think they should have compared with other people.
   a. Mass society theory
   b. Relative deprivation theory

   c. Resource mobilization theory
   d. New social movements theory

5. Which of the following is NOT one of four stages of a social movement?
   a. Emergence
   b. Conflict
   c. Institutionalization
   d. Decline

6. **True or False** During the organization stage of social movements, the movement becomes more bureaucratic.

7. **True or False** Gossip and urban legends are types of rumors.

8. **True or False** Mobs typically last longer than riots.

9. **True or False** Environmentalism is an example of the new social movements theory.

10. **True or False** Technological advances are available to most low-income people.

11. What conditions are necessary to produce social movements? What draws people into a social movement? If you're involved in a social movement now, which one and why?

12. What, personally, have been some of the benefits and costs of technology in your life? Ilustrate your answer with specific examples.

1. a 2. b 3. b 4. b 5. b 6. False 7. True 8. False 9. True 10. False

TABLE 16.1	FIVE TYPES OF SOCIAL MOVEMENTS	
Movement	**Goal**	**Examples**
Alternative	Change some people in a specific way	Alcoholics Anonymous, transcendental meditation
Redemptive	Change some people, but completely	Jehovah's Witnesses, born-again Christians
Reformative	Change everyone, but in specific ways	Gay rights advocates, Mothers Against Drunk Driving (MADD)
Resistance	Preserve status quo by blocking or undoing change	Antiabortion groups, white supremacists
Revolutionary	Change everyone completely	Right-wing militia groups, Communism, Islamic State in Iraq and Syria (ISIS)